T0201341

FOURTH EDITION

Matter & Interactions

VOLUME I
Modern Mechanics

VOLUME II
Electric and Magnetic Interactions

RUTH W. CHABAY
BRUCE A. SHERWOOD

North Carolina State University

VICE PRESIDENT & PUBLISHER Petra Recter
EXECUTIVE EDITOR Jessica Fiorillo
ASSOCIATE EDITOR Aly Rentrop
EDITORIAL ASSISTANT Amanda Rillo
SENIOR MARKETING MANAGER Kristy Ruff
MARKETING ASSISTANT Claudine Scrivanich
SENIOR CONTENT MANAGER Kevin Holm
SENIOR PRODUCTION EDITOR Elizabeth Swain
DESIGN DIRECTOR Harry Nolan
TEXT/COVER DESIGN Maureen Eide
COVER IMAGE Ruth Chabay

Cover Description: The cover image is a snapshot from a VPython program that models the motion of a mass-spring system in 3D (see Computational Problem P70 in Chapter 4).

This book was set in 10/12 Times Ten Roman in LaTex by MPS and printed and bound by Quad/Graphics. The cover was printed by Quad/Graphics.

This book is printed on acid-free paper.

Copyright © 2015, 2011, 2007 John Wiley & Sons, Inc. All rights reserved.

No part of this publication may be reproduced, stored in a retrieval system or transmitted in any form or by any means, electronic, mechanical, photocopying, recording, scanning, or otherwise, except as permitted under Sections 107 or 108 of the 1976 United States Copyright Act, without either the prior written permission of the Publisher, or authorization through payment of the appropriate per-copy fee to the Copyright Clearance Center, Inc., 222 Rosewood Drive, Danvers, MA 01923, website www.copyright.com. Requests to the Publisher for permission should be addressed to the Permissions Department, John Wiley & Sons, Inc., 111 River Street, Hoboken, NJ 07030-5774, (201)748-6011, fax (201)748-6008, website http://www.wiley.com/go/permissions.

Evaluation copies are provided to qualified academics and professionals for review purposes only, for use in their courses during the next academic year. These copies are licensed and may not be sold or transferred to a third party. Upon completion of the review period, please return the evaluation copy to Wiley. Return instructions and a free of charge return shipping label are available at www.wiley.com/go/returnlabel. Outside of the United States, please contact your local representative.

Complete text ISBN 978-1-118-87586-5
Complete text binder version ISBN 978-1-118-91451-9

Volume 1 ISBN 978-1-118-91449-6
Volume 1 Binder version ISBN 978-1-118-91452-6

Volume 2 ISBN 978-1-118-91450-2
Volume 2 Binder version ISBN 978-1-118-91453-3

Printed in the United States of America
SKY10035870_091522

Brief Contents

Contents

CHAPTER 23 *Electromagnetic Radiation* 939

The Supplements can be found at the web site, www.wiley.com/college/chabay

SUPPLEMENT S1 *Gases and Heat Engines*

SUPPLEMENT S3 *Waves*

SUPPLEMENT S2 *Semiconductor Devices*

Preface

TO THE STUDENT

This textbook emphasizes a 20th-century perspective on introductory physics. Contemporary physicists build models of the natural world that are based on a small set of fundamental physics principles and on an understanding of the microscopic structure of matter, and they apply these models to explain and predict a very broad range of physical phenomena. In order to involve students of introductory physics in the contemporary physics enterprise, this textbook emphasizes:

- Reasoning directly from a small number of fundamental physics principles, rather than from a large set of special-case equations.
- Integrating contemporary insights, such as atomic models of matter, quantized energy, and relativistic dynamics, throughout the curriculum.
- Engaging in the full process of creating and refining physical models (idealizing, making approximations, explicitly stating assumptions, and estimating quantities).
- Reasoning iteratively about the time-evolution of system behavior, both on paper and through the construction and application of computational models.

Because the physical world is 3-dimensional, we work in 3D throughout the text. Many students find the approach to 3D vectors used in this book easier than standard treatments of 2D vectors.

Textbook and Supplemental Resources

Modern Mechanics (Volume 1, Chapters 1–12) focuses on the atomic structure of matter and interactions between material objects. It emphasizes the wide applicability and utility of a small number of fundamental principles: the Momentum Principle, the Energy Principle, and the Angular Momentum Principle, and the Fundamental Assumption of Statistical Mechanics. We study how to explain and predict the behavior of systems as different as elementary particles, molecules, solid metals, and galaxies.

Electric and Magnetic Interactions (Volume 2, Chapters 13–23) emphasizes the somewhat more abstract concepts of electric and magnetic fields and extends the study of the atomic structure of matter to include the role of electrons. The principles of electricity and magnetism are the foundation for much of today's technology, from cell phones to medical imaging.

Additional resources for students are freely available at this site:

www.wiley.com/college/chabay

The web resources include several supplements. A copy of Chapter 1 is provided for students who are currently using Volume 2 but whose previous physics course did not use Volume 1. This chapter introduces 3D vectors and vector algebra, and includes an introduction to computational modeling in VPython, which is used throughout the textbook.

Supplement S1 treats the kinetic theory of gases and heat engines, and can be used by students who have completed Chapter 12 on Entropy. Supplement S2 explains the basic principles of PN junctions in semiconductor devices, and can be used by students who have completed Chapter 21: Patterns of Field in Space. Supplement S3 includes a more mathematically sophisticated treatment of mechanical and electromagnetic waves and wave phenomena, and can be used by students who have completed Chapter 23 on Electromagnetic Radiation.

Answers to odd-numbered problems may be found at the end of the book.

The new Student Solutions Manual is available for purchase as a printed supplement and contains fully worked solutions for a subset of end of chapter problems.

Prerequisites

This book is intended for introductory calculus-based college physics courses taken by science and engineering students. It requires a basic knowledge of derivatives and integrals, which can be obtained by studying calculus concurrently.

Modeling

Matter & Interactions places a major emphasis on constructing and using physical models. A central aspect of science is the modeling of complex real-world phenomena. A physical model is based on what we believe to be fundamental principles; its intent is to predict or explain the most important aspects of an actual situation. Modeling necessarily involves making approximations and simplifying assumptions that make it possible to analyze a system in detail.

Computational Modeling

Computational modeling is now as important as theory and experiment in contemporary science and engineering. We introduce you to serious computer modeling right away to help you build a strong foundation in the use of this important tool.

In this course you will construct simple computational models based on fundamental physics principles. You do not need any prior programming experience–this course will teach you the small number of computational concepts you will need. Using VPython, a computational environment based on the Python programming language, you will find that after less than an hour you can write a simple computational model that produces a navigable 3D animation as a side effect of your physics code.

Computational modeling allows us to analyze complex systems that would otherwise require very sophisticated mathematics or that could not be analyzed at all without a computer. Numerical calculations based on the Momentum Principle give us the opportunity to watch the dynamical evolution of the behavior of a system. Simple models frequently need to be refined and extended. This can be done straightforwardly with a computer model but is often impossible with a purely analytical (non-numerical) model.

VPython is free, and runs on Windows, MacOS, and Linux. Instructions in Chapter 1 tell you how to install it on your own computer, and how to find a set of instructional videos that will help you learn to use VPython.

Questions

As you read the text, you will frequently come to a question that looks like this:

> QUESTION What should I do when I encounter a question in the text?

A question invites you to stop and think, to make a prediction, to carry out a step in a derivation or analysis, or to apply a principle. These questions are answered in the following paragraphs, but it is important that you make a serious effort to answer the questions on your own before reading further. Be honest in comparing your answers to those in the text. Paying attention to surprising or counterintuitive results can be a useful learning strategy.

Checkpoints

Checkpoints at the end of some sections ask you to apply new concepts or techniques. These may involve qualitative reasoning or simple calculations. You should complete these checkpoints when you come to them, before reading further. The goal of a checkpoint is to help you consolidate your understanding of the material you have just read, and to make sure you are ready to continue reading. Answers to checkpoints are found at the end of each chapter.

Conventions Used in Diagrams

→	Force
⇒	Component of force
→	Velocity
→	Momentum
⇒	Electric field
⇒	Component of electric field
→	Magnetic field
⇒	Component of magnetic field
⊳	Position
⇒	Angular momentum
→	Torque
⊢------⊣	Distance
⌇	A path

The conventions most commonly used to represent vectors and scalars in diagrams in this text are shown in the margin. In equations and text, a vector will be written with an arrow above it: $\vec{p}$.

TO THE INSTRUCTOR

The approach to introductory physics in this textbook differs significantly from that in most textbooks. Key emphases of the approach include:

- Starting from fundamental principles rather than secondary formulas
- Atomic-level description and analysis
- Modeling the real world through idealizations and approximations
- Computational modeling of physical systems
- Unification of mechanics and thermal physics
- Unification of electrostatics and circuits
- The use of 3D vectors throughout

Web Resources for Instructors

Instructor resources are available at this web site:

www.wiley.com/college/chabay

Resources on this site include lecture-demo software, textbook figures, clicker questions, test questions, lab activities including experiments and computational modeling, a computational modeling guide, and a full solutions manual. Contact your Wiley representative for information about this site.

Electronic versions of the homework problems are available in WebAssign:

www.webassign.net

Some instructor resources are available through WebAssign as well.

Other information may be found on the authors' *Matter & Interactions* web site:

matterandinteractions.org

Also on the authors' website are reprints of published articles about *Matter & Interactions*, including these:

- Chabay, R. & Sherwood, B. (1999). Bringing atoms into first-year physics. *American Journal of Physics* 67, 1045–1050.
- Chabay, R. W. & Sherwood, B. (2004). Modern mechanics. *American Journal of Physics* 72, 439–445.
- Chabay, R. W. & Sherwood, B. (2006). Restructuring Introductory E&M. *American Journal of Physics* 74, 329–336.
- Chabay, R. & Sherwood, B. (2008) Computational physics in the introductory calculus-based course. *American Journal of Physics* 76(4&5), 307–313.
- Beichner, R., Chabay, R., & Sherwood, B. (2010) Labs for the Matter & Interactions curriculum. *American Journal of Physics* 78(5), 456–460.

Computational Homework Problems

Some important homework problems require the student to write a simple computer program. The textbook and associated instructional videos teach VPython, which is based on the Python programming language, and which generates real-time 3D animations as a side effect of simple physics code written by students. Such animations provide powerfully motivating and instructive visualizations of fields and motions. VPython supports true vector computations, which encourages students to begin thinking about vectors as much more than mere components. VPython can be obtained at no cost for Windows, Macintosh, and Linux at vpython.org.

In the instructor resources section of matterandinteractions.org is "A Brief Guide to Computational Modeling in Matter & Interactions" which explains how to incorporate computation into the curriculum in a way that is easy for instructors to manage and which is entirely accessible to students with no prior programming experience. There you will also find a growing list of advanced computational physics textbooks that use VPython, which means that introducing students to Python and VPython in the introductory physics course can be of direct utility in later courses. Python itself is now widely used in technical fields.

Desktop Experiment Kit for Volume 2

On the authors' web site mentioned above is information about a desktop experiment kit for E&M that is distributed by PASCO. The simple equipment in this kit allows students to make key observations of electrostatic, circuit, and magnetic phenomena, tightly integrated with the theory (www.pasco.com, search for EM-8675). Several chapters contain optional experiments that can be done with this kit. This does not preclude having other, more complex laboratory experiences associated with the curriculum. For example, one such lab that we use deals with Faraday's law and requires signal generators, large coils, and oscilloscopes. You may have lab experiments already in place that will go well with this textbook.

What's New in the 4th Edition

The 4th edition of this text includes the following major new features:

- Increased support for computational modeling throughout, including sample code.
- Discussion throughout the text contrasting iterative and analytical problem solutions.
- Many new computational modeling problems (small and large).
- Improved discussion throughout the text of the contrast between models of a system as a point particle and as an extended system.
- An improved discussion of the Momentum Principle throughout Volume 1, emphasizing that the future momentum depends on two elements: the momentum now, and the impulse applied.
- Improved treatment of polarization surface charge in electrostatics (Chapter 14) and circuits (Chapter 18) based on the results of detailed 3D computational models.
- A more extensive set of problems at the end of each chapter, with improved indication of difficulty level.

In order to reduce cost and weight, some materials that have seen little use by instructors have been moved to the Wiley web site (www.wiley.com/college/chabay) where they are freely available. These materials include Supplement S1 (Chapter 13 in the 3rd Edition: kinetic theory of gases, thermal processes, and heat engines), Supplement S2 on PN junctions (formerly an optional section in Chapter 22 in the 3rd Edition), and Supplement S3 (a significantly extended version of Chapter 25 in the 3rd Edition: electromagnetic interference and diffraction, wave-particle duality, and a new section on mechanical waves and the wave equation).

Additional changes in the 4th Edition include:

- In Chapter 5, improved treatment of curving motion and an added section on the dynamics of multiobject systems.
- An improved sequence of topics in Chapter 6, with an explicit discussion of the role of energy in computational models, and an improved treatment of path independence, highlighting its limitation to point particles.
- A new section in Chapter 7 on the effect of the choice of reference frame on the form of the Energy Principle, and explicit instruction on how to model several kinds of friction in a computational model.
- In Chapter 8, discussion of the lifetime of excited states and on the probabilistic nature of energy transitions.
- In Chapter 9, now renamed "Translational, Rotational, and Vibrational Energy," improved treatment of the energetics of deformable systems.
- In Chapter 11, analysis of a physical pendulum.
- A detailed discussion in Chapter 16 of how to calculate potential difference by numerical path integration.
- An improved treatment of motional emf in the case of a bar dragged along rails (Chapter 20).

Suggestions for Condensed Courses

In a large course for engineering and science students with three 50-minute lectures and one 110-minute small-group studio lab per week, or in a studio format with five 50-minute sessions per week, it is possible to complete most but not all of the mechanics and E&M material in two 15-week semesters. In an honors course, or a course for physics majors, it is possible to do almost everything. You may be able to go further or deeper if your course has a weekly recitation session in addition to lecture and lab.

What can be omitted if there is not enough time to do everything? In mechanics, the one thing we feel should not be omitted is the introduction to entropy in terms of the statistical mechanics of the Einstein solid (Chapter 12). This is a climax of the integration of mechanics and thermal physics. One approach to deciding what mechanics topics can be omitted is to be guided by what foundation is required for Chapter 12. See other detailed suggestions below.

In E&M, one should not omit electromagnetic radiation and its effects on matter (Chapter 23). This is the climax of the whole E&M enterprise. One way to decide what E&M topics can be omitted is to be guided by what foundation is required for Chapter 23. See other detailed suggestions below.

Any starred section (*) can safely be omitted. Material in these sections is not referenced in later work. In addition, the following sections may be omitted:

Chapter 3 (The Fundamental Interactions): The section on determinism may be omitted.

Chapter 4 (Contact Interactions): Buoyancy and pressure may be omitted (one can return to these topics by using Supplement S1 on gases).

Chapter 7 (Internal Energy): If you are pressed for time, you might choose to omit the second half of the chapter on energy dissipation, beginning with Section 7.10.

Chapter 9 (Translational, Rotational, and Vibrational Energy): The formalism of finding the center of mass may be skipped, because the important applications have obvious locations of the center of mass. Although they are very instructive, it is possible to omit the sections contrasting point-particle with extended system models; you may also omit the analysis of sliding friction.

Chapter 10 (Collisions): A good candidate for omission is the analysis of collisions in the center-of-mass frame. Since there is a basic introduction to collisions in Chapter 3 (before energy is introduced), one could omit all of Chapter 10. On the other hand, the combined use of the Momentum Principle and the Energy Principle can illuminate both fundamental principles.

Chapter 11 (Angular Momentum): The main content of this chapter should not be omitted, as it introduces the third fundamental principle of mechanics, the Angular Momentum Principle. One might choose to omit most applications involving nonzero torque.

Chapter 12 (Entropy: Limits on the Possible): The second half of this chapter, on the Boltzmann distribution, may be omitted if necessary.

Chapter 15 (Electric Field of Distributed Charges): It is important that students acquire a good working knowledge of the patterns of electric field around some standard charged objects (rod, ring, disk, capacitor, sphere). If however they themselves are to acquire significant expertise in setting up physical integrals, they need extensive practice, and you might decide that the amount of time necessary for acquiring this expertise is not an appropriate use of the available course time.

Chapter 16 (Electric Potential): The section on dielectric constant can be omitted if necessary.

Chapter 17 (Magnetic Field): In the sections on the atomic structure of magnets, you might choose to discuss only the first part, in which one finds that the magnetic moment of a bar magnet is consistent with an atomic model. Omitting the remaining sections on spin and domains will not cause significant difficulties later.

Chapter 19 (Circuit Elements): The sections on series and parallel resistors and on internal resistance, meters, quantitative analysis of RC circuits, and multiloop circuits can be omitted. Physics and engineering students who need to analyze complex multiloop circuits will later take specialized courses on the topic; in the introductory physics course the emphasis should be on giving all students a good grounding in the fundamental mechanisms underlying circuit behavior.

Chapter 20 (Magnetic Force): We recommend discussing Alice and Bob and Einstein, but it is safe to omit the sections on relativistic field transformations. However, students often express high interest in the relationship between electric fields and magnetic fields, and here is an opportunity to satisfy some aspects of their curiosity. Motors and generators may be omitted or downplayed. The case study on sparks in air can be omitted, because nothing later depends critically on this topic, though it provides an introductory-level example of a phenomenon where an intuitively appealing model fails utterly, while a different model predicts several key features of the phenomenon. Another possibility is to discuss sparks near the end of the course, because it can be a useful review of many aspects of E&M.

Chapter 22 (Faraday's Law): Though it can safely be omitted, we recommend retaining the section on superconductors, because students are curious about this topic. The section on inductance may be omitted.

Chapter 23 (Electromagnetic Radiation): The treatment of geometrical optics may be omitted.

Acknowledgments

We owe much to the unusual working environment provided by the Department of Physics and the former Center for Innovation in Learning at Carnegie Mellon, which made it possible during the 1990s to carry out the research and development leading to the first edition of this textbook in 2002. We are grateful for the open-minded attitude of our colleagues in the Carnegie Mellon physics department toward curriculum innovations.

We are grateful to the support of our colleagues Robert Beichner and John Risley in the Physics Education Research and Development group at North Carolina State University, and to other colleagues in the NCSU physics department.

We thank Fred Reif for emphasizing the role of the three fundamental principles of mechanics, and for his view on the reciprocity of electric and gravitational forces. We thank Robert Bauman, Gregg Franklin, and Curtis Meyer for helping us think deeply about energy.

Much of Chapter 12 on quantum statistical mechanics is based on an article by Thomas A. Moore and Daniel V. Schroeder, "A different approach to introducing statistical mechanics," *American Journal of Physics*, vol. 65, pp. 26–36 (January 1997). We have benefited from many stimulating conversations with Thomas Moore, author of another introductory textbook that takes a contemporary view of physics, *Six Ideas that Shaped Physics*. Michael Weissman and Robert Swendsen provided particularly helpful critiques on some aspects of our implementation of Chapter 12.

We thank Hermann Haertel for opening our eyes to the fundamental mechanisms of electric circuits. Robert Morse, Priscilla Laws, and Mel Steinberg stimulated our thinking about desktop experiments. Bat-Sheva Eylon offered important guidance at an early stage. Ray Sorensen provided deep analytical critiques that influenced our thinking in several important areas. Randall Feenstra taught us about semiconductor junctions. Thomas Moore showed us a useful way to present the differential form of Maxwell's equations. Fred Reif helped us devise an assessment of student learning of basic E&M concepts. Uri Ganiel suggested the high-voltage circuit used to demonstrate the reality of surface charge. The unusual light bulb circuits at the end of Chapter 22 are based on an article by P. C. Peters, "The role of induced emf's in simple circuits," *American Journal of Physics* 52, 1984, 208–211. Thomas Ferguson gave us unusually detailed and useful feedback on the E&M chapters. Discussions with John Jewett about energy transfers were helpful. We thank Seth Chabay for help with Latin.

We thank David Andersen, David Scherer, and Jonathan Brandmeyer for the development of tools that enabled us and our students to write associated software.

The research of Matthew Kohlmyer, Sean Weatherford, and Brandon Lunk on student engagement with computational modeling has made major contributions to our instruction on computational modeling. Lin Ding developed an energy assessment instrument congruent with the goals of this curriculum.

We thank our colleagues David Brown, Krishna Chowdary, Laura Clarke, John Denker, Norman Derby, Ernst-Ludwig Florin, Thomas Foster, Jon D.H. Gaffney, Chris Gould, Mark Haugan, Joe Heafner, Robert Hilborn, Eric Hill, Andrew Hirsch, Leonardo Hsu, Barry Luokkala, Sara Majetich, Jonathan Mitschele, Arjendu Pattanayak, Jeff Polak, Prabha Ramakrishnan, Vidhya Ramachandran, Richard Roth, Michael Schatz, Robert Swendsen, Aaron Titus, Michael Weissman, and Hugh Young.

We thank a group of reviewers assembled by the publisher, who gave us useful critiques on the second edition of this textbook: Kelvin Chu, Michael Dubson, Tom Furtak, David Goldberg, Javed Iqbal, Shawn Jackson, Craig Ogilvie, Michael Politano, Norris Preyer, Rex Ramsier, Tycho Sleator, Robert Swendsen, Larry Weinstein, and Michael Weissman. We also thank the group who offered useful critiques on the third edition: Alex Small, Bereket Behane, Craig Wiegert, Galen Pickett, Ian Affleck, Jeffrey Bindel, Jeremy King, Paula Heron, and Surenda Singh.

We have benefited greatly from the support and advice of Stuart Johnson and Jessica Fiorillo of John Wiley & Sons. Elizabeth Swain of John Wiley & Sons was exceptionally skilled in managing the project. Helen Walden did a superb job of copyediting; any remaining errors are ours.

This project was supported, in part, by the National Science Foundation (grants MDR-8953367, USE-9156105, DUE-9554843, DUE-9972420, DUE-0320608, DUE-0237132, and DUE-0618504). We are grateful to the National Science Foundation and its reviewers for their long-term support of this challenging project. Opinions expressed are those of the authors, and not necessarily those of the Foundation.

How the Figures Were Made

Almost all of the figures in this book were produced by us (for the third edition the Aptara studio created the figures that show human figures, and the studio added full color to our two-color versions from the second edition). Our main tool was Adobe Illustrator. The many 3D computer-generated images were made using VPython, with optional processing in POV-Ray using a module written by Ruth Chabay to generate a POV-Ray scene description file corresponding to a VPython scene, followed by editing in Adobe Photoshop before exporting to Illustrator. We used TeXstudio for editing LaTeX, with a package due in part to the work of Aptara. All the computer work was done on Windows computers.

Ruth Chabay and Bruce Sherwood
Santa Fe, New Mexico, July 2014

Biographical Background

Ruth Chabay earned a Ph.D in physical chemistry from the University of Illinois at Urbana-Champaign; her undergraduate degree was in chemistry from the University of Chicago. She is Professor Emerita in the Department of Physics at North Carolina State University and was Weston Visiting Professor, Department of Science Teaching, at the Weizmann Institute of Science in Rehovot, Israel. She has also taught at the University of Illinois at Urbana-Champaign and Carnegie Mellon University. She is a Fellow of the American Physical Society.

Bruce Sherwood's Ph.D is in experimental particle physics from the University of Chicago; his undergraduate degree was in engineering science from Purdue University, after which he studied physics for one year at the University of Padua, Italy. He is Professor Emeritus in the Department of Physics at North Carolina State University. He has also taught at Caltech, the University of Illinois at Urbana-Champaign, and Carnegie Mellon University. He is a Fellow of the American Physical Society and of the American Association for the Advancement of Science.

Chabay and Sherwood have been joint recipients of several educational awards. At Carnegie Mellon University they received the Ashkin Award for Teaching in the Mellon College of Science in 1999 and the Teaching Award of the National Society of Collegiate Scholars in 2001. At North Carolina State University they received the Margaret Cox Award for excellence in teaching and learning with technology in 2005. In 2014 the American Association of Physics Teachers presented them with the David Halliday and Robert Resnick Award for Excellence in Undergraduate Physics Teaching.

CHAPTER
13

Electric Field

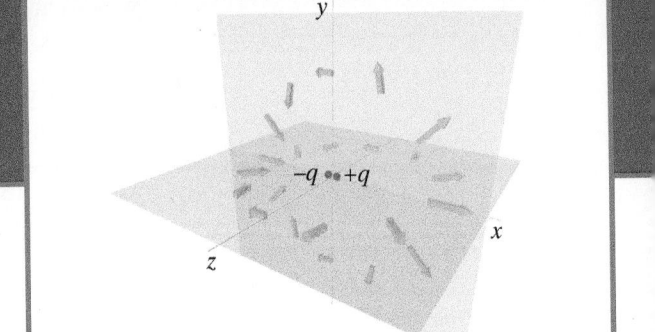

OBJECTIVES

After studying this chapter you should be able to

- Mathematically relate electric field and force.
- Calculate the 3D electric field at a particular location due to a collection of point charges.
- Explain the approximations made in deriving expressions for the electric field of a dipole, and use these approximate expressions appropriately.
- Graphically represent the magnitude and direction of the electric field of a dipole with arrows, at locations in a plane containing the dipole.
- Create a computational model to compute and display the electric field of a collection of point charges in 3D, and predict the motion of a charged particle that interacts with this field.

13.1 NEW CONCEPTS

Two important new ideas will form the core of our study of electric and magnetic interactions. The first is the concept of electric and magnetic fields. This concept is more abstract than the concept of force, which we used extensively in our study of modern mechanics. The reason we want to incorporate the idea of "field" into our models of the world is that this concept turns out to be a very powerful one, which allows us to explain and predict important phenomena that would otherwise be inaccessible to us.

The second important idea is a more sophisticated and complex model of matter. In our previous study of mechanics and thermal physics it was usually adequate to model a solid as an array of electrically neutral microscopic masses (atoms) connected by springs (chemical bonds). As we consider electric and magnetic interactions in more depth, we will find that we need to consider the individual charged particles—electrons and nuclei—that make up ordinary matter.

The material in this chapter lays the foundation for all succeeding chapters, so it is worth taking time to understand it thoroughly. In addition, if you did not use volume 1 of this textbook in your previous study of physics, you should work through the summary from Chapter 1 on 3D vectors, vector notation, and computational modeling available at no charge on the student website, www.wiley.com/college/chabay.

13.2 ELECTRIC CHARGE AND FORCE

In this section we briefly review concepts familiar to you from your previous studies.

Point Charges

There are two kinds of electric charge, which are called positive and negative. Particles with like charges repel each other (two positive or two negative particles); particles with unlike charges (positive and negative) attract each other. By "point particle" we mean an object whose radius is very small compared to the distance between it and all other objects of interest, so we can treat the object as if all its charge and mass were concentrated at a single mathematical point. Small particles such as protons and electrons can almost always be considered to be point particles.

The Coulomb Force Law for Point Particles

The electromagnetic interaction is one of the four fundamental physical interactions (see Chapter 3, The Fundamental Interactions). The electric force law, called Coulomb's law, describes the magnitude of the electric force between two point-like electrically charged particles:

$$|\vec{F}| = F = \frac{1}{4\pi\varepsilon_0}\frac{|Q_1 Q_2|}{r^2}$$

where Q_1 and Q_2 are the magnitudes of the electric charge of objects 1 and 2, and r is the distance between the objects. As indicated in Figure 13.1:

- The electric force acts along a line between two point-like objects.
- Like charges repel; unlike charges attract.
- Two charged objects interact even if they are some distance apart.

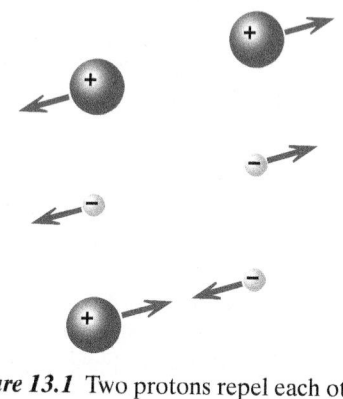

Figure 13.1 Two protons repel each other; two electrons repel each other; a proton and an electron attract each other.

Units and Constants

The SI unit of electric charge is the coulomb, abbreviated C. The charge of one proton is 1.6×10^{-19} C in SI units. The constant e is often used to represent this amount of positive charge: $e = +1.6 \times 10^{-19}$ C. An electron has a charge of $-e$.

The constant $1/4\pi\varepsilon_0$ has the value 9×10^9 N·m²/C². We write this constant in this way, instead of using the letter k, for two reasons. Since the letter k is often used for other quantities, we avoid confusion by not using it here. Also, the constant $\varepsilon_0 = 8.85 \times 10^{-12}$ C²/N·m² will appear by itself in important situations.

Charged Particles

There are many microscopic particles, some of which are electrically charged and hence interact with each other through the electric interaction. The characteristics of some charged particles are shown in the table below.

Particle	Mass	Charge	Radius
Electron	9×10^{-31} kg	$-e$ (-1.6×10^{-19} C)	? (too small to measure)
Positron	9×10^{-31} kg	$+e$ ($+1.6 \times 10^{-19}$ C)	?
Proton	1.7×10^{-27} kg	$+e$	$\sim 1 \times 10^{-15}$ m
Antiproton	1.7×10^{-27} kg	$-e$	$\sim 1 \times 10^{-15}$ m
Muon	1.88×10^{-28} kg	$+e$ (μ^+) or $-e$ (μ^-)	?
Pion	2.48×10^{-28} kg	$+e$ (π^+) or $-e$ (π^-)	$\sim 1 \times 10^{-15}$ m

This is not an exhaustive list; you may also have learned about other charged particles, such as the W and Δ particles, and other particles that are short-lived and not commonly encountered in everyday circumstances. Positrons and antiprotons are *antimatter*. A positron is an antielectron; if a positron and an electron encounter each other they will annihilate, releasing all their energy as high-energy photons. Antimatter is therefore not found in ordinary matter.

Ordinary matter is composed of protons, electrons, and neutrons (which have about the same size and mass as protons but are uncharged). However, some other charged particles do play a role in everyday processes, as we will see, for example, in our study of sparks in a later chapter.

Size and Structure of Atoms

As you have learned in chemistry or from other sources, matter is made of tiny atoms. In a solid metal, atoms are arranged in a regular three-dimensional array, called a lattice (Figure 13.2). A cubic centimeter of solid metal in which atoms are packed right next to each other contains around 1×10^{23} atoms, which is an astronomically large number.

A neutral atom has equal numbers of protons and electrons. The protons and neutrons are all found in the nucleus at the center of the atom. The electrons are spread out in a cloud surrounding the nucleus.

Each atom consists of a cloud of electrons continually in motion around a central "nucleus" made of protons and neutrons. If we imagine taking a snapshot of an iron atom, with a nucleus of 26 protons and 30 neutrons surrounded by 26 moving electrons, it might look something like Figure 13.3. In this figure the nucleus is hardly visible, because it is much smaller than the electron cloud, whose radius is on the order of 1×10^{-10} m.

The nucleus of the iron atom, depicted in Figure 13.4, has a radius of roughly 4×10^{-15} m, about 25,000 times smaller than the tiny electron cloud. If an iron atom were the size of a football field, the nucleus would have a radius of only 4 mm! Yet almost all of the mass of an atom is in the nucleus, because the mass of a proton or neutron is about 2000 times larger than the mass of an electron.

> **Checkpoint 1** What is the approximate radius of the electron cloud of a typical atom? Which of the following charged particles are constituents of ordinary matter? Protons, positrons, electrons, antiprotons, muons

13.3 THE CONCEPT OF "ELECTRIC FIELD"

Consider the following thought experiment. Having evacuated the air from the room (to avoid collisions with air molecules), you hold a proton in front of you and release it. There are no other objects nearby. You observe that the proton begins to move downward, picking up speed (accelerating) at the rate of 9.8 m/s each second (Figure 13.5). Recall that at speeds much less than the speed of light,

$$\frac{d\vec{p}}{dt} \approx m\frac{d\vec{v}}{dt} = m\vec{a}$$

QUESTION What do you think is responsible for this change in the velocity of the proton?

You probably inferred that the gravitational interaction of the Earth and the proton caused the downward acceleration of the proton. That is a reasonable explanation for this observation.

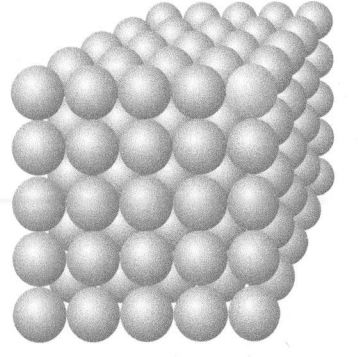

Figure 13.2 A metal lattice. The radius of a single atom is on the order of 1×10^{-10} m, and in a cube that is 1 cm on a side, there are on the order of 1×10^{23} atoms!

Figure 13.3 The electron cloud of an iron atom. The radius of the electron cloud is approximately 1×10^{-10} m. On this scale the tiny nucleus, located at the center of the cloud, would not actually be visible.

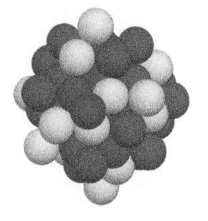

Figure 13.4 The nucleus of a iron atom contains 26 protons and 30 neutrons. Its radius is approximately 4×10^{-15} m.

Figure 13.5 Acceleration of a proton.

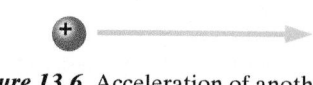

Figure 13.6 Acceleration of another proton at a later time.

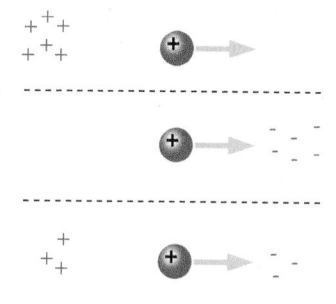

Figure 13.7 Three possible arrangements of charged particles that might be responsible for the observed high acceleration of a proton.

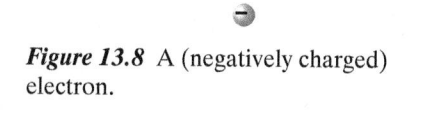

Figure 13.8 A (negatively charged) electron.

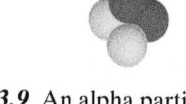

Figure 13.9 An alpha particle with two (positively charged) protons and two (uncharged) neutrons.

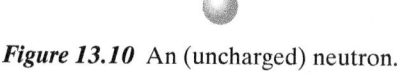

Figure 13.10 An (uncharged) neutron.

Now suppose that at a later time, you release another proton in the same location. This time you observe that the proton begins to move to the right, picking up speed at a rate of 1×10^{11} m/s^2 (Figure 13.6).

QUESTION What might be responsible for this change in the velocity of the proton?

This acceleration cannot be due to a gravitational interaction of the proton with the Earth, since the magnitude of the effect is too large, and the direction is not appropriate. Could it be due to a gravitational interaction with a nearby black hole? No, because if there were a black hole very nearby we would not be here to observe anything! Could it be an interaction via the strong (nuclear) force? No, because the strong force is a very short range force, and there are no other objects near enough.

It is, however, plausible that the interaction causing the acceleration of the proton could be an electric interaction, since electric interactions can have large effects and can occur over rather large distances.

QUESTION What charged objects might be responsible for this interaction, and where might they be?

There are many possible configurations of charged particles that might produce the observed effect. As indicated in Figure 13.7, there might be positively charged objects to your left. Alternatively, there could be negative charges to your right. Perhaps there are both—you can't draw a definite conclusion from a single observation. There are many possible arrangements of charges in space that could produce the observed effect.

(If, however, you made several observations of the proton over some time period, you might note a change in the proton's acceleration. For example, if you noticed that the acceleration of the proton increased as it moved to the right, you might suspect that there was a negative charge to your right, which would have an increasingly large effect on the proton as it moved closer to the charge.)

QUESTION On the basis of your observations so far, can you predict what you would observe if you released an electron instead of a proton at the same location (Figure 13.8)?

An electron would accelerate to the left rather than to the right, since it has a negative charge. Its acceleration would be greater than 1×10^{11} m/s^2, because the mass of the electron is much less than the mass of the proton.

QUESTION What would happen if you placed an alpha particle (a helium nucleus, with charge $+2e$) at the observation location (Figure 13.9)?

Since the alpha particle has a positive charge, it would accelerate to the right as the proton did. However, because the charge of the alpha particle is twice the charge of the proton, and its mass is about four times that of a proton, the magnitude of its acceleration would be one-half of 1×10^{11} m/s^2.

QUESTION If you released a neutron at the observation location, what would you expect to observe (Figure 13.10)?

Since the neutron has no electric charge, it would experience no electric force, and should simply accelerate toward the Earth at 9.8 m/s^2.

QUESTION Finally, suppose we do not put any particle at the observation location. Is there anything there?

Since we know that if we were to put a charged particle at that location, it would experience a force, it seems that in a certain sense there is something there, waiting for a charged particle to interact with. This "virtual force" is called "electric field." The electric field created by a charge is present throughout space at all times, whether or not there is another charge around to feel its effects. The electric field created by a charge penetrates through matter. The field permeates the neighboring space, lying in wait to affect anything brought into its web of interaction.

We define the electric field at a location by the following equation:

DEFINITION OF ELECTRIC FIELD

$$\vec{F}_2 = q_2 \vec{E}_1$$

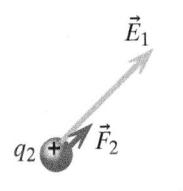

This equation says that the force on particle 2 is determined by the charge of particle 2 and by the electric field $\vec{E}_1$ made by all other charged particles in the vicinity, as shown in Figure 13.11.

How might we measure the electric field at a given location in space? We can't see it, but we can measure it indirectly. If we place a charge at that location in space, we can measure the force on the charge due to its interaction with the electric field at that location. We can determine the magnitude and direction of $\vec{E}$ by measuring a force on a known charge q:

Figure 13.11 A particle with charge q_2 experiences a force $\vec{F}_2$ because of its interaction with the electric field $\vec{E}_1$ produced by all other charged particles in the vicinity.

$$\vec{E} = \vec{F}/q$$

Electric field therefore has units of newtons per coulomb. Note that it doesn't matter whether the affected charge q is positive or negative. If $\vec{E}$ points north, a positive charge would experience a force to the north, while a negative charge would experience a force to the south (Figure 13.12); dividing this force by a negative q would still yield a vector $\vec{E}$ pointing north. Though we will usually write $\vec{E}$ for the value of the electric field at a particular location and time, this is really shorthand for $\vec{E}(x,y,z,t)$, since the electric field has a value at all locations and times.

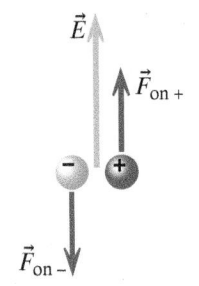

Figure 13.12 A positive charge experiences a force in the direction of the electric field at its location, while a negative charge experiences a force in the opposite direction.

Drawing Fields: Tail of Arrow at Observation Location

In diagrams we will use arrows to represent the electric field (a vector) at a specific observation location. By convention, we will almost always draw such an arrow with its tail at the location where the field is measured (the "observation location"). The length of the arrow is proportional to the magnitude of the field.

EXAMPLE

Electric Force and Electric Field

The charge of an alpha particle (a helium nucleus, consisting of two protons and two neutrons) is $2e = 2(1.6 \times 10^{-19}\,\text{C})$. An alpha particle at a particular location experiences a force of $\langle 0, -9.6 \times 10^{-17}, 0 \rangle\,\text{N}$. What is the electric field at that location? If the alpha particle were removed and an electron were placed at that location, what force would the electron experience?

Solution

To find the field we divide the force on the alpha particle by the charge of the alpha particle:

$$\vec{E} = \frac{\vec{F}}{q}$$

$$\vec{E} = \frac{\langle 0, -9.6 \times 10^{-17}, 0 \rangle\,\text{N}}{2(1.6 \times 10^{-19}\,\text{C})} = \langle 0, -300, 0 \rangle\,\text{N/C}$$

To find the force on the electron, we multiply the electric field by the charge on the electron:

$$\vec{F} = (-e)\vec{E}_1$$
$$\vec{F} = (-1.6 \times 10^{-19}\,\text{C})\langle 0, -300, 0 \rangle\;\text{N/C} = \langle 0, 4.8 \times 10^{-17}, 0 \rangle\;\text{N}$$

Since the electron's charge is negative, the force on the electron is in a direction opposite to the direction of the electric field. Note that we did not need to know the source of the electric field in order to calculate this force.

Checkpoint 2 What force would a proton experience if placed at the same location (see the preceding example)? What force would a neutron at the same location experience?

No "Self-Force"

It is important to note that a point charge is not affected by its *own* electric field. A point charge does not exert a force on itself! In the next section we will see that mathematically this is reassuring, because at the location of a point charge its own electric field would be infinite ($1/0^2$). Physically, this makes sense too, since after all, the charge can't start itself moving, nor is there any way to decide in what direction it should go.

When we use the electric field concept with point charges we always talk about a charge q_1 (the *source* charge) making a field $\vec{E}_1$, and a different charge q_2 in a different place being affected by that field with a force $\vec{F}_{\text{on }q_2} = q_2\vec{E}_1$.

The Physical Concept of "Field"

The word "field" has a special meaning in mathematical physics. A field is a physical quantity that has a value at every location in space. Its value at every location can be a scalar or a vector.

For example, the temperature in a room is a scalar field. At every location in the room, the temperature has a value, which we could write as $T(x,y,z)$, or as $T(x,y,z,t)$ if it were changing with time. The air flow in the room is a vector field. At every location in the room, air flows in a particular direction with a particular speed. Electric field is a vector field; at every location in space surrounding a charge the electric field has a magnitude and a direction.

QUESTION Think of another example of a quantity that is a field.

The field concept is also used with gravitation. Instead of saying that the Earth exerts a force on a falling object, we can say that the mass of the Earth creates a "gravitational field" surrounding the Earth, and any object near the Earth is acted upon by the gravitational field at that location (Figure 13.13). Gravitational field has units of newtons per kilogram. At a location near the Earth's surface we can say that there is a gravitational field $\vec{g}$ pointing downward (that is, toward the center of the Earth) of magnitude $g = 9.8$ newtons per kilogram.

A falling object of mass $m = 10\,\text{kg}$ experiences a gravitational force of magnitude $mg = (10\,\text{kg})(9.8\,\text{N/kg}) = 98$ newtons. If the object is moving at a speed slow compared to the speed of light, we can calculate its acceleration, $a = F/m = mg/m = 9.8\,\text{m/s}^2$.

Of course, farther away from the surface of the Earth, the gravitational field of the Earth is smaller in magnitude, just as the electric field of a charge decreases in magnitude as the distance from the charge increases.

Figure 13.13 Gravitational field of the Earth, in a plane cutting through the center of the Earth.

The field concept also applies to magnetism. As we will see in a later chapter, electric currents, including electric currents inside the Earth, create "magnetic fields" that affect compass needles.

QUESTION Find an example of a quantity that is *not* a field.

In contrast, many quantities are not fields. The position and velocity at one instant of a car going around a race track are not fields. The distance between the Earth and the Sun is not a field. The rate of change of the momentum of the Moon is not a field.

13.4 THE ELECTRIC FIELD OF A POINT CHARGE

By asking what expression for $\vec{E}_1$ satisfies the equation $\vec{F}_2 = q_2 \vec{E}_1$ for the force on one point charge by another, we can find an algebraic expression for the electric field at a location in space called the "observation location"—the location where we detect or measure the field—due to a charged particle q_1 (the "source charge") at the source location. The electric field at the observation location (marked by an "×" in Figure 13.14) is:

ELECTRIC FIELD OF A POINT CHARGE

$$\vec{E}_1 = \frac{1}{4\pi\varepsilon_0} \frac{q_1}{|\vec{r}|^2} \hat{r}$$

where q_1 is the source charge, $\vec{r}$ is the relative position vector giving the position of the observation location relative to the source charge, and $\hat{r}$ is the unit vector in the direction of $\vec{r}$. $|\vec{r}|$, the magnitude of $\vec{r}$, is the distance from the source location to the observation location.

This equation for electric field of a point charge may look complex, but we can make sense of it by taking it apart and looking at one piece at a time.

Direction of Electric Field

The direction of the electric field at the observation location depends on both the direction of $\hat{r}$ and the sign of the source charge q_1, which may be positive or negative (Figure 13.15).

If the source charge q_1 is positive, the quantity $q_1\hat{r}$ is in the direction of $\hat{r}$, pointing away from the source charge (Figure 13.14).

If the source charge q_1 is negative, the direction of $q_1\hat{r}$ is in the direction of $-\hat{r}$, pointing toward the source charge (Figure 13.16).

Distance Dependence of Electric Field

As in the gravitational force law and the electric force law, the square of the distance appears in the denominator (Figure 13.17). This means that the electric field of a point charge depends very strongly on distance. For example, if you double the distance between the source location and the observation location, the only thing that changes is the denominator, which gets four times bigger, so the electric field is only 1/4 as large as before.

QUESTION If you move the source charge five times farther away from the observation location, how does the electric field change?

The magnitude of the electric field decreases by a factor of 25.

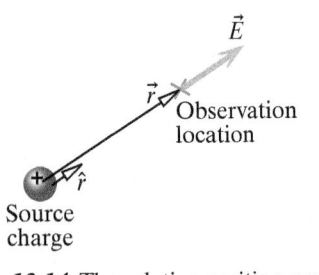

Figure 13.14 The relative position vector $\vec{r}$ extends from the source charge to the observation location, marked by "×". $\hat{r}$ is a unit vector with the same direction as the vector $\vec{r}$. Here the source charge is positive, so the electric field points away from the source charge.

$$\vec{E}_1 = \frac{1}{4\pi\varepsilon_0} \frac{q_1}{|\vec{r}|^2} \hat{r}$$

Figure 13.15 The direction of the electric field depends on both the unit vector $\hat{r}$ and the sign of the source charge.

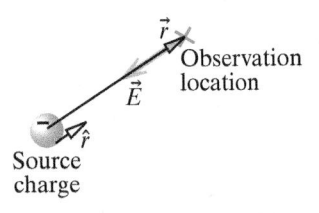

Figure 13.16 Here the source charge is negative, so the electric field points toward the source charge, in the direction of $-\hat{r}$.

$$\vec{E}_1 = \frac{1}{4\pi\varepsilon_0} \frac{q_1}{|\vec{r}|^2} \hat{r}$$

Figure 13.17 The magnitude of the electric field is inversely proportional to the square of the distance from source charge to observation location.

$$\vec{E}_1 = \frac{1}{4\pi\varepsilon_0} \frac{q_1}{|\vec{r}|^2}\hat{r}$$

Figure 13.18 The magnitude of the electric field is proportional to the magnitude of the source charge.

Magnitude of Charge

Because q_1, the charge of the source particle, appears in the numerator (Figure 13.18), the larger the charge of the source particle, the larger the magnitude of the electric field at any location in space.

> QUESTION   At a particular observation location, the magnitude of the electric field due to a point charge is found to be 10 N/C. If the source charge were replaced by a different particle whose charge was seven times larger, how would this change the electric field at the observation location?

The electric field would now have a magnitude of 70 N/C.

The Constant $\dfrac{1}{4\pi\varepsilon_0} = 9 \times 10^9 \ \text{N} \cdot \text{m}^2/\text{C}^2$

This may seem an odd way to write a single constant; this is done because it is based on the constant $\varepsilon_0 = 8.85 \times 10^{-12}\,\text{C}^2/\text{N}\cdot\text{m}^2$, which is in a sense a more fundamental constant, as we will see in a later chapter.

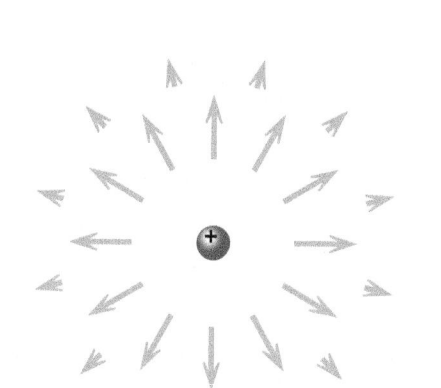

Figure 13.19 Electric field of a positive point charge, at various observation locations in a plane containing the charge.

Magnitude of Electric Field

The scalar quantity

$$\frac{1}{4\pi\varepsilon_0} \frac{q_1}{|\vec{r}|^2}$$

is *not* necessarily equal to the magnitude of the electric field. Because the charge q_1 can be negative or positive, the expression above may be negative or positive. The magnitude of a vector is a positive quantity. To get the magnitude of a field it is necessary to take the absolute value of the signed scalar quantity above. (Remember that the sign of q_1 helps determine the direction of the field.)

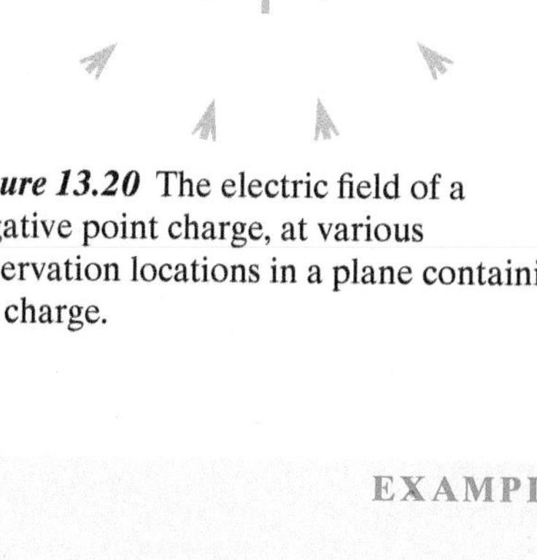

Patterns of Electric Field Near Point Charges

The electric field of a point charge is spherically symmetric. That is, the magnitude of the field at any location a distance r from the source charge is the same, and at each location the field points either toward the source charge or away from it, depending on the sign of the source charge. Figure 13.19 shows the electric field due to a positive source charge, at various locations in a plane containing the charge. The tail of each arrow is placed at the location where the electric field was measured. Notice that at every observation location the electric field points radially away from the positive source charge, and that the magnitude of the field decreases rapidly with distance from the charge.

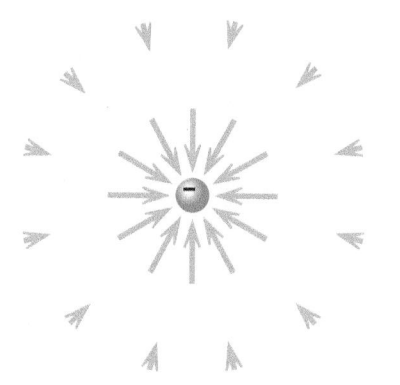

Figure 13.20 The electric field of a negative point charge, at various observation locations in a plane containing the charge.

Figure 13.20 shows the electric field due to a negative source charge. At each observation location, the electric field points radially inward toward the negative source charge (in the direction of $-\hat{r}$), so a positively charged particle would experience a force toward the negative charge. The magnitude of the field decreases rapidly with distance from the charge. We will frequently use two-dimensional diagrams because they are easy to draw, but it is important to remember the 3D character of electric fields in space.

EXAMPLE   **Field of a Particle**

A particle with charge $+2\,\text{nC}$ (a nanocoulomb is $1 \times 10^{-9}\,\text{C}$) is located at the origin. What is the electric field due to this particle at a location $\langle -0.2, -0.2, -0.2 \rangle$ m?

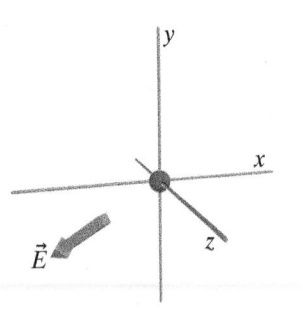

Figure 13.21 The electric field calculated in the example. This orientation of coordinate axes ($+x$ to the right, $+y$ up, and $+z$ out of the page, toward you) is the standard orientation we will use in this book.

$$\vec{r} = \langle \text{observation location} \rangle - \langle \text{source location} \rangle$$
$$= \langle -0.2, -0.2, -0.2 \rangle \text{ m} - \langle 0,0,0 \rangle \text{m}$$
$$\vec{r} = \langle -0.2, -0.2, -0.2 \rangle \text{m}$$
$$|\vec{r}| = \sqrt{(-0.2)^2 + (-0.2)^2 + (-0.2)^2} \text{m} = 0.35 \text{ m}$$
$$\hat{r} = \frac{\vec{r}}{|\vec{r}|} = \frac{\langle -0.2, -0.2, -0.2 \rangle \text{ m}}{0.35 \text{ m}} = \langle -0.57, -0.57, -0.57 \rangle$$
$$\frac{1}{4\pi\varepsilon_0} \frac{q}{r^2} = \left(9 \times 10^9 \frac{\text{N m}^2}{\text{C}^2} \right) \left(\frac{2 \times 10^{-9} \text{C}}{0.35^2 \text{ m}^2} \right) = 147 \frac{\text{N}}{\text{C}}$$

$$\vec{E} = \frac{1}{4\pi\varepsilon_0} \frac{q}{r^2} \hat{r}$$
$$= \left(147 \frac{\text{N}}{\text{C}} \right) \langle -0.57, -0.57, -0.57 \rangle$$
$$= \langle -84, -84, -84 \rangle \frac{\text{N}}{\text{C}}$$

The electric field points away from the positively charged particle, as shown in Figure 13.21.

EXAMPLE   **Source Location**

The electric field at location $\langle -0.13, 0.14, 0 \rangle$ m is found to be $\langle 6.48 \times 10^3, -8.64 \times 10^3, 0 \rangle$ N/C. The only charged particle in the surroundings has charge $-3\,\text{nC}$. What is the location of this particle?

Solution   Draw a diagram (Figure 13.22) to find the relative position of the source and the observation location. Remember that the electric field of a negatively charged particle points toward the particle. Find the magnitude of the electric field, and use this to determine the distance from the source to the observation location:

$$|\vec{E}| = \sqrt{(6.48 \times 10^3)^2 + (-8.64 \times 10^3)^2} \,\text{N/C} = 1.08 \times 10^4 \,\text{N/C}$$
$$|\vec{E}| = \frac{1}{4\pi\varepsilon_0} \frac{q}{|\vec{r}|^2}$$
$$|\vec{r}| = \sqrt{\frac{1}{4\pi\varepsilon_0} \frac{q}{|\vec{E}|}}$$
$$= \sqrt{9 \times 10^9 \,\text{Nm}^2/\text{C}^2 \frac{3 \times 10^{-9} \,\text{C}}{1.08 \times 10^4 \,\text{N/C}}}$$
$$= 0.05\,\text{m}$$

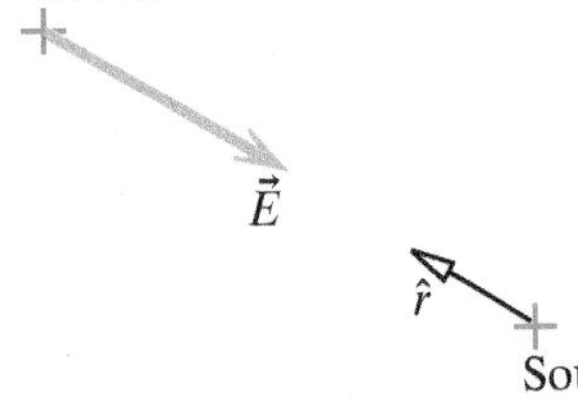

Figure 13.22 Draw this diagram, using the fact that the electric field due to a negative charge points toward the source charge.

Find the unit vector $\hat{r}$. From the diagram, note that in this case (because the source charge is negative) $\hat{r} = -\hat{E}$:

$$\hat{E} = \frac{\vec{E}}{|\vec{E}|}$$
$$= \frac{\langle 6.48 \times 10^3, -8.64 \times 10^3, 0 \rangle \,\text{N/C}}{1.08 \times 10^4 \,\text{N/C}}$$
$$= \langle 0.6, -0.8, 0 \rangle$$
$$\hat{r} = -\hat{E} = \langle -0.6, 0.8, 0 \rangle$$

Find $\vec{r}$, and use this to find the source location, where location 2 is the observation location and location 1 is the source location:

$$\vec{r} = |\vec{r}|\hat{r}$$
$$= (0.05\,\text{m})\langle-0.6,0.8,0\rangle = \langle-0.03,0.04,0\rangle\ \text{m}$$
$$\vec{r} = \vec{r}_2 - \vec{r}_1$$
$$\vec{r}_1 = \vec{r}_2 - \vec{r} = \langle-0.13,0.14,0\rangle - \langle-0.03,0.04,0\rangle$$
$$= \langle-0.10,0.10,0\rangle\ \text{m}$$

Checkpoint 3 A particle with charge $+1\,\text{nC}$ (a nanocoulomb is $1 \times 10^{-9}\,\text{C}$) is located at the origin. What is the electric field due to this particle at a location $\langle0.1,0,0\rangle$ m?

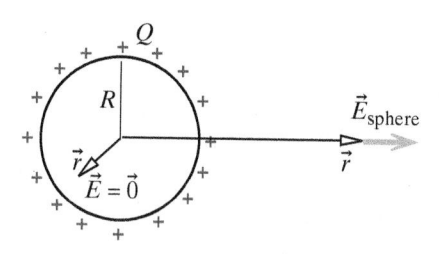

Figure 13.23 The electric field of a uniformly charged sphere, at locations outside and inside the sphere.

The Electric Field of a Uniformly Charged Sphere

In Chapter 15 we will apply the superposition principle to calculate the electric field of macroscopic objects with charge spread out over their surfaces. One of the results of such a calculation is sufficiently useful that we present it now, without going through the calculations involved.

As shown in Figure 13.23, a spherical object of radius R with charge Q uniformly spread out over its surface produces an electric field with the following distance dependence, if $\vec{r}$ is a vector from the center of the sphere to the observation location:

$$\vec{E}_{\text{sphere}} = \frac{1}{4\pi\varepsilon_0}\frac{Q}{r^2}\hat{r} \quad \text{for } r > R \text{ (outside the sphere)}$$
$$\vec{E}_{\text{sphere}} = \vec{0} \quad \text{for } r < R \text{ (inside the sphere)}$$

In other words, at locations outside the charged sphere, the electric field due to the sphere is the same as if all its charge were located at its center. Inside the sphere, the electric field due to all the charge on the surface of the sphere adds up to zero!

Informally, we may say that a uniformly charged sphere "acts like" a point charge, at locations outside the sphere. This is not only because the electric field due to the sphere is the same as the electric field of a point charge located at the center of the sphere, but also because the charged sphere responds to applied electric fields (due to other charges) in the same way as would a point charge located at its center.

Checkpoint 4 A sphere with radius 1 cm has a charge of $5 \times 10^{-9}\,\text{C}$ spread out uniformly over its surface. What is the magnitude of the electric field due to the sphere at a location 4 cm from the center of the sphere?

13.5 SUPERPOSITION OF ELECTRIC FIELDS

An important property of electric field is a consequence of the superposition principle for electric forces: the net electric field due to two or more charges is the vector sum of each field due to each individual charge. Superposition as a vector sum holds true for forces in general, so it also holds true for fields, because field is force per unit charge.

We state the superposition principle here in terms of electric field, in a form intended to emphasize an important aspect of the principle:

THE SUPERPOSITION PRINCIPLE

The net electric field at a location in space is the vector sum of the individual electric fields contributed by all charged particles located elsewhere.

The electric field contributed by a charged particle is unaffected by the presence of other charged particles.

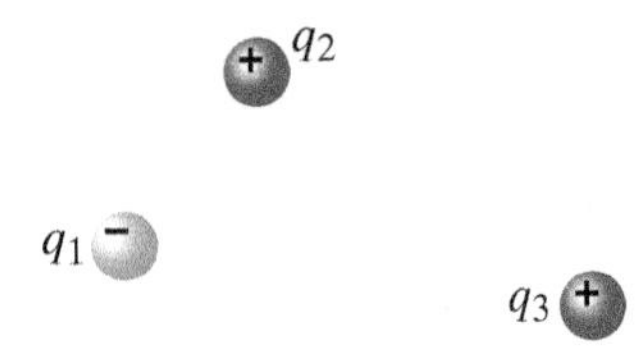

This may seem obvious, but it actually is a bit subtle. The presence of other particles does not affect the fundamental electric interaction between each pair of particles! The interaction doesn't get "used up"; the interaction between one particle and another is unaffected by the presence of a third particle.

A significant consequence of the superposition principle is that matter cannot "block" electric fields; electric fields "penetrate" through matter.

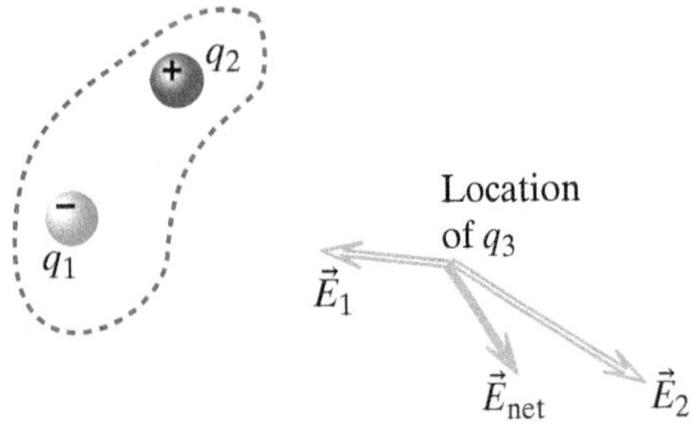

Figure 13.24 Three interacting charged particles.

Applying the Superposition Principle

Consider three charges q_1, q_2, and q_3 that interact with each other, as shown in Figure 13.24. We could calculate the net force that q_1 and q_2 exert on q_3 by using Coulomb's law and vector addition.

However, let's use the field concept to find the force on q_3 due to the other two charges. We pretend that q_3 isn't there, and we find the electric field $\vec{E}_{net}$ due to q_1 and q_2, at the location where q_3 would be. We can then calculate the force on q_3 as $q_3\vec{E}_{net}$. In Figure 13.25 the scheme is shown graphically. $\vec{E}_{net}$ is the vector sum of $\vec{E}_1$ (due to q_1) and $\vec{E}_2$ (due to q_2), at the location of q_3.

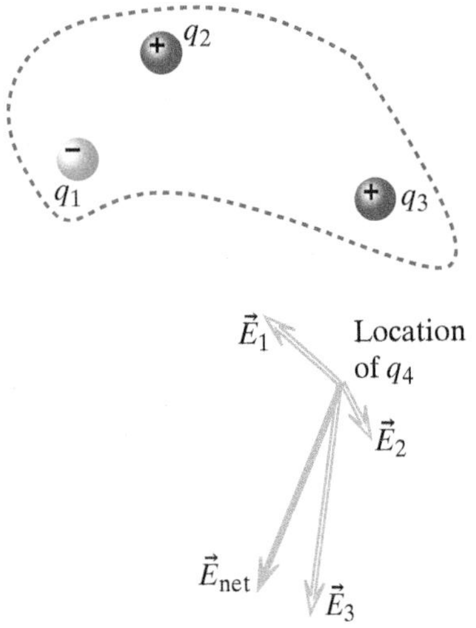

Figure 13.25 Electric field at the location of q_3.

Advantage of Electric Field in Calculations

We could just as easily have calculated the force on q_3 by using Coulomb's law directly, without using the more abstract concept of electric field. What advantage is there to using electric field? One of the advantages is that once you have calculated the field $\vec{E}_{net}$ due to q_1 and q_2, you can quickly calculate the force $Q\vec{E}_{net}$ on any amount of charge Q placed at that location, not just the force on q_3. The "force per unit charge" idea simplifies these kinds of calculations.

If we are interested in the effect that all three charges would have on a fourth charge q_4 placed at some location, we would include all three charges as the source of the electric field at that location, but exclude q_4, as shown in Figure 13.26.

Knowing the electric field $\vec{E}_{net}$ at a particular location, we know what force would act on a charge q_4 placed at that location: it would simply be $q_4\vec{E}_{net}$.

Figure 13.26 Electric field at the location of q_4.

EXAMPLE   **Electric Field and Force Due to Two Charges**

A small object with charge $Q_1 = 6\,\mathrm{nC}$ is located at the origin. A second small object with charge $Q_2 = -5\,\mathrm{nC}$ is located at $\langle 0.05, 0.08, 0\rangle$ m. What is the net electric field at location A $\langle -0.04, 0.08, 0\rangle$ m due to Q_1 and Q_2? If a small object with a charge of $Q_3 = -3\,\mathrm{nC}$ were placed at location A, what would be the force on this object?

Solution   Diagram: Figure 13.27

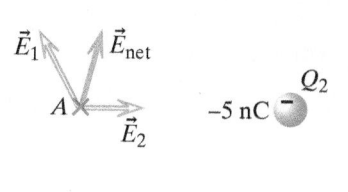

Q_1

6 nC

Figure 13.27 Two point charges make an electric field at location A.

$$\vec{r}_1 = \langle -0.04, 0.08, 0 \rangle \text{ m} - \langle 0,0,0 \rangle \text{ m} = \langle -0.04, 0.08, 0 \rangle \text{ m}$$

$$|\vec{r}_1| = \sqrt{(-0.04)^2 + (0.08)^2} = 0.0894 \text{ m}$$

$$\hat{r}_1 = \frac{\vec{r}_1}{|\vec{r}_1|} = \frac{\langle -0.04, 0.08, 0 \rangle \text{ m}}{(0.0894 \text{ m})} = \langle -0.447, 0.894, 0 \rangle$$

$$\vec{E}_1 = \frac{1}{4\pi\varepsilon_0} \frac{Q_1}{|\vec{r}_1|^2} \hat{r}_1$$

$$= \left(9 \times 10^9 \frac{\text{Nm}^2}{\text{C}^2} \right) \frac{(6 \times 10^{-9} \text{C})}{(0.0894 \text{ m})^2} \langle -0.447, 0.894, 0 \rangle$$

$$\vec{E}_1 = \langle -3.02 \times 10^3, 6.04 \times 10^3, 0 \rangle \text{ N/C}$$

$$\vec{r}_2 = \langle -0.04, 0.08, 0 \rangle \text{ m} - \langle 0.05, 0.08, 0 \rangle \text{ m} = \langle -0.09, 0, 0 \rangle \text{ m}$$

$$|\vec{r}_2| = \sqrt{(-0.09)^2} = 0.09 \text{ m}$$

$$\hat{r}_2 = \frac{\vec{r}_2}{|\vec{r}_2|} = \frac{\langle -0.09, 0, 0 \rangle \text{ m}}{(0.09 \text{ m})} = \langle -1, 0, 0 \rangle$$

$$\vec{E}_2 = \frac{1}{4\pi\varepsilon_0} \frac{Q_2}{|\vec{r}_2|^2} \hat{r}_2$$

$$= \left(9 \times 10^9 \frac{\text{Nm}^2}{\text{C}^2} \right) \frac{(-5 \times 10^{-9} \text{C})}{(0.09 \text{ m})^2} \langle -1, 0, 0 \rangle$$

$$\vec{E}_2 = \langle 5.56 \times 10^3, 0, 0 \rangle \text{ N/C}$$

$$\vec{E}_{\text{net}} = \vec{E}_1 + \vec{E}_2$$

$$= \langle -3.02 \times 10^3, 6.04 \times 10^3, 0 \rangle \text{ N/C} + \langle 5.56 \times 10^3, 0, 0 \rangle \text{ N/C}$$

$$\vec{E}_{\text{net}} = \langle 2.54 \times 10^3, 6.04 \times 10^3, 0 \rangle \text{ N/C}$$

In Figure 13.28 the charge Q_3 has been placed at the observation location.

$$\vec{F}_3 = Q_3 \vec{E}_{\text{net}}$$

$$= (-3 \times 10^{-9} \text{C}) \langle 2.54 \times 10^3, 6.04 \times 10^3, 0 \rangle \text{ N/C}$$

$$= \langle -7.51 \times 10^{-6}, -1.81 \times 10^{-5}, 0 \rangle \text{ N}$$

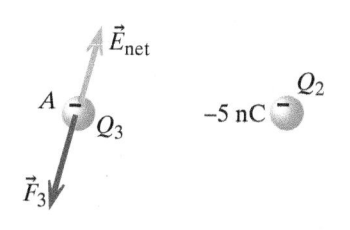

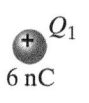

Q_1

6 nC

Figure 13.28 When a negative charge is placed at location A, it experiences a force in a direction opposite to the net field at that location.

The individual contributions and the net field are shown in Figure 13.27. When Q_3 is placed at location A, it experiences a force in the direction shown in Figure 13.28. The force on Q_3 is in a direction opposite to the direction of the net electric field at A, because Q_3 is negative.

13.6 THE ELECTRIC FIELD OF A DIPOLE

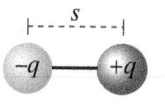

Figure 13.29 An electric dipole.

Neutral matter contains both positive and negative charges (protons and electrons). The simplest piece of neutral matter that we can analyze in detail is an "electric dipole" consisting of two equally but oppositely charged point-like objects, separated by a distance s, as shown in Figure 13.29. Dipoles occur frequently in nature, so this analysis has practical applications. For example, a single molecule of HCl is an electric dipole: the H end is somewhat positive and the Cl end is somewhat negative. Moreover, the analysis of a single dipole provides a basis for analyzing more complicated forms of matter.

By applying the superposition principle we can calculate the electric field due to a dipole at any location in space. We will be particularly interested in the electric field in two important locations: at a location along the axis of the dipole and at a location along the perpendicular bisector of the axis.

Along the Dipole Axis

Figure 13.30 shows a horizontal dipole with charge $+q$ and $-q$, and separation s (here q is a positive number, so $+q$ is a positive charge, and $-q$ is a negative charge). Its center is at the origin. We will apply the superposition principle to find the electric field $\vec{E}_1$ at an observation location on the axis of the dipole (in this case, the x axis), due to the dipole.

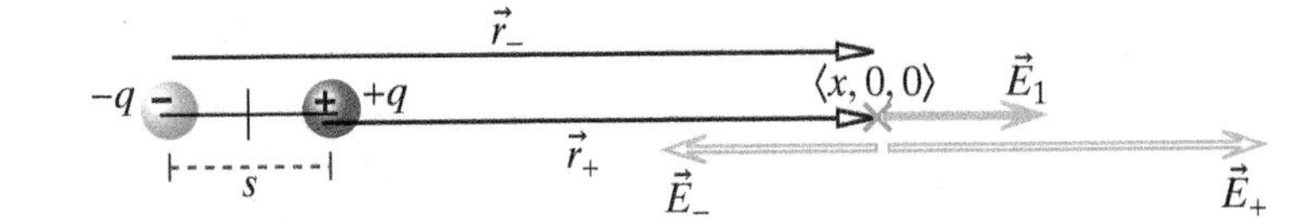

Figure 13.30 Electric field of a dipole at a location on the dipole axis (in the case shown, this is the x axis).

For the orientation shown, we can see that the net field $\vec{E}_1$ will be in the $+x$ direction, because at this location the field $\vec{E}_+$ of the closer $+q$ charge is larger than the field $\vec{E}_-$ of the more distant $-q$ charge. Note that if the dipole were oriented with its negative end on the right, $\vec{E}_1$ would point in the opposite direction. For the orientation shown, the electric field $\vec{E}_+$ due only to the positive end of the dipole axis is:

$$\vec{r}_+ = \langle x,0,0 \rangle - \left\langle \frac{s}{2},0,0 \right\rangle = \left\langle x - \frac{s}{2},0,0 \right\rangle$$

$$|\vec{r}_+| = \sqrt{\left(x - \frac{s}{2}\right)^2 + 0^2 + 0^2} = \left(x - \frac{s}{2}\right)$$

$$\hat{r}_+ = \frac{\left\langle x - \frac{s}{2},0,0 \right\rangle}{\left(x - \frac{s}{2}\right)} = \langle 1,0,0 \rangle$$

$$\vec{E}_+ = \frac{1}{4\pi\varepsilon_0} \frac{q}{\left(x - \frac{s}{2}\right)^2} \langle 1,0,0 \rangle = \left\langle \frac{1}{4\pi\varepsilon_0} \frac{q}{\left(x - \frac{s}{2}\right)^2},0,0 \right\rangle$$

Similarly, the electric field $\vec{E}_-$ due only to the negative end of the dipole is:

$$\vec{r}_- = \langle x,0,0 \rangle - \left\langle \frac{-s}{2},0,0 \right\rangle = \left\langle x + \frac{s}{2},0,0 \right\rangle$$

$$|\vec{r}_-| = x + \frac{s}{2}$$

$$\hat{r}_- = \langle 1,0,0 \rangle$$

$$\vec{E}_- = \frac{1}{4\pi\varepsilon_0} \frac{-q}{\left(x + \frac{s}{2}\right)^2} \langle 1,0,0 \rangle = \left\langle \frac{1}{4\pi\varepsilon_0} \frac{-q}{\left(x + \frac{s}{2}\right)^2},0,0 \right\rangle$$

We add the two contributions to get the net field at the observation location:

$$\vec{E}_1 = \vec{E}_+ + \vec{E}_- = \left\langle \left(\frac{1}{4\pi\varepsilon_0} \frac{q}{\left(x - \frac{s}{2}\right)^2} + \frac{1}{4\pi\varepsilon_0} \frac{-q}{\left(x + \frac{s}{2}\right)^2} \right),0,0 \right\rangle$$

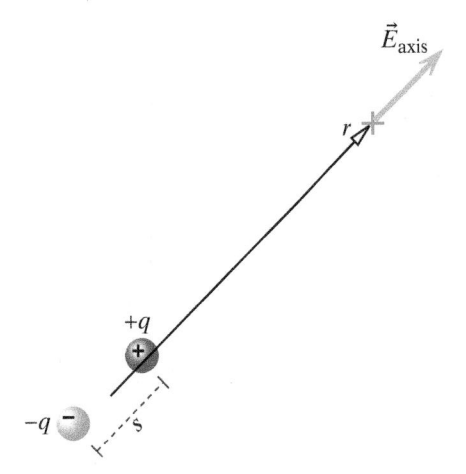

Figure 13.31 The electric field due to a dipole, at a location on the axis of the dipole, for a dipole with an arbitrary orientation.

Because the dipole could be oriented along any axis (see Figure 13.31), we can make this result more general by changing the x to an r, representing the distance from the center of the dipole to an observation location on the axis of the dipole, however the dipole may be oriented. Because the direction of the electric field at a location on the dipole axis depends on the orientation of the dipole, it is more useful to write an algebraic expression for the magnitude of the dipole, and to figure out the direction by looking at the particular situation.

$$|\vec{E}_{\text{on axis}}| = \left| \frac{1}{4\pi\varepsilon_0} \frac{q}{\left(r - \frac{s}{2}\right)^2} + \frac{1}{4\pi\varepsilon_0} \frac{-q}{\left(r + \frac{s}{2}\right)^2} \right|$$

$$= \frac{1}{4\pi\varepsilon_0} q \left(\frac{1}{\left(r - \frac{s}{2}\right)^2} - \frac{1}{\left(r + \frac{s}{2}\right)^2} \right)$$

$$= \frac{1}{4\pi\varepsilon_0} \frac{2qsr}{\left(r - \frac{s}{2}\right)^2 \left(r + \frac{s}{2}\right)^2}$$

QUESTION If the dipole in Figure 13.31 were rotated through an angle of 180°, so the negative end was closer to the observation location, what would be the direction of the electric field at the observation location?

If the dipole were rotated, the electric field would now point toward the dipole.

Approximation: Far from the Dipole

This is still a fairly messy expression, and it does not give us much information about how fast the magnitude of the field decreases as we get farther away from the dipole. Let's try to simplify it. If we are far from the dipole, as is normally the case when we interact with molecular dipoles, then the dipole separation s is very small compared to the distance r from the dipole, so:

$$\text{If } r \gg s, \text{ then } \left(r - \frac{s}{2}\right)^2 \approx \left(r + \frac{s}{2}\right)^2 \approx r^2$$

Using this approximation, we can simplify the expression:

$$|\vec{E}_{\text{on axis}}| \approx \frac{1}{4\pi\varepsilon_0} \frac{2qsr}{r^4} = \frac{1}{4\pi\varepsilon_0} \frac{2qs}{r^3}$$

ELECTRIC FIELD OF A DIPOLE, ON THE DIPOLE AXIS

$$|\vec{E}_{\text{axis}}| \approx \frac{1}{4\pi\varepsilon_0} \frac{2qs}{r^3} \quad \text{at a location on the dipole axis, if } r \gg s$$

q is the magnitude of one of the charges.
s is the separation between the charges.
r is the distance from the center of the dipole to the observation location.
The direction is determined by looking at the particular situation.

Since a dipole can have any orientation, we give only the magnitude of the field. For a particular dipole you will need to determine the direction of the field at a particular observation location by looking at the orientation of the dipole.

Is This Result Reasonable?

QUESTION Should we have expected the magnitude of the electric field of the dipole to be proportional to $1/r^2$?

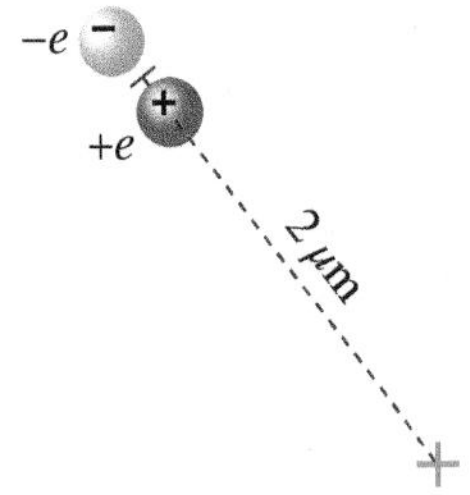

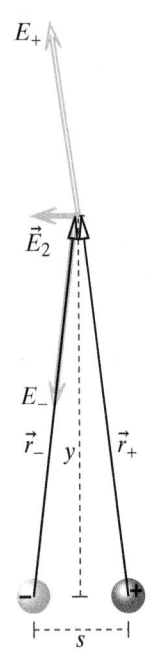

Figure 13.32 An HCl molecule.

The electric field of a single point charge is proportional to $1/r^2$, but when we add the electric fields contributed by more than one charge, the result may have quite a different distance dependence.

QUESTION Does our result have the correct units?

The units of electric field come out to N/C, as they should.

$$\left(\frac{\text{N}\cdot\text{m}^2}{\text{C}^2}\right)\frac{(\text{C})(\text{m})}{(\text{m}^3)} = \frac{\text{N}}{\text{C}}$$

EXAMPLE   **HCl Molecule**

An HCl molecule is a dipole, which can be considered to be a particle of charge $+e$ (a H$^+$ ion) and a particle of charge $-e$ (a Cl$^-$ ion) separated by a distance of about 1×10^{-10} m. What are the magnitude and direction of the electric field due to the HCl molecule at the location indicated in Figure 13.32, which is 2 μm (2×10^{-6} m) from the center of the molecule, on the molecular axis?

Solution

$$|\vec{E}| \approx \frac{1}{4\pi\varepsilon_0}\frac{2qs}{r^3} = \left(9 \times 10^9 \frac{\text{N}\cdot\text{m}^2}{\text{C}^2}\right)\frac{2(1.6\times10^{-19}\,\text{C})(1\times10^{-10}\,\text{m})}{(2\times10^{-6}\,\text{m})^3}$$

$$= 3.6 \times 10^{-2}\,\frac{\text{N}}{\text{C}}$$

Because the positive end of the dipole is closer to the observation location than the negative end, the electric field points away from the dipole, along the dipole axis.

Along the Perpendicular Axis

Next consider the electric field $\vec{E}_2$ at an observation location on the axis perpendicular to the axis of the dipole. For a dipole aligned as shown in Figure 13.33, this is the y axis. First we need to find $\vec{r}_+$ and $\vec{r}_-$, the vectors from each source charge to the observation location.

$$\vec{r}_+ = \langle 0,y,0\rangle - \left\langle \frac{s}{2},0,0\right\rangle = \left\langle -\frac{s}{2},y,0\right\rangle$$

$$\vec{r}_- = \langle 0,y,0\rangle - \left\langle -\frac{s}{2},0,0\right\rangle = \left\langle \frac{s}{2},y,0\right\rangle$$

The magnitudes of $\vec{r}_+$ and $\vec{r}_-$ are equal, because each charge is the same distance from the observation location.

$$|\vec{r}_+| = \sqrt{\left(-\frac{s}{2}\right)^2 + y^2 + 0^2} = \sqrt{\left(\frac{s}{2}\right)^2 + y^2}$$

$$|\vec{r}_-| = \sqrt{\left(\frac{s}{2}\right)^2 + y^2}$$

$$\hat{r}_+ = \frac{\left\langle -\frac{s}{2},y,0\right\rangle}{\left[\left(\frac{s}{2}\right)^2 + y^2\right]^{1/2}} \quad \text{and} \quad \hat{r}_- = \frac{\left\langle \frac{s}{2},y,0\right\rangle}{\left[\left(\frac{s}{2}\right)^2 + y^2\right]^{1/2}}$$

Figure 13.33 Electric field of a dipole at a location on the perpendicular axis.

$$\vec{E}_+ = \frac{1}{4\pi\varepsilon_0} \frac{q}{\left[\left(\frac{s}{2}\right)^2 + y^2\right]} \frac{\left\langle -\frac{s}{2}, y, 0 \right\rangle}{\left[\left(\frac{s}{2}\right)^2 + y^2\right]^{1/2}}$$

$$\vec{E}_- = \frac{1}{4\pi\varepsilon_0} \frac{-q}{\left[\left(\frac{s}{2}\right)^2 + y^2\right]} \frac{\left\langle \frac{s}{2}, y, 0 \right\rangle}{\left[\left(\frac{s}{2}\right)^2 + y^2\right]^{1/2}}$$

The net field $\vec{E}_2$ is the vector sum of $\vec{E}_+$ and $\vec{E}_-$:

$$\vec{E}_2 = \vec{E}_+ + \vec{E}_- = \frac{1}{4\pi\varepsilon_0} \frac{qs}{\left[\left(\frac{s}{2}\right)^2 + y^2\right]^{3/2}} \langle -1, 0, 0 \rangle$$

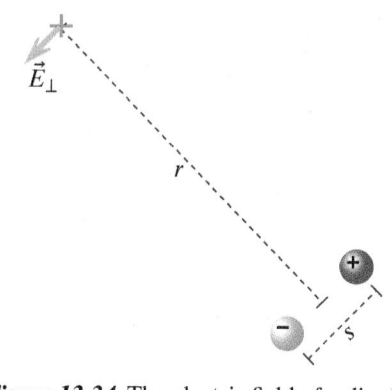

Figure 13.34 The electric field of a dipole at a location on the axis perpendicular to the dipole.

The y components of the two fields cancel, and the net field $\vec{E}_2$ is horizontal to the *left*, as shown in Figure 13.33, which may come as a bit of a surprise.

Again, since in general a dipole could have any orientation (for example, see Figure 13.34), it is appropriate to replace the y with an r representing the distance from the center of the dipole to a location on the axis perpendicular to the dipole, and to give only the magnitude of the electric field at this location.

$$|\vec{E}_\perp| = \frac{1}{4\pi\varepsilon_0} \frac{qs}{\left[\left(\frac{s}{2}\right)^2 + r^2\right]^{3/2}}$$

If we are far from the dipole, as is normally the case when we interact with molecular dipoles, the magnitude of the field can be further simplified. You should be able to show that in this important case ($r \gg s$), by making an appropriate approximation, we obtain:

ELECTRIC FIELD OF A DIPOLE, ON THE ⊥ AXIS

$$|\vec{E}_\perp| \approx \frac{1}{4\pi\varepsilon_0} \frac{qs}{r^3} \quad \text{on a location on the perpendicular axis, if } r \gg s$$

q is the magnitude of one of the charges.
s is the separation between the charges.
r is the distance from the center of the dipole to the observation location.
The direction is determined by looking at the particular situation.

Note that the magnitude of the electric field at a location along the y axis is half of the magnitude of the electric field at a location the same distance away on the x axis.

Along the Other Perpendicular Axis

> QUESTION What would be the electric field of a dipole at a location on the other perpendicular axis?

By looking at the symmetry of the system, we can conclude by inspection that the electric field along both perpendicular axes should have exactly the same distance dependence.

Other Locations

By applying the superposition principle we can calculate the electric field of a dipole at any location, though we may not get a simple algebraic expression.

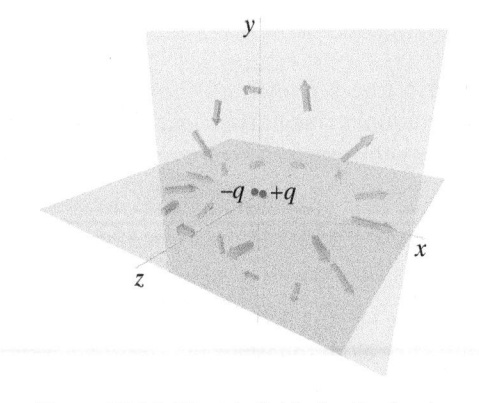

Figure 13.35 Electric field of a dipole at locations in two planes. Included are locations on the dipole axis and the perpendicular axis, as well as other locations.

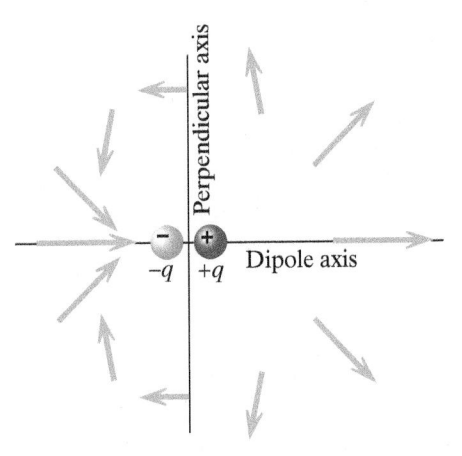

Figure 13.36 Electric field of a dipole, at locations in a plane containing the dipole.

Figure 13.35 shows the electric field of a dipole at locations in two perpendicular planes containing the dipole, calculated numerically. Figure 13.36 shows only locations in the vertical plane, displaying the symmetry of the field pattern. Problem P68 is a computer exercise that allows you to visualize the electric field of a dipole at any location.

The Importance of Approximations

By considering the common situation where $r \gg s$, we were able to come up with a simple algebraic expression for the electric field of a dipole at locations along the x and y axes, which gives us a clear insight into the distance dependence of the dipole field. If we had not made this simplification, we would have been left with a messy expression, and it would not have been easy to see that the electric field of a dipole falls off like $1/r^3$. (This turns out to be true for locations not on the axes as well, but the algebra required to demonstrate this is more complicated.)

> QUESTION   How would we calculate the electric field of a dipole at a location very near the dipole, where $r \approx s$?

Since the approximation we made in the calculations above is no longer valid, we would have to go back to adding the fields of the two charges exactly, as we did in the first step of our work above.

> **Checkpoint 5**   A dipole is located at the origin, and is composed of charged particles with charge $+e$ and $-e$, separated by a distance 2×10^{-10} m along the x axis. **(a)** Calculate the magnitude of the electric field due to this dipole at location $\langle 0, 2 \times 10^{-8}, 0 \rangle$ m. **(b)** Calculate the magnitude of the electric field due to this dipole at location $\langle 2 \times 10^{-8}, 0, 0 \rangle$ m.

Interaction of a Point Charge and a Dipole

Since we have analytical expressions for the electric field of a dipole at locations along the x, y, or z axes, we can calculate the force exerted on a point charge that interacts with the field of the dipole. Figure 13.37 shows a dipole acting on a point charge $+Q$ that is a distance $d \gg s$ from the center of the dipole.

EXAMPLE   **Force of a Dipole on a Point Charge**

What are the direction and the magnitude of the force exerted by the dipole on the point charge shown in Figure 13.37?

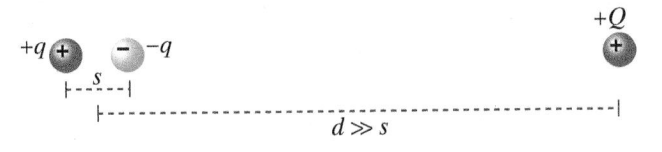

Figure 13.37 A dipole and a point charge.

Solution   The electric field due to the dipole at the location of the point charge points in the $-x$ direction (Figure 13.39). The dipole exerts a force on Q toward the dipole:

$$\vec{F} = Q\vec{E}_{\text{dipole}} = Q \left\langle -\frac{1}{4\pi\varepsilon_0}\frac{2qs}{d^3}, 0, 0 \right\rangle$$

> QUESTION   Does the direction of this electric force make sense?

Yes. The negative end of the dipole is closer to the location of interest, so its contribution to the net electric field is larger than the contribution from the positive end of the dipole.

QUESTION What are the magnitude and direction of the force exerted on the dipole by the point charge?

By the principle of reciprocity of electric forces (Newton's third law), the force exerted by Q on the dipole must be equal in magnitude and opposite in direction to the force on the point charge by the dipole.

QUESTION How is this possible? Since the particles making up the dipole have equal and opposite charges, why isn't the net force on the dipole zero?

The key lies in the distance dependence of the electric field of the point charge. Although the two ends of the dipole are very close together, the positive end is very slightly farther away from the point charge. Hence, the magnitude of the point charge's electric field is slightly less at this location, and the resulting repulsive force on the positive end of the dipole is slightly less than the attractive force on the negative end of the dipole. The net force is small, but attractive (Figure 13.38).

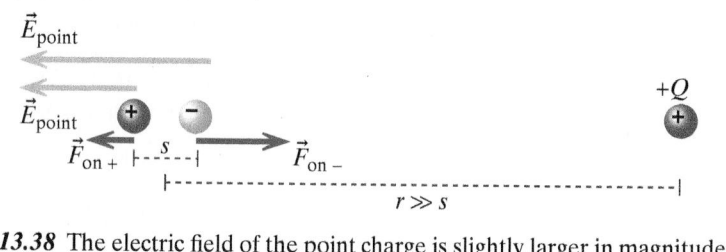

Figure 13.38 The electric field of the point charge is slightly larger in magnitude at the location of the negative end of the dipole. Hence, the net force on the dipole is to the right. The arrows representing field and force vectors are offset slightly from the location of the charges for clarity.

Checkpoint 6 Consider the situation in Figure 13.39. **(a)** If we double the distance d, by what factor is the force on the point charge due to the dipole reduced? **(b)** How would the magnitude of the force change if the point charge had a charge of $+3Q$? **(c)** If the charge of the point charge were $-2Q$, how would the force change?

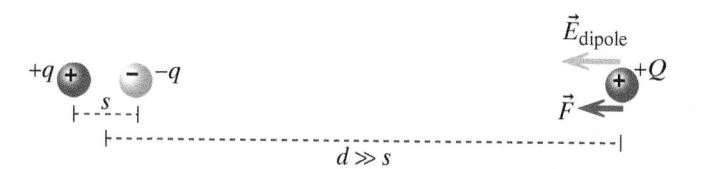

Figure 13.39 The electric field of the dipole at the location of the point charge, and the force on the point charge due to this field.

Electric Dipole Moment

The electric field of a dipole is proportional to the product qs, called the "electric dipole moment" and denoted by p.

$$p = qs \quad \text{(electric dipole moment)}$$

Far away from a dipole, the same electric field could be due to a small q and a large s, or a large q and a small s. The only thing that matters is the product qs. Moreover, this is the quantity that is measurable for molecules such as HCl and H_2O that are permanent dipoles, not the individual values of q (the amount of charge on an end) and s (the separation of the charges).

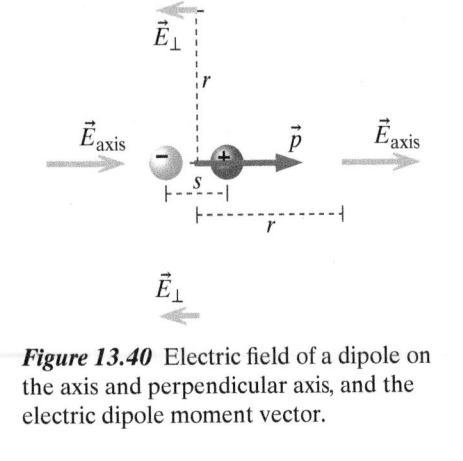

Figure 13.40 Electric field of a dipole on the axis and perpendicular axis, and the electric dipole moment vector.

We can rewrite our expressions for the electric field of a dipole in terms of the dipole moment p:

$$|\vec{E}_{\text{axis}}| \approx \frac{1}{4\pi\varepsilon_0}\frac{2p}{r^3} \quad \text{at a location on the dipole axis, if } r \gg s$$

$$|\vec{E}_\perp| \approx \frac{1}{4\pi\varepsilon_0}\frac{p}{r^3} \quad \text{at a location on the perpendicular axis, if } r \gg s$$

Dipole Moment as a Vector

The dipole moment can be defined as a vector $\vec{p}$ that points from the negative charge to the positive charge, with a magnitude $p = qs$ (Figure 13.40). Note that the electric field along the axis of the dipole (outside the dipole) points in the same direction as the dipole moment $\vec{p}$, which is one of the useful properties of the vector $\vec{p}$.

In Chapter 17 we will see a similar pattern of magnetic field around a magnetic dipole characterized by a magnetic dipole moment that is a vector.

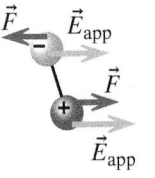

Figure 13.41 A dipole in a uniform electric field would experience no net force. It would experience a torque about its center of mass (in this example $\vec{\tau}_{\text{CM}}$ is out of the page, and the dipole would begin to rotate counterclockwise).

A Dipole in a Uniform Electric Field

We will see in Chapter 15 that it is possible to arrange a collection of point charges in a configuration that produces an electric field nearly uniform in magnitude and direction within a particular region. What would be the force on a dipole in such a uniform electric field?

The net force on the dipole would be zero. We see in Figure 13.41 that the forces on the two ends would be equal in magnitude and opposite in direction. However, there would be a torque on the dipole about its center of mass, and the dipole would begin to rotate.

Checkpoint 7 What would be the equilibrium position of the dipole shown in Figure 13.41? In this equilibrium position, compare the orientation of the vector dipole moment with the direction of the applied electric field.

EXAMPLE

Dipole and Charged Ball

A dipole consisting of point charges $+4\,\text{nC}$ and $-4\,\text{nC}$ separated by a distance of 2 mm is centered at location $\langle 0.03, 0.15, 0\rangle$ m as shown in Figure 13.42 (the size of the dipole is exaggerated in the diagram). A hollow plastic ball with radius 3 cm and charge $-0.2\,\text{nC}$ distributed uniformly over its surface is centered at location $\langle 0.11, 0.15, 0\rangle$ m. What is the net electric field at location C, $\langle 0.03, 0.04, 0\rangle$ m? $(1\,\text{nC} = 1 \text{ nanocoulomb} = 1 \times 10^{-9}\,\text{C}.)$

Solution

Apply the superposition principle:

$$\vec{E}_{\text{net}} = \vec{E}_{\text{dipole}} + \vec{E}_{\text{ball}}$$

The distance from the center of the dipole to the observation location is

$$|\vec{r}_d| = |\langle 0.03, 0.04, 0\rangle - \langle 0.03, 0.15, 0\rangle| = 0.11 \text{ m}$$

Because location C is reasonably far from the dipole, along the dipole axis, we can approximate the magnitude of the dipole field as:

Dipole

Charged ball

$\times C$

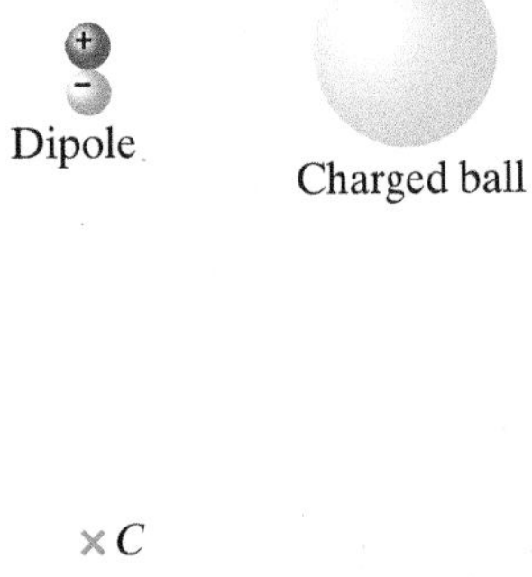

Figure 13.42 A dipole and a charged ball. Dipole size is exaggerated (not to scale).

$$|\vec{E}| \approx \frac{1}{4\pi\varepsilon_0}\frac{2qs}{r^3} = \left(9\times 10^9\ \frac{\text{N}\cdot\text{m}^2}{\text{C}^2}\right)\frac{2(4\times 10^{-9}\,\text{C})(2\times 10^{-3}\,\text{m})}{(0.11\,\text{m})^3} = 108\ \frac{\text{N}}{\text{C}}$$

By looking at the diagram we see that the direction of the dipole field is $\langle 0,1,0 \rangle$ (along the positive y axis, because the negative end is closer to location C). The electric field at location C due to the dipole is:

$$\vec{E}_{\text{dipole}} = \langle 0, 108, 0 \rangle \, \frac{N}{C}$$

The electric field due to the ball is the same as if the ball were a point charge located at its center. The vector $\vec{r}$ from the center of ball to location C is:

$$\vec{r} = \langle 0.03, 0.04, 0 \rangle \text{ m} - \langle 0.11, 0.15, 0 \rangle \text{ m} = \langle -0.08, -0.11, 0 \rangle \text{ m}$$

The magnitude of $\vec{r}$ is $|\vec{r}| = \sqrt{(-0.08)^2 + (-0.11)^2 + 0^2}$ m $= 0.136$ m, so the unit vector $\hat{r}$ is:

$$\hat{r} = \frac{\vec{r}}{|\vec{r}|} = \frac{\langle -0.08, -0.11, 0 \rangle \text{ m}}{0.136 \text{ m}} = \langle -0.588, -0.809, 0 \rangle$$

The electric field due to the charged ball is then

$$\vec{E}_{\text{ball}} = \frac{1}{4\pi\varepsilon_0} \frac{Q}{|\vec{r}|^2}\hat{r} = \left(9 \times 10^9 \, \frac{N \cdot m^2}{C^2} \right) \frac{(-2 \times 10^{-10}\,C)}{(0.136 \text{ m})^2} \langle -0.588, -0.809, 0 \rangle$$

$$= \langle 57.2, 78.7, 0 \rangle \, \frac{N}{C}$$

The net electric field is:

$$\vec{E}_{\text{net}} = \vec{E}_{\text{dipole}} + \vec{E}_{\text{ball}}$$

$$= \langle 0, 108, 0 \rangle \, \frac{N}{C} + \langle 57.2, 78.7, 0 \rangle \, \frac{N}{C}$$

$$= \langle 57.2, 187, 0 \rangle \, \frac{N}{C}$$

As shown in Figure 13.43, both the x and y components of $\vec{E}_{\text{ball}}$ are positive, which is correct, since the ball is negatively charged. Arrows representing each contribution to the electric field, and the net electric field, are shown in Figure 13.43.

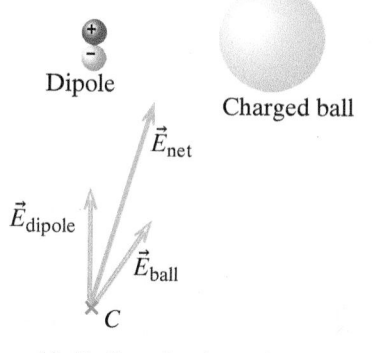

Figure 13.43 Contributions of dipole and ball, and net electric field at location C.

13.7 CHOICE OF SYSTEM

In our previous study of mechanics, when we analyzed multiparticle systems in terms of momentum or energy we had to decide what objects to include in our "system" and what objects to consider as external to the system. Similarly, when we use the field concept to calculate a force rather than using Coulomb's law directly, we split the Universe into two parts:

- the charges that are the sources of the field, and
- the charge that is affected by that field.

This is particularly useful when the sources of the field are numerous and fixed in position. We calculate the electric field due to such sources throughout a given region of space, and then we can predict what would happen to a charged particle that enters that region. (It is of course true that the moving charged particle exerts forces on all the other charges, but often we're not interested in those forces, or the other charges are not free to move.)

For example, in an oscilloscope, charges on metal plates are the sources of an electric field that affects the trajectory of single electrons (Figure 13.44). The electric field has nearly the same magnitude and direction everywhere within a box (this can be achieved with a suitable configuration of positive charges below and negative charges above). The trajectory of an electron

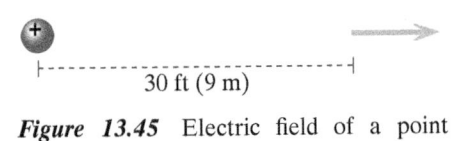

Figure 13.44 Path of an electron through a region of uniform electric field.

outside this box, where the field is nearly zero, is a straight line. Upon entering the box (through a small hole) the electron experiences a downward force (it has negative charge, hence the force on it is opposite to the field) and has a trajectory as shown.

> **Checkpoint 8** If the uniform upward-pointing electric field depicted in Figure 13.44 has a magnitude of 5000 N/C, what is the magnitude of the force on the electron while it is in the box? If a different particle experiences a force of 1.6×10^{-15} N when passing through this region, what is the charge of the particle?

13.8   IS ELECTRIC FIELD REAL?

Reflecting on what we have done so far, we can give two reasons for why we introduced the concept of electric field:

1. Once we know the electric field $\vec{E}$ at some location, we know the electric force $\vec{F} = q\vec{E}$ acting on *any* charge q that we place there.
2. In terms of electric field, we can describe the electric properties of matter, independent of how the applied electric field is produced.

Later we'll see that if the electric field in air exceeds about 3×10^6 N/C, air becomes a conductor, no matter how this electric field is produced. This is an example of the value of the electric field concept for parameterizing the behavior of matter. It would be very awkward to describe the properties of matter if we didn't split the problem into two parts using the field concept, in which we talk about the field made by other charges and the effect that a field has on a particular kind of matter.

Even with these advantages of the field concept, it may seem that it doesn't matter whether we think of electric field as merely a calculational convenience or as something real. However, Einstein's special theory of relativity, which has been experimentally verified in a wide variety of situations, rules out instantaneous action at a distance and implies that the field concept is necessary, not merely convenient, as we will show next.

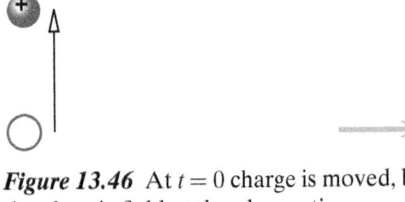

Figure 13.45 Electric field of a point charge.

Retardation: Electric Field and the Speed of Light

Special relativity predicts that nothing can move faster than the speed of light, not even information, and no one has ever observed a violation of this prediction. The speed of light is $c = 3 \times 10^8$ m/s, or about 1 ft (30 cm) per nanosecond, where a nanosecond is 1×10^{-9} s.

Consider a charge that is 30 ft away (about 9 m), as shown in Figure 13.45. It takes light 30 nanoseconds (30 ns) to travel from the charge to you.

Suppose that the charge is suddenly moved to a new location, as shown in Figure 13.46. You cannot observe change in the electric field until 30 ns have elapsed! If the field could change instantaneously, that would provide a mechanism for sending signals faster than the speed of light.

For a while (30 ns), the electric field has some kind of reality and existence independent of the original source of the field. If you place a positive charge at your observation location during this 30-ns interval, you will see it accelerate to the right, in the direction of the old electric field that is still valid.

At the end of the 30-ns delay, the field finally changes to correspond to the place to which the charge was moved 30 ns ago, as shown in Figure 13.47. You *see* the new position of the charge at the same instant that you notice the

Figure 13.46 At $t = 0$ charge is moved, but the electric field at the observation location does not change for 30 ns.

Figure 13.47 At $t = 30$ ns, the field at the observation location finally changes.

change in the electric field, because of course light itself travels at the speed of light.

What If the Source Charges Vanish?

An even more dramatic example concerns the electric field made by a remote electric dipole. Suppose that an electron and a positron are separated by a small distance s, making a dipole, and creating an electric field throughout space. The electron and positron may suddenly come together and react, annihilating each other and releasing a large amount of energy in the form of high-energy photons (gamma rays), which have no electric charge:

$$e^- + e^+ \rightarrow \gamma + \gamma$$

At a time shortly after the annihilation has occurred, there are no longer any charged particles in the vicinity. However, if you monitor the electric field in this region, you will keep detecting the (former) dipole's field for a while even though the dipole no longer exists! If you are a distance r away from the original location of the dipole, you will still detect the dipole's electric field for a time r/c, which is the same as the time it would take light to travel from that location to your position.

Coulomb's Law Is Correct Only at Low Speeds

This interesting behavior, called "retardation," means that Coulomb's law is not completely correct, since the equation

$$\vec{F} = \frac{1}{4\pi\varepsilon_0} \frac{q_1 q_2}{r^2} \hat{r}$$

doesn't contain time t or speed of light c. Coulomb's law is an approximation that is quite accurate as long as charges are moving slowly. However, when charges move with speeds that are a significant fraction of the speed of light, Coulomb's law is not adequate. Similarly, the equation for the electric field of a point charge,

$$\vec{E} = \frac{1}{4\pi\varepsilon_0} \frac{q}{r^2} \hat{r}$$

is only an approximation, valid if the charge is moving at a speed that is small compared with the speed of light.

Later we will consider the fields of moving charges, which include magnetic fields as well as electric fields. We will also consider the fields of *accelerated* charges, which produce a very special kind of electric field—a component of the electromagnetic field that makes up radio and television waves and light.

Evidently the electric field isn't just a calculational device. Space can be altered by the presence of an electric field, even when the source charges that made the field are gone.

Relativistic Electric Field

It turns out that for a charged particle moving at high speed in the $+x$ direction, the x component of electric field at any location is unchanged. However, a factor of

$$\gamma = \frac{1}{\sqrt{1 - v^2/c^2}}$$

appears in the y and z components of electric field, and they are much larger, as shown in Figure 13.48. The fields of relativistic particles are discussed further in Chapter 20.

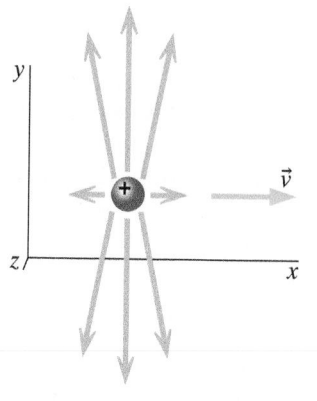

Figure 13.48 The pattern of electric field of a positively charged particle moving at high speed in the $+x$ direction is no longer spherically symmetric.

13.9 COMPUTATIONAL MODELING OF ELECTRIC FIELDS

■ In this section it is assumed that you have done at least some of the computational modeling activities in Chapters 1–12. If computational modeling in VPython is new to you, you will need to work through the computational part of the Chapter 1 supplement that is available on the Wiley student website before reading this section or doing the computational problems at the end of the chapter.

One of the goals of computational modeling in this and subsequent chapters is simply to calculate and visualize patterns of electric and magnetic fields in 3D space. Recognizing these patterns and understanding how they relate to source charge distributions will be important throughout the rest of this book.

The calculation of the electric field of a single point charge at a single observation location is similar to the calculation of gravitational force we have done before. In writing a program to tell a computer how to calculate an electric field, we use exactly the same organization that we use when doing the calculation with a calculator. Let's assume that our program creates an object named `particle_1` at some appropriate location in 3D space, and sets the variable `q1` equal to the charge of `particle_1`. The program also identifies the observation location by creating a vector called `r_obs`. Let's also assume that the program has defined a constant `oofpez` (standing for One-Over-Four-Pi-Epsilon-Zero) equal to $1/(4\pi\epsilon_0)$. A translation of the mathematical expressions (shown in red) into code (shown in blue) for calculating the electric field due to `particle_1` at the location `r_obs` into VPython code might look like this:

$\vec{r} = \vec{r}_{\text{obs}} - \vec{r}_1$	`r = r_obs - particle_1.pos`
$\hat{r} = \vec{r}/|\vec{r}|$	`rhat = r / mag(r)`
$\vec{E} = \left(\dfrac{1}{4\pi\epsilon_0}\dfrac{q_1}{|\vec{r}|^2}\right)\hat{r}$	`E = (oofpez * q1 / mag(r)**2) * rhat`

Instead of just printing the value of the electric field at a location, we typically want to use an arrow to display the field. Assuming that the program has defined an appropriate scale factor `sf` to scale the arrow so both the arrow and the sphere representing the source charge can be seen at the same time, we could use the following statement to do this (by convention, arrows representing electric field are orange throughout this book):

```
ea = arrow(pos=r_obs, color=color.orange, axis=sf*E)
```

QUESTION Figure 13.49 shows the display from a VPython program that calculates and displays the electric field at a single observation location, due to a single source charge. What is the sign of the source charge?

Since the electric field at the observation location points toward the source charge, the source charge must be negative.

QUESTION In the program used to generate Figure 13.49 the distance between the source charge and the observation location was 1.4×10^{-8} m, and the magnitude of the electric field was $|\vec{E}| = 7.2 \times 10^6$ N/C. Approximately what was the value of the scale factor `sf` used to adjust the length of the arrow?

The arrow is about half as long as the distance between source and observation location, so the scale factor must have been about $0.7 \times 10^{-8}/7.2 \times 10^6 \approx 1 \times 10^{-15}$.

Figure 13.49 Display generated by a VPython program to calculate the electric field due to one charge at a single observation location.

Finding the Net Field Due to Several Charges

In Chapter 15 we will concentrate on finding the electric field at one or more observation locations due to a large number of point charges. In that chapter we will introduce a new computational concept called a "list" that will facilitate this process. In the current chapter, we will stick to applying the computational concepts we have already covered. (If you already know how to use Python lists, you are welcome to use them in the problems at the end of the chapter.)

In Section 13.6 we learned how to calculate the electric field of a dipole at a particular location by adding the fields contributed by the two charges in the dipole. We were able to find analytical expressions for the magnitude of the electric field at any location on the axis of the dipole, or on the perpendicular axis. Finding an analytical expression for the magnitude and direction of the field of a dipole at observation locations not on an axis is much more difficult, but this is an easy problem to solve computationally.

If we call the two objects making up the dipole `particle_1` and `particle_2`, we can find the net field at an observation location `r_obs` this way:

```
E_net = vector(0,0,0)   ## initialize E_net

r = r_obs - particle_1.pos
rhat = r / mag(r)
E = (oofpez * q1 / mag(r)**2) * rhat
E_net = E_net + E

r = r_obs - particle_2.pos
rhat = r / mag(r)
E = (oofpez * q2 / mag(r)**2) * rhat
E_net = E_net + E

ea = arrow(pos=r_obs, axis=sf*E_net, color=color.orange)
```

In the code segment above, notice that the variable names `r` and `E` were reused. This is okay, because after we update `E_net` we no longer need the values stored in these variables, and we can set them to new values. Reusing variable names makes it possible to put calculations like these inside `while` loops, which will allow us to avoid repeating lines of code as is done in the segment above.

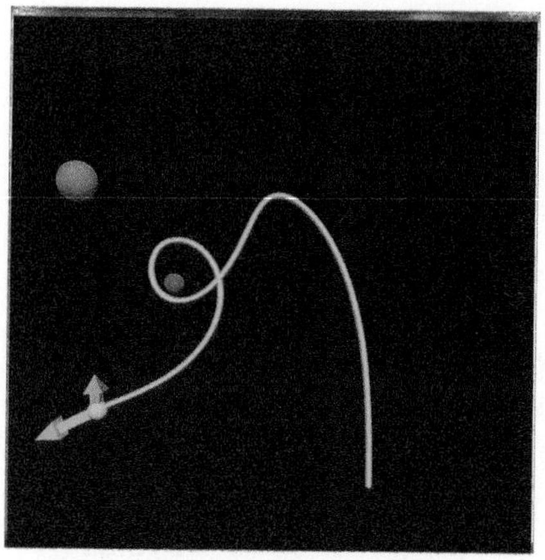

QUESTION In the code above, why does the following instruction appear twice?

```
E_net = E_net + E
```

This is an "update" instruction of the kind we've used with momentum and position. We incrementally update the value of `E_net` by sequentially adding to it the electric field of each particle.

Predicting Particle Motion

We can predict the motion of charged particles by iteratively updating momentum and position, as illustrated in Figure 13.50. Such a computational model differs from what we have done previously only in that we must calculate the net electric field at the particle's location in order to get the net force on the particle. The basic approach is this:

- Calculate the net electric field $\vec{E}_{net}$ at the location of the moving particle.
- Calculate the net force $\vec{F}_{net} = q\vec{E}_{net}$ on the moving particle.
- Update the momentum of the particle: $\vec{p}_f = \vec{p}_i + \vec{F}_{net} \Delta t$.
- Update the position: $\vec{r}_f = \vec{r}_i + \vec{v}_{avg} \Delta t$.

(The bracket to the left of the four bullets is labeled "Repeat".)

As we did in Chapters 2–11, we use the approximation $\vec{v}_{avg} \approx \vec{p}_f/m$ as long as the speed of the particle is small compared to the speed of light.

The computational loop in a program that moves a charged particle named `mp` (for "moving particle") with momentum `p_mp`, charge `q_mp`, and mass `m_mp`, in a region near a single point charge, might look like this:

Figure 13.50 Display generated by a VPython program to predict the motion of a charged particle that is affected by the electric field due to two unequal source charges. The orange arrow represents the net electric field at the current location of the moving particle. The green arrow represents the particle's momentum.

```
while t < tmax:
    rate(100)
    r = mp.pos - source.pos
    rhat = r / mag(r)
    E = (oofpez * q_source / mag(r)**2) * rhat
    F = q_mp * E
    p_mp = p_mp + F * deltat
    mp.pos = mp.pos + (p_mp/m_mp) * deltat
    t = t + deltat
```

QUESTION Why is the calculation of the electric field done inside
the computational loop?

Any quantity that changes in magnitude or direction must be recalculated in
each time step. The moving particle's position changes during each iteration,
and the electric field at its new position is different from the electric field at its
previous position. We need the new value of the electric field to find the new
force on the particle.

Many Observation Locations

The electric field of even a single point charge has a value at every location
in 3D space. Visualizing these patterns in 3D is straightforward in VPython.
Here is a small program that calculates and visualizes the electric field of a
single point charge at several locations on a circle. The display generated by
the program is shown in Figure 13.51.

```
from visual import *
oofpez = 9e9 ## One-Over-Four-Pi-Epsilon-Zero
qe = 1.6e-19
sf = 3e-16   ## scale factor
source = sphere(pos=vector(-0.5e-8,0,0), radius=1e-9,
                color=color.red)

q_source = qe
thetamax = 2*pi
dtheta = pi/6
R = 1e-8  ## radius of circle

theta = 0
while theta < thetamax:
    rate(500)
    r_obs = R * vector(cos(theta),cos(pi/2-theta),0)
    r = r_obs - source.pos
    rhat = r/mag(r)
    E = (oofpez * q_source / mag(r)**2) * rhat
    arrow(pos=r_obs, color=color.orange, axis=sf*E,
          shaftwidth = source.radius)
    theta = theta + dtheta
```

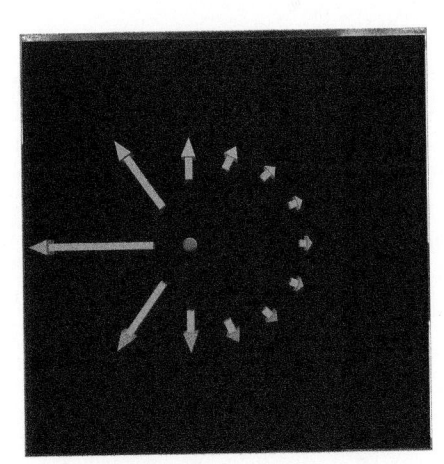

Figure 13.51 Display generated by the
VPython program shown in the text at
the right.

The extra argument in the constructor for the arrows is needed to force
VPython to make the shafts of the arrows all the same size. By default VPython
scales the width of the arrow shaft to the length of the arrow, but this can make
it difficult to compare arrows.

QUESTION What is the physical significance of the variable `theta`
used in the `while` loop, and how is it related to the observation
locations?

`theta` is the angle between the $+x$ axis and the relative position vector
$\vec{r}$. Direction cosines are used to construct the unit vector $\hat{r}$, which is then
multiplied by the radius of the desired circle to give the observation location.

QUESTION Why aren't the lengths of the orange arrows in Figure 13.51 all the same?

The distance from the source charge to the observation locations varies, so the magnitude of the electric field also varies. The center of the circle of observation locations is not at the location of the source charge.

SUMMARY

At a location where there is an electric field $\vec{E}_1$, a point particle with charge q_2 experiences a force $\vec{F}_2 = q_2\vec{E}_1$ (Figure 13.52).

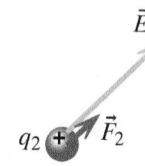

Figure 13.52

A point particle with charge q_1 makes an electric field at all locations in space (Figure 13.53) except its own location:

$$\vec{E}_1 = \frac{1}{4\pi\varepsilon_0}\frac{q_1}{|\vec{r}|^2}\hat{r}$$

where q_1 is the source charge, $\vec{r}$ is the relative position vector giving the position of the observation location relative to the source charge, $\hat{r}$ is the unit vector in the direction of $\vec{r}$, and $|\vec{r}|$, the magnitude of $\vec{r}$, is the distance from the source location to the observation location.

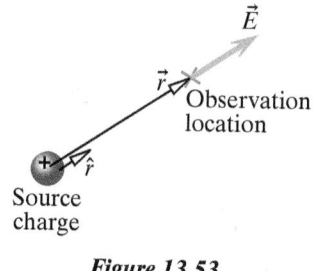

Figure 13.53

A point charge is not affected by its own electric field.

The source location is the location of the charge(s) that create an electric field.

The observation location is the location at which the electric field is detected (Figure 13.54).

Figure 13.54 Electric field at selected observation locations in a plane containing a positive or negative point charge.

The Superposition Principle

The net electric field at a particular location due to two or more charges is the vector sum of each field due to each individual charge. Each individual contribution is unaffected by the presence of the other charges. Thus, matter cannot "block" electric fields.

Retardation

Changes in electric fields (due to changes in source charge distribution) propagate through space at the speed of light. Therefore the electric field at a distant point does not change instantaneously when the source charge distribution changes.

Electric Dipole

An electric dipole consists of two point particles having equal and opposite charges $+q$ and $-q$, separated by a distance s (Figure 13.55). The electric dipole moment of a dipole is defined as $p \equiv qs$. The electric dipole moment vector $\vec{p}$ points from the negative charge toward the positive charge.

Electric field at a location on the dipole axis:

$$|\vec{E}_{\text{axis}}| \approx \frac{1}{4\pi\varepsilon_0}\frac{2qs}{r^3} \quad \text{for } r \gg s$$

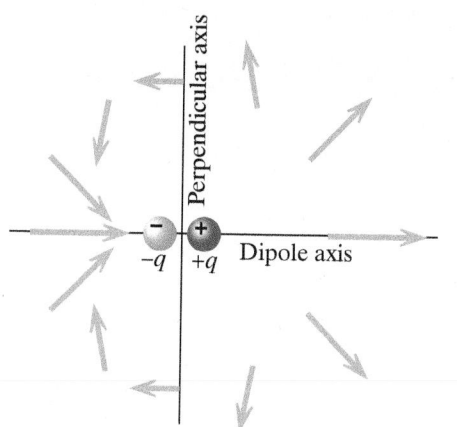

Figure 13.55 The pattern of electric field around an electric dipole.

The direction is determined by looking at the particular situation.

Electric field at a location on the perpendicular axis:

$$|\vec{E}_\perp| \approx \frac{1}{4\pi\varepsilon_0}\frac{qs}{r^3} \quad \text{for } r \gg s$$

The direction is determined by looking at the particular situation.

For a spherical object with a total charge Q distributed uniformly over its surface (a "uniform spherical shell of charge"):

at a location inside the shell, $|\vec{E}| = 0$

at a location outside the shell, $\vec{E} = \dfrac{1}{4\pi\varepsilon_0}\dfrac{Q}{|\vec{r}|^2}\hat{r}$

($\vec{r}$ extends from the center of sphere to the observation location; Q is the total charge on the sphere).

Constants:

$$\frac{1}{4\pi\varepsilon_0} = 9 \times 10^9\ \frac{\text{N}\cdot\text{m}^2}{\text{C}^2} \quad\text{and}\quad \varepsilon_0 = 8.85\times 10^{-12}\ \frac{\text{C}^2}{\text{N}\cdot\text{m}^2}$$

QUESTIONS

Q1 What is the relationship between the terms "field" and "force"? What are their units?

Q2 You are the captain of a spaceship. You need to measure the electric field at a specified location P in space outside your ship. You send a crew member outside with a meter stick, a stopwatch, and a small ball of known mass M and net charge $+Q$ (held by insulating strings while being carried). **(a)** Write down the instructions you will give to the crew member, explaining what observations to make. **(b)** Explain how you will analyze the data that the crew member brings you to determine the magnitude and direction of the electric field at location P.

Q3 Criticize the following statement: "A proton can never be at rest, because it makes a very large electric field near itself that accelerates it."

Q4 Draw a diagram showing two separated point charges placed in such a way that the electric field is zero somewhere, and indicate that position. Explain your reasons.

Q5 At location A there is an electric field in the direction shown by the orange arrow in Figure 13.56. This electric field is due to charged particles that are not shown in the diagram. **(a)** If a proton is placed at location A, which of the arrows $(a–h)$ best indicates the direction of the electric force on the proton? **(b)** If the proton is removed and an electron is placed at location A, which of the arrows in Figure 13.56 best indicates the direction of the electric force on the electron?

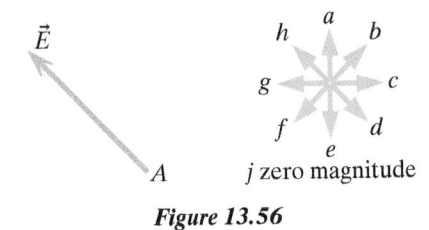

Figure 13.56

Q6 We found that the force exerted on a distant charged object by a dipole is given by

$$F_{\text{on } Q \text{ by dipole}} \approx Q\left(\frac{1}{4\pi\varepsilon_0}\frac{2qs}{r^3}\right)$$

In this equation, what is the meaning of the symbols q, Q, s, and r?

Q7 At a given instant in time, three charged objects are located near each other, as shown in Figure 13.57. Explain why the equation

$$F_{\text{on } Q \text{ by dipole}} \approx Q\left(\frac{1}{4\pi\varepsilon_0}\frac{2qs}{r^3}\right)$$

cannot be used to calculate the electric force on the ball of charge $+Q$.

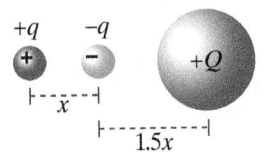

Figure 13.57

Q8 Where could you place one positive charge and one negative charge to produce the pattern of electric field shown in Figure 13.58? (As usual, each electric field vector is drawn with its tail at the location where the electric field was measured.) Briefly explain your choices.

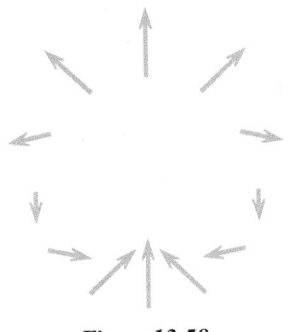

Figure 13.58

Q9 Consider Figure 13.59. Assume that the dipole is fixed in position. **(a)** What is the direction of the electric field at location A due to the dipole? **(b)** At location B? **(c)** If an electron were placed at location A, in which direction would it begin to move? **(d)** If a proton were placed at location B, in which direction would it begin to move? **(e)** Now suppose that an electron is placed at location A and held there, while the dipole is free to move. When the dipole is released, in what direction will it begin to move?

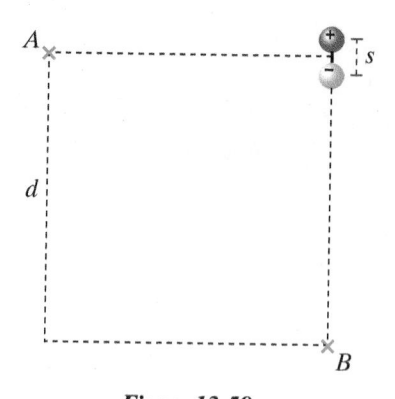

Figure 13.59

Q10 Which of these statements about a dipole are correct? Select all that are true. (1) At a distance d from a dipole, where $d \gg s$ (the separation between the charges), the magnitude of the electric field due to the dipole is proportional to $1/d^3$. (2) A dipole consists of two particles whose charges are equal in magnitude but opposite in sign. (3) The net electric field due to a dipole is zero, since the contribution of the negative charge cancels out the contribution of the positive charge. (4) At a distance d from a dipole, where $d \gg s$ (the separation between the charges), the magnitude of the electric field due to the dipole is proportional to $1/d^2$. (5) The electric field at any location in space, due to a dipole, is the vector sum of the electric field due to the positive charge and the electric field due to the negative charge.

Q11 If we triple the distance d, by what factor is the force on the point charge due to the dipole in Figure 13.60 reduced? (Note that the factor is smaller than one if the force is reduced and larger than one if the force is increased.)

Q12 If the charge of the point charge in Figure 13.60 were $-9Q$ (instead of Q): **(a)** By what factor would the magnitude of the

force on the point charge due to the dipole change? Express your answer as the ratio (magnitude of new force / magnitude of Fv). **(b)** Would the direction of the force change?

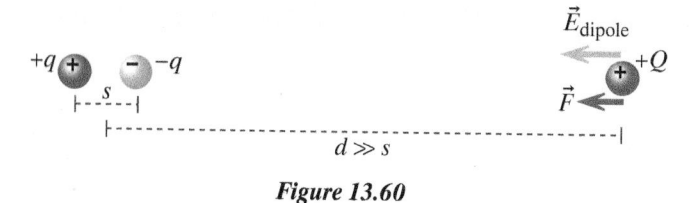

Figure 13.60

Q13 The distance between the dipole and the point charge in Figure 13.60 is d. If the distance between them were changed to $0.5d$, by what factor would the force on the point charge due to the dipole change? Express your answer as the ratio (magnitude of new force / magnitude of Fv).

Q14 Draw a diagram like the one in Figure 13.61.

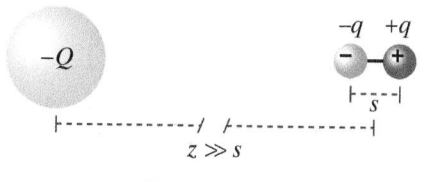

Figure 13.61

On your diagram, draw vectors showing: **(a)** the electric field of the dipole at the location of the negatively charged ball, **(b)** the net force on the ball due to the dipole, **(c)** the electric field of the ball at the center of the dipole, **(d)** the net force on the dipole due to the ball.

Q15 If the distance between the ball and the dipole in Figure 13.61 were doubled, what change would there be in the force on the ball due to the dipole?

PROBLEMS

Section 13.3

•**P16** An electron in a region in which there is an electric field experiences a force of magnitude 3.8×10^{-16} N. What is the magnitude of the electric field at the location of the electron?

•**P17** The electric field at a particular location is measured to be $\langle 0, -280, 0 \rangle$ N/C. What force would a positron experience if placed at this particular location?

•**P18** An electron in a region in which there is an electric field experiences a force of magnitude 3.7×10^{-16} N. What is the magnitude of the electric field at the location of the electron?

•**P19** If the particle in Figure 13.62 is a proton, and the electric field $\vec{E}_1$ has the value $\langle 2 \times 10^4, 2 \times 10^4, 0 \rangle$ N/C, what is the force $\vec{F}_2$ on the proton?

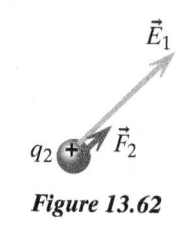

Figure 13.62

•**P20** An electron in a region in which there is an electric field experiences a force of $\langle 8.0 \times 10^{-17}, -3.2 \times 10^{-16}, -4.8 \times 10^{-16} \rangle$ N. What is the electric field at the location of the electron?

••**P21** In the region shown in Figure 13.63 there is an electric field due to a point charge located at the center of the dashed circle. The arrows indicate the magnitude and direction of the electric field at the locations shown.

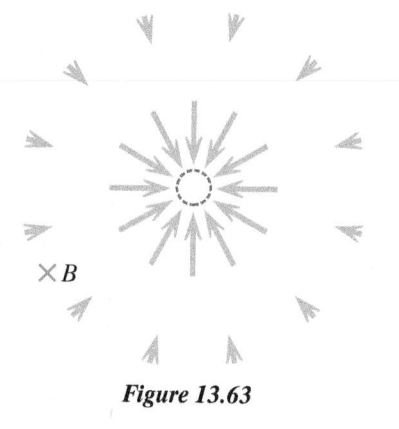

Figure 13.63

(a) What is the sign of the source charge? (b) Now a particle whose charge is -7×10^{-9} C is placed at location B. What is the direction of the electric force on the -7×10^{-9} C charge? (c) The electric field at location B has the value $\langle 2000, 2000, 0 \rangle$ N/C. What is the unit vector in the direction of $\vec{E}$ at this location? (d) What is the electric force on the -7×10^{-9} C charge? (e) What is the unit vector in the direction of this electric force?

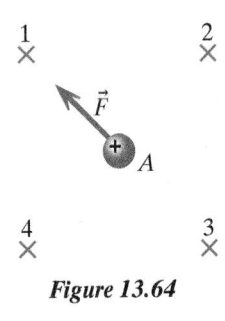

Figure 13.64

••**P22** In the region shown in Figure 13.64 there is an electric field due to charged objects not shown in the diagram. A tiny glass ball with a charge of 5×10^{-9} C placed at location A experiences a force of $\langle 4 \times 10^{-5}, -4 \times 10^5, 0 \rangle$ N, as shown in the figure. (a) Which arrow in Figure 13.65 best indicates the direction of the electric field at location A? (b) What is the electric field at location A? (c) What is the magnitude of this electric field? (d) Now the glass ball is moved very far away. A tiny plastic ball with charge -6×10^{-9} C is placed at location A. Which arrow in Figure 13.65 best indicates the direction of the electric force on the negatively charged plastic ball? (e) What is the force on the negative plastic ball? (f) You discover that the source of the electric field at location A is a negatively charged particle. Which of the numbered locations in Figure 13.64 shows the location of this negatively charged particle, relative to location A?

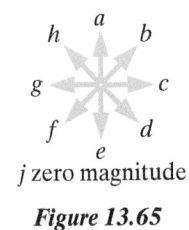

Figure 13.65

••**P23** An electron is observed to accelerate in the $+z$ direction with an acceleration of 1.6×10^{16} m/s^2. Explain how to use the definition of electric field to determine the electric field at this location, and give the direction and magnitude of the field.

••**P24** An object falling in a vacuum near a planet has a charge of -4×10^{-8} C and a mass of 0.3 kg. In this region of space there is an electric field $\langle 2 \times 10^7, 0, 0 \rangle$ N/C and a gravitational field $\langle 0, 5, 0 \rangle$ N/kg. What is the net force acting on the object?

••**P25** A proton is observed to have an instantaneous acceleration of 9×10^{11} m/s^2. What is the magnitude of the electric field at the proton's location?

Section 13.4

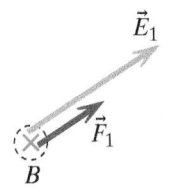

Figure 13.66

•**P26** In Figure 13.66 a proton at location A makes an electric field $\vec{E}_1$ at location B. A different proton, placed at location B, experiences a force $\vec{F}_1$.

Now the proton at B is removed and replaced by a lithium nucleus, containing three protons and four neutrons. (a) Now what is the value of the electric field at location B due to the proton? (b) What is the force on the lithium nucleus? (c) The lithium nucleus is removed, and an electron is placed at location B. Now what is the value of the electric field at location B due to the proton? (d) What is the magnitude of the force on the electron? (e) Which arrow in Figure 13.65 best indicates the direction of the force on the electron due to the electric field?

•**P27** You want to calculate the electric field at location $\langle 0.5, -0.1, -0.5 \rangle$ m, due to a particle with charge +9 nC located at $\langle -0.6, -0.7, -0.2 \rangle$ m. (a) What is the source location? (b) What is the observation location? (c) What is the vector $\vec{r}$ that points from the source location to the observation location? (d) What is $|\vec{r}|$? (e) What is the vector $\hat{r}$? (f) What is the value of $\dfrac{1}{4\pi\varepsilon_0}\dfrac{q}{|\vec{r}|^2}$? (g) Finally, what is the electric field, expressed as a vector?

•**P28** A particle with charge +5 nC (a nanocoulomb is 1×10^{-9} C) is located at the origin. What is the electric field due to this particle at a location $\langle 0.4, 0, 0 \rangle$ m?

•**P29** What is the electric field at a location $\langle -0.1, -0.1, 0 \rangle$ m, due to a particle with charge +4 nC located at the origin?

•**P30** In a hydrogen atom in its ground state, the electron is on average a distance of about 0.5×10^{-10} m from the proton. What is the magnitude of the electric field due to the proton at this distance from the proton?

•**P31** A sphere with radius 1 cm has a charge of 2×10^{-9} C spread out uniformly over its surface. What is the magnitude of the electric field due to the sphere at a location 4 cm from the center of the sphere?

•**P32** A sphere with radius 2 cm is placed at a location near a point charge. The sphere has a charge of -9×10^{-10} C spread uniformly over its surface. The electric field due to the point charge has a magnitude of 470 N/C at the center of the sphere. What is the magnitude of the force on the sphere due to the point charge?

•**P33** What are the magnitude and direction of the electric field $\vec{E}$ at location $\langle 20, 0, 0 \rangle$ cm if there is a negative point charge of 1 nC (1×10^{-9} C) at location $\langle 40, 0, 0 \rangle$ cm? Include units.

•**P34** A sphere with radius 2 cm is placed at a location near a point charge. The sphere has a charge of -8×10^{-10} C spread uniformly over its surface. The electric field due to the point charge has a magnitude of 500 N/C at the center of the sphere. What is the magnitude of the force on the sphere due to the point charge?

••P35 An electron is located at $\langle 0.8, 0.7, -0.8 \rangle$ m. You need to find the electric field at location $\langle 0.5, 1, -0.5 \rangle$ m, due to the electron. **(a)** What is the source location? **(b)** What is the observation location? **(c)** What is the vector $\vec{r}$? **(d)** What is $|\vec{r}|$? **(e)** What is the vector $\hat{r}$? **(f)** What is the value of $\dfrac{1}{4\pi\varepsilon_0} \dfrac{q}{|\vec{r}|^2}$? **(g)** Finally, what is the electric field at the observation location, expressed as a vector?

••P36 A charged particle located at the origin creates an electric field of $\langle -1.2 \times 10^3, 0, 0 \rangle$ N/C at a location $\langle 0.12, 0, 0 \rangle$ m. What is the particle's charge?

••P37 At a particular location in the room there is an electric field = $\langle 1000, 0, 0 \rangle$ N/C. Where would you place a single negative point particle of charge 1 μC in order to produce this electric field?

••P38 The electric field at a location C points north, and the magnitude is 1×10^6 N/C. Give numerical answers to the following questions: **(a)** Where relative to C should you place a single proton to produce this field? **(b)** Where relative to C should you place a single electron to produce this field? **(c)** Where should you place a proton and an electron, at equal distances from C, to produce this field?

••P39 You want to create an electric field = $\langle 0, 4104, 0 \rangle$ N/C at location $\langle 0, 0, 0 \rangle$. **(a)** Where would you place a proton to produce this field at the origin? **(b)** Instead of a proton, where would you place an electron to produce this field at the origin? (*Hint*: This problem will be much easier if you draw a diagram.)

••P40 A π^- ("pi-minus") particle, which has charge $-e$, is at location $\langle 7 \times 10^{-9}, -4 \times 10^{-9}, -5 \times 10^{-9} \rangle$ m. **(a)** What is the electric field at location $\langle -5 \times 10^{-9}, 5 \times 10^{-9}, 4 \times 10^{-9} \rangle$ m, due to the π^- particle? **(b)** At a particular moment an antiproton (same mass as the proton, charge $-e$) is at the observation location. At this moment what is the force on the antiproton, due to the π^-?

••P41 What is the electric field at a location $\vec{b} = \langle -0.1, -0.1, 0 \rangle$ m, due to a particle with charge $+3$ nC located at the origin?

••P42 At a particular location in the room there is an electric field $\vec{E} = \langle 1000, 0, 0 \rangle$ N/C. Figure out where to place a single positive point particle, and how much charge it should have, in order to produce this electric field (there are many possible answers!). Do the same for a single negatively charged point particle. Be sure to draw diagrams to explain the geometry of the situation.

••P43 Where must an electron be to create an electric field of $\langle 0, 160, 0 \rangle$ N/C at a location in space? Calculate its displacement from the observation location and show its location on a diagram.

••P44 The electric field at a location C points west, and the magnitude is 2×10^6 N/C. Give numerical answers to the following questions: **(a)** Where relative to C should you place a single proton to produce this field? **(b)** Where relative to C should you place a single electron to produce this field? **(c)** Where should you place a proton and an electron, at equal distances from C, to produce this field?

••P45 A lithium nucleus consisting of three protons and four neutrons accelerates to the right due to electric forces, and the initial magnitude of the acceleration is 3×10^{13} m/s/s. **(a)** What is the direction of the electric field that acts on the lithium nucleus? **(b)** What is the magnitude of the electric field that acts on the lithium nucleus? Be quantitative (that is, give a number). **(c)** If

this acceleration is due solely to a single helium nucleus (two protons and two neutrons), where is the helium nucleus initially located? Be quantitative (that is, give a number).

Section 13.5

••P46 **(a)** On a clear and carefully drawn diagram, place a helium nucleus (consisting of two protons and two neutrons) and a proton in such a way that the electric field due to these charges is zero at a location marked $\times$, a distance 1×10^{-10} m from the helium nucleus. Explain briefly but carefully, and use diagrams to help in the explanation. Be quantitative about the relative distances. **(b)** On a clear and carefully drawn diagram, place a helium nucleus and an electron in such a way that the electric field due to these charges is zero at a location marked $\times$. Explain briefly but carefully, and use diagrams to help in the explanation. Be quantitative about the relative distances.

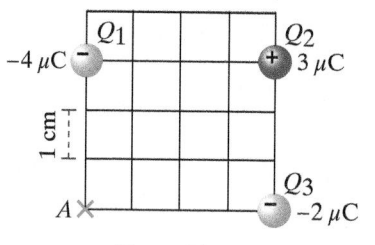

Figure 13.67

••P47 At a particular moment, three charged particles are located as shown in Figure 13.67. $Q_1 = -4\,\mu$C, $Q_2 = +3\,\mu$C, and $Q_3 = -2\,\mu$C. Your answers to the following questions should be vectors. (Recall that $1\,\mu$C $= 1 \times 10^{-6}$ C.) **(a)** Find the electric field at the location of Q_3, due to Q_1. **(b)** Find the electric field at the location of Q_3, due to Q_2. **(c)** Find the net electric field at the location of Q_3. **(d)** Find the net force on Q_3. **(e)** Find the electric field at location A due to Q_1. **(f)** Find the electric field at location A due to Q_2. **(g)** Find the electric field at location A due to Q_3. **(h)** What is the net electric field at location A? **(i)** If a particle with charge -3 nC were placed at location A, what would be the force on this particle?

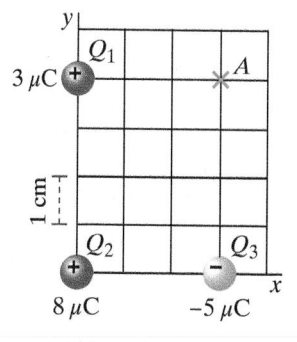

Figure 13.68

••P48 At a particular moment, one negative and two positive charges are located as shown in Figure 13.68. $Q_1 = +3\,\mu$C, $Q_2 = +8\,\mu$C, and $Q_3 = -5\,\mu$C. Your answers to each part of this problem should be vectors. (Recall that $1\,\mu$C $= 1 \times 10^{-6}$ C.) **(a)** Find the electric field at the location of Q_1, due to Q_2 and Q_3. **(b)** Use the electric field you calculated in part (a) to find the force on Q_1. **(c)** Find the electric field at location A, due to all three charges. **(d)** An alpha particle (He^{2+}, containing two protons and two neutrons) is released from rest at location A.

Use your answer from the previous part to determine the initial acceleration of the alpha particle.

••P49 An Fe^{3+} ion is located 400 nm (400×10^{-9} m, about 4000 atomic diameters) from a Cl^- ion, as shown in Figure 13.69 (ions not shown to scale).

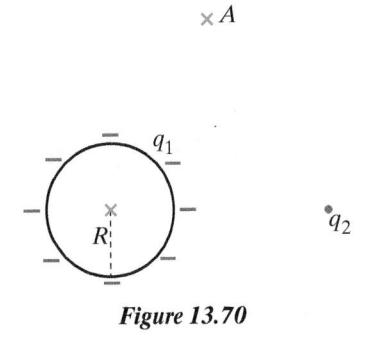

Figure 13.69

(a) Determine the magnitude and direction of the electric field $\vec{E}_A$ at location A, 100 nm to the left of the Cl^- ion. **(b)** Determine the magnitude and direction of the electric field $\vec{E}_B$ at location B, 100 nm to the right of the Cl^- ion. **(c)** If an electron is placed at location A, what are the magnitude and direction of the force on the electron?

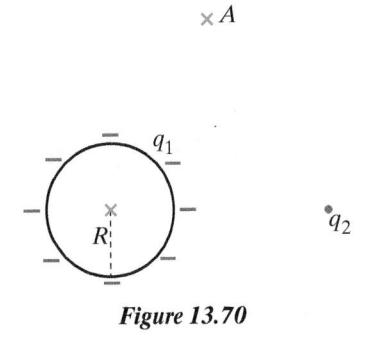

Figure 13.70

••P50 A hollow ball with radius $R = 2$ cm has a charge of $q_1 = -3$ nC spread uniformly over its surface, as shown in Figure 13.70. The center of the ball is at $\langle -3,0,0 \rangle$ cm. A point charge of $q_2 = 5$ nC is located at $\langle 4,0,0 \rangle$ cm. **(a)** What is the net electric field at location A, when $\vec{r}_A = \langle 0,6,0 \rangle$ cm? **(b)** Draw an arrow representing the net electric field at that location. Make sure that the arrow you drew makes sense.

••P51 Three nested hollow spheres have the same center. The innermost sphere has a radius of 2 cm and carries a uniformly distributed charge of 6 nC ($1 \text{ nC} = 1 \times 10^{-9}$ C). The middle sphere has a radius of 5 cm and carries a uniformly distributed charge of -4 nC. The outermost sphere has a radius of 10 cm and carries a uniformly distributed charge of 8 nC. **(a)** What is the magnitude of the electric field at a distance of 1 cm from the center? **(b)** What is the magnitude of the electric field at a distance of 4 cm from the center? **(c)** What is the magnitude of the electric field at a distance of 9 cm from the center?

Section 13.6

•P52 A dipole is located at the origin and is composed of charged particles with charge $+e$ and $-e$, separated by a distance 6×10^{-10} m along the x axis. The charge $+e$ is on the $+x$ axis. Calculate the electric field due to this dipole at a location $\langle 0, -5 \times 10^{-8}, 0 \rangle$ m.

•P53 A dipole is located at the origin and is composed of charged particles with charge $+e$ and $-e$, separated by a distance 2×10^{-10} m along the x axis. Calculate the magnitude of the electric field due to this dipole at a location $\langle 0, 3 \times 10^{-8}, 0 \rangle$ m.

•P54 The dipole moment of the HF (hydrogen fluoride) molecule has been measured to be 6.3×10^{-30} C·m. If we model

the dipole as having charges of $+e$ and $-e$ separated by a distance s, what is s? Is this plausible?

••P55 A dipole is located at the origin and is composed of charged particles with charge $+e$ and $-e$, separated by a distance 6×10^{-10} m along the y axis. The $+e$ charge is on the $-y$ axis. Calculate the force on a proton due to this dipole at a location $\langle 0, 4 \times 10^{-8}, 0 \rangle$ m.

••P56 A dipole is located at the origin and is composed of charged particles with charge $+2e$ and $-2e$, separated by a distance 2×10^{-10} m along the y axis. The $+2e$ charge is on the $+y$ axis. Calculate the force on a proton at a location $\langle 0, 0, 3 \times 10^{-8} \rangle$ m due to this dipole.

•P57 A dipole consists of two charges $+6$ nC and -6 nC, held apart by a rod of length 3 mm, as shown in Figure 13.71. **(a)** What is the magnitude of the electric field due to the dipole at location A, 5 cm from the center of the dipole? **(b)** What is the magnitude of the electric field due to the dipole at location B, 5 cm from the center of the dipole?

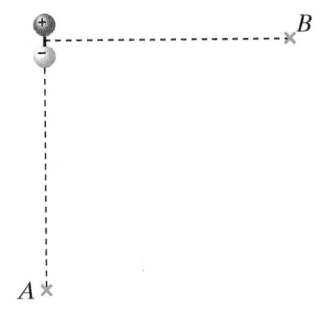

Figure 13.71

••P58 A dipole is centered at the origin and is composed of charged particles with charge $+2e$ and $-2e$, separated by a distance 7×10^{-10} m along the y axis. The $+2e$ charge is on the $-y$ axis, and the $-2e$ charge is on the $+y$ axis. **(a)** A proton is located at $\langle 0, 3 \times 10^{-8}, 0 \rangle$ m. What is the force on the proton due to the dipole? **(b)** An electron is located at $\langle -3 \times 10^{-8}, 0, 0 \rangle$ m. What is the force on the electron due to the dipole? (*Hint:* Make a diagram. One approach is to calculate magnitudes, and get directions from your diagram.)

••P59 Two dipoles are oriented as shown in Figure 13.72. Each dipole consists of two charges $+q$ and $-q$, held apart by a rod of length s, and the center of each dipole is a distance d from location A. If $q = 2$ nC, $s = 1$ mm, and $d = 8$ cm, what is the electric field at location A? (*Hint:* Draw a diagram and show the direction of each dipole's contribution to the electric field on the diagram.)

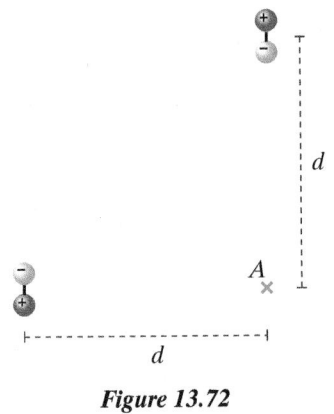

Figure 13.72

••**P60** Two dipoles are oriented as shown in Figure 13.73. Each dipole consists of charges held apart by a short rod (not shown to scale). What is the electric field at location A? Start by drawing a diagram that shows the direction of each dipole's contribution to the electric field at location A.

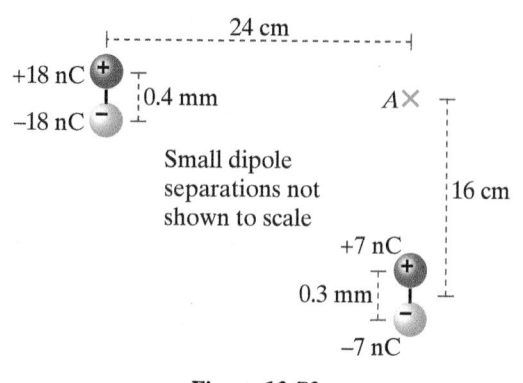

Figure 13.73

••**P61** A charge of $+1\,\text{nC}$ ($1 \times 10^{-9}\,\text{C}$) and a dipole with charges $+q$ and $-q$ separated by 0.3 mm contribute a net field at location A that is zero, as shown in Figure 13.74.

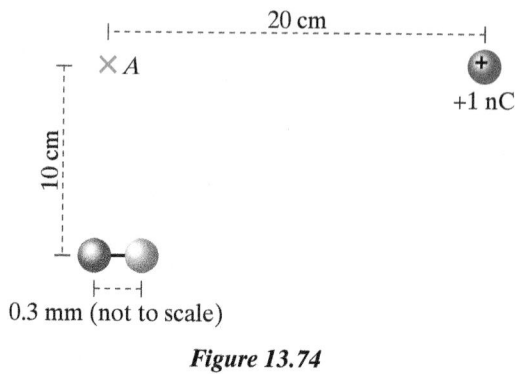

Figure 13.74

(a) Which end of the dipole is positively charged? **(b)** How large is the charge q?

•••**P62** A water molecule is asymmetrical, with one end positively charged and the other negatively charged. It has a dipole moment whose magnitude is measured to be 6.2×10^{-30} C·m. If the dipole moment is oriented perpendicular to an electric field whose magnitude is 4×10^5 N/m, what is the magnitude of the torque on the water molecule? Also, show that the vector torque is equal to $\vec{p} \times \vec{E}$, where $\vec{p}$ is the dipole moment.

•••**P63** Two identical permanent dipoles, each consisting of charges $+q$ and $-q$ separated by a distance s, are aligned along the x axis, a distance r from each other, where $r \gg s$ (Figure 13.75). Show all of the steps in your work, and briefly explain each step. **(a)** Draw a diagram showing all individual forces acting on each particle, and draw heavier vectors showing the net force on each dipole. **(b)** Show that the magnitude of the net force exerted on one dipole by the other dipole is this:

$$F \approx \frac{1}{4\pi\varepsilon_0} \frac{6q^2 s^2}{r^4}$$

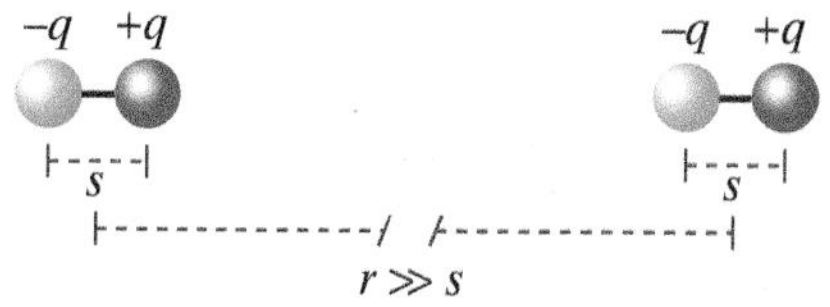

Figure 13.75

Section 13.8

••**P64** You make repeated measurements of the electric field $\vec{E}$ due to a distant charge, and you find it is constant in magnitude and direction. At time $t = 0$ your partner moves the charge. The electric field doesn't change for a while, but at time $t = 45$ ns you observe a sudden change. How far away was the charge originally?

COMPUTATIONAL PROBLEMS

If you did not learn how to create 3D computational models using VPython in your previous physics course, you should work through the introductory materials from Chapter 1 that are available at www.wiley.com/college/chabay.

To install the free 3D programming environment VPython, go to vpython.org and (carefully) follow the instructions for your operating system (Windows, MacOS, or Linux). Note the instructions given there on how to zoom and rotate the "camera" when viewing a 3D scene you have created.

To use an `arrow` object to visualize an electric field, it is usually necessary to scale the length of the arrow in order to make it fit on the screen with the objects that produce or experience that field. Watch VPython Instructional Video 5: Scalefactors, at vpython.org/video05.html to learn how to do this. In general it may be helpful to print the magnitude of an electric field using `print(mag(E))` in order to decide on an appropriate scale factor.

More detailed and extended versions of some of these computational modeling problems may be found in the lab

activities included in the *Matter & Interactions, 4th Edition,* resources for instructors.

•**P65** In Section 13.9 there is a program to calculate the electric field of a single point charge at multiple observation locations. **(a)** Study this program and make sure you can explain every line of code. **(b)** Modify the program so that the magnitude of the electric field at each observation location is the same. (There is more than one way to do this.) **(c)** Add an additional circle of observation locations in the yz plane, centered on the point charge.

••**P66** A particle with a charge of $+3$ nC is located at $\langle -0.04, 0, 0 \rangle$ m.   **(a)** Calculate the electric field at location $\langle -0.04, 0, 0.05 \rangle$ m due to this particle, and create an arrow to visualize the field at the observation location. Try a scale factor of about 2×10^{-6}. **(b)** Add an arrow representing the electric field at location $\langle -0.04, 0, -0.05 \rangle$ due to this particle. **(c)** Add two more arrows, each representing the electric field at a location 0.05 m from the particle in the $\pm y$ direction. **(d)** Add two more arrows, each representing the electric field at a location 0.05 m from the particle in the $\pm x$ direction.

••**P67** Three charged spheres lie in the xz plane. Sphere a has a charge of $+2$ nC, and is located at $\langle -0.03, 0, 0.03 \rangle$ m. Sphere b has a charge of $+4$ nC, and is located at $\langle 0.03, 0, 0.03 \rangle$ m. Sphere c has a charge of -2 nC, and is located at $\langle 0, 0, -0.011 \rangle$ m. Write a program to calculate and display (using an arrow) the electric field of these three charged spheres at location $\langle 0, 0.25, 0 \rangle$ m.

••**P68** The following code creates two objects representing a dipole oriented along the y axis. **(a)** Extend the program to calculate and display (using arrows) the electric field due to the dipole at 12 equally spaced observation locations located on a circle of radius 0.5 nm in the xy plane, centered on the dipole. **(b)** Add a second circle of arrows representing the electric field at observation locations in the yz plane.

```
from visual import *
scene.width = scene.height = 800
oofpez = 9e9
qe = 1.6e-19
sf = 5e-20
source_01 = sphere(pos=vector(0,0.1e-9,0),
                color=color.red,
                radius=0.5e-10)
q_01 = +qe
source_02 = sphere(pos=vector(0,-0.1e-9,0),
                color=color.blue,
                radius=0.5e-10)
q_02 = -qe
```

••**P69** The following skeleton program creates objects representing a stationary source charge and a moving antiproton. **(a)** Complete the program so that the antiproton is affected by the net electric field at its current location. **(b)** Add two arrows, and use them to visualize the momentum of the antiproton, and the electric field at the location of the antiproton. These arrows should move with the antiproton.

```
from visual import *
scene.width = scene.height = 760
scene.range = 3e-8
oofpez = 9e9
qe = 1.6e-19
source = sphere(pos=vector(-0.5e-8,0,0),
                radius=5e-10,
                color=color.red)
```

```
source_q = qe
##antiproton
ap = sphere(pos=vector(0.5e-8,-1e-8,0),
                radius=5e-10,
                color=color.cyan,
                make_trail = True)
ap_q = -qe
ap_m = 1.7e-27
ap_p = ap_m * vector(0,5e3,0)
deltat = 1e-15
t=0

while t < 8e-12:
    rate(500)
    ## add your code here
    ap.pos = ap.pos + (ap_p/ap_m) * deltat
    t = t + deltat
```

•••**P70** Start with the program you wrote in Problem P68 to calculate and display the electric field of a dipole. **(a)** Place a proton at location $\langle 0.3 \times 10^{-9}, 0, 0 \rangle$ m, and release it from rest. Compute and display the trajectory of the proton as it moves under the influence of the electric field of the dipole. You may wish to start with $\Delta t = 1 \times 10^{-17}$ s. **(b)** Simultaneously compute and plot a graph showing the potential energy U, kinetic energy K, and $(K + U)$ vs. time for the entire system (dipole + proton). Your graph will be more useful as a computational diagnostic tool if you do not include the potential energy associated with the interaction of the pair of charges making up the dipole, which does not change. **(c)** Explain the shape of the K and U graphs.

•••**P71** Write a computer program to calculate and plot a graph of the magnitude of the electric field of the dipole from Problem P68 at locations on the y axis as a function of distance from the center of the dipole. Vary y from 0.2 nm $(0.2 \times 10^{-9}$ m) to 0.5 nm from the center of the dipole. Do the calculation two different ways, and put both plots (in different colors) on the same axes: **(a)** Calculate the electric field exactly as the superposition of the fields due to the individual charges. **(b)** Calculate the electric field using the approximate equation for the dipole field derived in Section 13.6. **(c)** Comment on the validity of the approximate equation. How close to the dipole (compared to s, the dipole separation) do you have to get before the approximate equation no longer gives good results? What is your criterion for "good results"?

A N S W E R S   T O   C H E C K P O I N T S

1 $\approx 1 \times 10^{-10}$ m; protons and electrons
2 $\langle 0, -4.8 \times 10^{-17}, 0 \rangle$ N; $\langle 0, 0, 0 \rangle$ N
3 $\langle 900, 0, 0 \rangle$ N/C
4 2.8×10^4 N/C
5 (a) 3.6×10^4 N/C; **(b)** 7.2×10^4 N/C

6 (a) 1/8; **(b)** three times bigger; **(c)** two times bigger, opposite direction
7 Its electric dipole moment would point horizontally, in the same direction as the applied electric field, with the $+$ charge to the right.
8 3.2×10^{-19} C

Electric Fields and Matter

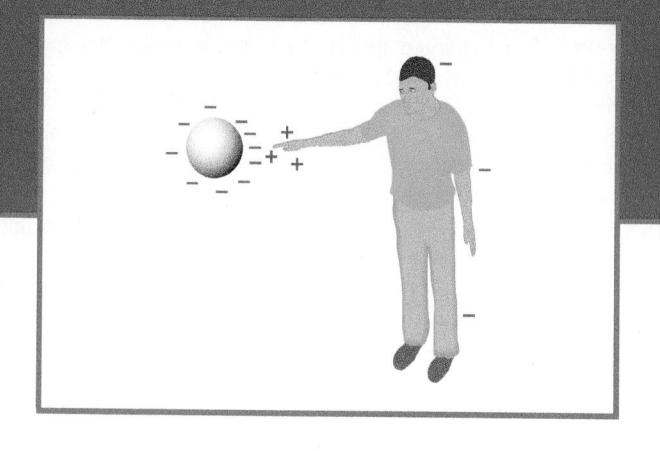

OBJECTIVES

After studying this chapter you should be able to

- Explain the difference between a conductor and an insulator, and graphically depict the polarization of each when subjected to an external electric field.
- Explain why a charged object is attracted to a neutral object.
- Calculate the drift speed of a charged particle in a conductor.
- Explain why the net electric field inside a conductor at equilibrium must be zero.
- Mathematically relate polarizability, induced dipole moment, and applied electric field.

14.1 CHARGED PARTICLES IN MATTER

Since ordinary matter is composed of charged particles, electric fields can affect matter. In order to understand the effect of electric fields on matter, in this chapter we will extend our microscopic model of matter to include the fact that matter contains charged particles: protons and electrons.

Net Charge

DEFINITION OF NET CHARGE

The net charge of an object is the sum of the charges of all of its constituent particles. An object with a net charge of zero is called "neutral." An object with a nonzero net charge (either positive or negative) is called "charged."

Elementary particles such as protons and electrons are electrically charged. If a proton and an electron combine to form a hydrogen atom, however, the hydrogen atom is electrically "neutral"—its net charge is the sum of the charges of its constituent particles, which in this case is zero:

$$(+e) + (-e) = 0$$

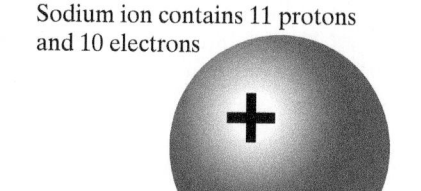

Sodium ion contains 11 protons and 10 electrons

Lost electron

Figure 14.1 A sodium ion Na^+ consists of a sodium atom that has lost an electron.

A sodium atom has 11 protons in its nucleus and 11 electrons surrounding the nucleus, so it has a net charge of zero and is electrically neutral. However, a sodium atom can lose an electron, becoming a sodium ion, Na^+.

QUESTION What is the net charge of a sodium ion, Na^+ (Figure 14.1)?

A sodium ion has 11 protons and 10 electrons, so its net charge is

$$(+11e) + (-10e) = +e = +1.6 \times 10^{-19} \, C$$

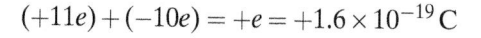

Ordinary matter is electrically neutral. However, it is possible to remove or add charged particles, giving an object a nonzero net charge.

QUESTION   Is a dipole electrically neutral?

A dipole is neutral, because the sum of its constituent charges is zero:

$$(+q)+(-q)=0$$

Evidently even a neutral object can make a nonzero electric field in the surrounding space.

Conservation of Charge

In an extremely wide variety of experiments, no one has ever observed a change in the net charge of the universe. These results are summarized by the fundamental principle called "conservation of charge": if the net charge of a system changes, the net charge of the surroundings must change by the opposite amount. For example, if your comb acquires negative charge, your hair acquires an equal amount of positive charge:

CONSERVATION OF CHARGE

The net charge of a system plus its surroundings cannot change.

Conservation of charge is a fundamental principle, because it applies to every system in every situation.

Consider the annihilation reaction between an electron and a positron:

$$e^- + e^+ \rightarrow \gamma + \gamma$$

QUESTION   In this reaction an electron and a positron are destroyed, creating two high-energy photons (called gamma rays). Does this reaction violate the principle of conservation of charge?

No. The net charge of the system (electron plus positron) was initially zero; the charge of the two photons is also zero. Even though charged particles were destroyed, the net charge of the system did not change.

> **Checkpoint 1** A carbon atom is composed of 6 protons, 6 neutrons, and 6 electrons. What is the net charge of this atom? A neutral chlorine atom contains 17 protons and 17 electrons. When a chlorine atom gains one extra electron, it becomes a chloride ion. What is the net charge of a chloride ion?

Conductors and Insulators

All materials are made of atoms that contain electrons and protons. However, at the microscopic level there can be differences in structure that lead to very different behavior when macroscopic objects are exposed to electric fields. In this chapter we will examine two different kinds of materials: conductors and insulators. (There are other classes of technologically important materials, such as semiconductors and superconductors, which we will discuss briefly in later chapters.)

Conductors: Some materials contain charged particles that can move easily through the material. These materials are called conductors. Most metals, such as copper, silver, iron, aluminum, and gold, are excellent conductors because they contain highly mobile electrons, as we will discuss later. Aqueous salt solutions are also conductors, because there are mobile positive and negative ions. An example is ordinary salt water, containing positive sodium ions (Na^+) and negative chloride ions (Cl^-).

Insulators: In many materials the electrons are tightly bound to the atoms, and there are no charged particles that can move through the material. These materials are called insulators because they can electrically "insulate" one charged object from another. You are familiar with many insulating materials, such as rubber, plastics, wood, paper, and glass. A related experiment is EXP11.

DEFINITION OF "CONDUCTOR" AND "INSULATOR"

A conductor contains mobile charged particles that can move throughout the material.

An insulator has no mobile charged particles.

14.2 HOW OBJECTS BECOME CHARGED

Many of the interactions we observe in our everyday lives are electric in nature. By being systematic in observing the behavior of simple systems and by thinking carefully in our analysis of this behavior, we can uncover some deep questions about the interaction of ordinary matter with electric fields.

In the section "Basic Experiments" at the end of this chapter are some simple experiments that you can do to make the issues vivid. All you need is a roll of "invisible" tape such as Scotch® brand Magic™ Tape or a generic brand of frosted tape (Figure 14.2). When you pull a long piece of invisible tape off a roll, it often curls up or sticks to your hand, and this is due to electric interactions between the electrically charged tape and your hand. We encourage you to do these simple experiments as you study this chapter.

In experimenting with charged objects such as invisible tape we find the following:

- Objects can have a charge that is positive, negative, or zero.
- Like charges repel (Figure 14.3), unlike charges attract.
- The electric force
  - acts along a line between the charges,
  - decreases rapidly as the distance between the charges increases, and
  - is proportional to the amounts of both charges.

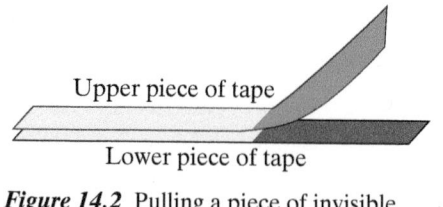

Figure 14.2 Pulling a piece of invisible tape off of another piece charges the tapes electrically. See Experiment EXP1.

Upper piece of tape

Lower piece of tape

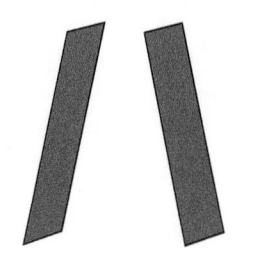

Figure 14.3 Two pieces of invisible tape with like charges repel each other.

Charging by Contact

A charged object that has a net negative charge has more electrons than protons. A positively charged object has fewer electrons than protons.

> QUESTION Normally most ordinary objects appear to be neutral. How can they become significantly charged?

It has been known for a long time that if you rub a glass rod with silk, the glass rod becomes positively charged and the silk negatively charged (Figure 14.4). If you rub a clear plastic object such as a pen through your hair (or with fur, wool, or even cotton), the plastic ends up having a negative charge and so repels electrons (Figure 14.5). A similar process occurs when you separate one piece of invisible tape from another. Many objects acquire a nonzero net charge through contact with other objects. See the related experiments EXP1–EXP4.

There are a variety of possible mechanisms for charging by contact. Large organic molecules in the plastic or your hair could break at their weakest bond in such a way that negative ions (negatively charged fragments) are deposited on the plastic and/or positive ions (positively charged fragments) are deposited on your hair. Electrons can move from one object to another,

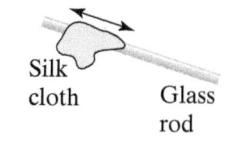

Silk cloth

Glass rod

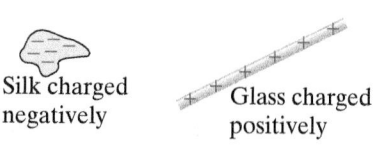

Silk charged negatively

Glass charged positively

Figure 14.4 Rub silk on glass and the glass becomes positively charged.

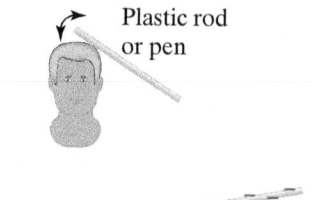

Plastic rod or pen

Hair charged positively

Plastic charged negatively

Figure 14.5 Rub a plastic pen on your hair or shirt and the pen becomes negatively charged.

especially if the objects are metals, as we will discuss later. It is known that rubbing is not essential to transferring charge from one object to another: mere contact can be sufficient. However, rubbing produces many points of contact, which may facilitate transfer.

Protons Are Not Removed from Nuclei

One thing is certain: you cannot remove bare nuclei from inside the surface atoms or remove protons from inside the nuclei of the surface atoms by rubbing. The amount of energy required to do this would be enormous. Removing protons would amount to transmuting one element into another! The nucleus is buried deep inside the atom, and the protons are bound tightly in the nucleus. A much smaller force is required to remove one electron from an atom, or to break a chemical bond and transfer an entire ion to another object. Therefore the only things that can be transferred by contact are positive or negative ions, or electrons.

We can make an approximate comparison of the energy required to charge an object by different mechanisms: breaking a bond, removing an electron from an atom, or removing a proton from a nucleus. We saw in Chapter 8 (Energy Quantization) that the energy required to ionize a hydrogen atom (that is, to move the electron very far away from the nucleus) was about 14 eV (recall that $1 \text{ eV} = 1.6 \times 10^{-19} \text{ J}$), so we can estimate the energy required to remove an electron from any atom as several eV. Similarly, the energy required to break one of the oxygen–hydrogen bonds in water is about 450 kJ/mol, which is 4.6 eV per molecular bond.

In contrast, as we saw in Chapter 6 (The Energy Principle), an input of 2.2 MeV (2.2×10^6 eV) is required to break apart the nucleus of a deuterium atom, which consists of one proton and one neutron, so we can estimate the energy required to remove a proton from a nucleus as millions of electron-volts.

Mechanism	Energy Required
Break chemical bond	several eV
Remove electron	several eV
Remove proton from nucleus	several *million* eV

It is clear that either breaking bonds or removing single electrons is a possible mechanism for charging a macroscopic object by contact, but removing protons from the atomic nuclei is not.

The Location of Charge Transfer

Suppose that two neutral strips of invisible tape are stuck together, then pulled quickly apart, as shown in Figure 14.6. Both tapes will become charged as a result of this process. (The sign of the charge on each tape depends on the chemical composition of the tape and the glue, which varies from brand to brand.)

QUESTION If the upper tape ends up being negatively charged after the tapes are pulled apart, what is the charge of the lower tape?

The principle of conservation of charge requires that the net charge of this isolated system remain unchanged. Since the tapes started out neutral, the lower tape must now be positively charged, and the absolute value of the charge of each tape must be the same. For example, if the charge of the upper tape is -1 nC, the charge of the lower tape must be $+1$ nC (1 nC $= 1 \times 10^{-9}$ C).

QUESTION After the tapes are separated, where are the excess charges on the negative upper tape located? Where are the excess charges on the positive lower tape located?

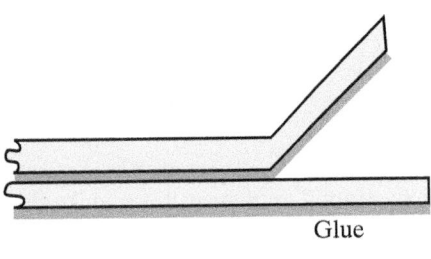

Glue

Figure 14.6 Two initially neutral tapes being pulled apart and becoming charged. Where are the regions with excess + and − charges?

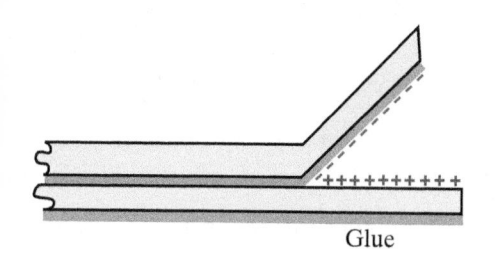

Figure 14.7 In this case the upper tape becomes negatively charged, so there must be excess electrons or negative ions on the part of its lower surface that is no longer touching the lower tape. The corresponding part of the lower tape has a deficiency of electrons, so there are now positive ions on it.

The tapes interact only along the bottom surface of the upper tape and the top surface of the lower tape, and because plastic is a good insulator the charges cannot move to a different location. Any excess charges must end up on one of these surfaces, as shown in Figure 14.7. The parts of the tapes that have been separated now have an excess or a deficiency of negative charge. Electrons or negative ions could have moved from the lower tape to the upper tape, or positive ions could have moved from the upper tape to the lower tape.

How Much Charge Is on a Charged Object?

Objects like invisible tapes, plastic pens, Ping-Pong balls, balloons, and glass rods can be electrically charged by rubbing with an appropriate material, ripping apart, or other similar contact. It would be useful to know approximately how much excess charge is on an ordinary small object like these when it is charged. Even knowing an approximate order of magnitude would be useful—is it closer to 10 C, 0.1 C, or 1×10^{-10} C?

EXAMPLE

Approximate Amount of Charge on a Piece of Tape

We can make a rough estimate of the amount of excess charge on a piece of tape by doing a simple experiment, illustrated in Figure 14.8. Prepare a pair of oppositely charged invisible tapes by pulling apart a pair of initially neutral tapes, so that the magnitude of charge on the two tapes is the same. If we suspend the positively charged tape between two books and lower the negatively charged tape toward it, when the negative tape gets close enough we will see that the positive tape is lifted upward. If we know the mass of the tape, we can estimate the electric force on the tape, and from this and the distance between the tapes, estimate the charge on the tape.

Solution

Momentum Principle:

$$\frac{dp_y}{dt} = F_{\text{electric}} - F_{\text{grav}}$$

$$0 = Q_B E_{A,y} - mg$$

In Figure 14.8, estimate the electric field due to A by approximating A as a point charge (a very rough approximation):

$$E_{A,y} \approx \frac{1}{4\pi\varepsilon_0} \frac{Q_A}{d^2}$$

so

$$Q_B \frac{1}{4\pi\varepsilon_0} \frac{Q_A}{d^2} \approx mg$$

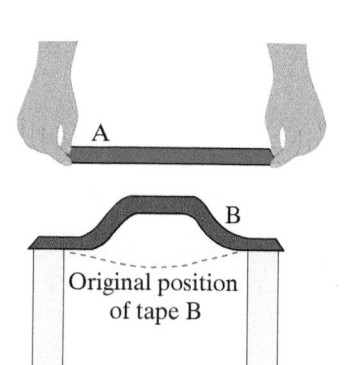

Figure 14.8 Tapes A and B are oppositely charged. When A gets sufficiently close, the electrical force on B is sufficient to lift it up.

If the 20-cm-long piece of tape has a mass of about 0.15 g, and the bottom tape starts to be lifted when the top tape is about 2.5 cm away, then, since the magnitude of the charge on the tapes is the same,

$$Q \approx \sqrt{\frac{(1.5 \times 10^{-4}\ \text{kg})(9.8\ \text{N/kg})(0.025\ \text{m})^2}{\left(9 \times 10^9 \dfrac{\text{N} \cdot \text{m}^2}{\text{C}^2}\right)}} \approx 1 \times 10^{-8}\,\text{C} = 10\ \text{nC}$$

This is a rough estimate, but it turns out to be a reasonable one—a small object charged by rubbing usually has a charge on the order of 10 nC. In Chapter 15 we'll see how to calculate the electric field of an object that is not point-like but has charge spread over its entire surface. Using these techniques, we find that the electric field at a location very close to the surface of the charged tape is around 2×10^5 N/C. Since the electric field required to ionize the air itself is about 3×10^6 N/C, the field near the surface of the charged tape is pretty large.

Experiment EXP6 guides you through a more careful version of this experiment. Your results may be somewhat different, depending on the design of the experiment (there are other possible geometries), the length of your tapes, and other factors.

Fraction of Surface Atoms with Excess Charge

Assuming that the net charge on the negative tape is distributed uniformly over its surface, we can estimate the fraction of those surface atoms that have gained an excess electron or negative ion.

EXAMPLE   **Fraction of Surface Atoms with Excess Charge**

What fraction of the atoms on the charged surface of the tape have gained or lost charge? (Assume that an atom gains or loses at most one electron charge.)

Solution   Surface area of a 1-cm-wide, 20-cm-long tape:

$$A = (0.2 \text{ m})(0.01 \text{ m}) = 2 \times 10^{-3} \text{ m}^2$$

Approximate cross-sectional area occupied by an atom whose radius is approximately 1×10^{-10} m:

$$A_{\text{atom}} \approx (2 \times 10^{-10} \text{ m})^2 = 4 \times 10^{-20} \text{ m}^2$$

Number of atoms on the surface:

$$\frac{A}{A_{\text{atom}}} = \frac{2 \times 10^{-3} \text{ m}^2}{4 \times 10^{-20} \text{ m}^2} = 5 \times 10^{16} \text{ atoms}$$

Number of excess electrons (or ions):

$$\frac{10 \times 10^{-9} \text{ C}}{1.6 \times 10^{-19} \text{ C}} = 6.25 \times 10^{10}$$

Fraction of surface atoms with excess charge:

$$\frac{6.25 \times 10^{10}}{5 \times 10^{16}} \approx 1 \times 10^{-6}$$

Thus only about one in a million atoms on the surface of the tape has acquired an excess electron or lost an electron—a small fraction.

At the macroscopic level, the charge on the surface of the tape may appear to be distributed quite uniformly. At the atomic level, though, we see that the atoms with excess charge are sprinkled quite sparsely over the surface.

14.3 POLARIZATION OF ATOMS

A positively charged object, like a positive piece of invisible tape, is attracted not only to a negative tape, but also to your hand, your desk, your book, and every other nearby neutral object (Figure 14.9). The same is true for a negatively charged object, such as a negative piece of invisible tape. Since

$$\vec{F} = q\vec{E}$$

this implies that your neutral hand must make a nonzero electric field. It is not obvious how or why this can happen.

Figure 14.9 A charged tape, whether positive or negative, is attracted to your neutral hand.

QUESTION   Why are charged objects attracted to neutral objects?

The attraction of both positively and negatively charged invisible tape to your hand, and to many other neutral objects, is deeply mysterious. The net charge of a neutral object is zero, so your neutral hand should not make an electric field that could act on a charged tape, nor should your neutral hand experience a force due to the electric field made by a charged tape. Nothing in our statement of the properties of electric interactions allows us to explain this attraction!

The Structure of an Atom

An external charge can cause a shift in the position of the charges that make up a neutral atom or molecule. To see this clearly we need to look in more detail at the structure of atoms. We'll consider a hydrogen atom because it is the simplest atom, but the effects we discuss occur with other atoms as well.

In Figure 14.10 we show a special kind of picture of a hydrogen atom, based on quantum mechanics, the theory that describes the detailed structure of atoms. A hydrogen atom consists of an electron and a nucleus normally consisting of one proton (and no neutrons). The lightweight electron doesn't follow a well-defined orbit around the heavy nucleus the way the Earth does around the Sun. Rather, there is only a probability for finding the electron in any particular place.

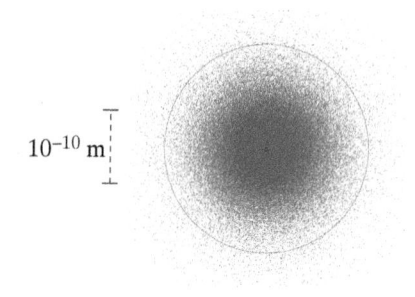

Figure 14.10 shows this probability graphically. You can think of the picture as a multiple exposure. For each exposure, the position of the electron at that time is shown as a dot. Because the electron is most likely to be found near the nucleus, that part of the multiple exposure is so dark you can't see the individual dots. The electron is seldom found a long way from the nucleus, so as you get farther and farther from the nucleus the density of dots gets less and less.

We call this probability distribution the "electron cloud." In hydrogen the cloud consists of just one electron, but in other atoms the electron cloud is made up of many electrons. The average location of the electron is in the center, at the same location as the nucleus. You're just as likely to find the electron to the right of the nucleus as to the left of the nucleus.

It is impossible to show the nucleus accurately on this scale. Although the mass of a proton is 2000 times the mass of an electron, the radius of the proton, about 1×10^{-15} m, is only about 1/100,000 as big as the radius of the electron cloud, which is itself only about 1×10^{-10} m! We used an oversize red dot to mark the position of the tiny nucleus in Figure 14.10.

In the following exercise, remember that in the previous chapter we pointed out that the electric field produced by a uniformly distributed sphere of charge, outside the sphere, is the same as though all the charge were located at the center of the sphere (this will be discussed in more detail in Chapter 15).

> **Checkpoint 2** A student asked, "Since the positive nucleus of the atom is hidden inside a negative electron cloud, why doesn't all matter appear to be negatively charged?" Explain to the student the flaw in this reasoning.

Polarization of Atoms

If the electron cloud in an atom could be considered to be spherically uniform and always centered on the nucleus, a neutral atom would have no interaction with an external charge. If the electron cloud were centered on the nucleus, the electric field produced by the N electrons would exactly cancel the field produced by the N protons. However, the electron cloud doesn't always stay centered, as we'll see next.

In an atom the electron cloud is not rigidly connected to the nucleus. The electron cloud and the nucleus can move relative to each other. If an external

Figure 14.10 A quantum-mechanical view of a hydrogen atom. The picture is a two-dimensional slice through a three-dimensional spherical distribution. Each dot represents the location of the electron at the time of a multiple-exposure photo. The tiny nucleus is shown as a red dot at the center of the electron cloud (actual nuclear radius is only about 1×10^{-15} m). A circle is drawn through regions of constant density.

10^{-10} m

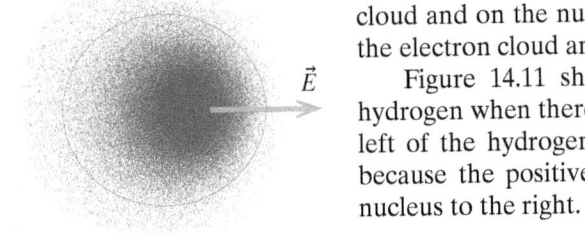

Figure 14.11 A positive charge makes an electric field that shifts the electron cloud of the hydrogen atom to the left (and shifts the hydrogen nucleus to the right). An ellipse is drawn through regions of constant density. It is now more probable that the electron will be found to the left of the nucleus than to the right.

charge is nearby, it creates an electric field, which exerts forces on the electron cloud and on the nucleus. Under the influence of this "applied" electric field the electron cloud and the nucleus shift position relative to each other.

Figure 14.11 shows the probability distribution or electron cloud for hydrogen when there is an external positive charge located somewhere to the left of the hydrogen atom. You can see that the cloud has been distorted, because the positive charge attracts the electron to the left and repels the nucleus to the right.

Average Location of the Electron

The average location of the electron is now not at the center where the nucleus is located, but is displaced somewhat to the left of the nucleus. That is, each time you take a snapshot, you're more likely to find the electron to the left of the nucleus than to the right of the nucleus.

The hydrogen atom isn't immediately torn apart, because the attraction between the nucleus and the electron is stronger than the forces exerted by the distant external charge. However, if the external charge gets *very* close the hydrogen atom may break up or react with the external charge. If the external charge were a proton, it could combine with the hydrogen atom to form ionized molecular hydrogen (H_2^+).

You can see in Figure 14.11 that the outer regions of the cloud are affected the most by the external charge. This is because in the outer regions the electron is farther from the nucleus and can be influenced more by the external positive charge. In an atom containing several electrons, the outer electrons are affected the most. The picture is deliberately exaggerated to show the effect: unless the polarization is caused by charges only a few atomic diameters away, the shift in the electron cloud is normally too small to represent accurately in a drawing.

An atom is said to be "polarized" when its electron cloud has been shifted by the influence of an external charge so that the electron cloud is not centered on the nucleus.

Diagrams of Polarized Atoms or Molecules

For most purposes we can approximate the charge distribution of the polarized atom as consisting of an approximately spherical negative cloud whose center is displaced from the positive nucleus (Figure 14.12). A uniform spherical charge distribution acts as if it were a point charge located at the center of the sphere, both in the sense that it makes an electric field outside the sphere identical to the electric field of a point charge and that it responds to applied fields as though it were a point charge. It is therefore reasonable to model a polarized atom as a dipole, consisting of two opposite point charges separated by a small distance.

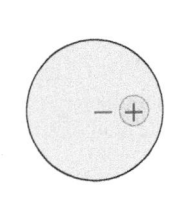

Figure 14.12 We can approximate a polarized atom as a roughly spherical electron cloud whose center is displaced from the positive nucleus.

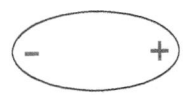

Figure 14.13 A simplified representation of a polarized atom or molecule.

To simplify drawing a polarized atom or molecule and to emphasize its most important aspects, we will usually represent it as an exaggeratedly elongated blob, with + and − at the ends (Figure 14.13).

Induced Dipoles Are Created by Applied Electric Fields

Figure 14.12 shows quite clearly that a polarized atom or molecule is a dipole, since there are two opposite charges separated by a distance. However, the polarized atom or molecule is not a permanent dipole. If the applied electric field is removed (for example, by removing the charges making that field), the electron cloud will shift back to its original position, and there will no longer be any charge separation. We call the polarized atom or molecule an "induced" dipole, because the dipole was induced (caused) to form by the presence of an applied electric field.

An "induced dipole" is created when a neutral object is polarized by an applied electric field. The induced dipole will vanish if the applied field is removed.

A "permanent dipole" consists of two opposite charges separated by a fixed distance, such as HCl or H_2O molecules, or the dipole that can be constructed out of + and − tapes (Experiment EXP8).

Polarizability

It has been found experimentally that for almost all materials, the amount of polarization induced (that is, the dipole moment $\vec{p}$ of the polarized atoms or molecules) is directly proportional to the magnitude of the applied electric field. This result can be written like this:

$$\vec{p} = \alpha \vec{E}$$

The constant α is called the "polarizability" of a particular material. The polarizability of many materials has been measured experimentally, and these experimental values may be found in reference volumes. Recall that the magnitude of the electric dipole moment of a dipole is

$$|\vec{p}| = qs$$

and the direction is from the negative charge toward the positive charge (see Chapter 13).

Checkpoint 3 A typical atomic polarizability is 1×10^{-40} C·m/(N/C). If the q in $p = qs$ is equal to the proton charge e, what charge separation s could you produce in a typical atom by applying a large field of 3×10^6 N/C, which is large enough to cause a spark in air?

A Neutral Atom and a Point Charge

In the previous chapter we found that since the electric field of a dipole was proportional to $1/r^3$, the force exerted by a dipole on a point charge was also proportional to $1/r^3$. Because of the reciprocity of the electric force, the force on the dipole by the point charge was therefore also proportional to $1/r^3$. Let us extend this analysis by considering the case of a point charge q_1 and a neutral atom.

Even though the entire process happens very quickly, it is instructive to analyze it as if it occurred in several steps. (Of course, the process is not instantaneous, since information about changes in electric field takes a finite time to propagate to distant locations.)

Step 1: At the location of the atom there is an electric field $\vec{E}_1$ due to the point charge (Figure 14.14). This electric field affects both the nucleus and the electron cloud, both of which, due to their spherical symmetry, can be modeled as point charges. The force on the electron cloud and the force on the nucleus are in opposite directions. Since the electron cloud and the nucleus can move relative to each other, they shift in opposite directions, until a new equilibrium position is reached.

The atom is now polarized, with dipole moment $\vec{p}_2 = \alpha \vec{E}_1$ proportional to the applied electric field $\vec{E}_1$.

Figure 14.14 At the location of the atom there is an electric field $\vec{E}_1$ due to the point charge.

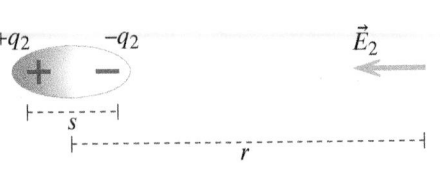

Figure 14.15 The polarized atom makes an electric field $\vec{E}_2$ at the location of the point charge.

Step 2: The polarized atom now has a dipole moment $p = q_2 s$. The atom, which is now an induced dipole, makes an electric field $\vec{E}_2$ at the location of the point charge (Figure 14.15). We can write an expression for the magnitude of $\vec{E}_2$:

$$|\vec{E}_2| = E_2 = \frac{1}{4\pi\varepsilon_0} \frac{2p}{r^3} = \frac{1}{4\pi\varepsilon_0} \frac{2\alpha E_1}{r^3}$$

Since we know $\vec{E}_1$, the electric field of the point charge at the location of the dipole, we can put that into our equation:

$$E_2 = \frac{1}{4\pi\varepsilon_0} \frac{2\alpha}{r^3} E_1 = \left(\frac{1}{4\pi\varepsilon_0} \frac{2\alpha}{r^3} \right) \left(\frac{1}{4\pi\varepsilon_0} \frac{q_1}{r^2} \right)$$

$$= \left(\frac{1}{4\pi\varepsilon_0} \right)^2 \left(\frac{2\alpha q_1}{r^5} \right)$$

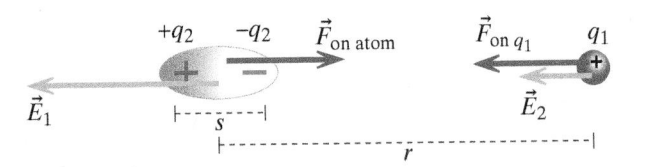

Figure 14.16 The force on the point charge due to the electric field of the polarized atom is equal in magnitude to the force on the polarized atom due to the electric field of the point charge.

QUESTION **Step 3:** What is the force exerted on the point charge by the induced dipole (Figure 14.16)?

$$\vec{F}_1 = q_1 \vec{E}_2 = q_1 \left(\frac{1}{4\pi\varepsilon_0} \right)^2 \left(\frac{2\alpha q_1}{r^5} \right) \hat{r} = \left(\frac{1}{4\pi\varepsilon_0} \right)^2 \left(\frac{2\alpha q_1^2}{r^5} \right) \hat{r}$$

We find that the force on the point charge by the polarized atom is proportional to $1/r^5$.

QUESTION What is the force on the neutral atom by the point charge?

Because of the reciprocity of the electric interaction (Newton's third law), the force on the neutral atom by the point charge is equal in magnitude and opposite in direction to the force on the point charge by the neutral atom:

$$\vec{F}_2 = -\vec{F}_1 = -\left(\frac{1}{4\pi\varepsilon_0} \right)^2 \left(\frac{2\alpha q_1^2}{r^5} \right) \hat{r}$$

so the force on a (polarized) neutral atom by a point charge also is proportional to $1/r^5$.

EXAMPLE

A Proton Attracts a Helium Atom

A proton and a helium atom are at rest, 20 nm apart. The polarizability of a helium atom is $\alpha = 2.3 \times 10^{-41}$ C·m/(N/C). **(a)** What is the initial acceleration of the helium atom? **(b)** What would be the initial acceleration of the helium atom if it were 100 nm from the proton?

Solution

(a) The mass of a helium atom is

$$m = \frac{4 \times 10^{-3} \text{ kg/mole}}{6.02 \times 10^{23} \text{ atoms/mole}} = 6.64 \times 10^{-27} \text{ kg}$$

The initial acceleration of the helium atom is

$$a = \frac{F}{m} = \left(\frac{1}{4\pi\varepsilon_0}\right)^2 \left(\frac{2\alpha q_1^2}{mr^5}\right)$$

$$= \left(9 \times 10^9 \frac{\text{N} \cdot \text{m}^2}{\text{C}^2}\right)^2 \left(\frac{2(2.3 \times 10^{-41} \text{ C} \cdot \text{m/(N/C)})(1.6 \times 10^{-19} \text{ C})^2}{(6.64 \times 10^{-27} \text{ kg})(20 \times 10^{-9} \text{ m})^5}\right)$$

$$= 4.5 \times 10^6 \text{ m/s}^2$$

This is an extremely high acceleration compared to $g = 9.8$ m/s^2.

(b) Since the force is proportional to $1/r^5$, increasing the distance r by a factor of 5 means that the force and acceleration are reduced by a large factor of $5^5 = 3125$. The acceleration is now only $(4.5 \times 10^6 \text{ m/s}^2)/5^5 = 1400$ m/s^2.

Checkpoint 4 If the distance between a neutral atom and a point charge is doubled, by what factor does the force on the atom by the point charge change?

Interaction of Charged Objects and Neutral Matter

We are now in a position to explain why both positively and negatively charged objects (such as + and − tapes) are strongly attracted to neutral matter.

QUESTION Try to explain in detail what happens when a positively charged tape is brought near your hand. This is a complex process; consider all the interactions involved.

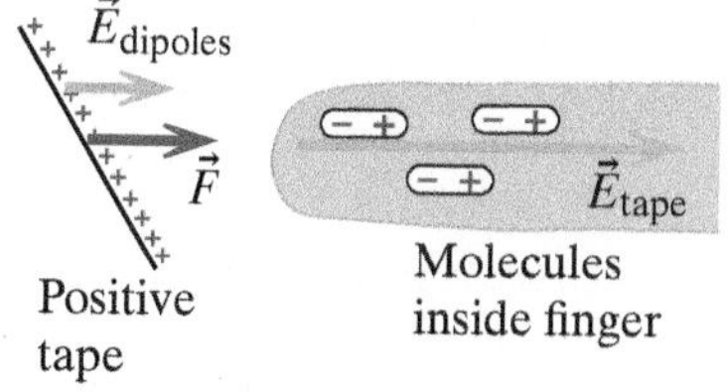

In considering the interactions of fields and matter, the following scheme is useful. (1) Identify any sources of electric fields. (2) Identify any charges at other locations that can be affected by these fields. (3) Redistribution of the affected charges may create an electric field at the location of the original source charges: are they affected?

The positively charged tape makes an electric field, which points away from the tape. This electric field is present inside your hand, and affects atoms, molecules, and ions inside your hand. Figure 14.17 shows the polarization caused inside your finger by the electric field of the tape. The induced dipoles in your finger create an electric field at the location of the tape, which attracts the tape. You should be able to construct a diagram like the one in Figure 14.17 illustrating what happens when a negatively charged tape interacts with your finger.

You may have noticed that the attraction between your neutral hand and a hanging charged invisible tape changes much more rapidly with distance $(1/r^5)$ than does the interaction between two charged tapes $(1/r^2)$. A related experiment is EXP8.

Figure 14.17 The electric field of a positively charged tape polarizes your finger. The induced dipoles in your finger create an electric field at the location of the tape, which attracts the tape.

Checkpoint 5 Explain in detail, including diagrams, what happens when a negatively charged tape is brought near your finger.

Determining the Charge of an Object

Suppose that you have a negatively charged tape hanging from the desk, and you rub a wooden pencil on a wool sweater and bring it near the tape.

> QUESTION If the tape swings toward the pencil, does this show that the pencil had been charged positively by rubbing it on the wool?

Not necessarily. Even if the pencil is uncharged, the charged tape will polarize the pencil and be attracted by the induced dipoles.

> QUESTION Can a charged object repel a neutral object? Why or why not? Draw diagrams to help you make your point.

Polarization always brings the unlike-sign charge closer, yielding a net attraction. Repulsion of an induced dipole can't happen. Therefore repulsion is the better test of whether an object is charged.

Electric Field Penetrates Intervening Matter

The superposition principle states that the presence of matter does not affect the electric field produced by a charged object. Intervening matter does not "screen" or "shield" the electric field, just as your desk does not "screen" or "shield" your book from the gravitational field of the Earth. See Figure 14.18; a related experiment is EXP10.

You may have already observed one case of electric field passing through intervening matter. You see the same interaction between a charged invisible tape and your hand, or another tape, when approaching either side of a hanging tape, despite the charges being on just one side of the tape. The charges are initially on either the slick side or the sticky side (depending on whether it is the upper or lower tape in a pair of tapes), and one can show that the charges can't move through the tape to the other side (Experiment EXP11).

Intervening Matter and Superposition

The fact that an electric field acts through intervening matter is another example of the superposition principle. It is true that the repulsion between two like-charged pieces of tape is weaker when a piece of paper is in the way (Experiment EXP10), but when viewed in terms of the superposition principle this reduction is not due to the paper partially "blocking" the field of the other tape. Rather, we say that the net field is due to the superposition of two fields: the *same* field that you would have had without the paper intervening, plus another field due to the induced dipoles in the paper.

At this time we can't prove that this view is correct and that there is no blocking of electric field. However, we will find repeatedly that the superposition principle makes the right predictions for a broad range of phenomena and offers a simpler explanation than any kind of hypothetical blocking effect.

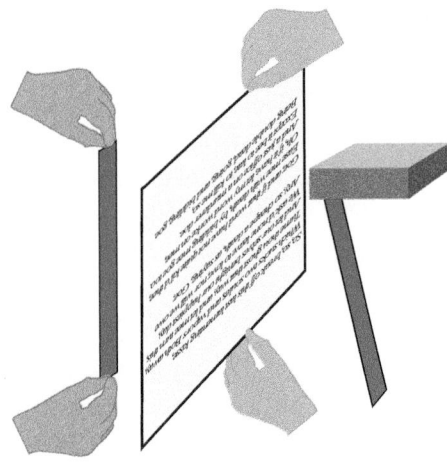

Figure 14.18 Two like charged tapes repel each other even if another object such as a piece of paper is placed between them. Since the tapes are also attracted to the paper, the net effect may be small.

14.4 POLARIZATION OF INSULATORS

In insulators, all of the electrons are firmly bound to the atoms or molecules making up the material. We have seen that an individual atom or molecule can be polarized by an applied electric field, producing an induced dipole of atomic or molecular dimensions. The electrons in an atom or molecule of an insulator shift position slightly, but remain bound to the molecule—no charged particles

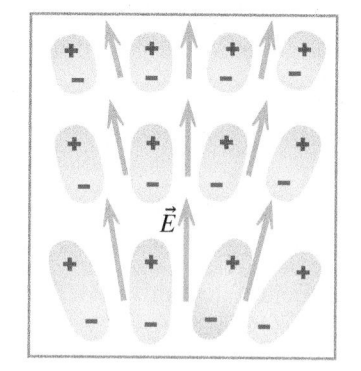

Figure 14.19 A block of insulating material (plastic, glass, etc.) polarized by an applied electric field. The molecules are not shown to scale!

can move more than about one atomic diameter, or 1×10^{-10} m (most move much less than this distance; see Checkpoint 3).

In Figure 14.19 we show a solid block of insulating material, each of whose molecules has been polarized by an applied electric field (that is, an electric field made by external charges—in this case a single positive charge). The molecules are of course not shown to scale! This is an example of "induced polarization"—the electric field has induced the normally unpolarized insulator to become polarized. In each molecule the electrons have moved a very short distance, and the molecules themselves are not free to move. However, the net effect can be very large because there are many molecules in the insulator to be affected. Note that the polarized molecules align with the electric field that is polarizing them, and that the stronger the electric field the larger the "stretch" of the induced dipole.

Polarization Happens Very Rapidly

Because the electron cloud is displaced only a tiny distance when an atom or molecule polarizes, this process happens extremely rapidly. The process can take much less than a nanosecond to complete.

Diagrams Showing Polarization of Insulators

In diagrams of insulators we show polarized molecules exaggerated in size, to indicate that individual molecules in an insulator polarize, but the electrons remain bound to the molecule. We show the extent of polarization by the degree to which the molecule is "stretched." Keep this diagrammatic convention in mind, and compare it to diagrams of polarized conductors in the following sections.

Charge on or in an Insulator

Since there are no mobile charged particles in an insulator, excess charges stay where they are. Excess charge can be located in the interior of an insulator, or can be bound to a particular spot on the surface without spreading out along that surface (Figure 14.20).

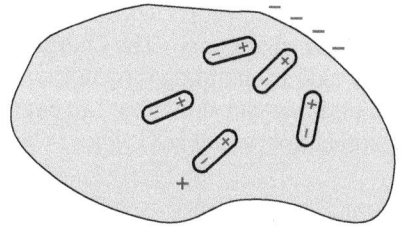

Figure 14.20 In an insulator, charge can occur in patches on the surface, and there can be excess charge inside.

Low-Density Approximation

When atoms or molecules in an insulator polarize in response to an electric field created by external charged objects, the polarized molecules themselves make electric fields that affect neighboring molecules. Because the effect of polarized molecules on each other is typically small compared to the effect of the original applied field, we will neglect this when discussing polarized insulators.

In formal terms, when an electric field $E_{applied}$ is applied to a dense material (a solid or a liquid), the induced dipole moment of one of the atoms or molecules in the material isn't simply $p = \alpha E_{applied}$, but is really $p = \alpha |\vec{E}_{applied} + \vec{E}_{dipoles}|$, where $E_{dipoles}$ is the additional electric field at the location of one of the molecules, due to all the other induced dipoles in the material. In this text we make the simplifying assumption of low density and assume that $E_{dipoles}$ is small compared to $E_{applied}$. This is good enough for our purposes, but accurate measurements of polarizability must take this effect into account.

14.5 POLARIZATION OF CONDUCTORS

As we stated earlier, a conductor has some kind of charged particles that can move freely throughout the material. In contrast to an insulator, where electrons and nuclei can move only very small distances (around 1×10^{-10} m,

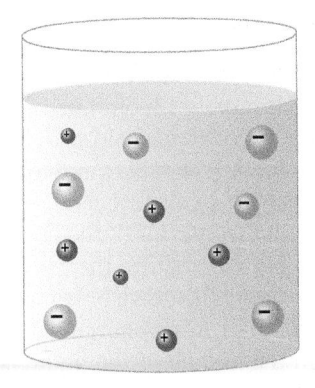

Figure 14.21 A beaker containing an ionic solution (salt water).

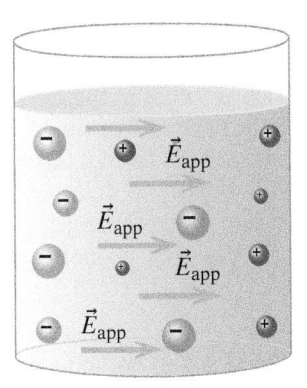

Figure 14.22 Under the influence of an applied electric field (labeled $\vec{E}_{app}$), the liquid polarizes. There is a slight excess ion concentration at the two sides of the beaker. The electric field due to the redistributed ions is not shown.

or much less), the charged particles in a conductor are free to move large distances.

Ionic Solutions

Ionic solutions are conductors, such as a solution of sodium chloride (table salt) in water. In salt water, the mobile charged particles are Na^+ ions and Cl^- ions (Figure 14.21; there are also very small concentrations of H^+ and OH^- ions, which are not shown).

> QUESTION What happens when an electric field is created in the region of the beaker?

When an electric field is applied to a conductor, the mobile charged particles begin to move in the direction of the force exerted on them by the field. However, as the charges move, they begin to pile up in one location, creating a concentration of charge that itself creates an electric field in the region occupied by the remaining mobile charges. The net electric field in the region is the superposition of the applied (external) field and the electric field created by the relocated charges in the material. Figure 14.22 is a diagram of the polarization that occurs in the salt water. The ions (charged atoms or molecules) are in constant motion, so the actual situation isn't simple. Moreover, the interior of the liquid is full of positive and negative ions; there's just a slight excess concentration of ions near the sides of the beaker.

Drift Speed and Applied Electric Field

If a beaker of salt water is placed in a region where there is an electric field (due to charges outside the beaker), a sodium ion or a chloride ion will experience an electric force, and will begin to move in the direction of the force. However, even if the force remains constant, the ion will not keep accelerating, because it will collide with water molecules or with other ions. In effect, there is a kind of friction at the microscopic level.

To keep the ions in a salt solution moving at a constant speed, a constant electric field must be applied to the solution. The speed at which mobile charges (in this case, sodium or chloride ions) move through a conductor is called the *drift speed*. Drift speed is directly proportional to the net electric field at the location of the charge. The proportionality constant is called the *mobility* of the mobile charges.

DRIFT SPEED

$$\bar{v} = uE_{net}$$

$\bar{v}$ is the average drift speed of a mobile charge.

u is the mobility of the charge. The units of mobility are $\dfrac{m/s}{N/C}$.

E_{net} is the magnitude of the net electric field at the location of the mobile charge.

As implied by this equation, if the net electric field at the location of a mobile charge in a conductor is zero, the charge will stop moving.

> **Checkpoint 6** An electric field of magnitude 190 N/C is applied to a solution containing chloride ions. The mobility of chloride ions in solution is 7.91×10^{-8} (m/s)/(N/C). What is the average drift speed of the chloride ions in the solution?

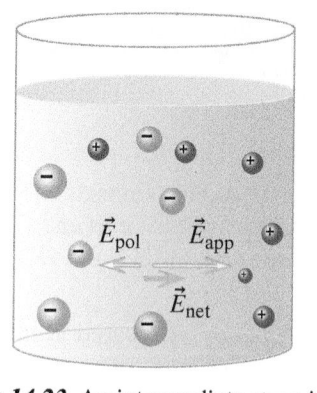

Figure 14.23 An intermediate stage in the polarization process, before polarization is complete.

The Polarization Process in an Ionic Solution

Polarization occurs very rapidly, but it is not instantaneous. Let's "slow down time" so we can talk about the process of polarization; we'll operate on a time scale of attoseconds (1×10^{-18} seconds!). To simplify our analysis, we'll imagine that we are able temporarily to "freeze" the ions in the salt water, and to release them after we have brought charges nearby to apply an electric field.

Consider the net electric field at a location in the interior of the liquid, at a time attoseconds after polarization has begun, but long before the process has finished. The electric field E_{app} due to external charges is shown in Figure 14.23, and also a smaller electric field E_{pol} due to the polarization charges present at this time. The net electric field at a location in the middle of the liquid is now smaller than it was before polarization began.

> QUESTION Will the polarization of the salt water increase beyond what it is now?

At this instant the net electric field in the solution still has magnitude greater than zero, so ions in the solution will still experience forces in the direction to increase the polarization. Because $E_{net} \neq 0$, the drift speed $\bar{v}$ of the ions is not zero. More ions will pile up at the sides of the beaker, and the net electric field in the interior will be further weakened.

Eventually the conductor will reach equilibrium on the microscopic level. Equilibrium at the microscopic level means that there is no net motion of mobile charges in any direction:

EQUILIBRIUM INSIDE A CONDUCTOR

When a conductor is in equilibrium at the microscopic level:

$$\bar{v} = 0$$

The average drift speed of the mobile charges inside the conductor is zero. There is no net flow of charges in any direction.

> QUESTION How weak does the net electric field inside the conductor get? In the final state of equilibrium (when there is no further increase in polarization), how big is the net electric field in the interior of the liquid?

An Example of a "Proof by Contradiction"

You may have correctly deduced that in the final state the net electric field in the conductor goes to zero at equilibrium. A rigorous way to reason about this using formal logic is to construct a "proof by contradiction." In a proof by contradiction, we assume the opposite of what we want to prove, then, making valid logical deductions from this assumption, show that we reach a conclusion that is impossible or contradictory. We therefore conclude that the original assumption was wrong, and its opposite must be true.

1. Assume that in equilibrium the magnitude of the net electric field in the interior of an ionic solution is nonzero.
2. Since $E_{net} > 0$, all mobile ions in the solution will experience a nonzero force. Since $u > 0$, the average drift speed $\bar{v} > 0$, and all ions move in the direction of the force.
3. Since $\bar{v} > 0$, there is a net flow of charges. Therefore the system cannot be in equilibrium, because by definition in equilibrium $\bar{v} = 0$ and there is no net flow of charges. This result, that $\bar{v} > 0$, contradicts our original assumption (point 1 above) that the ionic solution is in equilibrium.

4. Because we have reached a contradiction, we must conclude that the original assumption (that the net electric field in the solution may be nonzero in equilibrium) is wrong. Thus, we conclude that the net electric field in an ionic solution in equilibrium must be zero.

This reasoning holds true for any conductor, including not only ionic solutions, but solid metal objects as well.

Superposition

Note that the electric field inside the liquid is zero, not because of any "blocking" of fields due to external charges, but by the superposition of two effects: the effect of the external charges and the effect of the polarization charges. This is another example of the superposition principle in action.

It is not true that the net electric field in a solution is zero at *all* times. While the ionic solution is in the process of polarizing, it is not in equilibrium; there is a nonzero electric field, and hence a nonzero force on an ion in the liquid, as you saw above. If electrodes are placed in the ionic solution and connected to a battery, the battery prevents the system from reaching equilibrium. In such a case (no equilibrium), there can be a field continuously acting on ions inside the liquid, resulting in continuous shifting of the ions through the liquid, constituting an electric current.

Since there are a very large number of ions in the solution, none of them has to move very far during the polarization process. Even a tiny shift leads to the buildup of an electric field large enough to cancel out the applied electric field.

Polarization of Salt Water in the Body

Your own body consists mainly of salt water, including the blood and the insides of cells. Look again at the diagrams in which you focused on the way an external charge polarizes individual molecules inside your finger. An additional effect is the polarization of the salt water inside your finger. As shown in Figure 14.22, there will be a shift of Na^+ and Cl^- ions in the blood and tissues. This shift may be a larger effect than the molecular polarization. It is a bit unsettling to realize that a charged tape or comb messes with the inside of your body!

14.6   CHARGE MOTION IN METALS

You probably know that metals are very good electrical conductors. In almost all metals, the mobile charged particles are electrons.

The Mobile Electron Sea

The atoms in a solid piece of metal are arranged in a regular 3D geometric array, called a "lattice" (Figure 14.24). The inner electrons of each metal atom are bound to the nucleus. Some of the outer electrons participate in chemical bonds between atoms (the "springs" in the ball-and-spring model of a solid). However, some of the outer electrons (usually one electron per atom) join a "sea" of mobile electrons that are free to move throughout the entire macroscopic piece of solid metal (Figure 14.25). In a sense, the entire hunk of metal is like one giant molecule, in which some of the electrons are spread out over the entire crystal. The electrons are not completely free; they are bound to the metal as a whole and are difficult to remove from the metal. (For example, electrons do not drip out when you shake a piece of metal!) Metals are excellent conductors because of the presence of these mobile electrons.

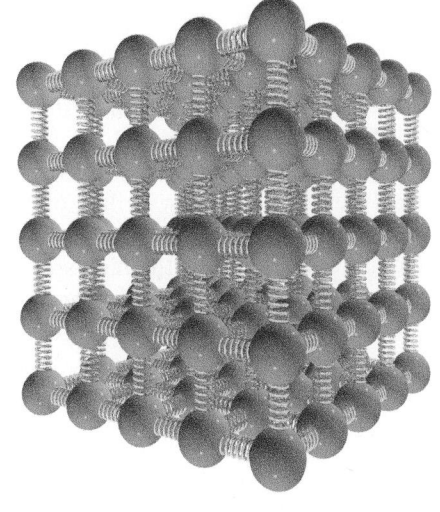

Figure 14.24 The ball-and-spring model of a solid.

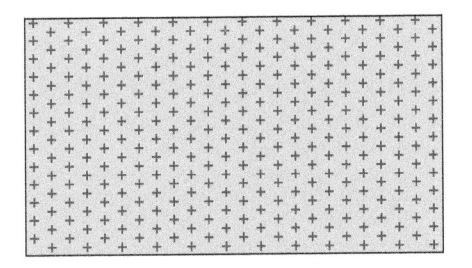

Figure 14.25 A 2D slice of an unpolarized metal: uniform mobile-electron sea (blue), positive atomic cores (red "+" symbols).

No Net Interaction between Mobile Electrons

Although the roaming electrons repel each other strongly, this repulsion between electrons is neutralized on the average by the attractions exerted by the positive atomic cores (a "core" is a neutral atom minus its roaming electron, so it has a charge of $+e$). The effect is that on average, the net electric field inside a piece of metal in equilibrium is zero.

Because of this, in some ways the mobile electrons look like an ideal gas: they move in a region free from the electric field, so they appear not to interact with each other or with the atomic cores. In fact, in some simple models of electron motion the mobile electron sea is treated as an ideal gas.

The Drude Model

In a simple classical model of electron motion (called the "Drude model" after the physicist who first proposed it), a mobile electron in the metal, under the influence of the electric field inside the metal, does accelerate and gain energy, but then it loses that energy by colliding with the lattice of atomic cores, which is vibrating because of its own thermal energy and acquires more thermal energy due to the collisions of the electrons with the lattice. After a collision, an electron again gets accelerated, and again collides. This process is what makes the metal filament in a light bulb get hot. Figure 14.26 shows a graph of this start–stop motion for a single electron.

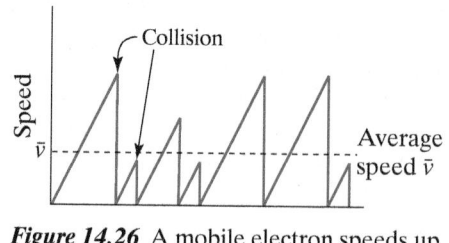

Figure 14.26 A mobile electron speeds up under the influence of the electric field inside the wire, then collides with an atomic core and loses energy.

Drift Speed and Electron Mobility

The average speed of an electron in this start–stop motion is called the "drift" speed $\bar{v}$, and we say that the electron "drifts" through the metal. Actually, the slow drift motion is superimposed on high-speed motion of the electrons in all directions inside the metal, much as the wind is a slow drift motion superimposed on the high-speed motion of air molecules in all directions. A full treatment of electrons in a metal, including the reason for the high-speed motion in all directions, requires quantum mechanics, but the simple classical Drude model allows us to understand most of the important aspects of circuits on a microscopic level.

We express the momentum principle

$$\frac{\Delta \vec{p}}{\Delta t} = \vec{F}_{\text{net}}$$

in a form involving finite time steps for momentum in the direction of $\vec{F}_{\text{net}}$,

$$\Delta p = F_{\text{net}} \Delta t = e E_{\text{net}} \Delta t$$

where E_{net} is the magnitude of the net electric field inside the wire, and Δt is the time between collisions. If we make the simplifying assumption that the electron loses all its momentum during each collision, we have

$$\Delta p = p - 0 = e E_{\text{net}} \Delta t$$

The speed of the electron (of mass m_e) at the time of collision turns out to be small compared to the speed of light, so we have

$$v = \frac{p}{m_e} = \frac{e E_{\text{net}} \Delta t}{m_e}$$

However, the time between collisions is not the same for all electrons. Some experience longer times between collisions, some shorter times. To get an average, drift speed $\bar{v}$ for all electrons at a particular instant, we need the average time $\overline{\Delta t}$ between collisions:

$$\bar{v} = \frac{eE_{\text{net}}\,\overline{\Delta t}}{m_e}$$

$\overline{\Delta t}$, the average time between collisions of the electrons with the atomic cores, is determined by the high-speed random motion of the electrons and by the temperature of the metal. (At a higher temperature the thermal motion of the atomic cores is greater, and the average time between collisions is reduced, leading to a smaller drift speed for the same field E.)

QUESTION Is drift speed directly proportional to the magnitude of the electric field?

Assuming that increasing the electric field does not result in a significant change in temperature, then doubling the electric field E doubles the drift speed $\bar{v}$ attained in that time; hence the drift speed is directly proportional to the electric field. As mentioned earlier, the proportionality factor is called the electron "mobility" and is denoted by u (or by μ in some books).

$$\bar{v} = uE_{\text{net}}$$

Evidently,

$$u = \frac{e}{m_e}\overline{\Delta t}$$

Different metals have different electron mobilities. The higher the mobility, the higher the drift speed for a given electric field. The direction of the drift velocity of a mobile (negatively charged) electron is opposite to the direction of the electric field.

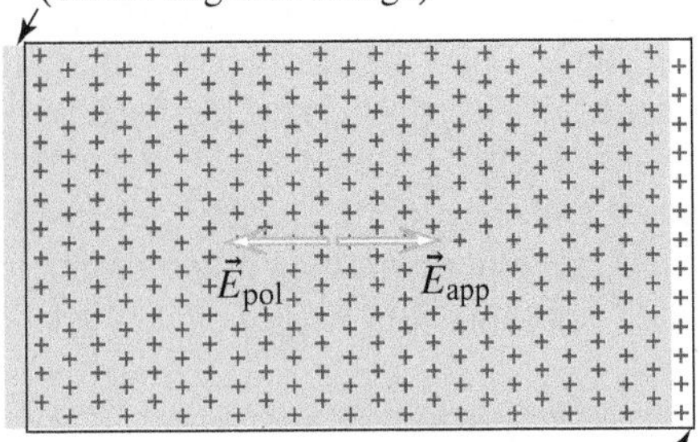

Excess of electrons
(excess negative charge)

Deficiency of electrons
(excess positive charge)

Figure 14.27 Polarized metal: mobile electron sea shifted left relative to the positive atomic cores, under the influence of an applied electric field. There is an excess of electrons on the left side and a deficiency of electrons on the right. These charges contribute to the net electric field inside the metal.

> **Checkpoint 7** The mobility of the mobile electrons in copper is 4.5×10^{-3} (m/s)/ (N/C). How large an electric field would be required to give the mobile electrons in a block of copper a drift speed of 1×10^{-3} m/s?

Polarization Happens Very Quickly

When an electric field (due to some external charges) is applied to a metal, the metal polarizes. We can describe the polarization of a metal as shifting the entire mobile electron sea relative to the fixed positive cores. In Figure 14.27, in response to an applied field, electrons have piled up on the left, creating a very thin negatively charged layer near the surface. There is a corresponding deficiency of electrons on the right, creating a very thin positively charged layer near the surface.

The shift in the electron sea is extremely small, much less than an atomic diameter! (See Checkpoint 3.) It is not necessary for electrons at one end of the block to move to the other end. Just displace the entire electron sea slightly and you have lots of excess electrons on one surface. Because the electron sea has to move only a tiny distance, this displacement can happen very rapidly—it can take much less than a nanosecond.

Diagrams Showing Polarization of Metals

Figure 14.28 shows the polarization of a neutral metal block caused by a positive point charge to the left of the block, as calculated in a computational model of the polarization process. Blue represents negative charge and red represents positive charge. The block is still neutral, but negative charge (excess electron density) has accumulated on the surface of the left side of the block, leaving positive charge (deficiency of electrons) on the right side of the block. Note that excess charge tends to concentrate along the edges and on the corners of the block. As expected, the charge on the block's surface is more dense where the polarizing field is greater, closer to the point charge.

The computational model used to generate this image does quantitatively what we did qualitatively in reasoning about the polarization of ionic solutions and metals. In Chapter 18, after introducing additional concepts that are involved, we will describe this algorithm in more detail.

Figure 14.28 The polarization of a neutral block of metal by a positive point charge. Shades of blue represent negative charge; red represents positive charge.

In Figure 14.29 we depict a polarized metal in a simplified way that is both easy to draw and easy to interpret at a glance.

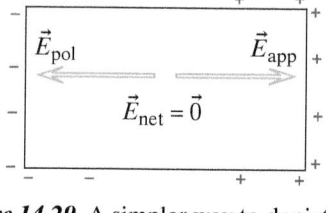

Figure 14.29 A simpler way to depict the polarization of a polarized metal. Excess charges are drawn outside the boundary lines, to indicate that they are on the surface.

- We show − and + signs outside the surfaces to indicate which surfaces have thin layers of negative charge (electron excess) or positive charge (electron deficiency), as a result of shifts in the mobile electron sea. To avoid ambiguity we draw + and − signs just outside the surface of a metal object to indicate that the excess charge (excess or deficiency of electrons) is on the surface of the object.
- We do not show the positive atomic cores and the mobile electron sea inside the metal, because the interior is all neutral. The diagram is much easier to interpret if we do not clutter up neutral regions with charges that must be counted to see whether they balance.

Compare these conventions to the convention we used earlier to show how the individual atoms or molecules polarize in an insulator.

Polarized and/or Charged

Take care to use technical terms precisely. The metal block shown in Figure 14.29 is *polarized*. It is *not* charged; its net charge is still zero.

On the other hand, a charged object can also be polarized. The positively charged metal block depicted schematically in Figure 14.30 is also polarized. "Polarized" and "charged" are not synonyms.

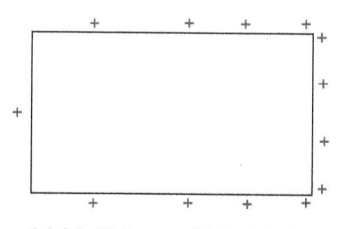

Figure 14.30 This metal block is both charged and polarized.

Net Electric Field in a Metal Goes to Zero in Equilibrium

The reasoning process that we went through when considering the polarization of ionic solutions applies equally well to metals, or to any conductor. We used proof by contradiction to demonstrate that in equilibrium the net electric field inside a conductor must be zero (because if it were not zero, mobile charged particles would move under the influence of the field, and the system would not be in equilibrium).

It is intriguing that it is possible for mobile charges to rearrange themselves in such a way that the net electric field is zero not just at one single location, but also at every location inside the metal. It would be a very difficult problem for us to calculate exactly where to place charged particles to make a net field of zero inside a metal object, but in fact the many mobile charges do rearrange in just such a way as to accomplish this. It can be shown that it is because of the $1/r^2$ distance dependence of the electric field that this is possible—if the exponent were not exactly 2.0, the world would be quite different.

When equilibrium is reached in a metal, things are essentially unchanged in the interior of the metal. There is no excess charge—we still have a uniform sea of electrons filling the space around the positive atomic core. The net electric field inside the metal, which is the sum of the applied field and the field due to the charge buildup on the edges of the metal, is still zero.

$$\vec{E}_{net} = \vec{E}_{app} + \vec{E}_{pol} = 0 \quad \text{in a conductor at equilibrium}$$

At the surfaces there is some excess charge, so we can represent a polarized metal as having thin layers of charge on its surfaces but being unpolarized in the interior, unlike an insulator.

The shifting of the mobile electron sea in metals is a much larger effect than occurs in insulators, where the polarization is limited by the fact that all the electrons, including the outermost ones, are bound to the atoms, unlike the situation in metals. A polarized insulator is a collection of tiny (molecule-sized) dipoles, whereas a polarized metal forms one giant dipole.

E Is Not Always Zero Inside a Metal

Do not overgeneralize our previous conclusions. It is not true that the net electric field in a metal is zero at all times. While the metal is in the process of polarizing, the metal is not in equilibrium, and there is a nonzero electric field inside the metal, creating a nonzero force on electrons in the electron sea, as you saw above. In an electric circuit, the battery prevents the system from reaching equilibrium. In such a nonequilibrium situation, there can be an electric field inside the metal, and hence a force continuously acting on electrons in the mobile electron sea, resulting in continuous shifting of the electron sea around the closed circuit, constituting an electric current.

Excess Charges on Conductors

Another important property of metals (and of the $1/r^2$ property of the electric interaction, as we will see when we study Gauss's law in a later chapter) is that any excess charges on a piece of metal, or any conductor, are always found on an outer or inner surface. This makes intuitive sense, since any excess charges will repel each other and will end up as far apart as possible—on the surface of the conductor. Any multiatom region in the interior of the conductor has a net charge of zero. Moreover, the mutual repulsion among any excess charges makes the mobile electron sea redistribute itself in such a way that charge appears almost immediately all over the surface (Figure 14.31).

Here is a summary of the behavior of conductors vs. insulators:

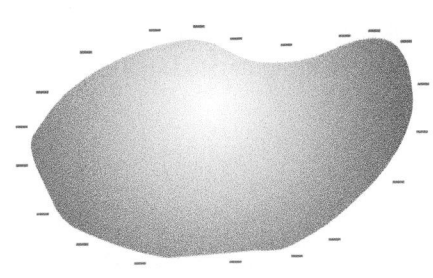

Figure 14.31 In a metal, charge is spread all over the surface (not necessarily uniformly), and there is no excess charge inside.

	Conductor	Insulator
Mobile charges	Yes	No
Polarization	Entire sea of mobile charges moves	Individual atoms or molecules polarize
Equilibrium	$\vec{E}_{net} = \vec{0}$ inside	$\vec{E}_{net}$ nonzero inside
Excess charge	Spread over surface	In patches anywhere

Plastic rod

Metal ball

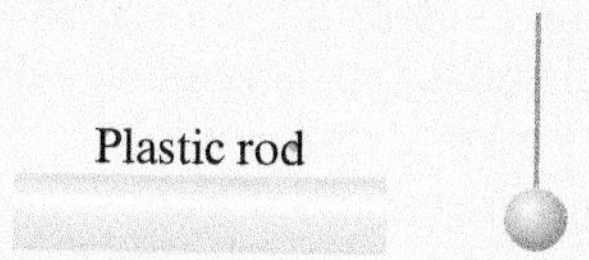

Figure 14.32 The plastic rod and ball are initially uncharged.

Plastic rod

This end was rubbed with wool

Metal ball

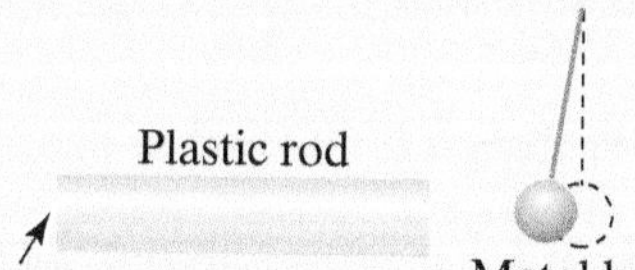

Figure 14.33 The left end of the plastic rod was rubbed with wool and became negatively charged.

EXAMPLE

Plastic and Metal Rods

A lightweight (conducting) metal ball hangs from a thread, to the right of an (insulating) plastic rod. Both are initially uncharged (Figure 14.32).

(a) You rub the left end of the plastic rod with wool, depositing charged molecular fragments whose total (negative) charge is that of 1×10^9 electrons. You observe that the ball moves toward the rod, as shown in Figure 14.33.

Explain. Show all excess charged particles, polarization, and so on, clearly in a diagram. Make it clear whether charged particles that you show are on the surface of an object or inside it.

(b) You perform a similar experiment with a (conducting) metal rod. You touch the left end of the rod with a charged metal object, depositing 1×10^9 excess electrons on the left end. You then remove the object. You see the ball deflect more than it did with the plastic rod in part (a), as shown in Figure 14.34.

Explain. Show all excess charged particles, polarization, and so on, clearly in a diagram. Make it clear whether charged particles that you show are on the surface of an object or inside it.

Metal rod

This end touched a charged metal object

Metal ball

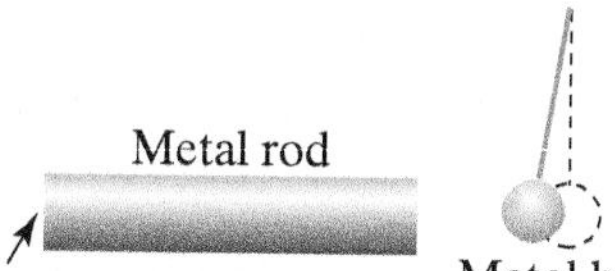

Figure 14.34 Replace the plastic rod with a neutral metal rod, then touch the left end with a negatively charged metal object. The ball deflects more than it did with the plastic rod.

Solution

(a) The plastic rod is an insulator, so the excess charge remains on the left-end surface of the rod. This charge polarizes the molecules inside the rod. The original charge plus the polarized molecules make a field that polarizes the neutral metal ball, as shown in Figure 14.35. The excess charge on the polarized metal ball is on the surface of the ball, because the ball is a conductor. The interior of the ball is neutral.

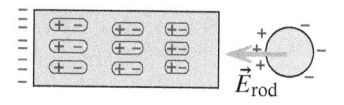

Figure 14.35

The field due to the plastic rod (charge on end plus polarized molecules) exerts a net force to the left on the polarized metal ball (Figure 14.36).

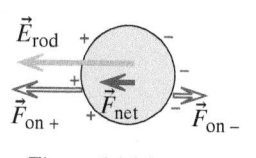

Figure 14.36

(b) The excess negative charge spreads all over the surface of the metal rod and ball, which are conductors. This excess negative charge polarizes the metal ball. The polarized metal ball in turn polarizes the negatively charged metal rod somewhat, as shown in Figure 14.37.

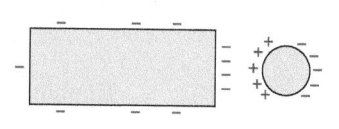

Figure 14.37

Polarization of the ball is greater in this case, because more of the original charge is closer to the ball (Figure 14.38). The net force on the ball is greater in this case than it was with the plastic rod.

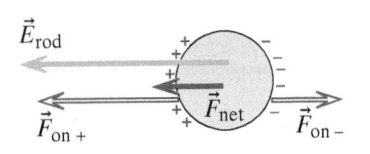

Figure 14.38

EXAMPLE   **A Ball and a Wire**

The center of a small spherical metal ball of radius R, carrying a negative charge $-Q$, is located a distance r from the center of a short, thin, neutral copper wire of length L (Figure 14.39). The ball and the wire are held in position by threads that are not shown. If $R = 5\,\text{mm}$, $Q = 1 \times 10^{-9}\,\text{C}$, $r = 10\,\text{cm}$, and $L = 4\,\text{mm}$, calculate the force that the ball exerts on the wire.

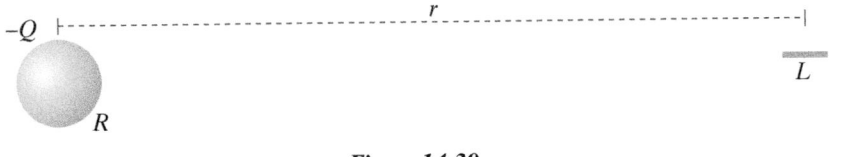

Figure 14.39

Solution   The ball makes a field that polarizes the wire as shown in Figure 14.40. The polarized wire in turn makes a field that polarizes the ball, but let's assume that we can neglect this tiny effect, so we can model the ball as a point charge. The polarized wire will be attracted by the ball, and the ball will be attracted by the polarized wire.

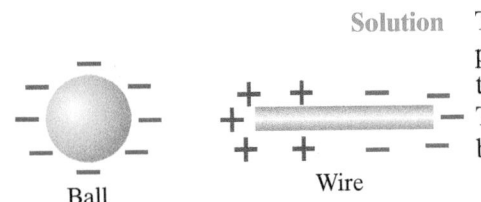

Figure 14.40 The ball polarizes the wire. Neglect the effect of the wire on the ball.

$F_{\text{on ball}} = F_{\text{on wire}}$ (reciprocity of electric forces)

$F_{\text{on ball}} = Q_{\text{ball}} E_{\text{wire}}$ (field made by polarized wire affects ball)

$$E_{\text{wire}} \approx \frac{1}{4\pi\varepsilon_0} \frac{2qL}{r^3}$$ (model wire as dipole with charges $+q$ and $-q$ at ends)

We need to find the charge q on one end of the wire. We know that at any location inside the metal wire, $\vec{E}_{\text{net}} = 0$ in equilibrium (Figure 14.41). Consider a location in the center of the wire, and model the wire as though there were $+q$ and $-q$ on the ends, a distance L apart, ignoring the small amount of charge on the rest of the wire (here we can't use dipole equations; we're between the charges).

Figure 14.41 The net electric field inside the wire must be zero in equilibrium.

$$E_{\text{net},x} = E_{\text{ball},x} + E_{+\text{end},x} + E_{-\text{end},x}$$

$$0 = -\frac{1}{4\pi\varepsilon_0} \frac{Q}{r^2} + \frac{1}{4\pi\varepsilon_0} \frac{q}{(L/2)^2} + \frac{1}{4\pi\varepsilon_0} \frac{q}{(L/2)^2}$$

$$q \approx \frac{Q}{8} \left(\frac{L}{r} \right)^2$$

At a location outside the wire we model the wire as a dipole, so the force on the ball (which is equal in magnitude to the force on the wire) is this:

$$F \approx Q \left(\frac{1}{4\pi\varepsilon_0} \frac{2qL}{r^3} \right) = \frac{1}{4\pi\varepsilon_0} \frac{Q^2}{4} \left(\frac{L}{r} \right)^2 \frac{L}{r^3} = \frac{1}{4\pi\varepsilon_0} \frac{Q^2 L^3}{4r^5}$$

Not too surprisingly, we find a force proportional to $1/r^5$.

Now we can calculate numerical values:

$$q \approx \frac{(1 \times 10^{-9}\,\mathrm{C})}{8} \left(\frac{4 \times 10^{-3}\,\mathrm{m}}{0.1\,\mathrm{m}}\right)^2 = 2 \times 10^{-13}\,\mathrm{C}$$

which is a very small charge. This justifies our assumption that the polarized wire won't polarize the ball to any significant extent. The force is tiny:

$$F = \left(9 \times 10^9 \frac{\mathrm{N \cdot m^2}}{\mathrm{C^2}}\right) \frac{(1 \times 10^{-9}\,\mathrm{C})^2 (4 \times 10^{-3}\,\mathrm{m})^3}{4(0.1\,\mathrm{m})^5} = 1.4 \times 10^{-11}\,\mathrm{N}$$

If we double r, there is $1/4$ as much q and $1/32$ as much force. If we double Q, there is 2 times as much q and 4 times as much force.

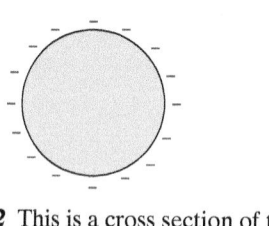

Figure 14.42 This is a cross section of the metal ball.

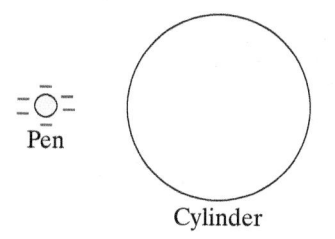

Figure 14.43 This is a cross section of the metal or plastic cylinder.

Checkpoint 8 (a) An object can be both charged *and* polarized. On a negatively charged metal ball, the charge is spread uniformly all over the surface (Figure 14.42). If a positive charge is brought near, the charged ball will polarize. If any of the following quantities is zero, state this explicitly. (1) Draw the approximate final charge distribution on the ball. (2) At the center, draw the electric field due to the external positive charge. (3) At the center, draw the electric field due to the charge on the surface of the ball. (4) At the center, draw the net electric field.

(b) Next, consider a negatively charged plastic pen that is brought near a neutral solid metal cylinder (Figure 14.43). If any of the following quantities is zero, state this explicitly. (1) Show the approximate charge distribution for the metal cylinder. (2) Draw a vector representing the net force exerted by the pen on the metal cylinder, and explain your force vector briefly but completely, including all relevant interactions. (3) At the center, draw the electric field due to the external negative charge. (4) At the center, draw the electric field due to the charge on the surface of the ball. (5) At the center, draw the net electric field.

(c) Replace the solid metal cylinder with a solid plastic cylinder. (1) Show the approximate charge distribution for the plastic cylinder. (2) Draw a vector representing the net force exerted by the pen on the plastic cylinder. (3) Explain your force vector briefly but completely, including all relevant interactions.

14.7 CHARGE TRANSFER

We have seen that objects made of insulators can often acquire a nonzero net charge if they are rubbed by another insulator. You can charge initially neutral pieces of invisible tape by stripping them off other pieces of tape. You can charge an initially neutral pen by rubbing it on your hair. In both these cases, some kind of charged particle is added to or removed from a surface that was originally neutral. In Section 14.2 we discussed possible mechanisms for the transfer of charge between one insulating object and another, including the transfer of positive or negative ions and the transfer of electrons.

Since a conducting object contains mobile charged particles, the process of charging or discharging a conductor involves a flow of charged particles from one conductor to another. For example, electrons from the mobile electron sea in a metal object can move into the mobile electron sea of a different metal object if the objects come into contact with each other.

Although you may not have previously thought of yourself as a conductor, your own body plays an interesting role in some kinds of charging or discharging phenomena. In the following discussion you will see why.

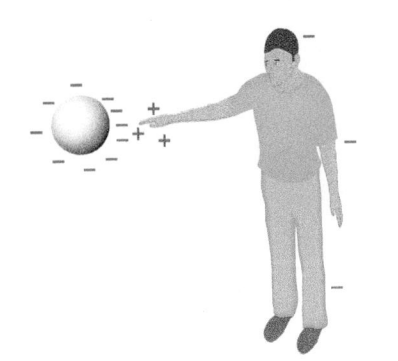

Figure 14.44 The metal is charged, and the person is uncharged but slightly polarized.

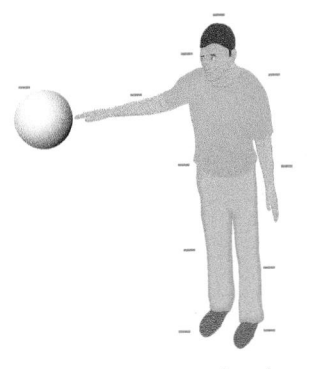

Figure 14.45 The net negative charge is distributed over a much larger area, nearly neutralizing the metal.

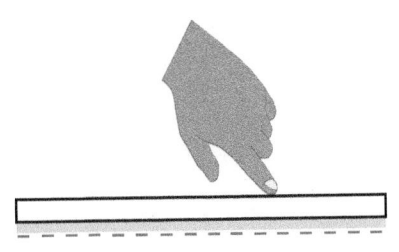

Figure 14.46 You run your finger along the slick side of the tape, and the tape seems to become neutralized.

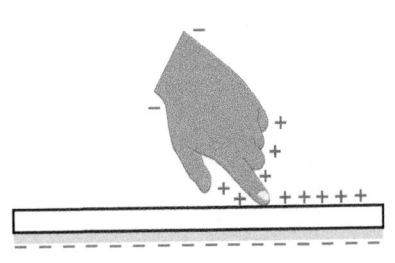

Figure 14.47 Positive ions from the salt solution on your skin are attracted to the negatively charged tape.

Discharging by Contact

If you exercise on a hot day, you sweat, and your body becomes covered with a layer of salt water. Even in a cool place, when you are not moving, there is usually a thin layer of salt water covering your skin. As we saw earlier, salt water is a conductor, so you have a conducting film all over the surface of your skin.

When you approach a negatively charged surface, your body polarizes as shown in Figure 14.44. The polarization includes not only atomic or molecular polarization but also polarization of the blood and sweat, which are salt solutions.

When you touch the charged object, the negatively charged object attracts positive Na^+ ions in the film of salt water on the skin. The Na^+ ions pick up an electron, partially neutralizing the excess negative charge of the object (Figure 14.45). The body acquires a net negative charge. (The Na atom can react with the water to form NaOH and hydrogen!) In the case of a small piece of metal, on which charge is free to redistribute itself, this process nearly neutralizes the metal, because the original net amount of charge is now spread out over the much larger area of metal plus human body.

Similarly, a positive metal surface would attract negative Cl^- ions from your skin, which give up an electron to the metal. The body acquires a net positive charge. (Chlorine can be emitted in tiny quantities!)

Grounding

Touching a small charged object is a pretty effective way to discharge the object, even though you're wearing shoes with insulating soles. An even better way to discharge a conducting object is to "ground" it by making a good connection to the earth or ground (typically through a water pipe that goes into the ground). Earth is a rather good conductor due to the presence of water containing ions. Grounding spreads charge throughout a huge region, neutralizing an object essentially completely.

Discharging an Insulator

You can easily discharge a charged metal foil by briefly touching it anywhere, because it is a conductor. It is more difficult to discharge a charged strip of invisible tape, which is an insulator and does not allow charge to move through the tape.

To discharge a charged piece of invisible tape, it turns out that one may simply rub one's fingers across the slick side of the tape, as shown in Figure 14.46. After doing this, one finds that the tape no longer interacts with neutral objects like your hand and appears to be neutral itself.

QUESTION How is it possible to discharge a tape by rubbing the slick side even when it was the sticky side that got charged? Tape is an insulator, so charges can't move through the tape.

There are mobile charges on your skin. Positive ions from the salt solution on the skin are attracted to the negatively charged tape and are deposited on its slick surface, so the tape becomes neutral (net charge becomes zero) as shown in Figure 14.47. The + charges on the top and the − charges on the bottom actually make the tape into a dipole (and there are induced dipoles inside the tape), but these dipoles exert much weaker forces on other objects than the negatively charged tape did. Thus the tape acts like ordinary neutral matter. A related experiment is EXP12.

Charging by Induction

It is possible to make use of the polarizability of a conductor to make it acquire a net charge, without actually touching a charged object. The process, called "charging by induction," is illustrated in the sequence in Figure 14.48. In this example, a piece of neutral aluminum foil hangs from a neutral insulating tape. You charge a plastic pen by rubbing it on wool, and bring it near the left side of the neutral foil. A related experiment is EXP13.

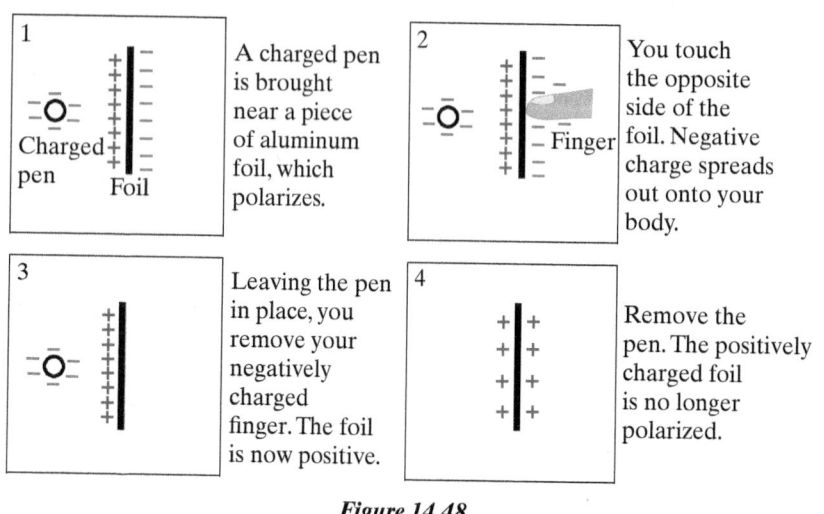

Figure 14.48

Effect of Humidity on Tapes

Isolated atoms are always symmetrical and unpolarized unless an external charge shifts the electron cloud and makes an induced dipole. However, some molecules are permanently polarized even in the absence of an external charge, and this leads to important physical and chemical effects.

For example, water molecules are permanently polarized. The water molecule (H_2O) is not spherically symmetrical but has both hydrogen atoms off to one side of the oxygen atom. In Figure 14.49 the δ^+ and δ^- symbols are used to indicate that slight shifts of the electron clouds to the right leave the right side of the molecule a bit negative and the left side a bit positive, so the water molecule is a permanent dipole.

Many of water's unusual chemical and physical properties are due to this structure. In particular, the charged ends can bind to ions, which is why many chemicals dissolve well in water.

When water molecules in the air strike a surface they sometimes become attached to the surface, probably because the charged ends bind to the surface. A film of water builds up on all surfaces. Pure water is a very poor conductor but does contain small amounts of mobile H^+ and OH^- ions. More important, the water dissolves surface contaminants such as salt, and the impure water provides an effective path for charges to spread onto neighboring objects. After a while a charged surface loses its original charge, so experiments with charged objects work better when the humidity is low. A related experiment is EXP14.

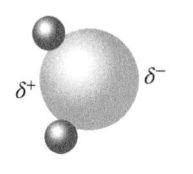

Figure 14.49 A water molecule is a permanent dipole.

14.8 PRACTICAL ISSUES IN MEASURING ELECTRIC FIELD

There are practical issues in measuring the electric field by placing a charge somewhere and measuring the force on it, if the charge affected by the field significantly alters the original distribution of source charges. Consider a negatively charged metal sphere (Figure 14.50). The electric field due to the sphere points radially inward.

If we place a particle with very little charge q near this charged sphere (a single proton, for example), it hardly alters the distribution of charge on the metal sphere. The small force on the small charge is $q\vec{E}$, where $\vec{E}$ is the electric field we calculated in the absence of the small additional charge q.

However, if we place a particle with a big charge Q near the sphere, the sphere polarizes to a significant extent (Figure 14.51). We show the electric field due solely to the new charge distribution on the sphere (we don't show the large additional contribution to the net electric field due to Q). Clearly, the force on Q is not simply Q times the *original* $\vec{E}$, but Q times a significantly larger field.

With these effects in mind, we need to qualify our previous method for measuring electric field, in which we measure the force exerted on a charge q and determine the force per unit charge:

$$\vec{E} = \vec{F}/q$$

This procedure is valid only if q is small enough not to disturb significantly the arrangement of other charges that create $\vec{E}$.

Since no object can have a charge smaller than e (the charge of a proton), sometimes it is not possible to find a charge small enough that it doesn't disturb the arrangement of source charges. In this case, we can't measure the electric field without changing the field!

On the other hand, if we know the locations of the source charges, we can calculate the electric field at a location, by applying the superposition principle and adding up the contributions of all the point charges that are the sources of the field:

$$\vec{E}_{\text{net}} = \frac{1}{4\pi\varepsilon_0}\left(\frac{q_1}{r_1^2}\hat{r}_1 + \frac{q_2}{r_2^2}\hat{r}_2 + \frac{q_3}{r_3^2}\hat{r}_3 + \cdots\right)$$

If even the smallest possible charge e would disturb this arrangement of source charges, we can't use the calculated field to predict the force that would act on a charge placed at this location. However, we could use the calculated field to predict the polarization of a neutral atom placed at that location, because a neutral atom, even if (slightly) polarized, would disturb the existing arrangement of source charges much less than a charged object would. To turn this around, placing a neutral atom at a location and observing its polarization could offer a less intrusive way to measure an electric field at that location.

Another way to improve the measurement would be to measure $\vec{F}/q$ for a positive q and also measure $\vec{F}/q$ for a negative q, and average the results. A negative charge would polarize the sphere in Figure 14.50 in such a way as to reduce rather than increase the value of $\vec{F}/q$, by pushing the negative charges on the sphere farther away.

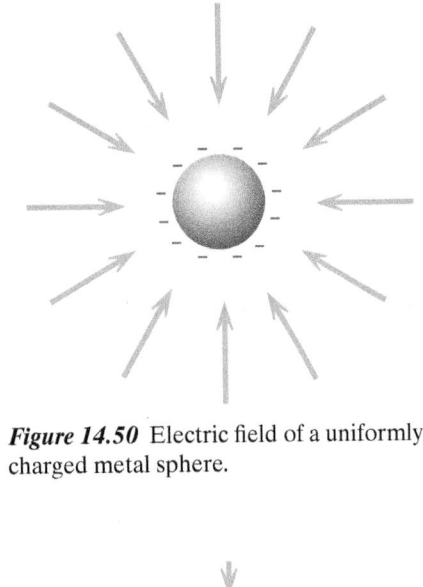

Figure 14.50 Electric field of a uniformly charged metal sphere.

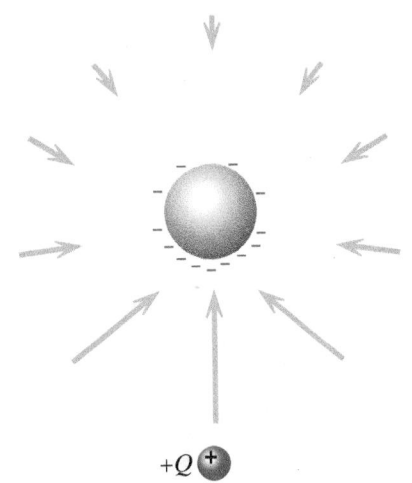

Figure 14.51 Electric field due to the polarized sphere.

SUMMARY

Net Charge
The net charge of an object is the sum of the charges of all of its constituent particles. An object with a net charge of zero is called "neutral."

Conservation of Charge
The net charge of a system plus the net charge of its surroundings cannot change.

Polarization of an atom or molecule produces an induced dipole

$$p = qs = \alpha E_{\text{applied}}$$

(where p is the dipole moment and α is the atomic polarizability).

Insulator
An insulator has no mobile charges.

Conductor

A conductor contains mobile charges that can move through the material.

$$\bar{v} = uE_{\text{net}}$$

The average drift speed $\bar{v}$ of a mobile charged particle in a conductor is directly proportional to the magnitude of the net electric field inside the material. The proportionality constant u is called the "mobility" and has units of (m/s)/(N/C). Different materials have different mobilities.

Metal

A metal is a conductor. It has a mobile electron sea, spread throughout the object, like an ideal gas.

	Conductor	Insulator
Mobile charges	Yes	No
Polarization	Entire sea of mobile charges moves	Individual atoms or molecules polarize
Equilibrium	$\vec{E}_{\text{net}} = \vec{0}$ inside	$\vec{E}_{\text{net}}$ nonzero inside
Excess charge	Spread over surface	In patches anywhere

Force between a point charge and a neutral atom is proportional to $1/r^5$:

$$F_{\text{on pt}} = q_{\text{pt}}E_{\text{ind. dipole}} = \left(\frac{1}{4\pi\varepsilon_0}\right)^2 \left(\frac{2\alpha q_1^2}{r^5}\right)$$

EXPERIMENTS

Is Invisible Tape Electrically Charged?

When you pull a long piece of invisible tape off a roll, it often curls up or sticks to your hand. We will do some simple experiments with invisible tape.

Obtain a roll of invisible tape, such as Scotch® brand Magic™ Tape or a generic brand. It must be the kind of tape that almost disappears when you smooth it down on a surface, not ordinary cellophane tape.

Our first task is to determine whether or not a piece of invisible tape might be electrically charged.

> QUESTION How can we decide whether a piece of invisible tape is electrically charged?

If an object has a net electric charge, it should create an electric field in the surrounding space. Another charged object placed nearby should therefore experience an electric force. If we observe a change in an object's momentum, we can conclude that a force acts on the object.

We know that the electric field of a point charge has these characteristics:

- The magnitude of $\vec{E}$ is proportional to the amount of charge.
- The magnitude of $\vec{E}$ decreases with distance from the charge.
- The direction of $\vec{E}$ is directly away from or toward the source charge.

Therefore, since $\vec{F}_2 = q_2\vec{E}_1$, the electric force on object 2 should have the same properties. In addition, we should be able to observe both attraction and repulsion, since charges of different sign will be affected differently by a particular field.

We will observe the interactions of two pieces of invisible tape, and see whether they meet the criteria listed above.

Preparing a U Tape

Use a strip of tape about 20 cm long (about 8 in, about as long as this paper is wide). Shorter pieces are not flexible enough, and longer pieces are difficult to handle. Fold under one end of the strip to make a nonsticky handle, as shown in Figure 14.52.

Figure 14.52

HOW TO PREPARE A U TAPE

- Stick a strip of tape with a handle down onto a smooth flat surface such as a desk. This is a "base" tape.
- Smooth this base tape down with your thumb or fingertips. This base tape provides a standard surface to work from. (Without this base tape, you get different effects on different kinds of surfaces.)
- Stick another tape with a handle down on top of the base tape, as shown in Figure 14.53.

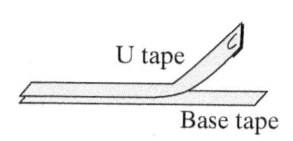

Figure 14.53

- Smooth the upper tape down well with your thumb or fingertips.
- Write U (for Upper) on the handle of the upper tape.
- With a quick motion, pull the U tape up and off the base tape, leaving the base tape stuck to the desk.
- Hang the U tape vertically from the edge of the desk, and bring your hand near the hanging tape, as shown in Figure 14.54.

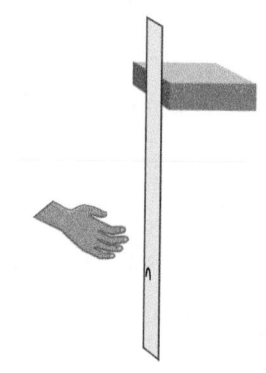

Figure 14.54

- If the tape is in good condition and the room is not too humid, you should find that there is an attraction between the hanging strip of tape and your hand when you get close to the tape. If there is no attraction, remake the U tape.

Experimental technique: try to handle the tapes only by their ends while you are doing an experiment.

EXP1 Interaction of Two U Tapes

(a) If U tapes are electrically charged, how would you expect two U tapes to interact with each other? Would you expect them to repel each other, attract each other, or not to interact at all? Make a prediction, and briefly state a reason.

(b) Make two U ("upper") tapes by following the procedure detailed above. Make sure that both tapes interact with your hand. Hang one on the edge of a desk. Bring the second U tape near the hanging U tape. Since the hanging tape is attracted to your hands, try to keep your hands out of the way. For example, you might approach the vertically hanging tape with the other tape oriented horizontally, held by two hands at its ends. What happens?

You should have seen the two U tapes repel each other. If you did not observe repulsion, try remaking the U tapes (or making new ones, both from the same roll of invisible tape). It is important to see this effect before continuing further.

Making a Tape Not Interact
You may have already discovered that if you handle a U tape too much, it no longer repels another U tape. Next we will learn a systematic way for making this happen.

- Make sure that you have an active U tape, which is attracted to your hand.
- Holding onto the bottom of the U tape, slowly rub your fingers or thumb back and forth along the *slick* side of the tape (Figure 14.55).
- You should find that the U tape no longer interacts with your hand. If it still does, repeat the process.

This is a little odd; if the U tape was originally electrically charged, the charges would presumably have been on its sticky side. However, by running a finger along the other side (the slick side) we have apparently "neutralized" it—it now appears uncharged. It will be a while before we can explain this peculiar effect, but now we have a useful way to neutralize a U tape.

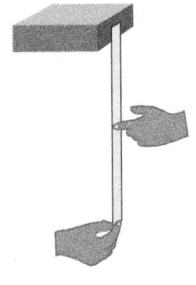

Figure 14.55

EXP2 Is This an Electric Interaction?
To decide whether the interaction between two U tapes (Figure 14.56) is or is not an electric interaction, we will see whether it obeys the criteria for an electric interaction. (As is done throughout the scientific community, it is important to compare your results to the results of other experimenters.)

Figure 14.56

(a) Does the force act along a line connecting the two tapes? Think of a way to determine whether or not the force between two tapes acts along a line drawn from one object to the other, and do the experiment. What did you find? (What would you see if this were not the case?)

(b) Does the force decrease rapidly as the distance between the tapes increases? How can you determine this?

(c) Is the force proportional to the amounts of both charges? Design and carry out an experiment to test this. One way to vary the amount of charge on a tape is to neutralize part of one of the tapes, by running your finger along the length of the slick side of the tape, being careful that your finger touches only a portion of the width of the tape. What do you observe?

The real world is messy! You may have noted several difficulties in making your measurements. For example, both of the tapes are attracted to your hand, and also repel each other. If you tried to use a ruler to measure the distance between the tapes, you might have found that the tapes are attracted to the ruler, too.

Unlike Charges
So far we have observed that two U tapes repel each other, that the force acts along a line between the tapes, that the strength of the repulsion decreases as the tapes get farther away from each other, and that the strength of the interaction depends on the amount of charge on the tape. These observations are consistent with the hypothesis that the U tapes are electrically charged and that all U tapes have like electric charge.

> **QUESTION** How could you prepare a tape that might have an electric charge unlike the charge of a U tape? Think of a plan before reading further.

Perhaps you reasoned along these lines: We don't know how the U tape became charged, but if the tapes started out neutral, maybe the U tape pulled some charged particles off of the lower tape (or vice versa). So now the lower tape should have an equal amount of charge, of the opposite sign.

Making an L Tape
Here is a reproducible procedure for making an L tape, whose charge is unlike the charge of a U tape:

HOW TO PREPARE AN L TAPE

- Stick a strip of tape with a handle down onto a base tape, smooth this tape down thoroughly with your thumb or fingertips, and write L (for Lower) on the handle of this tape.
- Stick another tape with a handle down on top of the L tape, and write U (for Upper) on the handle of this tape. Smooth the upper tape down well with your thumb or fingertips.

▪ You now have three layers of tape on the desk: a base tape, an L tape, and a U tape (Figure 14.57).

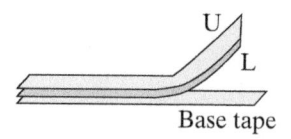

Figure 14.57

▪ Slowly lift the L tape off the base tape, bringing the U tape along with it (and leaving the bottom base tape stuck to the desk). Hang the double layer of tape vertically from the edge of the desk and see whether there is attraction between it and your hand (Figure 14.58). If so, get rid of these interactions (hold the bottom of the tape and slowly rub the slick side with your fingers or thumb).

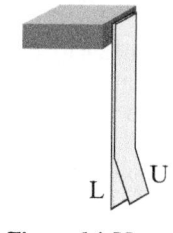

Figure 14.58

▪ Check that the tape pair is no longer attracted to your hand. This is important!

▪ Hold onto the bottom tab of the L tape and quickly pull the U tape up and off (Figure 14.59). Hang the U tape vertically from the edge of the desk, not too close to the L tape!

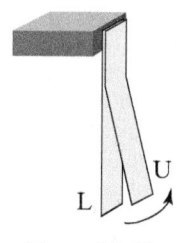

Figure 14.59

Repeating exactly the same procedure, make another pair of tapes so that you have at least two U tapes and two L tapes. Before separating the tapes from each other, always remember to make sure that the tapes are not attracted to your hand.

QUESTION An important step in preparing an L tape is to neutralize the L/U tape pair before separating the two tapes. Considering the principle of charge conservation, why is this step important? What can go wrong if this step is omitted?

The principle of conservation of charge states that if the pair has a total charge of zero before separation, the two tapes will have a total charge of zero after separation: one tape will have a charge of $+q$ and the other a charge of $-q$. However, if the total charge before separation is nonzero and positive (say), the separated tapes could both have positive charge, as long as their individual charges add up to the original amount.

EXP3 Observations of L and U Tapes
You should now have two L tapes and two U tapes. Make sure that both the U tapes and the L tapes are active (attracted to your hand).

 (a) If an L tape is indeed electrically charged, and its charge is unlike the charge on a U tape, what interaction would you *predict* between an L tape and a U tape?

 (b) What interaction do you *observe* between an L tape and a U tape?

 (c) What interaction would you *predict* between two L tapes?

 (d) What interaction do you *observe* between two L tapes?

 (e) Is the pattern of interactions consistent with the statement: "Like charges repel; unlike charges attract"?

A U Tape and an L Tape: Distance Dependence of Attraction
If U and L tapes are electrically charged, then we would expect the strength of the attractive interaction to decrease as the distance between the tapes increases. Make the same sort of observations you made with two U tapes.

EXP4 Distance Dependence of Force between U and L Tapes
Move a U tape very slowly toward a hanging L tape. Observe the deflections of the tapes from the vertical, at several distances (for example, the distance at which you first see attraction, half that distance, etc.). The deflections of the tapes away from the vertical is a measure of the strength of the interaction.

 (a) Does the force decrease rapidly as the distance between the tapes increases?

 (b) Why is this measurement more difficult with a U and an L tape than with two U tapes?

Summary and Conclusions: U and L Tapes
Let's summarize the observations and try to conclude, at least tentatively, whether U and L tapes are electrically charged. Presumably you have observed the following:

▪ Objects can have a charge that is positive, negative, or zero.
▪ Like charges repel, unlike charges attract.
▪ The electric force
  ▪ acts along a line between the charges,
  ▪ decreases rapidly as the distance between the charges increases, and
  ▪ is proportional to the amounts of both charges.

Our observations of U and L tapes seem to be consistent with a description of the electric interactions between charged objects. We tentatively conclude that U and L tapes are electrically charged, and have unlike charges.

How a Plastic Comb or Pen Becomes Charged
Charged objects, such as invisible tape, are negatively charged if they have more electrons than protons, and positively charged if they have fewer electrons than protons. Are U tapes positively or negatively charged? How can we tell? Charging an object in a standard manner gives us a "litmus test."

It is known that if you rub a glass rod with silk, the glass rod becomes positively charged and the silk negatively charged. Likewise, if you rub a clear plastic object such as a pen through your hair (or with fur, wool, or even cotton), the plastic ends up having a negative charge and so repels electrons. A similar process occurs when you separate one tape from another.

See Section 14.2 for a discussion of how objects become charged through rubbing or contact.

EXP5 Determining the Charge on U and L Tapes

Prepare a U tape and an L tape, and hang them from your desk. Test them with your hand to make sure they are both charged. Rub a plastic pen or comb on your hair (clear plastic seems to charge best), or on a piece of cotton or wool, and bring it close to each tape. You should observe that one of the tapes is repelled by the pen, and one is attracted to it.

Knowing that the plastic is negatively charged, what can you conclude about the sign of the electric charge on U tapes? On L tapes?

These results may be reversed if you try a different brand of "invisible tape." Be sure to compare your results with those of other students. Make sure you all agree on the assignment of "+" and "−" labels to your tapes. (If other groups are using different brands of tape, you may disagree on whether U tapes or L tapes are positive, but the electric interactions between your + tapes and their + tapes should be repulsive!)

In any of your experiments, did you find any objects, other than tapes or a charged comb or pen, that repelled a U or L tape? If so, those objects must have been charged. List these objects and whether the charge was + or −.

Amount of Charge on a Tape

We have concluded that U tapes and L tapes are electrically charged, but we have no idea how much charge is on one of the charged tapes—we don't even know an approximate order of magnitude for this quantity. Even a rough measurement of the amount of charge on a tape would be useful, because it would give us a feel for the amount of charge there might be on an ordinary object that is observed to interact electrically with other objects. Therefore the following experiment is an important one.

EXP6 Amount of Excess Charge on a Tape

In this problem you will design and carry out an experiment to determine the approximate number of excess electron charges on the surface of a negatively charged tape.

Initial Estimates

Since we do not know what order of magnitude to expect for our answer, it is important to put upper and lower bounds on reasonable answers.

(a) What is the smallest amount of excess charge that a tape could possibly have?

(b) What is the largest amount of excess charge a tape could possibly have?

Design and Perform an Experiment

A centimeter ruler is printed on the inside back cover of this textbook. A piece of half-inch-wide (1.2 cm) invisible tape, 20 cm long (8 in), has a mass of about 0.16 g.

(c) Make a clear and understandable diagram of your experimental setup, indicating each quantity you measured. Report all measurements you made.

Analyze the Results

(d) Clearly present your physical analysis of your data. Make an appropriate diagram, labeling all vector quantities. Reason from fundamental physics principles. Explicitly report any simplifying assumptions or approximations

you have made in your analysis. Report two quantities:

- The amount of charge on a tape, in coulombs
- The number of excess electrons to which this charge corresponds

Present your analysis clearly. Your reasoning must be clear to a reader.

(e) Estimate whether the true amount of excess charge is larger or smaller than the value you calculated from your experimental data. Explain your reasoning briefly.

Is This a Lot of Charge?

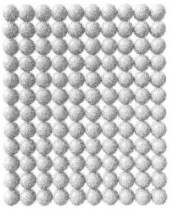

Figure 14.60

(f) What fraction of the molecules on the surface of the tape have gained an extra electronic charge? To estimate this, you may assume that molecules in the tape are arranged in a square array, as indicated in Figure 14.60, and that the diameter of a molecule in the tape is about 3×10^{-10} m. Does your answer suggest that it is a common event or a rare event for a molecule to gain an extra electron?

(g) If the electric field at a location in air exceeds 3×10^6 N/C, the air will become ionized and a spark will be triggered. In Chapter 15 we will see that the electric field in a region very close to a uniformly charged disk or plate depends approximately only on the charge Q per unit area A:

$$E = \frac{1}{e\varepsilon_0}\left(\frac{Q}{A}\right)$$

Use this model (or make a different but justifiable simplifying assumption) to calculate the magnitude of the electric field at a location in the air very close to your tape (less than 1 mm from the surface of the tape). How does it compare to the electric field needed to trigger a spark in the air?

Interaction of Charges and Neutral Matter

We have focused on the interactions of U and L tapes with other U and L tapes. Let's look more broadly at the interactions of charged tapes with other objects.

EXP7 Interactions of U and L Tapes with Other Objects

Which other objects (paper, metal, plastic, etc.) have an attractive interaction with a hanging U or L tape, and which objects have a repulsive interaction? Which objects have no interaction? Record the objects you try and the interactions observed.

The attraction of both U and L tapes to your hand, and to many other objects, is deeply mysterious. The net charge of a neutral object is 0, so your neutral hand should not make an electric field that could act on a charged tape; nor should your neutral hand experience a force due to the electric field made by a charged tape. Nothing in our statement of the properties of electric interactions allows us to explain this attraction!

Observing Interactions with Dipoles
You can also use charged tapes to observe the behavior of a dipole.

> QUESTION Consider the forces that a positive charge Q exerts on the charges making up a dipole (Figure 14.61) and describe the main features of the resulting motion of the dipole.

Figure 14.61

There is a twist (torque) that tends to align the dipole along the line connecting the charge Q and the center of the dipole, with the negative end of the dipole closer to the single positive charge. The dipole has a nonzero net force acting on it that makes it move toward the positive charge.

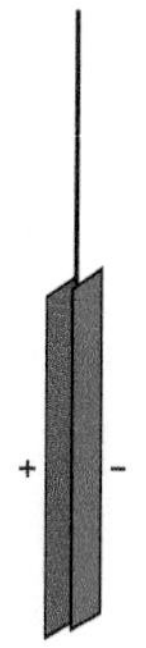

Figure 14.62

EXP8 An Electric "Compass"
Make a tall dipole and observe the motion. Take a + tape and a − tape and stick them together, overlapping them only enough to hold them together. Avoid discharging the tapes with too much handling. Hang the combination from a thread or a hair, as in Figure 14.62.

Now approach the tapes with a charged object and admire how sensitive a charge detector you have made! Slowly move the charged object all around the dipole and observe how the dipole tracks the object.

If you draw an appropriately labeled arrow on the tape, you have an electric "compass" that points in the direction of electric field.

EXP9 Observing Attraction of Like-Charged Objects (!)
Because of the very rapid $1/r^5$ increase in the attraction to neutral matter at short distances, it is sometimes the case that at short distances the attractive effect can actually overcome the repulsion between like-charged objects. Specifically, it could be that

$$N\left(\frac{1}{4\pi\varepsilon_0}\right)^2 \frac{2\alpha q_1^2}{r^5} > \frac{1}{4\pi\varepsilon_0} \frac{q_1 q_2}{r^2}$$

where q_1 and q_2 are the excess charges on the surfaces of the two tapes, and N is the total number of neutral atoms in tape 2 (since each neutral atom participates in the attraction.) As can be seen by dividing the inequality by q_1, you can enhance the effect by making q_1 be significantly larger than q_2, so you might wish to partially discharge one of the tapes.

Try it! Hold one of the U tapes horizontally, with its slick side facing away from you and toward the slick side of the hanging tape. Move toward the hanging tape and check that the hanging tape is repelled as you approach. Then move close enough so that the tapes touch each other (a partner may have to hold the bottom of the hanging tape in order to be able to get very close).

You may be able to detect some slight attraction when the tapes are very close together or touch each other, despite the fact that the tapes repel at longer distances. Do you see such an effect? The effect is quite easy to see in the interaction between a highly charged Van de Graaff generator and a charged tape.

EXP10 Interaction through a Piece of Paper
Have a partner hold a piece of paper close to, but not touching, a hanging U tape. Bring another U tape toward the hanging tape from the other side of the paper, holding both ends of this tape so that it can't swing (Figure 14.63).

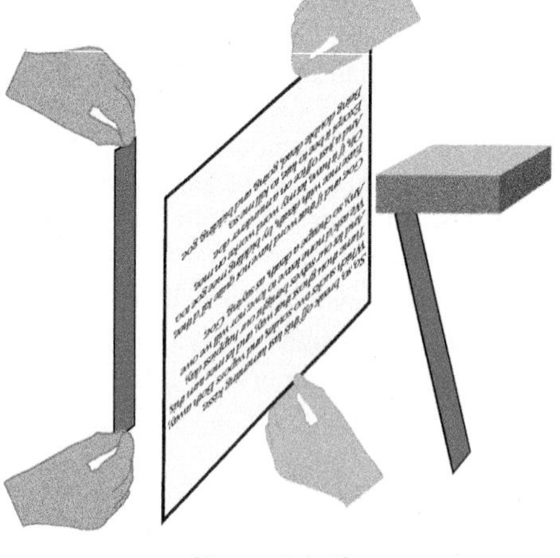

Figure 14.63

Can you observe repulsion occurring right through the intervening paper? This is difficult, because the paper attracts the hanging tape, which masks the repulsion due to the other tape. You can heighten the sensitivity of the experiment by moving the tape rhythmically toward and away from the hanging tape, as though you were pushing a swing. This lets you build up a sizable swing in the hanging tape even though the repulsive force is quite small, because you are adding up lots of small interactions. Using rhythmic movements, are you able to observe repulsion through the intervening paper?

The effect is especially hard to observe if you have weak repulsion due to high humidity. The farther away you can detect repulsion, the better, because the competing attraction falls off rapidly with distance. Under good conditions of low humidity, when tapes remain strongly charged and repulsion is observable with the tapes quite far apart from each other, it is possible to see repulsion with the paper in place, showing that electric field does go right through intervening matter. You have seen evidence of this when you observed attraction between a tape and your hand even when you approached the slick side of the tape.

EXP11 Is Tape a Conductor or an Insulator?
Prepare a hanging tape that has the top half charged and the bottom half uncharged. After a second or two, check to see whether the bottom half of the tape has become charged.

Based on this observation, is tape a conductor or an insulator? That is, are the charges free or bound? Explain fully and rigorously how your observations justify your conclusion. (*Hint:* Draw a diagram showing what effect the charges on the upper half of the tape have on each other and on charges inside the tape, and reason through what will happen if any of these charges are free to move.)

EXP12 Discharging a Tape
The previous exercise suggests that a key element in neutralizing a tape is the salt solution on the surface of your finger. Design one or more experiments you can do to confirm or reject this explanation of discharging a tape.

EXP13 Charging by Induction
Hang a short piece of aluminum foil (about the width of the tape and half as long as your thumb) from a tape, with another piece of tape added to the bottom of the foil as a handle (Figure 14.64).

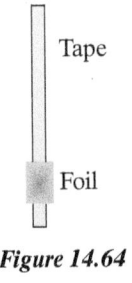

Tape

Foil

Figure 14.64

Now carry out the following operations *exactly* as specified:
1. Make sure that the tape and foil are uncharged (touch the foil, and rub the slick side of the tape).
2. Have a partner hold onto the bottom tape to keep the foil from moving.
3. Bring a charged plastic pen or comb very close to the foil, *but don't touch the foil with the plastic.*
4. *While holding the plastic near the foil*, tap the *back* of the foil with your finger.
5. Move your finger away from the foil, *then* move the plastic away from the foil.

You should find that the metal foil is now strongly charged. This process is called "charging by induction." Now touch the charged aluminum foil with your finger and observe that this discharges the foil, as predicted by our earlier discussion.

Complete a "comic strip" of diagrams (Figure 14.65) illustrating the charging by induction process you carried out. Make sure you have the sign of the charges right. In each diagram show charge distributions, polarization, movement of charges, and so on. For each frame, explain briefly what happens. Remember that excess charges on the metal foil can only be on the surface.

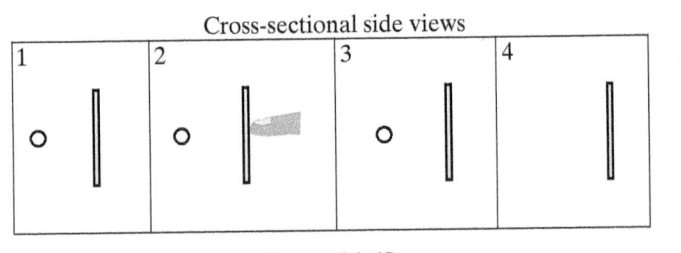

Cross-sectional side views

Figure 14.65

This process is called charging by induction because the entire piece of foil becomes an induced dipole when it is polarized by the external charge. Charging by induction makes it possible to charge a metal without touching the external charge to the metal.

Explain the process of discharging the metal foil by touching it, using the same kind of time-sequence "comic strip" diagrams you used previously. Illustrate the important aspects of each step in the process. Include any changes to your body as well as to the foil. Be precise in your use of words.

EXP14 A Water Film as a Conductor
Prepare a hanging tape that has the top half charged and the bottom half uncharged. Let it hang while you do other work, but check every few minutes to see what has happened in the two halves. What do you predict will happen to the state of charge in the two halves? What do you observe over a period of many minutes? (If the room is very dry or very wet you may not be able to see this effect.)

Suppose you were to breathe heavily through your mouth onto the slick and sticky sides of a short section in the middle of a long charged tape. Your breath is very moist. What do you predict you would find immediately afterward? Try the experiment—what do you observe? (Repeat if you see no effect.)

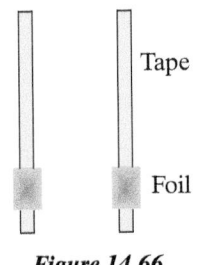

Tape

Foil

Figure 14.66

EXP15 Transferring Charge by Contact
Make two identical hanging foil arrangements (Figure 14.66), and charge one of the aluminum foils by induction. Discharge the other foil by touching it with your finger. Charge a tape or pen or comb, and note the approximate strength of the interaction between it and the charged foil.

Next make the two foils touch each other, being careful not to touch either foil with your fingers. Note the approximate strength of the interaction that there is now between the plastic and each foil.

Compared with the situation before the two foils touched, what sign of charge, and roughly how much, is there now on each foil? Discuss this fully with your partners, and convince yourselves that you understand the process. Make a written explanation, including appropriate diagrams.

What would you expect to happen if one piece of foil were much larger than the other?

QUESTIONS

Q1 Criticize the following statement: "Since an atom's electron cloud is spherical, the effect of the electrons cancels the effect of the nucleus, so a neutral atom can't interact with a charged object." ("Criticize" means to explain why the given statement is inadequate or incorrect, as well as to correct it.)

Q2 Criticize the following statement: "A positive charge attracts neutral plastic by polarizing the molecules and then attracting the negative side of the molecules." ("Criticize" means to explain why the given statement is inadequate or incorrect, as well as to correct it.)

Q3 Jill stuck a piece of invisible tape down onto another piece of tape. Then she yanked the upper tape off the lower tape, and she found that this upper tape strongly repelled other upper tapes and was charged positive. Jack ran his thumb along the slick (upper) side of the upper tape, and the tape no longer repelled other upper tapes. Jill and Jack explained this by saying that Jack rubbed some protons out of the carbon nuclei in the tape. Give a critique of their explanation. If Jill's and Jack's explanation is deficient, give a physically possible explanation for why the upper tape no longer repelled other upper tapes. Include explanatory diagrams.

Q4 Atom A is easier to polarize than atom B. Which atom, A or B, would experience a greater attraction to a point charge a distance r away? Explain your reasoning.

Q5 Is the following statement true or false? If true, what principle makes it true? If false, give a counterexample or say why. See Figure 14.67.

"The electric field E_{point} at the center of an induced dipole, due to the point charge, is equal in magnitude and opposite in direction to the electric field E_{dipole} at the location of the point charge, due to the induced dipole."

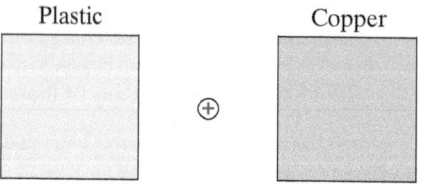

Point charge   Induced dipole

Figure 14.67

Q6 Explain briefly why the attraction between a point charge and a dipole has a different distance dependence for induced dipoles $(1/r^5)$ than for permanent dipoles $(1/r^3)$. (You need not explain either situation in full detail: just explain why there is this difference in their behavior.)

Q7 A large positive charge pulls on a distant electron. How does the *net* force on the electron change if a slab of glass is inserted between the large positive charge and the electron? Does the net force get bigger, smaller, or stay the same? Explain, using only labeled diagrams.

(Be sure to show *all* the forces on the electron before determining the net force on the electron, not just the force exerted by the large positive charge. Remember that the part of the net force on the electron contributed by the large positive charge does not change when the glass is inserted: the electric interaction extends through matter.)

Q8 Explain briefly why repulsion is a better test for the sign of a charged object than attraction is.

Q9 Carbon tetrachloride (CCl_4) is a liquid whose molecules are symmetrical and so are not permanent dipoles, unlike water

molecules. Explain briefly how the effect of an external charge on a beaker of water (H_2O) differs from its effect on a beaker of CCl_4. (*Hint:* Consider the behavior of the permanent dipole you made out of U and L tapes.)

Q10 A positive charge is located between a neutral block of plastic and a neutral block of copper (Figure 14.68). Draw the approximate charge distribution for this situation.

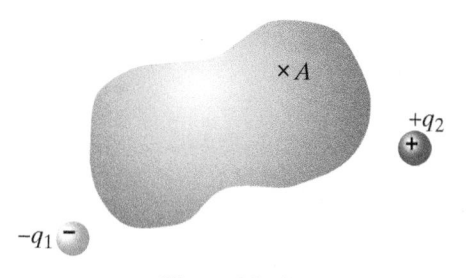

Figure 14.68

Q11 Make a table showing the major differences in the electric properties of plastic, salt water, and copper. Include diagrams showing polarization by an external charge.

Q12 Figure 14.69 shows a neutral, solid piece of metal placed near two point charges. Copy this diagram. **(a)** On your diagram, show the polarization of the piece of metal. **(b)** Then, at location A *inside* the solid piece of metal, carefully draw and label three vectors: (1) $\vec{E}_1$, the electric field due to $-q_1$; (2) $\vec{E}_2$, the electric field due to $+q_2$; (3) $\vec{E}_3$, the electric field due to all of the charges on the metal. **(c)** Explain briefly why you drew the vectors the way you did.

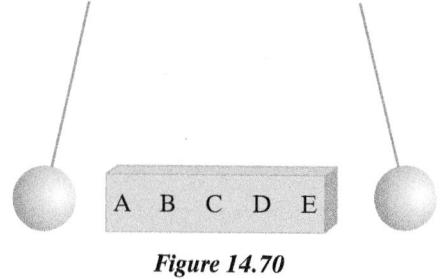

Figure 14.69

Q13 Two identical metal spheres are suspended from insulating threads. One is charged with excess electrons, and the other is neutral. When the two spheres are brought near each other, they swing toward each other and touch, then swing away from each other. **(a)** Explain in detail why both these swings happen. In your explanation, include clear diagrams showing charge distributions, including the final charge distribution. **(b)** Next the spheres are moved away from each other. Then a block of plastic is placed between them as shown in Figure 14.70. The original positions of the spheres are indicated, before the plastic is placed between them. Sketch the new positions of the spheres. Explain, including charge distributions on the spheres.

A B C D E

Figure 14.70

(c) Show the polarization of a molecule inside the plastic at points A, B, C, D, and E. Explain briefly.

Q14 A small glass ball is rubbed all over with a small silk cloth and acquires a charge of +5 nC. The silk cloth and the glass ball are placed 30 cm apart.

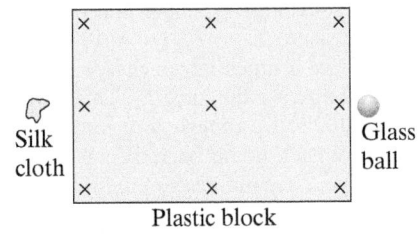

Figure 14.71

(a) On a diagram like that shown in Figure 14.71, draw the electric field vectors qualitatively at the locations marked ×. Pay careful attention to directions and to relative magnitudes. Use dashed lines to explain your reasoning graphically, and draw the final electric field vectors with solid lines.
(b) Next, a neutral block of copper is placed between the silk and the glass.

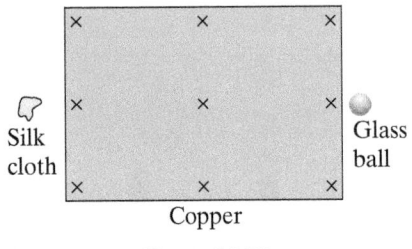

Figure 14.72

On a diagram like that shown in Figure 14.72, carefully show the approximate charge distribution for the copper block and the electric field vectors inside the copper at the locations marked ×.
(c) The copper block is replaced by a neutral block of plastic. Carefully show the approximate molecular polarization of the plastic block at the locations marked × in Figure 14.73.

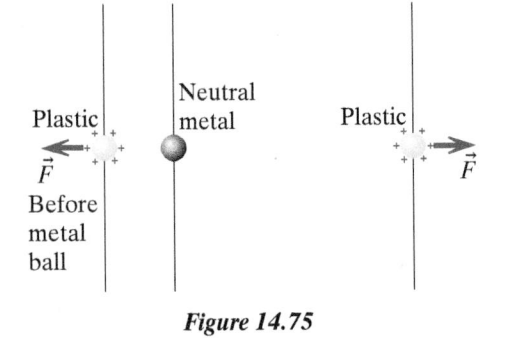

Figure 14.73

(d) Even if you have to state your result as an inequality, make as quantitative a statement as you can about the electric field at the location of the glass ball and the net force on the ball when the plastic block is in place compared to when there is no block. Explain briefly.
Q15 A student said, "When you touch a charged piece of metal, the metal is no longer charged: all the charge on the metal is neutralized." As a practical matter, this is nearly correct, but it

isn't exactly right. What's wrong with saying that all the charge on the metal is neutralized?

Q16 You are wearing shoes with thick rubber soles. You briefly touch a negatively charged metal sphere. Afterward, the sphere seems to have little or no charge. Why? Explain in detail.

Q17 Criticize the following statement: "When you rub your finger along the slick side of a U tape, the excess charges flow onto your finger, and this discharges the tape." Draw diagrams illustrating a more plausible explanation.

Q18 Can you charge a piece of plastic by induction? Explain, using diagrams. Compare with the amount of charging obtained when you charge a piece of metal by induction.

Q19 As shown in Figure 14.74(a), an electroscope consists of a steel ball connected to a steel rod, with very thin gold foil leaves connected to the bottom of the rod (in good electric contact with the rod). The bottom of the electroscope is enclosed in a glass jar and held in place by a rubber stopper.

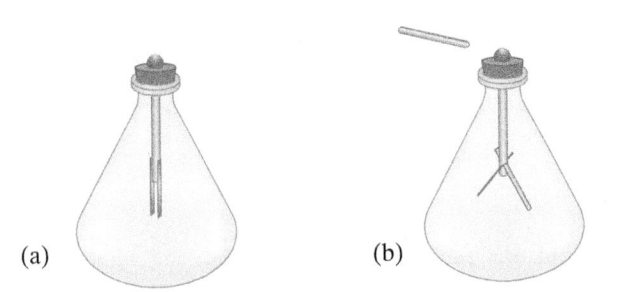

Figure 14.74

(a) The electroscope is brought near to but not touching a positively charged glass rod as shown in Figure 14.74(b). The foil leaves are observed to spread apart. Explain why in detail, using as many diagrams as necessary.
(b) The electroscope is moved far away from the glass rod and the steel ball is touched momentarily to a metal block. The foil leaves spread apart and stay spread apart when the electroscope is moved away from the block. As the electroscope is moved close to but not touching the positively charged glass rod, the foil leaves move closer together. Is the metal block positive, negative, or neutral? How do you know? Explain.

Q20 Two plastic balls are charged equally and positively and held in place by insulating threads, as shown in Figure 14.75.

Figure 14.75

They repel each other with an electric force of magnitude F. Then an uncharged metal ball is held in place by insulating threads between the balls, closer to the left ball. State what change (if any) there is in the net electric force on the left ball and on the net electric force on the right ball. Show relevant force vectors. Also show the charge distribution on the metal ball. Explain briefly but completely.

Q21 You take two invisible tapes of some unknown brand, stick them together, and discharge the pair before pulling them apart and hanging them from the edge of your desk. When you bring an uncharged plastic pen within 10 cm of either the U tape or the L tape you see a slight attraction. Next you rub the pen through your hair, which is known to charge the pen negatively. Now you find that if you bring the charged pen within 8 cm of the L tape you see a slight repulsion, and if you bring the pen within 12 cm of the U tape you see a slight attraction. Briefly explain all of your observations.

Q22 You have three metal blocks marked A, B, and C, sitting on insulating stands. Block A is charged $+$, but blocks B and C are neutral (Figure 14.76).

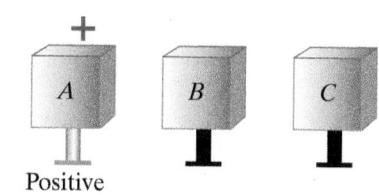

Positive

Figure 14.76

Without using any additional equipment and without altering the amount of charge on block A, explain how you could make block B be charged $+$ and block C be charged $-$. Explain your procedure in detail, including diagrams of the charge distributions at each step in the process.

Q23 You have two identical neutral metal spheres labeled A and B, mounted on insulating posts, and you have a plastic pen that charges negatively when you rub it on your hair (Figure 14.77).

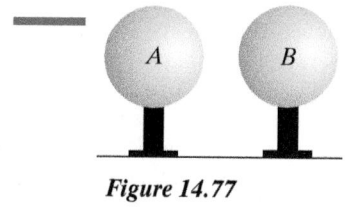

Figure 14.77

(a) $(+ \text{ and } -)$ Explain in detail, including diagrams, what operations you would carry out to give sphere A some positive charge and sphere B an equal amount of negative charge.

(b) $(+ \text{ and } +)$ Explain in detail, including diagrams, what operations you would carry out on the neutral spheres to give sphere A some positive charge and sphere B an equal amount of positive charge (the spheres are initially uncharged).

Q24 Here is a variant of "charging by induction." Place two uncharged metal objects so as to touch each other, one behind the other. Call them *front* object and *back* object. While you hold a charged comb in front of the *front* object, your partner moves away the *back* object (handling it through an insulator so as not to discharge it). Now you move the comb away. Explain this process. Use only labeled diagrams in your explanation (no prose!).

Q25 Metal sphere A is charged negatively and then brought near an uncharged metal sphere B (Figure 14.78). Both spheres rest on insulating supports, and the humidity is very low.

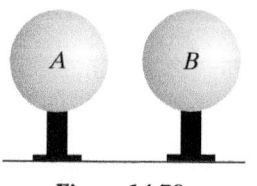

Figure 14.78

(a) Use $+$'s and $-$'s to show the approximate distribution of charges on the two spheres. (*Hint:* Think hard about *both* spheres, not just B.)

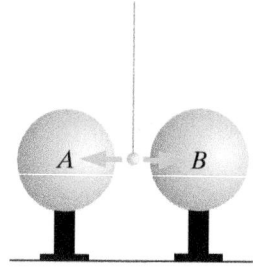

Figure 14.79

(b) A small, lightweight hollow metal ball, initially uncharged, is suspended from a string and hung between the two spheres (Figure 14.79). It is observed that the ball swings rapidly back and forth hitting one sphere and then the other. This goes on for 5 seconds, but then the ball stops swinging and hangs between the two spheres. Explain in detail, step by step, why the ball swings back and forth and why it finally stops swinging. Your explanation must include good physics diagrams.

Q26 Suppose that you try to measure the electric field $\vec{E}$ at a location by placing a charge Q_1 there and observing the force $\vec{F}_1$, so that you measure $E_1 = F_1/Q_1 = 1000$ N/C. Then you remove Q_1 and place a much larger charge $Q_2 = 30Q_1$ at the same location, and observe the force $\vec{F}_2$. This time you measure $E_2 = F_2/Q_2 = 1100$ N/C, though you expected to measure 1000 N/C again. What's going on here? Why didn't you get $E = 1000$ N/C in your second measurement? Sketch a possible situation that would lead to these measurements.

PROBLEMS

Section 14.1

•P27 You rub a plastic comb through your hair and it now carries a charge of -4×10^{-10} C. What is the charge on your hair?

•P28 Many heavy nuclei are "alpha emitters": they emit an alpha particle, which is the historical name for the nucleus of a helium atom, which contains two protons. For example, a

thorium nucleus containing 90 protons is an alpha-emitter. What element does thorium turn into as a result of emitting an alpha particle?

Section 14.2

•P29 Which of the following could be reasonable explanations for how a piece of invisible tape gets charged? Select all that apply. (1) Protons are pulled out of nuclei in one tape and transferred to another tape. (2) Charged molecular fragments are broken off one tape and transferred to another. (3) Electrons are pulled out of molecules in one tape and transferred to another tape. (4) Neutrons are pulled out of nuclei in one tape and transferred to another tape.

•**P30** You rub a clear plastic pen with wool, and observe that a strip of invisible tape is attracted to the pen. Assuming that the pen has a net negative charge, which of the following could be true? Select all that apply. (1) The tape might be negatively charged. (2) The tape might be positively charged. (3) The tape might be uncharged. (4) There is not enough information to conclude anything.

•**P31** Which observation provides evidence that two objects have the same sign charge? (a) The two objects repel each other. (b) The two objects attract each other. (c) The two objects do not interact at all. (d) The strength of the interaction between the two objects depends on distance.

Section 14.3

•**P32** Which statements about a neutral atom are correct? Select all that apply. (1) A neutral atom is composed of both positively and negatively charged particles. (2) The positively charged particles in the nucleus are positrons. (3) The electrons are attracted to the positively charged nucleus. (4) Positively charged protons are located in the tiny, massive nucleus. (5) The radius of the electron cloud is twice as large as the radius of the nucleus. (6) The negatively charged electrons are spread out in a "cloud" around the nucleus.

•**P33** If the distance between a neutral atom and a point charge is tripled, by what factor does the force on the atom by the point charge change? Express your answer as a ratio: new force/original force.

•**P34** There is a region where an electric field points to the right, due to charged particles somewhere. A neutral carbon atom is placed inside this region. Draw a diagram of the situation, and use it to answer the following question: Which of the following statements are correct? Select all that apply. (1) Because the net charge of the carbon atom is zero, it cannot be affected by an electric field. (2) The electron cloud in the carbon atom shifts to the left. (3) The neutral carbon atom polarizes and becomes a dipole. (4) The nucleus of the carbon atom shifts to the left.

•**P35** A charged particle with charge q_1 is a distance r from a neutral atom, as shown in Figure 14.80.

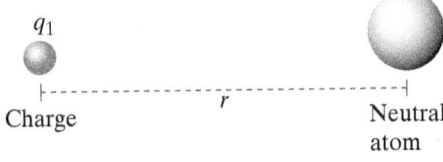

Figure 14.80

(1) If q_1 is negative, which diagram (1–10) in Figure 14.81 best shows the charge distribution in the neutral atom in this situation?

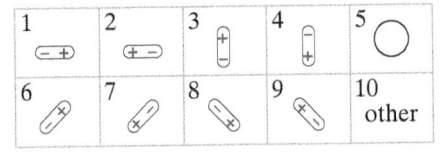

Figure 14.81

(2) Which of the arrows (a–j) in Figure 14.82 best indicates the direction of the electric field at the location of the charged particle, made by the polarized neutral atom?

Figure 14.82

(3) Which of the arrows (a–j) in Figure 14.82 best indicates the direction of the force on the charged particle, due to the polarized neutral atom? (4) Which of the arrows (a–j) in Figure 14.82 best indicates the direction of the force on the polarized neutral atom, due to the charged particle?

•**P36** A charged piece of invisible tape is brought near your hand, as shown in Figure 14.83. Your hand is initially neutral.

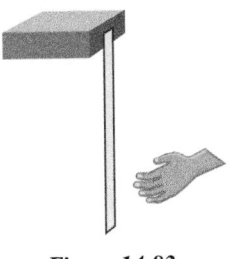

Figure 14.83

(a) If the tape is negatively charged, which of the diagrams 1–10 in Figure 14.81 best shows the polarization of a neutral molecule in your hand? (b) Which arrow in Figure 14.82 best indicates the direction (a–j) of the electric field at the location of the tape due to the large number of polarized molecules in your hand? (c) Which arrow in Figure 14.82 best indicates the direction (a–j) of the force on the tape due to the polarized molecules in your hand? (d) Which arrow in Figure 14.82 best indicates the direction (a–j) of the force on your hand due to the charged tape?

••**P37** An electron and a neutral carbon atom are initially 1×10^{-6} m apart (about 10 000 atomic diameters), and there are no other particles in the vicinity. The polarizability of a carbon atom has been measured to be $\alpha = 1.96 \times 10^{-40}$ C · m/(N/C). (a) Calculate the initial magnitude and direction of the acceleration of the electron. Explain your steps clearly. Pay particular attention to clearly defining your algebraic symbols. Don't put numbers into your calculation until the very end. (b) If the electron and carbon atom were initially twice as far apart, how much smaller would the initial acceleration of the electron be?

••**P38** In Problem P37, replace "electron" with "water molecule" and repeat the analysis. A water molecule has a permanent dipole moment whose magnitude is 6.2×10^{-30} C · m, which is much larger than the induced dipole for this situation. Assume that the dipole moment of the water molecule points toward the carbon atom.

•••**P39** In Figure 14.84 there is a permanent dipole on the left with dipole moment $\mu_1 = Qs_1$ and a neutral atom on the right with polarizability α, so that it becomes an induced dipole with dipole moment $\mu_2 = qs_2 = \alpha E_1$, where E_1 is the magnitude of the electric field produced by the permanent dipole. Show that the force the permanent dipole exerts on the neutral atom is

$$F \approx \left(\frac{1}{4\pi\varepsilon_0}\right)^2 \frac{12\alpha\mu_1^2}{r^7}$$

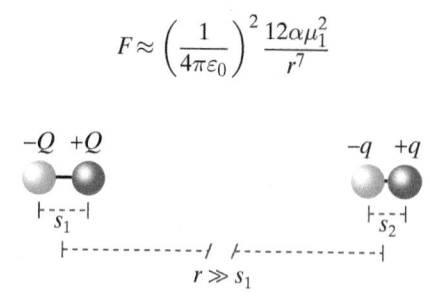

Figure 14.84

Hint: It is convenient to use the "binomial expansion" that you may have learned in calculus, that $(1+\epsilon)^n \approx 1+n\epsilon$ is $\epsilon \ll 1$. Note that n can be negative.

•••P40 Try rubbing a plastic pen through your hair, and you'll find that you can pick up a tiny scrap of paper when the pen is about one centimeter above the paper. From this simple experiment you can estimate how much an atom in the paper is polarized by the pen! You will need to make several assumptions and approximations. Hints may be found at the end of the chapter. **(a)** Suppose that the center of the outer electron cloud ($q = -4e$) of a carbon atom shifts a distance s when the atom is polarized by the pen. Calculate s algebraically in terms of the charge Q on the pen. **(b)** Assume that the pen carries about as much charge Q as we typically find on a piece of charged invisible tape. Evaluate s numerically. How does this compare with the size of an atom or a nucleus? **(c)** Calculate the polarizability α of a carbon atom. Compare your answer to the measured value of $1.96 \times 10^{-40}\,\text{C} \cdot \text{m}/(\text{N/C})$ (T. M. Miller and B. Bederson, "Atomic and molecular polarizabilities: a review of recent advances," *Advances in Atomic and Molecular Physics*, 13, 1–55, 1977). **(d)** Carefully list all assumptions and approximations you made.

Section 14.4

•P41 A solid plastic ball has negative charge uniformly spread over its surface. Which of the diagrams in Figure 14.85 best shows the polarization of molecules inside the ball?

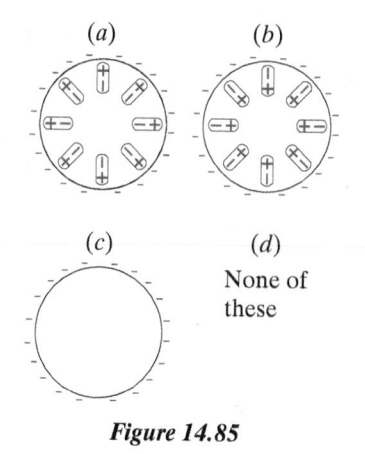

Figure 14.85

•P42 A dipole consisting of two oppositely charged balls connected by a wooden stick is located as shown in Figure 14.86. A block of plastic is located nearby, as shown. Locations B, C,

and D all lie on a line perpendicular to the axis of the dipole, passing through the midpoint of the dipole.

Before selecting answers to the following questions, draw your own diagram of this situation, showing all the fields and charge distributions requested.

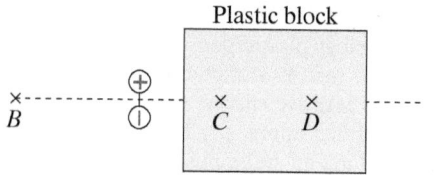

Figure 14.86

Answer the following questions by selecting either a direction (a–j) from Figure 14.82 or an orientation of a polarized molecule (1–10) from the diagrams in Figure 14.81.
(a) Which of the arrows (a–j) best indicates the direction of the electric field at location C due only to the dipole? **(b)** Which of the arrows (a–j) best indicates the direction of the electric field at location D due only to the dipole? **(c)** Which of the diagrams (1–10) best indicates the polarization of a molecule of plastic at location C? **(d)** Which of the diagrams (1–10) best indicates the polarization of a molecule of plastic at location D? **(e)** Which of the following statements is correct? (1) A molecule located at C would not be polarized at all. (2) The polarization of a molecule located at D would be the same as the polarization of a molecule located at C. (3) A molecule located at D would be polarized more than a molecule located at C. (4) A molecule located at D would be polarized less than a molecule located at C. **(f)** Which of the arrows (a–j) best indicates the direction of the electric field at location B due only to the dipole? **(g)** Which of the arrows (a–j) best indicates the direction of the electric field at location B due only to the plastic block? The magnitude of the electric field at B due to the plastic is less than the magnitude of the electric field at B due to the dipole. **(h)** Which of the arrows (a–j) best indicates the direction of the net electric field at location B? **(i)** Which of the following statements is correct? (1) The electric field at B due only to the dipole would be larger if the plastic block were not there. (2) The electric field at B due only to the dipole would be the same if the plastic block were not there. (3) The electric field at B due only to the dipole would be smaller if the plastic block were not there. (4) The electric field at B due only to the dipole would be zero if the plastic block were not there.

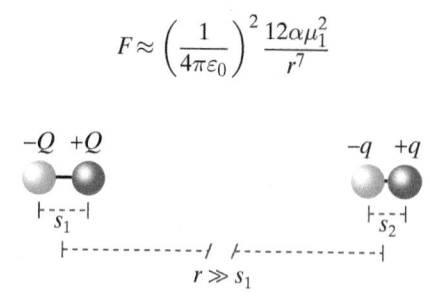

Figure 14.87

(j) Using the diagrammatic conventions discussed in the text, a student drew the diagram in Figure 14.87 to help answer the questions asked above. Which of the following statements about the student's diagram are true? Check all that apply. (1) The direction of polarization of the plastic block is wrong. (2) The diagram is correct; this is just a different way of drawing the polarization. (3) The diagram shows mobile charges; this is wrong because an insulator does not have mobile charged particles.

Section 14.5

•P43 The mobility of Na⁺ ions in water is 5.2×10^{-8} (m/s)/(N/C). If an electric field of 2400 N/C is maintained in the fluid, what is the drift speed of the sodium ions?

•P44 An electric field is applied to a solution containing bromide ions. As a result, the ions move through the solution with an average drift speed of 3.7×10^{-7} m/s. The mobility of bromide ions in solution is 8.1×10^{-8} (m/s)/(N/C). What is the magnitude of the net electric field inside the solution?

Section 14.6

•P45 Which of the following are true? Check all that apply. (1) If the net electric field at a particular location inside a piece of metal is zero, the metal is not in equilibrium. (2) The net electric field inside a block of metal is zero under all circumstances. (3) The net electric field at any location inside a block of copper is zero if the copper block is in equilibrium. (4) The electric field from an external charge cannot penetrate to the center of a block of iron. (5) In equilibrium, there is a net flow of mobile charged particles inside a conductor.

•P46 Which of the following are true? Select all that apply. (1) In equilibrium, there is no net flow of mobile charged particles inside a conductor. (2) The electric field from an external charge cannot penetrate to the center of a block of iron. (3) The net electric field inside a block of aluminum is zero under all circumstances. (4) If the net electric field at a particular location inside a piece of metal is not zero, the metal is not in equilibrium. (5) The net electric field at any location inside a block of copper is zero if the copper block is in equilibrium.

•P47 A negatively charged iron block is placed in a region where there is an electric field downward (in the $-y$ direction) due to charges not shown. Which of the diagrams (a–f) in Figure 14.88 best describes the charge distribution in and/or on the iron block?

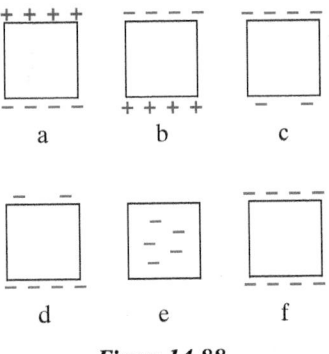

Figure 14.88

•P48 Two small, negatively charged plastic spheres are placed near a neutral iron block, as shown in Figure 14.89. Which arrow (a–j) in Figure 14.89 best indicates the direction of the net electric field at location A?

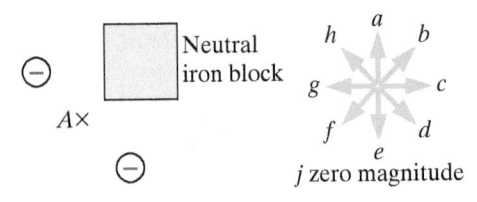

Figure 14.89

•P49 A neutral copper block is polarized as shown in Figure 14.90, due to an electric field made by external charges (not shown). Which arrow (a–j) in Figure 14.90 best indicates the direction of the net electric field at location B, which is inside the copper block?

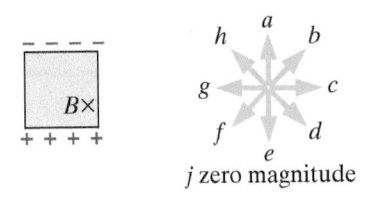

Figure 14.90

•P50 (a) Which of the diagrams (A–F) in Figure 14.91 correctly displays the polarization of a *metal* sphere by an electric field that points to the left, using the conventions discussed in this chapter? (b) Which of the diagrams (A–F) in Figure 14.91 correctly displays the polarization of a *plastic* sphere by an electric field that points to the left, using the conventions discussed in this chapter?

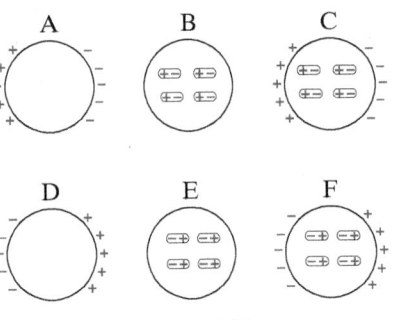

Figure 14.91

•P51 You place a neutral block of nickel near a small glass sphere that has a charge of 2×10^{-8} C uniformly distributed over its surface, as shown in Figure 14.92.

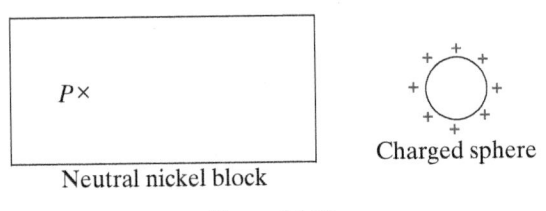

Figure 14.92

(a) About how long do you have to wait to make sure that the mobile electron sea inside the nickel block has reached equilibrium? (1) Less than a nanosecond (1×10^{-9} s), (2) Several hours, (3) About 1 s, (4) About 10 min (b) In equilibrium, what is the average drift speed of the mobile electrons inside the nickel block? (1) About 1×10^5 m/s, (2) About 1×10^{-5} m/s, (3) 0 m/s (c) In the equation $\bar{v} = uE$, what is the meaning of the symbol u? (1) The density of mobile electrons inside the metal, in electrons/m³, (2) The mobility of an electron inside the metal, in (m/s)/(N/C), (3) The time it takes a block of metal to reach equilibrium, in seconds

•P52 This question focuses on reasoning about equilibrium inside the nickel block shown in Figure 14.92. Start with these premises:

▪ The definition of equilibrium inside a conductor and
▪ The relationship between average drift speed and electric field in a conductor

to reason about which situations are *possible* inside the nickel block at equilibrium. Some of the situations listed below are possible, some are ruled out by one premise, and some are ruled out by two premises. If a situation is ruled out by two premises, choose both.

Case 1: $\bar{v} = 0$ and $E_{net} = 0$ (1) Possible, (2) Not possible by definition of equilibrium, (3) Not possible because $\bar{v} = uE_{net}$
Case 2: $\bar{v} = 0$ and $E_{net} > 0$ (1) Possible, (2) Not possible by definition of equilibrium, (3) Not possible because $\bar{v} = uE_{net}$
Case 3: $\bar{v} > 0$ and $E_{net} = 0$ (1) Possible, (2) Not possible by definition of equilibrium, (3) Not possible because $\bar{v} = uE_{net}$
Case 4: $\bar{v} > 0$ and $E_{net} > 0$ (1) Possible, (2) Not possible by definition of equilibrium, (3) Not possible because $\bar{v} = uE_{net}$
Now that you have considered each case, in equilibrium, which one is the only situation that is physically possible? (1) Case 1, (2) Case 2, (3) Case 3, (4) Case 4

•P53 A positively charged sphere is placed near a neutral block of nickel, as shown in Figure 14.92. **(a)** Which of the diagrams in Figure 14.93 best represents the equilibrium distribution of charge on the neutral nickel block?

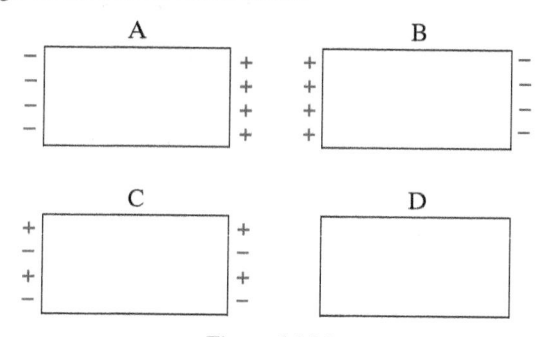

Figure 14.93

(b) At location P inside the nickel block the electric field due to the charged sphere is $\langle -625,0,0 \rangle$ N/C. At equilibrium, which of the following statements must be true? (1) It is not possible to determine the electric field at location P due only to charges on the surface of the nickel block. (2) The electric field at location P due only to charges on the surface of the nickel block is $\langle 0,0,0 \rangle$ N/C. (3) Because the net electric field at location P is $\langle 0,0,0 \rangle$ N/C, the field at P due only to charges on the surface of the polarized nickel block must be $\langle 625,0,0 \rangle$ N/C.

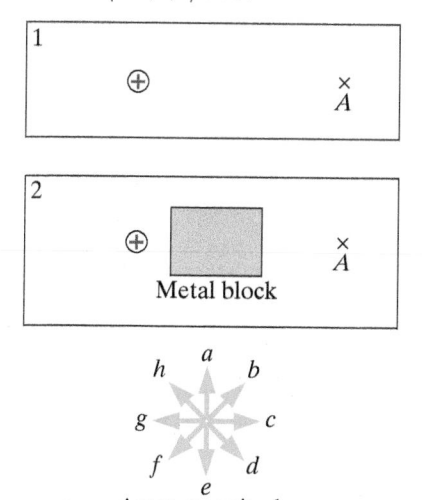

Figure 14.94

•P54 **(a)** The positively charged particle shown in diagram 1 in Figure 14.94 creates an electric field $\vec{E}_p$ at location A. Which of the arrows (*a–j*) in Figure 14.94 best indicates the direction of $\vec{E}_p$ at location A? **(b)** Now a block of metal is placed in the location shown in diagram 2 in Figure 14.94. Which of the arrows (*a–j*) in Figure 14.94 best indicates the direction of the electric field $\vec{E}_m$ at location A due only to the charges in and/or on the metal block? **(c)** $|\vec{E}_p|$ is greater than $|\vec{E}_m|$. With the metal block still in place, which of the arrows (*a–j*) in Figure 14.94 best indicates the direction of the net electric field at location A? **(d)** With the metal block still in place, which of the following statements about the magnitude of $\vec{E}_p$, the field due only to the charged particle, is correct? (1) $|\vec{E}_p|$ is less than it was originally, because the block is in the way. (2) $|\vec{E}_p|$ is the same as it was originally, without the block. (3) $|\vec{E}_p|$ is zero, because the electric field due to the particle can't go through the block. **(e)** With the metal block still in place, how does the magnitude of $\vec{E}_{net}$ at location A compare to the magnitude of $\vec{E}_p$? **(f)** Which of the arrows (*a–j*) in Figure 14.94 best indicates the direction of the net electric field at the center of the metal block (inside the metal)?

•P55 In a particular metal, the mobility of the mobile electrons is 0.0077 (m/s)/(N/C). At a particular moment the net electric field everywhere inside a cube of this metal is 0.053 N/C in the $+x$ direction. What is the average drift speed of the mobile electrons in the metal at this instant?

••P56 A neutral solid metal sphere of radius 0.1 m is at the origin, polarized by a point charge of 6×10^{-8} C at location $\langle -0.3,0,0 \rangle$ m. At location $\langle 0,0.07,0 \rangle$ m, what is the electric field contributed by the polarization charges on the surface of the metal sphere? How do you know?

••P57 A point charge of 3×10^{-9} C is located at the origin. **(a)** What is the magnitude of the electric field at location $\langle 0.2,0,0 \rangle$ m? **(b)** Next, a short, straight, thin copper wire 3 mm long is placed along the x axis with its center at location $\langle 0.1,0,0 \rangle$ m. What is the approximate change in the magnitude of the electric field at location $\langle 0.2,0,0 \rangle$ m? **(c)** Does the magnitude of the electric field at location $\langle 0.2,0,0 \rangle$ m increase or decrease as a result of placing the copper wire between this location and the point charge? **(d)** Does the copper metal block the electric field contributed by the point charge?

••P58 A metal ball with diameter of a half a centimeter and hanging from an insulating thread is charged up with 1×10^{10} excess electrons. An initially uncharged identical metal ball hanging from an insulating thread is brought in contact with the first ball, then moved away, and they hang so that the distance *between their centers* is 20 cm. **(a)** Calculate the electric force one ball exerts on the other, and state whether it is attractive or repulsive. If you have to make any simplifying assumptions, state them explicitly and justify them. **(b)** Now the balls are moved so that as they hang, the distance *between their centers* is only 5 cm. Naively one would expect the force that one ball exerts on the other to increase by a factor of $4^2 = 16$, but in real life the increase is a bit less than a factor of 16. Explain why, including a diagram. (Nothing but the distance between centers is changed—the charge on each ball is unchanged, and no other objects are around.)

••P59 A thin, hollow spherical plastic shell of radius R carries a uniformly distributed negative charge $-Q$. A slice through the plastic shell is shown in Figure 14.95.

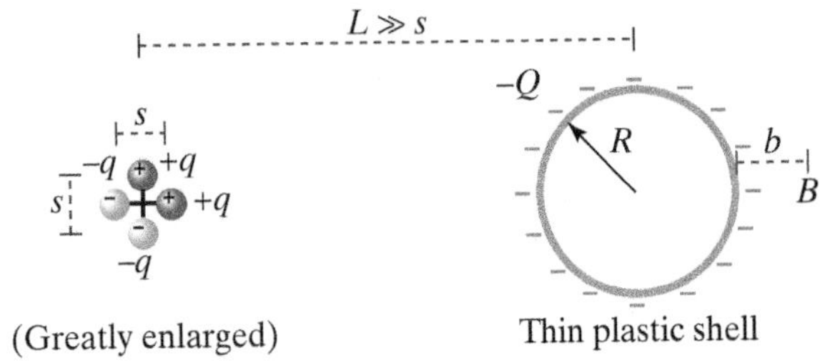

Figure 14.95

(Greatly enlarged)

Thin plastic shell

To the left of the spherical shell are four charges packed closely together as shown (the distance s is shown greatly enlarged for clarity). The distance from the center of the four charges to the center of the plastic shell is L, which is much larger than s ($L \gg s$). Remember that a uniformly charged sphere makes an electric field as though all the charge were concentrated at the center of the sphere. **(a)** Calculate the x and y components of the electric field at location B, a distance b to the right of the outer surface of the plastic shell. Explain briefly, including showing the electric field on a diagram. Your results should not contain any symbols other than the given quantities R, Q, q, s, L, and b (and fundamental constants). You need not simplify the final algebraic results except for taking into account the fact that $L \gg s$. **(b)** What simplifying assumption did you have to make in part (a)? **(c)** The plastic shell is removed and replaced by an uncharged metal ball, as in Figure 14.96.

At location A inside the metal ball, a distance b to the left of the outer surface of the ball, accurately draw and label the electric field $\vec{E}_{ball}$ due to the ball charges and the electric field $\vec{E}_4$ of the four charges. Explain briefly.

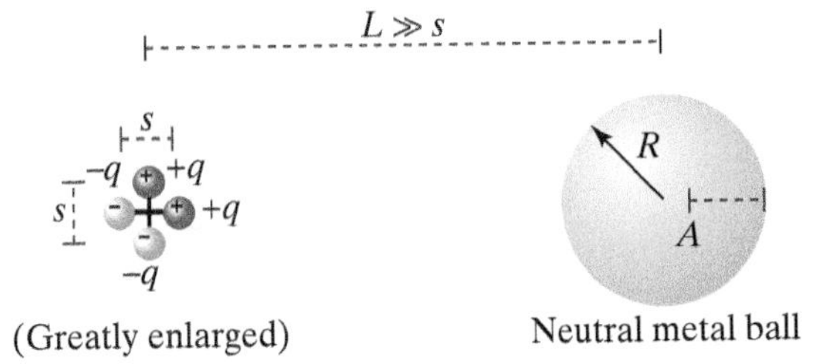

Figure 14.96

(Greatly enlarged)

Neutral metal ball

(d) Show the distribution of ball charges. **(e)** Calculate the x and y components of the net electric field at location A.

••P60 A very thin spherical plastic shell of radius 15 cm carries a uniformly distributed negative charge of $-8\,nC$ ($-8 \times 10^{-9}\,C$) on its outer surface (so it makes an electric field as though all the charge were concentrated at the center of the sphere). An uncharged solid metal block is placed nearby. The block is 10 cm thick, and it is 10 cm away from the surface of the sphere. See Figure 14.97. **(a)** Sketch the approximate charge distribution of the neutral solid metal block.

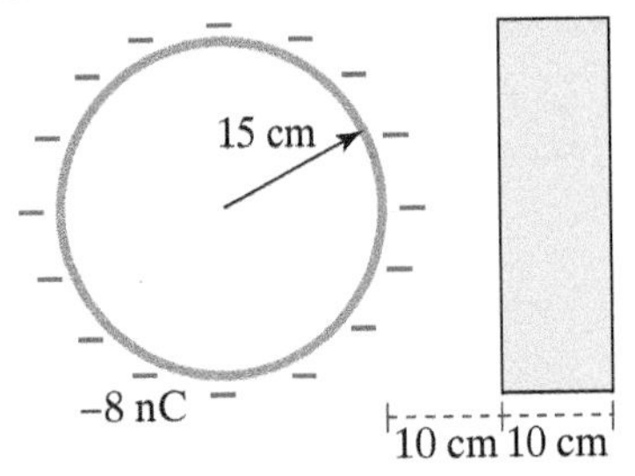

15 cm

−8 nC

10 cm 10 cm

Figure 14.97

(b) Draw the electric field vector at the center of the metal block that is due solely to the charge distribution you sketched (that is, excluding the contributions of the sphere). **(c)** Calculate the magnitude of the electric field vector you drew. Explain briefly. If you must make any approximations, state what they are.

Section 14.7

•P61 You run your finger along the slick side of a positively charged tape, and then observe that the tape is no longer attracted to your hand. Which of the following are not plausible explanations for this observation? Check all that apply. (1) Sodium ions (Na^+) from the salt water on your skin move onto the tape, leaving the tape with a zero (or very small) net charge. (2) Electrons from the mobile electron sea in your hand move onto the tape, leaving the tape with a zero (or very small) net charge. (3) Chloride ions (Cl^-) from the salt water on your skin move onto the tape, leaving the tape with a zero (or very small) net charge. (4) Protons are pulled out of the nuclei of atoms in the tape and move onto your finger.

•P62 You observe that a negatively charged plastic pen repels a charged piece of invisible tape. You then observe that the same piece of tape is repelled when brought near a metal sphere. You are wearing rubber-soled shoes, and you touch the metal sphere with your hand. After you touch the metal sphere, you observe that the tape is attracted to the metal sphere. Which of the following statements could be true? Check all that apply. (1) Electrons from the sphere traveled through your body into the Earth. (2) Electrons from the sphere moved into the salt water on your skin, where they reacted with sodium ions. (3) After you touched it, the metal sphere was very nearly neutral. (4) Chloride ions from the salt water on your hand moved onto the sphere. (5) The excess negative charge from the sphere spread out all over your body. (6) Electrons from your hand moved onto the sphere. (7) Sodium ions from the salt water on your hand moved onto the sphere.

•P63 Blocks A and B are identical metal blocks. Initially block A is neutral, and block B has a net charge of 5 nC. Using insulating handles, the blocks are moved so they touch each other. After touching for a few seconds, the blocks are separated (again using insulating handles). **(a)** What is the final charge of block A? **(b)** What happened while the blocks were in contact with each other? (1) Protons moved from block B to block A. (2) Positrons moved from block B to block A. (3) Electrons moved from block A to block B. (4) Both protons and electrons moved. (5) No charged particles moved.

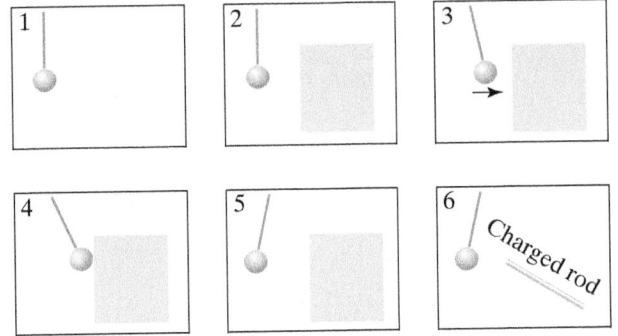

Figure 14.98

••P64 The diagrams in Figure 14.98 show a sequence of events involving a small lightweight aluminum ball that is suspended from a cotton thread. In order to get enough information, you will need to read through the entire sequence of events described below before beginning to answer the questions. Before trying to select answers, you will need to draw your own diagrams showing the charge state of each object in each situation. **(a)** A small, lightweight aluminum ball hangs from a cotton thread. You touch the ball briefly with your fingers, then release it (Diagram 1 in Figure 14.98). Which of the diagrams in Figure 14.99 best shows the distribution of charge in and/or on the ball at this moment, using the diagrammatic conventions discussed in this chapter? **(b)** A block of metal that is known to be charged is now moved near the ball (Diagram 2 in Figure 14.98). The ball starts to swing toward the block of metal, as shown in Diagram 3 in Figure 14.98. Remember to read through the whole sequence before answering this question: Which of the diagrams in Figure 14.99 best shows the distribution of charge in and/or on the ball at this moment? **(c)** The ball briefly touches the charged metal block (Diagram 4 in Figure 14.98). Then the ball swings away from the block and hangs motionless at an angle, as shown in Diagram 5 in Figure 14.98. Which of the diagrams in Figure 14.99 best shows the distribution of charge in and/or on the ball at this moment? **(d)** Finally, the block is moved far away. A negatively charged rod is brought near the ball. The ball is repelled by the charged rod, as shown in Diagram 6 in Figure 14.98. Which of the diagrams in

Figure 14.99 best shows the distribution of charge in and/or on the ball at this moment?

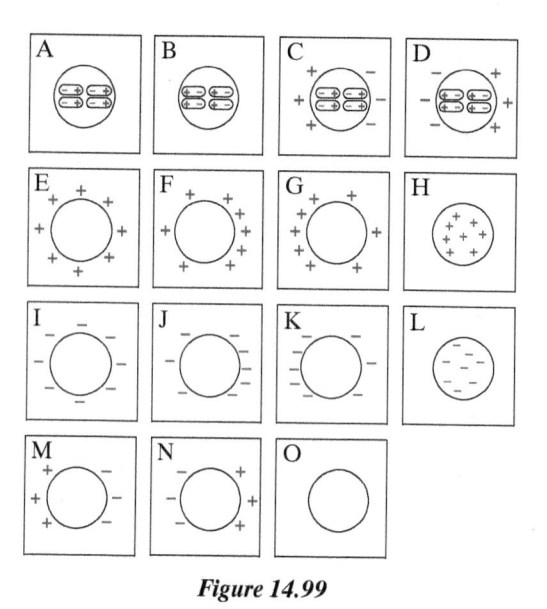

Figure 14.99

ANSWERS TO CHECKPOINTS

1 0; $-e = -1.6 \times 10^{-19}$ C
2 The student has forgotten to consider the superposition principle. Electric interactions go right through matter, so the effect of the positive nucleus is not blocked by the surrounding electron cloud. There are exactly as many protons in the nucleus as there are electrons, and normally the electron cloud is centered on the nucleus, so the net effect is zero.
3 About 2×10^{-15} m, about the diameter of a proton!
4 $2^{-5} = 1/32$
5 Negative tape is attracted to finger:

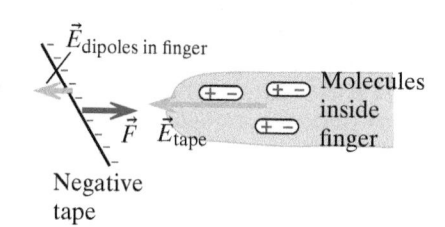

Figure 14.100

6 1.5×10^{-5} m/s
7 0.22 N/C
8 (a) Note shift of charge distribution; it is no longer uniform:

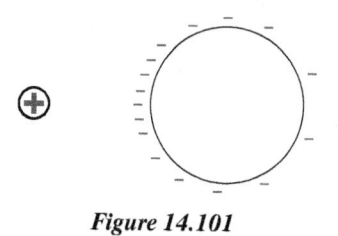

Figure 14.101

The net field is zero:

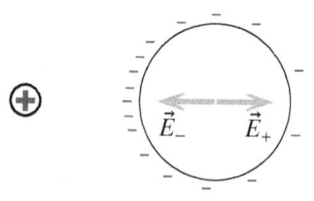

Figure 14.102

(b) Negative pen polarizes the neutral metal cylinder by shifting the electron sea; + charges are closer than − charges, so the pen exerts a net attraction on the cylinder.

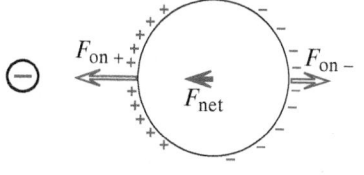

Figure 14.103

The net field is zero:

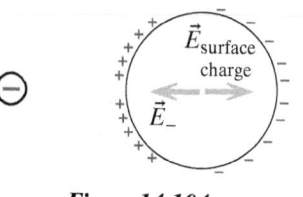

Figure 14.104

(c) Negative pen polarizes the neutral plastic cylinder by polarizing the molecules; + charges are closer than − charges, so the pen exerts a net attraction on the cylinder.

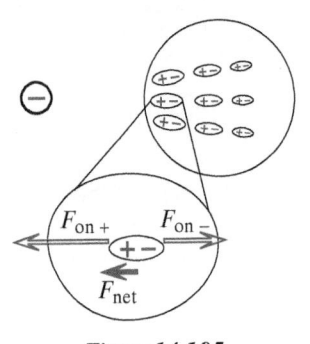

Figure 14.105

Hints for Problem P40, Polarizability of a carbon atom What must be the force on a single carbon atom in the paper at the moment the paper is lifted by the pen? You know how to calculate the force on a point charge due to a dipole. How does this relate to the force on the dipole by the point charge? In this problem, is there something you can model as a dipole and something else you can model as a point charge? Note that the dipole moment ($p = qs$) of a polarized atom or molecule is directly proportional to the applied electric field. In this case the charged pen is generating the applied electric field.

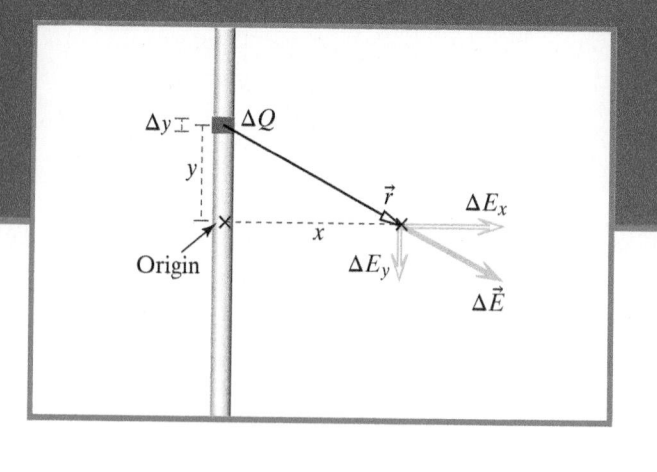

CHAPTER
15

Electric Field of Distributed Charges

OBJECTIVES

After studying this chapter you should be able to

- Set up an integral to calculate the electric field at any location due to a continuous charge distribution.
- Correctly use approximate analytical expressions for the electric field of various charge distributions.
- Numerically (computationally) calculate the electric field at a single location due to a continuous charge distribution.
- Numerically (computationally) calculate the electric field at many observation locations due to a charge distribution.

In many real-world situations, charge is spread out over the surface of a macroscopic object. This chapter focuses on mathematical techniques for adding up the contributions to the electric field of very large numbers of point charges distributed over large areas. The most general technique is to divide the charge distribution into a large but finite number of pieces, approximate each piece by a point charge, and use a computer to add up the contributions ("numerical integration"). This chapter introduces an approach to setting up and carrying out such a computation. Such a calculation can be done by hand, but it is most easily done using a computer. Section 15.9 introduces a new computational concept that greatly facilitates the calculation and display of electric fields in 3D.

In a few special but important cases we can get an analytical solution by using an integral to add up the contributions. We are able to do this for some locations near a charged rod, ring, disk, capacitor, and sphere. A major advantage of the analytical approach is that the result allows us to see how the field varies with distance from the charge distribution, which is important in many applications. These results turn out to be useful because it is often possible to model ordinary objects as combinations of spheres, rods, rings, and disks, and thereby to estimate their electric fields.

Whether we solve a problem analytically or numerically, a key step in the process is coming up with an algebraic description of the location of an object or part of an object in 3D space. Drawing and labeling diagrams can save time and minimize frustration in this process.

15.1 A UNIFORMLY CHARGED THIN ROD

As an example of how to calculate the electric field due to large numbers of charges, we'll consider the electric field of a uniformly charged thin rod (Figure 15.1). The rod might, for example, be a glass rod that was rubbed all over with silk, giving it a nearly uniform distribution of positive charge on its

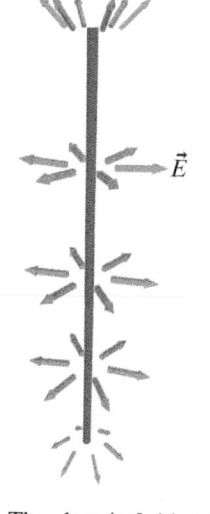

Figure 15.1 The electric field at locations near a positively charged rod. Near the middle of the rod the electric field vectors lie in a plane perpendicular to the rod.

surface. We'll consider a thin rod of length L and total positive charge Q. To illustrate the process, we will pick an observation location on the midplane of the rod (the plane perpendicular to the rod, passing through the midpoint of the rod); this simplifies the algebra we will need to do.

> QUESTION Before launching into a calculation, think about the pattern of electric field you would expect to observe around a rod. Viewed end-on, what would it look like?

The field of a point charge was spherically symmetric, since a point charge itself has spherical symmetry. Since a rod is shaped like a cylinder, we should expect cylindrical symmetry in the electric field of a uniformly charged rod. Near a positively charged rod, the electric field might look like Figure 15.1. If we took a slice perpendicular to the rod, near the middle of the rod, the pattern of field might look like Figure 15.2.

The process of finding the electric field due to charge distributed over a macroscopic object has four steps:

1. Divide the charged object into small pieces. Make a diagram and draw the electric field $\Delta \vec{E}$ contributed by one of the pieces.
2. Choose an origin and axes. Write an algebraic expression for the electric field $\Delta \vec{E}$ due to one piece.
3. Add up the contributions of all pieces, either numerically or symbolically.
4. Check that the result is physically correct.

We will carry out this process to find the electric field of a uniformly charged thin rod, at a location on the midplane.

Step 1: Divide the Distribution into Pieces; Draw $\Delta \vec{E}$

To apply the superposition principle, we imagine cutting up the thin rod into very short sections each with positive charge ΔQ, as shown in Figure 15.3. The Greek capital letter delta (Δ) denotes a small portion of something or a change in something. Here ΔQ is a small portion of the total charge Q of the rod, which contributes $\Delta \vec{E}$ to the net field at the observation location. We can treat the piece ΔQ as though it were a point charge, which should be a fairly good approximation as long as its size is small compared to the distance to the observation location.

In Figure 15.3 we have picked a representative piece of the rod that is not at a "special" location (in this case, neither at an end nor at the middle), and drawn its contribution $\Delta \vec{E}$ to the net field at our chosen observation location.

Assumptions

We assume that the rod is so thin that we can ignore the thickness of the rod. We choose the piece of charge ΔQ to be small enough that it can be modeled as a point charge.

Step 2: Write an Expression for the Electric Field Due to One Piece

We will approximate each piece of the rod as a point charge. Therefore the expression for $\Delta \vec{E}$ will have the familiar form of the equation for the electric field of a point charge. The challenge is to write this expression in terms of our chosen origin and axes, and to figure out a general expression for the amount of charge on each piece.

In order to write an algebraic expression for the contribution to the electric field of one representative piece of the rod, we need to pick an origin and axes. It is possible to put the origin anywhere, but some choices make the algebra

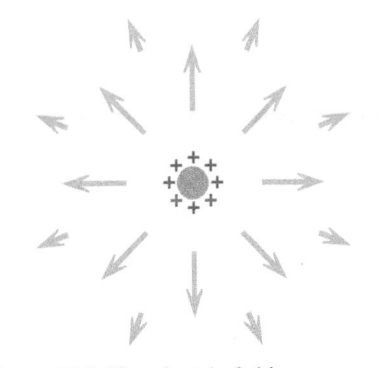

Figure 15.2 The electric field near a charged rod should be cylindrically symmetric. This is a representation of the field in a plane perpendicular to the rod, near the middle of the rod.

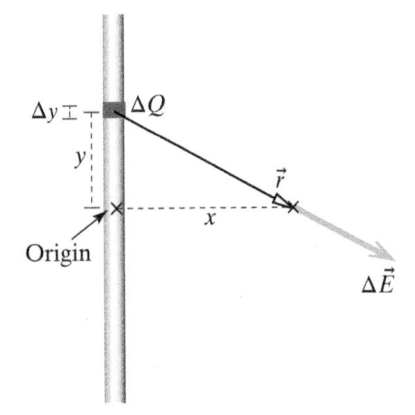

Figure 15.3 $\Delta \vec{E}$ is the contribution to the total field at location $\langle x, 0, 0 \rangle$ made by a small piece of the rod of length Δy.

easier than others. Here we put the origin at the center of the rod; the x axis extends to the right, and the y axis extends up.

The location of one piece of the rod (shown in Figure 15.4) depends on y.

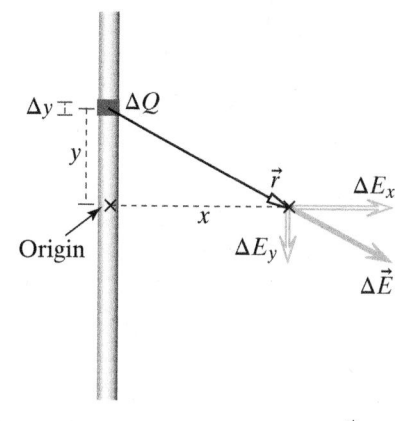

Figure 15.4 The components of $\Delta\vec{E}$ may be calculated by multiplying the scalar quantity by the components of $\hat{r}$.

QUESTION What variables should remain in our answer?

The coordinates of the observation location should remain, but we will sum (integrate) over all pieces of the rod, so the coordinates of the piece of the rod should not remain. However, in the expression we want to integrate it is okay to have variables representing the location of the rod segment; they are called "integration variables" and will disappear after we do the sum.

A piece of the rod is a distance y from the origin. The length of the piece is Δy, a small increment in the integration variable y, shown in Figure 15.4. Our invented variable y will not appear in the final result. It is important to understand that if there is no Δ *something* in the expression (such as Δy), we can't do a sum or evaluate an integral.

The vector $\vec{r}$ points from the source (the representative little piece of charge ΔQ) to the observation location. We can read the components of $\vec{r}$ from the diagram in Figure 15.4:

$$\vec{r} = \langle\text{obs. loc.}\rangle - \langle\text{source}\rangle = \langle x,0,0\rangle - \langle 0,y,0\rangle$$
$$= \langle x,-y,0\rangle$$

Note that $-y$, the y component of $\vec{r}$, is negative (as it should be).

$$|\vec{r}| = [x^2 + (-y)^2]^{1/2}$$
$$\hat{r} = \frac{\vec{r}}{|\vec{r}|} = \frac{\langle x,-y,0\rangle}{[x^2 + (-y)^2]^{1/2}}$$

The scalar part of the field, due to this piece, is:

$$\frac{1}{4\pi\varepsilon_0}\frac{\Delta Q}{r^2} = \frac{1}{4\pi\varepsilon_0}\frac{\Delta Q}{[x^2 + (-y)^2]}$$

We can now write the vector $\Delta\vec{E}$ (Figure 15.3):

$$\Delta\vec{E} = \frac{1}{4\pi\varepsilon_0}\frac{\Delta Q}{[x^2 + (-y)^2]}\frac{\langle x,-y,0\rangle}{[x^2 + (-y)^2]^{1/2}}$$
$$\Delta\vec{E} = \frac{1}{4\pi\varepsilon_0}\frac{\Delta Q}{(x^2 + y^2)^{3/2}}\langle x,-y,0\rangle$$

The components of $\Delta\vec{E}$ (Figure 15.4) are then:

$$\Delta E_x = \frac{1}{4\pi\varepsilon_0}\frac{x\Delta Q}{(x^2 + y^2)^{3/2}}$$
$$\Delta E_y = \frac{1}{4\pi\varepsilon_0}\frac{-y\Delta Q}{(x^2 + y^2)^{3/2}}$$
$$\Delta E_z = 0$$

ΔQ and the Integration Variable

QUESTION Which things are constants and which things are variables in the expressions for ΔE_x and ΔE_y?

You should have identified y as a variable, since it differs for each small piece. We need to rewrite the expression in a form in which it is easy to add up all the ΔE_x's and ΔE_y's due to all the pieces of the rod. This means expressing everything in terms of one integration variable, related to the coordinates of the piece; in this case y is the integration variable.

In particular, we need to express the charge ΔQ in terms of the integration variable y. The rod is uniformly charged with a total charge Q (positive or negative), so the amount of charge on a section of length Δy is equal to

$$\Delta Q = \left(\frac{\Delta y}{L}\right) Q$$

since $\Delta y/L$ is the fraction of the whole rod represented by Δy. Alternatively, there is a linear charge density of Q/L in coulombs/meter, so on a length of Δy m there is an amount of charge $\Delta Q = (Q/L)\Delta y$. The important thing is to realize that ΔQ is a small fraction of the total charge Q, and that you must express ΔQ in terms of Δy.

Algebraic Expression for $\Delta \vec{E}$

Putting it all together, we have an expression for the x component:

$$\Delta E_x = \frac{1}{4\pi\varepsilon_0}\frac{Q}{L}\frac{x}{(x^2+y^2)^{3/2}}\Delta y$$

A similar approach yields an expression for the y component:

$$\Delta E_y = \frac{1}{4\pi\varepsilon_0}\frac{Q}{L}\frac{-y}{(x^2+y^2)^{3/2}}\Delta y$$

A Note on the Symbol Δ

We have used the Greek letter Δ (capital delta) in two ways in setting up this problem. ΔE or ΔQ refers to a small contribution to a total quantity. Δy refers to a change in the integration variable y, which determines the location of the piece of the rod currently under consideration. You will need to use the symbol Δ in both these ways when setting up problems like this one.

Step 3: Add Up the Contributions of All the Pieces

Before performing the actual summation, we should think about what we expect the answer to be. Since we have chosen an observation location on the midplane of the rod, we should expect the y component of the final result to be zero, as indicated in Figure 15.5.

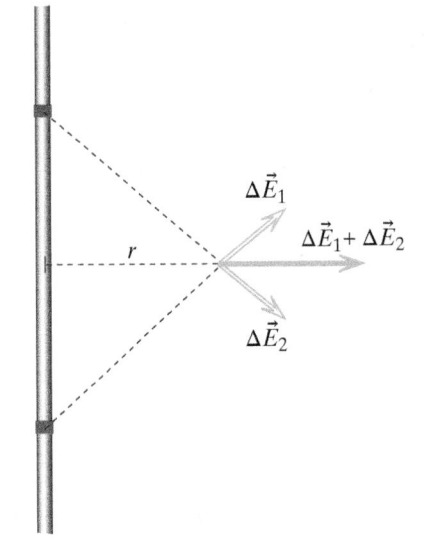

Figure 15.5 Along a line in the midplane only the x component of the electric field is nonzero, because the y contributions cancel.

x components:

Each ΔQ contributes ΔE_x to the net field. If we number each piece 1, 2, 3, and so on, we have

$$\Delta E_x = \Delta E_{x_1} + \Delta E_{x_2} + \Delta E_{x_3} + \cdots$$

It is standard practice to write such sums in a more compact form using a Σ (the Greek capital sigma stands for "Summation"):

$$E_x = \sum \Delta E_x = \sum \frac{1}{4\pi\varepsilon_0}\frac{Q}{L}\frac{x}{(x^2+y^2)^{3/2}}\Delta y$$

y components:

$$E_y = \sum \Delta E_y = \sum \frac{1}{4\pi\varepsilon_0}\frac{Q}{L}\frac{-y}{(x^2+y^2)^{3/2}}\Delta y$$

At this point we have to decide how to add up these terms. Each term in this summation is different. Although we can choose to make all the pieces the same length Δy and take Δy out of the sum, y itself is different for every piece along the rod. How can we add up all the contributions of all the pieces?

Numerical Summation

One way to add up all these different contributions would be to divide the rod into 10 slices as shown in Figure 15.6, calculate the contribution of each of those 10 slices (using our equation with $\Delta y = L/10$), and add up these 10 numbers. Of course that would be only an approximate result, because each slice isn't really a point particle, but it might be good enough for many purposes. Let's do this calculation for the case of $L = 1$ m, $\Delta y = 0.1$ m, $Q = 1$ nC, and $x = 0.05$ m. For each slice we'll take y to be at the center of the slice, and we'll use a calculator to evaluate each term in the summation. The results are shown in Figure 15.7, including the sum for all pieces. As expected, we see that the sum of the y components is zero.

Accuracy

It would be possible to get a more accurate value by cutting the rod into more slices, but it would be tedious to do the calculations by hand. We wrote a little computer program to do these calculations for various numbers of slices, and the results of these calculations are shown in Figure 15.8.

Judging from the computations summarized in Figure 15.8, our calculation with just 10 slices was not very accurate (though it might be good enough for some purposes), while taking more than 50 slices makes almost no difference. The accuracy depends on how good an approximation it is to consider one slice as a point charge. With 50 slices, each slice Δy is 1/50th of a meter long (0.02 m), and the nearest slice is 0.05 m away from the observation location. Apparently the 0.02-m slices are adequately approximated by point charges at their centers, since using smaller 0.01-m slices (100 slices) gives practically the same result.

How do we know that our computer summations are correct? For 10 slices they agree with the calculator results. For larger and larger numbers of slices the computer results approach a constant value, which is expected. Later we will discuss additional ways to check such work.

Summation as an Integral

A major disadvantage of adding up the contributions numerically is that we don't get an analytical (algebraic) form for the electric field. This means that we can't easily answer such questions as, "How does the electric field vary with r near a rod?" For a point charge we know that the field goes like $1/r^2$, and far from a permanent dipole it goes like $1/r^3$. Is there a way to do the summation to get an algebraic rather than a numerical answer for the rod? That is what integral calculus was invented to do.

The key idea of integral calculus applied to problems like ours is to imagine taking not 50 or 100 slices but an infinite number of infinitesimal slices. We let the number of slices $N = L/(\Delta y)$ increase without bound, and the corresponding slice length $\Delta y = L/N$ decreases without bound. We take the limit as Δy gets arbitrarily small:

$$E_x = \lim_{\Delta y \to 0} \frac{1}{4\pi\varepsilon_0} \frac{Q}{L} x \sum \frac{1}{(x^2 + y^2)^{3/2}} \Delta y$$

This limit is called a "definite integral" and is written like this:

$$E_x = \frac{1}{4\pi\varepsilon_0} \frac{Q}{L} x \int_{-L/2}^{+L/2} \frac{1}{(x^2 + y^2)^{3/2}} dy$$

The integral sign $\int$ is a distorted S standing for Summation, just as the Greek sigma Σ stands for summation. It is important that in this context you think of an integral as a sum of many contributions.

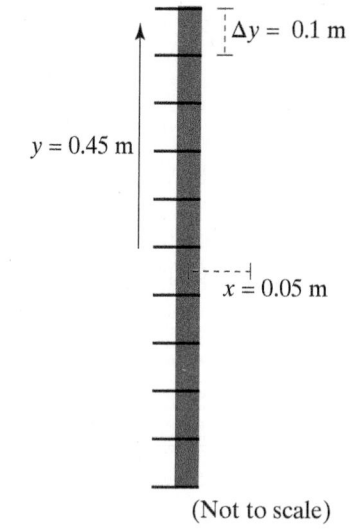

$\Delta y = 0.1$ m

$y = 0.45$ m

$x = 0.05$ m

(Not to scale)

Figure 15.6 Division of the rod into 10 equal slices.

y	E_x	E_y
+0.45	0.484	−4.363
+0.35	1.018	−7.127
+0.25	2.715	−13.577
+0.15	11.384	−34.153
+0.05	127.279	−127.279
−0.05	127.279	127.279
−0.15	11.384	34.153
−0.25	2.715	13.577
−0.35	1.018	7.127
−0.45	0.484	4.363
Sum	285.764	0.000

Figure 15.7 Contributions to $\vec{E}$.

Number of Slices	E_x
10	285.764
20	353.764
50	358.214
100	358.214

Figure 15.8 Effect of number of slices.

The integration variable y ranges from $y = -L/2$ (the bottom of the rod) to $y = +L/2$ (the top of the rod), so these are the limits on the definite integral. If you have drawn an appropriately labeled diagram, you should be able to read the correct limits of the integration variable off the diagram. This is another of the benefits of drawing a well-labeled physics diagram.

Our small length Δy has now changed into dy, an "infinitesimal" increment in y that *must* appear in the integrand. This is why Δy *must* appear in the algebraic expression for the contribution of one piece of the charge distribution.

Evaluating the Integral

Most of the physics in this problem went into setting up the integral. Evaluating the integral is simply mathematics. In some cases it is easy to evaluate the integral. In this particular case, the integral is not a very simple one, but if you would like to exercise your skill at integration, give it a try! Otherwise, it can be found in tables of integrals in mathematical handbooks and in some calculus textbooks. Some calculators can do such integrals, or one can use a tool such as Maple or Mathematica. Looking up the result in a table of integrals, we get the following:

$$E_x = \frac{1}{4\pi\varepsilon_0}\left(\frac{Q}{L}x\right)\left[\frac{y}{x^2\sqrt{x^2+y^2}}\right]_{-L/2}^{+L/2}$$

$$E_x = \frac{1}{4\pi\varepsilon_0}\left[\frac{Q}{x\sqrt{x^2+(L/2)^2}}\right]$$

Note that as we expected this result does not contain the integration variable y, which was simply a variable referring to the coordinates of one piece of the rod and was necessary in setting up the summation.

The y component of the electric field can also be found by integration. In this situation the result comes out to zero, as expected.

$$E_y = \frac{1}{4\pi\varepsilon_0}\frac{Q}{L}\int_{-L/2}^{+L/2}\frac{-y}{(x^2+y^2)^{3/2}}dy$$

$$E_y = 0$$

Replace x with r

Because the rod, and its associated electric field, are cylindrically symmetric, the axis we called the x axis could have been rotated around the rod by any angle, and we would have obtained the same answer. To indicate this, we replace x with r in our result (you may recognize this as converting to cylindrical coordinates):

**ELECTRIC FIELD OF A
UNIFORMLY CHARGED THIN ROD**

$$E = \frac{1}{4\pi\varepsilon_0}\left[\frac{Q}{r\sqrt{r^2+(L/2)^2}}\right]$$

at a location a distance r from the midpoint of the rod along a line perpendicular to the rod. Q is the total charge on the rod, and L is the length of the rod. The direction is either radially away from or toward the rod (depending on the sign of Q). See Figure 15.9.

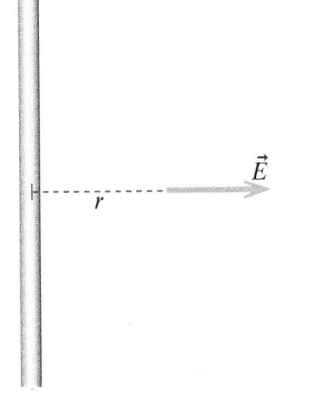

Figure 15.9 Electric field of a uniformly charged thin rod, near the midpoint of the rod.

Step 4: Check the Result

Because there are many opportunities to make mistakes in this procedure, it is extremely important to check the result in as many ways as you can. Different checks provide information about different kinds of possible errors.

Direction: First, is the direction qualitatively correct? We have the electric field pointing straight away from the midpoint of the rod, which is correct, given the symmetry of the situation. The vertical component of the electric field should indeed be zero.

Units: Second, do we have the right units? The units should be the same as the units of the expression for the electric field for a single point particle:

$$\frac{1}{4\pi\varepsilon_0}\frac{Q}{r^2}$$

We easily verify that our answer does have the right units, since

$$\frac{1}{r\sqrt{r^2+(L/2)^2}} \qquad \text{has the same units as} \qquad \frac{1}{r^2}$$

QUESTION Next, let's try a special case for which we already know the answer. If r is very much larger than L, the distant rod looks almost like a point charge, so the net field ought to look like the field of a point charge. Does it?

If $r \gg L$, then $\dfrac{1}{r\sqrt{r^2+(L/2)^2}} \approx \dfrac{1}{r\sqrt{r^2+0}} = \dfrac{1}{r^2}$.

Special case, $L \ll r$: For another special case, suppose that r is fairly near the rod, but the rod is so short as to look almost like a point charge. In that case $L \ll r$, and we again get $E \rightarrow \dfrac{1}{4\pi\varepsilon_0}\dfrac{Q}{r^2}$, as we should.

Compare to numerical calculation: Earlier we carried out a numerical calculation for the case of $Q = 1 \times 10^{-9}$ C, $L = 1$ m, and $r = 0.05$ m, and we found that if we used 50 or more slices the result was $E_x = 358.214$ N/C. The analytical solution gives

$$E = \frac{1}{4\pi\varepsilon_0}\frac{Q}{r\sqrt{r^2+(L/2)^2}}$$

$$= \frac{1}{4\pi\varepsilon_0}\left(\frac{1\times 10^{-9}}{(0.05)\sqrt{0.05^2+0.5^2}}\right)$$

$$= 358.213 \text{ N/C}$$

which agrees very well with the numerically calculated value.

Checking a Numerical Solution

Similar techniques can be used to check a numerical integration done on a computer. For example, set the length of the rod to be very short or the distance to be very large, and the numerical integration should give a result equal to what you calculate by hand for a point charge.

> **Checkpoint 1** If the total charge on a thin rod of length 0.4 m is 2.5 nC, what is the magnitude of the electric field at a location 1 cm from the midpoint of the rod, perpendicular to the rod?

Special Case: A Very Long Rod

A very important special case, which we will refer to often, is the case of a rod that is very long (or alternatively a rod very close to the observation location). In either case $L \gg r$.

QUESTION Before reading further, try to show that if $L \gg r$, the electric field is approximately:

$$\frac{1}{4\pi\varepsilon_0}\frac{2(Q/L)}{r}$$

$$E = \frac{1}{4\pi\varepsilon_0}\left[\frac{Q}{r\sqrt{r^2+(L/2)^2}}\right]$$

If $L \gg r$, then $\dfrac{1}{r\sqrt{r^2+(L/2)^2}} \approx \dfrac{1}{r\sqrt{(L/2)^2}} = \dfrac{1}{r(L/2)}$, so

$$E \approx \frac{1}{4\pi\varepsilon_0}\frac{2(Q/L)}{r}$$

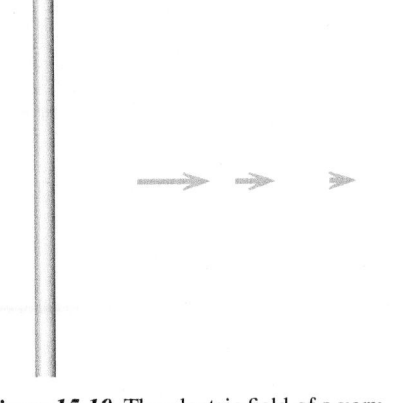

Figure 15.10 The electric field of a very long uniformly charged positive rod, shown at locations along a line very near the center of the rod. Only a small portion of the rod is visible.

This distance dependence is illustrated in Figure 15.10. This also holds for a short rod if we are very close, so that $L \gg r$.

APPROXIMATE ELECTRIC FIELD: VERY LONG ROD

$$E \approx \frac{1}{4\pi\varepsilon_0}\frac{2(Q/L)}{r} \quad \text{if } L \gg r$$

at a location a perpendicular distance $r \ll L$ from the rod, as long as the observation location is not too near the ends of the rod. Q is the total charge on the rod, and L is the length of the rod. The direction is either radially away from or toward the rod (depending on the sign of Q).

Although we proved this result only for observation locations a perpendicular distance $r \ll L$ from the center of the rod, the result is actually a good approximation all along the rod as long as you're not too near the ends of the rod. Figure 15.11, which is the result of an accurate numerical integration, shows that $\vec{E}$ hardly varies along the central region of the rod. You can also see this in a three-dimensional numerical integration shown in Figure 15.12.

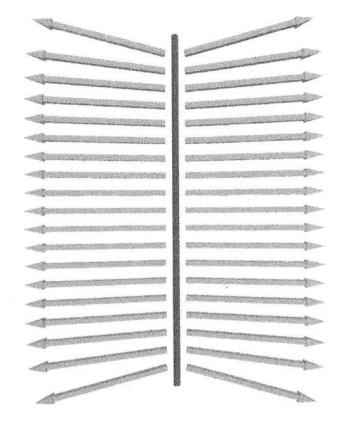

Figure 15.11 A numerical integration of the electric field near a uniformly positively charged rod.

15.2 PROCEDURE FOR CALCULATING ELECTRIC FIELD

Here is a summary of the procedure for calculating electric field:

Step 1: Cut Up the Charge Distribution into Pieces and Draw $\Delta\vec{E}$

- Divide the charge distribution into pieces whose field is known. In particular, very small pieces can be approximated by point particles.
- Pick a representative piece, and at the location of interest draw a vector $\Delta\vec{E}$ showing the contribution to the electric field of this representative piece. Drawing this vector helps you figure out the direction of the net field at the location of interest.

Step 2: Write an Expression for the Electric Field Due to One Piece

- Pick an origin for your coordinate system, and show it on your diagram.
- Draw the vector $\vec{r}$ from the source piece to the observation location. Write algebraic expressions for $\vec{r}$ and $\hat{r}$ (a unit vector in the direction of $\vec{r}$).
- Write an algebraic expression for the magnitude $|\Delta\vec{E}|$ contributed by the representative piece. Multiply by $\hat{r}$ to get a vector $\Delta\vec{E}$, from which you can read the components ΔE_x, ΔE_y, and ΔE_z. Your expressions should contain one or more "integration variables" related to the coordinates of the piece.

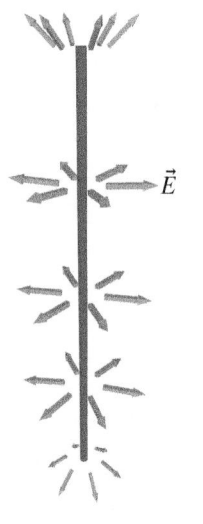

Figure 15.12 A numerical integration in three dimensions of the electric field of a uniformly positively charged rod.

- Write the amount of charge on the piece, Δq, in terms of your variables.
- If a representative piece is small in size, your algebraic expressions should include small increments of the integration variable. For example, if your integration variable is y, your expressions must be proportional to Δy.

Step 3: Add Up the Contributions of All the Pieces

- The net field is the sum of the contributions of all the pieces. To write the sum as a definite integral, you must include limits given by the range of the integration variable. If the integral can be done symbolically, do it. If not, choose a finite number of pieces and do the sum with a calculator or a computer.

Step 4: Check the Result

- Check that the direction of the net field is qualitatively correct.
- Check the units of your result, which should be newtons per coulomb.
- Look at special cases. For example, if the net charge is nonzero, your result should reduce to the field of a point charge when you are very far away. For a numerical integration on a computer, check that the computation gives the correct numerical result for special cases that can be calculated by hand.

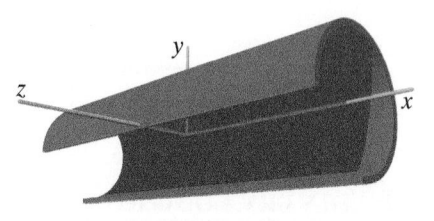

Figure 15.13 A charged hollow 3/4 cylinder. The observation location is at the center of the cylinder.

EXAMPLE

A Hollow Cylinder

A hollow 3/4 cylinder (Figure 15.13) carries a positive charge Q spread uniformly over its surface. The radius of the cylinder is R, and its length is L, where $L \gg R$. Calculate the electric field at the center of the cylinder.

Solution

(1) Since we know the electric field near the center of a long rod, we could consider the cylinder to be made up of many thin charged rods, as shown in Figure 15.14. We will place the origin at the observation location, as shown. Each rod contributes $\Delta \vec{E}$ as shown in Figure 15.15.

(2) The location of the center of a particular rod can be described in terms of the angle θ, defined as shown in Figure 15.15.

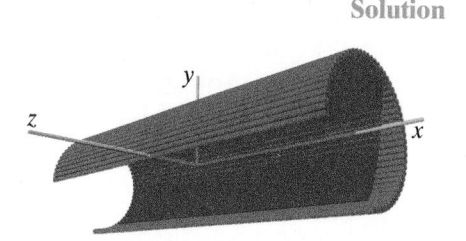

Figure 15.14 We can consider the cylinder to be composed of a large number of very thin charged rods.

$$\vec{r} = \langle \text{obs. location} \rangle - \langle \text{source} \rangle = \langle 0,0,0 \rangle - \langle 0, R\sin\theta, R\cos\theta \rangle$$
$$= \langle 0, -R\sin\theta, -R\cos\theta \rangle$$
$$r = \sqrt{(-R\cos\theta)^2 + (-R\sin\theta)^2} = R$$
$$\hat{r} = \frac{\vec{r}}{r} = \langle 0, -\sin\theta, -\cos\theta \rangle$$

The amount of charge on one rod is given by:

$$dQ = Q\left(\frac{d\theta}{\theta_{\text{total}}}\right) = \frac{Qd\theta}{(3/2)\pi}$$

where $d\theta$ is the angular width of one rod, and $\theta_{\text{total}} = (3/2)\pi$ is the angular extent of the cylinder.

The contribution of this rod to the net electric field is:

$$\Delta \vec{E} = |\Delta \vec{E}|\hat{r} \approx \frac{1}{4\pi\varepsilon_0}\frac{2dQ}{L}\frac{1}{r}\langle 0, -\sin\theta, -\cos\theta \rangle$$
$$= \frac{1}{4\pi\varepsilon_0}\frac{2}{LR}\frac{Qd\theta}{(3/2)\pi}\langle 0, -\sin\theta, -\cos\theta \rangle$$

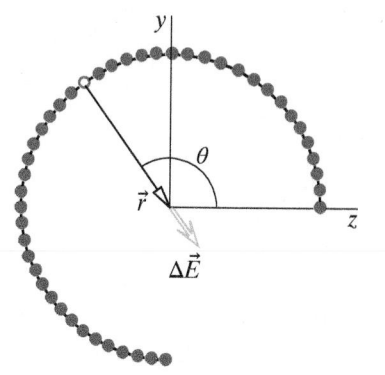

Figure 15.15 An end view of the cylinder, looking along the x axis in the $+x$ direction.

because in this case $r = R$; the center of every rod is the same distance R from the observation location.

(3) We need to consider the x, y, and z components of electric field separately. To add up the contributions of all of the rods, we can integrate:

$$E_x = \int_0^{\frac{3}{2}\pi} \frac{1}{4\pi\varepsilon_0} \frac{2Q}{LR(3/2)\pi} \cdot 0\, d\theta = 0$$

$$E_y = \int_0^{\frac{3}{2}\pi} \frac{1}{4\pi\varepsilon_0} \frac{2Q}{LR(3/2)\pi} (-\sin\theta)\, d\theta = \frac{1}{4\pi\varepsilon_0} \frac{2Q}{LR(3/2)\pi} \cos\theta \Big|_0^{3\pi/2}$$

$$= -\frac{1}{4\pi\varepsilon_0} \frac{4Q}{3\pi LR}$$

$$E_z = \int_0^{\frac{3}{2}\pi} \frac{1}{4\pi\varepsilon_0} \frac{2Q}{LR(3/2)\pi} (-\cos\theta)\, d\theta = \frac{-1}{4\pi\varepsilon_0} \frac{2Q}{LR(3/2)\pi} \sin\theta \Big|_0^{3\pi/2}$$

$$= \frac{1}{4\pi\varepsilon_0} \frac{4Q}{3\pi LR}$$

so the net electric field is:

$$\vec{E} = \left\langle 0, -\frac{1}{4\pi\varepsilon_0} \frac{4Q}{3\pi LR}, \frac{1}{4\pi\varepsilon_0} \frac{4Q}{3\pi LR} \right\rangle$$

(4) The direction is reasonable, since the y component of $\vec{E}$ should be negative, and the z component should be positive.

Units are $(\text{N} \cdot \text{m}^2/\text{C}^2)(\text{C}/(\text{m} \cdot \text{m})) = \text{N/C}$, which is correct.

Special case: If this were a complete hollow cylinder, with $\theta_{\text{total}} = 2\pi$, the limits of integration would be 0 to 2π, and both E_y and E_z would be zero. The zero field makes sense, because there is just as much charge above the center as below, or to the $+z$ or $-z$ side of the center. Later when we introduce Gauss's law it will be possible to prove that inside a long uniformly charged cylindrical tube the contributions of all the charges cancel out, and everywhere inside the tube (not too near the ends) the electric field is zero, not just at the center.

15.3 A UNIFORMLY CHARGED THIN RING

Next we'll calculate the electric field of a uniformly charged thin ring. We'll encounter rings of charge later when we study electric circuits. We'll also use the results for a ring to find the electric field of a disk later in this chapter.

Field of a Ring, at a Point on the Axis

We'll calculate the electric field due to a uniformly charged ring of radius R and total positive charge q. We'll do only the easiest case—the field at a location along the axis of the ring, which is a line going through the center and perpendicular to the ring. Finding the electric field at other locations is harder.

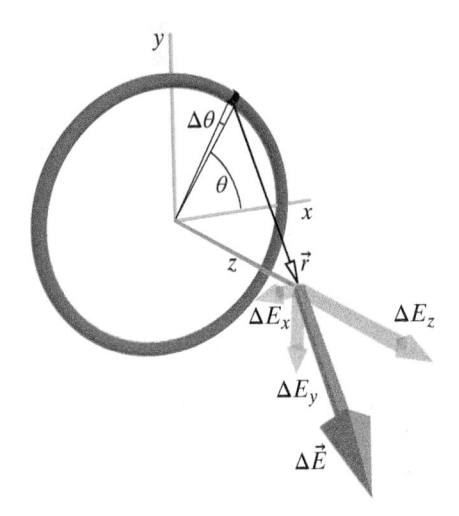

Figure 15.16 The contribution $\Delta\vec{E}$ to the total electric field of a charged ring, made by a segment of the ring of angular size $\Delta\theta$. We have chosen the center of the ring for the origin of our coordinate system. The x, y, and z components of $\Delta\vec{E}$ are indicated.

Step 1: Cut Up the Charge Distribution into Pieces and Draw $\Delta\vec{E}$

In Figure 15.16 a piece has angular length $\Delta\theta$.

Step 2: Write an Expression for the Electric Field Due to One Piece

Origin: Center of the ring. Axes shown in Figure 15.16.
Location of piece: Described by θ, where $\theta = 0$ is along the x axis.

Integration variable: θ

$$\vec{r} = \langle \text{obs. loc.} \rangle - \langle \text{source} \rangle = \langle 0,0,z \rangle - \langle R\cos\theta, R\sin\theta, 0 \rangle$$
$$= \langle -R\cos\theta, -R\sin\theta, z \rangle$$

$$|\vec{r}| = \sqrt{(-R\cos\theta)^2 + (-R\sin\theta)^2 + z^2} = (R^2 + z^2)^{1/2}$$

$$\hat{r} = \frac{\vec{r}}{|\vec{r}|} = \frac{\langle -R\cos\theta, -R\sin\theta, z \rangle}{(R^2 + z^2)^{1/2}}$$

$$\frac{1}{4\pi\varepsilon_0} \frac{\Delta q}{r^2} = \frac{1}{4\pi\varepsilon_0} \frac{\Delta q}{(R^2 + z^2)}$$

$$\Delta q = q\left(\frac{\Delta\theta}{2\pi}\right) \quad \text{(There are } 2\pi \text{ radians in the complete ring.)}$$

$$\Delta\vec{E} = (\Delta E)\hat{r} = \left(\frac{1}{4\pi\varepsilon_0} \frac{q\left(\dfrac{\Delta\theta}{2\pi}\right)}{(R^2 + z^2)}\right) \frac{\langle -R\cos\theta, -R\sin\theta, z \rangle}{(R^2 + z^2)^{1/2}}$$

$$\Delta\vec{E} = \frac{1}{4\pi\varepsilon_0} \frac{q}{2\pi} \frac{\Delta\theta}{(R^2 + z^2)^{3/2}} \langle -R\cos\theta, -R\sin\theta, z \rangle$$

Components: From the symmetry of the situation, we see that the x components will sum to zero, as will the y components. The z components will not cancel, so we need ΔE_z:

$$\Delta E_z = \frac{1}{4\pi\varepsilon_0} \frac{q}{2\pi} \frac{z}{(R^2 + z^2)^{3/2}} \Delta\theta$$

$$dE_z = \frac{1}{4\pi\varepsilon_0} \frac{q}{2\pi} \frac{z}{(R^2 + z^2)^{3/2}} d\theta$$

Step 3: Add Up the Contributions of All the Pieces

In this case we can evaluate the integral (sum) analytically:

$$E_z = \int_0^{2\pi} dE_z$$

$$= \int_0^{2\pi} \frac{1}{4\pi\varepsilon_0} \frac{q}{2\pi} \frac{z}{(R^2 + z^2)^{3/2}} d\theta$$

$$= \frac{1}{4\pi\varepsilon_0} \frac{q}{2\pi} \frac{z}{(R^2 + z^2)^{3/2}} \int_0^{2\pi} d\theta$$

ELECTRIC FIELD OF A UNIFORMLY CHARGED THIN RING

$$E = \frac{1}{4\pi\varepsilon_0} \frac{qz}{(R^2 + z^2)^{3/2}}$$

along the axis, a distance z from the center of the ring, for a ring of radius R and charge q (Figure 15.17). The direction is parallel or antiparallel to the axis, depending on the sign of q.

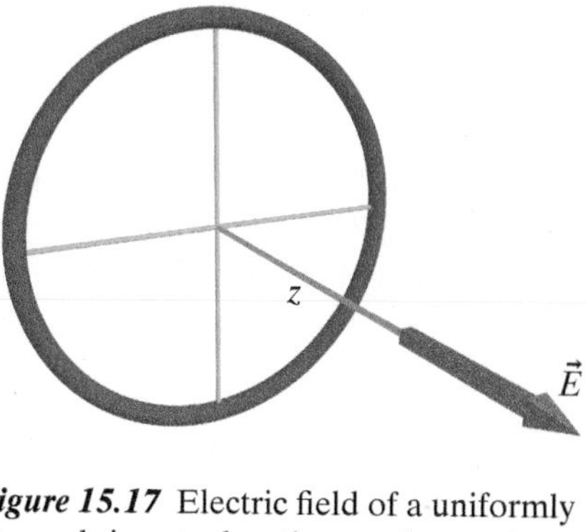

Figure 15.17 Electric field of a uniformly charged ring at a location on the z axis.

Step 4: Check the Result

Direction: Correct, by symmetry.

Units: $\left(\dfrac{\text{N} \cdot \text{m}^2}{\text{C}^2}\right)\left(\dfrac{\text{C} \cdot \text{m}}{(\text{m}^2)^{3/2}}\right) = \dfrac{\text{N}}{\text{C}}$, which is correct.

Special case: Exact center of the ring:

$$z = 0 \Rightarrow E = 0$$

This is correct, since all contributions to E will cancel at this location.

Special case: $z \gg R$:

$$(R^2 + z^2)^{3/2} \approx (z^2)^{3/2} = z^3, \text{ so } E \propto z/z^3 = 1/z^2$$

This is correct; at locations far from the ring, the ring should look like a point charge.

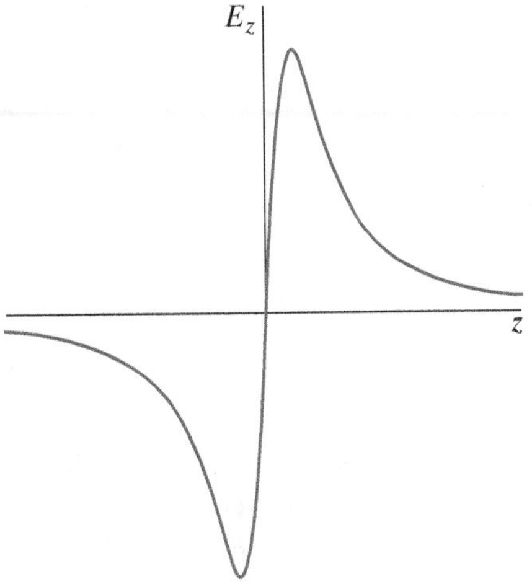

Distance Dependence of the Electric Field of a Ring

Since the field is zero at the center of the ring but falls off like $1/z^2$ far from the ring, the field must first increase, then decrease, with distance from the ring. With a graphing calculator you can easily show that a plot of E_z vs. z looks like Figure 15.18.

Note that we have calculated the field of a ring only along the axis perpendicular to the ring (Figure 15.19). As shown in Figure 15.20, the pattern of the field at other locations is much more complex.

Figure 15.20 is the result of a numerical integration, in which the ring was cut up into short segments and the summation was made at many observation locations. We calculated the field analytically along the z axis not just because this is one of the rare cases where an analytical solution is possible, but also because this result will be useful to us later, including in the next section.

Figure 15.18 The z component of electric field of a uniformly charged ring along the z axis (perpendicular to the ring).

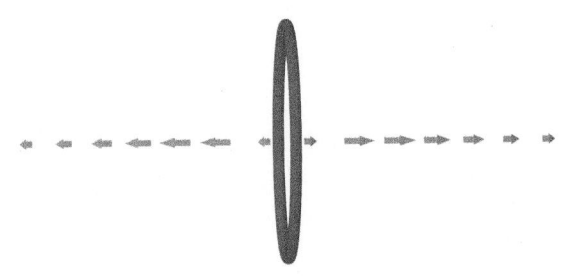

Figure 15.19 The electric field of a ring along the axis, varying with distance. As the graph in Figure 15.18 indicates, as the distance from the ring increases the magnitude of the field first increases, then decreases.

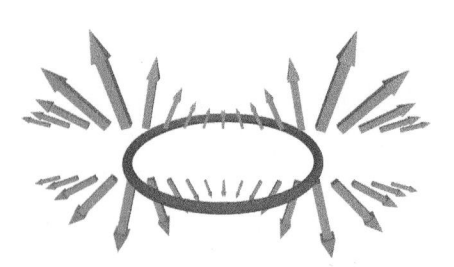

Figure 15.20 The electric field of a charged ring at locations across the ring, a short distance out of the plane of the ring.

Checkpoint 2 Two rings of radius 5 cm are 20 cm apart and concentric with a common horizontal axis. The ring on the left carries a uniformly distributed charge of +35 nC, and the ring on the right carries a uniformly distributed charge of −35 nC. **(a)** What are the magnitude and direction of the electric field on the axis, halfway between the two rings? **(b)** If a charge of −5 nC were placed midway between the rings, what would be the magnitude and direction of the force exerted on this charge by the rings? **(c)** What are the magnitude and direction of the electric field midway between the rings if both rings carry a charge of +35 nC?

15.4 A UNIFORMLY CHARGED DISK

One reason that a uniformly charged disk is important is that two oppositely charged metal disks can form a "capacitor," a device that is important in electric circuits. Before discussing metal disks, we will consider a glass disk that has been rubbed with silk in such a way as to deposit a uniform density of positive charge all over the surface.

Figure 15.21 A uniformly charged disk viewed edge-on. The electric field is shown at locations across the diameter of the disk. Note the uniformity of the field near the central region of the disk for a given distance from the disk.

Field Along the Axis of a Uniformly Charged Disk

The electric field of a uniformly charged disk of course varies in both magnitude and direction at observation locations near the disk, as illustrated in Figure 15.21, which shows the computed pattern of electric field at many locations near a uniformly charged disk (done by numerical integration, with the surface of the disk divided into small areas). Note, however, that near the center of the disk the field is quite uniform for a given distance from the disk. Even near the edge of the disk the magnitude of the electric field isn't very different, though the direction is no longer nearly perpendicular to the disk.

Again, we'll pick an easy location for an analytical solution—the field at any location along the axis of the disk, which is a line going through the center and perpendicular to the disk, as shown in Figure 15.22. This is a more useful choice than one might expect, since the field turns out to be nearly uniform in regions far from the edge of the disk as can be seen in Figure 15.21; our result will be applicable to a variety of situations.

We consider a disk of radius R, with a total charge Q uniformly distributed over the front surface of the disk.

Step 1: Cut Up the Charge Distribution into Pieces; Draw $\Delta \vec{E}$

Use thin concentric rings as pieces, as shown in Figure 15.22, since we already know the electric field of a uniform ring. Approximate each ring as having some average radius r.

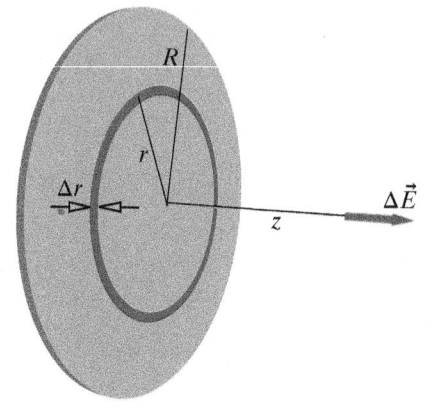

Figure 15.22 A ring of width Δr makes a contribution ΔE to the total electric field.

Step 2: Write an Expression for the $\Delta \vec{E}$ Due to One Piece

Origin: Center of ring.
Location of piece: Given by radius r of ring.
Integration variable: r

$$\vec{r} = \langle 0, 0, z \rangle$$

$$|\vec{r}| = z$$

$$\hat{r} = \frac{\vec{r}}{|\vec{r}|} = \langle 0, 0, 1 \rangle$$

$$\Delta \vec{E} = \frac{1}{4\pi\varepsilon_0} \frac{(\Delta q)z}{(r^2 + z^2)^{3/2}} \langle 0, 0, 1 \rangle$$

Δq *in terms of variables:* (See Figure 15.23.)

$$\Delta q = Q \frac{(\text{area of ring})}{(\text{area of disk})} = Q \frac{2\pi r \Delta r}{\pi R^2}$$

$$\Delta \vec{E} = \frac{1}{4\pi\varepsilon_0} \frac{\left(Q \dfrac{2\pi r \Delta r}{\pi R^2} \right) z}{(r^2 + z^2)^{3/2}} \langle 0, 0, 1 \rangle$$

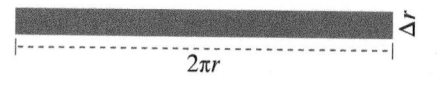

Figure 15.23 A ring of radius r and thickness Δr cut and rolled out straight.

Note that both Δq and r will be different for each piece.

Components to calculate: Only ΔE_z is nonzero.

$$\Delta E_z = \frac{1}{2\varepsilon_0} \left(\frac{Q}{\pi R^2} \right) \frac{z}{(r^2 + z^2)^{3/2}} r \Delta r$$

or, for infinitesimally thin rings:

$$dE_z = \frac{1}{2\varepsilon_0} \left(\frac{Q}{\pi R^2} \right) \frac{z}{(r^2 + z^2)^{3/2}} r \, dr$$

Step 3: Sum All Contributions

$$E_z = \int_0^R \frac{1}{2\varepsilon_0} \left(\frac{Q}{\pi R^2}\right) \frac{z}{(r^2 + z^2)^{3/2}} r \, dr$$

Many of these quantities have the same values for different values of r, and these can be taken out of the integral as common factors:

$$E_z = \frac{1}{2\varepsilon_0} \left(\frac{Q}{\pi R^2}\right) z \int_0^R \frac{r}{(r^2 + z^2)^{3/2}} dr$$

This particular integral can be done by a change of variables, letting $u = (r^2 + z^2)$. You can work it out yourself, look up the result in a table of integrals, or use a symbolic math package or calculator to evaluate it. The result is:

$$E_z = \frac{1}{2\varepsilon_0} \left(\frac{Q}{\pi R^2}\right) \left[1 - \frac{z}{(R^2 + z^2)^{1/2}}\right]$$

for a uniformly charged disk of charge Q and radius R, at locations along the axis of the disk. This is often written in terms of the area A of the disk ($A = \pi R^2$):

$$E_z = \frac{(Q/A)}{2\varepsilon_0} \left[1 - \frac{z}{(R^2 + z^2)^{1/2}}\right]$$

Step 4: Check

Direction: Away from the disk if Q is positive, as expected.
Special location: $0 \ll z \ll R$ (very close to the disk, but not touching it. See Figure 15.25.)

$$E \approx \frac{Q/A}{2\varepsilon_0} \left[1 - \frac{z}{R}\right]$$

If z/R is extremely small, $[1 - z/R]$ reduces to 1, and

$$E \approx \frac{Q/A}{2\varepsilon_0}$$

Interestingly, this field is nearly independent of distance! This result is approximately true near any large uniformly charged plate, not just a circular one.

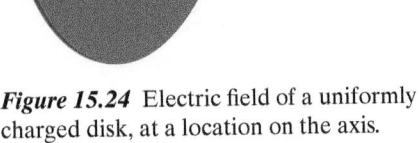

Figure 15.24 Electric field of a uniformly charged disk, at a location on the axis.

ELECTRIC FIELD OF A UNIFORMLY CHARGED DISK

$$E = \frac{(Q/A)}{2\varepsilon_0} \left[1 - \frac{z}{(R^2 + z^2)^{1/2}}\right]$$

at a location along the axis, a distance z from the disk (Figure 15.24). The direction is perpendicular to the surface of the disk. Q is the total charge on the disk, A is the area of the surface of the disk, and R is the radius of the disk.

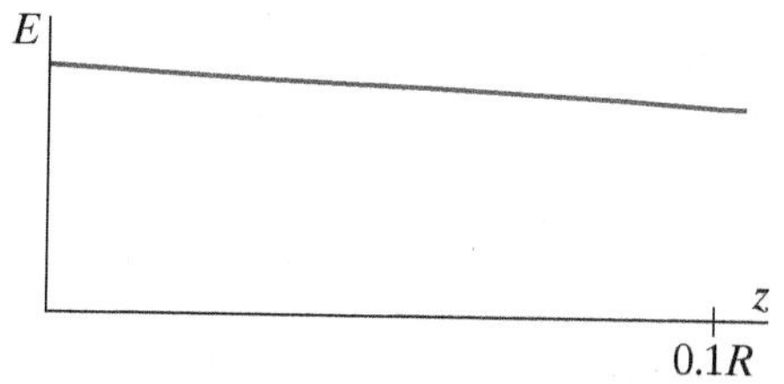

Figure 15.25 Magnitude of electric field along the axis of a disk, for $z < 0.1R$.

APPROXIMATE ELECTRIC FIELD OF A UNIFORMLY CHARGED DISK

$$E \approx \frac{Q/A}{2\varepsilon_0}\left[1 - \frac{z}{R}\right]$$

or

$$E \approx \frac{Q/A}{2\varepsilon_0}$$

at a location a perpendicular distance $z \ll R$ from the disk, as long as the observation location is not too near the edge of the disk (Figure 15.25). The direction is perpendicular to the surface of the disk. Q is the total charge on the disk, A is the area of the surface of the disk, and R is the radius of the disk.

Although we proved this result only for observation locations a perpendicular distance $z \ll R$ from the center of the disk, the result is actually a good approximation as long as you're not too near the edge of the disk, as can be seen in Figure 15.21, which is the result of an accurate numerical integration.

EXAMPLE   **A Rod and a Disk**

In Figure 15.26 a thin plastic disk of radius 0.6 m is uniformly charged with $Q_{disk} = -3 \times 10^{-7}$ C and is attached to a thin glass rod of length 2.4 m that is uniformly charged with $Q_{rod} = 5 \times 10^{-8}$ C. The center of the rod and the center of disk are at the origin. The rod lies along the x axis and the disk lies in the yz plane. What is the (vector) electric field at location $\langle 0.02, 0.01, 0 \rangle$ m?

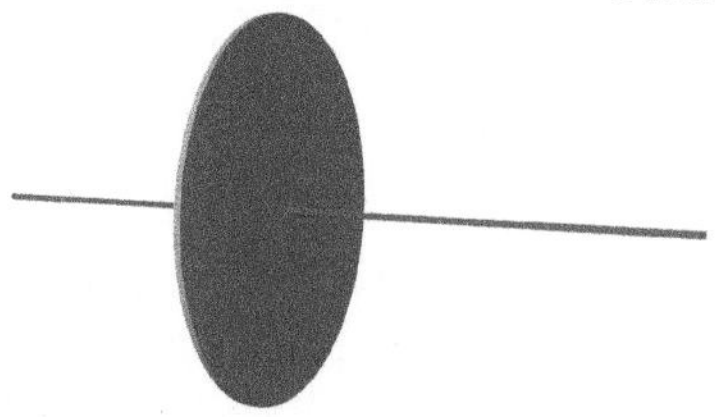

Solution   According to the superposition principle $\vec{E}$ is the vector sum of the electric field $\vec{E}_{disk}$ contributed by the disk and the electric field $\vec{E}_{rod}$ contributed by the rod:

$$\vec{E} = \vec{E}_{disk} + \vec{E}_{rod}$$

Because the observation location is close to the disk (0.02 m $\ll$ 0.6 m) and far from the edge of the disk, we can approximate the field contributed by the disk:

Figure 15.26 A thin positively charged glass rod (red) passes through a thin negatively charged circular plastic disk (blue).

$$E_{disk} \approx \frac{Q/A}{2\varepsilon_0}$$

$$= \frac{(|-3 \times 10^{-7}|\text{C})/(\pi(0.6\text{ m})^2)}{2(8.85 \times 10^{-12}\,\text{C}^2/\,\text{N}\cdot\text{m}^2)}$$

$$= 1.49 \times 10^4\,\text{N/C}$$

Because the disk is negatively charged, $\vec{E}_{disk}$ is in the $-x$ direction, so

$$\vec{E}_{disk} \approx \langle -1.49 \times 10^4, 0, 0 \rangle\,\text{N/C}$$

Because the observation location is close to the rod (0.01 m $\ll$ 2.4 m) and far from the ends of the rod, we can approximate the field contributed by the rod:

$$E_{rod} \approx \frac{1}{4\pi\varepsilon_0}\frac{2Q/L}{r}$$

$$= (9 \times 10^9\,\text{N}\cdot\text{m}^2/\text{C}^2)\frac{2(5 \times 10^{-8}\,\text{C})/2.4\text{ m}}{0.01\text{ m}}$$

$$= 3.75 \times 10^4\,\text{N/C}$$

Because the rod is positively charged, $\vec{E}_{rod}$ is in the $+y$ direction, so

$$\vec{E}_{rod} \approx \langle 0, 3.75 \times 10^4, 0 \rangle \, \text{N/C}$$

Therefore the net electric field is

$$\vec{E} = \vec{E}_{disk} + \vec{E}_{rod}$$
$$\approx \langle -1.49 \times 10^4, 3.75 \times 10^4, 0 \rangle \, \text{N/C}$$

The disk does not "block" the contribution to the field by the left half of the rod, on the other side of the disk from the observation location. As the superposition principle stated, the field $\vec{E}_{rod}$ contributed by the rod is unaffected by the presence of the disk, and the field $\vec{E}_{disk}$ contributed by the disk is unaffected by the presence of the rod.

The disk and rod were "thin," so they didn't contain much matter to be polarized. If these objects were "thick," the net electric field might have a significant additional contribution from induced dipoles in the disk and rod.

Checkpoint 3 Suppose that the radius of a disk is $R = 20\,\text{cm}$, and the total charge distributed uniformly all over the disk is $Q = 6 \times 10^{-6}\,\text{C}$. Use the exact result to calculate the electric field 1 mm from the center of the disk, and also 3 mm from the center of the disk. Does the field decrease significantly?

15.5   TWO UNIFORMLY CHARGED DISKS: A CAPACITOR

Consider two uniformly charged metal disks placed very near each other (separation or gap distance s), carrying charges of $-Q$ and $+Q$ (Figure 15.27). This arrangement is called a capacitor, and we will work with such devices later in electric circuits.

A single charged metal disk cannot have a truly uniform charge density, because the mobile charges tend to push each other to the edge of the disk. Nevertheless, our results for disks made of insulating material are approximately correct for metal disks, especially in this two-disk configuration if the disks are very close together ($s \ll R$). Due to the attraction by the neighboring disk, almost all of the charge is nearly uniformly distributed on the inner surfaces of the disks, with very little charge on the outer surfaces of the disks.

We'll calculate the strength of the electric field at locations near the center of the disks, both inside and outside the capacitor.

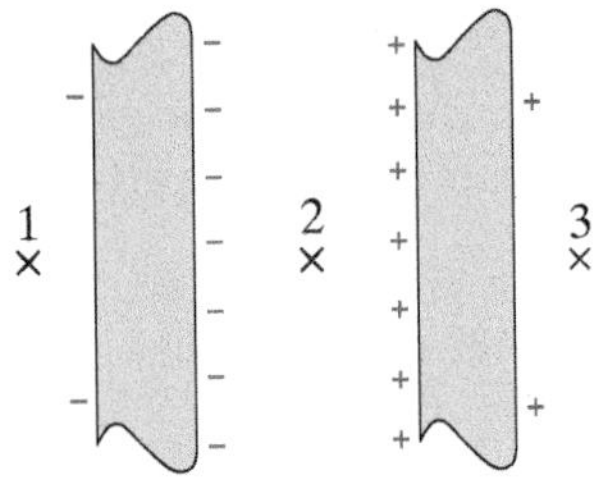

Step 1: Cut Up the Charge Distribution into Pieces and Draw $\Delta\vec{E}$

We know the electric field made by a single uniformly charged disk, so we can use this as a "piece." Consider the locations labeled 1, 2, and 3 in Figure 15.28, which shows a side view of the region near the center of the disks, blown up so there is room to draw. The disks are very close together ($s \ll R$), and extend up and down beyond the boundaries of the drawing.

QUESTION Before reading farther, predict the direction and relative magnitude (that is, whether the net field will be large or small) of $\vec{E}_{net}$ at locations 1, 2, and 3 in Figure 15.28.

At locations 1 and 3, the contributions of the two disks are in opposite directions, and nearly equal, so the net field is very small. However, the

Figure 15.27 The two charged metal plates of a capacitor are separated by a very small gap s (exaggerated in this figure).

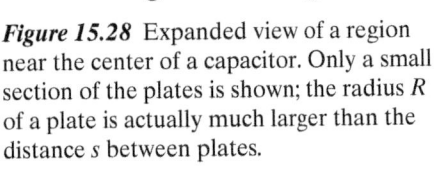

Figure 15.28 Expanded view of a region near the center of a capacitor. Only a small section of the plates is shown; the radius R of a plate is actually much larger than the distance s between plates.

distances to the two disks are slightly different, so $E_+ \neq E_-$, and the net field is to the right at locations 1 and 3 (Figure 15.29). At location 2, both disks contribute $\Delta \vec{E}$ in the same direction, so $\vec{E}_{net}$ is large, and to the left.

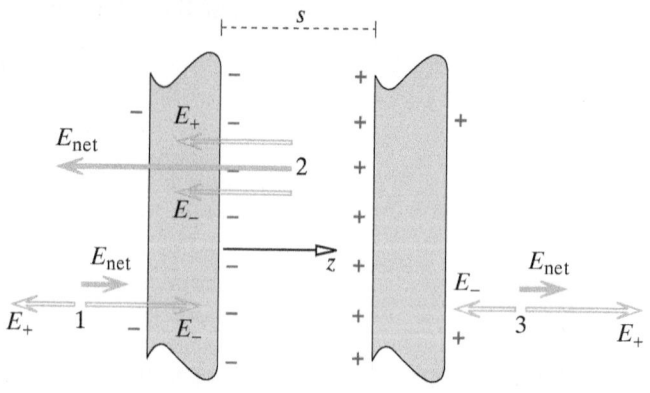

Figure 15.29 Electric field at three locations near the center of a capacitor. As shown, we choose an origin on the inner face of the left disk, with $+z$ running to the right.

Step 2: Write an Expression for the Electric Field Due to One Piece

We will write the expressions in terms of the horizontal coordinate z, so the expressions will be valid at all locations (1, 2, and 3).

Origin: At surface of left plate; z axis runs to right (Figure 15.29).
Location of a piece given by: Pieces are at $z = 0$ and $z = s$.
Distance from piece to observation location: z (for the left plate) or $s - z$ (for the right plate).
Δq in terms of variables: $+Q$ or $-Q$ (all the charge on a disk).
Assumptions: Uniform charge density on inner surface of each disk (ignore the small charge on the outer surfaces of the disks); $(s \ll R)$; locations 1 and 3 are very near the disks $(z \ll R)$.

$$A = \pi R^2 \quad \text{(the area of one disk)}$$

Contribution of each piece:

$$E_- \approx \frac{Q/A}{2\varepsilon_0} \left[1 - \frac{z}{R} \right] \text{ to the left}$$

$$E_+ \approx \frac{Q/A}{2\varepsilon_0} \left[1 - \frac{s-z}{R} \right] \text{ to the left}$$

Step 3: Add Up the Contributions of All the Pieces

Location 2: At a distance z from the negative plate (and a distance $s - z$ from the positive plate), the electric field is to the left, with magnitude given by the following:

$$
\begin{aligned}
E_2 &\approx \frac{Q/A}{2\varepsilon_0} \left[1 - \frac{z}{R} \right] + \frac{Q/A}{2\varepsilon_0} \left[1 - \frac{s-z}{R} \right] \\
&\approx \frac{Q/A}{2\varepsilon_0} \left[1 - \frac{z}{R} + 1 - \frac{s-z}{R} \right] \\
&\approx \frac{Q/A}{\varepsilon_0} \left[1 - \frac{s/2}{R} \right] \\
&\approx \frac{Q/A}{\varepsilon_0}
\end{aligned}
$$

The electric field between the plates is essentially twice the field due to one plate. The magnitude hardly depends on z, so the field is remarkably

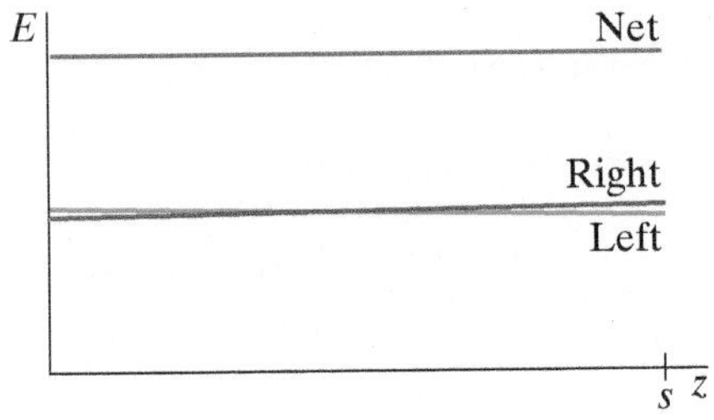

Figure 15.30 E at a location inside a capacitor, showing the contributions from each plate, and the net field, as a function of location inside the capacitor.

uniform. As you move away from the negative plate, the very slightly smaller contribution of the negative plate is nearly compensated by the very slightly larger contribution of the positive plate, as shown in Figure 15.30.

Location 3: $\vec{E}_3$ points to the right (see Figure 15.29) and its magnitude is the difference of the magnitudes of the fields of the two disks:

$$E_3 \approx \frac{Q/A}{2\varepsilon_0}\left[1 - \frac{z-s}{R}\right] - \frac{Q/A}{2\varepsilon_0}\left[1 - \frac{z}{R}\right] = \frac{Q/A}{2\varepsilon_0}\left(\frac{s}{R}\right)$$

Location 1: $\vec{E}_1$ also points to the right, and

$$E_1 \approx \frac{Q/A}{2\varepsilon_0}\left[1 - \frac{z}{R}\right] - \frac{Q/A}{2\varepsilon_0}\left[1 - \frac{z+s}{R}\right] = \frac{Q/A}{2\varepsilon_0}\left(\frac{s}{R}\right)$$

The electric field at locations outside the capacitor is called the "fringe field." The fringe field is very small compared to the field inside the capacitor. Inside the gap, the fields of the two disks are in the same direction, but outside the gap the fields of the two disks are in opposite directions.

> QUESTION Calculate the ratio of the outside field to the inside field, and show that this ratio is very small if $s \ll R$.

The result is $s/(2R)$, which is small if $s \ll R$. This is the fringe field at a location outside the capacitor but very close to it. (If, however, you are very far away, so $z \gg R$, the capacitor looks like an electric dipole, and the fringe field falls off like $1/z^3$.)

Step 4: Check the Result

Units: Inside the capacitor

$$\frac{C/m^2}{N \cdot m^2/C^2} = \frac{N}{C}$$

Units: Fringe field

$$\left(\frac{C/m^2}{N \cdot m^2/C^2}\right)\left(\frac{m}{m}\right) = \frac{N}{C}$$

Applicability

Because we have considered locations far from the edges of the plates, our results apply not only to circular capacitors but also to capacitors with rectangular or other shapes of plates, as long as the same conditions are satisfied — the plate separation s must be much smaller than the width or height of a plate.

Although we cannot prove it at this time (we will need Gauss's law — see Chapter 21), it is also true that the magnitude and direction of the electric field are practically the same everywhere in the gap, not just near the center of the plates. Only if you get near the outer edges of the plates does the electric field vary much from its center value.

ELECTRIC FIELD OF A CAPACITOR

$$E \approx \frac{Q/A}{\varepsilon_0}$$

near the center of a two-plate capacitor (each plate has area A, one plate has charge $+Q$, other plate has charge $-Q$; separation s between plates is very small compared to the radius of a plate). The direction is perpendicular to the plates.

Fringe field (just outside the plates, near center of disk):

$$E_{\text{fringe}} \approx \frac{Q/A}{2\varepsilon_0}\left(\frac{s}{R}\right)$$

Checkpoint 4 If the magnitude of the electric field in air exceeds roughly 3×10^6 N/C, the air breaks down and a spark forms. For a two-disk capacitor of radius 50 cm with a gap of 1 mm, what is the maximum charge (plus and minus) that can be placed on the disks without a spark forming (which would permit charge to flow from one disk to the other)? Under these conditions, what is the strength of the fringe field just outside the center of the capacitor?

15.6 A SPHERICAL SHELL OF CHARGE

As mentioned in Chapter 13, a sphere with charge spread uniformly over its surface produces a surprisingly simple pattern of electric field. For brevity, we will use the phrase "a uniform spherical shell" to refer to "a sphere with charge spread uniformly over its surface," since the charge itself forms a thin, shell-like spherical layer. A spherical object of radius R with charge Q uniformly spread out over its surface produces an electric field with the following distance dependence, if $\vec{r}$ is a vector from the center of the sphere to the observation location (Figure 15.31):

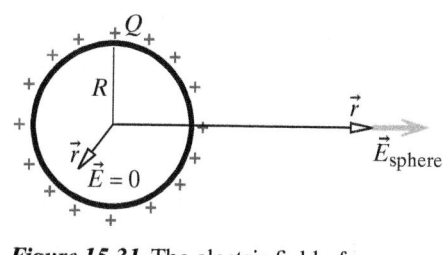

Figure 15.31 The electric field of a uniformly charged sphere (shown in cross section), at locations outside and inside the sphere.

ELECTRIC FIELD OF A UNIFORMLY CHARGED SPHERICAL SHELL

$$\vec{E}_{\text{sphere}} = \frac{1}{4\pi\varepsilon_0} \frac{Q}{r^2} \hat{r} \quad \text{for } r > R \text{ (outside the sphere)}$$

$$\vec{E}_{\text{sphere}} = 0 \quad \text{for } r < R \text{ (inside the sphere)}$$

Q is the total charge on the surface of the sphere, and R is the radius of the sphere.

This can be shown by setting up and evaluating an integral, as we did in the case of the rod, the ring, and the disk, following the procedure that we used in those cases. The geometry involved is somewhat complex, and the details of the process are given in Section 15.10. In later chapters we will see alternative ways to prove this, one involving Gauss's law and one involving electric potential.

In this section we will try to develop a qualitative understanding of how this surprising result can be true.

Outside a Uniform Spherical Shell of Charge

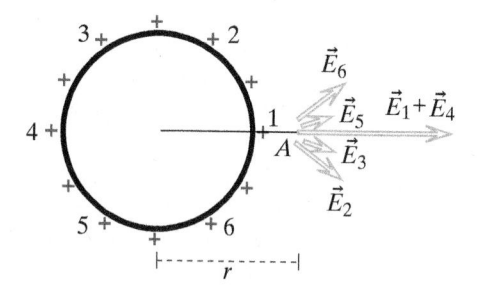

Figure 15.32 Contributions to the net electric field outside a charged sphere.

Let's divide the sphere into six regions, as shown in Figure 15.32. $\vec{E}_2$ means "the electric field due to the charges in region 2 of the spherical shell." As long as we are rather far from a region of distributed charge we can approximate the electric field of that region as being due to a point particle with the total charge of the region.

Direction

Consider the field at location A outside the shell, a distance r from the center of the sphere. By symmetry, the net electric field is horizontal and to the right, because the vertical component of the net field cancels for pairs of fields such as $\vec{E}_2$ and $\vec{E}_6$, or $\vec{E}_3$ and $\vec{E}_5$. It is clear that the net field (the superposition of the contributions of all the source charges) is radially outward from the center of the sphere.

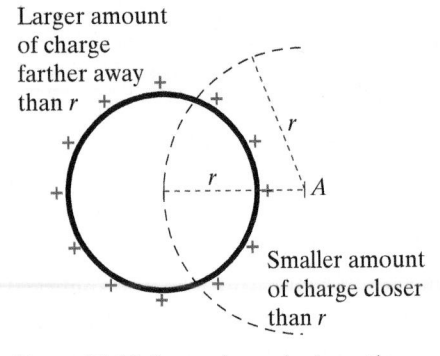

Larger amount of charge farther away than r

Smaller amount of charge closer than r

Figure 15.33 Some charge is closer than r to point A, but more of the charge is farther away.

Magnitude

It is somewhat surprising that the magnitude of the net field turns out to be

$$E = \frac{1}{4\pi\varepsilon_0} \frac{Q}{r^2}$$

where Q is the total charge on the sphere and r is the distance to location A from the center of the sphere (not from the surface of the sphere!).

In Figure 15.33, notice that the distance from location A to some regions is less than r, while the distance from location A to other regions is greater than r. Because of the $1/r^2$ distance dependence for point charges, a smaller amount of charge closer than r makes as big a contribution to the net field as the larger amount of charge that is farther away than r.

It is an extraordinary aspect of the $1/r^2$ behavior of Coulomb's law that the net effect of all these charges comes out to be simply equivalent to placing one point charge Q at the center of the sphere. This is not an obvious result!

Force on a Uniformly Charged Sphere

A uniformly charged spherical shell not only makes a field outside that looks as though it were made by a point charge, but the shell also reacts to outside charges as though it were a point charge.

Consider the interaction between a uniform spherical shell and an outside point charge. The shell exerts a force on the point charge as though both were point charges. By the reciprocity principle for electric forces (Newton's third law), the real point charge must exert an equal and opposite force on the shell, and so it exerts a force on the shell as though the shell were a point charge. Hence from the outside a uniform spherical shell looks just like a point charge, both as a source of electric field and when it reacts to external fields.

Inside a Uniform Spherical Shell of Charge

Now consider a location B inside the shell, as indicated in Figure 15.34. A small number of charges near location B (region 1) make a large contribution to the left, while a large number of charges far away from location B (regions 3, 4, and 5) each make small contributions to the right. Because the surface area of a portion of the sphere is proportional to r^2 while the electric field contributed by this region is proportional to $1/r^2$, these contributions exactly cancel each other. The electric field at location B due to charges on the surface of the sphere turns out to be exactly zero. Although intuitively plausible, this is not an obvious result.

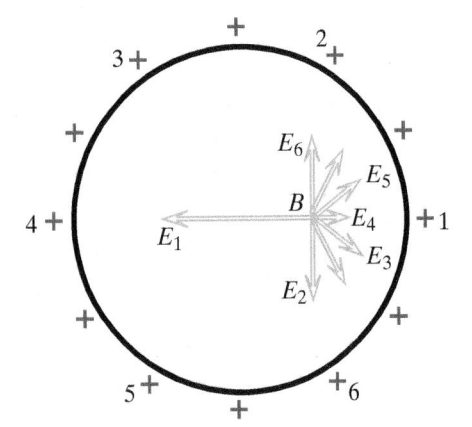

Figure 15.34 Contributions to the net electric field at a location inside a uniformly charged sphere add up to zero.

Warning: E Is Not Always Zero Inside a Charged Sphere

Do not overgeneralize this result. Other charges in the Universe may make a nonzero electric field inside the shell. It is only the electric field contributed by charges uniformly distributed on the surface of the sphere that is zero inside the sphere.

Implications

The fact that the electric field due to uniform charges on the shell is zero at any location inside the shell has interesting implications. For example, if the uniformly charged shell is filled with plastic, the charges on the surface do not polarize the molecules in the plastic, because the surface charges contribute zero field. Even if the location of a molecule is very close to the charged surface of the sphere, E inside the sphere is still zero at that location. (Charges external to the sphere may contribute to a nonzero electric field inside the sphere, however.)

Since a solid metal sphere is a conductor, any excess charge on a metal sphere arranges itself on the outer surface. Inside the metal there is no excess charge, and there is no field, because inside a metal in equilibrium the electric field is zero. The field inside a metal sphere in equilibrium is zero even if external charges polarize the metal and make the charge on the outer surface have a nonuniform distribution.

What Is the Electric Field Right at the Surface?

The electric field is zero just inside the surface of a uniformly charged spherical shell, but the field is nonzero just outside the surface. You might wonder what the field is right at the surface. The physical reality is that the electric field is highly variable in direction and magnitude right at the surface, because the individual excess point charges (electrons and ions) on the surface are many atomic diameters away from each other. In Chapter 14 we found that typically only about one in a million atoms on a surface has an extra charge. It is only when you get some thousands of atomic diameters away from the surface (about 1×10^{-7} m) that it is appropriate to consider the surface charge as approximately a uniform, continuous sheet of charge. As long as you are this far away from the surface, the field outside the surface looks like the field of a point charge at the center of the sphere, and the field inside the surface is zero.

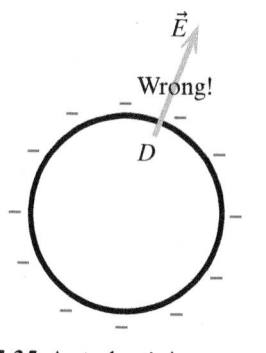

Figure 15.35 A student's incorrect prediction of the electric field inside a uniformly charged sphere.

> **Checkpoint 5** What is wrong with Figure 15.35 and this associated incorrect student explanation? "The electric field at location D inside the uniformly charged sphere points in the direction shown, because the charges closest to this location have the largest effect." (Spheres provide the most common exception to the normally useful rule that the nearest charges usually make the largest contribution to the electric field.)

15.7 A SOLID SPHERE CHARGED THROUGHOUT ITS VOLUME

Until this point, all of the charge distributions we have considered involved charge distributed over the surface of an object. There are some cases, however, in which charge can be distributed throughout an object. One such case is the nucleus of an atom, which is composed of protons and neutrons packed into a sphere of radius on the order of 10^{-15} m. We can treat the charge density inside the nucleus as approximately uniform. Another such case is the electron cloud in an atom, in which the negative charge of the electrons appears to be distributed throughout a spherical region of radius on the order of 10^{-10} m (although not necessarily uniformly).

We will consider a solid sphere with radius R, having charge Q uniformly distributed throughout its volume.

Cut Up the Charge Distribution into Pieces and Draw $\Delta \vec{E}$

We can model the sphere as a series of concentric spherical shells.

Outside a Solid Sphere of Charge

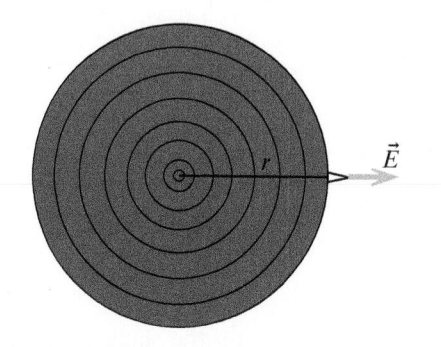

Figure 15.36 A charged solid sphere modeled as a series of concentric spherical shells, all charged.

At a location outside the sphere, and hence outside all the shells (Figure 15.36), each shell looks like a point charge at the center of the sphere. Hence, outside the sphere:

$$E = \frac{1}{4\pi\varepsilon_0} \frac{Q}{r^2} \quad \text{for } r > R$$

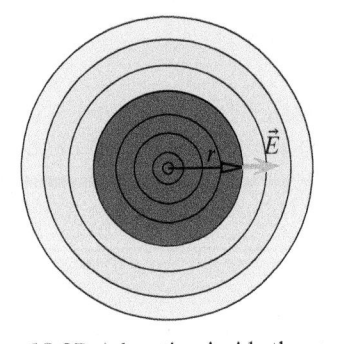

Figure 15.37 A location inside the gray shells, but outside the red shells. At this location only the inner shells make a nonzero contribution to the electric field.

Inside a Solid Sphere of Charge

At a location inside the sphere, we are inside some of the spherical shells (gray shells in Figure 15.37); the contribution of these shells to $\vec{E}_{net}$ is therefore zero.

We are outside the rest of the shells (red shells in Figure 15.37); each of these shells looks like a point charge at the center of the sphere. To get $\vec{E}_{net}$ at this location we must add the contributions of all the inner shells. Since all the inner shells together will contribute

$$E = \frac{1}{4\pi\varepsilon_0}\frac{\Delta Q}{r^2}$$

our only remaining task is to find ΔQ.

$$\Delta Q = Q\frac{(\text{volume of inner shells})}{(\text{volume of sphere})} = Q\frac{\frac{4}{3}\pi r^3}{\frac{4}{3}\pi R^3}$$

Assembling the result:

$$E = \frac{1}{4\pi\varepsilon_0}\frac{Q}{r^2}\frac{\frac{4}{3}\pi r^3}{\frac{4}{3}\pi R^3} = \frac{1}{4\pi\varepsilon_0}\frac{Q}{R^3}r \quad \text{for } r < R$$

so the electric field inside the sphere is directly proportional to r.

QUESTION Does this make sense?

Yes. While inside the sphere, as r increases, the amount of charge inside the observation location increases. The larger volume ($V \propto r^3$) more than compensates for the $1/r^2$ field dependence. By symmetry, $E = 0$ at $r = 0$.

In the special case that $r = R$, we have $E = \frac{1}{4\pi\varepsilon_0}\frac{Q}{R^2}$, which is correct.

15.8 INFINITESIMALS AND INTEGRALS IN SCIENCE

In pure mathematics, an infinitesimal quantity is conceived to be arbitrarily close to zero. In science, however, this may not make sense. For example, in the case of the uniformly charged thin rod it makes no physical sense to choose dz to be smaller than an atomic diameter. It would enormously complicate the summation if we had to take into account the detailed distribution of charged particles inside each atom. All that is required is that dz be small compared to the significant dimensions of the situation (r and L). We want each slice dz to be an adequate approximation of a point charge, and $dz = 0.02$ m may be good enough! We can consider the mathematical infinitesimal dz to be an idealization of what a scientist means by an infinitesimal quantity, which is a quantity that is "small enough" within the desired and possible precision of measurements and analysis.

Oddly, the result of doing an integral is approximate, not exact, because at the microscopic level the actual charge distribution is not in fact uniformly distributed. In Chapter 14 we found that on a charged piece of tape the individual electrons or ions were quite far from each other on an atomic scale. However, if you are many atomic diameters away from a "uniformly" charged rod, the mathematical integral gives an excellent approximation to the actual electric field.

There is another important sense in which our analysis is approximate. If a rod is an insulator, and has been charged by rubbing it with a cloth, the charge distribution will be only approximately uniform, even on a macroscopic scale, since charge transfer cannot be accurately controlled. If the rod is a metal, charge will tend to pile up at the ends rather than being uniformly distributed. Nevertheless, modeling a charged rod as a uniformly charged rod in practice can be a useful model. We just need to remember that our analytical solution is approximate, not exact.

15.9 3D NUMERICAL INTEGRATION WITH A COMPUTER

In Section 15.1 we found an algebraic expression for the electric field of a uniformly charged thin rod, at observation locations on the midplane of the rod. However, this expression is not valid for observation locations that are not near the midplane. We can easily write a computer program to calculate the electric field of the rod at any observation location.

A Thin Rod: One Observation Location

■ The use of Python lists to facilitate calculations involving many charges and observation locations is introduced in two VPython videos: VPython Video 7: Lists, Part 1 (http://vpython.org/video07.html), and VPython Video 8: Lists, Part 2 (http://vpython.org/video08.html). Watch these videos and do the challenge activities at the end of each video before reading this section.

We'll start by reproducing the numerical calculation for an observation location in the midplane, approximating the rod by a collection of 10 point charges. We did this before by hand, with results that are summarized in Figure 15.7. We'll start with this observation location so we can check our result by comparing it to the analytical result.

The following code creates a Python list of point charges represented by spheres, each corresponding to one slice of the rod (see the note about lists in the margin of this page). We'll use the values used to generate Figure 15.7: a total charge of 1 nC spread uniformly over a thin rod of length 1 m, and an observation location of $\vec{r}_{obs} = \langle 0.05, 0, 0 \rangle$ m.

```
from visual import *
scene.height = 800
oofpez = 9e9
Qtot = 1e-9
L = 1
N = 10
dy = L/N
## create a list of point charges
slices = []  ## an empty list
i = 0
y0 = -L/2 + dy/2   ## center of bottom slice
while i < N:
    a = sphere(pos=vector(0,y0+i*dy,0), radius=dy/2,
               color=color.red, q=Qtot/N)
    slices.append(a)  ## add sphere to list
    i = i + 1
```

QUESTION A sphere has only three attributes: `pos`, `radius`, and `color`. Why is an attribute `q` added in the instruction that creates a sphere named `a`?

As explained in VPython Video 7, in Python it is legal to make up new attributes for objects. This is done here for convenience: we need the charge of each slice, and it is easier to store this information in a new attribute than to make a separate list of the charges of every slice. In this case, we happen to know that each slice has the same charge, but in a more complex situation this might not be true (for example, if the rod were not uniformly charged), so this is a very convenient way of keeping track of the charge of each slice.

QUESTION Why is `y0` set to `-L/2 + dy/2`?

The thin rod is aligned along the *y* axis, and centered at the origin. The bottom end of the rod is at $\langle -L/2, 0, 0 \rangle$. However, the center of the bottom slice is not at the end of the rod. The height of a slice is `dy`, so the center of this slice must be a distance `dy/2` up from the bottom of the rod.

QUESTION Why is the *y* coordinate calculated as `y0+i*dy`?

Before entering the loop we set y0 to represent the *y* coordinate of the center of the bottom slice, and the first time through the loop i is zero, so y0+i*dy is equal to y0. The second time through the loop i is 1, so the *y* coordinate is y0+dy, which is the *y* coordinate of the next higher slice, and so on.

To make sure that the positions of the spheres representing the point charges in our model of the rod are correct, we can print their positions. Recall that the first element of a list is element 0. As explained in the VPython videos on lists, we can refer to the 0th object in a list named slices by the name slices[0]; if the object is a sphere, we can refer to its position attribute by adding .pos as usual (slices[0].pos). In the code below the variable i is used for the list index.

```
## print index and position
i = 0
while i < len(slices):
    print(i, slices[i].pos)
    i = i + 1
```

Figure 15.38 shows the 10 spheres created by the program. The output from the print() statement is shown below. These positions agree with the *y* values in Figure 15.7, although the order in which they are listed is different because we created the bottom sphere first.

Figure 15.38 A uniformly charged rod modeled as ten point charges.

```
0 <0, -0.45, 0>
1 <0, -0.35, 0>
2 <0, -0.25, 0>
3 <0, -0.15, 0>
4 <0, -0.05, 0>
5 <0, 0.05, 0>
6 <0, 0.15, 0>
7 <0, 0.25, 0>
8 <0, 0.35, 0>
9 <0, 0.45, 0>
```

To calculate the net electric field at the observation location, we can write a loop to add up the contributions of each slice:

```
## calculate E at observation location
r_obs = vector(0.05,0,0)   ## to test calculation
E_net = vector(0,0,0)
i = 0
while i < N:
    rate(100)
    r = r_obs - slices[i].pos
    rhat = r/mag(r)
    E = (oofpez * slices[i].q / mag(r)**2) * rhat
    E_net = E_net + E
    i = i + 1
print(E_net)
```

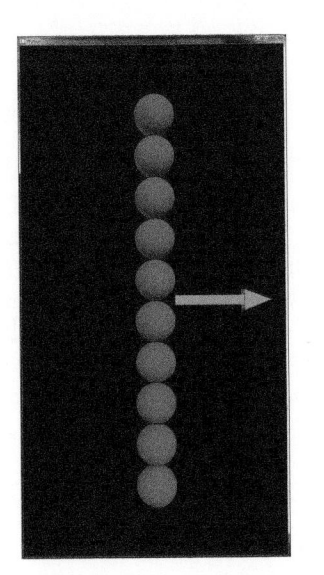

Figure 15.39 The electric field at a location in the midplane, calculated by modeling a uniformly charged rod as 10 point charges.

The net field calculated by this program comes out to $\langle 285.764, 1.43885 \times 10^{-13}, 0 \rangle$ N/C. In Figure 15.39 an arrow has been added to display the electric field at the observation location.

QUESTION Why isn't the *y* component of $\vec{E}_{net}$ zero, as expected?

There are slight rounding errors in any calculation. Since the calculated *y* component of the net field is 15 orders of magnitude smaller than the *x* component, it is sufficiently close to zero for our purposes. More seriously, the magnitude of the electric field given by this calculation is significantly smaller than the value obtained from the analytical expression, which is 385.213 N/C. As discussed in Section 15.1, increasing the number of point charges in the model of the rod will initially increase the accuracy of the calculation. However, at some point adding more slices does not produce a noticeable change in the result.

In Problem P63 you will improve the accuracy of this calculation by increasing the number of point charges used to model the rod, visualize the result with an arrow, and then use the resulting program to calculate and display the electric field due to the rod at locations that are not near the midplane.

There is a somewhat more elegant way to go through the elements of a list within a loop. In the following version of the code we've been discussing, the name this_slice is repeatedly recycled; during the first iteration it refers to the 0th element of the list, during the second iteration it refers to the 1st element of the list, and so on. Either version of the loop is acceptable—both accomplish the same thing.

```
## calculate E at observation location
r_obs = vector(0.05,0,0)  ## to test calculation
E_net = vector(0,0,0)
for this_slice in slices:
    rate(100)
    r = r_obs - this_slice.pos
    rhat = r/mag(r)
    E = (oofpez * this_slice.q / mag(r)**2) * rhat
    E_net = E_net + E
print(E_net)
```

A Thin Rod: Multiple Observation Locations

To see the 3D pattern of electric field associated with a distribution of source charges, we need to calculate and display the field at many observation locations. We can easily do this by making a list of observation locations.

As an example, let's create a list of arrows at observation locations lying on a circle around the rod. Initially the arrows will have zero length (although if we wanted to make sure their positions were correct, we could give them a nonzero axis, pointing in an arbitrary direction).

```
## create a list of arrows
observation = []
yobs = 0.4
dtheta = pi/6
theta = 0
R = 0.15 ## radius of circle
while theta < 2*pi:
    a = arrow(pos=vector(R*cos(theta), yobs, R*sin(theta)),
              color=color.orange,
              axis=vector(0,0,0))
    observation.append(a)
    theta = theta + dtheta
```

We now have two lists: a list of source charges and a list of arrows whose positions are observation locations. For each observation location, we need to add up the field contributed by all source charges. This means we need two nested loops, and therefore two levels of indentation. In the outer loop, which uses the index j, the next observation location is chosen. In the inner loop, which uses the index i, the field at the chosen observation location is calculated.

```
## calculate E at observation location
sf = 0.002  ## arrow scale factor
j = 0
## outer loop
while j < len(observation):
    rate(500)
    earrow = observation[j]
    ## add E of all slices for this obs. loc.
    i = 0
    E_net = vector(0,0,0)
    ## inner loop
    while i < N:
        r = earrow.pos - slices[i].pos
        rhat = r/mag(r)
        E = (oofpez * slices[i].q / mag(r)**2) * rhat
        E_net = E_net + E
        i = i + 1
        ## end of inner loop
    earrow.axis = sf*E_net
    j = j + 1
    ## end of outer loop
```

The result of this calculation is shown in Figure 15.40. In Problems P66–P67 you will add additional observation locations, or create a different pattern of observation locations.

Figure 15.40 The calculated electric field at many observation locations.

15.10 *INTEGRATING THE SPHERICAL SHELL

In this section, we'll sketch the proof that a uniformly charged spherical shell looks from the outside like a point charge but on the inside has a zero electric field. Two quite different proofs are given in Chapters 16 and 21.

Step 1: Divide the Sphere into Pieces

Divide the spherical shell into rings of charge, each delimited by the angle θ and the angle $\theta + \Delta\theta$, and carrying an amount of charge ΔQ (Figure 15.41). Each ring contributes ΔE at an observation point a distance r from the center of the spherical shell.

Step 2: Write an Expression for ΔE

Origin: At the center of the sphere.

Use spherical coordinates (r, θ, ϕ). As shown in Figure 15.42, θ is an angle measured from the r (horizontal) axis; ϕ refers to rotation about the r (horizontal) axis.

Location of one piece (ring): Given by angle θ.

Components: There is only a horizontal component.

Distance from center of one ring to observation location: $d = (r - R\cos\theta)$

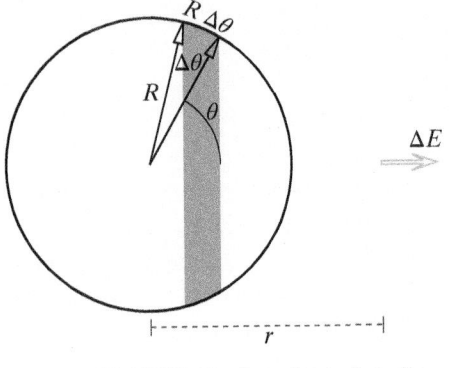

Figure 15.41 Divide the spherical shell (shown here in cross section) into rings, each carrying a (variable) amount of charge ΔQ.

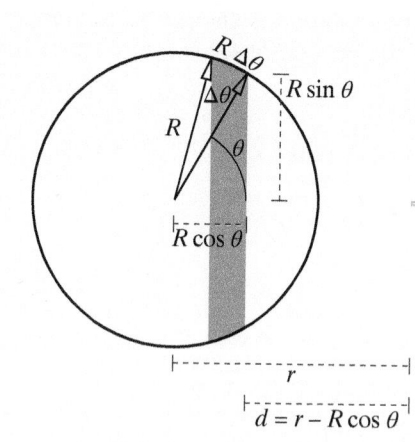

Figure 15.42 The sphere may be divided into ring-shaped segments.

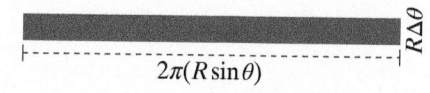

Figure 15.43 We can calculate the area of the ring of charge by "unrolling" it.

Amount of charge on each ring:

$$\Delta Q = Q \frac{\text{(surface area of ring)}}{\text{(surface area of sphere)}} = Q \left(\frac{2\pi(R\sin\theta)(R\Delta\theta)}{4\pi R^2} \right)$$

As before, we calculate the surface area ΔA of the ring by laying the ring out flat (Figure 15.43), noting that the radius of the ring is $R\sin\theta$ and its width is $R\Delta\theta$ (since arc length is radius times angle, with angle measured in radians).
Integration variable: θ
Magnitude of ΔE:

$$\Delta E = \frac{1}{4\pi\varepsilon_0} \frac{(\Delta Q)d}{(d^2 + (R\sin\theta)^2)^{3/2}}$$

$$= \frac{1}{4\pi\varepsilon_0} \frac{(r - R\cos\theta)}{[(r - R\cos\theta)^2 + (R\sin\theta)^2]^{3/2}} Q \left(\frac{2\pi(R\sin\theta)}{4\pi R^2} \right) (R\Delta\theta)$$

Step 3: Add Up Contributions of All the Pieces

$$E = \frac{1}{4\pi\varepsilon_0} \frac{Q}{2} \int_0^\pi \frac{(r - R\cos\theta)}{[(r - R\cos\theta)^2 + (R\sin\theta)^2]^{3/2}} \sin\theta \, d\theta$$

The limits on the integral are determined by the fact that if we let θ range between 0 and π radians (180°), we add up rings that account for the entire surface of the spherical shell.

Evaluating the integral is simply math (in this case, it is rather difficult math). The results are:
Outside the shell $(r > R)$:

$$E = \frac{1}{4\pi\varepsilon_0} \frac{Q}{r^2} \quad \text{for } r > R$$

as though all the charge were concentrated into a point at the center of the spherical shell.
Inside the shell $(r < R)$: $E = 0$ for $r < R$.

SUMMARY

A procedure for calculating the electric field due to a distribution of electric charges, by applying the superposition principle.

Step 1: Divide the charge distribution into pieces and draw $\Delta\vec{E}$.
Choose pieces whose electric field you know.

Step 2: Write an expression for the electric field due to one piece.
This involves selecting an origin, finding $\vec{r}$ and $\hat{r}$, describing the location of one piece in terms of an integration variable, expressing Δq in terms of your variables, and writing expressions for ΔE_x, ΔE_y, and ΔE_z.

Step 3: Add up the contributions of all the pieces.
Sometimes the sum can be written as a definite integral that can be symbolically evaluated; otherwise you must calculate a finite sum numerically.

Step 4: Check the result.
Checks should include a qualitative check of field direction, a check of the units of the answer, and checks of special cases.

Useful results:

Rod
Electric field of a uniformly charged thin rod, at a location a distance r from the midpoint of the rod along a line perpendicular to the rod (Figure 15.44). Q is the total

charge on the rod, and L is the length of the rod. Direction: either radially away from or toward the rod (depending on the sign of Q).

$$E = \frac{1}{4\pi\varepsilon_0} \frac{Q}{r\sqrt{r^2 + (L/2)^2}}$$

Approximate electric field of a rod:

$$E \approx \frac{1}{4\pi\varepsilon_0} \frac{2(Q/L)}{r} \quad (\text{if } L \gg r)$$

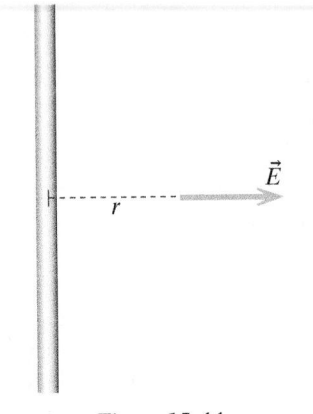

Figure 15.44

Ring

Electric field of a thin uniformly charged ring at a location along the axis, a distance z from the center of the ring, for a ring of radius R and charge q (Figure 15.45). Direction: parallel or antiparallel to the axis, depending on the sign of q.

$$E = \frac{1}{4\pi\varepsilon_0} \frac{qz}{(R^2 + z^2)^{3/2}}$$

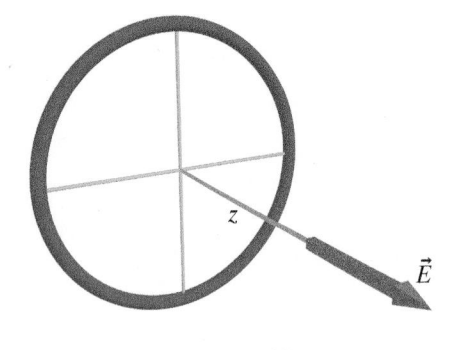

Figure 15.45

Disk

Electric field of a uniformly charged disk of radius R, a distance z from the center of the disk along a line perpendicular to the disk (Figure 15.46):

$$E = \frac{(Q/A)}{2\varepsilon_0} \left[1 - \frac{z}{(R^2 + z^2)^{1/2}}\right]$$

where $A = \pi R^2$.

Approximation: close to the disk ($z \ll R$):

$$E \approx \frac{Q/A}{2\varepsilon_0} \left[1 - \frac{z}{R}\right]$$

Approximation: extremely close to the disk ($z \ll R$):

$$E \approx \frac{Q/A}{2\varepsilon_0}$$

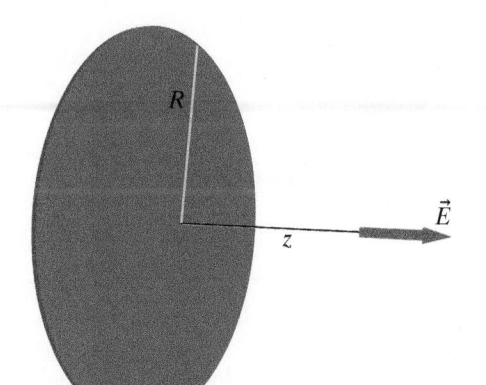

Figure 15.46

Capacitor

Electric field at a location inside a capacitor (Figure 15.47):

$$E \approx \frac{Q/A}{\varepsilon_0}$$

Near center of a two-plate capacitor (each plate has area A, one plate has charge $+Q$, other plate has charge $-Q$; separation between plates is very small compared to the radius of a plate). Direction: perpendicular to the plates. Fringe field just outside the plates:

$$E_{\text{fringe}} \approx \frac{Q/A}{2\varepsilon_0} \left(\frac{s}{R}\right)$$

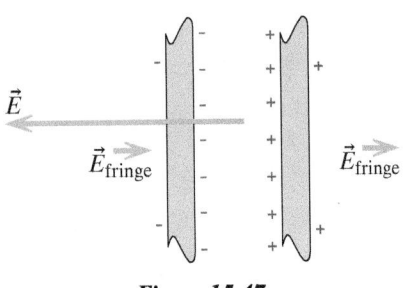

Figure 15.47

Spherical shell

Inside a uniformly charged thin spherical shell:

$$E = 0 \quad \text{due to charges on the shell}$$

Outside a uniformly charged thin spherical shell:

$$E = \frac{1}{4\pi\varepsilon_0}\frac{Q}{r^2} \quad \text{due to charges on the shell}$$

where r is the distance from the center of the sphere to the observation location (Figure 15.48):

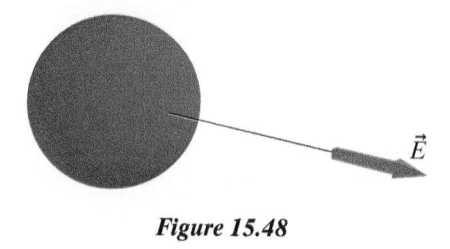

Figure 15.48

Sphere charged throughout its volume

Inside a solid sphere of radius R, charged throughout its volume, at a location a distance r from the center of the sphere:

$$E = \frac{1}{4\pi\varepsilon_0}\frac{Q}{R^3}r$$

Warning: Do not confuse this result with the previous one. The words "solid" and "shell" refer to the location of the charge (on the surface or throughout the object). If a plastic ball is solid but has charge uniformly distributed over its surface and none inside, the electric field inside the plastic, due to the charge on the surface, is zero—this is an example of a spherical shell of charge.

QUESTIONS

Q1 Consider setting up an integral to find an algebraic expression for the electric field of a uniformly charged rod of length L, at a location on the midplane. If we choose an origin at the center of the rod, what are the limits of integration?

Q2 A rod is 2.5 m long. Its charge is -2×10^{-7} C. The observation location is 4 cm from the rod, in the midplane. In the expression

$$E = \frac{1}{4\pi\varepsilon_0}\frac{Q}{r\sqrt{r^2 + (L/2)^2}}$$

what is r in meters?

Q3 Graph the magnitude of the full expression for the field E of a rod along the midplane vs. r. Does E fall off monotonically (with distance)?

Q4 A rod with uniformly distributed charge 2×10^{-8} C is 50 cm long. We need to calculate E at a distance of 1 cm from the midpoint of the rod. Which equation for the electric field of a rod should we use? (1) Exact, (2) Approximate, (3) Either exact or approximate, (4) Neither—we have to do it numerically, (5) Neither—we need to integrate

Q5 A student claimed that the equation for the electric field outside a cube of edge length L, carrying a uniformly distributed charge Q, at a distance x from the center of the cube, was

$$E = \frac{Q}{\varepsilon_0}\frac{L}{x^{1/2}}$$

Explain how you know that this cannot be the right equation.

Q6 A student claimed that the equation for the electric field outside a cube of edge length L, carrying a uniformly distributed charge Q, at a distance x from the center of the cube, was

$$\frac{1}{4\pi\varepsilon_0}\frac{50QL}{x^3}$$

Explain how you know that this cannot be the right equation.

Q7 When calculating the electric field of an object with electric charge distributed approximately uniformly over its surface, what is the order in which you should do the following operations? (1) Check the direction and units. (2) Write an expression for the electric field due to one point-like piece of the object. (3) Divide up the object into small pieces of a shape whose field is known. (4) Sum the vector contributions of all the pieces.

Q8 Explain briefly how knowing the electric field of a ring helps in calculating the field of a disk.

Q9 Consider the algebraic expression for the electric field of a uniformly charged ring, at a location on the axis of the ring. Q is the charge on the entire ring, and ΔQ is the charge on one piece of the ring. $\Delta\theta$ is the angle subtended by one piece of the ring (or, alternatively, Δr is the arc length of one piece). What is ΔQ, expressed in terms of given constants and an integration variable? What are the integration limits?

Q10 The rod in Figure 15.49 carries a uniformly distributed positive charge. Which arrow (a–h) best represents the direction of the electric field at the observation location marked with a red X?

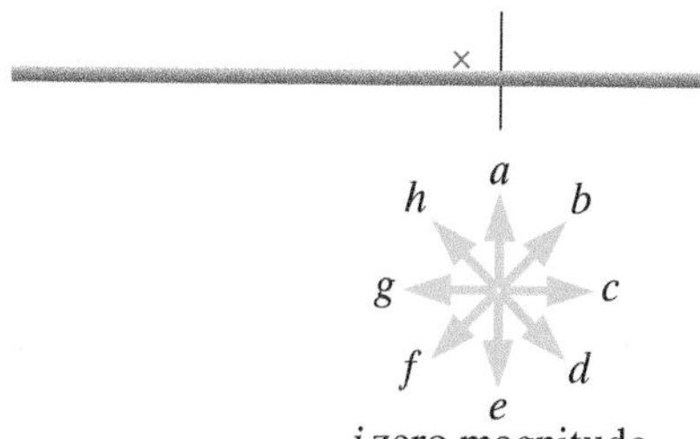

Figure 15.49

Q11 By thinking about the physical situation, predict the magnitude of the electric field at the center of a uniformly charged ring of radius R carrying a charge $+Q$. Then use the equation derived in the text to confirm this result.

Q12 Define "fringe field."

Q13 Explain qualitatively how it is possible for the electric field at locations near the center of a uniformly charged disk not to vary with distance away from the disk.

Q14 Coulomb's law says that electric field falls off like $1/z^2$. How can E for a uniformly charged disk depend on $[1 - z/R]$, or be independent of distance?

Q15 Two very thin circular plastic sheets are close to each other and carry equal and opposite uniform charge. Explain briefly why the field between the sheets is much larger than the field outside. Illustrate your argument on a diagram.

Q16 The electric field inside a capacitor is shown on the left in Figure 15.50. Which option (1–5) best represents the electric field at location A?

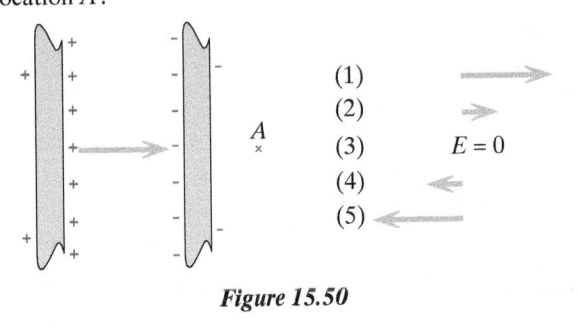

(1)

(2)

(3) $E = 0$

(4)

(5)

Figure 15.50

Q17 A solid spherical plastic ball was rubbed with wool in such a way that it acquired a uniform negative charge all over the surface. Make a sketch showing the polarization of molecules inside the plastic ball, and explain briefly.

Q18 A student said, "The electric field inside a uniformly charged sphere is always zero." Describe a situation where this is not true.

Q19 Give an example of a configuration of charges that yields an electric field or force whose magnitude varies approximately with distance as specified: (1) Field independent of distance, (2) Field proportional to $1/r$, (3) Field proportional to $1/r^2$, (4) Field proportional to $1/r^3$, (5) Force (not field) that is proportional to $1/r^5$.

PROBLEMS

Section 15.1

•P20 If the total charge on a uniformly charged rod of length 0.4 m is 2.2 nC, what is the magnitude of the electric field at a location 3 cm from the midpoint of the rod?

•P21 A thin rod lies on the x axis with one end at $-A$ and the other end at A, as shown in Figure 15.51. A charge of $-Q$ is spread uniformly over the surface of the rod. We want to set up an integral to find the electric field at location $\langle 0, y, 0 \rangle$ due to the rod. Following the procedure discussed in this chapter, we have cut up the rod into small segments, each of which can be considered as a point charge. We have selected a typical piece, shown in red on the diagram.

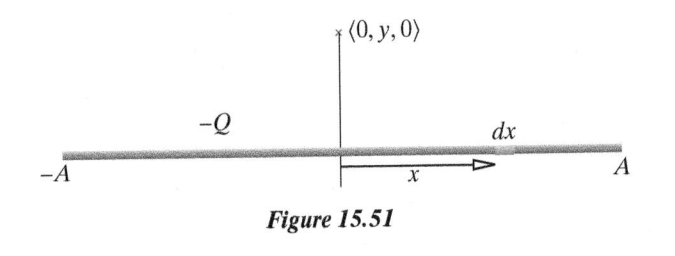

Figure 15.51

Answer using the variables x, y, dx, A, Q as appropriate. Remember that the rod has charge $-Q$.
(a) In terms of the symbolic quantities given above and on the diagram, what is the charge per unit length of the rod? **(b)** What is the amount of charge dQ on the small piece of length dx? **(c)** What is the vector from this source to the observation location? **(d)** What is the distance from this source to the observation location? **(e)** When we set up an integral to find the electric field at the observation location due to the entire rod, what will be the integration variable?

•P22 A plastic rod 1.7 m long is rubbed all over with wool, and acquires a charge of -2×10^{-8} C (Figure 15.52). We choose the center of the rod to be the origin of our coordinate system, with the x axis extending to the right, the y axis extending up, and the z axis out of the page. In order to calculate the electric field at location $A = \langle 0.7, 0, 0 \rangle$ m, we divide the rod into eight pieces, and approximate each piece as a point charge located at the center of the piece.

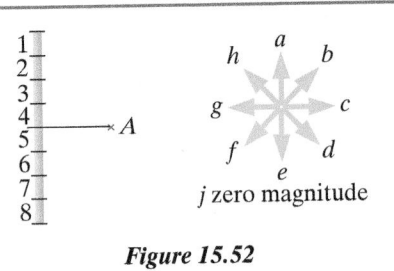

Figure 15.52

(a) What is the length of one of these pieces? **(b)** What is the location of the center of piece number 3? **(c)** How much charge is on piece number 3? (Remember that the charge is negative.) **(d)** Approximating piece 3 as a point charge, what is the electric field at location A due only to piece 3? **(e)** To get the net electric field at location A, we would need to calculate $\Delta \vec{E}$ due to each of the eight pieces, and add up these contributions. If we did that, which arrow (a–h) would best represent the direction of the net electric field at location A?

•P23 A clear plastic pen 12 cm long is rubbed all over with wool, and acquires a negative charge of -2 nC. You want to figure out the electric field a distance of 18 mm from the pen, near the middle of the pen. **(a)** You decide to model the pen as a rod consisting of a series of five segments, each of which you will consider to be approximately point-like. What is the length of each segment in meters? **(b)** What is the amount of charge ΔQ on each of the five segments? **(c)** In general, if the rod has a length L and total charge Q, and you divide the rod into N segments, what is the amount of charge ΔQ on each piece? **(d)** If the length of each segment is dL, write a symbolic expression for the number of pieces N in terms of the length of the rod L and the length of one piece dL. **(e)** Now write a symbolic expression for the amount of charge on each piece in terms of the length of the rod and the length of a small piece.

•P24 A thin glass rod of length 80 cm is rubbed all over with wool and acquires a charge of 60 nC, distributed uniformly over its surface. Calculate the magnitude of the electric field due to the rod at a location 7 cm from the midpoint of the rod. Do the calculation two ways, first using the exact equation for a rod of any length, and second using the approximate equation for a long rod.

•••**P25** A water molecule is a permanent dipole with a known dipole moment p ($=qs$). There is a water molecule in the air a very short distance x from the midpoint of a long glass rod of length L carrying a uniformly distributed positive charge $+Q$ (Figure 15.53). The axis of the dipole is perpendicular to the rod. Note that $s \ll x \ll L$. (The charged rod induces an increase in the dipole moment, but the induced portion of the dipole moment is completely negligible compared to p. It is convenient to use the "binomial expansion" that you may have learned in calculus, that $(1+\epsilon)^n \approx 1+n\epsilon$ is $\epsilon \ll 1$. Note that n can be negative.)

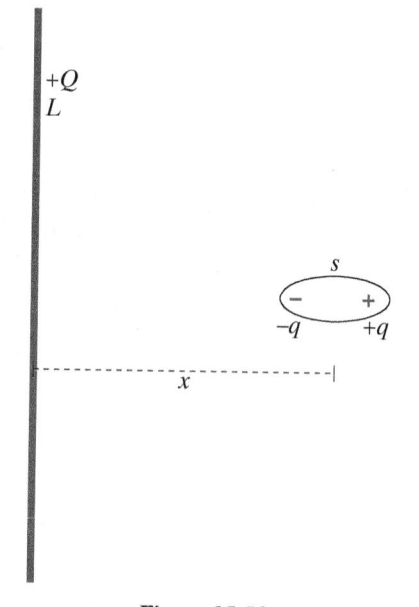

Figure 15.53

(a) Find the magnitude and direction of the electric force acting on the water molecule. Your final result for the magnitude of the force must be expressed in terms of Q, L, x, and p. You can use q and s in your calculations, but your final result must not include q or s, since it is only their product $p = qs$ that is measurable for a water molecule. Explain your work carefully, including appropriate diagrams. **(b)** If the electric field of the rod has the magnitude 1×10^6 N/C at the location of the water molecule, 1 cm from the rod, what is the magnitude of the horizontal component of the acceleration of the water molecule? The measured dipole moment for H_2O is 6.2×10^{-30} C·m, and the mass of one mole is 18 g $(1+1+16)$. Be sure to check units in your calculation!

•••**P26** An electrostatic dust precipitator that is installed in a factory smokestack includes a straight metal wire of length $L = 0.8$ m that is charged approximately uniformly with a total charge $Q = 0.4 \times 10^{-7}$ C. A speck of coal dust (which is mostly carbon) is near the wire, far from both ends of the wire; the distance from the wire to the speck is $d = 1.5$ cm. Carbon has an atomic mass of 12 (6 protons and 6 neutrons in the nucleus). A careful measurement of the polarizability of a carbon atom gives the value

$$\alpha = 1.96 \times 10^{-40} \frac{\text{C} \cdot \text{m}}{\text{N/C}}$$

(a) Calculate the initial acceleration of the speck of coal dust, neglecting gravity. Explain your steps clearly. Your answer must be expressed in terms of Q, L, d, and α. You can use other quantities in your calculations, but your final result must not include them. Don't put numbers into your calculation until the very end, but then show the numerical calculation that you carry out on your calculator. It is convenient to use the "binomial expansion" that you may have learned in calculus, that $(1+\epsilon)^n \approx 1+n\epsilon$ is $\epsilon \ll 1$. Note that n can be negative. **(b)** If the speck of coal dust were initially twice as far from the charged wire, how much smaller would be the initial acceleration of the speck?

Section 15.2

••**P27** Consider a thin plastic rod bent into an arc of radius R and angle α (Figure 15.54). The rod carries a uniformly distributed negative charge $-Q$.

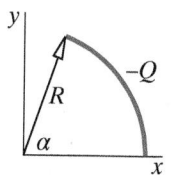

Figure 15.54

What are the components E_x and E_y of the electric field at the origin? Follow the standard four steps: **(a)** Use a diagram to explain how you will cut up the charged rod, and draw the $\Delta \vec{E}$ contributed by a representative piece. **(b)** Express algebraically the contribution each piece makes to the x and y components of the electric field. Be sure to show your integration variable and its origin on your drawing. (*Hint:* An arc of radius R and angle $\Delta\theta$ measured in radians has a length $R\Delta\theta$.) **(c)** Write the summation as an integral, and simplify the integral as much as possible. State explicitly the range of your integration variable. Evaluate the integral. **(d)** Show that your result is reasonable. Apply as many tests as you can think of.

••**P28** A strip of invisible tape 0.12 m long by 0.013 m wide is charged uniformly with a total net charge of 3 nC (nano = 1×10^{-9}) and is suspended horizontally, so it lies along the x axis, with its center at the origin, as shown in Figure 15.55. Calculate the approximate electric field at location $\langle 0, 0.03, 0 \rangle$ m (location A) due to the strip of tape. Do this by dividing the strip into three equal sections, as shown in Figure 15.55, and approximating each section as a point charge.

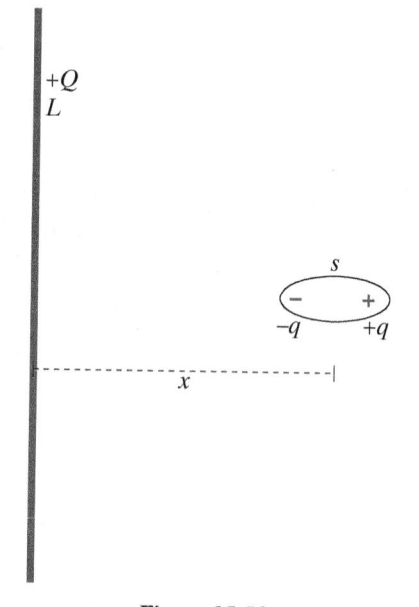

Figure 15.55

(a) What is the approximate electric field at A due to piece 1?
(b) What is the approximate electric field at A due to piece 2?
(c) What is the approximate electric field at A due to piece 3?
(d) What is the approximate net electric field at A? **(e)** What could you do to improve the accuracy of your calculation?

••**P29** Consider a thin glass rod of length L lying along the x axis with one end at the origin. The rod carries a uniformly distributed positive charge Q.

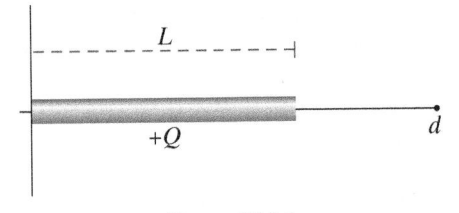

Figure 15.56

At a location $d > L$, on the x axis to the right of the rod in Figure 15.56, what is the electric field due to the rod? Follow the standard four steps. **(a)** Use a diagram to explain how you will cut up the charged rod, and draw the $\Delta \vec{E}$ contributed by a representative piece. **(b)** Express algebraically the contribution each piece makes to the electric field. Be sure to show your integration variable and its origin on your drawing. **(c)** Write the summation as an integral, and simplify the integral as much as possible. State explicitly the range of your integration variable. Evaluate the integral. **(d)** Show that your result is reasonable. Apply as many tests as you can think of.

••**P30** Consider a thin plastic rod bent into a semicircular arc of radius R with center at the origin (Figure 15.57). The rod carries a uniformly distributed negative charge $-Q$.

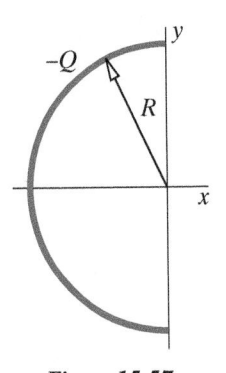

Figure 15.57

(a) Determine the electric field $\vec{E}$ at the origin contributed by the rod. Include carefully labeled diagrams, and be sure to check your result. **(b)** An ion with charge $-2e$ and mass M is placed at rest at the origin. After a *very* short time Δt the ion has moved only a very short distance but has acquired some momentum $\vec{p}$. Calculate $\vec{p}$.

Section 15.3

•**P31** Two rings of radius 5 cm are 24 cm apart and concentric with a common horizontal x axis. The ring on the left carries a uniformly distributed charge of $+31$ nC, and the ring on the right carries a uniformly distributed charge of -31 nC. **(a)** What are the magnitude and direction of the electric field on the x axis, halfway between the two rings? **(b)** If a charge of -9 nC were placed midway between the rings, what would be the force exerted on this charge by the rings?

•**P32** Two rings of radius 4 cm are 12 cm apart and concentric with a common horizontal x axis. The ring on the left carries a uniformly distributed charge of $+40$ nC, and the ring on the right carries a uniformly distributed charge of -40 nC. **(a)** What is the

electric field due to the right ring at a location midway between the two rings? **(b)** What is the electric field due to the left ring at a location midway between the two rings? **(c)** What is the net electric field at a location midway between the two rings? **(d)** If a charge of -2 nC were placed midway between the rings, what would be the force exerted on this charge by the rings?

•**P33** Two rings of radius 2 cm are 20 cm apart and concentric with a common horizontal x axis. What is the magnitude of the electric field midway between the rings if both rings carry a charge of $+35$ nC?

•••**P34** A thin-walled hollow circular glass tube, open at both ends, has a radius R and length L. The axis of the tube lies along the x axis, with the left end at the origin (Figure 15.58). The outer sides are rubbed with silk and acquire a net positive charge Q distributed uniformly. Determine the electric field at a location on the x axis, a distance w from the origin. Carry out all steps, including checking your result. Explain each step. (You may have to refer to a table of integrals.)

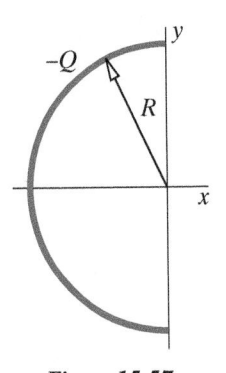

Figure 15.58

Section 15.4

•**P35** Suppose that the radius of a disk is 21 cm, and the total charge distributed uniformly all over the disk is 5.0×10^{-6} C. **(a)** Use the exact result to calculate the electric field 1 mm from the center of the disk. **(b)** Use the exact result to calculate the electric field 3 mm from the center of the disk. **(c)** Does the field decrease significantly?

•**P36** For a disk of radius $R = 20$ cm and $Q = 6 \times 10^{-6}$ C, calculate the electric field 2 mm from the center of the disk using all three equations:

$$E = \frac{(Q/A)}{2\varepsilon_0}\left[1 - \frac{z}{(R^2 + z^2)^{1/2}}\right],$$

$$E \approx \frac{Q/A}{2\varepsilon_0}\left[1 - \frac{z}{R}\right], \quad \text{and} \quad E \approx \frac{Q/A}{2\varepsilon_0}.$$

How good are the approximate equations at this distance? For the same disk, calculate E at a distance of 5 cm (50 mm) using all three equations. How good are the approximate equations at this distance?

•**P37** A disk of radius 16 cm has a total charge 4×10^{-6} C distributed uniformly all over the disk. **(a)** Using the exact equation, what is the electric field 1 mm from the center of the disk? **(b)** Using the same exact equation, find the electric field 3 mm from the center of the disk. **(c)** What is the percent difference between these two numbers?

•**P38** For a disk of radius 20 cm with uniformly distributed charge 7×10^{-6} C, calculate the magnitude of the electric field on the axis of the disk, 5 mm from the center of the disk, using each of the following equations:

(a) $E = \frac{(Q/A)}{2\varepsilon_0}\left[1 - \frac{z}{(R^2 + z^2)^{1/2}}\right],$

(b) $E \approx \dfrac{Q/A}{2\varepsilon_0}\left[1 - \tfrac{z}{R}\right]$,

(c) $E \approx \dfrac{Q/A}{2\varepsilon_0}$.

(d) How good are the approximate equations at this distance?
(e) At what distance does the least accurate approximation for the electric field give a result that is closest to the most accurate: at a distance $R/2$, close to the disk, at a distance R, or far from the disk?

••P39 A large, thin plastic disk with radius $R = 1.5$ m carries a uniformly distributed charge of $-Q = -3 \times 10^{-5}$ C as shown in Figure 15.59. A circular piece of aluminum foil is placed $d = 3$ mm from the disk, parallel to the disk. The foil has a radius of $r = 2$ cm and a thickness $t = 1$ mm.

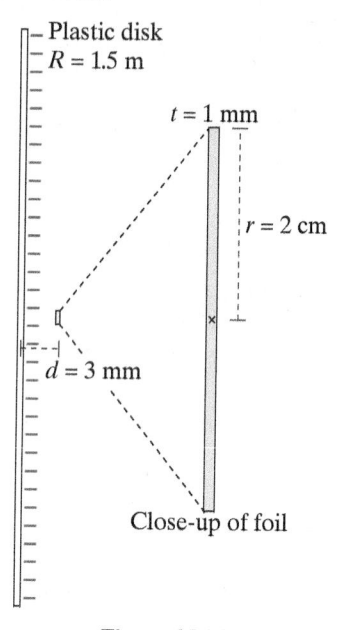

Figure 15.59

(a) Show the charge distribution on the close-up of the foil.
(b) Calculate the magnitude and direction of the electric field at location $\times$ at the center of the foil, inside the foil.
(c) Calculate the magnitude q of the charge on the left circular face of the foil.

••P40 A thin circular sheet of glass of diameter 3 m is rubbed with a cloth on one surface and becomes charged uniformly. A chloride ion (a chlorine atom that has gained one extra electron) passes near the glass sheet. When the chloride ion is near the center of the sheet, at a location 0.8 mm from the sheet, it experiences an electric force of 5×10^{-15} N, toward the glass sheet. It will be useful to you to draw a diagram on paper, showing field vectors, force vectors, and charges, before answering the following questions about this situation.

Which of the following statements about this situation are correct? Select all that apply. (1) The electric field that acts on the chloride ion is due to the charge on the glass sheet and to the charge on the chloride ion. (2) The electric field of the glass sheet is equal to the electric field of the chloride ion. (3) The charged disk is the source of the electric field that causes the force on the chloride ion. (4) The net electric field at the location of the chloride ion is zero. (5) The force on the chloride ion is equal to the electric field of the glass sheet.

In addition to an exact equation for the electric field of a disk, the text derives two approximate equations. In the current situation we want an answer that is correct to three significant figures. Which of the following is correct? We should not use an approximation if we have enough information to do an exact calculation. (1) $R \gg z$, so it is adequate to use the most approximate equation here. (2) z is nearly equal to R, so we have to use the exact equation. (3) $z \ll R$, so we can't use an approximation.

How much charge is on the surface of the glass disk? Give the amount, including sign and correct units.

Section 15.5

•P41 If the magnitude of the electric field in air exceeds roughly 3×10^6 N/C, the air breaks down and a spark forms. For a two-disk capacitor of radius 47 cm with a gap of 1 mm, what is the maximum charge (plus and minus) that can be placed on the disks without a spark forming (which would permit charge to flow from one disk to the other)?

•P42 Consider a capacitor made of two rectangular metal plates of length L and width W, with a very small gap s between the plates. There is a charge $+Q$ on one plate and a charge $-Q$ on the other. Assume that the electric field is nearly uniform throughout the gap region and negligibly small outside. Calculate the attractive force that one plate exerts on the other. Remember that one of the plates doesn't exert a net force on itself.

•P43 A capacitor consists of two large metal disks of radius 1.1 m placed parallel to each other, a distance of 1.2 mm apart. The capacitor is charged up to have an increasing amount of charge $+Q$ on one disk and $-Q$ on the other. At about what value of Q does a spark appear between the disks?

••P44 If the magnitude of the electric field in air exceeds roughly 3×10^6 N/C, the air breaks down and a spark forms. For a two-disk capacitor of radius 51 cm with a gap of 2 mm, if the electric field inside is just high enough that a spark occurs, what is the strength of the fringe field just outside the center of the capacitor?

••P45 In a cathode-ray tube, an electron travels in a vacuum and enters a region between two deflection plates where there is an upward electric field of magnitude 1×10^5 N/C (Figure 15.60).

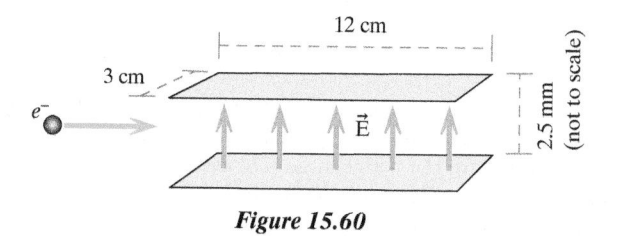

Figure 15.60

(a) Sketch the trajectory of the electron, continuing on well past the deflection plates (the electron is going fast enough that it does not strike the plates).
(b) Calculate the acceleration of the electron while it is between the deflection plates.
(c) The deflection plates measure 12 cm by 3 cm, and the gap between them is 2.5 mm. The plates are charged equally and oppositely. What are the magnitude and sign of the charge on the upper plate?

••P46 In Figure 15.61 are two uniformly charged disks of radius R that are very close to each other (gap $\ll R$). The disk on the left has a charge of $-Q_{\text{left}}$ and the disk on the right has a charge of $+Q_{\text{right}}$ (Q_{right} is greater than Q_{left}). A uniformly charged thin rod of length L lies at the edge of the disks, parallel to the axis of the disks and centered on the gap. The rod has a charge of $+Q_{rod}$.

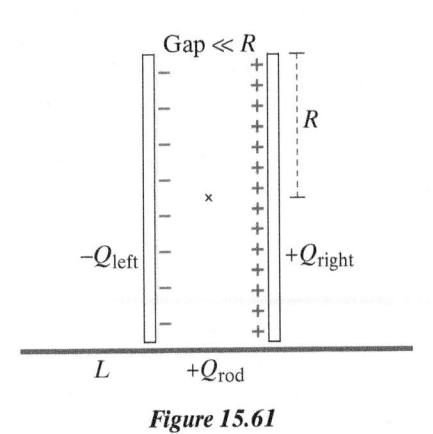

Figure 15.61

(a) Calculate the magnitude and direction of the electric field at the point marked $\times$ at the center of the gap region, and explain briefly, including showing the electric field on a diagram. Your results must not contain any symbols other than the given quantities R, Q_{left}, Q_{right}, L, and Q_{rod} (and fundamental constants), unless you define intermediate results in terms of the given quantities. **(b)** If an electron is placed at the center of the gap region, what are the magnitude and direction of the electric force that acts on the electron?

••P47 A capacitor made of two parallel uniformly charged circular metal disks carries a charge of $+Q$ and $-Q$ on the inner surfaces of the plates and very small amounts of charge $+q$ and $-q$ on the outer surfaces of the plates. Each plate has a radius R and thickness t, and the gap distance between the plates is s. How much charge q is on the outside surface of the positive disk, in terms of Q?

•••P48 This is a challenging problem that requires you to construct a model of a real physical situation, to make idealizations, approximations, and assumptions, and to work through a detailed analysis based on physical principles. Be sure to allow yourself enough time to think it through. You may need to measure, estimate, or look up various quantities.

A clear plastic ball-point pen is rubbed thoroughly with wool. The charged plastic pen is held above a small, uncharged disk-shaped piece of aluminum foil, smaller than a hole in a sheet of three-ring binder paper. **(a)** Make a clear physics diagram of the situation, showing charges, fields, forces, and distances. Refer to this physics diagram in your analysis. **(b)** Starting from fundamental physical principles, predict quantitatively how close you must move the pen to the foil in order to pick up the foil (that is, predict an actual numerical distance). State explicitly all approximations and assumptions that you make. **(c)** Try the experiment and compare your observation to your prediction.

Section 15.6

•P49 A solid metal ball of radius 1.5 cm bearing a charge of -17 nC is located near a solid plastic ball of radius 2 cm bearing a uniformly distributed charge of $+7$ nC (Figure 15.62) on its outer surface. The distance between the centers of the balls is 9 cm. **(a)** Show the approximate charge distribution in and on each ball. **(b)** What is the electric field at the center of the metal ball due only to the charges on the plastic ball? **(c)** What is the net electric field at the center of the metal ball? **(d)** What is the electric field

at the center of the metal ball due only to the charges on the surface of the metal ball?

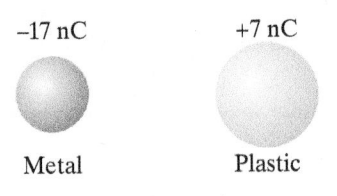

Metal   Plastic

Figure 15.62

•P50 You stand at location A, a distance d from the origin, and hold a small charged ball. You find that the electric force on the ball is 0.08 N. You move to location B, a distance $2d$ from the origin, and find the electric force on the ball to be 0.04 N. What object located at the origin might be the source of the field? (1) A point charge, (2) A dipole, (3) A uniformly charged rod, (4) A uniformly charged ring, (5) A uniformly charged disk, (6) A capacitor, (7) A uniformly charged hollow sphere, (8) None of the above

If the force at B were 0.0799 N, what would be your answer? If the force at B were 0.01 N, what would be your answer? If the force at B were 0.02 N, what would be your answer?

•P51 A thin plastic spherical shell of radius 5 cm has a uniformly distributed charge of -25 nC on its outer surface. A concentric thin plastic spherical shell of radius 8 cm has a uniformly distributed charge of $+64$ nC on its outer surface. Find the magnitude and direction of the electric field at distances of 3 cm, 7 cm, and 10 cm from the center. See Figure 15.63.

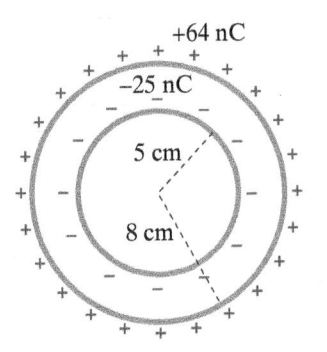

Figure 15.63

•P52 A hollow ball of radius 7 cm, made of very thin glass, is rubbed all over with a silk cloth and acquires a negative charge of -9×10^{-8} C that is uniformly distributed all over its surface. Location A in Figure 15.64 is inside the sphere, 1 cm from the surface. Location B in Figure 15.64 is outside the sphere, 2 cm from the surface. There are no other charged objects nearby.

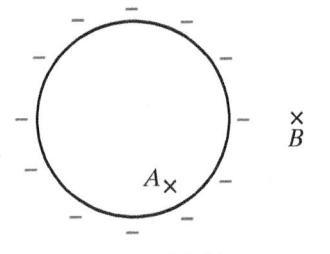

Figure 15.64

Which of the following statements about E_{ball}, the magnitude of the electric field due to the ball, are correct? Select all that apply. **(a)** At location A, E_{ball} is 0 N/C. **(b)** All of the charges on the surface of the sphere contribute to E_{ball} at location A. **(c)** A hydrogen atom at location A would polarize because it is close to the negative charges on the surface of the sphere. What is E_{ball} at location B?

••P53 A thin hollow spherical glass shell of radius 0.17 m carries a uniformly distributed positive charge $+6 \times 10^{-9}$ C, as shown in Figure 15.65. To the right of it is a horizontal permanent dipole with charges $+3 \times 10^{-11}$ and -3×10^{-11} C separated by a distance 2×10^{-5} m (the dipole is shown greatly enlarged for clarity). The dipole is fixed in position and is not free to rotate. The distance from the center of the glass shell to the center of the dipole is 0.6 m.

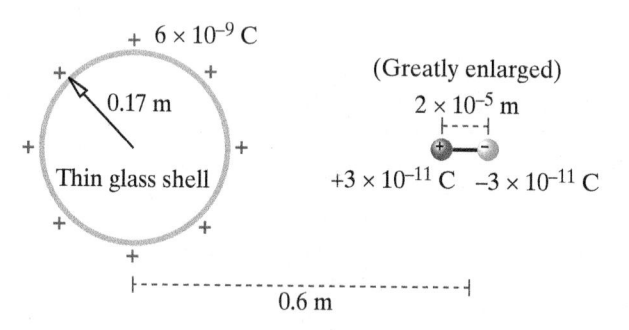

Figure 15.65

(a) Calculate the net electric field at the center of the glass shell. **(b)** If the sphere were a solid metal ball with a charge $+6 \times 10^{-9}$ C, what would be the net electric field at its center? **(c)** Draw the approximate charge distribution in and/or on the metal sphere.

••P54 Two large, thin, charged plastic circular plates each of radius R are placed a short distance s apart; s is much smaller than the dimensions of a plate (Figure 15.66). The right-hand plate has a positive charge of $+Q$ evenly distributed over its inner surface (Q is a positive number). The left-hand plate has a negative charge of $-2Q$ evenly distributed over its inner surface. A very thin plastic spherical shell of radius r is placed midway between the plates (and shown in cross section). It has a uniformly distributed positive charge of $+q$. You can ignore the contributions to the electric field due to the polarization of the thin plastic shell and the thin plastic plates.

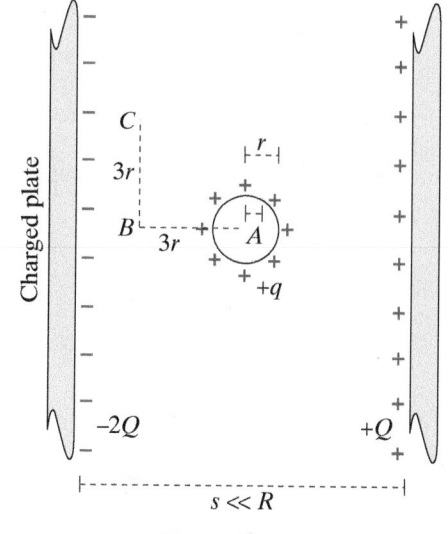

Figure 15.66

(a) Calculate the x and y components of the electric field at location A, a horizontal distance $r/2$ to the right of the center of the sphere. **(b)** Calculate the x and y components of the electric field at location B, a horizontal distance $3r$ to the left of the center of the sphere. **(c)** Calculate the x and y components of the electric field at location C, a horizontal distance $3r$ to the left and a vertical distance $3r$ above the center of the sphere.

••P55 A glass sphere carrying a uniformly distributed charge of $+Q$ is surrounded by an initially neutral spherical plastic shell (Figure 15.67).

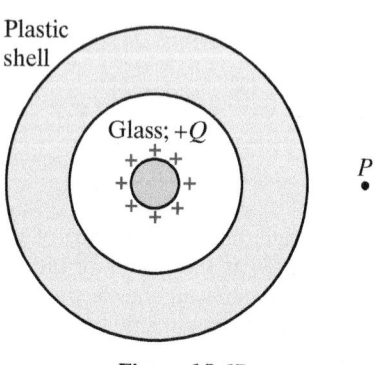

Figure 15.67

(a) Qualitatively, indicate the polarization of the plastic. **(b)** Qualitatively, indicate the polarization of the inner glass sphere. Explain briefly. **(c)** Is the electric field at location P outside the plastic shell larger, smaller, or the same as it would be if the plastic weren't there? Explain briefly. **(d)** Now suppose that the glass sphere carrying a uniform charge of $+Q$ is surrounded by an initially neutral metal shell (Figure 15.68). Qualitatively, indicate the polarization of the metal.

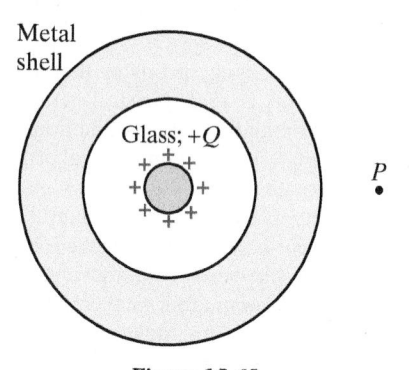

Figure 15.68

(e) Now be *quantitative* about the polarization of the metal sphere and prove your assertions. **(f)** Is the electric field at location P outside the metal shell larger, smaller, or the same as it would be if the metal shell weren't there? Explain briefly.

••P56 Breakdown field strength for air is roughly 3×10^6 N/C. If the electric field is greater than this value, the air becomes a conductor. **(a)** There is a limit to the amount of charge that you can put on a metal sphere in air. If you slightly exceed this limit, why would breakdown occur, and why would the breakdown occur very near the surface of the sphere, rather than somewhere else? **(b)** How much excess charge can you put on a metal sphere of 10-cm radius without causing breakdown in the neighboring air, which would discharge the sphere? **(c)** How much excess charge can you put on a metal sphere of only 1-mm radius? These results hint at the reason why a highly charged piece of metal tends to spark at places where the radius of curvature is small, or at places where there are sharp points.

••**P57** Two thin plastic spherical shells (shown in cross section in Figure 15.69) are uniformly charged. The center of the larger

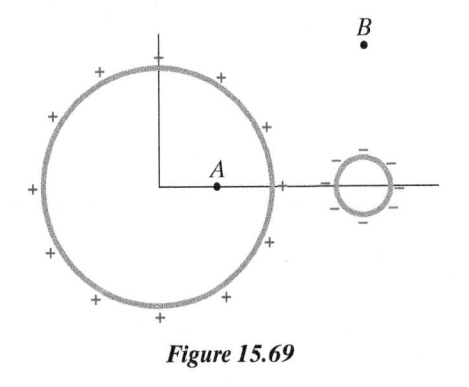

Figure 15.69

sphere is at $(0,0)$; it has a radius of 12 cm and a uniform positive charge of $+4 \times 10^{-9}$ C. The center of the smaller sphere is at $(25\text{ cm}, 0)$; it has a radius of 3 cm and a uniform negative charge of -1×10^{-9} C. **(a)** What are the components $E_{A,x}$ and $E_{A,y}$ of the electric field $\Delta\vec{E}$ at location A (6 cm to the right of the center of the large sphere)? Neglect the small contribution of the polarized molecules in the plastic, because the shells are very thin and don't contain much matter. **(b)** What are the components $E_{B,x}$ and $E_{B,y}$ of the electric field at location B (15 cm above the center of the small sphere)? Neglect the small contribution of the polarized molecules in the plastic, because the shells are very thin and don't contain much matter. **(c)** What are the components F_x and F_y of the force on an electron placed at location B?

••**P58** A solid plastic sphere of radius R_1 has a charge $-Q_1$ on its surface (Figure 15.70). A concentric spherical metal shell of inner radius R_2 and outer radius R_3 carries a charge Q_2 on the inner surface and a charge Q_3 on the outer surface. Q_1, Q_2, and Q_3 are positive numbers, and the total charge $Q_2 + Q_3$ on the metal shell is greater than Q_1.

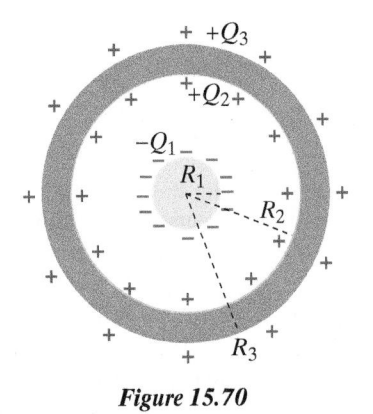

Figure 15.70

At an observation location a distance r from the center, determine the magnitude and direction of the electric field in the following regions, and explain briefly in each case. For parts (a)–(d), be sure to give both the direction and the magnitude of the electric field, and explain briefly: **(a)** $r < R_1$ (inside the plastic sphere), **(b)** $R_1 < r < R_2$ (in the air gap), **(c)** $R_2 < r < R_3$ (in the metal), **(d)** $r > R_3$ (outside the metal). **(e)** Suppose $-Q_1 = -5$ nC.

What is Q_2? Explain fully on the basis of fundamental principles. **(f)** What can you say about the molecular polarization in the plastic? Explain briefly. Include a drawing if appropriate.

••**P59** A small, thin, hollow spherical glass shell of radius R carries a uniformly distributed positive charge $+Q$. Below it is a horizontal permanent dipole with charges $+q$ and $-q$ separated by a distance s (s is shown greatly enlarged in Figure 15.71 for clarity). The dipole is fixed in position and is not free to rotate. The distance from the center of the glass shell to the center of the dipole is L.

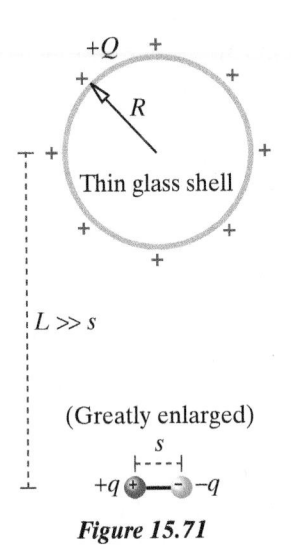

Figure 15.71

(a) Calculate the magnitude and direction of the electric field at the center of the glass shell, and explain briefly, including showing the electric field on the diagram. Your results must not contain any symbols other than the given quantities R, Q, q, s, and L (and fundamental constants), unless you define intermediate results in terms of the given quantities. What simplifying assumption do you have to make? **(b)** If the upper sphere were a solid metal ball with a charge $+Q$, what would be the magnitude and direction of the electric field at its center? Explain briefly. Show the distribution of charges everywhere, and at the center of the ball accurately draw and label the electric field due to the ball charges $\vec{E}_{\text{ball}}$ and the electric field $\vec{E}_{\text{dipole}}$ of the dipole.

••**P60** A very thin glass rod 4 m long is rubbed all over with a silk cloth (Figure 15.72). It gains a uniformly distributed charge $+1.3 \times 10^{-6}$ C. Two small spherical rubber balloons of radius 1.2 cm are rubbed all over with wool. They each gain a uniformly distributed charge of -2×10^{-8} C. The balloons are near the midpoint of the glass rod, with their centers 3 cm from the rod. The balloons are 2 cm apart (4.4 cm between centers).

Length 4 m (drawing not to scale)

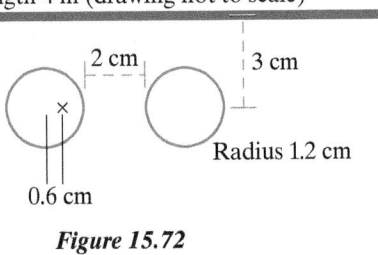

Figure 15.72

(a) Find the magnitude of the electric field at the location marked by the ×, 0.6 cm to the right of the center of the left balloon. Also calculate the angle the electric field makes with the horizontal. Show all your work, including showing vectors. State any approximations you are making. **(b)** Next, a neutral hydrogen atom is placed at that same location (marked by the ×). Draw a diagram showing the effect on the hydrogen atom while it is at that location. The polarizability of atomic hydrogen has been measured to be $\alpha = 7.4 \times 10^{-41}$ C·m/(N/C). What is the distance between the center of the proton and the center of the electron cloud in the hydrogen atom?

••P61 A very long thin wire of length L carries a total charge $+Q$, nearly uniformly distributed along the wire (Figure 15.73). The center of a small neutral metal ball of radius r is located a distance d from the wire, where $d \ll L$. Determine the approximate magnitude of the force that the rod exerts on the ball. Show labeled physics information on the diagram. Explain how you model the situation.

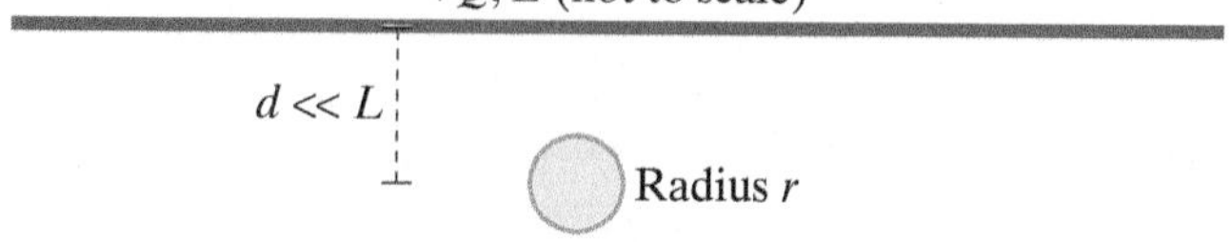

$+Q, L$ (not to scale)

$d \ll L$

Radius r

Figure 15.73

Section 15.7

•••P62 A simplified model of a hydrogen atom is that the electron cloud is a sphere of radius R with uniform charge density and total charge $-e$. (The actual charge density in the ground state is nonuniform.) See Figure 15.74.

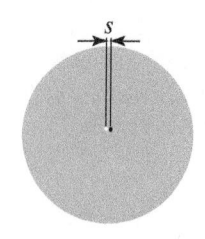

Figure 15.74

(a) For the uniform-density model, calculate the polarizability α of atomic hydrogen in terms of R. Consider the case where the magnitude E of the applied electric field is much smaller than the electric field required to ionize the atom. Suggestions for your analysis: Imagine that the hydrogen atom is inside a capacitor whose uniform field polarizes but does not accelerate the atom. Consider forces on the proton in the equilibrium situation, where the proton is displaced a distance s from the center of the electron cloud ($s \ll R$ in the diagram). **(b)** For a hydrogen atom, R can be taken as roughly 1×10^{-10} m (the Bohr model of the H atom gives $R = 0.5 \times 10^{-10}$ m). Calculate a numerical value for the polarizability α of atomic hydrogen. For comparison, the measured polarizability of a hydrogen atom is $\alpha = 7.4 \times 10^{-41}$ C·m(N/C); see the note below. **(c)** If the magnitude E of the applied electric field is 1×10^6 N/C, use the measured value of α to calculate the shift s shown in Figure 15.74. **(d)** For some purposes it is useful to model an atom as though the nucleus and electron cloud were connected by a spring. Use the measured value of α to calculate the effective spring stiffness k_s for atomic hydrogen. For comparison, measurements of Young's modulus show that the effective spring stiffness of the *interatomic* force in solid aluminum is about 16 N/m. **(e)** If α were twice as large, what would k_s be?

Note: Quantum-mechanical calculations agree with the experimental measurement of α reported in T. M. Miller and B. Bederson, "Atomic and molecular polarizabilities: a review of recent advances," *Advances in Atomic and Molecular Physics* **13**, 1–55, 1977. They use cgs units, so their value is $1/(4\pi\varepsilon_0)$ greater than the value given here.

COMPUTATIONAL PROBLEMS

More detailed and extended versions of some of these computational modeling problems may be found in the lab activities included in the *Matter & Interactions* resources for instructors.

•P63 Start with the program in Section 15.9 that calculates the electric field of a uniformly charged rod at a single location on the midplane. **(a)** Improve the accuracy of the calculation by increasing the number of point charges used to model the rod. Determine experimentally the minimum number of point charges required to get an accurate answer, and explain your criterion for accuracy. **(b)** Use your improved program to calculate the electric field of the rod at location $\langle 0.05, -0.35, 0 \rangle$ m, and display your result with an arrow. Report the value of the electric field you calculated.

•P64 Write a program to model a long thin rod of length 0.5 m with a net charge of -2 nC, lying on the x axis and centered at the origin. **(a)** Calculate the electric field of the rod at

location $\langle -0.4, 0.3, 0.3 \rangle$ m, and display the field with an arrow. **(b)** Determine experimentally the minimum number of point charges required to give an accurate result, and explain how you determined this.

••P65 A strip of invisible tape 12 cm by 2 cm is charged uniformly with a total net charge of 4×10^{-8} and hangs vertically. Write a program to calculate and display both the tape and the electric field due to this strip of tape, at a location a perpendicular distance of 3 cm from the center of the tape. Use the same scale factor to display the arrow representing the electric field in all parts of this problem. **(a)** Divide the strip into three sections, each 4 cm high by 2 cm wide, and use numerical summation to calculate the magnitude of the electric field at a perpendicular distance of 3 cm from the center of the tape. **(b)** Divide the strip into eight sections, each 3 cm high by 1 cm wide, and use numerical summation to calculate the magnitude of the electric field at a perpendicular distance of 3 cm from the center of the tape. Compare the result to your previous result. **(c)** Divide the tape into a large number of small sections. Compare your result to your previous results.

•P66 Start with the program in Section 15.9 that calculates the electric field of a uniformly charged rod, modeled as 10 point charges, at multiple observation locations in a circle around the rod. **(a)** Add more arrows to the list named "observation," so that the electric field of the rod is displayed with 5 rings of arrows centered on the rod, at different locations. **(b)** Increase the number of point charges used to model the rod to increase the accuracy of the calculation. Explain your criteria for determining an appropriate number of point charges.

••P67 Write a computer program to compute and display the electric field of a thin rod of length 2 m, aligned along the z axis and centered at the origin. A total charge of -60 nC is uniformly distributed along the surface of the rod. Start by modeling the rod as a collection of 5 point charges. **(a)** Increase the number of point charges until you are satisfied with the accuracy of the calculation. Explain the criteria you used to make this decision. **(b)** Display the electric field at locations around the rod with arrows arranged in 6 rings centered on the rod, at different z locations along the rod. Scale the arrows appropriately to fit the display.

•••P68 Four thin rods, each of length 0.4 m, form a square loop in the yz plane, as shown in yellow in Figure 15.75. A charge of 3 nC is uniformly distributed over the surface of each rod (the net charge of the loop is 12 nC). Write a program to calculate and display the electric field of the loop at 12 observation locations: six that lie on a vertical line that crosses the x axis at $x = -0.1$ m, and six that lie on a vertical line that crosses the x axis at $x = +0.1$ m. These lines are shown in cyan in Figure 15.75.

•••P69 Model a uniformly charged ring as a set of point charges placed on a circle. Place the ring in the yz plane so that its axis is along the x axis. Give the ring a total charge of $Q = 50 \times 10^{-9}$ C and a radius of $R = 0.1$ m. **(a)** Start an electron from rest at location $\langle 0.15, 0, 0 \rangle$ m, and model its motion. In each step you will need to calculate the net electric field at the location of the electron. Describe the motion you observe. **(b)** Vary `deltat` and the number of point charges used to model the uniformly charged ring to make sure your calculation is accurate. **(c)** Experiment with different initial locations that do not lie on the axis of the ring. You may also wish to experiment with giving the electron some initial momentum. Make a screen shot of the most interesting trajectory you find. Figure 15.76 is an example of one such trajectory.

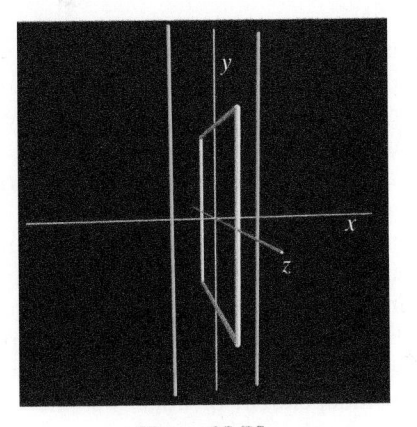

Figure 15.75

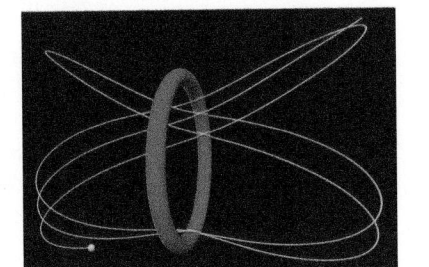

Figure 15.76

ANSWERS TO CHECKPOINTS

1 1.12×10^4 N/C

2 (a) 4.51×10^4 N/C to the right; **(b)** 2.25×10^{-4} N to the left; **(c)** zero

3 2.684×10^6 N/C; 2.657×10^6 N/C

4 About 2.1×10^{-5} C; about 3000 N/C

5 Although the other charges are much farther away, there are a lot more of them, and in a sphere the two effects exactly cancel each other. The electric field at location D is zero.

Electric Potential

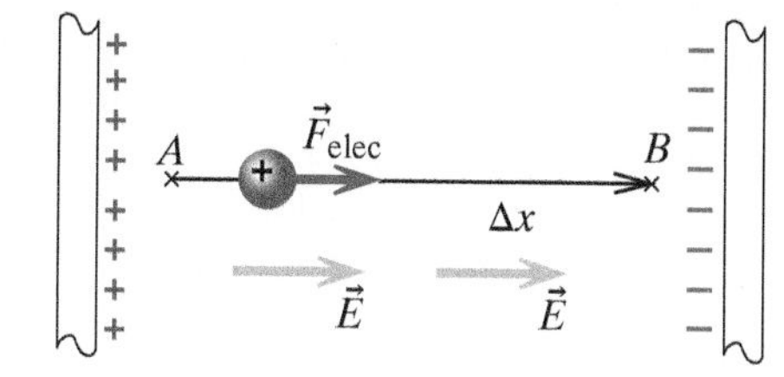

OBJECTIVES

After studying this chapter you should be able to

- Mathematically relate electric field, electric potential energy, and electric potential
- Calculate differences in electric potential in and around conductors and insulators

In analyzing mechanical systems, both at the macroscopic and the microscopic level, we found that it was frequently important to consider not only forces and momenta, but also work and energy, in trying to model the behavior of a physical system. Similarly, to complement our use of the concept of electric field and electric force, we need the concept of electric potential. Electric potential is defined as electric potential energy per unit charge.

The concept of electric potential is useful for some of the same reasons that the concept of electric field is useful. It allows us to reason about energy in a range of situations without having to worry about the details of some particular distribution of point charges. Electric potential has practical importance, in part because batteries and electric generators maintain a potential difference across themselves that is nearly independent of what is connected to them. The concept also provides significant theoretical power, enabling us to draw conclusions about a surprising range of issues, including, for example, what patterns of electric field in space are possible, and the magnitude and direction of the average electric field inside an insulator due to the polarization of molecules in the insulator.

16.1 A REVIEW OF POTENTIAL ENERGY

This section is a review of the concept of potential energy, which was first introduced in Chapter 6. If this concept is very familiar to you, you may wish to skip ahead to Section 16.2, in which we apply the concept to systems of interacting charged particles similar to those that we will typically consider in this volume.

Energy of a Single Particle

The energy of a single particle with charge q_1 (Figure 16.1) consists solely of its particle energy. The particle energy is the sum of the rest energy mc^2 of the particle and its kinetic energy K.

$$\text{Particle energy} = mc^2 + K$$

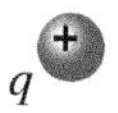

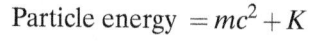

Figure 16.1 No electric potential energy is associated with a single charged particle.

q

Most of the processes we will study in this volume are low-energy processes that do not involve significant changes in rest energy. (In contrast, processes such as fission and fusion, in which atomic nuclei combine or split apart to form different elements, involve the release of very large amounts of energy. These are the subject of some of the applications in Chapter 6.)

Kinetic energy is energy associated with motion. The kinetic energy of a particle moving at low speed may be approximated as

APPROXIMATE KINETIC ENERGY OF A PARTICLE

$$K \approx \frac{1}{2}mv^2 \quad \text{for } v \ll c$$

Recall that energy is a scalar quantity, not a vector, so it depends only on the particle's speed but not the direction of its motion. (For a discussion of the relativistically correct form for particle energy and kinetic energy, see Chapter 6.)

A Reminder about the Symbol Δ

In this chapter we will be discussing changes in energy. Remember that the symbol Δ (uppercase Greek delta) indicates a change in a quantity, calculated by subtracting the initial value of the quantity from its final value.

$$\Delta K = K_{\text{final}} - K_{\text{initial}}$$

EXAMPLE   **Accelerated Electron**

An electron is traveling at a speed of 6×10^4 m/s. After passing through a region in which there is an electric field, its speed is 2×10^4 m/s. (Note that these speeds are small compared to the speed of light.) What is the change in kinetic energy of the electron?

Solution

$$K_{\text{initial}} \approx \frac{1}{2}m_e v_i^2 = \frac{1}{2}(9 \times 10^{-31}\text{ kg})(6 \times 10^4\text{ m/s})^2 = 1.6 \times 10^{-21}\text{ J}$$

$$K_{\text{final}} \approx \frac{1}{2}m_e v_f^2 = \frac{1}{2}(9 \times 10^{-31}\text{ kg})(2 \times 10^4\text{ m/s})^2 = 1.8 \times 10^{-22}\text{ J}$$

$$\Delta K = 1.8 \times 10^{-22}\text{ J} - 1.6 \times 10^{-21}\text{ J} = -1.42 \times 10^{-21}\text{ J}$$

Since the electric force slowed down the electron, it makes sense that $K_{\text{final}} < K_{\text{initial}}$, the electron's kinetic energy decreased, and $\Delta K < 0$.

Checkpoint 1 A proton initially travels at a speed of 3000 m/s. After it passes through a region in which there is an electric field, the proton's speed is 5000 m/s. **(a)** What is the initial kinetic energy of the proton? **(b)** What is the final kinetic energy of the proton? **(c)** What is the change in kinetic energy of the proton?

A Single Particle Does Not Have Potential Energy

When analyzing changes in energy, we must choose a *system* consisting of one or more objects to analyze. In choosing the system, we must decide what objects to include in the system; all objects not included in the system are part of the surroundings. We are free to choose whatever system we find convenient or interesting—the laws of physics apply to any choice of system. However, the choice of system determines the forms of energy we need to consider.

One possible choice of system is to choose a single particle as the system and classify all other particles as part of the surroundings. In this case, the system can have only the two kinds of energy mentioned above: rest energy and kinetic energy. Potential energy is associated with interactions between particles in a system. A single particle does not have potential energy.

The kinetic energy of a single particle can be changed if positive or negative work is done on the particle by its surroundings.

CHANGE IN ENERGY OF A SINGLE-PARTICLE SYSTEM

$$\Delta K_{\text{sys}} = W_{\text{surr}}$$

Work

As discussed in Chapter 6, work is a scalar quantity, with units of joules. The work done on a particle is simply the product of the force exerted on the particle times the displacement through which the force acts. Because force is a vector and displacement is a vector, work is a sum:

$$W = F_x \Delta x + F_y \Delta y + F_z \Delta z$$

This sum can be written in a compact form as the "dot product" of two vectors.

$$\vec{A} \cdot \vec{B} = (A_x B_x + A_y B_y + A_z B_z) \quad \text{(a scalar quantity)}$$

so work can be written as

$$W = \vec{F} \cdot \Delta \vec{l} = \langle F_x, F_y, F_z \rangle \cdot \langle \Delta x, \Delta y, \Delta z \rangle = F_x \Delta x + F_y \Delta y + F_z \Delta z$$

where $\langle \Delta x, \Delta y, \Delta z \rangle = \Delta \vec{l}$ denotes the displacement of the particle.

It is occasionally useful to know another, equivalent, evaluation equation for a vector dot product:

$$\vec{A} \cdot \vec{B} = |\vec{A}||\vec{B}| \cos \theta$$

where θ is the angle between the two vectors.

EXAMPLE **Work Done by Electric Force**

A dust particle with charge 2×10^{-11} C moves from $\langle 0.1, -0.3, 0.4 \rangle$ m to $\langle 0.2, -0.3, -0.2 \rangle$ m. In this region there is an electric field of $\langle 2000, 0, 4000 \rangle$ N/C. How much work is done on the dust particle by the electric force?

Solution
$$\Delta \vec{l} = \langle 0.2, -0.3, -0.2 \rangle \text{ m} - \langle 0.1, -0.3, 0.4 \rangle \text{ m} = \langle 0.1, 0, -0.6 \rangle \text{ m}$$

$$W = [(2 \times 10^{-11})(2000)(0.1) + (2 \times 10^{-11})(0)(0) + (2 \times 10^{-11})(4000)(-0.6)] \text{ J}$$

$$= -4.4 \times 10^{-8} \text{ J}$$

The work done on the particle was negative, so the kinetic energy of the particle decreased.

Potential Energy Is Associated with Pairs of Interacting Objects

A different choice is to include two or more particles in the system. In a system consisting of two or more interacting particles, there can be energy associated with the interactions of particles inside the system. If the rest energy of the particles does not change (that is, if the identity of the particles does not change), then the kinetic energy of the system can be changed either by work done by forces exerted by objects in the surroundings (W_{surr}) or by work done by forces due to interactions between particles inside the system (internal work, W_{int}). (In general, there can also be other kinds of transfer of energy into the system, including transfers due to a temperature difference between the system and the surroundings (Q). In this volume, we will deal primarily with situations in which there are no energy transfers other than mechanical work W, and we will omit these other terms from the Energy Principle equation.)

$$\Delta K_{sys} = W_{surr} + W_{int}$$

We can rearrange this equation to put all terms associated only with particles inside the system on the left side of the equation:

Energy change in system = external energy input

$$\Delta K_{sys} + (-W_{int}) = W_{surr}$$

The term $-W_{int}$, the negative of the work done by internal forces, is called the change in the *potential energy U* of the system of particles:

DEFINITION: CHANGE IN POTENTIAL ENERGY OF A SYSTEM

$$\Delta U = -W_{int}$$

Therefore, the energy principle for a multiparticle system in which the particles do not change identity can be written:

ENERGY PRINCIPLE FOR A MULTIPARTICLE SYSTEM

$$\Delta K_{sys} + \Delta U = W_{surr}$$

for a situation in which $\Delta(mc^2) = 0$ and other energy transfers such as Q are zero.

The preceding equation, expressed in words, says that for a system of two or more particles, the change in kinetic energy of the system plus the change in the potential energy of the system is equal to the amount of work done on the system by external forces, assuming that there are no other (nonwork) kinds of energy transfers into the system.

16.2 SYSTEMS OF CHARGED OBJECTS

In most cases of interest in this volume, we will choose a system that includes all interacting charged objects. In some situations we will need to consider the effect of an external force, since a force may be required to move a charged particle in a direction opposite to the electric force acting on it. However, in many simple cases, we can neglect external forces. Although the Earth exerts a gravitational force on moving charged particles, this gravitational force is typically very small compared to the electric forces acting on the particles, and we can usually neglect it. In many cases, therefore, the energy principle for the systems in which we are interested will simplify to this:

$$\Delta K_{sys} + \Delta U = 0 \quad \text{(zero work done by surroundings)}$$

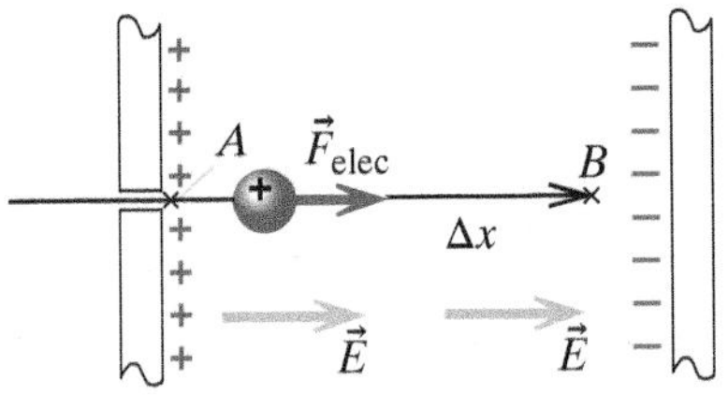

Decrease in U, Increase in K

As an example, consider a system consisting of two uniformly charged plates (a capacitor) and a single proton, as shown in Figure 16.2. Suppose that the

Figure 16.2 A proton traveling to the right enters a capacitor through a tiny hole in the left plate. Inside the capacitor it travels from location A to location B, in a region of nearly uniform electric field. The electric field outside the capacitor is negligible.

proton is initially traveling to the right, and enters the capacitor through a small hole in the left plate. While the proton is inside the capacitor, an electric force (an internal force, due to the electric field inside the capacitor) acts on it, doing work on the proton.

The energy principle, applied to this system, is

$$\Delta K_{proton} + \Delta U_{electric} = 0$$

QUESTION Will the kinetic energy of the proton increase or decrease?

Because the force on the proton acts in the *same* direction as the displacement of the proton, the kinetic energy of the proton increases. When it reaches location B near the negative plate, the kinetic energy of the proton–capacitor system will have increased by an amount equal to the work done on the proton by the electric field inside the capacitor, which is due to all the charged particles on the plates of the capacitor. The speed of the proton will be greater.

QUESTION Will the electric potential energy of the system increase or decrease?

Since the kinetic energy will increase, the electric potential energy must decrease.

QUESTION How do we calculate the change in the electric potential energy of this system?

The change in the electric potential energy of this system is equal to the negative of the internal work: in this case, the work done on the proton by the uniform electric field inside the capacitor.

$$\Delta U_{electric} = -W_{int} = -(F_x \Delta x + F_y \Delta y + F_z \Delta z)$$
$$= -(eE_x \Delta x + eE_y \Delta y + eE_z \Delta z)$$

In this case, since $\vec{E} = (E_x, 0, 0)$ and $\Delta\vec{l} = \langle \Delta x, 0, 0 \rangle$,

$$\Delta U_{electric} = -(eE_x \Delta x + 0 + 0)$$
$$= -eE_x \Delta x$$

From Figure 16.2 we can see that $\Delta x = x_f - x_i$ is a positive quantity, and E_x is also positive, so the change in the potential energy of the system $-eE_x \Delta x$ is negative, as predicted. For this system and this process, we find that

$$\Delta K = -\Delta U_{electric} > 0$$

and the kinetic energy of the proton increases.

EXAMPLE **ΔU and ΔK**

For concreteness, find ΔU and ΔK if the magnitude of the electric field inside the capacitor is 2×10^3 N/C, and the distance traveled by the proton is 4 mm.

Solution Since for the proton both the displacement and the force are in the x direction:

$$\Delta U_{electric} = -(1.6 \times 10^{-19} C)(2 \times 10^3 \, N/C)(0.004 \, m) = -1.3 \times 10^{-18} \, J$$
$$\Delta K = -\Delta U_{electric} = +1.3 \times 10^{-18} \, J$$

Since the potential energy of the system is less in the final state than it was initially, the kinetic energy of the system must have increased; the proton is moving faster when it reaches location B, as we would expect.

Increase in U, Decrease in K

Suppose that instead of a proton, an electron is initially traveling to the right, and enters the capacitor through the tiny hole in the left plate. The electron moves to location B, as shown in Figure 16.3.

QUESTION Will the kinetic energy of the electron increase or decrease?

Because the electric force acts in a direction *opposite* to the motion of the electron, the kinetic energy of the electron will decrease as it moves toward location B. The speed of the electron will be lower at location B.

QUESTION Will the electric potential energy of the system increase or decrease?

As in the previous example, there are no significant energy inputs to the system from external sources. Therefore, applying the energy principle to this system and this process:

$$\Delta K_{\text{electron}} + \Delta U_{\text{electric}} = 0$$

Since kinetic energy will decrease, electric potential energy must increase.

QUESTION What is the change in the electric potential energy of this system?

The change in the electric potential energy of this system is equal to the negative of the internal work: in this case, the work done on the electron by the uniform electric field inside the capacitor. In the simple situation illustrated in Figure 16.3, the only nonzero component of the electric field is the x component, so

$$\Delta U_{\text{electric}} = -(F_x \Delta x + 0 + 0)$$
$$= -((-e)E_x \Delta x + 0 + 0)$$
$$= e E_x \Delta x$$

From Figure 16.3 we can see that $\Delta x = x_f - x_i$ is a positive quantity, and E_x is also positive, so the change in the potential energy of the system $e E_x \Delta x$ is positive. Since the electric force on the electron (due to the field inside the capacitor) is in a direction *opposite* to the displacement of the electron, the work done by this internal force is negative, and the change in potential energy is positive.

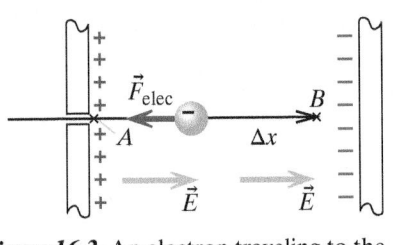

Figure 16.3 An electron traveling to the right enters a capacitor through a tiny hole in the left plate. Inside the capacitor it travels from location A to location B, in a region of nearly uniform electric field. The electric field outside the capacitor is negligible.

EXAMPLE **ΔU and ΔK**

As before, suppose that the magnitude of the electric field inside the capacitor is 2×10^3 N/C; the distance traveled by the electron is 4 mm.

Solution For the electron the direction of the force is opposite to the displacement:

$$\Delta U_{\text{electric}} = -(-1.6 \times 10^{-19} \text{C})(2 \times 10^3 \text{ N/C})(0.004 \text{ m}) = +1.3 \times 10^{-18} \text{ J}$$
$$\Delta K = -\Delta U_{\text{electric}} = -1.3 \times 10^{-18} \text{ J}$$

The kinetic energy of the electron decreases, and the potential energy of the system increases.

EXAMPLE **Proton Moving in a Uniform Electric Field**

A proton moves from location A to location B in a region of uniform electric field, as shown in Figure 16.4. If the magnitude of the electric field inside the capacitor shown is 190 N/C and the distance from location A to location B is 1.5 cm, what is the change in electric potential energy of the system during this process? What is the change in kinetic energy of the proton during this process? If the proton is initially at rest, what is its speed when it reaches location B?

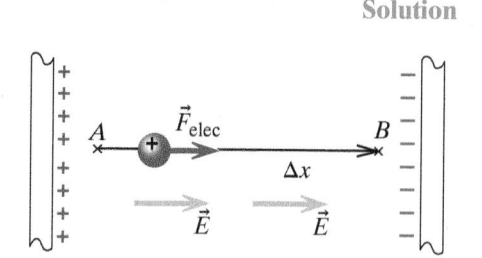

Figure 16.4 A proton moves from location A to location B through a region of uniform electric field.

Solution   Initial location: A
Final location: B

Change in electric potential energy:

$$\Delta U = -W_{int} = -eE_x \Delta x$$
$$= -(1.6 \times 10^{-19} \text{C})(190 \text{N/C})(0.015 \text{ m})$$
$$= -4.6 \times 10^{-19} \text{ J}$$

Change in kinetic energy of proton:

$$\Delta K + \Delta U = 0$$
$$\Delta K = -\Delta U = +4.6 \times 10^{-19} \text{ J}$$

Final speed of proton:

$$K_f - 0 = \frac{1}{2} m v_f^2 - 0$$
$$v_f = \sqrt{\frac{2K_f}{m}} = \sqrt{\frac{2(4.6 \times 10^{-19} \text{ J})}{(1.7 \times 10^{-27} \text{ kg})}} = 2.3 \times 10^4 \frac{\text{m}}{\text{s}}$$

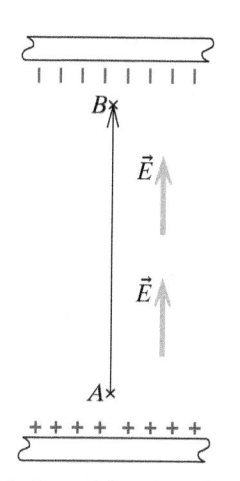

Figure 16.5 A particle moves from location A to location B in a region in which there is a uniform electric field.

Checkpoint 2   A proton moves from location A to location B in a region of uniform electric field, as shown in Figure 16.5. **(a)** If the magnitude of the electric field inside the capacitor in Figure 16.5 is 3500 N/C, and the distance between location A and location B is 3 mm, what is the change in electric potential energy of the system (proton + plates) during this process? **(b)** What is the change in the kinetic energy of the proton during this process? **(c)** If the proton is initially at rest, what is its speed when it reaches location B? **(d)** How do the answers to (a) and (b) change if the proton is replaced by an electron?

16.3   POTENTIAL DIFFERENCE IN A UNIFORM FIELD

When working with electric forces, we found it quite useful to be able to calculate the electric field at a particular location without having to worry about what kind of charged particle might come along to be affected by this field. When we know the electric field $\vec{E}_1$ at a location, to find the electric force on whatever particle happens to be at that location, we simply multiply $\vec{E}_1$ by the charge of the particle, q_2:

$$\vec{F}_2 = q_2 \vec{E}_1$$

Likewise, when calculating changes in potential energy of a system, it would be useful to find a quantity that is independent of the charge of the particle that might move through a region. Then we could simply multiply this quantity by the charge of whatever particle happened to move through the region, to get the change in potential energy of the system. We would like to be able to write:

$$\Delta U = q \text{ (something)}$$

In the two cases discussed in the previous section (proton or electron moving in a uniform field), many things about the system were the same: the capacitor, the electric field in the region, and the path taken by the particle between the initial location A and final location B. Let's rewrite the change in electric potential energy in each case as the product of the charge of the displaced particle and another quantity:

If the particle is a proton: $\qquad \Delta U_{\text{electric}} = -eE_x\Delta x = (+e)(-E_x\Delta x)$

If the particle is an electron: $\qquad \Delta U_{\text{electric}} = +eE_x\Delta x = (-e)(-E_x\Delta x)$

The quantity that is the same in both cases is $(-E_x\Delta x)$. This quantity is called the "difference of the electric potential between locations A and B," and is given the symbol ΔV.

POTENTIAL DIFFERENCE

$$\Delta U_{\text{electric}} = q\Delta V \quad \text{and therefore} \quad \Delta V = \frac{\Delta U_{\text{electric}}}{q}$$

Like energy, electric potential is a scalar quantity. The dimensions of a difference in electric potential ΔV are joules/coulomb, or "difference of energy per unit charge." This is such an important quantity that it has its own name:

$$1\frac{\text{joule}}{\text{coulomb}} = 1 \text{ volt}$$

Units of Field and of Energy

Note that since $\Delta V = -(E_x\Delta x + E_y\Delta y + E_z\Delta z)$, and the units of ΔV are volts, then the units of electric field must be volts per meter. This is in fact equivalent to newtons per coulomb, and can be used interchangeably.

We can also now understand the origin of the energy unit "electron volt," or eV. One eV is equal to 1.6×10^{-19} J. If an electron moves through a potential difference of one volt, there is a change in the electric potential energy whose magnitude is

$$|\Delta U| = (e)(1 \text{ volt}) = (1.6 \times 10^{-19}\text{C})(1\,\text{J/C}) = 1.6 \times 10^{-19} \text{ J}$$

The electron volt is a convenient unit for measuring energies of atomic processes in physics and chemistry. Note that an electron volt is a unit of energy, not potential (potential is measured in volts, not electron volts).

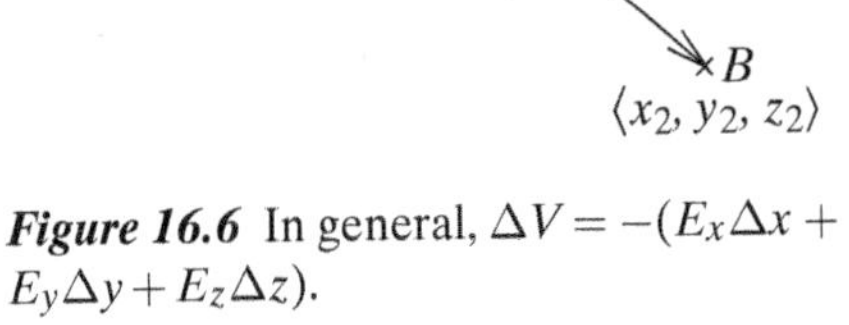

Path Not Parallel to Electric Field

The path between two locations does not have to be parallel to the electric field, as shown in Figure 16.6. The general three-dimensional expression for change in electric potential in a uniform field is:

$$\Delta V = -(E_x\Delta x + E_y\Delta y + E_z\Delta z)$$

Note that this quantity can be positive or negative, since each component of the electric field can be positive or negative, and each component of the displacement can also be positive or negative.

The quantity $(E_x\Delta x + E_y\Delta y + E_z\Delta z)$ can be written as the dot product of the electric field and the displacement vector $\Delta\vec{l} = \langle \Delta x, \Delta y, \Delta z \rangle$:

POTENTIAL DIFFERENCE IN A UNIFORM FIELD

$$\Delta V = -(E_x\Delta x + E_y\Delta y + E_z\Delta z)$$

or, using dot product notation:

$$\Delta V = -\vec{E} \bullet \Delta\vec{l} = -\langle E_x, E_y, E_z \rangle \bullet \langle \Delta x, \Delta y, \Delta z \rangle$$

$\langle x_1, y_1, z_1 \rangle$

A

$\vec{E}$

$\Delta\vec{l} = \langle \Delta x, \Delta y, \Delta z \rangle$

$\vec{E}$

B

$\langle x_2, y_2, z_2 \rangle$

Figure 16.6 In general, $\Delta V = -(E_x\Delta x + E_y\Delta y + E_z\Delta z)$.

EXAMPLE   **Potential and Potential Energy**

Figure 16.7 shows two locations, A and B, in a region of uniform electric field. For a path starting at A and going to B, calculate the following quantities: **(a)** the difference in electric potential, **(b)** the potential energy change for the system when a proton moves along this path, and **(c)** the potential energy change for the system when an electron moves along this path.

Solution

$\langle -0.4, 0, 0 \rangle$ m $\langle 0.2, 0, 0 \rangle$ m

A× B×

$\vec{E} = \langle 500, 0, 0 \rangle$ N/C

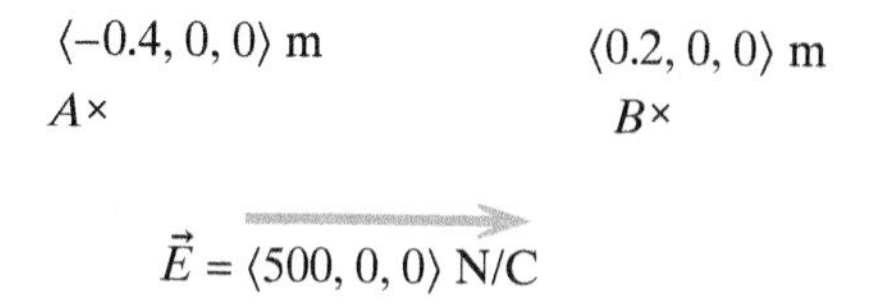

Figure 16.7 A region of uniform electric field.

(a) Potential difference:
Initial location: A
Final location: B

$$\Delta\vec{l} = \langle 0.2,0,0 \rangle \text{ m} - \langle -0.4,0,0 \rangle \text{ m} = \langle 0.6,0,0 \rangle \text{ m}$$
$$\Delta V = -(E_x\Delta x + E_y\Delta y + E_z\Delta z)$$
$$= -\left(500\frac{\text{N}}{\text{C}} \cdot 0.6 \text{ m} + 0\cdot0 + 0\cdot0 \right)$$
$$= -300 \text{ V}$$

(b) For a proton:

$$\Delta U = (+e)(\Delta V) = (1.6\times10^{-19}\text{C})(-300\text{V}) = -4.8\times10^{-17}\text{ J}$$

(c) For an electron:

$$\Delta U = (-e)(\Delta V) = (-1.6\times10^{-19}\text{C})(-300\text{V}) = 4.8\times10^{-17}\text{ J}$$

EXAMPLE   **Field and Potential**

Suppose that in a certain region of space (Figure 16.8) there is a nearly uniform electric field of magnitude 100 N/C (that is, the electric field is the same in magnitude and direction throughout this region). If you move 2 m at an angle of 30° to this electric field, what is the change in potential?

Solution

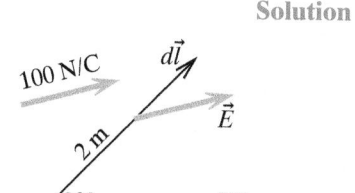

100 N/C $d\vec{l}$

$\vec{E}$

2 m

30°

Figure 16.8 Moving at a 30° angle to the electric field.

$$\Delta V \approx -\vec{E}\cdot\Delta\vec{l} = -E\Delta l\cos\theta$$
$$\Delta V = -(100\,\text{N/C})(2\,\text{m})\cos(30°) = -173\text{ V}$$

Checkpoint 3 For a path starting at B and going to A (Figure 16.9), calculate **(a)** the change in electric potential, **(b)** the potential energy change for the system when a proton moves from B to A, and **(c)** the potential energy change for the system when an electron moves from B to A. For a path starting at B and going to C, calculate **(d)** the change in electric potential, **(e)** the potential energy change for the system when a proton moves from B to C, and **(f)** the potential energy change for the system when an electron moves from B to C.

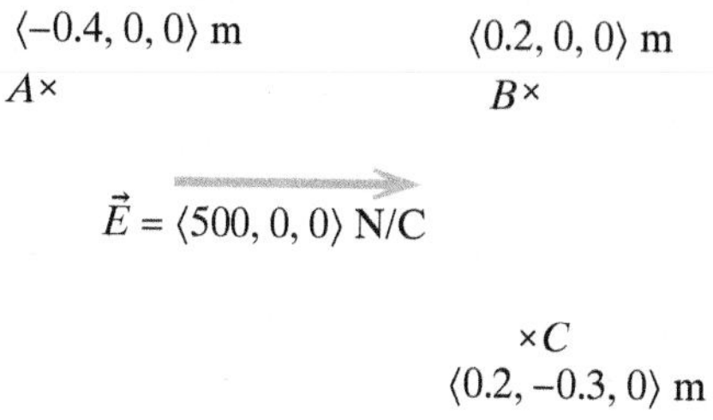

$\langle -0.4, 0, 0 \rangle$ m $\langle 0.2, 0, 0 \rangle$ m

A× B×

$\vec{E} = \langle 500, 0, 0 \rangle$ N/C

×C
$\langle 0.2, -0.3, 0 \rangle$ m

Figure 16.9 A region of uniform electric field (Checkpoint 3).

Calculating Field from Potential Difference

We have just seen how to calculate potential difference ΔV from the electric field $\vec{E}$. Conversely, we can calculate electric field from potential difference. Moving in the x direction, we have $\Delta V = -E_x\Delta x$, or

$$E_x = -\frac{\Delta V}{\Delta x}$$

EXAMPLE   **Calculating Field from Potential Difference**

A capacitor has large plates that are 1 mm apart, and the potential difference from one plate to the other is 50 V. What is the magnitude of the electric field between the plates?

Solution   Choose the x axis to be perpendicular to the plates. Then

$$|E| = \left| -\frac{50\ \text{V}}{1 \times 10^{-3}\ \text{m}} \right| = 5 \times 10^4\ \text{V/m}$$

Gradient of Potential

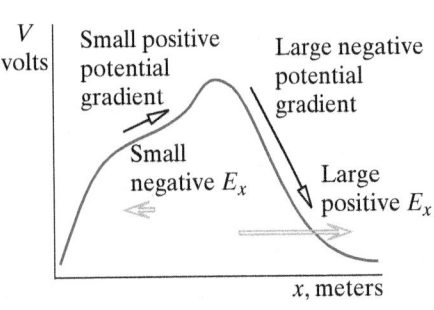

Figure 16.10 The negative slope of potential vs. x is the x component of the electric field.

In the limit of very short distance, E_x is given by a derivative:

$$E_x = -\lim_{\Delta x \to 0} \frac{\Delta V}{\Delta x} = -\frac{dV}{dx}$$

This is called the "gradient" of the potential, by analogy with the gradient of a hill, where a steep hill might have a "15% grade." Figure 16.10 emphasizes that a large gradient of potential (rapid change of potential with position) corresponds to a large electric field.

ELECTRIC FIELD IS THE NEGATIVE GRADIENT OF THE POTENTIAL

$$E_x = -\frac{dV}{dx}, \qquad E_y = -\frac{dV}{dy}, \qquad E_z = -\frac{dV}{dz}$$

More technically, the gradient should be written using partial derivatives, like this:

$$E_x = -\frac{\partial V}{\partial x}, \qquad E_y = -\frac{\partial V}{\partial y}, \qquad E_z = -\frac{\partial V}{\partial z}$$

The meaning of $\partial V/\partial x$ is simply "take the derivative of V with respect to x while holding y and z constant." The fact that $E_x = -\partial V/\partial x$ is closely related to the fact that (as we saw in Chapter 6 on energy) $F_x = -\partial U/\partial x$, because $E = F/q$ and $V = U/q$.

The gradient is often written in the form $\vec{E} = -\vec{\nabla} V$, where the inverted delta (nabla) represents taking the x, y, and z partial derivatives of the potential to find the x, y, and z components of the electric field.

Checkpoint 4 Suppose that the potential difference in going from location $\langle 2.00, 3.50, 4.00 \rangle$ m to location $\langle 2.00, 3.52, 4.00 \rangle$ m is 3 V. What is the approximate value of E_y in this region? Include the appropriate sign.

16.4 SIGN OF POTENTIAL DIFFERENCE

The potential difference ΔV can be positive or negative, and the sign is extremely important, since the sign determines whether a particular charged particle will gain or lose energy in moving from one place to another. An increase in potential energy $q\Delta V$ is associated with a decrease in kinetic energy, whereas a decrease in potential energy $q\Delta V$ is associated with an increase in kinetic energy.

Direction of Path Same as Direction of Electric Field: ΔV Is Negative

If the direction of the path from initial location to final location is the same as the direction of the electric field, the potential difference is negative. Mathematically, this is a result of the minus sign in the definition of potential difference.

In Figure 16.11, the path starting at A and ending at B travels in the same direction as the electric field in the region, so

$$\Delta x = (0.6 - 0.4)\ \text{m} = 0.2\ \text{m}$$

$$\Delta V = -E_x \Delta x = -\left(200\,\frac{\text{V}}{\text{m}}\right)(0.2\ \text{m}) = -40\ \text{V}$$

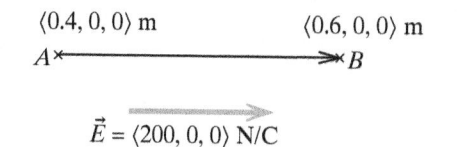

$(0.4, 0, 0)$ m $\qquad\qquad (0.6, 0, 0)$ m

A ⟶ B

$\vec{E} = \langle 200, 0, 0 \rangle$ N/C

Figure 16.11 Direction of path same as direction of electric field.

Direction of Path Opposite to Direction of Electric Field: ΔV Is Positive

On the other hand, in Figure 16.12, the path starting at B and ending at A travels opposite to the electric field in the region, and

$$\Delta V = -E_x \Delta x = -\left(200\,\frac{\text{V}}{\text{m}}\right)(-0.2\ \text{m}) = +40\ \text{V}$$

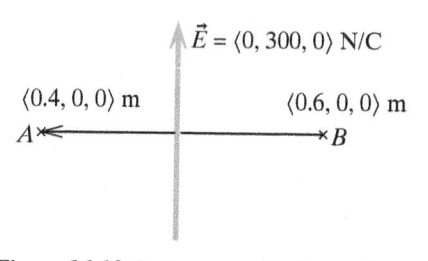

$(0.4, 0, 0)$ m $\qquad\qquad (0.6, 0, 0)$ m

A ⟵ B

$\vec{E} = \langle 200, 0, 0 \rangle$ N/C

Figure 16.12 Direction of path opposite to direction of electric field.

Path Perpendicular to the Electric Field

QUESTION What if a path is perpendicular to the electric field?

Consider the situation in Figure 16.13, where the path starting at B and ending at A is perpendicular to the electric field. Then

$$\Delta V = -(E_x \Delta x + E_y \Delta y) = -\left(\left(0\,\frac{\text{V}}{\text{m}}\right)(-0.2\ \text{m}) + \left(300\,\frac{\text{V}}{\text{m}}\right)(0\ \text{m})\right) = 0\ \text{V}$$

$\vec{E} = \langle 0, 300, 0 \rangle$ N/C

$(0.4, 0, 0)$ m $\qquad\qquad (0.6, 0, 0)$ m

A ⟵ B

Figure 16.13 Path perpendicular to the direction of the electric field.

Along a path whose direction is perpendicular to the electric field in the region, the potential does not change! Clearly, if a charged particle traveled along this path, an external force would be required to counteract the electric force. However, neither the external force nor the electric force would do any work on the particle, because the displacement in the direction of these forces would be zero. Thus, the kinetic energy of the particle would not change, and the potential energy of the system would not change.

Summary: Direction of the Path Relative to the Electric Field

It is easy to make sign errors in formal calculations using potential, and we need an independent check on the sign. The following rules are extremely useful:

SIGN OF ΔV AND DIRECTION OF PATH RELATIVE TO $\vec{E}$

Path going in the direction of $\vec{E}$: Potential is decreasing $(\Delta V < 0)$

Path going opposite to $\vec{E}$: Potential is increasing $(\Delta V > 0)$

Path perpendicular to $\vec{E}$: Potential does not change $(\Delta V = 0)$

Indicating the Path Direction

To show the direction of the path, we use the following notation. As usual, note that the symbol Δ indicates "final − initial."

$\Delta V = V_B - V_A$ indicates a path starting at A and ending at B.
$\Delta V = V_A - V_B$ indicates a path starting at B and ending at A.

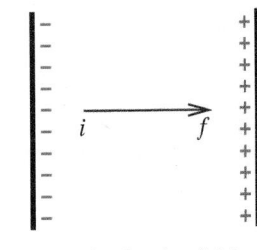

Figure 16.14 Checkpoint 5 (a).

Figure 16.15 Checkpoint 5(b).

(We will see later that the symbol V_A actually has a meaning, and denotes the "potential at location A," which is defined as the potential difference along a path starting infinitely far away and traveling to location A.)

Because you can determine the sign of a potential difference on purely physical grounds, you should never get the wrong sign! In the following exercises you are asked to give the correct sign for the potential difference between two locations. Use the rule given above.

Checkpoint 5 **(a)** In Figure 16.14, what is the direction of the electric field? Is $\Delta V = V_f - V_i$ positive or negative? **(b)** In Figure 16.15, what is the direction of the electric field? Is $\Delta V = V_f - V_i$ positive or negative?

16.5 POTENTIAL DIFFERENCE IN A NONUNIFORM FIELD

Two Adjacent Regions with Different Fields

In the previous sections we considered only situations in which the electric field is uniform within a region. However, often the real world is more complex than this. We need to be able to calculate changes in electric potential and electric potential energy in regions of space in which the electric field is not uniform.

Consider a situation in which the path from the initial location to the final location of interest passes through two regions of uniform but different electric field. In region 1 the electric field is $\vec{E}_1$, and in region 2 the electric field is $\vec{E}_2$, as shown in Figure 16.16. What would be the change in electric potential energy of a particle traveling along a path from location A to location B? (Assume that there is a tiny hole in the middle plate, so a particle can pass through it.)

In order to calculate ΔV in this situation, we need to divide the path into two pieces. Each displacement vector $\Delta \vec{l}$ must be small enough that the electric field is uniform in the region through which it passes. Essentially, we have to invent a new point, which we can call C, on the boundary between the two regions, as shown in Figure 16.17.

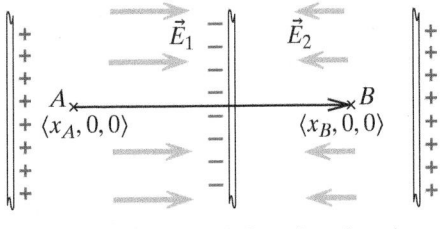

Figure 16.16 The path from location A to location B passes through two regions in which the field is different.

Now we can calculate the difference in potential along the two segments of the path:

From location A to location C: $\Delta V_1 = -E_{1x}(x_C - x_A)$
From location C to location B: $\Delta V_2 = -E_{2x}(x_B - x_C)$
From location A to location B:

$$\Delta V = \Delta V_1 + \Delta V_2 = -E_{1x}(x_C - x_A) - E_{2x}(x_B - x_C)$$

In general, we need to add up all components (x, y, and z), so we can write the general equation as a sum:

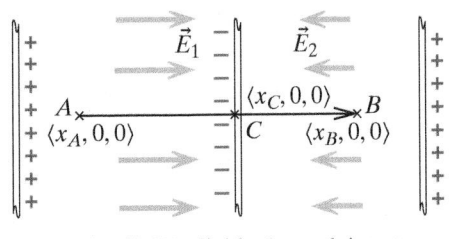

Figure 16.17 We divide the path into two segments by inventing a point C. Each segment of the path now passes through a region of uniform electric field.

POTENTIAL DIFFERENCE ACROSS SEVERAL REGIONS

$$\Delta V = \sum_i [(-E_{i,x}\Delta x_i) + (-E_{i,y}\Delta y_i) + (-E_{i,z}\Delta z_i)] = -\sum \vec{E} \bullet \Delta \vec{l}$$

EXAMPLE **Potential Difference Across Two Regions**

Suppose that from $x = 0$ to $x = 3$ the electric field is uniform and given by $\vec{E} = \langle 400, 0, 0 \rangle$ N/C, and that from $x = 3$ to $x = 5$ the electric field is uniform and given by $\vec{E} = \langle 1000, 500, 0 \rangle$ N/C (see Figure 16.18). What is the potential difference $\Delta V = V_C - V_A$?

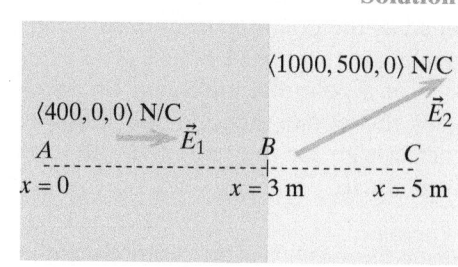

Figure 16.18 The electric field differs in the two regions.

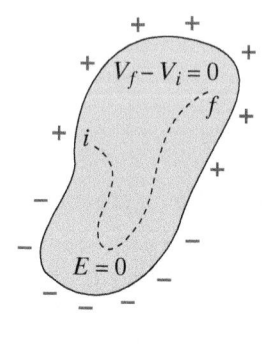

Figure 16.19 Potential difference is zero inside a metal in equilibrium.

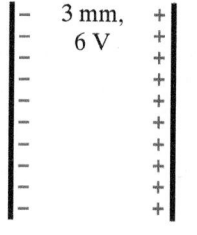

Figure 16.20 A capacitor with large plates and small gap.

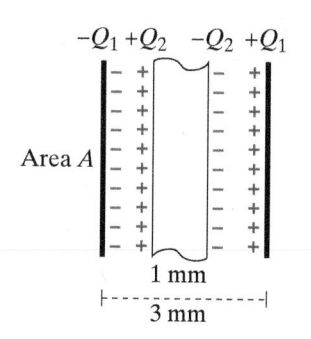

Figure 16.21 Insert a metal slab in the middle of the capacitor gap.

Solution Split the path into two parts, A to B and B to C. In each part the electric field is constant in magnitude and direction. We have the following:

$$\Delta V = -\sum \vec{E} \cdot \Delta \vec{l}$$

$$\Delta V = -\langle 400, 0, 0\rangle \frac{\text{V}}{\text{m}} \cdot \langle 3, 0, 0\rangle \,\text{m} - \langle 1000, 500, 0\rangle \frac{\text{V}}{\text{m}} \cdot \langle 2, 0, 0\rangle \,\text{m}$$

$$\Delta V = (-1200 - 2000)\,\text{V} = -3200\,\text{V}$$

Checkpoint 6 In the earlier Figure 16.17, location A is at $\langle 0.5, 0, 0\rangle$ m, location C is $\langle 1.3, 0, 0\rangle$ m, and location B is $\langle 1.7, 0, 0\rangle$ m. $\vec{E}_1 = \langle 650, 0, 0\rangle$ N/C and $\vec{E}_2 = \langle -350, 0, 0\rangle$ N/C. Calculate the following quantities: **(a)** ΔV along a path going from A to B, and **(b)** ΔV along a path going from B to A.

A Conductor in Equilibrium

In Chapter 14 we proved that inside a conductor, such as a metal object, in equilibrium, the net electric field is zero (because otherwise mobile charges would shift until they contribute a field large enough to cancel the applied field). Therefore it must be the case that in equilibrium the net electric field is zero at all locations along any path through the conductor. This means that the potential difference is zero between two locations inside the metal (Figure 16.19).

In a conductor in equilibrium the difference in potential between any two locations is zero, because $E = 0$ inside the conductor.

$$\Delta V = -(E_x \Delta x + E_y \Delta y + E_z \Delta z) = 0, \quad \text{so } V = \text{constant}$$

We'll use this fact, in combination with the approach developed in this section, to examine a situation involving the potential difference across multiple regions, one of which is inside a metal object. Suppose that a capacitor with large plates and a small gap of 3 mm initially has a potential difference of 6 V from one plate to the other (Figure 16.20).

> **QUESTION** What are the direction and magnitude of the electric field in the gap?

The direction of the electric field in the gap is toward the left, away from the positive plate and toward the negative plate. The magnitude can be found from noting that the potential difference $\Delta V = Es = 6$ V, so that $E = (6\,\text{V})/(0.003\,\text{m}) = 2000$ V/m.

> **QUESTION** Would inserting a metal slab into the gap change ΔV?

Suppose we insert into the center of the gap a 1-mm-thick metal slab with the same area as the capacitor plates, as shown in Figure 16.21. We are careful not to touch the charged capacitor plates as we insert the metal slab. We want to calculate the new potential difference between the two outer plates.

First we need to understand the new pattern of electric field that comes about when the metal slab is inserted. The charges on the outer plates are $-Q_1$ and $+Q_1$, both distributed approximately uniformly over a large plate area A. The metal slab of course polarizes and has charges of $+Q_2$ and $-Q_2$ on its surfaces, as indicated in Figure 16.21.

The electric field inside a capacitor is approximately $E = (Q/A)/\varepsilon_0$, where Q is the charge on one plate and A is the area of the plate, if the plate separation s is small compared to the size of the plates.

> **QUESTION** Since the electric field inside the metal slab must be zero, we can conclude that Q_2 is equal to Q_1. Why?

The plates and the surfaces of the slab have the same area A. The outer charges $-Q_1$ and $+Q_1$ produce an electric field in the metal slab $E_1 = (Q_1/A)/\varepsilon_0$ to the left. The inner charges $+Q_2$ and $-Q_2$ are arranged like the charges on a capacitor, so they produce an electric field in the metal slab $E_2 = (Q_2/A)/\varepsilon_0$ to the right. The sum of these two contributions must be zero, because the electric field inside a metal in equilibrium must be zero. Hence Q_2 is equal to Q_1.

As a result, the left pair of charges produces a field like that of a capacitor, and the right pair of charges also produces a field like that of a capacitor. The effect is that after inserting the metal slab, the electric field is still approximately 2000 V/m in the air gaps but is now zero inside the metal slab. (The fringe fields are small if the gap is small.)

> QUESTION Now that we know the electric field everywhere, what are the potential differences across each of the three regions between the plates (air gap, metal slab, air gap)?

The electric field in the air gap is essentially unchanged, so

$$\Delta V_{\text{left}} = \Delta V_{\text{right}} = (2000 \text{ V/ m})(0.001 \text{ m}) = 2 \text{ V}$$

ΔV inside the metal slab must be zero, because $E = 0$ inside the slab. The potential difference between the plates of the capacitor is now:

$$\Delta V = 2 \text{ V} + 0 \text{ V} + 2 \text{ V} = 4 \text{ V}$$

not 6 V as it was originally. (Note that there is no such thing as "conservation of potential"; the potential difference changed when we inserted the slab.)

When you know that the electric field inside a metal must be zero, you know that the sum of all the contributions to that electric field must be zero. This is a powerful tool for reasoning about fields and charges in and on metals in equilibrium.

> **Checkpoint 7** In Figure 16.22, location A is inside a charged metal block, and location B is outside the block. The metal block sits on an insulating surface and is not in contact with any other object. The electric field outside the block is $\vec{E}_2 = \langle -50, 0, 0 \rangle$ V/m. **(a)** Calculate ΔV along a path going from A to B. **(b)** Calculate ΔV along a path going from B to A.

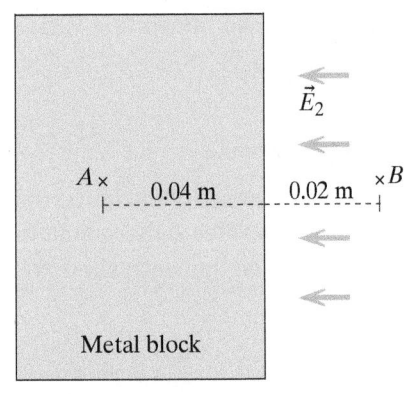

Figure 16.22 Location A is inside a charged metal block. Location B is outside the block.

ΔV in a Region of Varying Electric Field

In some situations, the electric field in a region varies continuously. For example, the electric field due to a single point charge is different in magnitude and direction at every location in space. To find a change in electric potential between two locations in such a region, we need to take into account the varying electric field.

In the previous chapter we evaluated the electric field produced by distributed charges (rod, ring, disk, sphere) by adding up the contributions of pieces of the distribution. For example, to find the electric field due to a uniformly charged rod we divided the rod into short pieces, approximated the field of each short piece by the field of a point charge, and added up all these contributions either numerically or, where possible, analytically, by evaluating an integral. The more pieces, and the shorter each piece, the more accurate was the result, and in the case of an integral there were an infinite number of infinitesimally short pieces of the rod.

We can use a similar approach to evaluate electric potential differences between two locations in a region of varying electric field. The basic idea is to choose a path from the initial location to the final location, break that path into short pieces, evaluate ΔV for each piece, add up all these ΔV's, and check the result. Here are details of these steps:

Step 1: Choose a Path and Divide It into Pieces $\Delta \vec{l}$

- Choose a path from the initial location i to the final location f. Although the potential difference is independent of the path from i to f, we need to choose a specific path in order to carry out the calculation, and it needs to be a path along which we are able to calculate the electric field $\vec{E}$ at every location along the path.
- Divide the path into many short pieces, where each piece is represented as a vector $\Delta \vec{l}$.
- At a couple of locations somewhere near the middle of the path, sketch the electric field $\vec{E}$ to visualize and guide the calculation. (Remember that the result depends on $\vec{E}$ all along the path—don't focus on the endpoints.)

Step 2: Write an Expression for $\Delta V = -\vec{E} \bullet \Delta \vec{l}$ of One Piece

- Each piece of the path contributes an amount $\Delta V = -\vec{E} \bullet \Delta \vec{l}$ to the potential difference. $\vec{E}$ has a different value at each location on the path. To calculate each $\Delta V = -\vec{E} \bullet \Delta \vec{l}$ we will use the value of $\vec{E}$ midway along $\Delta \vec{l}$. If we are doing a numerical calculation, we (or a computer) must calculate $\vec{E}$ for each step. If we are setting up an expression that we can try to integrate analytically, our expression for $\vec{E}$ should depend on an integration variable related to position along the path.

Step 3: Add Up the Contributions of All the Pieces

- Add up the ΔV contributions to get $V_f - V_i$. The summation might be done numerically, or in some cases it may be possible to evaluate the summation as the integral $V_f - V_i = - \int_i^f \vec{E} \bullet d\vec{l}$, where we imagine that each short piece $\Delta \vec{l}$ becomes an infinitesimal length $d\vec{l}$.

Step 4: Check the Result

- Check that the result makes sense. Most importantly, does the overall sign make physical sense? Does the result have the correct units of volts (joules per coulomb)? Does the result give the right value in special cases where the correct result is known?

Potential Difference Near a Point Charge

To see how this scheme works, we'll carry out these steps to calculate the potential difference between two locations near a point charge of amount $Q = 3 \times 10^{-9}$ C. The initial location is a distance $r_i = 2 \times 10^{-3}$ m from the point charge, and the final location is a distance $r_f = 8 \times 10^{-3}$ m from the point charge. First we will solve the problem numerically; second we'll solve the problem analytically by integrating symbolically.

(1) Choose a Path and Cut it into Pieces $\Delta \vec{l}$: In Figure 16.23 we want to calculate the potential difference in going from location i to location f, that is, $V_f - V_i$. By sketching $\vec{E}$ at various places, we see that along the path $\vec{E}$ is parallel to the path. We'll divide the path into five equal lengths, each of length $(8 \times 10^{-3}$ m $- 2 \times 10^{-3}$ m$)/5 = 1.2 \times 10^{-3}$ m, and we'll calculate $\Delta V = -\vec{E} \bullet \Delta \vec{l}$ numerically for each of the five pieces.

(2) Write an expression for $\Delta V = -\vec{E} \bullet \Delta \vec{l}$: The dot product is simple: $\Delta V = -E \Delta l \cos 0 = -E \Delta l$, where E is the usual field of a point charge. Therefore

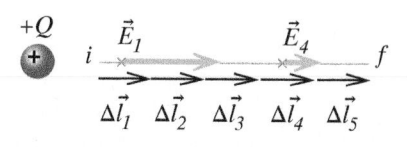

Figure 16.23 The path from i to f has been divided into five steps. The electric field is drawn at the center of the first and fourth steps. The arrows representing each $\Delta \vec{l}$ have been offset downward slightly from the path for clarity.

the potential difference of each piece is the following, where we can write $\Delta l = \Delta r$:

$$\Delta V = \frac{1}{4\pi\varepsilon_0}\frac{Q}{r^2}\Delta r$$

Although we've divided the path into five short pieces, within each piece the electric field does vary with distance from the point charge. In order to use something like the average value of E for a piece, we'll let r be the distance from the point charge to the middle of the piece. We could of course improve the accuracy of our calculation by dividing the path into more than five pieces.

(3) Add Up the Contributions of All the Pieces: In Figure 16.24 are the calculations in which we calculate the ΔV for each piece and add them up to find $V_f - V_i$.

It would be possible to get a more accurate value by cutting the path into more pieces, but it would be tedious to do the calculations by hand. We wrote a little computer program to do these calculations for various numbers of pieces, and the results of these calculations are shown in Figure 16.25.

(4) Check the Result: The first check to make is whether the sign of the potential difference makes physical sense. Referring back to Figure 16.23, we see that $\vec{E}$ is in the same direction as $\Delta\vec{l}$ along the path, so the sign of the potential difference along each piece, $-E\Delta l\cos 0 = -E\Delta l$, is negative, and the sum of all these ΔVs is negative. Note too that a positive charge placed at r_i would be repelled and would head "downhill" to lower potential, which means that the potential difference is negative. So the negative sign is correct. You should also check your own calculations to make sure that the units are correct.

r, m	ΔV, volts
$2.6E{-}3$	$-4.793E3$
$3.8E{-}3$	$-2.244E3$
$5.0E{-}3$	$-12960E3$
$6.2E{-}3$	$-8.429E2$
$7.4E{-}3$	$-5.917E2$
Sum	$-9.767E3$

Figure 16.24 Numerical integration for $V_f - V_i$.

# pieces	$V_f - V_i$, volts
5	$-9.767E3$
10	$-10.028E3$
15	$-10.081E3$
20	$-10.100E3$
100	$-10.124E3$

Figure 16.25 Effect of number of pieces.

Analytical Integration

In step 2 above we developed an expression for the potential difference along a short piece of the path. We expect that as we divide the path into more and more shorter and shorter pieces our calculation should be more accurate, and in fact in Figure 16.25 we see that for 15, 20, or 100 pieces there is very little change in the calculated potential difference (they're all about -10.1×10^3 V), which suggests that the true value is in fact about -10.1×10^3 V.

In step 3 (adding up all the contributions to the potential difference), let's go to the limit of an infinite number of pieces, each piece of infinitesimal length. This means that $\Delta r \to dr$ and the sum becomes an integral:

$$
\begin{aligned}
V_f - V_i &= -\int_{r_i}^{r_f} \frac{1}{4\pi\varepsilon_0}\frac{Q}{r^2}dr \\
&= -\frac{Q}{4\pi\epsilon_0}\left[-\frac{1}{r}\right]_{r_i}^{r_f} \\
&= \frac{1}{4\pi\varepsilon_0}\frac{Q}{r_f} - \frac{1}{4\pi\varepsilon_0}\frac{Q}{r_i}
\end{aligned}
$$

When we evaluate this expression for the specified values of Q, r_i, and r_f, we find that $V_f - V_i = -10.125 \times 10^3$ V. Compare with the numerical integration with 100 pieces, which gave $V_f - V_i = -10.124 \times 10^3$ V. Note that if r increases, as it does in this example, the quantity $1/r$ decreases and $1/r_f$ is smaller than $1/r_i$, so moving to larger r implies a negative potential difference.

Note the important result that near a point charge the potential difference along a radial path can be written like this:

$$V_f - V_i = \Delta\left(\frac{1}{4\pi\varepsilon_0}\frac{Q}{r}\right)$$

We'll say more about this later in the chapter.

Potential Difference Due to a Proton

A proton is located at the origin. Location C is 1×10^{-10} m from the proton, and location D is 2×10^{-8} m from the proton, along a line radially outward. **(a)** What is the potential difference $V_D - V_C$? **(b)** How much work would be required to move an electron from location C to location D?

Solution **(a)** Potential difference:

$$\Delta V = -\int_C^D E_x dx = -\int_{1 \times 10^{-10}}^{2 \times 10^{-8}} \left(\frac{1}{4\pi\varepsilon_0} \frac{Q}{x^2} \right) dx$$

$$\Delta V = \frac{1}{4\pi\varepsilon_0} (1.6 \times 10^{-19} \text{C}) \left(\frac{1}{2 \times 10^{-8} \text{ m}} - \frac{1}{1 \times 10^{-10} \text{ m}} \right)$$

$$= -14.3 \text{ V}$$

(b) Work to move an electron: The least amount of work will be required if we don't change the kinetic energy of the proton.

$$\cancel{\Delta K} + \Delta U = W_{\text{ext}}$$

$$0 + (-e)(\Delta V) = (-1.6 \times 10^{-19} \text{C})(-14.3 \text{ V}) = 2.3 \times 10^{-18} \text{ J}$$

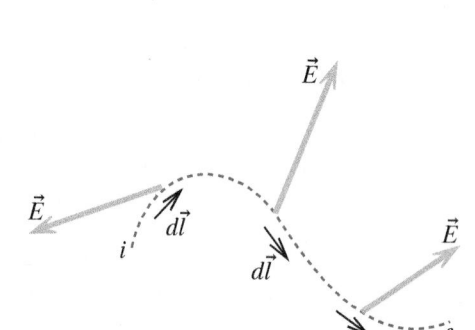

Figure 16.26 The potential difference for each step along this path is $-\vec{E} \bullet \Delta \vec{l}$. The potential difference $\Delta V = V_f - V_i$ is the sum of all these contributions, which becomes an integral for infinitesimal steps.

$\langle -0.4, 0, 0 \rangle$ m $\langle 0.2, 0, 0 \rangle$ m
$A \times$ $B \times$

$\vec{E} = \langle 500, 0, 0 \rangle$ N/C

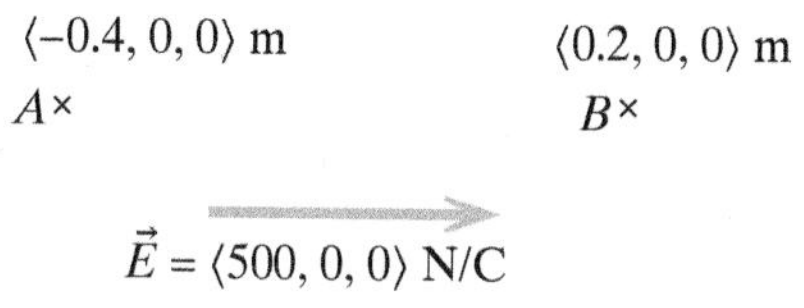

Figure 16.27 A region of uniform electric field.

General Definition of Potential Difference

The most general definition of potential difference is the integral expression, which can be used in any situation (see Figure 16.26).

GENERAL DEFINITION OF POTENTIAL DIFFERENCE

$$\Delta V = -\int_i^f \vec{E} \bullet d\vec{l}$$

This general expression is correct for regions in which the field is uniform, as well as for regions in which the field varies. For example, in the situation shown in Figure 16.27, where the field is uniform,

$$\Delta V = -\int_i^f \vec{E} \bullet d\vec{l} = -\int_{-0.4}^{0.2} (500 \text{ V/m}) dx = -500 \text{ V/m} \int_{-0.4}^{0.2} dx = -300 \text{ V}$$

The result of the integration is the same as we got in the example involving uniform field in Section 16.3.

Complex Paths

Using $V_f - V_i = \Delta V = -\int_i^f \vec{E} \bullet d\vec{l}$ to calculate the potential difference along a path that is not straight, along which the electric field is varying in magnitude and direction (such as that shown in Figure 16.26), can be mathematically challenging. In many cases the best way to do this is to do it numerically, using a computer. In this text we will not ask you to do arbitrarily complex integrations of this kind analytically.

To establish the basic principles we will typically consider relatively simple situations. Often the electric field is uniform (same magnitude and direction) all along a straight path, which makes the calculation very simple. Another relatively simple situation is one in which the magnitude varies but not the direction along a straight path. A third simple situation that we will encounter is one in which we move along a curved path, but the electric field is always in the same direction as this curved path, and often constant in magnitude, which occurs in electric circuits. In each of these simple cases it can be quite easy to evaluate the potential difference, as we saw in the preceding exercises.

EXAMPLE

A Disk and a Spherical Shell

In Figure 16.28 a thin spherical shell made of plastic carries a uniformly distributed negative charge $-Q_1$ and a thin circular disk made of glass carries a uniformly distributed positive charge $+Q_2$. The radius R_1 of the plastic spherical shell is very small compared to the large radius R_2 of the glass disk. The distance from the surface of the spherical shell to the glass disk is d, and d is much smaller than R_2. Find the potential difference $V_2 - V_1$. Location 1 is at the center of the plastic sphere, and location 2 is just outside the glass disk. State what approximations or simplifying assumptions you make.

Solution

Initial location: 1
Final location: 2
Path: Straight line from 1 to 2 (see Figure 16.29). $\vec{E}_{net} = \vec{E}_{shell} + \vec{E}_{disk}$ according to the Superposition Principle, so

$$-\int_1^2 \vec{E}_{net} \bullet d\vec{l} = -\int_1^2 \vec{E}_{shell} \bullet d\vec{l} - \int_1^2 \vec{E}_{disk} \bullet d\vec{l}$$

This allows us to find ΔV due to the shell and ΔV due to the disk separately, then add them. Alternatively, we could add the two electric field contributions, then integrate $\vec{E}_{net} \bullet d\vec{l}$ along a path from 1 to 2.

Simplifying assumption: Neglect the polarization of the plastic and glass, because there is little matter in the thin shell and thin disk, so the field of the polarized molecules is negligible compared to the contributions of $-Q_1$ and $+Q_2$.

ΔV due to shell: Put origin at center of shell.
Inside the shell: $V_{surface\ of\ shell} - V_1 = 0$ because $E_{shell} = 0$ inside the shell.
To the right of the shell:

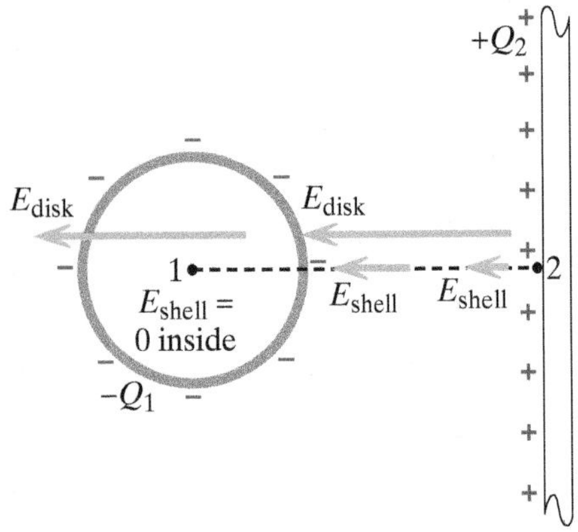

$$\vec{E}_{shell} = \left\langle \frac{1}{4\pi\varepsilon_0}\frac{-Q_1}{x^2},0,0 \right\rangle, \Delta\vec{l} = \langle dx,0,0 \rangle$$

$$V_2 - V_{surface\ of\ shell} = -\int_{R_1}^{R_1+d} \frac{1}{4\pi\varepsilon_0}\frac{-Q_1}{x^2}dx$$

$$= \frac{1}{4\pi\varepsilon_0}\left(\frac{-Q_1}{R_1+d} - \frac{-Q_1}{R_1} \right)$$

Check the sign: As we move toward the disk, we're moving opposite to the field, so the potential should increase. Result agrees, since $+Q_1/R_1$ is the larger term.

ΔV due to disk: an approximation is that since $d \ll R_2$ and $R_1 \ll R_2$,

$$\vec{E}_{disk} \approx \left\langle \frac{-Q_2/(\pi R_2^2)}{2\varepsilon_0},0,0 \right\rangle$$

We could set up an integral, placing the origin at the center of the disk. However, since $\vec{E}_{disk}$ is approximately uniform, we recognize that the result of integrating will be

$$V_2 - V_1 = \frac{Q_2/(\pi R_2^2)}{2\varepsilon_0}(R_1+d)$$

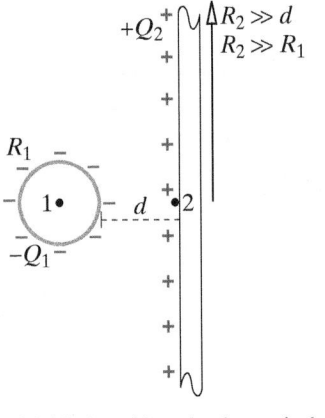

Figure 16.28 A uniformly charged plastic shell and a uniformly charged glass disk.

Figure 16.29 Electric field contributions of the shell and the disk, shown at locations along the path.

Check the sign: As we move toward the disk, we're moving opposite to the field, so the potential should increase. The result agrees.

ΔV due to both shell and disk:

$$V_2 - V_1 = \frac{1}{4\pi\varepsilon_0}\left(\frac{-Q_1}{R_1+d} - \frac{-Q_1}{R_1}\right) + \frac{Q_2/(\pi R_2^2)}{2\varepsilon_0}(R_1+d)$$

16.6 PATH INDEPENDENCE

In Chapter 6 we showed that potential energy differences depend only on the initial and final states of a system and are independent of path: different path, different process, but same change of state. Similarly, the potential difference $\Delta V = V_B - V_A$ between two locations A and B does not depend on the path taken between the locations.

To calculate the potential difference between two locations based on the electric field in the region, it is necessary to choose a path. We are free, however, to choose a path that simplifies our calculations.

A Simple Example of Two Different Paths

To see how the potential difference between two locations can be the same along different paths, we'll consider two different paths through a capacitor. Suppose you move from the positive plate of a capacitor to the negative plate, moving at an angle to the electric field $\vec{E}$ (Figure 16.30).

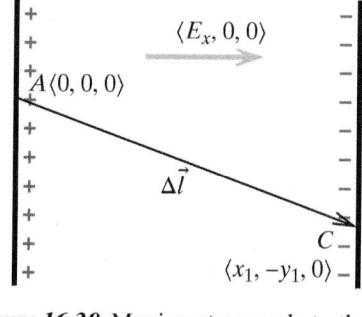

Figure 16.30 Moving at an angle to the electric field.

QUESTION Calculate the potential difference in going from A to C: $\Delta V = V_C - V_A = ?$

Since the electric field is the same in magnitude and direction all along the path, we can write

$$\Delta V = -(E_x\Delta x + E_y\Delta y + E_z\Delta z)$$
$$= -E_x(x_1 - 0) + 0(-y_1 - 0) + 0\cdot 0 = -E_x x_1$$

Suppose instead that we move along the two-part path shown in Figure 16.31.

QUESTION Now $\Delta V = V_C - V_A = ?$

Along the path from A to B,

$$V_B - V_A = -(E_x\Delta x + E_y\Delta y + E_z\Delta z)$$
$$= -E_x(x_1 - 0) + 0\cdot 0 + 0\cdot 0 = -E_x x_1$$

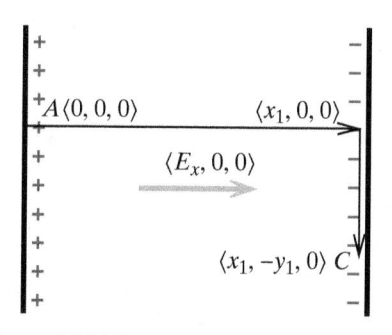

Figure 16.31 Moving at right angles to the electric field.

Along the path from B to C,

$$V_C - V_B = -(E_x\Delta x + E_y\Delta y + E_z\Delta z)$$
$$= -E_x(0) + 0(-y_1 - 0) + 0\cdot 0 = 0$$

Therefore we again find that

$$V_C - V_A = (V_C - V_B) + (V_B - V_A) = -E_x x_1$$

which is what we found for the direct path from A to C.

QUESTION Does it make sense that $V_C - V_B = 0$?

Yes. Just as no work is required to move at right angles to a force, the electric potential does not change along a path perpendicular to the electric field. The potential difference doesn't depend on the angle of our path in Figure 16.30 because the sideways components of our steps don't affect ΔV.

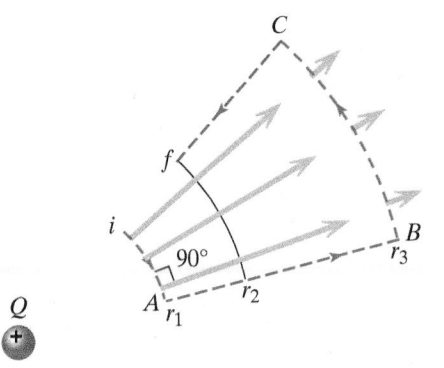

Figure 16.32 What is the potential difference along this complicated path?

Two Different Paths Near a Point Charge

Along a path straight away from a stationary charge Q we know this:

$$\Delta V = \frac{1}{4\pi\varepsilon_0} Q \left[\frac{1}{r_f} - \frac{1}{r_i}\right] \quad \text{near a positive or negative point charge } Q$$

It is instructive to work out the potential difference along a different path.

QUESTION In the situation shown in Figure 16.32, calculate the potential difference for each branch of the path from location i to A to B to C to location f, in terms of Q and the radii r_1, r_2, and r_3, then add these up to get the potential difference along this path. As a first step, draw the electric field at various locations along the path.

From i to A, $\vec{E} \perp \Delta\vec{l}$, so $\Delta V_1 = 0$.

From A to B, $\Delta V_2 = \frac{1}{4\pi\varepsilon_0} Q \left[\frac{1}{r_3} - \frac{1}{r_1}\right]$.

From B to C, $\vec{E} \perp \Delta\vec{l}$, so $\Delta V_3 = 0$.

From C to f, $\Delta V_4 = \frac{1}{4\pi\varepsilon_0} Q \left[\frac{1}{r_2} - \frac{1}{r_3}\right]$.

Adding up these ΔV's, we get:

$$V_f - V_i = \Delta V_1 + \Delta V_2 + \Delta V_3 + \Delta V_4$$

$$= 0 + \frac{1}{4\pi\varepsilon_0} Q \left[\frac{1}{r_3} - \frac{1}{r_1}\right] + 0 + \frac{1}{4\pi\varepsilon_0} Q \left[\frac{1}{r_2} - \frac{1}{r_3}\right]$$

$$= \frac{1}{4\pi\varepsilon_0} Q \left[\frac{1}{r_2} - \frac{1}{r_1}\right]$$

This is exactly the same result that you would get by going directly in a straight line from i to f. We see here that it is only when $d\vec{l}$ has a radial component that there is a nonzero contribution to ΔV. Along the curved arcs the electric field is perpendicular to the path, so the arcs contribute zero to the potential difference. Going along the straight lines is equivalent to just going from r_1 to r_2. Evidently it doesn't matter what path we take. We say that potential difference is "path independent."

An Arbitrary Path Near a Point Charge

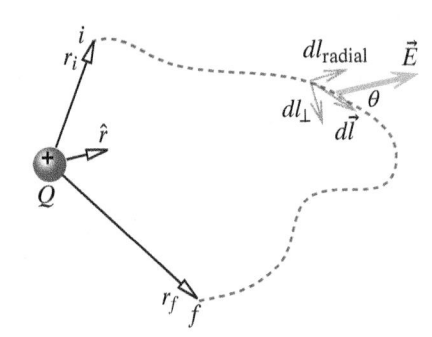

Figure 16.33 Calculating the potential difference along an arbitrary path.

We can see more generally how this works out for any path near a stationary point charge. Follow an arbitrary path from initial location i to final location f (Figure 16.33). At each location along the path draw the component dl_{radial} of $d\vec{l}$ in the direction of $\vec{E}$, pointing away from the charge Q.

At every location along the path we have

$$-\vec{E} \bullet d\vec{l} = -E \, dl \cos\theta$$

However, $dl\cos\theta$ is the radial component of $d\vec{l}$ and is equal to dl_{radial}, the change in distance from the charge. Therefore along this arbitrary path we are simply integrating $-E \, dl_{radial}$, so

$$\Delta V = \frac{1}{4\pi\varepsilon_0} Q \left[\frac{1}{r_f} - \frac{1}{r_i}\right]$$

All that matters are the initial and final distances from the charge, not the path we happen to follow. This is consistent with our earlier result for potential

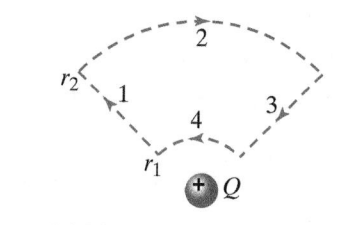

Figure 16.34 Checkpoint 8.

difference $\Delta V = (1/(4\pi\varepsilon_0))(Q/r)$ along a straight path between location i and location f. What is interesting is seeing how it worked out that the integral of $-\vec{E} \bullet d\vec{l}$ didn't depend on what path we took.

> **Checkpoint 8** Calculate the potential difference along the closed path consisting of two radial segments and two circular segments centered on the charge Q (Figure 16.34). Show that the four ΔV's add up to zero. It is helpful to draw electric field vectors at several locations on each path segment to help keep track of signs.

Round-Trip Potential Difference Is Zero

An important consequence of path independence is that the integral of $-\vec{E} \bullet d\vec{l}$ along a round-trip path gives a potential difference of zero, as we can now show (Figure 16.35). Move along path 1 (A-B-C), and then back to location A along path 2 (C-D-A).

QUESTION Suppose that the potential difference along path 1 is -25 V. What would be the potential difference in going backward along path 2 (A-D-C)? Forward along path 2 (C-D-A)?

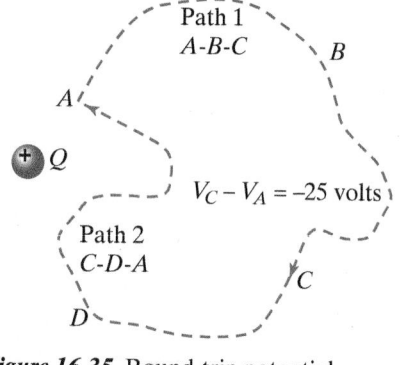

Figure 16.35 Round-trip potential difference on a path from A back to A again is zero.

Going backward along path 2 is the same as going forward along path 1, so the potential difference along the path (A-D-C) must be -25 V. Going in the opposite direction, along path 2 (C-D-A), reverses the sign of the parallel component of the electric field, so the potential difference along path 2 must be $+25$ V.

QUESTION If we walk all the way around the closed path, A-B-C-D-A, what is the potential difference?

The round-trip potential difference is zero (-25 V plus $+25$ V). Note that this result is the direct consequence of the path independence of electric potential. It is also not surprising, since we do expect $V_A - V_A = 0$!

> Potential difference due to a stationary point charge
> is independent of the path.
> Potential difference along a round trip is zero.

Since potential difference can be calculated as an integral involving the electric field, and the electric field is the sum of all the contributions of many stationary point charges, potential difference is the sum of the contributions of all the point charges. The potential difference integral is independent of the path for one point charge, and since all charge distributions are made up of atomic point charges, the round-trip potential difference must be zero for any configuration of stationary charges whatsoever.

PATH INDEPENDENCE

ΔV is always independent of the path taken between two locations.
For a round-trip path, ΔV is zero.

Conservation of Energy

The fact that $\Delta V = 0$ for a round trip along any path is fundamentally related to the principle of Conservation of Energy. We can use a proof by contradiction to show this connection.

Let's assume that it is possible to fasten down a collection of stationary charged particles in such a configuration that a round trip along the path shown in Figure 16.35 gives a potential difference greater than zero. If this is possible, then an electron could travel around this path many times, and $\Delta U = -e\Delta V$

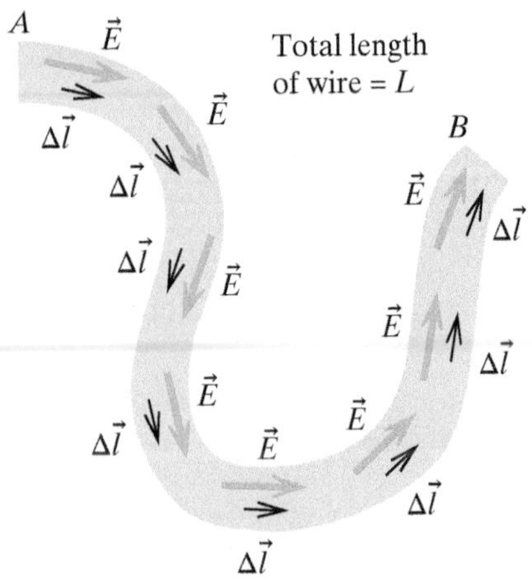

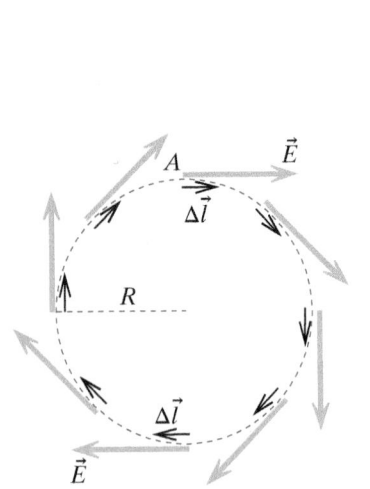

Figure 16.36 Inside a wire in a circuit, there can be an electric field that has uniform magnitude E but follows the direction of the wire. The wire's length is L.

of the system would be negative each time. Therefore the kinetic energy of the system would increase each time, despite the fact that no external energy inputs were made. We would have invented a perpetual motion machine! Because this outcome would violate the Energy Principle, we can conclude that our assumption (that it was possible for $\Delta V_{\text{round trip}}$ to be nonzero) must have been incorrect.

These are extremely important general results, and in the future we will often have occasion to refer to these properties of electric potential due to charges. Path independence is important because when we use potential we are reasoning about the relationship between initial and final states, independent of the intervening process. We are free to analyze the most convenient process that takes the system from the initial state to the final state.

Application: A Metal Not in Equilibrium

During the (extremely) brief time when a metal is polarizing due to an external charge, there is a nonzero electric field inside the metal. Before equilibrium is reached, the electric field in the metal is not zero.

As we will see in detail in a later chapter, in an electric circuit the battery maintains a nonzero electric field inside the wires, as a kind of continuing polarization process that doesn't lead to equilibrium. The electric field continually pushes the sea of mobile electrons through the wires. The electric field is not zero in this situation, which is not equilibrium.

> QUESTION In Figure 16.36 there is a nonzero electric field of uniform magnitude E throughout the interior of a wire of length L, and the direction of the electric field follows the direction of the wire. What is the potential difference $V_B - V_A$?

In this case, since for each step $\vec{E} \| \Delta \vec{l}$, the potential difference is simply $-EL$, the sum of all the $-\vec{E} \cdot \Delta \vec{l} = -E\Delta l$ terms along the wire. Thus if a metal is not in equilibrium, there can be a nonzero potential difference inside the metal. However, in a circuit a thick copper wire may have such a small electric field in it that there is very little potential difference from one end to the other, in which case the potential is almost (but not quite) constant in the wire.

> **Checkpoint 9** In a circuit there is a copper wire 40 cm long with a potential difference from one end to the other of 0.01 V. What is the magnitude of the electric field inside the wire?

Impossible Patterns of Electric Field

We can use the fact that the potential difference over a round-trip path must be zero to reason about particular patterns of electric field. Surprisingly, we can actually conclude that certain patterns of electric field can never be created by a collection of stationary charged particles. For example, consider the pattern of electric field shown in Figure 16.37, where the electric field is tangent to the circle at every point on the circle.

> QUESTION Try to think of an arrangement of stationary charges that could produce the pattern of electric field shown in Figure 16.37.

Figure 16.37 Is it possible to produce this pattern of electric field ($\vec{E}$ tangent to a circle at every point on the circle) with some arrangement of point charges?

We can demonstrate very simply that it is impossible to produce this pattern of electric field by any arrangement of stationary point charges! Consider a path that starts at location A and goes clockwise around the circle, ending back at location A. Since at every location along the path $\vec{E}$ is parallel to $d\vec{l}$, the round-trip potential difference along this path is:

$$\Delta V = -\int \vec{E} \cdot d\vec{l} = -(2\pi R)E$$

However, we showed previously that for a single point charge, and hence for any assemblage of point charges, $\Delta V = 0$ for a round trip. Therefore this "curly" pattern of electric field must be *impossible* to produce by arranging any number of stationary point charges!

Important Limitation

In the next chapter we will study magnetic fields, which can be produced by moving charges. We will see in Chapter 22 on Faraday's Law that in the presence of time-varying magnetic fields the round-trip integral of the electric field can be nonzero. In this chapter we have assumed that there were no time-varying magnetic fields.

16.7 THE POTENTIAL AT ONE LOCATION

Usually we are interested in the potential difference between two locations; for example, $\Delta V = V_B - V_A$. However, it is sometimes useful to consider the potential at only one location.

The potential at location A is defined as the potential difference between a location infinitely far away from all charged particles, and the location of interest:

POTENTIAL AT ONE LOCATION

$$V_A = V_A - V_\infty$$

The potential at location A, sometimes referred to as "the potential relative to infinity," is the potential difference between location A and a location infinitely far away from all charged particles.

Of course, this equation makes sense only if V_∞, the potential at infinity, is zero, and by convention it is defined to be so. This is consistent with the fact that the potential energy of a system of two charged particles that are infinitely far from each other must be zero.

Since potential has a value at every location in space, potential (but not potential difference) is itself a field, but it is a scalar field, not a vector field.

Potential at a Location Near a Point Charge

As an example, we will find the potential at a location a distance r from a single point particle of charge q. To do this, we will find the potential difference between this location and a location infinitely far away.

$$V_r = V_r - V_\infty = -\int_\infty^r \frac{1}{4\pi\varepsilon_0} \frac{q}{x^2} dx = \frac{1}{4\pi\varepsilon_0} \frac{q}{x}\Big|_\infty^r = \frac{1}{4\pi\varepsilon_0} \frac{q}{r}$$

QUESTION How does the sign of this result depend on the sign of the charged particle?

If $q < 0$, $V_r < 0$, but if $q > 0$, $V_r > 0$. Figure 16.38 is a graph of potential as a function of distance from a positive or negative point charge.

Potential and Potential Energy at One Location

If we know the value of the potential at location A, we can use this to find what the potential energy of the system would be if we placed a charged object at that location:

$$U_A = qV_A$$

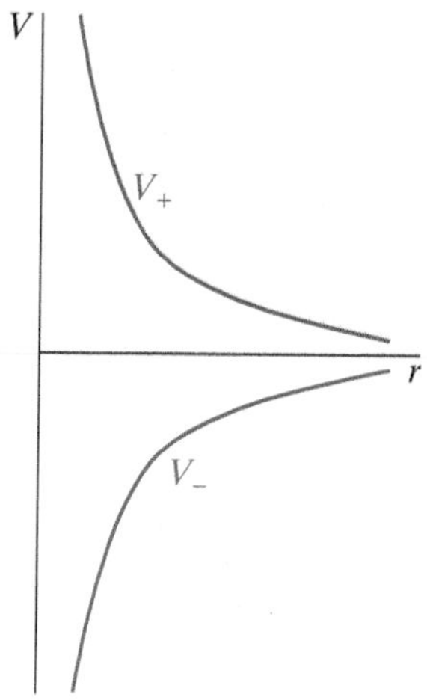

Figure 16.38 Potential as a function of distance from a positive (red) or negative (blue) point charge.

Using our result for V we get the familiar expression for the electric potential energy of two particles with charges q_1 and q_2 separated by a distance r:

$$U = q_1 \left(\frac{1}{4\pi\varepsilon_0} \frac{q_2}{r} \right) = \frac{1}{4\pi\varepsilon_0} \frac{q_1 q_2}{r}$$

EXAMPLE

Potential at a Location Near a Sphere

A metal sphere of radius 3 mm has a charge of -2×10^{-9} C. What is the potential at location C, which is 5 mm from the center of the sphere?

Solution

$$V_C = V_C - V_\infty = \frac{1}{4\pi\varepsilon_0} Q \left(\frac{1}{x_C} - \frac{1}{\infty} \right)$$

$$= \left(9 \times 10^9 \, \frac{\text{N} \cdot \text{m}^2}{\text{C}^2} \right) (-2 \times 10^{-9}\text{C}) \left(\frac{1}{0.005 \text{ m}} - \frac{1}{\infty} \right)$$

$$= -3600 \text{ V}$$

We can calculate the potential due to a charge distribution in two ways: either by dividing the distribution into point-like pieces and adding the potential due to each piece, or by calculating $-\int \vec{E} \bullet d\vec{l}$ along a path from infinity to the location of interest.

EXAMPLE

Potential Along the Axis of a Ring

Use superposition to calculate the potential at a location z from the center of a thin ring, along the axis, by adding the potential due to each piece of the ring. The ring carries a total charge Q. The total charge Q is made up of point charges, each of charge q, and each of these point charges is a distance $\sqrt{z^2 + R^2}$ from the location of interest (Figure 16.39).

Solution

The potential contributed by any one of these point charges q is

$$V = \frac{1}{4\pi\varepsilon_0} \frac{q}{\sqrt{z^2 + R^2}}$$

for one point charge q.

Adding up the potential contributed by all of the point charges, we have the following result:

$$V_{\text{ring}} = \sum \frac{1}{4\pi\varepsilon_0} \frac{q}{\sqrt{z^2 + R^2}}$$

$$V_{\text{ring}} = \frac{1}{4\pi\varepsilon_0} \frac{1}{\sqrt{z^2 + R^2}} \sum q$$

$$V_{\text{ring}} = \frac{1}{4\pi\varepsilon_0} \frac{Q}{\sqrt{z^2 + R^2}}$$

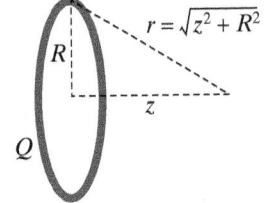

Figure 16.39 The potential due to a uniformly charged ring.

Integrating

This same result can be obtained by integrating the electric field of the ring along a path, using the expression for this field that we found in Chapter 15.

$$V(z) = -\int_\infty^z \frac{1}{4\pi\varepsilon_0} \frac{Qw}{(R^2 + w^2)^{3/2}} dw$$

Conversely, from the result for the potential of the ring we can determine the electric field of the ring by differentiation, since $E_z = -\partial V/\partial z$ (see Problem P85).

Checkpoint 10 Show that if you are very far from the ring ($z \gg R$), the potential is approximately equal to that of a point charge. (This is to be expected, because if you are very far away, the ring appears to be nearly a point.)

Potential Inside a Conductor at Equilibrium

In a previous section we noted that because the net electric field is zero inside a metal object in equilibrium, the potential difference between any two locations inside the metal must be zero. Since this is true, we can conclude that the potential at any location inside the metal must be the same as the potential at any other location (Figure 16.40).

In a conductor in equilibrium the potential is exactly the same everywhere inside the conductor, because $E = 0$ inside the conductor.

$$\Delta V = -(E_x \Delta x + E_y \Delta y + E_z \Delta z) = 0$$

so V is constant.

QUESTION Does this mean that the potential is zero at every location inside a metal in equilibrium?

No. The potential inside a metal object in equilibrium is constant (the same at every location), but it need not be zero. Charges on the surface of the metal object or on other objects may contribute to a nonzero but uniform potential inside the metal.

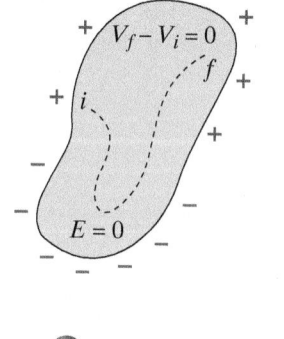

Figure 16.40 The potential is constant inside a metal in equilibrium.

EXAMPLE

The Potential Inside a Charged Metal Sphere

A solid metal sphere of radius R has a charge of Q uniformly distributed over its surface. What is the potential (relative to infinity) throughout the sphere?

Solution

At any distance $r > R$, the potential relative to infinity due to the charged sphere is $\dfrac{1}{4\pi\varepsilon_0}\dfrac{Q}{r}$, so the potential relative to infinity just outside the surface of the sphere is $\dfrac{1}{4\pi\varepsilon_0}\dfrac{Q}{R}$. Inside the sphere the electric field is zero everywhere, so the potential difference between any location inside the sphere and the surface is zero. Therefore the potential inside the sphere is $\dfrac{1}{4\pi\varepsilon_0}\dfrac{Q}{R}$ everywhere.

This situation is a good example of the indirect relationship between field and potential. Inside the sphere the electric field is zero everywhere but the electric potential is nonzero everywhere.

Avoiding a Common Confusion

A common error is to assume that the electric field at a location determines the potential at a location. In fact, the electric field at location A has very little to do with the potential at location A! Similarly, the electric field at A and the electric field at B have very little to do with the potential difference between locations A and B; it is the electric field in the intervening region that determines ΔV. As we have just seen, the electric field due to a charged spherical shell is zero inside the shell, but the potential inside the shell due to the charges on the sphere is nonzero.

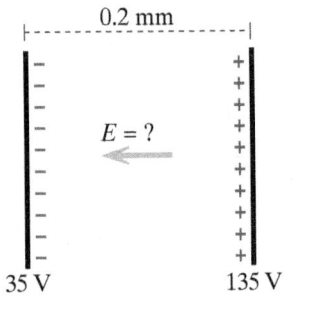

Figure 16.41 What is the magnitude of the electric field?

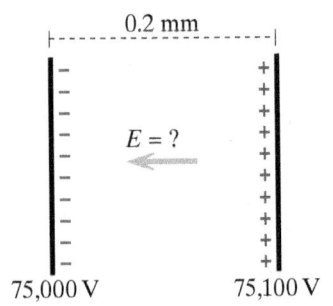

Figure 16.42 What is the magnitude of the electric field?

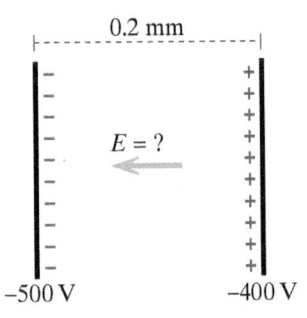

Figure 16.43 What is the magnitude of the electric field?

Shifting the Zero of Potential

As we saw in Chapter 6, electric potential energy (and therefore potential) must go to zero at large distances, to satisfy the requirements of relativity. Nevertheless, we are often interested only in differences of potential or potential energy, in which case we can choose a different zero location as a matter of convenience. This is exactly analogous to approximating change in gravitational potential energy as $\Delta(-GMm/r) \approx \Delta(mgy)$ near the surface of the Earth, and then choosing the zero of potential energy to be at or near the surface, rather than at infinity.

The electric field $\vec{E}$ is given by the gradient of the potential, not by the value of the potential. This means that only potential difference ΔV matters in determining the electric field, not the value V of the potential.

QUESTION Find the magnitude of the electric field in the 0.2-mm gap of a narrow-gap capacitor whose negative plate is at a potential of 35 V and whose positive plate is at a potential of 135 V (Figure 16.41).

The magnitude of the electric field in this case is

$$(100\ \text{V})/(0.2 \times 10^{-3}\ \text{m}) = 5 \times 10^5\ \text{V/m}$$

The direction is from the positive plate toward the negative plate (direction of decreasing potential).

QUESTION Suppose instead that the potential of the negative plate is 75,000 V and the potential of the positive plate is 75,100 V (Figure 16.42). Now what is the electric field in the 0.2-mm gap?

This makes no difference in the calculation, because it is only the 100-V difference in the potential that counts, not the particular values of the potential.

QUESTION Now suppose that the potential of the negative plate is −500 V and the potential of the positive plate is −400 V (Figure 16.43). What is the electric field in the 0.2-mm gap?

The magnitude of the electric field is again 5×10^5 V/m. These calculations show that what counts in calculating the electric field is the gradient of potential or potential difference per unit distance, not the value of the potential. This is exactly analogous to the fact that it is no harder to walk up from the 75th floor to the 76th floor of a building than it is to walk up from the 1st floor to the 2nd floor. To put it another way, adding a constant V_0 to all the potential values changes nothing as far as potential difference and electric field are concerned:

$$(V_f + V_0) - (V_i + V_0) = V_f - V_i$$

Reflection: Potential and Potential Difference

We have seen that there are two different ways to find the potential at a particular location:

- Add up the contributions of all point charges at all other locations:

$$V_A = \sum \frac{1}{4\pi\varepsilon_0} \frac{q_i}{r_i}$$

- Travel along a path from a point very far away to the location of interest, adding up $-\vec{E} \cdot d\vec{l}$ at each step:

$$V_A = -\int_\infty^A \vec{E} \cdot d\vec{l}$$

One way to think of the meaning of the potential at location A is to consider how much work per unit charge you would have to do to move a charged particle from a location very far away to location A. This work would of course depend on the electric field in the region through which you moved the charge.

We have also seen that there are two ways to find the potential difference between two locations A and B:

■ Subtract the potential at the initial location A from the potential at the final location B:

$$\Delta V = V_B - V_A$$

■ Travel along a path from A to B, adding up $-\vec{E} \bullet d\vec{l}$ at each step:

$$\Delta V = V_B - V_A = -\int_A^B \vec{E} \bullet d\vec{l}$$

Again, note that the potential difference between two locations depends on the electric field in the region between the two locations. If you were to move a charge from location A to location B, the electric field in the intervening region would determine how much work per unit charge you would have to do.

16.8 COMPUTING POTENTIAL DIFFERENCES

A computer can be useful in calculating potential differences in two situations: first, if the locations and charges of the sources of electric field are known, and second, if we have information about the electric field in the region of interest—either an analytical expression or a sufficient number of measurements.

A Known Charge Distribution

If the distribution of source charges is known, the easiest approach is to model each source as a point charge or collection of point charges. From Section 16.7, we have the following results:

■ The potential relative to infinity at location A, due to a single point charge q, is $\dfrac{1}{4\pi\epsilon_0}\dfrac{q}{r_A}$.

■ The potential at location A due to many point charges is simply the sum of the individual contributions of each point charge.
■ The potential difference $V_B - V_A$ is independent of the path taken from A to B.

A program to compute the potential difference $V_B - V_A$, therefore, has a relatively straightforward structure:

■ Make a list of source charges and locations.
■ Add up the contributions of each source charge to the potential at A.
■ Add up the contributions of each source charge to the potential at B.
■ Find $V_B - V_A$.

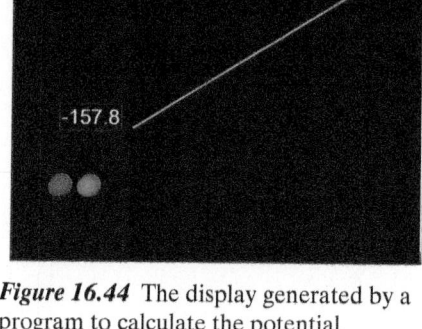

Figure 16.44 The display generated by a program to calculate the potential difference between two locations near a dipole, not on the dipole axis.

The program below does this calculation for a path near a dipole. Neither point A nor point B lies on the dipole axis. The display generated by the program (modified to use label objects to display the potential at each endpoint) is shown in Figure 16.44.

```
from visual import *
scene.width = scene.height = 800
oofpez = 9e9
```

```
s = 0.01
qd = 2e-9
## source charges
s1 = sphere(pos=(-s/2,0,0), radius=0.003,
            color=color.red, q=qd)
s2 = sphere(pos=(s/2,0,0), radius=0.003,
            color=color.blue, q=-qd)
sources = [s1,s2]
## path
A = vector(0.02,0.02,0)
B = vector(0.12,0.08,0)
path = [A,B]
curve(pos=path, color=color.green)   ## draw path
## potentials
VA = 0
VB = 0
i = 0
while i < len(sources):
    rate(500)
    r_A = A - sources[i].pos
    VA = VA + oofpez*sources[i].q/mag(r_A)
    r_B = B - sources[i].pos
    VB = VB + oofpez*sources[i].q/mag(r_B)
    i = i + 1
print('VA=', VA, 'VB=', VB, 'VB-VA=', VB-VA)
```

Problems P88–91 ask you to modify this program to calculate potential differences near other charge distributions such as rings and rods.

A Known Electric Field

It is sometimes the case that the spatial dependence of electric field in a region of space is known, even though we may not know exactly the location and value of the source charges associated with the field. In such a case, we can write a program to follow the procedure used in Section 16.5 to calculate a potential difference numerically. The structure of such a program is discussed in optional Section 16.13.

16.9   POTENTIAL DIFFERENCE IN AN INSULATOR

Reasoning from the definition of equilibrium and the existence of a sea of mobile electrons in a metal, we were able to conclude that in equilibrium the net electric field everywhere inside a metal is zero. Given this, we were able to conclude that the potential difference between any two locations inside a metal object in equilibrium must be zero.

The situation inside a polarized insulator is more complex. An applied field, such as the uniform field inside a capacitor (Figure 16.45), polarizes the molecules in the insulator. These polarized molecules themselves contribute to the net field inside the material. Presumably, however, the electric field inside the plastic due to the polarized molecules varies depending on the observation location.

For example, consider locations A and B inside the polarized plastic shown in Figure 16.46, where we show electric field contributed solely by the induced dipoles in the polarized plastic. If we consider columns of polarized molecules to be similar to capacitors (consisting of two vertical sheets of charge), then location A is inside "capacitor" 1, while location B is between "capacitors" 2 and 3. At location A the dominant contribution by the dipoles will be from the

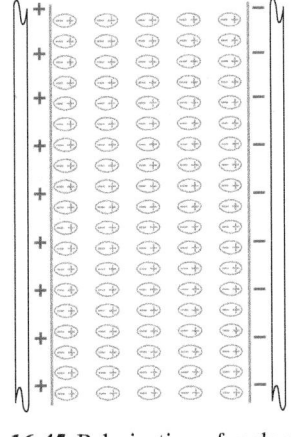

Figure 16.45 Polarization of molecules in a piece of plastic inside a capacitor.

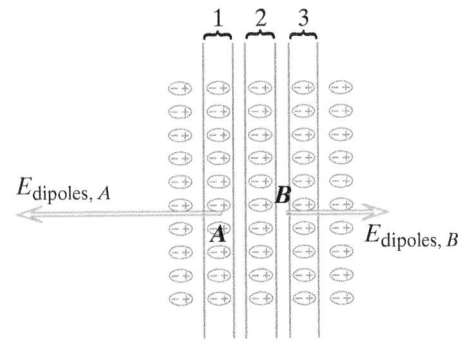

Figure 16.46 Electric field contributed solely by the induced dipoles in the plastic.

molecules in capacitor 1, and the electric field due to the dipoles will be large and to the left.

At location B, the fringe fields of the nearest two capacitors represent the dominant contributions by the dipoles, and the field due to the dipoles will point to the right and be smaller than the field due to the dipoles at A. If we follow a path from the left side of the plastic to the right side, $\vec{E}$ due to the dipoles will sometimes point in the direction of $d\vec{l}$ and sometimes opposite to the direction of $d\vec{l}$.

It would be useful to consider an average electric field due to the polarized molecules inside the plastic, but it isn't clear how we would calculate such a field. It is not even clear from simply inspecting the situation what the direction of this average field will be. We can construct a surprisingly simple argument based on potential to help us answer this question.

Round-Trip Potential Difference

Although the pattern of $E_{dipoles}$ is complex inside the material, it is much less complex at locations outside the plastic. Intuitively, you may be able to see that at locations outside the plastic the electric field due to the polarized plastic will have a similar pattern to the electric field near a single dipole.

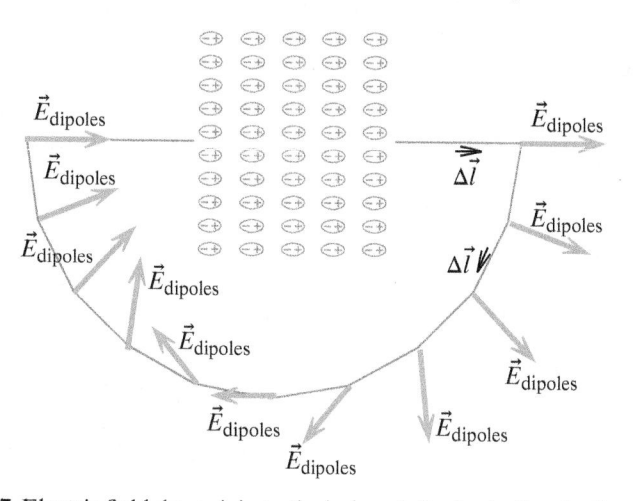

Figure 16.47 Electric field due solely to the induced dipoles in the plastic, along a path outside the plastic.

In Figure 16.47 the electric field due just to the polarized plastic is shown at locations along a path outside the plastic. If we travel clockwise around the path, it is clear from the diagram that $\vec{E} \bullet d\vec{l}$ is always a positive quantity: there is never a component of $\vec{E}$ opposite to $d\vec{l}$. Therefore $-\int \vec{E} \bullet d\vec{l} < 0$ along the path outside the plastic.

Since the molecular dipoles consist of stationary point charges, the round-trip path integral of the electric field (the round-trip potential difference) due to the molecular dipoles must sum to zero.

> QUESTION Explain why this implies that the average field inside the plastic must point to the left and cannot point to the right.

If the average field inside the plastic pointed to the right, the round-trip path integral of electric field in Figure 16.47 would be nonzero, which is impossible. A very general argument concerning the nature of potential lets us deduce the direction of the average field inside the plastic.

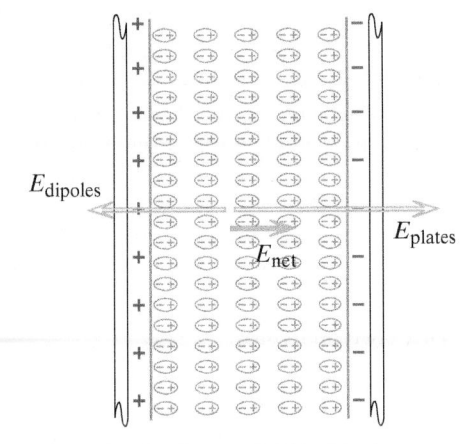

Figure 16.48 The net field inside the polarized insulator is smaller in magnitude than the applied field.

Material	K
Vacuum	1 (exactly)
Air	1.0006
Carbon tetrachloride	2.2
Typical plastic	5
Sodium chloride	6.1
Water	80 at 20 °C
Strontium titanate	310

Figure 16.49 Values of the dielectric constant of several materials.

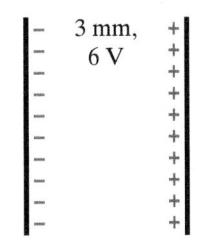

Figure 16.50 Capacitor before inserting a glass disk 1 mm in thickness between the plates.

Dielectric Constant and Net Field

The net electric field inside the plastic is the sum of the field due to the capacitor plates and the field due to the induced dipoles in the plastic, which, as we have just demonstrated, points in a direction opposite to the capacitor's field (Figure 16.48). The net field inside the plastic is therefore smaller than the field due to the capacitor alone. We can say informally that the electric field is "weakened" inside an insulator. Note carefully that the *net* field is in the same direction as the field made by the plates, but smaller in magnitude.

We define a constant K, called the "dielectric constant" (dielectric is another word for insulator), as the factor by which the net electric field is weakened inside an insulator:

$$\text{Inside an insulator:} \quad \vec{E}_{\text{insulator}} = \frac{\vec{E}_{\text{applied}}}{K}$$

For a capacitor, the applied field is $(Q/A)/\varepsilon_0$ and the net field inside a capacitor that is filled with an insulator is $(Q/A)/\varepsilon_0/K$. The dielectric constant is related to the atomic polarizability, but the details are beyond the scope of this introductory textbook. (The difficulty in making the connection is in accounting for the electric fields produced by the other polarized molecules.)

Note that the dielectric constant K is always bigger than 1, because polarization always weakens the net electric field inside the insulator. The easier it is to polarize a molecule, the bigger is the field-weakening effect and the bigger the value of K. Figure 16.49 gives representative values of the dielectric constant K for various insulators.

If there is no insulator in the gap, the potential difference across the capacitor is $|\Delta V| = Es$ (since $\vec{E}$ is uniform, and parallel to $d\vec{r}$). If the insulator fills the gap and we maintain the same charges $+Q$ and $-Q$ on the metal plates, both the electric field E and the potential difference ΔV are reduced by the same factor K:

$$\Delta V_{\text{insulator}} = \frac{\Delta V_{\text{vacuum}}}{K}$$

Since we can't actually get inside the insulator and measure the field there, we determine K for a particular insulator by measuring the effect on the potential difference ΔV between the plates for a fixed capacitor charge Q.

In summary, placing an insulator between the plates of a capacitor

- decreases electric field inside the insulator
- decreases potential difference across the insulator

If the insulator doesn't fill the gap, the electric field inside the insulator is reduced by the factor K, but the electric field at other places in the gap is hardly affected, because the electric field of the insulator in those places turns out to be small compared to the field made by the plates. (In our model of the insulator, consisting of capacitor-like sheets of plus and minus charges, the field made by the dipoles outside the insulator is a fringe field.) Outside the capacitor plates, however, the fringe field of the plates is small, and the small field contributed by the insulator significantly reduces the net field.

In Section 16.5 we examined the electric field and potential difference between the two metal plates of a capacitor, then inserted a metal slab. In the following Checkpoint we will start with the same device, but insert a glass slab.

Checkpoint 11 A capacitor with a 3-mm gap has a potential difference of 6 V (Figure 16.50). A disk of glass 1 mm thick, with area the same as the area of the metal plates, has a dielectric constant of 2.5. It is inserted in the middle of the gap between the metal plates. Now what is the potential difference of the two metal plates? (It helps to make a diagram showing the electric field along a path.)

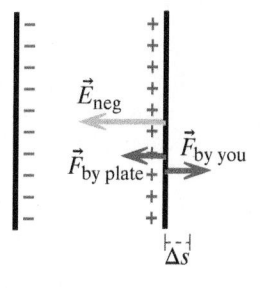

Figure 16.51 To move a capacitor plate you must exert a force equal to the force on the plate by the other plate.

16.10 ENERGY DENSITY AND ELECTRIC FIELD

Until now we have thought of energy (that is, electric potential energy) as associated with interacting charged particles. There is an alternative view, however, that considers energy to be stored in electric fields themselves. To see how this view works out quantitatively, we'll consider moving one plate of a capacitor.

The force that one capacitor plate exerts on the other (Figure 16.51) is equal to the charge Q on one plate times the field made by the other plate, which is half the total field in the gap:

$$E_{\text{one plate}} = \frac{(Q/A)}{2\varepsilon_0} \quad \text{(for very small gap)}$$

so the force on one plate is

$$F = Q\frac{(Q/A)}{2\varepsilon_0}$$

Suppose that you pull the positive plate of the capacitor slowly away from the negative plate, exerting a force only infinitesimally larger than the force exerted by the other plate. The work that you do in moving the plate a distance Δs goes into increasing the electric potential energy U:

$$\Delta U = W = Q\left(\frac{Q/A}{2\varepsilon_0}\right)\Delta s$$

We can rearrange this expression in the following way:

$$\Delta U = \frac{1}{2}\varepsilon_0\left(\frac{Q/A}{\varepsilon_0}\right)^2 A \; \Delta s$$

The expression in parentheses is the electric field E inside the capacitor. The quantity $A \; \Delta s$ is the change in the volume occupied by electric field inside the capacitor (length times width times increase in the plate separation). Therefore we can write

$$\frac{\Delta U}{\Delta(\text{volume})} = \frac{1}{2}\varepsilon_0 E^2$$

We ascribe an *energy density* (joules per cubic meter) to the electric field. By pulling the capacitor plates apart we increased the volume of space in which there is a sizable electric field. We say that the energy expended by us was converted into energy stored *in the electric field*.

This view is numerically equivalent to our earlier view of electric potential energy, but it turns out to be a more fundamental view. For example, the energy carried by electromagnetic radiation propagating through space far from any charges can best be expressed in terms of the energy density associated with the field.

FIELD ENERGY DENSITY

$$\frac{1}{2}\varepsilon_0 E^2 \; (\text{units are J/m}^3)$$

The energy density (joules per cubic meter) is stored in a region where there is an electric field of magnitude E.

Although we derived the field energy density for the particular case of a capacitor, the result is general. Anywhere there is electric field, there is energy density given by this equation.

Checkpoint 12 The energy density inside a certain capacitor is 10 J/m³. What is the magnitude of the electric field inside the capacitor? What is the energy density associated with an electric field of 3×10^6 V/m (large enough to initiate a spark)?

Figure 16.52 An electron and a positron are released from rest, some distance apart. We choose the electron as the system and everything else (including the positron) as the surroundings.

*An Electron and a Positron

In Chapter 6 we stated the principle of conservation of energy in this way: The change in energy of a system plus the change in energy of its surroundings must be zero. In the following example we will see that in order to understand energy conservation in quite a simple situation it is necessary to invoke the idea that energy is stored in fields.

Consider an electron and a positron that are released from rest some distance from each other (Figure 16.52). We will take the electron to be the system under consideration, so therefore the positron and everything else in the Universe are the "surroundings." Because of the attractive electric force between the particles, the electron accelerates toward the positron, gaining kinetic energy. By the principle of conservation of energy, the energy of the surroundings must therefore decrease. (Remember that the system can have no potential energy, since it consists of a single particle, and potential energy is a property of pairs of particles.)

QUESTION Does the energy of the positron decrease?

No, the energy of the positron also increases, since it accelerates toward the electron, gaining kinetic energy.

QUESTION Where is there a decrease of energy in the surroundings?

Evidently the energy stored in the fields surrounding the two particles must decrease. Clearly, the electric field at any location in space does change as the positions of the particles change. The electric field in the region between the particles gets larger, but the electric field everywhere else in space decreases (since E_{dipole} is proportional to s, the distance between the particles). It would be a somewhat daunting task to integrate E^2 over the volume of the Universe, with the additional complication that close to a charged particle E approaches infinity. However, we do not actually need to do this integral to figure out the change in energy of the electric field throughout space. Since

$$\Delta(\text{Field energy}) + \Delta K_{\text{positron}} + \Delta K_{\text{electron}} = 0$$

then

$$\Delta(\text{Field energy}) = -2(\Delta K_{\text{electron}})$$

In this example, the principle of conservation of energy leads us directly to the idea that energy must be stored in electric fields, since there is no other way to account for the decrease of energy in the surroundings.

If we had chosen the electron plus the positron as our system, we would have found that ΔU_{el} is equal to $-2(\Delta K_{\text{electron}})$. The change in potential energy for the two-particle system is the same as the change in the field energy. Evidently in a multiparticle system we can either consider a change in potential energy or a change in field energy (but not both); the quantities are equal.

The idea of energy stored in fields is a general one. It is not only electric fields that carry energy, but magnetic fields and gravitational fields as well.

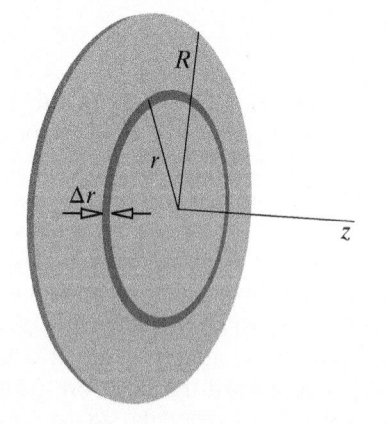

Figure 16.53 A ring of radius r and width Δr makes a contribution V_{ring} to the potential of the disk.

16.11 *POTENTIAL OF DISTRIBUTED CHARGES

Potential Along the Axis of a Uniformly Charged Disk

Consider a disk of radius R (area $A = \pi R^2$) with charge Q uniformly distributed over its surface. To calculate the potential due to this disk we follow a procedure similar to the one we used in finding the electric field of a charge distribution. We divide the disk into rings as we did in finding the electric field of a disk in Chapter 15 (Figure 16.53).

At location z along the axis of a disk, the potential contributed by one ring is given by the result found earlier:

$$V_{\text{ring}} = \frac{1}{4\pi\varepsilon_0} \frac{\Delta q}{(z^2 + r^2)^{1/2}} = \frac{1}{4\pi\varepsilon_0} \frac{Q}{A} \frac{2\pi r \Delta r}{(z^2 + r^2)^{1/2}}$$

We have used the result from the previous chapter for the charge on one of the rings, $\Delta q = (Q/A)2\pi r\Delta r$. Add up the contributions of all the rings:

$$V = \frac{1}{2\varepsilon_0} \frac{Q}{A} \int_0^R \frac{r\,dr}{(z^2 + r^2)^{1/2}}$$

$$V = \frac{1}{2\varepsilon_0} \frac{Q}{A} \left[(z^2 + r^2)^{1/2}\right]_0^R$$

$$V = \frac{1}{2\varepsilon_0} \frac{Q}{A} \left[(z^2 + R^2)^{1/2} - z\right]$$

The units for this result check: meters in the bracket, and m² in the A term in the denominator, so the units are those of $Q/(\varepsilon_0 R)$, which are indeed units of potential.

We can find the electric field from the negative gradient of the potential. We want the z component of electric field, and taking the derivative of the potential we find the following:

$$E = -\frac{\Delta V}{\Delta z} \rightarrow -\frac{dV}{dz} = \frac{(Q/A)}{2\varepsilon_0} \left[1 - \frac{z}{(z^2 + R^2)^{1/2}}\right]$$

This is the same result obtained in Chapter 15 by summing the contributions to the electric field.

Checkpoint 13 What is the potential at the center of a spherical shell of radius R that has a charge Q uniformly distributed over its surface? Don't do an integral of the electric field; just add up the contributions to the potential by the charges.

16.12 *INTEGRATING THE SPHERICAL SHELL

In this section, we'll use potential to prove that a uniformly charged spherical shell looks from the outside like a point charge but on the inside has a zero electric field. We did this by integrating the electric field directly in Chapter 15, and a very different kind of proof using Gauss's law is given in a later chapter.

Divide the spherical shell into rings of charge, each delimited by the angle θ and the angle $\theta + \Delta\theta$, and carrying an amount of charge ΔQ (Figure 16.54). The angle θ will be the integration variable used in summing up the contributions of the various rings.

Each ring contributes V_{ring} at an observation point a distance r from the center of the spherical shell. Each ring is a distance $(r - R\cos\theta)$ from

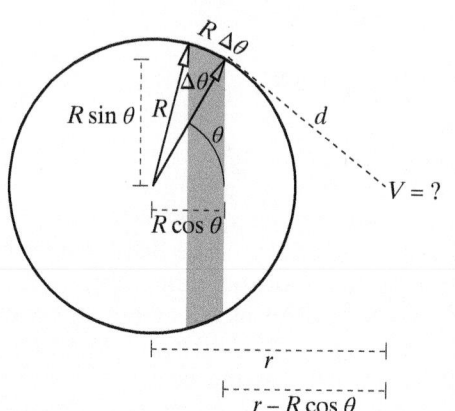

Figure 16.54 The sphere may be divided into ring-shaped segments.

the observation location (we will eventually choose $\Delta\theta$ so small that it makes no difference whether we measure to the edge of the ring or to the center of the ring). Therefore the distance d from the observation location to each charge on the ring is $\sqrt{[(r-R\cos\theta)^2+(R\sin\theta)^2]}$:

$$V_{\text{ring}} = \frac{1}{4\pi\varepsilon_0} \frac{\Delta Q}{[(r-R\cos\theta)^2+(R\sin\theta)^2]^{1/2}}$$

The total amount of charge on each ring is

$$\Delta Q = \left(\frac{Q}{4\pi R^2}\right)\Delta A$$

where ΔA is the surface area of the ring, since the total charge Q is spread uniformly all over the spherical surface, whose total area is $4\pi R^2$. To calculate the surface area ΔA of the ring, as before we lay the ring out flat (Figure 16.55), noting that the radius of the ring is $R\sin\theta$ and its width is $R\,\Delta\theta$ (since arc length is radius times angle, with angle measured in radians).

Putting these elements together, we have for the potential due to the ring

$$V_{\text{ring}} = \frac{1}{4\pi\varepsilon_0} \frac{1}{[(r-R\cos\theta)^2+(R\sin\theta)^2]^{1/2}} \left(\frac{Q}{4\pi R^2}\right) \times 2\pi(R\sin\theta)R\Delta\theta$$

We can add up all these contributions in the form of a definite integral:

$$V = \frac{1}{4\pi\varepsilon_0} \frac{Q}{2} \int_0^\pi \frac{\sin\theta d\theta}{[(r-R\cos\theta)^2+(R\sin\theta)^2]^{1/2}}$$

The limits on the integral are determined by the fact that if we let θ range between 0 and π rad (180°), we add up rings that account for the entire surface of the spherical shell.

A change of variables lets us evaluate this integral. Let

$$u = (r-R\cos\theta)^2+(R\sin\theta)^2$$
$$u = r^2 - 2rR\cos\theta + (R\cos\theta)^2 + (R\sin\theta)^2$$

However, $(R\cos\theta)^2 + (R\sin\theta)^2 = R^2$, from the Pythagorean theorem. Therefore

$$u = r^2 - 2rR\cos\theta + R^2$$

$$du = 2rR\sin\theta d\theta \quad \text{and} \quad \sin\theta d\theta = \frac{1}{2rR}du$$

$$V = \frac{1}{4\pi\varepsilon_0} \frac{Q}{2} \frac{1}{2rR} \int_{\theta=0}^{\theta=\pi} \frac{du}{u^{1/2}}$$

$$V = \frac{1}{4\pi\varepsilon_0} \frac{Q}{2} \frac{1}{2rR} \left[2\sqrt{(r-R\cos\theta)^2+(R\sin\theta)^2}\right]_0^\pi$$

$$V = \frac{1}{4\pi\varepsilon_0} \frac{Q}{2rR} \left[\sqrt{(r+R)^2} - \sqrt{(r-R)^2}\right]$$

$$V = \frac{1}{4\pi\varepsilon_0} \frac{Q}{2rR} [(r+R) \mp (r-R)]$$

This gives two results, depending on which sign is taken for the square root. Take the $-$ sign, which turns out to correspond to a location outside the shell:

$$V = \frac{1}{4\pi\varepsilon_0} \frac{Q}{r} \quad \text{(outside the shell; } r > R)$$

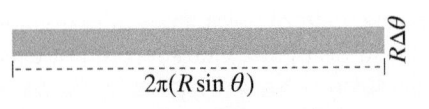

$2\pi(R\sin\theta)$

$R\Delta\theta$

Figure 16.55 We can calculate the area of the ring of charge by "unrolling" it.

This says that the potential outside a uniformly charged spherical shell is exactly the same as the potential due to a point charge located at the center of the shell, as though the shell were collapsed to a point. Since the electric field is the negative gradient of the potential, the electric field outside the shell is the same as if the shell were collapsed to its center.

If we take the + sign, we have

$$V = \frac{1}{4\pi\varepsilon_0} \frac{Q}{R} \quad \text{(inside the shell; } r < R)$$

This says that the potential inside the shell is constant, and equal to the potential just outside the shell. A constant potential means that the electric field inside the shell is zero. Note that for locations inside the spherical shell, some of the rings give a field to the left, and some give a field to the right. It turns out that these contributions exactly cancel each other.

It is the $1/r^2$ behavior of the electric field of point charges that leads to these unusual results. If the electric field of a point charge were not exactly proportional to $1/r^2$, the electric field outside a uniform spherical shell would not look like the field of a point charge at the center, and the electric field would not be zero inside the shell.

16.13 *NUMERICAL INTEGRATION ALONG A PATH

The key issue in writing a program to calculate $-\int_A^B \vec{E} \cdot \Delta\vec{l}$ is specifying the path. One way to do this is to find the vector $\vec{B} - \vec{A}$ from the initial to the final location, and then to divide this vector by the number of steps you want to take along the path to get $\Delta\vec{l}$. For example:

```
A = vector(0.15,0,0)
B = vector(0.20,0.35,0)
Nsteps = 10
deltal = (B-A)/Nsteps
```

Each iteration of the loop corresponds to one step along the path. For each step, we need to find $-\vec{E} \cdot \Delta\vec{l}$, using the value of $\vec{E}$ in the middle of the step. We then need to advance the position one step along the path. For example:

```
deltaV_BA = 0
L = vector(0.15,0,0)  ## initial position
i = 0
while i < Nsteps:
    rate(10) ## slow rate to watch the process
    E = ## calculate E at location L + deltal/2
    ## display E and deltal as arrows
    deltaV = -dot(E,deltal)
    deltaV_BA = deltaV_BA + deltaV
    L = L + deltal
    i = i + 1
```

If the full path is more complicated than a single straight line, it may be necessary to divide the path into several segments.

SUMMARY

Potential difference

$$\Delta V = V_f - V_i = -\int_i^f \vec{E} \bullet d\vec{l}$$

along a path from i to f

$$= -\int_i^f E_x dx - \int_i^f E_y dy - \int_i^f E_z dz$$

Units of electric potential are joules/coulomb, or volts.
Special case:
Regions of uniform electric field:

$$\Delta V = \sum(-\vec{E}\bullet\Delta\vec{l}) = \sum(-E_x\Delta x + -E_y\Delta y + -E_z\Delta z)$$

Change in electric potential energy for a particle of charge q moved through a potential difference ΔV:

$$\Delta U_{electric} = q\Delta V$$

Sign of ΔV and direction of path relative to $\vec{E}$:
Path going in the direction of $\vec{E}$: $\Delta V < 0$
Path going opposite to $\vec{E}$: $\Delta V > 0$
Path perpendicular to $\vec{E}$: $\Delta V = 0$

Point charge: $\Delta V = \Delta\left(\dfrac{1}{4\pi\varepsilon_0}\dfrac{Q}{r}\right)$

If the electric field is known along a path from location A to location B, the potential difference can be calculated as a path integral of the electric field:

Step 1: Choose a path and cut it into pieces $\Delta\vec{l}$.
Step 2: Write an expression for $\Delta V = -\vec{E}\bullet\Delta\vec{l}$ of one piece.
Step 3: Add up the contributions of all the pieces.
Step 4: Check the result, especially the sign.

If you can calculate the potential relative to infinity at locations A and B, it is not necessary to integrate the electric field along a path.

Path independence
Change in potential between two locations is independent of path.
For a round-trip path, $\Delta V = 0$.

Conductor in equilibrium

$\Delta V = -\int_i^f \vec{E} \bullet d\vec{l} = 0$ along any path, so potential is uniform (constant everywhere) throughout a metal in equilibrium.

Field is gradient of potential

$$E_x = \frac{\partial V}{\partial x}, \quad E_y = \frac{\partial V}{\partial y}, \quad E_z = \frac{\partial V}{\partial z}$$

Dielectric constant
Inside an insulator $\vec{E}_{insulator} = \dfrac{\vec{E}_{applied}}{K}$; K is the dielectric constant; $K > 1$, so

$$\Delta V_{insulator} = \frac{\Delta V_{vacuum}}{K}$$

In summary, placing an insulator between the plates of a capacitor

- decreases the electric field inside the insulator
- decreases the potential difference across the insulator

Field energy density

$$\frac{1}{2}\varepsilon_0 E^2 \quad (\text{units are J/m}^3)$$

QUESTIONS

Q1 Which statements about potential energy are correct? Select all that apply. (1) Potential energy can be either positive or negative. (2) Potential energy is a property of a pair of objects. (3) When two interacting charged objects get very far away from each other, their potential energy approaches zero. (4) A single particle does not have potential energy. (5) A single isolated object can have potential energy. (6) Potential energy is always a positive number. (7) The potential energy of a pair of interacting charged objects increases as the distance between them increases.

Q2 The graph in Figure 16.56 shows the electric potential energy for a system of two interacting objects, as a function of the distance between the objects. What system(s) might this graph represent?

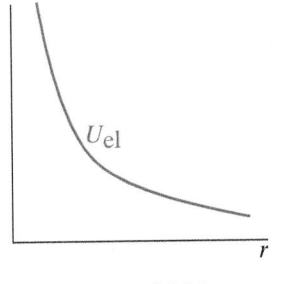

Figure 16.56

(1) Two protons, (2) Two sodium ions, (3) Two neutrons, (4) Two chloride ions, (5) Two electrons, (6) A proton and an electron, (7) A sodium ion and a chloride ion

Q3 The graph in Figure 16.57 shows the electric potential energy for a system of two interacting objects, as a function of the distance between the objects. What system(s) might this graph represent?

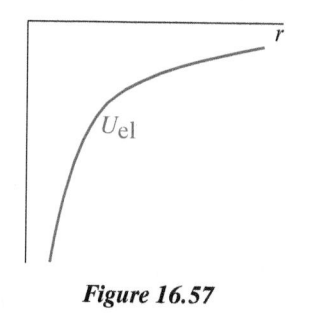

Figure 16.57

(1) Two protons, (2) Two sodium ions, (3) Two neutrons, (4) Two chloride ions, (5) Two electrons, (6) A proton and an electron, (7) A sodium ion and a chloride ion

Q4 What is the difference between electric potential energy and electric potential?

Q5 What are the units of electric potential energy, of electric potential, and of electric field?

Q6 In Figure 16.58, what is the direction of the electric field? Is $\Delta V = V_f - V_i$ positive or negative?

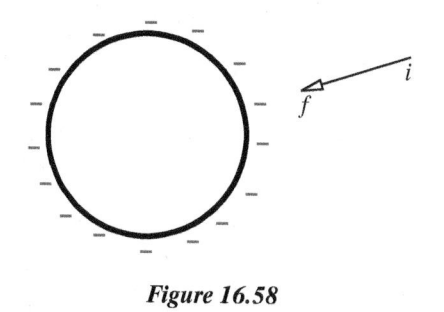

Figure 16.58

Q7 You travel along a path from location A to location B, moving in the same direction as the direction of the net electric field in that region. What is true of the potential difference $V_B - V_A$? (1) $V_B - V_A > 0$, (2) $V_B - V_A < 0$, (3) $V_B - V_A = 0$

Q8 You travel along a path from location A to location B, moving in a direction opposite to the direction of the net electric field in that region. What is true of the potential difference $V_B - V_A$? (1) $V_B - V_A > 0$, (2) $V_B - V_A < 0$, (3) $V_B - V_A = 0$

Q9 You travel along a path from location A to location B, moving in a direction perpendicular to the direction of the net electric field in that region. What is true of the potential difference $V_B - V_A$? (1) $V_B - V_A > 0$, (2) $V_B - V_A < 0$, (3) $V_B - V_A = 0$

Q10 Figure 16.59 shows several locations inside a capacitor. You need to calculate the potential difference $V_D - V_C$.

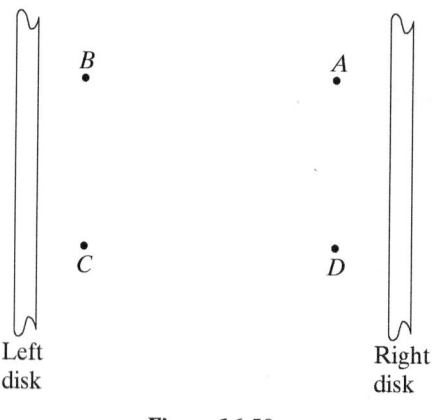

Figure 16.59

(a) What is the direction of the path ($+x$ or $-x$)? **(b)** If the charge on the right plate is negative and the charge on the left plate is positive, what is the sign of $V_D - V_C$?

Q11 Figure 16.60 shows a portion of a long, negatively charged rod. You need to calculate the potential difference $V_A - V_B$.

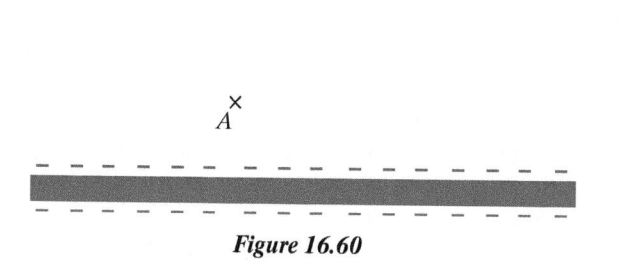

Figure 16.60

(a) What is the direction of the path ($+y$ or $-y$)? **(b)** What is the sign of $V_A - V_B$?

Q12 Location A is a distance d from a charged particle. Location B is a distance $2d$ from the particle. Which of the following statements are true? It may help to draw a diagram. (1) If the charge of the particle is negative, $V_B - V_A$ is negative. (2) If the charge of the particle is positive, $V_A < V_B$. (3) If $V_B < V_A$, we know that the particle must be positive. (4) $V_B < V_A$, regardless of the sign of the charge of the particle. (5) The sign of $(V_B - V_A)$ does not give us any information about the sign of the charge of the particle.

Q13 Locations $A = \langle a,0,0 \rangle$ and $B = \langle b,0,0 \rangle$ are on the $+x$ axis, as shown in Figure 16.61. Four possible expressions for the electric field along the x axis are given below. For each expression for the electric field, select the correct expression (1–8) for the potential difference $V_A - V_B$. In each case K is a numerical constant with appropriate units.

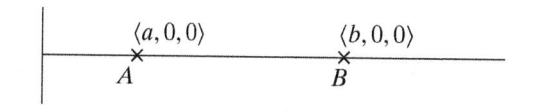

Figure 16.61

(a) $\vec{E} = \left\langle \dfrac{K}{x^2}, 0, 0 \right\rangle$, **(b)** $\vec{E} = \left\langle \dfrac{K}{x^3}, 0, 0 \right\rangle$, **(c)** $\vec{E} = \left\langle \dfrac{K}{x}, 0, 0 \right\rangle$,

(d) $\vec{E} = \langle Kx, 0, 0 \rangle$

1. $V_A - V_B = 0$

2. $V_A - V_B = K(a - b)$

3. $V_A - V_B = K\left(\dfrac{1}{a} - \dfrac{1}{b}\right)$

4. $V_A - V_B = K\left(\dfrac{1}{a^3}a - \dfrac{1}{b^3}b\right)$

5. $V_A - V_B = \frac{1}{2}K\left(b^2 - a^2\right)$

6. $V_A - V_B = K\ln\left(\dfrac{b}{a}\right)$

7. $V_A - V_B = K\left(a^2 - b^2\right)$

8. $V_A - V_B = \frac{1}{2}K\left(\dfrac{1}{a^2} - \dfrac{1}{b^2}\right)$

Q14 Figure 16.62 shows a pattern of electric field in which the electric field is horizontal throughout this region but is larger toward the top of the region and smaller toward the bottom. If this pattern of electric field can be produced by some arrangement of stationary charges, sketch such a distribution of charges. If this pattern of electric field cannot be produced by any arrangement of stationary charges, prove that it is impossible.

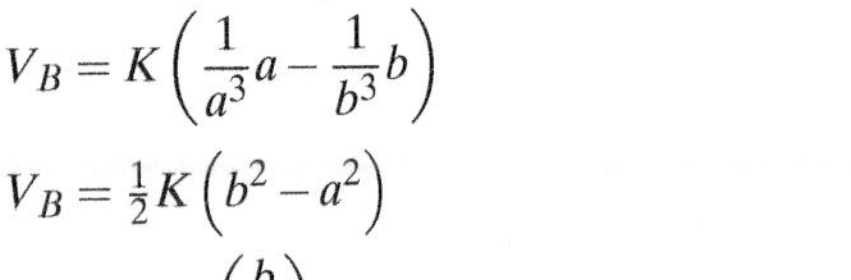

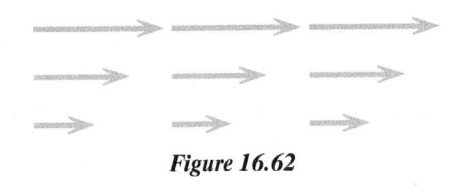

Figure 16.62

Q15 The graph in Figure 16.63 is a plot of electric potential versus distance from an object. Which of the following could be the object?
(1) A neutron, (2) A sodium ion (Na$^+$), (3) A chloride ion (Cl$^-$), (4) A proton, (5) An electron

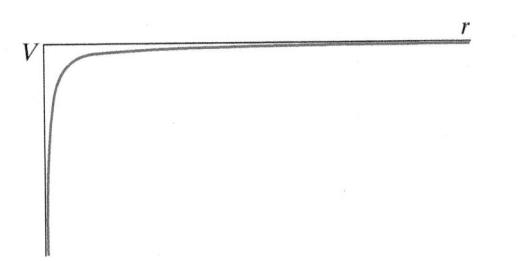

Figure 16.63

Q16 The graph in Figure 16.64 is a plot of electric potential versus distance from an object. Which of the following could be the object?

Figure 16.64

(1) A neutron, (2) A sodium ion (Na$^+$), (3) A chloride ion (Cl$^-$), (4) A proton, (5) An electron

Q17 A student said, "The electric field at the center of a charged spherical shell is zero, so the potential at that location is also zero." Explain to the student why this statement is incorrect.

Q18 For each of the following statements, say whether it is true or false and explain why it is true or false. Be complete in your explanation, but be brief. Pay particular attention to the distinction between potential V and potential difference ΔV. **(a)** The electric potential inside a metal in equilibrium is always zero. **(b)** If there is a constant large positive potential throughout a region, the electric field in that region is large. **(c)** If you get close enough to a negative point charge, the potential is negative, no matter what other charges are around. **(d)** Near a point charge, the potential difference between two points a distance L apart is $-EL$. **(e)** In a region where the electric field is varying, the potential difference between two points a distance L apart is $-(E_f - E_i)L$.

Q19 We discussed a method for measuring the dielectric constant by placing a slab of the material between the plates of a capacitor. Using this method, what would we get for the dielectric constant if we inserted a slab of metal (not quite touching the plates, of course)?

PROBLEMS

Section 16.1

•P20 What is the kinetic energy of a proton that is traveling at a speed of 3725 m/s?

•P21 If the kinetic energy of an electron is 4.4×10^{-18} J, what is the speed of the electron? You can use the approximate (nonrelativistic) equation here.

•P22 An electron passes through a region in which there is an electric field, and while it is in the region its kinetic energy decreases by 4×10^{-17} J. Initially the kinetic energy of the electron was 4.5×10^{-17} J. What is the final speed of the electron? (You can use the approximate (nonrelativistic) equation here.)

•P23 At a particular instant an electron is traveling with speed 6000 m/s. **(a)** What is the kinetic energy of the electron? **(b)** If a proton were traveling at the same speed (6000 m/s), what would be the kinetic energy of the proton?

•**P24** In Chapter 6 we saw that the electric potential energy of a system of two particles is given by the equation

$$U_{el} = \frac{1}{4\pi\varepsilon_0} \frac{q_1 q_2}{r}$$

(a) What is the electric potential energy of two protons separated by a distance of 9 nm (9×10^{-9} m)? **(b)** What is the electric potential energy of a proton and an electron separated by the same distance?

•**P25** A proton that initially is traveling at a speed of 300 m/s enters a region where there is an electric field. Under the influence of the electric field the proton slows down and comes to a stop. What is the change in kinetic energy of the proton?

Section 16.3

•**P26** You move from location i at $\langle 2,7,5 \rangle$ m to location f at $\langle 5,6,12 \rangle$ m. All along this path there is a nearly uniform electric field of $\langle 1000,200,-510 \rangle$ N/C. Calculate $V_f - V_i$, including sign and units.

•**P27** A capacitor with a gap of 1 mm has a potential difference from one plate to the other of 36 V. What is the magnitude of the electric field between the plates?

•**P28** An electron starts from rest in a vacuum, in a region of strong electric field. The electron moves through a potential difference of 35 V. **(a)** What is the kinetic energy of the electron in electron volts (eV)? **(b)** What would happen if the particle were a proton?

•**P29** Locations A, B, and C are in a region of uniform electric field, as shown in the diagram in Figure 16.65. Location A is

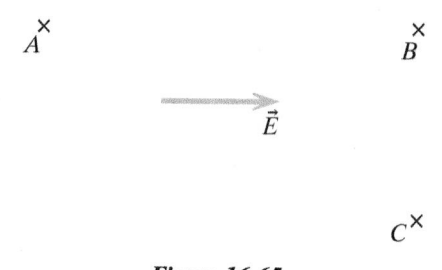

Figure 16.65

at $\langle -0.5,0,0 \rangle$ m. Location B is at $\langle 0.5,0,0 \rangle$ m. In the region the electric field has the value $\langle 750,0,0 \rangle$ N/C. For a path starting at B and ending at A, calculate: **(a)** the displacement vector $\Delta \vec{l}$, **(b)** the change in electric potential, **(c)** the potential energy change for the system when a proton moves from B to A, **(d)** the potential energy change for the system when an electron moves from B to A.

•**P30** Locations A, B, and C are in a region of uniform electric field, as shown in Figure 16.65. Location A is at $\langle -0.5,0,0 \rangle$ m. Location B is at $\langle 0.5,0,0 \rangle$ m. In the region the electric field has the value $\langle 750,0,0 \rangle$ N/C. **(a)** For a path starting at B and ending at C, calculate: (1) the displacement vector $\Delta \vec{l}$, (2) the change in electric potential, (3) the potential energy change for the system when a proton moves from B to C, (4) the potential energy change for the system when an electron moves from B to C, **(b)** Which of the following statements are true in this situation? Choose all that are correct. (1) the potential difference cannot be zero because the electric field is not zero along this path, (2) when a proton moves along this path, the electric force does zero net work on the proton, (3) $\Delta \vec{l}$ is perpendicular to $\vec{E}$.

•**P31** Locations A, B, and C are in a region of uniform electric field, as shown in Figure 16.66. Location A is at $\langle -0.3,0,0 \rangle$ m. Location B is at $\langle 0.4,0,0 \rangle$ m. In the region the electric field has the value $\langle -850,400,0 \rangle$ N/C. For a path starting at A and ending at B, calculate:

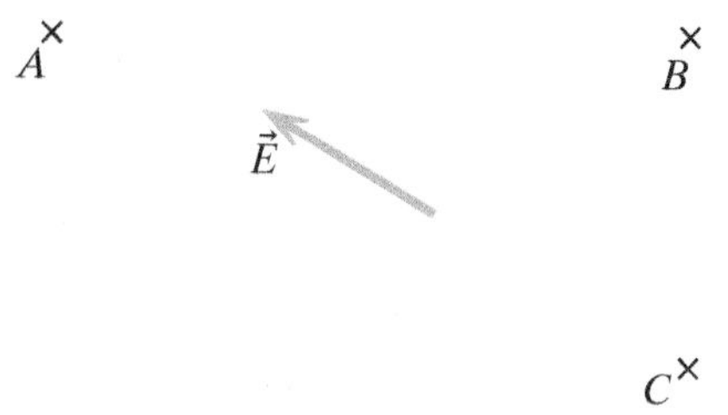

$A^\times$ $B^\times$

$\vec{E}$

$C^\times$

Figure 16.66

(a) the displacement vector $\Delta \vec{l}$, **(b)** the change in electric potential, **(c)** the potential energy change for the system when a proton moves from A to B, **(d)** the potential energy change for the system when an electron moves from A to B.

•**P32** Locations A, B, and C are in a region of uniform electric field, as shown in Figure 16.66. Location A is at $\langle -0.3,0,0 \rangle$ m. Location B is at $\langle 0.4,0,0 \rangle$ m. In the region the electric field has the value $\langle -850,400,0 \rangle$ N/C. For a path starting at C and ending at A, calculate: **(a)** the displacement vector $\Delta \vec{l}$, **(b)** the change in electric potential, **(c)** the potential energy change for the system when a proton moves from C to A, **(d)** the potential energy change for the system when an electron moves from C to A.

•**P33** Locations A and B are in a region of uniform electric field, as shown in Figure 16.67. Along a path from B to A, the change in potential is -2200 V. The distance from A to B is 0.28 m. What is the magnitude of the electric field in this region?

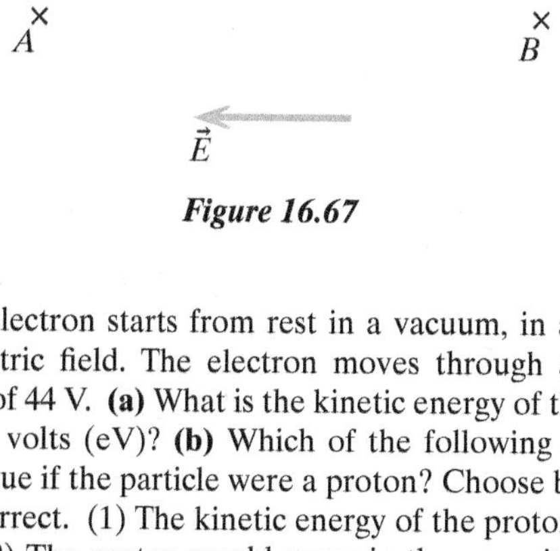

$A^\times$ $B^\times$

$\vec{E}$

Figure 16.67

•**P34** An electron starts from rest in a vacuum, in a region of strong electric field. The electron moves through a potential difference of 44 V. **(a)** What is the kinetic energy of the electron in electron volts (eV)? **(b)** Which of the following statements would be true if the particle were a proton? Choose both if they are both correct. (1) The kinetic energy of the proton would be negative. (2) The proton would move in the opposite direction from the electron.

•**P35** If the electric field exceeds about 3×10^6 N/C in air, a spark occurs. Approximately, what is the absolute value of the maximum possible potential difference between the plates of a capacitor whose gap is 3 mm, without causing a spark in the air between them?

•**P36** A capacitor with a gap of 2 mm has a potential difference from one plate to the other of 30 V. What is the magnitude of the electric field between the plates?

•**P37** You move from location i at $\langle 2,5,4 \rangle$ m to location f at $\langle 3,5,9 \rangle$ m. All along this path there is a nearly uniform electric field whose value is $\langle 1000, 200, -500 \rangle$ N/C. Calculate $\Delta V = V_f - V_i$, including sign and units.

•P38 In a particular region there is a uniform electric field of $\langle -760, 380, 0 \rangle$ V/m. Location A is $\langle 0.2, 0.1, 0 \rangle$ m, location B is $\langle 0.7, 0.1, 0 \rangle$, and location C is $\langle 0.7, -0.4, 0 \rangle$ m. (a) What is the change in potential along a path from B to A? (b) What is the change in potential along a path from A to C? (c) An alpha particle (two protons and two neutrons) moves from A to C. What is the change in potential energy of the system (alpha + source charges)?

••P39 The potential difference from one end of a 1-cm-long wire to the other in a circuit is $\Delta V = V_B - V_A = 1.5$ V, as shown in Figure 16.68. Which end of the wire is at the higher potential? What are the magnitude and direction of the electric field E inside the wire?

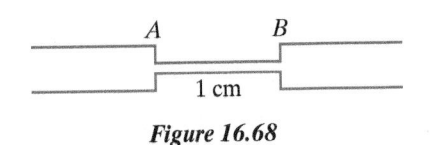

Figure 16.68

••P40 In the cathode ray tube found in old television sets, which contains a vacuum, electrons are boiled out of a very hot metal filament placed near a negative metal plate. These electrons start out nearly at rest and are accelerated toward a positive metal plate. They pass through a hole in the positive plate on their way toward the picture screen, as shown in the diagram in Figure 16.69. If the high-voltage supply in the television set maintains a potential difference of 15,000 V between the two plates, what speed do the electrons reach?

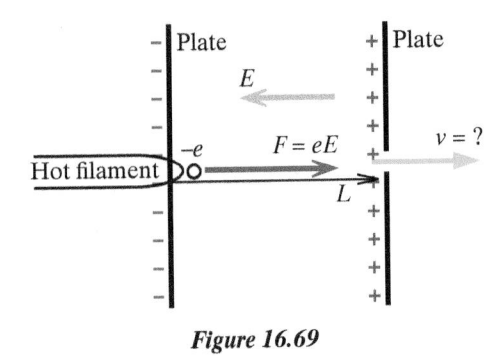

Figure 16.69

••P41 Two very large disks of radius R are carrying uniformly distributed charges Q_A and Q_B. The plates are parallel and 0.1 mm apart, as shown in Figure 16.70. The potential difference between the plates is $V_B - V_A = -10$ V. (a) What is the direction of the electric field between the disks? (b) Invent values of Q_A, Q_B, and R that would make $V_B - V_A = -10$ V.

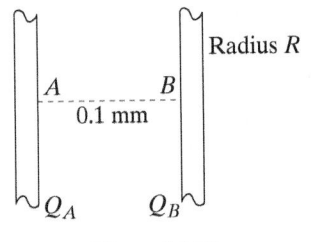

Figure 16.70

••P42 A particular oscilloscope, shown in Figure 16.71, has an 18,000-V power supply for accelerating electrons to a speed adequate to make the front phosphor-coated screen glow when the electrons hit it. Once the electron has emerged from the accelerating region, it coasts through a vacuum at nearly constant speed.

One can apply a potential difference of plus or minus 40 V across the deflection plates to steer the electron beam up or down on the screen to paint a display (other deflection plates not shown in the diagram are used to steer the beam horizontally).

Each of the two deflection plates is a thin metal plate of length $L = 8$ cm and width (into the diagram) 4 cm. The distance between the deflection plates is $s = 3$ mm. The distance from the deflection plates to the screen is $d = 30$ cm.

When there is a 40-V potential difference between the deflection plates, what is the deflection y of the electron beam where it hits the screen? An approximate treatment is fine, but state your assumptions. As is usually the case, it pays to carry out all of your calculations algebraically and only evaluate the final algebraic result numerically. Note the exaggerated vertical scale: the deflection is actually small compared to the distance to the screen.

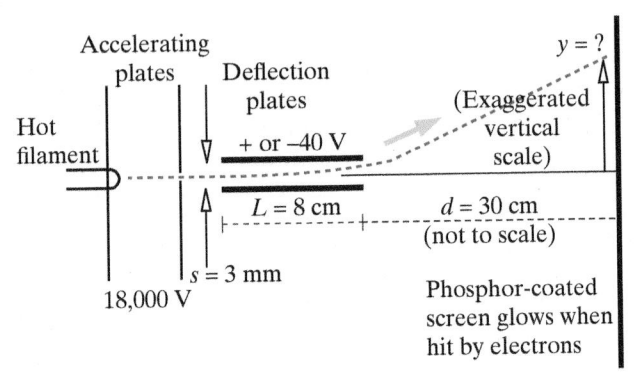

Figure 16.71

Section 16.5

•P43 Location C is 0.02 m from a small sphere that has a charge of 4 nC uniformly distributed on its surface. Location D is 0.06 m from the sphere. What is the change in potential along a path from C to D?

•P44 What is the change in potential ΔV in going from a location 3×10^{-10} m from a proton to a location 4×10^{-10} m from the proton?

•P45 An electron is initially at rest. It is moved from a location 4×10^{-10} m from a proton to a location 6×10^{-10} m from the proton. What is the change in electric potential energy of the system of proton and electron?

•P46 How much work is required to move a proton and an electron at rest a distance 3×10^{-8} m apart to be at rest a distance 7×10^{-8} m apart?

••P47 As shown in Figure 16.72, three large, thin, uniformly charged plates are arranged so that there are two adjacent regions of uniform electric field. The origin is at the center of the central plate. Location A is $\langle -0.4, 0, 0 \rangle$ m, and location B is $\langle 0.2, 0, 0 \rangle$ m. The electric field $\vec{E}_1$ has the value $\langle 725, 0, 0 \rangle$ V/m, and $\vec{E}_2$ is $\langle -425, 0, 0 \rangle$ V/m.

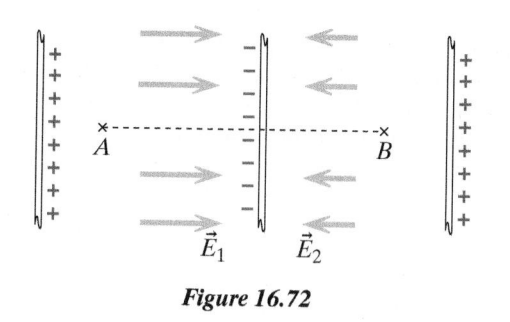

Figure 16.72

(a) Consider a path from A to B. Along this path, what is the change in electric potential? **(b)** What is the change in electric potential along a path from B to A? **(c)** There is a tiny hole in the central plate, so a moving particle can pass through the hole. If an electron moved from A to B along the path shown, what would be the change in its kinetic energy? **(d)** What is the minimum kinetic energy the electron must have at location A in order to ensure that it reaches location B?

••P48 If throughout a particular region of space the potential can be expressed as $V = 4xz + 2y - 5z$, what are the vector components of the electric field at location $\langle x, y, z \rangle$?

••P49 A dipole is oriented along the x axis. The dipole moment is $p\ (= qs)$. **(a)** Calculate exactly the potential V (relative to infinity) at a location $\langle x, 0, 0 \rangle$ on the x axis and at a location $\langle 0, y, 0 \rangle$ on the y axis, by superposition of the individual $1/r$ contributions to the potential. **(b)** What are the approximate values of V at the locations in part (a) if these locations are far from the dipole? **(c)** Using the approximate results of part (b), calculate the gradient of the potential along the x axis, and show that the negative gradient is equal to the x component E_x of the electric field. **(d)** Along the y axis, $dV/dy = 0$. Why isn't this equal to the magnitude of the electric field E along the y axis?

••P50 A capacitor consists of two large metal disks placed a distance s apart. The radius of each disk is R ($R \gg s$), and the thickness of each disk is t, as shown in Figure 16.73. The disk on the left has a net charge of $+Q$, and the disk on the right has a net charge of $-Q$. Calculate the potential difference $V_2 - V_1$, where location 1 is inside the left disk at its center, and location 2 is in the center of the air gap between the disks. Explain briefly.

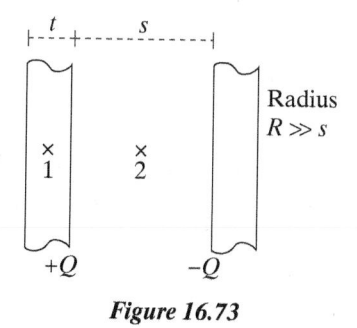

Figure 16.73

••P51 The diagram in Figure 16.74 shows three very large metal disks (seen edgewise), carrying charges as indicated. On each surface the charges are distributed approximately uniformly. Each disk has a very large radius R and a small thickness t. The distances between the disks are a and b, as shown; they also

are small compared to R. Calculate $V_2 - V_1$, and explain your calculation briefly.

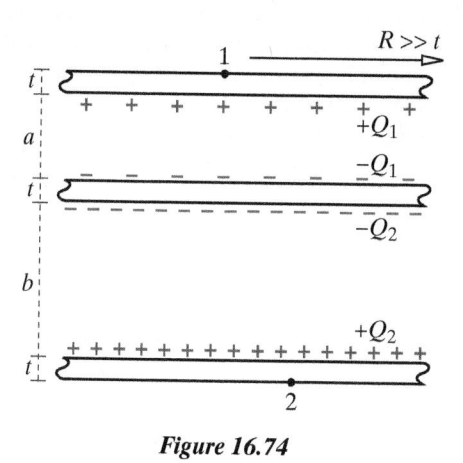

Figure 16.74

••P52 Three charged metal disks are arranged as shown in Figure 16.75 (cutaway view). The disks are held apart by insulating supports not shown in the diagram. Each disk has an area of $2.5\ \text{m}^2$ (this is the area of one flat surface of the disk). The charge $Q_1 = 5 \times 10^{-8}$ C and the charge $Q_2 = 4 \times 10^{-7}$ C.

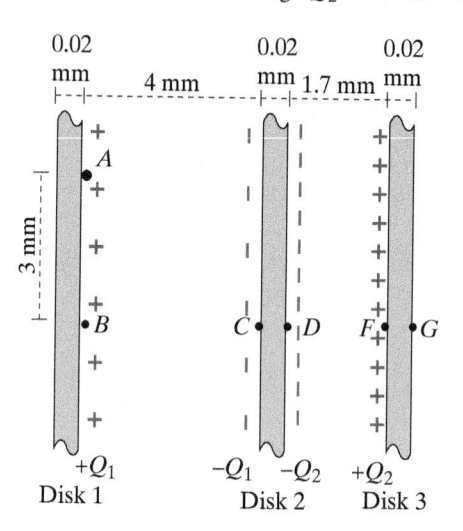

Figure 16.75

(a) What is the electric field (magnitude and direction) in the region between disks 1 and 2? **(b)** Which of the following statements are true? Choose all that apply. (1) Along a path from A to B, $\vec{E} \perp \Delta\vec{l}$. (2) $V_B - V_A = 0$. (3) $V_B - V_A = -\dfrac{Q_1/2.5}{\varepsilon_0} *$ (0.003) V. **(c)** To calculate $V_C - V_B$, where should the path start and where should it end? **(d)** Should $V_C - V_B$ be positive or negative? Why? (1) Positive, because $\Delta\vec{l}$ is opposite to the direction of $\vec{E}$. (2) Negative, because $\Delta\vec{l}$ is in the same direction as $\vec{E}$. (3) Zero, because $\Delta\vec{l} \perp \vec{E}$. **(e)** What is the potential difference $V_C - V_B$? **(f)** What is the potential difference $V_D - V_C$? **(g)** What is the potential difference $V_F - V_D$? **(h)** What is the potential difference $V_G - V_F$? **(i)** What is the potential difference $V_G - V_A$? **(j)** The charged disks have tiny holes that allow a particle to pass through them. An electron that is traveling at a fast speed approaches the plates from the left side. It travels along a path from A to G. Since no external work is done on the system of plates + electron, $\Delta K + \Delta U = W_{ext} = 0$. Consider the following states: initial, electron at location A; final, electron at location G. (1) What is the change in potential energy of the system? (2) What is the change in kinetic energy of the electron?

••P53 The long rod shown in Figure 16.76 has length L and carries a uniform charge $-Q$. Calculate the potential difference $V_A - V_C$. All of the distances are small compared to L. Explain your work carefully.

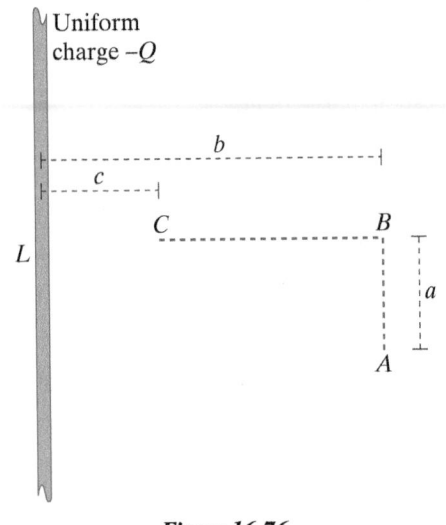

Figure 16.76

••P54 Three very large charged metal plates are arranged as shown in Figure 16.77. The radius of each plate is 4 m, and each plate is $w = 0.05$ mm thick. The separation d_1 is 6 mm, and the separation d_2 is 2 mm. Each plate has a tiny hole in it, so it is possible for a small charged particle to pass through all the plates.

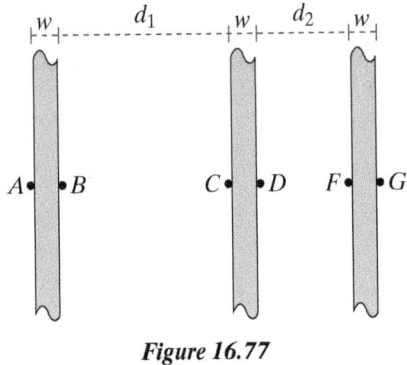

Figure 16.77

You are able to adjust the apparatus by varying the electric field in the region between location D and location F. You need to adjust this setting so that a fast-moving electron moving to the right, entering at location A, will have lost exactly 5.2×10^{-18} J of kinetic energy by the time it reaches location G. Using a voltmeter, you find that the potential difference $V_C - V_B = -16$ V. Based on this measurement, you adjust the electric field between D and F to the appropriate value. **(a)** Consider the system of (electron + plates). Neglecting the small amount of work done by the gravitational force on the electron, during this process (electron going from A to G), what is $\Delta K + \Delta U$? **(b)** What is the change in potential energy for the system during this process? **(c)** What is $V_G - V_A$? **(d)** What is $V_F - V_D$? **(e)** What is the electric field (magnitude and direction) in the region between locations D and F?

••P55 In the Van de Graaff generator shown in Figure 16.78, a rubber belt carries electrons up through a small hole in a large hollow spherical metal shell. The electrons come off the upper part of the belt and drift through a wire to the outer surface of the metal shell, so that the metal shell acquires a sizable negative charge, approximately uniformly distributed over the sphere. At a time when the sphere has acquired a sizable charge $-Q$, approximately how much work must be done by the motor to move one more electron from the base (a distance h below the sphere) to the upper pulley (located a distance $R/2$ from the center of the hollow sphere)? Explain your work, and state explicitly what approximations you had to make.

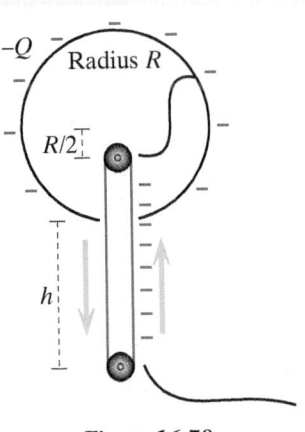

Figure 16.78

••P56 A thin spherical shell made of plastic carries a uniformly distributed negative charge $-Q_1$. As shown in Figure 16.79, two large thin disks made of glass carry uniformly distributed positive and negative charges $+Q_2$ and $-Q_2$. The radius R_1 of the plastic spherical shell is very small compared to the radius R_2 of the glass disks. The distance from the center of the spherical shell to the positive disk is d, and d is much smaller than R_2. **(a)** Find the potential difference $V_2 - V_1$ in terms of the given quantities (Q_1, Q_2, R_1, R_2, and d). Point 1 is at the center of the plastic sphere, and point 2 is just outside the sphere. **(b)** Find the potential difference $V_3 - V_2$. Point 2 is just below the sphere, and point 3 is right beside the positive glass disk.

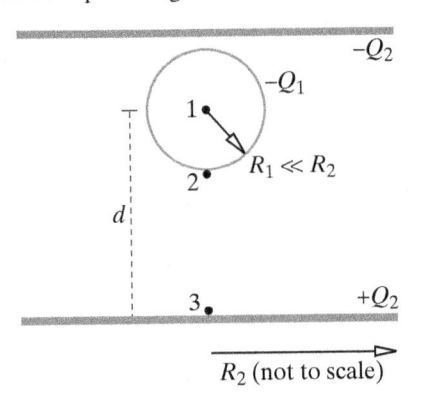

Figure 16.79

(c) Suppose that the plastic shell is replaced by a solid metal sphere with radius R_1 carrying charge $-Q_1$. State whether the absolute magnitudes of the potential differences would be greater than, less than, or the same as they were with the plastic shell in place. Explain briefly, including an appropriate diagram.

••**P57** A thin spherical shell of radius R_1 made of plastic carries a uniformly distributed negative charge $-Q_1$. A thin spherical shell of radius R_2 made of glass carries a uniformly distributed positive charge $+Q_2$. The distance between centers is L, as shown in Figure 16.80. **(a)** Find the potential difference $V_B - V_A$. Location A is at the center of the glass sphere, and location B is just outside the glass sphere. **(b)** Find the potential difference $V_C - V_B$. Location B is just outside the glass sphere, and location C is a distance d to the right of B. **(c)** Suppose the glass shell is replaced by a solid metal sphere with radius R_2 carrying charge $+Q_2$. Would the magnitude of the potential difference $V_B - V_A$ be greater than, less than, or the same as it was with the glass shell in place? Explain briefly, including an appropriate physics diagram.

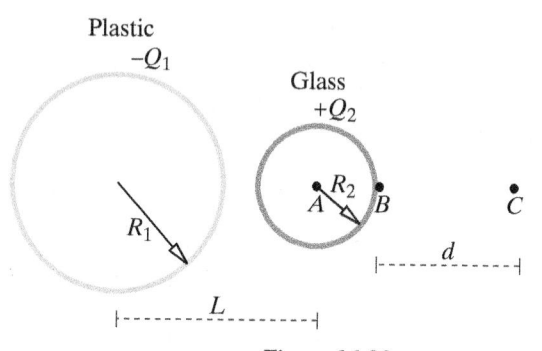

Plastic
$-Q_1$

Glass
$+Q_2$

R_2

A B

C

R_1

d

L

Figure 16.80

••**P58** As shown in Figure 16.81, a solid metal sphere of radius r_1 has a charge $+Q$. It is surrounded by a concentric spherical metal shell with inner radius r_2 and outer radius r_3 that has a charge $-Q$ on its inner surface and $+Q$ on its outer surface. In the diagram, point A is located at a distance r_4 from the center of the spheres. Points B and C are inside the metal shell, very near the outer and inner surfaces, respectively. Point E is just inside the surface of the solid sphere. Point D is halfway between C and E. Point F is a distance $r_1/2$ from the center. **(a)** Is each of the following potential differences greater than zero, equal to zero, or less than zero? Briefly explain why in terms of the directions of the electric field and of the path: (1) $V_B - V_A$, (2) $V_C - V_B$, (3) $V_D - V_C$, (4) $V_F - V_E$. **(b)** Calculate V_F, the potential at location F. Explain your work.

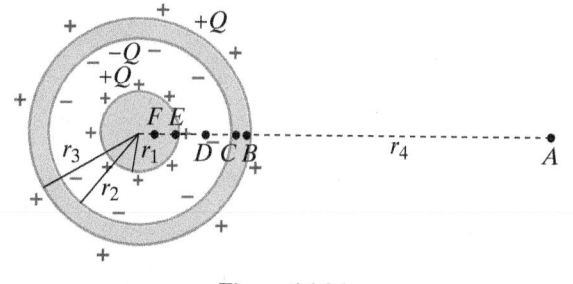

$+Q$

$-Q$

$+Q$

F E

r_3 r_1 D C B r_4 A

r_2

Figure 16.81

••**P59** In a region with an uniform electric field, you measure a potential difference of 74 V from the origin to a position of $\langle 0,0,10 \rangle$ m. Now we add a uniformly charged, thin spherical plastic shell centered at the origin. The spherical shell has a radius of 5 m and a charge of -3530 nC. Draw a diagram to help answer the following questions: **(a)** What is the potential difference from the origin to a position of $\langle 0,0,5 \rangle$ m (at the surface of the spherical

shell)? **(b)** What is the potential difference from the position of $\langle 0,0,5 \rangle$ m to a position of $\langle 0,0,10 \rangle$ m?

••**P60** A dipole is centered at the origin, with its axis along the y axis, so that at locations on the y axis, the electric field due to the dipole is given by

$$\vec{E} = \left\langle 0, \frac{1}{4\pi\varepsilon_0} \frac{2qs}{y^3}, 0 \right\rangle \frac{\text{V}}{\text{m}}$$

The charges making up the dipole are $+3$ nC and -3 nC, and the dipole separation is 2 mm (Figure 16.82). What is the potential difference along a path starting at location $\langle 0,0.03,0 \rangle$ m and ending at location $\langle 0,0.04,0 \rangle$ m?

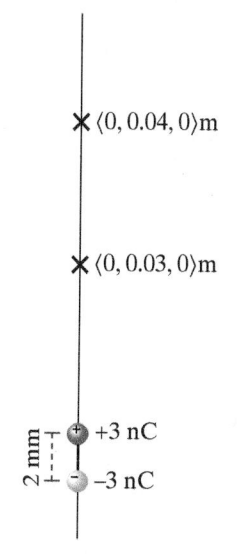

$\times \langle 0, 0.04, 0 \rangle$m

$\times \langle 0, 0.03, 0 \rangle$m

$+3$ nC

2 mm

-3 nC

Figure 16.82

••**P61** A thin spherical glass shell of radius R carries a uniformly distributed charge $+Q$, and a thin spherical plastic shell of radius R carries a uniformly distributed charge $-Q$. The surfaces of the spheres are a distance $L + 2d$ from each other, and locations A and B are a distance d from the surfaces of the spheres, as shown in Figure 16.83. Calculate the potential difference $V_B - V_A$.

$+Q$

R

A

B

$-Q$

R

d

L

d

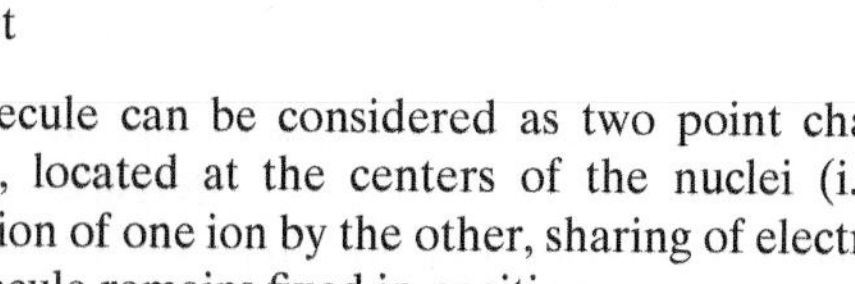

Figure 16.83

••**P62** An HCl molecule (Figure 16.84) in the gas phase has an internuclear separation s. For the purposes of this problem, assume that

- The molecule can be considered as two point charges, H^+ and Cl^-, located at the centers of the nuclei (i.e., ignore polarization of one ion by the other, sharing of electrons, etc.).
- The molecule remains fixed in position.
- The distances a and b shown in the diagram are much greater than shown ($a > b \gg s$).

(a) Draw vectors showing the relative magnitude and direction of the electric field at each of the three points indicated by a square box. Your drawing should be accurate enough to show clearly whether one vector is equal in magnitude, greater in magnitude, or less in magnitude than another vector, but need not be more accurate than that. **(b)** What is the potential difference $\Delta V = V_B - V_A$ in terms of the given quantities? **(c)** An electron passes point A traveling along the x axis in the negative x direction, toward the Cl$^-$. Its kinetic energy is K_A. What is its kinetic energy at point B?

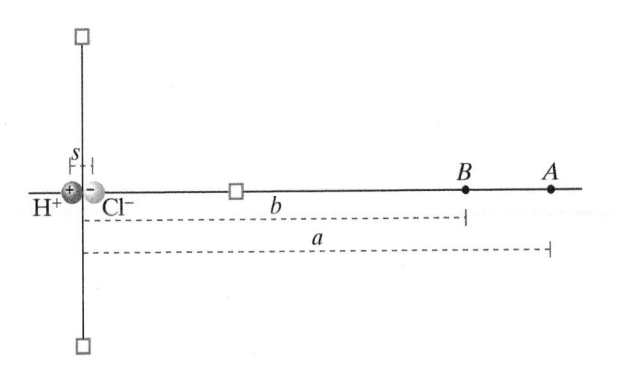

Figure 16.84

••P63 A very long, thin glass rod of length $2R$ carries a uniformly distributed charge $+q$, as shown in Figure 16.85. A very large plastic disk of radius R, carrying a uniformly distributed charge $-Q$, is located a distance d from the rod, where $d \ll R$. Calculate the potential difference $V_B - V_A$ from the center of the surface of the disk (location A) to a location a distance h from the center of the disk (location B). If you have to make any approximations, state what they are.

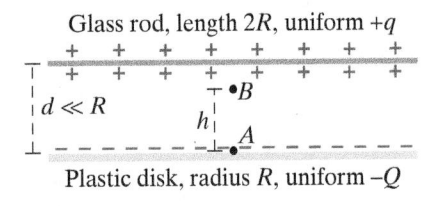

Glass rod, length $2R$, uniform $+q$

Plastic disk, radius R, uniform $-Q$

Figure 16.85

•••P64 A long thin metal wire with radius r and length L is surrounded by a concentric long narrow metal tube of radius R, where $R \ll L$, as shown in Figure 16.86. Insulating spokes hold the wire in the center of the tube and prevent electrical contact between the wire and the tube. A variable power supply is connected to the device as shown. There is a charge $+Q$ on the inner wire and a charge $-Q$ on the outer tube. As we will see when we study Gauss's law in a later chapter, the electric field inside the tube is contributed solely by the wire, and the field outside the wire is the same as though the wire were infinitely thin; the outer tube does not contribute as long as we are not near the ends of the tube. **(a)** In terms of the charge Q, length L, inner radius r, and outer radius R, what is the potential difference $V_{tube} - V_{wire}$ between the inner wire and the outer tube? Explain, and include checks on your answer. **(b)** The power-supply voltage is slowly increased until you see a glow in the air very near the inner wire. Calculate this power-supply voltage (give a numerical value), and explain your calculation. The length $L = 80$ cm, the inner radius $r = 0.7$ mm, and the outer radius $R = 3$ cm. This device is called a "Geiger–Müller tube" and was one of the first electronic particle detectors. The voltage is set just below the threshold for making the air glow near the wire. A charged particle that passes near the center wire can trigger breakdown in the air, leading to a large current that can be easily measured.

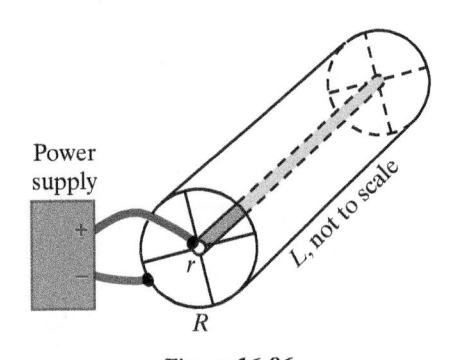

Figure 16.86

Section 16.6

•P65 In the region of space depicted in Figure 16.87 there are several stationary charged objects that are not shown in the diagram. Along a path $A = B = C = D$ you measure the following potential differences: $V_B - V_A = 12$ V; $V_C - V_B = -5$ V; $V_D - V_C = -15$ V. What is the potential difference $V_A - V_D$?

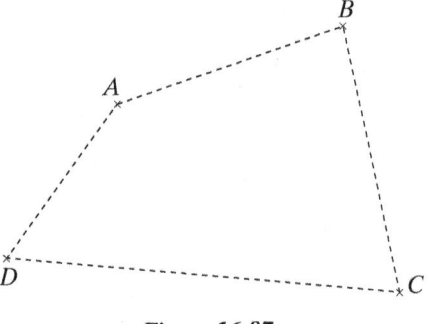

Figure 16.87

••P66 A capacitor consists of two charged disks of radius R separated by a distance s, where $R \gg s$. The magnitude of the charge on each disk is Q. Consider points A, B, C, and D inside the capacitor, as shown in Figure 16.88. **(a)** Show that $\Delta V = V_C - V_A$ is the same for these paths by evaluating ΔV along each path: (1) Path 1: $A = B = C$, (2) Path 2: $A = C$, (3) Path 3: $A = D = B = C$. **(b)** If $Q = 43$ μC, $R = 4.0$ m, $s_1 = 1.5$ mm, and $s_2 = 0.7$ mm, what is the value of $\Delta V = V_C - V_A$? **(c)** Choose two different paths from point A back to point A again, and show that $\Delta V = 0$ for a round trip along both of these paths.

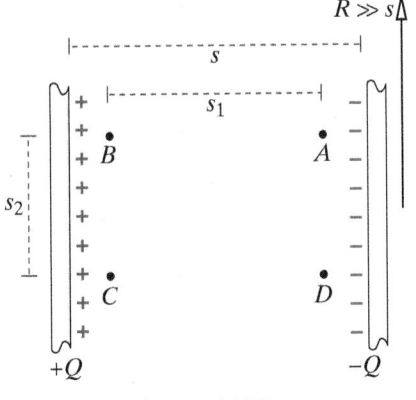

Figure 16.88

••P67 A uniform spherical shell of charge $+Q$ is centered at point B, as shown in Figure 16.89. Show that $\Delta V = V_C - V_A$ is independent of path by calculating ΔV for each of these two paths (actually do the integrals): **(a)** Path 1: $A = B = C$ (along a straight line through the shell). **(b)** Path 2: $A = D = C$ (along a circular arc around point B).

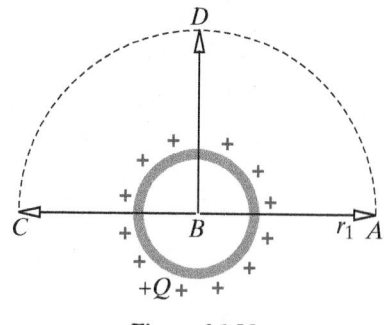

Figure 16.89

••P68 Four voltmeters are connected to a circuit as shown in Figure 16.90. As is usual with voltmeters, the reading on the voltmeter is positive if the negative lead (black wire, usually labeled COM) is connected to a location at lower potential, and the positive lead (red) is connected to a location at higher potential. The circuit contains two devices whose identity is unknown, and a rod (green) of length 9 cm made of a conducting material. At a particular moment, the readings observed on the voltmeters are: voltmeter A: -1.6 V; voltmeter B: -6 V; voltmeter C: -3.5 V. **(a)** At this moment, what is the reading on voltmeter D, both magnitude and sign? **(b)** What are the magnitude and direction of the electric field inside the rod?

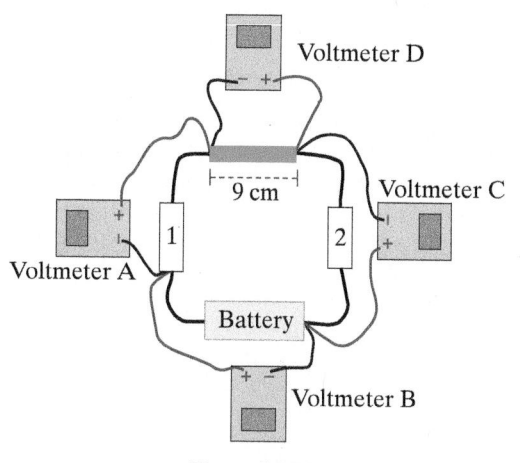

Figure 16.90

Section 16.7

•P69 A particle with charge $+q_1$ and a particle with charge $-q_2$ are located as shown in Figure 16.91. What is the potential (relative to infinity) at location A?

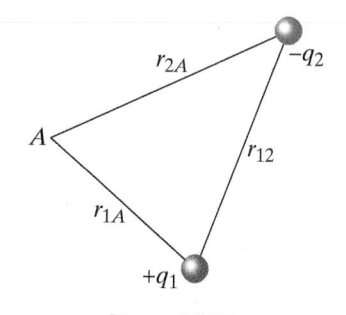

Figure 16.91

•P70 What is the electric potential at a location 2.5×10^{-9} m from an electron?

•P71 Calculate the potential at location A in Figure 16.92.

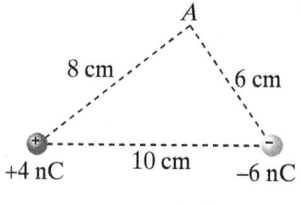

Figure 16.92

•P72 Two capacitors have the same size of plates and the same distance (2 mm) between plates. The potentials of the two plates in capacitor 1 are -10 V and $+10$ V. The potentials of the two plates in capacitor 2 are 350 V and 370 V. What is the electric field inside capacitor 1? Inside capacitor 2?

•P73 Suppose that the potential at $z = 3.15$ m is 37 V, and the potential at $z = 3.27$ m is -36 V. What is the approximate value of E_z in this region? Include the appropriate sign.

•P74 Four point charges, q_1, q_2, q_3, and q_4, are located at the corners of a square with side d, as shown in Figure 16.93. What is the potential (relative to infinity) at location A, at the center of the square? What is the potential at location B, in the middle of one side?

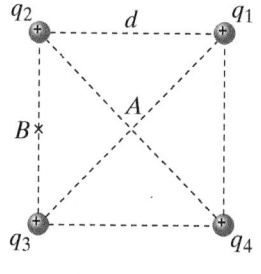

Figure 16.93

•P75 What is the potential (relative to infinity) at location B, a distance h from a ring of radius a with charge $-Q$, as shown in Figure 16.94?

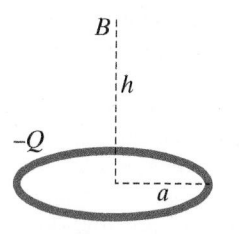

Figure 16.94

••P76 What is the maximum possible potential (relative to infinity) of a metal sphere of 10-cm radius in air? What is the maximum possible potential of a metal sphere of only 1-mm radius? These results hint at the reason why a highly charged piece of metal (with uniform potential throughout) tends to spark at places where the radius of curvature is small or at places where there are sharp points. Remember that the breakdown electric field strength for air is roughly 3×10^6 V/m.

••P77 A small solid glass ball of radius r is irradiated by a beam of positive ions and gains a charge $+q$ distributed uniformly throughout its volume. What is the potential (relative to infinity) at location A, inside the ball, a distance $r/2$ from the center of the ball?

••P78 A small metal sphere of radius r initially has a charge q_0. Then a long copper wire is connected from this small sphere to a distant, large, uncharged metal sphere of radius R. Calculate the final charge q on the small sphere and the final charge Q on the large sphere. You may neglect the small amount of charge on the wire. What other approximations did you make? (Think about potential ....)

••P79 A metal sphere of radius r_1 carries a positive charge of amount Q. A concentric spherical metal shell with inner radius r_2 and outer radius r_3 surrounds the inner sphere and carries a total positive charge of amount $4Q$, with some of this charge on the outer surface (at r_3) and some on the inner surface (at r_2). (a) How is the charge of $4Q$ distributed on the two surfaces of the outer shell? Prove this! (b) What is the potential (relative to infinity) just outside r_3, halfway between r_2 and r_3, just inside r_2, just outside r_1, and at the center?

••P80 A thin plastic spherical shell of radius a is rubbed all over with wool and gains a charge of $-Q$. What is the potential relative to infinity at location B, a distance $a/3$ from the center of the sphere, as shown in Figure 16.95?

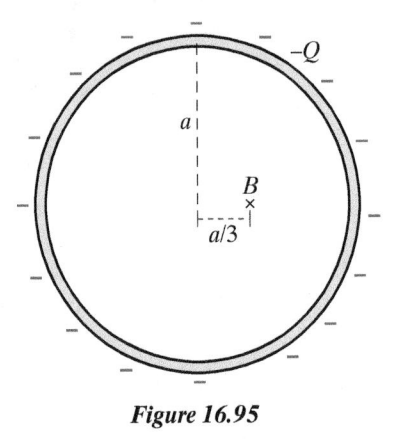

Figure 16.95

••P81 A solid plastic sphere of radius R has charge $-Q$ distributed uniformly over its surface. It is far from all other objects. (a) Sketch the molecular polarization in the interior of the sphere and explain briefly. (b) Calculate the potential relative to infinity at the center of the sphere.

•••P82 A nucleus contains Z protons that on average are uniformly distributed throughout a tiny sphere of radius R. (a) Calculate the potential (relative to infinity) at the center of the nucleus. Assume that there are no electrons or other charged particles in the vicinity of this bare nucleus. (b) Graph the potential as a function of distance from the center of the nucleus. (c) Suppose that in an accelerator experiment a positive pion is produced at rest at the center of a nucleus containing Z protons. The pion decays into a positive muon (essentially a heavy positron) and a neutrino. The muon has initial kinetic energy K_i. How much kinetic energy does the muon have by the time it has been repelled very far away from the nucleus? (The muon interacts with the nucleus only through Coulomb's law and is unaffected by nuclear forces. The massive nucleus hardly moves and gets negligible kinetic energy.) (d) If the nucleus is gold, with 79 protons, what is the numerical value of $K_f - K_i$ in electron volts?

Section 16.9

••P83 An isolated large-plate capacitor (not connected to anything) originally has a potential difference of 1000 V with an air gap of 2 mm. Then a plastic slab 1 mm thick, with dielectric constant 5, is inserted into the middle of the air gap as shown in Figure 16.96. Calculate the following potential differences, and explain your work.

$$V_1 - V_2 =? \qquad V_2 - V_3 =?$$
$$V_3 - V_4 =? \qquad V_1 - V_4 =?$$

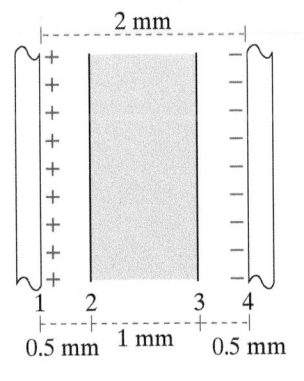

Figure 16.96

••P84 An isolated parallel-plate capacitor of area A_1 with an air gap of length s_1 is charged up to a potential difference ΔV_1. A second parallel-plate capacitor, initially uncharged, has an area A_2 and a gap of length s_2 filled with plastic whose dielectric constant is K. You connect a wire from the positive plate of the first capacitor to one of the plates of the second capacitor, and you connect another wire from the negative plate of the first capacitor to the other plate of the second capacitor. What is the final potential difference across the first capacitor?

Section 16.11

••P85 Refer to Section 16.7 for the potential relative to infinity along the axis of a ring of radius R carrying a charge Q. Let the axis of the ring be the z axis of the coordinate system, and determine E_z at any location z along the axis. (If the charge is nearly uniformly distributed around the ring, at these locations there is no E_x or E_y, due to the symmetry of the situation.) Compare with the result obtained in the previous chapter by integration.

••P86 A rod uniformly charged with charge $-q$ is bent into a semicircular arc of radius b, as shown in Figure 16.97. What is the potential relative to infinity at location A, at the center of the arc?

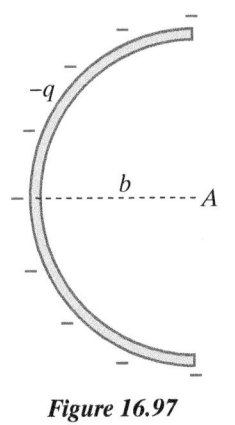

Figure 16.97

••P87 We know the magnitude of the electric field of a dipole at a location on the x axis and at a location on the y axis, if we are far from the dipole. **(a)** Find $\Delta V = V_B - V_A$ along a line perpendicular to the axis of the dipole shown in Figure 16.98. Do it two ways: from the superposition of V due to the two charges and from the integral of the electric field. **(b)** Find $\Delta V = V_C - V_D$ along the axis of the dipole, as shown in Figure 16.99. Include the correct sign. Do it two ways: from the superposition of V due to the two charges and from the integral of the electric field. **(c)** What is the change in potential energy ΔU in moving an electron from D to C?

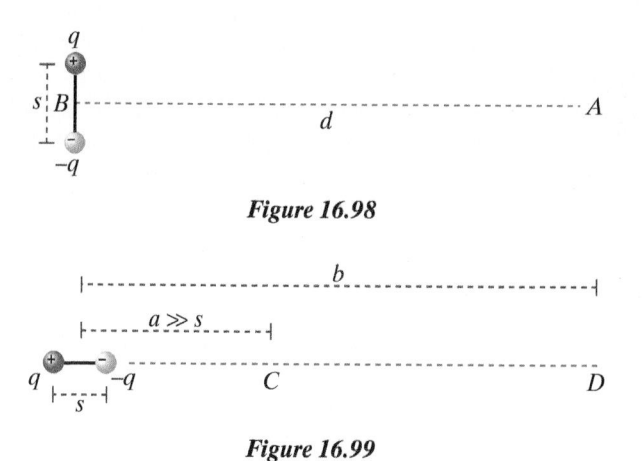

Figure 16.98

Figure 16.99

COMPUTATIONAL PROBLEMS

More detailed and extended versions of some computational modeling problems may be found in the lab activities included in the *Matter & Interactions* resources for instructors.

•P88 Modify the program from Section 16.8 to calculate the potential difference $V_C - V_D$ near the dipole, where $C = \langle -0.03, -0.02, 0 \rangle$ m and $D = \langle -0.05, -0.07, 0 \rangle$ m.

••P89 Replace the dipole in the program from Section 16.8 with a uniformly charged thin rod. The rod is centered at the origin, and its axis is aligned along the y axis. The rod is 0.05 m long, and its net charge is -2×10^{-8} C. Calculate the potential difference $V_B - V_A$, where $A = \langle 0.02, 0.02, 0 \rangle$ m, and $B = \langle 0.12, 0.08, 0 \rangle$ m. Explain how you decided how many point charges to use in modeling the rod.

••P90 Replace the dipole in the program from Section 16.8 with a uniformly charged thin ring. The ring is centered at the origin, and its axis is aligned with the z axis. The radius of the ring is 0.04 m, and its net charge is -3×10^{-8} C. Calculate the potential difference $V_B - V_A$, where $A = \langle 0.02, 0, 0.02 \rangle$ m, and

$B = \langle 0.12, 0, 0.08 \rangle$ m. Explain how you decided how many point charges to use in modeling the ring.

••P91 Consider a uniformly charged cylinder 0.3 m long, with a radius of 1 mm, carrying a total charge of 4×10^{-8} C. The cylinder lies along the x axis, with its center at the origin. Calculate the potential difference along a path going from $\langle 0, 0.05, 0 \rangle$ to $\langle 0.15, 0.05, 0 \rangle$. Explain how you decided how many point charges to use in modeling the cylinder.

•••P92 As in Problem P91, consider a uniformly charged cylinder 0.3 m long, with a radius of 1 mm, carrying a total charge of 4×10^{-8} C. The cylinder lies along the x axis, with its center at the origin. Compute the potential difference along a path from $\langle 0, 0.05, 0 \rangle$ to $\langle 0.15, 0.05, 0 \rangle$ by calculating and displaying (with arrows) $\vec{E}$ and $\Delta \vec{l}$ at many steps along the path and adding up the contributions $-\vec{E} \bullet \Delta \vec{l}$. The VPython function dot(A,B) gives the dot product of the two vectors A and B. See optional Section 16.13 for suggestions on how to structure the program.

ANSWERS TO CHECKPOINTS

1 (a) 7.65×10^{-21} J; **(b)** 2.12×10^{-20} J; **(c)** 1.36×10^{-20} J
2 (a) -1.68×10^{-18} J; **(b)** 1.68×10^{-18} J; **(c)** 4.45×10^4 m/s;
(d) 1.68×10^{-18} J; -1.68×10^{-18} J
3 (a) 300 V; **(b)** 4.8×10^{-17} J; **(c)** -4.8×10^{-17} J; **(d)** 0; **(e)** 0; **(f)** 0
4 -150 V/m
5 (a) To the left, positive; **(b)** To the right, negative
6 (a) -380 V; **(b)** $+380$ V
7 (a) $+1$ V; **(b)** -1 V
8 $\Delta V_1 = \dfrac{1}{4\pi\varepsilon_0}\left(\dfrac{Q}{r_2} - \dfrac{Q}{r_1}\right)$
$\Delta V_2 = 0$

$\Delta V_3 = \dfrac{1}{4\pi\varepsilon_0}\left(\dfrac{Q}{r_1} - \dfrac{Q}{r_2}\right)$
$\Delta V_4 = 0$
$\Delta V_{\text{round trip}} = 0$
9 0.025 V/m (a very small field)
10 If $z \gg R$, $\sqrt{z^2+R^2} \approx z$, and $V \approx \dfrac{1}{4\pi\varepsilon_0}\dfrac{Q}{z}$.
11 4.8 V
12 1.5×10^6 V/m; 39.8 J
13 Since each little charge dQ is a distance R from the center, the potential at the center is $\sum \dfrac{1}{4\pi\varepsilon_0}\dfrac{dQ}{R} = \dfrac{1}{4\pi\varepsilon_0}\dfrac{Q}{R}$.

Magnetic Field

$$\vec{B} = \frac{\mu_0}{4\pi} \frac{q\vec{v} \times \hat{r}}{r^2}$$

OBJECTIVES

After studying this chapter you should be able to

- Calculate the 3D magnetic field due to a single moving charge or to a current, and represent it graphically using arrows.
- Use a magnetic compass to determine the magnitude and direction of current in a wire.
- Relate the magnetic field of a bar magnet to its magnetic dipole moment.

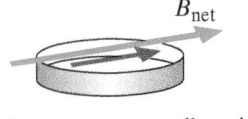

Figure 17.1 A compass needle points in the direction of the net magnetic field at its location.

Electric fields are not the only kind of field associated with charged particles. When a compass needle turns and points in a particular direction, we say that there is a "magnetic field" pointing in that direction, which forces the needle to line up with it (Figure 17.1). Initially we'll simply define magnetic field as whatever it is that is detected by a compass. The twist of a compass needle is an indicator of magnetic fields, just as the twist of a suspended electric dipole is an indicator of electric fields, as you saw in Chapter 14.

Magnetic fields are made by moving charges, so we will need to have some moving charges at hand. We will assemble some simple electric circuits in which currents can run steadily, providing a convenient source of moving charges. In this chapter we will study magnetic field, and we will also develop an atomic-level description of magnets. Later, in Chapter 20, we will study the forces that magnetic fields exert on moving charges.

There is a set of simple experiments related to the topics discussed in this chapter, which allow you to explore these phenomena yourself using equipment such as batteries, wires, flashlight bulbs, and a compass. See the Experiments section at the end of this chapter for further information.

17.1 ELECTRON CURRENT

Magnetic fields are associated with moving charges. A current in a wire provides a convenient source of moving charges and allows us to experiment with producing and detecting magnetic fields.

In equilibrium, there is no net motion of the sea of mobile electrons inside a metal. In the electric circuits we will construct in this chapter, we can arrange things so the electron sea does keep moving continuously. This continuous flow of electrons is called an "electron current," and it is an indication that the system is not in equilibrium. In order to be able to talk about what things affect electron current, we need a precise definition:

DEFINITION OF ELECTRON CURRENT

The electron current i is the number of electrons per second that enter a section of a conductor.

In an electric circuit with a steady current flowing, the electron current i is the same in every section of a wire of uniform thickness and composition (in the next chapter we'll see why). If we could count the number of electrons per second passing a particular point in a circuit (Figure 17.2), we could measure electron current directly. This is difficult to do, so we use indirect measurements to determine the magnitude of electron current in a wire. One such indirect measurement involves measuring the magnetic field created by the moving electrons.

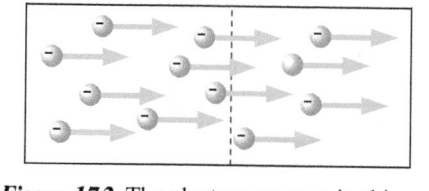

Figure 17.2 The electron current in this conductor is the number of electrons per second passing the dashed line.

As electrons drift through a wire, they collide with the atomic cores, and this "friction" heats the wire (and prevents the electrons from going faster and faster). Both the heating and the magnetic effects are proportional to the electron current—the number of electrons that enter the wire every second.

> **Checkpoint 1** If 1.8×10^{16} electrons enter a light bulb in 3 ms, what is the magnitude of the electron current at that point in the circuit? If the electron current at a particular location in a circuit is 9×10^{18} electrons/s, how many electrons pass that point in 10 min?

Simple Circuits

A simple electric circuit, involving a battery, wire, and a flashlight bulb, is a convenient means of producing a supply of moving electrons. We will refer to such circuits throughout this chapter. Despite the simplicity of the materials, the physics questions that can be investigated with such circuits are centrally important ones. If you have the appropriate equipment, you may wish to do these experiments yourself. See EXP1.

17.2 DETECTING MAGNETIC FIELDS

We can use a magnetic compass as a detector of magnetic fields. Just as the deflection of a charged piece of invisible tape or the twisting of a permanent electric dipole indicates the presence of an electric field, the twisting of a compass needle indicates the presence of a magnetic field. See EXP2.

> QUESTION How can you be sure that a compass needle is not simply responding to electric fields?

In observing the behavior of a compass, we find that:

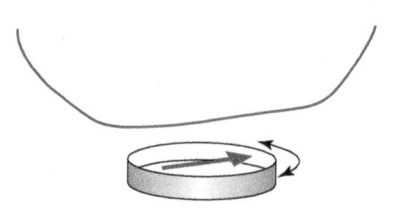

Figure 17.3 If a compass needle is originally pointing north, and a current-carrying wire aligned north–south is placed on the compass, the needle deflects. The deflection direction depends on whether the current runs northward or southward.

- The compass needle is affected by the proximity of objects made of iron or steel, even if these objects are electrically neutral (and therefore attract both positively and negatively charged tapes). It is also affected by the presence of nickel or cobalt, though these are less readily available.
- The compass needle is unaffected by objects made of most other elements, including aluminum, copper, zinc, and carbon, whereas charged tapes interact with these objects.
- If it is not near objects made of iron, the compass needle points toward the Earth's magnetic north pole, whereas neither electrically charged objects nor electric dipoles do this.

If you bring a current-carrying wire near the compass, something very interesting happens: the compass needle is deflected while current is running in the wire, but not while the wire is disconnected from the battery. This effect was discovered by accident by the Danish scientist Oersted in 1820 while doing

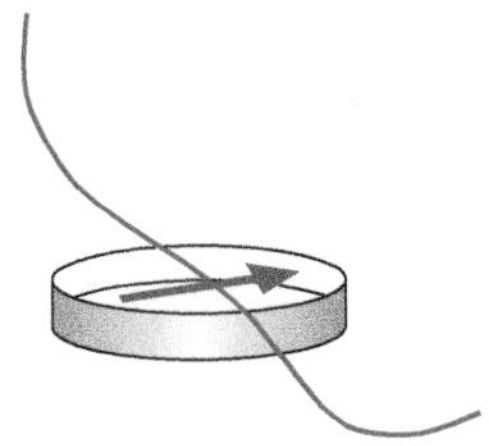

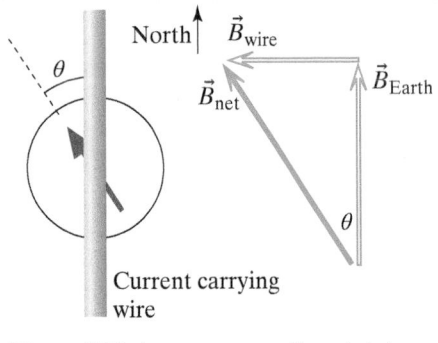

Figure 17.4 If a current-carrying wire aligned east–west is placed on a compass, the needle does not deflect.

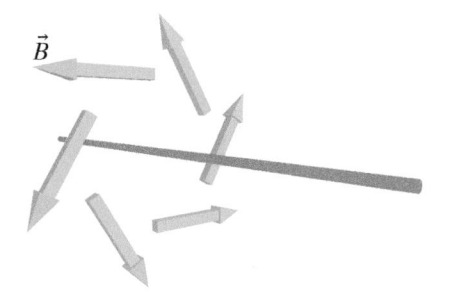

Figure 17.5 A compass needle points in the direction of the net magnetic field, which is the superposition of the magnetic field of the Earth and the magnetic field of the current-carrying wire that passes over the compass, aligned north–south.

Figure 17.6 The magnetic field made by a long straight current-carrying wire. The magnetic field is perpendicular to the wire everywhere.

a lecture demonstration in a physics class. (He must have been surprised!) The phenomenon is often called "the Oersted effect." The biggest effect occurs if the wire points north–south (Figure 17.3); the deflection direction depends on whether the current runs northward or southward. The compass needle doesn't deflect if the wire points east–west (Figure 17.4).

From such experiments, one can draw the following conclusions:

- The magnitude of the magnetic field produced by a current of moving electrons depends on the amount of current.
- A wire with no current running in it produces no magnetic field.
- The magnetic field due to the current appears to be perpendicular to the direction of the current.
- The direction of the magnetic field due to the current under the wire is opposite to the direction of the magnetic field due to the current above the wire.

A Model for the Observations

The moving electrons in a wire create a magnetic field at various locations in space, including at the location of the compass. The vector sum of the Earth's magnetic field $\vec{B}_{\text{Earth}}$ plus the magnetic field $\vec{B}_{\text{wire}}$ of a current-carrying wire makes a net magnetic field $\vec{B}_{\text{net}}$ (Figure 17.5). Since a compass needle points in the direction of the net magnetic field at its location, the needle turns away from its original north–south direction to align with the net field. Because the magnetic field made by the wire is perpendicular to the wire and is in opposite directions above and below the wire, the pattern of magnetic field made by the wire must look like Figure 17.6.

It happens (conveniently) to be the case that the magnitudes of $\vec{B}_{\text{Earth}}$ and $\vec{B}_{\text{wire}}$ are similar. If B_{Earth} were much larger than B_{wire}, then we would see almost no response from the compass.

> QUESTION What would we observe if B_{wire} were much larger than B_{Earth}?

If B_{wire} were much larger than B_{Earth}, then the compass deflection would be nearly 90°, since the net magnetic field would be primarily due to the magnetic field made by the current in the wire.

There is hardly any observable electric interaction of the current-carrying wire with nearby materials, because the wires have no net charge. (Actually, we'll see later that there are tiny amounts of excess charge on the surface, but too little to observe easily.)

EXAMPLE   **Compass Deflection**

A current-carrying wire is oriented north–south and laid on top of a compass. At the location of the compass needle the magnetic field due to the wire points west and has a magnitude of 3×10^{-6} T (tesla). The horizontal component of Earth's magnetic field has a magnitude of about 2×10^{-5} T. What compass deflection will you observe?

Solution   The compass needle will point in the direction of the net magnetic field, so from the diagram in Figure 17.7:

$$\theta = \tan^{-1}\left(\frac{B_{\text{wire}}}{B_{\text{Earth}}}\right) = \tan^{-1}\left(\frac{3 \times 10^{-6}\,\text{T}}{2 \times 10^{-5}\,\text{T}}\right) = 8.5° \text{ west}$$

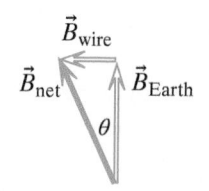

Figure 17.7 The compass points in the direction of the net magnetic field.

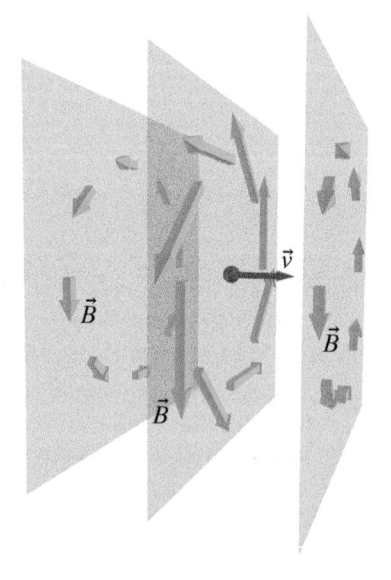

Figure 17.8 The magnetic field made by a moving positive charge, shown in three planes normal to $\vec{v}$.

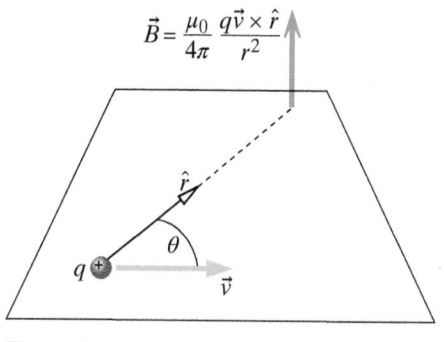

Figure 17.9 The Biot–Savart law involves a cross product, $\vec{v} \times \hat{r}$.

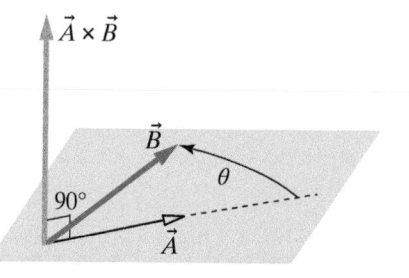

Figure 17.10 The cross product produces a vector that is perpendicular to the two original vectors.

Checkpoint 2 A current-carrying wire oriented north–south and laid over a compass deflects the compass 15° east. What are the magnitude and direction of the magnetic field made by the current? The horizontal component of Earth's magnetic field is about 2×10^{-5} T.

17.3 BIOT–SAVART LAW: SINGLE MOVING CHARGE

Careful experimentation has shown that a stationary point charge makes an electric field given by this equation, called Coulomb's law:

$$\vec{E} = \frac{1}{4\pi\varepsilon_0} \frac{q}{r^2} \hat{r}$$

Similarly, careful experimentation shows that a moving point charge not only makes an electric field but also makes a magnetic field (Figure 17.8), which curls around the moving charge. This curly pattern is characteristic of magnetic fields. (In contrast, we saw in Chapter 16 that it is impossible to produce a curly electric field by arranging stationary point charges.) Magnetic field is measured in "teslas" and its magnitude and direction are given by the Biot–Savart law (pronounced bee-oh sah-VAR):

THE BIOT–SAVART LAW FOR A SINGLE CHARGE

$$\vec{B} = \frac{\mu_0}{4\pi} \frac{q\vec{v} \times \hat{r}}{r^2}$$

where

$$\frac{\mu_0}{4\pi} = 1 \times 10^{-7} \frac{\text{T} \cdot \text{m}^2}{\text{C} \cdot \text{m/s}} \; exactly$$

$\vec{v}$ is the velocity of the point charge q, and $\hat{r}$ is a unit vector that points from the source charge toward the observation location.

The Vector Cross Product

The Biot–Savart law involves a cross product, $\vec{v} \times \hat{r}$ (Figure 17.9). We will summarize the discussion of cross products found in Chapter 11 on angular momentum.

In the mathematics of vectors, the cross product is an operation that combines two vectors to produce a third vector that is perpendicular to the plane defined by the original vectors (Figure 17.10). The cross product of two vectors $\vec{A}$ and $\vec{B}$ is written as $\vec{A} \times \vec{B}$. Its magnitude is

$$|\vec{A} \times \vec{B}| = |\vec{A}||\vec{B}| \sin\theta$$

where θ is the angle between the two vectors when they are placed tail to tail. θ is always less than or equal to 180°. The direction of $\vec{A} \times \vec{B}$ is given by a right-hand rule:

- Draw the vectors $\vec{A}$ and $\vec{B}$ so their tails are in the same location.
- Point the fingers of your right hand in the direction of $\vec{A}$ (Figure 17.11), then fold them through the angle θ toward $\vec{B}$ (Figure 17.12). You may need to turn your hand over to be able to do this.
- Stick your thumb out. Your thumb points in the direction of the cross product.

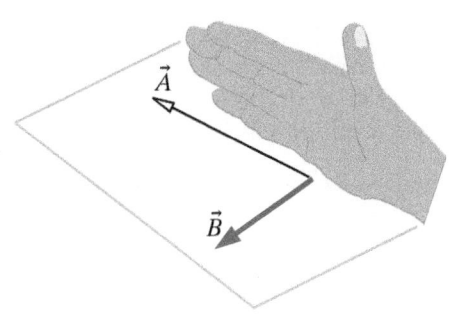

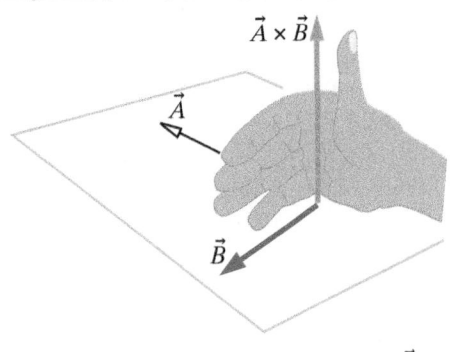

Figure 17.11 With the tails of $\vec{A}$ and $\vec{B}$ together, point your fingers along $\vec{A}$.

Figure 17.12 Fold your fingers toward $\vec{B}$. Your extended thumb points in the direction of $\vec{A} \times \vec{B}$.

In Figure 17.12, your fingers fold counterclockwise and $\vec{A} \times \vec{B}$ points out of the plane. On the other hand, if you were evaluating $\vec{B} \times \vec{A}$, in Figure 17.12 you would have had to turn your hand over, your fingers would have folded clockwise from $\vec{B}$ to $\vec{A}$, and your thumb would have pointed into the plane. In summary: fold counterclockwise and the cross product is out of the plane; fold clockwise and the cross product is into the plane.

Another way to determine the magnitude and direction of a cross product is to use the algebraic expression given below:

CROSS PRODUCT $\vec{A} \times \vec{B}$

$$\vec{A} \times \vec{B} = \langle A_y B_z - A_z B_y, A_z B_x - A_x B_z, A_x B_y - A_y B_x \rangle$$
$$|\vec{A} \times \vec{B}| = |\vec{A}||\vec{B}| \sin \theta$$

The direction of the vector is given by the right-hand rule.

For example,

$$\langle 2,3,4 \rangle \times \langle 5,6,7 \rangle = \langle 3 \cdot 7 - 4 \cdot 6, 4 \cdot 5 - 2 \cdot 7, 2 \cdot 6 - 3 \cdot 5 \rangle$$
$$= \langle -3, 6, -3 \rangle$$

This approach to calculating a cross product is particularly useful in computer calculations. Note the cyclic nature of the subscripts: xyz, yzx, zxy.

If you have studied determinants, you may recognize that the components of $\vec{A} \times \vec{B}$ can be thought of as the "minors" of a 3 by 3 determinant in which $\vec{A}$ and $\vec{B}$ are the second and third rows, and the first row is $\hat{\imath} = \langle 1,0,0 \rangle$, $\hat{\jmath} = \langle 0,1,0 \rangle$, $\hat{k} = \langle 0,0,1 \rangle$.

Two-Dimensional Projections

Because it is more difficult to sketch a situation in three dimensions, whenever possible we will work with two-dimensional projections onto a plane. If $\vec{A}$ and $\vec{B}$ lie in the xy plane, the cross-product vector $\vec{A} \times \vec{B}$ points in the $+z$ direction (out of the page, $\odot$) or in the $-z$ direction (into the page, $\otimes$). The symbol $\odot$ is supposed to suggest the tip of an arrow pointing toward you (cross product pointing toward you). The symbol $\otimes$ is supposed to suggest the feathers on the end of an arrow that is heading away from you (cross product pointing away from you).

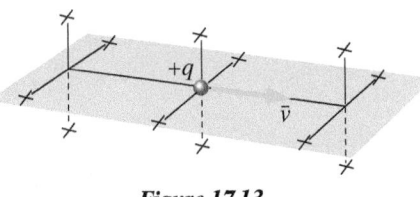

Figure 17.13

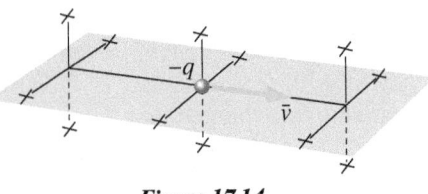

Figure 17.14

Work through the following checkpoint before going on, to make sure you understand how to find the direction of $\vec{B}$ using the right-hand rule.

Checkpoint 3 At the locations marked with $\times$ in Figure 17.13, determine the direction of the magnetic-field vectors due to a positive charge $+q$ moving with a velocity $\vec{v}$. **(a)** For each observation location, draw the unit vector $\hat{r}$ *from* the charge *to* that location, then consider the cross product $\vec{v} \times \hat{r}$. Pay attention not only to the directions of the magnetic field but also to the relative magnitudes of the vectors. **(b)** If the charge is negative $(-q)$ as in Figure 17.14, how does this change the pattern of magnetic field?

Sign of the Moving Charge

Assume that the moving charged particles in a current-carrying wire are negatively charged and move in the direction shown in Figure 17.15.

QUESTION Use the Biot–Savart law and the right-hand rule to predict the direction of the magnetic field at several locations around the wire (Figure 17.15).

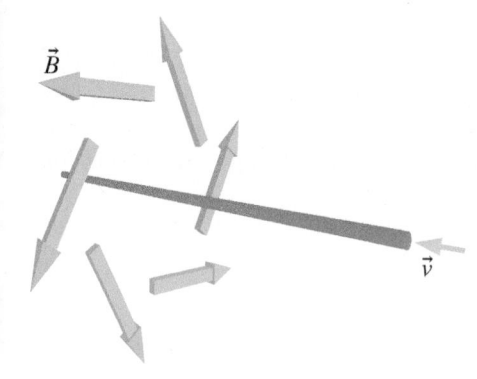

Figure 17.15 The magnetic field due to moving charges in the wire curls around the wire. Here the moving charges are assumed to be negative.

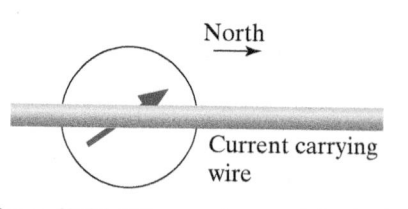

Figure 17.16 If the moving particles in the wire are electrons, in what direction are they moving? If we assume that the particles are positively charged, in which direction would they be moving?

Convince yourself that $q\vec{v} \times \hat{r}$ does correctly give the direction and pattern of magnetic field near a wire; be sure to take into account the negative sign of q. From this curly pattern you can see that the direction of compass deflection will be different if the wire is above or below the compass, which is observed.

> QUESTION Try the analysis again, but assume that the moving charges are positively charged particles moving in a direction opposite to that shown. Do you still predict the curly magnetic field as shown in Figure 17.15?

You should have concluded that for the purpose of predicting the direction of magnetic field, it does not matter whether we assume that the moving charges are negative and moving in a given direction or positive and moving in the opposite direction. It is common to describe current flow in terms of "conventional current," which means assuming that the moving charges are positive, even if this is known not to be the case in a particular situation.

> **Checkpoint 4** In Figure 17.16 a current-carrying wire lies on top of a compass. Judging from the deflection of the compass away from north, what is the direction of the electron current in the wire? If the current were due to the motion of positive charges, which way would the positive charges be moving?

17.4 RELATIVISTIC EFFECTS

Magnetic Field Depends on Your Frame of Reference

Electric fields are made by charges, whether at rest or moving. Magnetic fields seem to be made solely by moving charges—but consider the following odd "thought experiment."

Suppose that Jack sits in the classroom with a charged piece of invisible tape stuck to the edge of his desk. He can of course observe an electric field due to the charged tape, but he doesn't observe any magnetic field. His compass is unaffected by the charged tape, since those charges aren't moving:

$$\vec{B} = \frac{\mu_0}{4\pi} \frac{q\vec{v} \times \hat{r}}{r^2} = \vec{0} \quad \text{since } v = 0$$

Jill runs at high speed through the classroom past the charged tape, carrying her own very sensitive compass. She observes an electric field due to the charged tape. In addition, however, in her frame of reference the charged tape is moving, so she observes a small magnetic field due to the moving charges, which affects her compass! As far as Jack is concerned, the charged tape just makes an electric field, but apparently Jill sees a mixture of electric and magnetic fields made by what is for her a moving charged tape.

Up until now we have implied that electric fields and magnetic fields are fundamentally different, but this "thought experiment" shows that they are in fact closely related. Moreover, this connection raises questions about the Biot–Savart law:

$$\vec{B} = \frac{\mu_0}{4\pi} \frac{q\vec{v} \times \hat{r}}{r^2}$$

Just what velocity are we supposed to use in this equation? The velocity of the tape relative to Jack (which is zero) or the velocity of the tape relative to Jill (opposite her own velocity)?

The correct answer is that you use the velocities of the charges as you observe them in your frame of reference. Using these velocities in the

Biot–Savart law, you calculate the magnetic field; in your frame of reference you observe a magnetic field that agrees with your prediction. Observers in a different reference frame use the velocities observed in their frame to calculate the magnetic field using the Biot–Savart law, and in their frame of reference they observe a magnetic field that agrees with their prediction. You and they make different predictions and observe different magnetic fields, but both you and they find agreement between theory and experiments. Jack predicts that his compass won't deflect because the charged tape isn't moving, and that's what he observes. Jill predicts that her compass will deflect because she sees a moving charged tape, and she observes that her compass does deflect. (Actually, for Jill to see a large effect she would have to move at very high speed, because there isn't much charge on a charged tape.)

Evidently there is a deep connection between electric fields and magnetic fields, and this connection is made explicit in Einstein's special theory of relativity. Later in the text we will see further aspects of this connection.

Retardation

Remember that when you move a charge, the electric field of that charge at a distance from the charge doesn't change instantaneously. The electric field doesn't change until a time sufficient for light to reach the observation location, and you measure a change in the electric field at the same instant that you see the charge move.

The same retardation effect is observed with magnetic fields. If you suddenly change the current in a wire, the magnetic field at some distance from the source of the magnetic field stays the same until enough time has elapsed for light to reach the observation location. Thus magnetic field has some reality in its own right, independent of the moving charges that originally produced it.

The Biot–Savart law does not contain any reference to time, so it cannot be relativistically correct. Like Coulomb's law, the Biot–Savart law is only approximately correct and will give accurate results only if the speeds of the moving source charges are small compared to the speed of light.

17.5 ELECTRON CURRENT AND CONVENTIONAL CURRENT

An easy way to observe the magnetic fields made by moving charged particles is to initiate and sustain a current—a continual flow of charged particles in one direction. In order to do this, we need to find a way to produce and sustain an electric field inside a wire, since we know that if the electric field inside a metal becomes zero, the metal object will be in equilibrium, and no current will flow. How exactly it is possible to arrange charges in order to create such an electric field everywhere in a wire is the subject of Chapter 18. For the moment, we'll assume that we have somehow accomplished this by assembling a circuit from batteries, wires, and perhaps light bulbs, since evidently currents do flow in such circuits.

In order to apply the Biot–Savart law to predict the magnitude and direction of the magnetic field associated with this current, we need to know the number of moving charges making the field and how fast they are moving.

An Equation for Electron Current

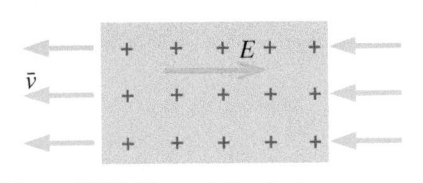

Figure 17.17 The mobile electron sea drifts to the left with a speed $\bar{v}$.

Consider a section of a metal wire through which the mobile electron sea is continuously shifting under the influence of a a nonzero electric field, as indicated in Figure 17.17. There can be no excess charge anywhere inside the wire. To every mobile electron there corresponds a singly charged positive atomic core—an atom minus one electron that has been released to roam freely in the mobile-electron sea. Averaged over a few atomic diameters, the

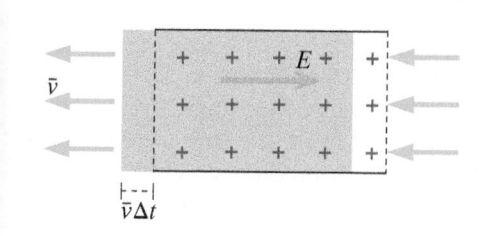

Figure 17.18 During a time Δt the electron sea shifts a distance $\bar{v}\Delta t$.

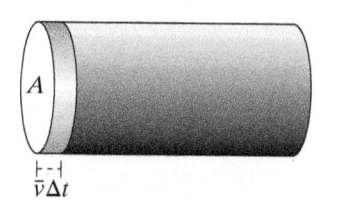

Figure 17.19 The volume of the electron sea that has flowed past a point in the wire during a time Δt is $A\bar{v}\Delta t$.

interior of the metal is neutral, and the repulsion between the mobile electrons is balanced out on the average by the attractions of the positive cores.

Suppose that in a metal wire the mobile electron sea as a whole has an average "drift speed" $\bar{v}$. Assume that the current is evenly distributed across the entire cross section of the wire (in a later chapter we will be able to show that this is true.) After a short time Δt, this section of the electron sea will drift to the left a distance $\bar{v}\Delta t$, as indicated in Figure 17.18.

If the cross-sectional area of this wire is A, the volume of the disk-shaped portion of the sea that has flowed out of the left end of this section of the wire (and into the adjoining section of wire) in the time Δt is $A\bar{v}\Delta t$ (Figure 17.19).

> QUESTION If the density of mobile electrons (that is, the number of mobile electrons per unit volume) everywhere in the metal is n, how many electrons are in this disk?

$$n \times \text{volume} = n(A\bar{v}\Delta t) = nA\bar{v}\Delta t$$

The number of electrons in this disk is the number of electrons that have flowed out of this section of the wire in the short time Δt.

> QUESTION What is the rate at which mobile electrons flow past the left end of this piece of the wire?

We divide by the time Δt to calculate the rate at which mobile electrons flow past the left end of this piece of the wire:

$$\frac{nA\bar{v}\Delta t}{\Delta t} = nA\bar{v}$$

This result, that the rate at which electrons pass some section of the wire is $nA\bar{v}$, is sufficiently important to be worth remembering (though it is even better to be able to re-do the simple derivation that gives this result).

ELECTRON CURRENT

The rate i at which electrons pass a section of a wire:

$$i = nA\bar{v}$$

n is the mobile electron density (the number of mobile electrons per unit volume), A is the cross-sectional area of the wire, and $\bar{v}$ is the average drift speed of the electrons. In SI units, the units of i are electrons per second.

In a circuit, electrons leave the negative end of the battery (the end that is marked $-$) and flow through the wire to the positive end of the battery. See EXP3.

EXAMPLE   **Drift Speed of Electrons**

Suppose that $i = 3.4 \times 10^{18}$ electrons/s are drifting through a copper wire. (This is a typical value for a simple circuit.) The diameter of the wire is 1 mm, and the density of mobile electrons in copper is $8.4 \times 10^{28} \text{m}^{-3}$. **(a)** What is the drift speed of the electrons? **(b)** About how many minutes would it take for an electron traveling at this speed to drift from one end to the other end of a wire 30 cm long, about one foot?

Solution   **(a)**

$$A = \pi \left(\frac{1 \times 10^{-3}\,\text{m}}{2} \right)^2 = 8 \times 10^{-7}\,\text{m}^2$$

$$\bar{v} = \frac{i}{nA} = \frac{3.4 \times 10^{18}\,\text{s}^{-1}}{(8.4 \times 10^{28}\,\text{m}^{-3})(8 \times 10^{-7}\,\text{m}^2)}$$

$$= 5.06 \times 10^{-5}\,\text{m/s}$$

(b)

$$t = \frac{0.3\,\text{m}}{5.06 \times 10^{-5}\,\text{m/s}} = 5930\,\text{s} = 99\,\text{min}$$

This result is puzzling: If the drift speed is so slow, how can a lamp light up as soon as you turn it on? We'll come back to this in Chapter 18.

Conventional Current

In most metals, current consists of drifting electrons. However, there are a few materials in which the moving charges act as though they were positive "holes" in the sea of mobile electrons. The positive holes act in every way like real positive particles. For example, holes drift in the direction of the electric field, and they have a charge of $+e$.

Electron current moves from the negative end of a battery through a circuit to the positive end of the battery; hole current would go in the opposite direction. Given this information, consider the following question:

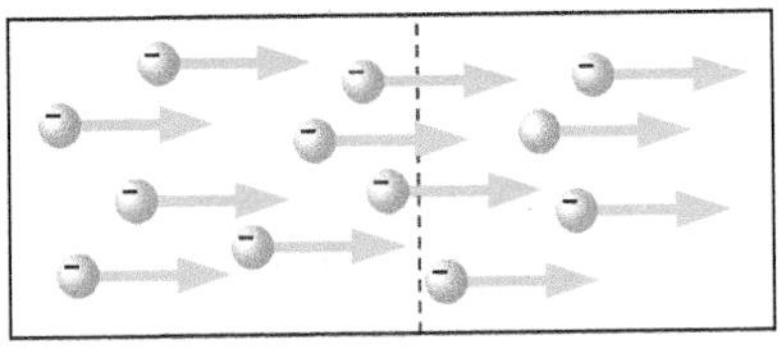

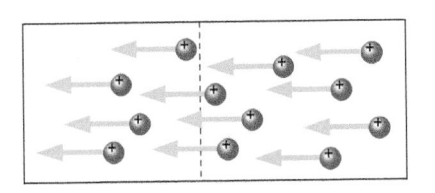

Figure 17.20 An electron current to the right moves negative charge from left to right.

QUESTION From observations of the direction of magnetic field around a copper wire, can one tell whether the current in copper consists of electrons, or holes?

Observations of the magnetic field due to a current-carrying wire are not sufficient to tell the difference, because if the sign of the moving charge is changed, the direction of the drift velocity is also changed: $(+e)(+\bar{v}) \rightarrow (-e)(-\bar{v})$ (Figures 17.20 and 17.21). Therefore the prediction of the Biot–Savart law is exactly the same in either case.

In copper, as in most metals, the current is due to moving electrons. The few metals that have hole current include aluminum and zinc. In the doped semiconductors important in electronics, p-type material (positive type) involves hole current, and n-type (negative type) involves electron current. In Chapter 20 we will study the Hall effect, which can be used to determine whether the current in a wire is due to electrons or to holes.

Figure 17.21 A conventional current to the left would move positive charge from right to left, which would have the same effect as moving negative charge from left to right, as in the previous figure.

Because most effects (other than the Hall effect) are the same for electron current and hole current, it is traditional to define "conventional current" to run in the direction of hole current, even if the actual current consists of moving electrons, in which case the conventional current runs in the opposite direction to the electrons (Figure 17.21). This simplifies calculations by eliminating the minus sign associated with electrons. (This sign ultimately goes back to a choice made by Benjamin Franklin when he arbitrarily assigned "positive" charge to be the charge we now know to be carried by protons.)

In addition, conventional current I is defined not as the number of holes passing some point per second but rather as the amount of charge (in coulombs) passing that point per second. This is the number of holes per second multiplied by the (positive) charge $|q|$ associated with one hole:

CONVENTIONAL CURRENT

$$I = |q|nA\bar{v} \quad \text{(coulombs per second, or amperes)}$$

1 C/s = 1 ampere (abbreviated A). The direction of conventional current is opposite to the direction of electron current. The moving charged particles are assumed to be positive. A battery is assumed to send conventional current out of the + end.

In a metal, $|q| = e$, but in an ionic solution the moving ions might have charges that are a multiple of e.

EXAMPLE

Electron Current and Conventional Current

In the preceding example an electron current of $i = 3.4 \times 10^{18}$ electrons/s ran through a copper wire. What was the conventional current in the wire?

Solution

$$I = |q|i = (1.6 \times 10^{-19}\,\text{C})(3.4 \times 10^{18}\,\text{s}^{-1}) = 0.544\,\text{A}$$

Electric Field and Current

A nonzero electric field is necessary to maintain a current in a conductor. In Chapter 14 we saw that the average drift speed in a metal is $\bar{v} = uE$, where u is the electron mobility and E is the electric field inside the metal (which is not in equilibrium). Therefore, the conventional current in a wire is

$$I = |q|nA\bar{v} = |q|nA(uE)$$

If we know the magnitude of the electric field, we can predict what the current will be in the wire. In Chapter 18 we will see how to maintain a nonzero electric field inside a metal.

> **Checkpoint 5** The electron mobility in copper is 4.5×10^{-3} (m/s)/(N/C), and the density of mobile electrons is $8.4 \times 10^{28}\,\text{m}^{-3}$. In a copper wire 1 mm in diameter there is an electric field whose magnitude is 0.05 N/C. What is the conventional current in this wire? (Note that a small electric field drives a sizable current in copper, which is a good conductor.)

17.6 THE BIOT–SAVART LAW FOR CURRENTS

We don't often observe the magnetic field of a single moving charge. Usually we are interested in the magnetic field produced by a large number of charges moving through a wire in a circuit. The superposition principle is valid for magnetic fields, so we need to add up the contributions of the individual charges to the magnetic field.

Let's calculate the magnetic field due to a bunch of moving positive charges contained in a small thin wire of length Δl and cross-sectional area A (Figure 17.22). If there are n moving charges per unit volume, in this small volume $A\,\Delta l$ there are $nA\,\Delta l$ moving charges. We will measure the magnetic field at a location far enough away from this small volume that each moving charge produces approximately the same magnetic field at that location.

QUESTION Show that the net magnetic field of all the moving charges in this small volume, far from the volume, is

$$\Delta\vec{B} = \frac{\mu_0}{4\pi}\frac{I\Delta\vec{l}\times\hat{r}}{r^2}$$

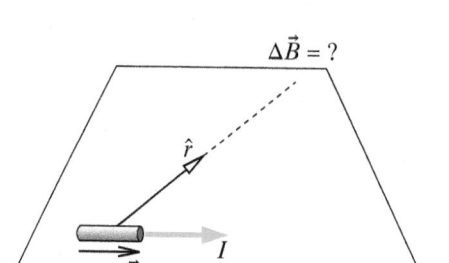

Figure 17.22 Magnetic field contributed by a short thin length Δl of current-carrying wire.

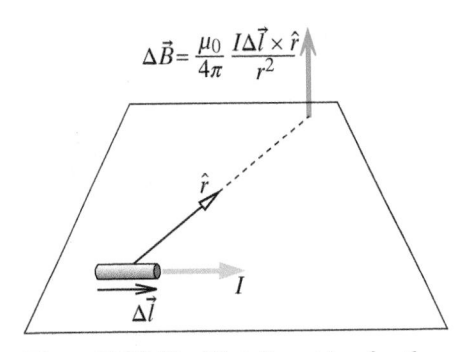

$$\Delta \vec{B} = \frac{\mu_0}{4\pi} \frac{I\Delta \vec{l} \times \hat{r}}{r^2}$$

Figure 17.23 The Biot–Savart law for the magnetic field contributed by a short thin length of wire carrying conventional current I. The magnetic field due to this current segment is perpendicular to the plane defined by $\Delta \vec{l}$ and $\hat{r}$.

where $\Delta \vec{l}$ is a vector with magnitude Δl (the length of the segment of wire) pointing in the direction of the conventional current I.

This results from the fact that the volume $A\,\Delta l$ contains $n(A\,\Delta l)$ moving charges, so the magnetic field is $n(A\,\Delta l)$ times as large as the magnetic field of one of the charges. Each moving charge has a charge q, and the conventional current is $I = |q|(nA\bar{v})$. Collecting terms, we find the result given above.

We have an alternative form of the Biot–Savart law for the magnetic field contributed by a short thin length of current-carrying wire (Figure 17.23):

THE BIOT–SAVART LAW FOR A SHORT THIN LENGTH OF WIRE

$$\Delta \vec{B} = \frac{\mu_0}{4\pi} \frac{I\Delta \vec{l} \times \hat{r}}{r^2}$$

where $\dfrac{\mu_0}{4\pi} = 1 \times 10^{-7} \dfrac{\text{T} \cdot \text{m}^2}{\text{C} \cdot \text{m/s}}$ exactly.

$\Delta \vec{l}$ is a vector in the direction of the conventional current I, and whose magnitude Δl is the length of the segment of wire. $\hat{r}$ is a unit vector that points from the wire segment toward the observation location.

QUESTION   Explain why this equation also gives the right results if the moving charges are (negative) electrons, as long as we interpret $\Delta \vec{l}$ as a vector in the direction of the conventional current.

The law works for moving electrons because although they have the opposite charge, they also move in the direction opposite to the direction of the conventional current. These two changes in sign cancel each other.

It is important to keep in mind that there is really only one Biot–Savart law, not two. Always remember that

$$\Delta \vec{B} = \frac{\mu_0}{4\pi} \frac{I\Delta \vec{l} \times \hat{r}}{r^2}$$

is simply the result of adding up the effects of many moving charges in a short thin length of wire, each of which contributes a magnetic field

$$\vec{B} = \frac{\mu_0}{4\pi} \frac{q\vec{v} \times \hat{r}}{r^2}$$

The key point is that there are $nA\,\Delta l$ electrons in a short length of wire, each moving with average speed $\bar{v}$, so that the sum of all the $q\bar{v}$ contributions is $(nA\,\Delta l)|q|\bar{v} = (|q|nA\bar{v})\Delta l = I\Delta l$.

17.7   THE MAGNETIC FIELD OF CURRENT DISTRIBUTIONS

In the next sections we will apply the Biot–Savart law to find the magnetic field of various distributions of currents in wires. We will use the same four-step approach we used to find the electric field of distributed charges in Chapter 15:

1. Cut up the current distribution into pieces; draw $\Delta \vec{B}$ for a representative piece.
2. Write an expression for the magnetic field due to one piece.
3. Add up the contributions of all the pieces.
4. Check the result.

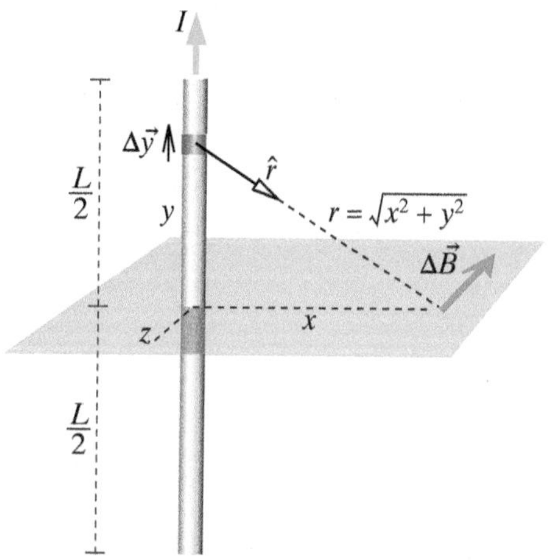

Figure 17.24 Cut the wire into short pieces and draw the magnetic field contributed by one of the pieces.

Applying the Biot–Savart Law: A Long Straight Wire

Using the procedure outlined above, we can calculate the magnetic field near a long straight wire. This will make it possible to predict the compass needle deflection that you observe when you bring a wire near your compass, and you can compare your prediction with your experimental observation.

The long straight wire is one of the few cases that we can calculate completely by hand. Except for a few special cases, calculating the magnetic field due to distributed currents is often best done by computer. The basic concepts you would use in a computer program are the same as we will use here, but the computer would carry out the tedious arithmetic involved in the summation step of the procedure.

Step 1: Cut Up the Distribution into Pieces and Draw $\Delta\vec{B}$

We will consider a wire of length L (Figure 17.24). We cut up the wire into very short sections each of length Δy, where Δy is a small portion of the total length. We will calculate the magnetic field a perpendicular distance x from the center of the wire. Using the right-hand rule, we can draw the direction of $\Delta\vec{B}$ for this "piece" of current.

Step 2: Write an Expression for the Magnetic Field Due to One Piece

We will place the origin at the center of the wire.

$$\vec{r} = \langle x,0,0\rangle - \langle 0,y,0\rangle = \langle x,-y,0\rangle$$
$$|\vec{r}| = [x^2 + (-y)^2]^{1/2} = [x^2 + y^2]^{1/2}$$
$$\hat{r} = \frac{\vec{r}}{r} = \frac{\langle x,-y,0\rangle}{[x^2+y^2]^{1/2}}$$

Since the location of one piece is specified by y, this will be the integration variable. We can express $\Delta\vec{l}$ in terms of y:

$$\Delta\vec{l} = \Delta y\langle 0,1,0\rangle$$
$$\Delta\vec{B} = \frac{\mu_0}{4\pi}\frac{I\Delta\vec{l}\times\hat{r}}{(x^2+y^2)} = \frac{\mu_0}{4\pi}\frac{I\Delta y\langle 0,1,0\rangle}{(x^2+y^2)} \times \frac{\langle x,-y,0\rangle}{[x^2+y^2]^{1/2}}$$

Evaluating the cross product, we get:

$$\langle 0,1,0\rangle \times \langle x,-y,0\rangle = \langle 0,0,-x\rangle$$

Our final expression for $\Delta\vec{B}$ is:

$$\Delta\vec{B} = -\frac{\mu_0}{4\pi}\frac{Ix\Delta y}{(x^2+y^2)^{3/2}}\langle 0,0,1\rangle$$

Every piece of the straight wire contributes some magnetic field in the $-z$ direction, so we need to calculate only the z component.

Step 3: Add Up the Contributions of All the Pieces

By letting $\Delta y \to 0$, we can calculate the magnitude B by writing the sum as an integral:

$$B = \frac{\mu_0}{4\pi}Ix\int_{y=-L/2}^{y=+L/2}\frac{dy}{(x^2+y^2)^{3/2}}$$

Fortunately, this integral can be found in standard tables of integrals:

$$B = \frac{\mu_0}{4\pi} Ix \left[\frac{y}{x^2 \sqrt{x^2 + y^2}} \right]_{y=-L/2}^{y=+L/2}$$

$$B = \frac{\mu_0}{4\pi} Ix \left[\frac{L/2}{x^2 \sqrt{x^2 + (L/2)^2}} - \frac{-L/2}{x^2 \sqrt{x^2 + (L/2)^2}} \right]$$

$$B = \frac{\mu_0}{4\pi} \frac{LI}{x \sqrt{x^2 + (L/2)^2}}$$

An extremely important case is that of a very long wire ($L \gg x$) or, equivalently, the magnetic field very near a short wire ($x \ll L$).

QUESTION Show that you get

$$B \approx \frac{\mu_0}{4\pi} \frac{2I}{x} \quad \text{if } L \gg x$$

If $L \gg x$, $\sqrt{x^2 + (L/2)^2} \approx L/2$, which leads to the stated result.

In summary, Figure 17.25 shows the pattern of magnetic field a distance r from the center of a straight wire of length L. Because of the symmetry of this field, it does not matter where we draw our x axis—the field will be the same all around the rod. To indicate this, we replace x in our equation with r, the radial distance from the rod.

Although the magnitude of the field is constant at constant r, the direction of the field is different at each angle, since the field curls around the wire. To express this as a vector equation we would have to include a cross product in the expression; it's simpler to get the direction using the right-hand rule.

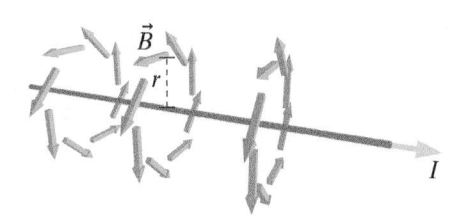

Figure 17.25 Magnetic field of a straight current-carrying wire, at selected locations a distance r radially outward from the wire.

MAGNETIC FIELD OF A STRAIGHT WIRE

$$B_{\text{wire}} = \frac{\mu_0}{4\pi} \frac{LI}{r \sqrt{r^2 + (L/2)^2}}$$

for length L, conventional current I, a perpendicular distance r from the center of the wire.

$$B_{\text{wire}} \approx \frac{\mu_0}{4\pi} \frac{2I}{r} \quad \text{if } L \gg r$$

Historically, the result for a very long straight wire was first obtained by the French physicists Biot and Savart. Their names have come to be associated with the more fundamental principle (the Biot–Savart law) that leads to this result. See EXP4.

Step 4: Check the Result

Direction: In Figure 17.25 the right-hand rule is consistent with the diagram.
Units:

$$\left(\frac{\text{T} \cdot \text{m}}{\text{A}} \right) \frac{(\text{m} \cdot \text{A})}{\text{m} \cdot \text{m}} = \text{T}$$

Far away ($r \gg L$):

$$B_{\text{wire}} \approx \frac{\mu_0}{4\pi} \frac{LI}{r^2} = \frac{\mu_0}{4\pi} \frac{|I \Delta \vec{l} \times \hat{r}|}{r^2}$$

which is indeed the magnetic field contributed by a short wire of length Δl.

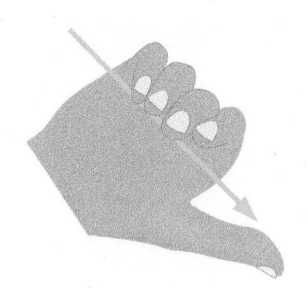

Figure 17.26 The thumb of the right hand points in the direction of conventional current flow I, and the fingers curl around in the direction of the magnetic field.

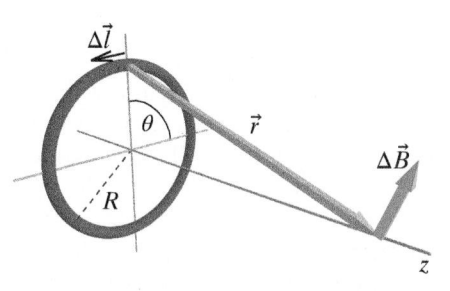

Figure 17.27 Cut the loop into short pieces and draw the magnetic field contributed by one of the pieces. Note that $d\vec{l} \perp \hat{r}$ for every piece of the loop. The loop is in the xy plane, with x to the right and y up. $\Delta\vec{B}$ is perpendicular to the plane defined by $\Delta\vec{l}$ and $\vec{r}$.

Another Right-Hand Rule

There is another right-hand rule that is often convenient to use with current-carrying wires. As you can see in Figure 17.26, if you grasp the wire in your right hand with your thumb pointing in the direction of conventional current, your fingers curl around in the direction of the magnetic field. Clearly this right-hand rule is consistent with the one we used for the Biot–Savart law.

17.8 A CIRCULAR LOOP OF WIRE

Next we'll calculate the magnetic field of a circular loop of wire that carries a conventional current I. We'll do only the easiest case—the magnetic field at any location along the axis of the loop, which is a line going through the center and perpendicular to the loop.

This calculation is important for two reasons. First, many scientific and technological applications of magnetism involve circular loops of current-carrying wire. Second, the calculation will also lead into an analysis of atomic current loops in your bar magnet. After calculating the magnetic field we will compare our predictions with experiments.

Step 1: Cut Up the Distribution into Pieces and Draw $\Delta\vec{B}$

See Figure 17.27. We cut up the loop into very short sections each of length Δl (a small portion of the total circumference $2\pi R$) (Figure 17.27; side view in Figure 17.28). Determine the direction of $\Delta\vec{B}$ with the right-hand rule. Note that the angle between $\Delta\vec{l}$ and $\hat{r}$ is 90° for every location on the ring.

Step 2: Write an Expression for the Magnetic Field Due to One Piece

To work out $\Delta\vec{B}$ for an arbitrary $\Delta\vec{l}$ will be algebraically messy, since

$$\Delta\vec{l} = \langle R\cos(\theta + d\theta), R\sin(\theta + d\theta), 0\rangle - \langle R\cos\theta, R\sin\theta, 0\rangle$$

and

$$\vec{r} = \langle 0, 0, z\rangle - \langle R\cos\theta, R\sin\theta, 0\rangle$$

However, we can simplify the problem considerably by noticing the symmetry of the situation. Because of the circular symmetry of the ring, the ΔB_x and ΔB_y contributed by one piece will be canceled by the contributions of a piece located on the other side of the loop.

Furthermore, the ΔB_z contributed by each piece of the ring will be exactly the same. This allows us to select one piece for which ΔB_z is easy to calculate, and use this value for every piece in the sum. We will select the piece shown in Figure 17.27, located on the y axis. We choose the origin to be the center of the loop.

$$\vec{r} = \langle 0, 0, z\rangle - \langle 0, R, 0\rangle = \langle 0, -R, z\rangle$$
$$|\vec{r}| = [R^2 + z^2]^{1/2}$$
$$\hat{r} = \frac{\vec{r}}{r} = \frac{\langle 0, -R, z\rangle}{[R^2 + z^2]^{1/2}}$$

The location of the piece depends on θ, so this will be the integration variable.

$$\Delta\vec{l} = \langle -R\,\Delta\theta, 0, 0\rangle$$
$$\Delta\vec{B} = \frac{\mu_0}{4\pi}\frac{I\Delta\vec{l}\times\hat{r}}{r^2} = \frac{\mu_0}{4\pi}I\frac{\langle -R\,\Delta\theta, 0, 0\rangle \times \langle 0, -R, z\rangle}{[R^2 + z^2]^{3/2}}$$

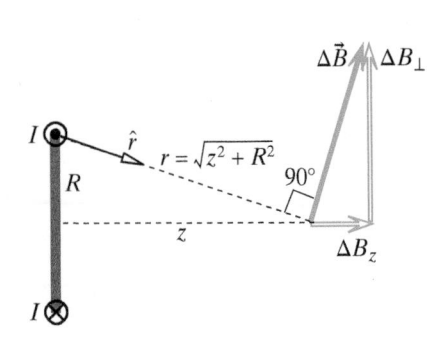

Figure 17.28 Side view of the current-carrying loop. Again note that $d\vec{l}\perp\hat{r}$ for every piece of the loop.

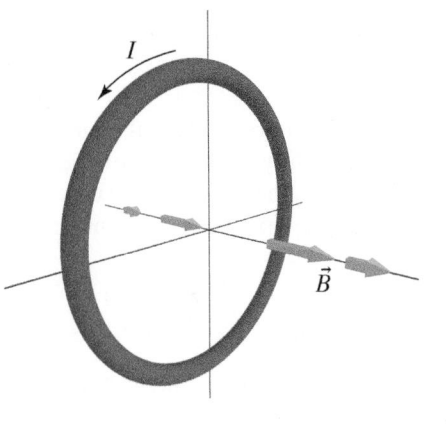

Figure 17.29 Magnetic field of a circular loop of current-carrying wire.

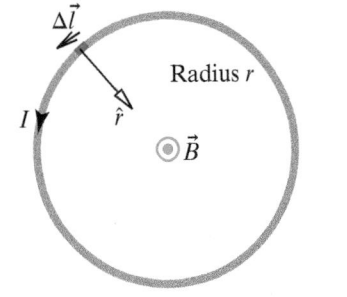

Figure 17.30 Magnetic field at the center of a loop of current-carrying wire.

Evaluating the cross product, we get:

$$\langle -R\,\Delta\theta, 0, 0\rangle \times \langle 0, -R, z\rangle = \langle 0, zR\,\Delta\theta, R^2\,\Delta\theta\rangle$$

We need only the z component, since the others will add up to zero:

$$\Delta B_z = \frac{\mu_0}{4\pi}\frac{IR^2\Delta\theta}{[R^2+z^2]^{3/2}}$$

In Figure 17.28 we show the component ΔB_z along the axis and the component $\Delta B_\perp$ perpendicular to the axis.

Step 3: Add Up the Contributions of All the Pieces

We can express the sum as an integral, where θ, which specifies the location of a piece of the ring, runs all the way around the ring. Remember that ΔB_z contributed by each piece is the same.

$$B_z = \int_0^{2\pi}\frac{\mu_0}{4\pi}\frac{IR^2\,d\theta}{(z^2+R^2)^{3/2}} = \frac{\mu_0}{4\pi}\frac{IR^2}{(z^2+R^2)^{3/2}}\int_0^{2\pi}d\theta$$

The integral $\int_0^{2\pi}d\theta$ is just 2π. Here is the result (Figure 17.29):

MAGNETIC FIELD OF A LOOP

$$B_{\text{loop}} = \frac{\mu_0}{4\pi}\frac{2\pi R^2 I}{(z^2+R^2)^{3/2}}$$

for a circular loop of radius R and conventional current I at a distance z from the center, along the axis. See EXP5.

Step 4: Check the Result

Units:

$$\left(\frac{\text{T}\cdot\text{m}}{\text{A}}\right)\frac{(\text{m}^2\cdot\text{A})}{(\text{m}^2)^{3/2}} = \text{T}$$

Direction: Check several pieces with the right-hand rule.
Special case: Center of the loop (Figure 17.30), where the magnetic field is especially easy to calculate from scratch. By the right-hand rule, $\Delta\vec{l}\times\hat{r}$ points out of the page for every piece of the loop, and its magnitude is simply Δl, since $\Delta\vec{l}\perp\hat{r}$. The magnitude of $\hat{r}$ is 1, and $\sin(90°)=1$.
We have this:

$$B = \sum\frac{\mu_0}{4\pi}\frac{I\Delta l}{R^2} = \frac{\mu_0}{4\pi}\frac{I(2\pi R)}{R^2} = \frac{\mu_0}{4\pi}\frac{2\pi I}{R}$$

because the sum of all the Δl's is just $2\pi R$, the circumference of the ring.

> QUESTION Show that the general equation for the magnetic field of a loop reduces to this result if you let $z=0$, which is another kind of check.

Qualitative Features of the Magnetic Field of a Loop of Wire

It is interesting to see what the magnetic field is *far* from the loop, along the axis of the loop.

QUESTION Show that if z is very much larger than the radius R, the magnetic field is approximately equal to

$$\frac{\mu_0}{4\pi}\frac{2\pi R^2 I}{z^3}$$

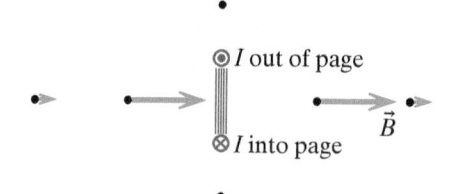

Figure 17.31 Predict the direction of the magnetic field of the coil, above and below the coil. The plane of the coil is perpendicular to the page.

The key to this result is that if $z \gg R$, $(z^2 + R^2)^{3/2} \approx (z^2)^{3/2} = z^3$. We see that the magnetic field of a circular loop falls off like $1/z^3$.

QUESTION Figure 17.31 shows the pattern of magnetic field along the axis of a coil containing many loops. What is the direction of the magnetic field at the points above and below the coil? (*Hint:* Apply the Biot–Savart law qualitatively to the near and far halves of the circular loop.)

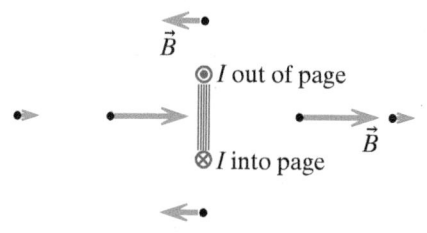

Figure 17.32 The magnetic field above and below the coil points in the opposite direction to the field along the axis.

Far above the coil in Figure 17.31, the closer upper part of the coil contributes a larger magnetic field to the left than does the slightly farther away lower part of the coil to the right. A detailed calculation shows that this is also true a short distance above the coil (Figure 17.32). Compare the pattern of magnetic field with the pattern of *electric* field around an electric dipole.

Magnetic Field at Other Locations Outside the Loop

The magnetic field at other locations outside the loop is more difficult to calculate analytically, but the magnetic field has a characteristic dipole pattern as shown in Figure 17.33, which is the result of a computer calculation that added up all the contributions of many short sections of the loop.

A Special Right-Hand Rule for Current Loops

There is another "right-hand rule" that is often used to get the direction of the magnetic field along the axis of a loop. Let the fingers of your right hand curl around in the direction of the conventional current, and your thumb will point in the direction of the magnetic field at any location on the axis.

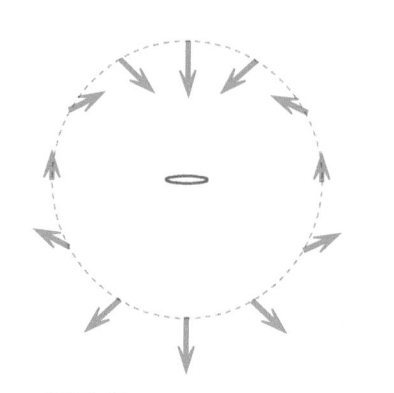

Figure 17.33 The magnetic field of a current loop (which lies in the xz plane, viewed edge-on), at locations outside the loop, in a plane containing the axis of the loop.

QUESTION Try using this right-hand rule to determine the direction of the magnetic field at the indicated observation location in Figure 17.34.

You should find that the magnetic field points down. This right-hand rule should of course give the same result as applying the more general right-hand rule to the cross product $\Delta\vec{l} \times \hat{r}$ and adding up the contributions of the various parts of the loop, as called for by the Biot–Savart law.

QUESTION On the diagram, consider $\Delta\vec{l} \times \hat{r}$ for two short pieces of the loop, on opposite sides of the loop. Show that the two pieces together contribute a magnetic field in the downward direction above the loop.

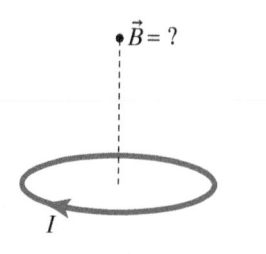

Figure 17.34 Curl the fingers of your right hand in the direction of the conventional current, and your thumb will point in the direction of the magnetic field.

Magnetic Field of Thin and Long Coils

A coil of wire with N turns like the one in Figure 17.35 produces a magnetic field that is approximately N times the magnetic field produced by one turn, as long as the coil is "thin"; that is, that all the turns are in nearly the same plane.

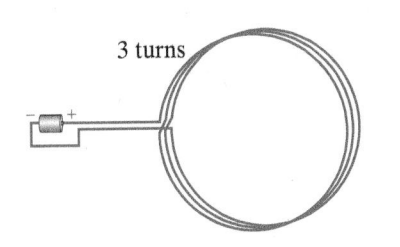

Figure 17.35 A thin coil containing three turns.

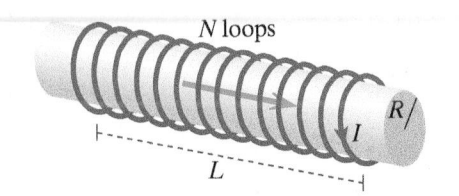

Figure 17.36 A solenoid is a long coil that produces a uniform magnetic field inside.

In contrast, a solenoid is a long coil made by winding a wire around a long tube (Figure 17.36). In Section 17.14 we show that the magnetic field along the centerline of a solenoid of length L has a surprisingly simple form, $B \approx \mu_0 NI/L$, which is nearly the same all along the centerline except near the ends. In Chapter 21 we'll see that the magnetic field is also nearly uniform across the cross section of the solenoid, so a solenoid is useful for producing a uniform magnetic field (constant in magnitude and direction) throughout the interior of the solenoid, not too near the ends.

17.9 COMPUTATION AND 3D VISUALIZATION

Although it is possible to understand something about patterns of electric field by considering only 2-D planar slices of space, this is not the case with magnetic fields. The fact that the vector cross-product occurs in the expression for calculating magnetic field is an indication of how three-dimensional patterns of magnetic field are.

It is easy to write a short VPython program that helps in visualizing both the spatial and temporal variation of the magnetic fields due to moving charged particles. We already have all of the computational tools we'll need to do this.

Recall from Chapter 11 on angular momentum that the VPython syntax for a vector cross product is

```
cross(v,rhat)
```

The organization of a program to calculate and display the magnetic field of a moving particle is straightforward:

- Set up a list of arrows at observation locations.
- Create a `while` loop to move a charged particle.
- For each time step Δt, loop through all observation locations, calculating the magnetic field at each one, and adjusting the axis of the arrow to represent the field.

Assuming that we have done the usual setup, including defining a constant `mzofp` (mu zero over four pi, equal to $\mu_0/4\pi$), and creating a list of arrows called `obs`, with their positions at the desired locations, the structure of the computational loop in such a program might look like this:

```
while t < tmax:
    source.pos = source.pos + v*deltat
    i = 0
    while i < len(obs):
        r = obs[i].pos - source.pos
        rhat = r/mag(r)
        B = mzofp * source.q * cross(v,rhat)/mag(r)**2
        obs[i].axis = sf*B
        i = i + 1
    t = t + deltat
```

Several of the problems at the end of the chapter involve visualizations of this kind, dealing with the magnetic field of a moving charge. A somewhat different approach is needed for calculating the magnetic field generated by current-carrying wires, where $q\vec{v} \times \vec{r}$ is replaced by $I\Delta\vec{l} \times \vec{r}$. The easiest way to approach these situations in the VPython context is to use a `curve` object, whose `pos` attribute is a list of points that are connected by straight lines. The ith vector piece $\Delta\vec{l}$ of a current-carrying wire represented as a `curve` object named `wire` is `dl = wire.pos[i+1] - wire.pos[i]`, and the center of the $\Delta\vec{l}$ vector is `wire.pos[i] + dl/2`. Make sure that the last value of `i` in your `while` loop references the next to the last element in the curve's position list, so that `wire.pos[i+1]` is a legal reference. Here is the basic idea:

```
# Create a curve object to represent a wire loop
wire = curve(radius=R/4, color=color.orange)
N = 40 # N dl's around the loop
dtheta = 2*pi/N
theta = 0
# Include one extra point for end of last dl
while theta <= 2*pi+dtheta/2:
    # Append another point to the curve
    wire.append(pos=R*vector(0, cos(theta), sin(theta)))
    theta = theta + dtheta

# Create a list named obs of observation locations
# ......

# Calculate B at all observation locations
iobs = 0
while iobs < len(obs):
    B = vector(0,0,0)
    i = 0
    while i < N:
        rate(1000)
        dl = wire.pos[i+1] - wire.pos[i]
        loc = wire.pos[i] + dl/2
        r = obs[iobs].pos-loc
        rhat = r/mag(r)
        B = B + mzofp*I*cross(dl,rhat)/mag(r)**2
        i = i + 1
    obs[iobs].axis = scalefactor*B
    iobs = iobs + 1
```

17.10 MAGNETIC DIPOLE MOMENT

Recall the equation for the electric field along the axis of an electric dipole, at a distance r far from the dipole:

$$E_{\text{axis}} \approx \frac{1}{4\pi\varepsilon_0}\frac{2p}{r^3}$$

where the electric dipole moment $p = qs$. Similarly, in the equation for the magnetic field along the axis of a current-carrying coil at a distance r far from the coil, we can write this:

$$B_{\text{axis}} \approx \frac{\mu_0}{4\pi}\frac{2\mu}{r^3}$$

where the magnetic dipole moment $\mu = IA$. (If there are N loops, $\mu = NIA$.) Here A is the area of the loop (πR^2 for circular loops). This equation for magnetic field is approximately valid even if the loop is not circular. The magnetic dipole moment $\vec{\mu}$ is considered to be a vector pointing in the direction of the magnetic field along the axis (Figure 17.37). This means that the direction of the magnetic dipole moment can be obtained by curling the fingers of your right hand in the direction of the conventional current, and your thumb points in the direction of the magnetic dipole moment.

We will see later that the concept of magnetic dipole moment also applies to magnets and provides a way of characterizing the strength of a magnet.

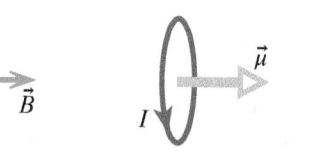

Figure 17.37 The magnetic dipole moment $\vec{\mu}$ is considered to be a vector pointing in the direction of the magnetic field along the axis.

Checkpoint 6 What is the magnetic dipole moment of a 3000-turn rectangular coil that measures 3 cm by 5 cm and carries a current of 2 A?

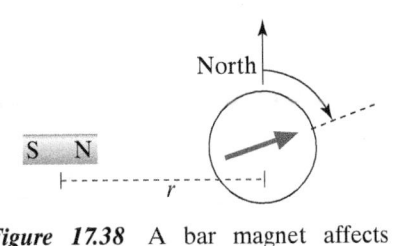

Twisting of a Magnetic Dipole

For reasons that we will discuss in a later chapter, the magnetic dipole moment vector $\vec{\mu}$ acts just like a compass needle. In an applied magnetic field, a current-carrying loop rotates so as to align the magnetic dipole moment $\vec{\mu}$ with the field. See EXP6.

17.11 THE MAGNETIC FIELD OF A BAR MAGNET

From its name, you might guess that a magnet also makes a magnetic field. We say that a magnetic field is present if we see a compass needle twist. If you bring a bar magnet near a compass, you see that a bar magnet does make a compass needle deflect (Figure 17.38). Two magnets interact with each other, attracting or repelling depending on their relative orientations.

Magnetic interactions have some similarity to electric interactions: one can observe both attraction and repulsion, the superposition principle is valid, and the interactions pass through matter. However, there are also some major differences. A bar magnet interacts only with iron or steel objects, whereas a charged invisible tape interacts with *all* objects. A magnet can be permanently magnetic, but a charged invisible tape or plastic pen loses its charge within a relatively short time. Two negatively charged objects always repel each other, whereas two magnets may repel or attract, depending on their orientations. The magnetic field around a current-carrying wire has a "curly" pattern that we don't see with electric fields. See EXP7 and EXP8.

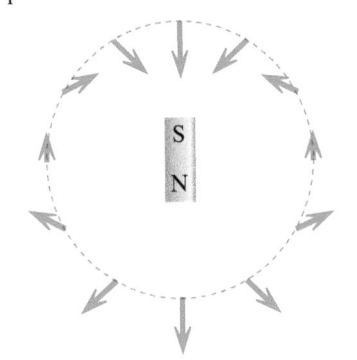

Figure 17.38 A bar magnet affects a compass.

Figure 17.39 A bar magnet is a magnetic dipole. The field pattern is that of a dipole, and the magnitude of the field is proportional to $1/r^3$.

A Bar Magnet Is a Magnetic Dipole

The pattern of directions of magnetic field around a bar magnet is very similar to the pattern of directions of the magnetic field around a current loop, and to the pattern of the electric field around a permanent electric dipole seen in Chapter 13 (Figure 17.39). Also, along the axis of a magnet or a current loop or a permanent electric dipole, the magnetic or electric field varies like $1/r^3$ as can be seen in Experiment EXP9. Moreover, the magnitude of the field off to the side of an electric dipole or a magnet is half as large as it is at the same distance along the axis. Because of these strong similarities, a magnet is often called a "magnetic dipole." In Problem P59 you can use experimental data to determine the magnetic dipole moment of a bar magnet. See EXP9–EXP11.

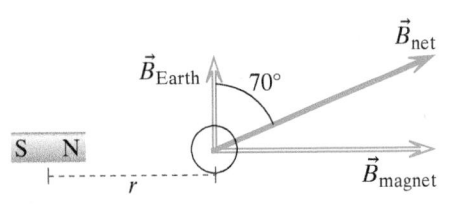

Figure 17.40 Determining the magnetic dipole moment of a bar magnet.

EXAMPLE

Determining the Magnetic Dipole Moment of a Magnet

A compass originally points north. A bar magnet whose mass is 69.5 g is aligned east–west and placed near a compass as shown in Figure 17.40. When the distance between the center of the magnet and the center of the compass is 23.3 cm, the compass deflects 70°. What is the magnetic dipole moment of the bar magnet?

Solution

$$\vec{B}_{net} = \vec{B}_{Earth} + \vec{B}_{magnet}$$
$$B_{magnet} = B_{Earth} \tan 70°$$

$$B_{\text{magnet}} = (2 \times 10^{-5}\,\text{T})\tan 70° = 5.5 \times 10^{-5}\,\text{T}$$

$$B_{\text{magnet}} = \frac{\mu_0}{4\pi}\frac{2\mu}{r^3}$$

$$\mu = \frac{B_{\text{m}}r^3}{2\dfrac{\mu_0}{4\pi}} = \frac{(5.5 \times 10^{-5}\,\text{T})(0.233\,\text{m})^3}{2\left(1 \times 10^{-7}\dfrac{\text{T}\cdot\text{m}}{\text{A}}\right)} = 3.5\,\text{A}\cdot\text{m}^2$$

The Magnetic Field of the Earth

The magnetic field of the Earth has a pattern that looks like that of a bar magnet (Figure 17.41). In the Northern Hemisphere, the Earth's magnetic field dips downward toward the Earth. A horizontally held compass is affected only by the horizontal component of the magnetic field, and this horizontal component gets smaller as you move closer to the magnetic poles.

For example, the horizontal component is smaller in Canada than it is in Mexico, despite the fact that the magnitude of the magnetic field is larger in Canada. The horizontal component is zero right at the magnetic poles, where the magnitude of the magnetic field is largest.

The Earth is a big magnet, but its type S pole is in the Arctic, and the N end of a compass needle points toward this type S pole.

The Earth's magnetic poles are not located exactly at the geographic poles. The S pole is located in northern Canada, about 1300 km (800 mi) from the geographic North Pole, and the N pole is on the Antarctic continent, about 1300 km from the geographic South Pole.

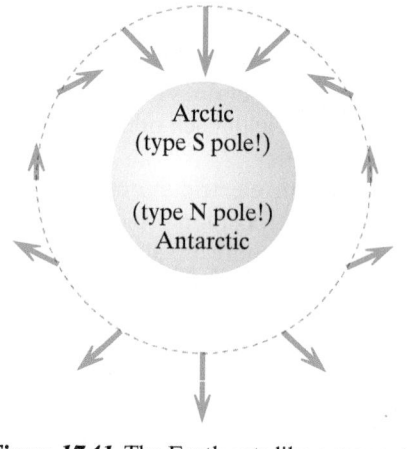

Figure 17.41 The Earth acts like a magnet. Its type N pole is in the Antarctic, and its type S pole is in the Arctic.

Dependence on Distance

The horizontal component of the Earth's magnetic field, which is the component that affects a horizontally held compass, is different at different latitudes, depending on the distance from the magnetic poles of the Earth. Here are measurements of the magnitude of the horizontal component of the Earth's magnetic field at a few selected locations in the United States:

Location	Horizontal Component of Earth's Magnetic Field
Maine	about 1.5×10^{-5} T
Much of the United States	about 2×10^{-5} T
Florida, Hawaii	about 3×10^{-5} T

Knowing the horizontal component of the Earth's magnetic field, we can use the deflection of the compass needle to measure the magnitude of the magnetic field of the magnet. We can then study both the distance dependence and the directional pattern of the magnetic field of the magnet.

Figure 17.42 No magnetic monopoles. This pattern of outward-going magnetic field has never been observed.

Magnetic Monopoles

There is one dramatic difference between electric and magnetic dipoles. The individual positive and negative electric charges (monopoles) making up an electric dipole can be separated from each other, and these point charges make outward-going or inward-going electric fields (which vary like $1/r^2$). One might expect a magnetic dipole to be made of positive and negative magnetic monopoles, but no one has ever found an individual magnetic monopole. Such a magnetic monopole would presumably make an outward-going or inward-going magnetic field (which would vary like $1/r^2$), but such a pattern of magnetic field has never been observed.

Figure 17.42 is an example of the sort of magnetic field pattern that many scientists have diligently looked for but not found. If you cut a magnet in two, you don't get two magnetic monopoles—you just get two magnets!

EXAMPLE

A Circuit in the Antarctic

In a research station in the Antarctic, a circuit containing a partial loop of wire, of radius 5 cm, lies on a table. A top view of the circuit (looking down on the table) is shown in Figure 17.43. A current of 5 A runs in the circuit in the direction shown, and other portions of the circuit are far away. You have a bar magnet with magnetic dipole moment 1.2 A·m². How far above location A, at the center of the loop, would you have to hold the bar magnet, and in what orientation, so that the net magnetic field at location A would be zero? The magnitude of the Earth's magnetic field in the Antarctic is about 6×10^{-5} T.

Solution

Orientation:
  $\vec{B}_{Earth}$ points out of the page (out of the ground, in the Antarctic).
  $\vec{B}_{circuit}$ points out of the page (out of the table).
  The bar magnet must be held with its north pole downward (into page), to make a magnetic field into the ground.
  Components of magnetic field out of page at location A:
  $B_{Earth} = 6 \times 10^{-5}$ T
  $B_{straight\ wires} = 0$ (The straight wires contribute nothing to the magnetic field at A, because for each wire $d\vec{l} \times \hat{r} = 0$.)

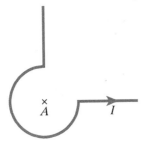

Figure 17.43 Top view of a circuit lying on a table (looking down at the table).

B due to 3/4 loop: $\vec{B} = \int d\vec{B} = \int \dfrac{\mu_0}{4\pi} \dfrac{I d\vec{l} \times \hat{r}}{r^2}$

Origin: Center of the loop
$|\vec{r}| = R$ (constant)

At all locations $d\vec{l} \perp \hat{r}$, so $|d\vec{l} \times \hat{r}| = dl \sin 90° = dl = R d\theta$, as shown in Figure 17.44. Therefore

$$|\vec{B}_{loop}| = \int_{\pi/2}^{2\pi} \frac{\mu_0}{4\pi} \frac{I(R d\theta)}{R^2} = \frac{\mu_0}{4\pi} \frac{IR}{R^2} \int_{\pi/2}^{2\pi} d\theta = \frac{\mu_0}{4\pi} \frac{I}{R} \theta \Big|_{\pi/2}^{2\pi} = \frac{\mu_0}{4\pi} \frac{3\pi I}{2R}$$

$$|\vec{B}_{loop}| = \left(1 \times 10^{-7} \frac{T \cdot m}{A}\right) \frac{(3\pi)(5\,A)}{2(0.05\,m)} = 4.7 \times 10^{-5}\,T$$

$$B_{Earth} + B_{3/4\ loop} = 6 \times 10^{-5}\,T + 4.7 \times 10^{-5}\,T = 1.06 \times 10^{-4}\,T$$

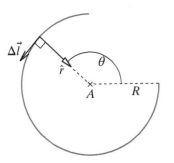

Figure 17.44 $d\vec{l} \perp \hat{r}$ for every step around the loop. The magnitude of r is constant and equal to the radius of the loop.

We need

$$B_{magnet} = 1.06 \times 10^{-4}\,T \approx \frac{\mu_0}{4\pi} \frac{2\mu}{z^3}$$

$$z = \left[\frac{\frac{\mu_0}{4\pi}(2\mu)}{B_{magnet}}\right]^{1/3} = \left[\frac{\left(1 \times 10^{-7} \frac{T \cdot m}{A}\right)(2)(1.2\,A \cdot m^2)}{(1.1 \times 10^{-4}\,T)}\right]^{1/3} = 0.13\,m$$

Thus we hold the magnet with its center about 13 cm above the circuit, with its north pole toward the ground. (We model the magnet as a dipole, located at the center of the magnet.)

17.12 THE ATOMIC STRUCTURE OF MAGNETS

We have seen that the magnetic field of a circular current loop looks suspiciously similar to the magnetic field of a magnet. The pattern of directions of the magnetic field looks the same, and it is even the case that for both magnets and current loops the magnitude of the field at large distances along the axis falls off like $1/r^3$. Both magnets and current loops can be described in terms of a magnetic dipole moment $\vec{\mu}$.

A Single Atom Can Be a Magnetic Dipole

Inside every atom there are moving charged particles. In an atom in a magnetic material, these subatomic "currents" may produce tiny magnetic dipole moments that add up to a nonzero magnetic dipole moment for each atom. (In contrast, in a nonmagnetic material, these tiny magnetic dipole moments would add up to zero.)

There are three possibilities for the source of the subatomic currents that might give a single atom a nonzero magnetic dipole moment:

- An electron orbiting around the nucleus (Figure 17.45)
- An electron spinning on its own axis (Figure 17.45)
- Rotational motion in the nucleus (protons and neutrons have spin about their own axes and can also orbit around inside the nucleus)

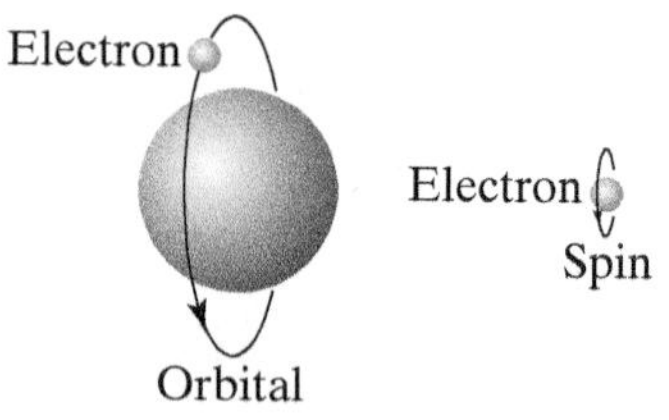

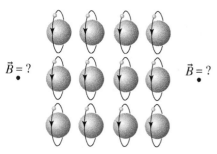

Figure 17.45 Electrons have spin as well as orbital motion that can contribute to the magnetic dipole moment.

In all of these situations the object has angular momentum, and the magnetic dipole moment turns out to be proportional to the angular momentum L of the particle:

$$\mu = (\text{factor})L$$

We can use a simple model to estimate this proportionality factor.

Magnetic Dipole Moment of an Orbiting Electron

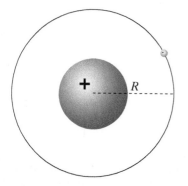

Figure 17.46 A simple model for a magnet might involve electrons in circular atomic orbits.

It is easiest to estimate this proportionality factor by considering the magnetic dipole moment of an orbiting electron, using the simple Bohr model of an atom to get an order of magnitude estimate. As indicated in Figure 17.46, let's suppose that each atom has one unpaired electron that orbits the nucleus along a circular path, at constant speed. Each orbiting electron can be considered to be a tiny current loop that makes a dipole magnetic field.

> QUESTION In Figure 17.46, what is the direction of the resulting magnetic field to the left and right of the atoms, and of the magnetic dipole moment of one atom? (Remember that electrons are negative.)

The magnetic field due to each orbiting electron would point to the left, so the magnetic dipole moment of each atom, and of the whole object, would point to the left.

For a current loop, we know that

$$\mu = I(\pi R^2)$$

> QUESTION What would be the current I for one orbiting electron (Figure 17.47)?

Figure 17.47 Simple model of an atom in which an outer electron orbits a positive inner core.

The current I is a measure of how much charge passes one location per second. The charge on the electron is $-e$, and the time it takes the electron to go around once is $T = 2\pi R/v$, where v is the speed of the electron, so

$$I = \frac{e}{\left(\dfrac{2\pi R}{v}\right)} = \frac{ev}{2\pi R}$$

Therefore the magnetic dipole moment of a single orbiting electron would be

$$\mu = I(\pi R^2) = \left(\frac{ev}{2\pi R}\right)(\pi R^2) = \frac{1}{2}eRv$$

Relating Magnetic Dipole Moment to Angular Momentum

The magnitude of the angular momentum of an electron in a circular orbit is

$$|\vec{L}| = |\vec{r} \times \vec{p}|$$

In a circular orbit $\vec{r} \perp \vec{p}$, so this reduces to

$$L = Rp \sin 90° = Rmv \quad (v \ll c)$$

We can express the magnetic dipole moment of the orbiting electron in terms of angular momentum by multiplying by 1:

$$\mu = \left(\frac{m}{m}\right)\left(\frac{1}{2}eRv\right) = \frac{1}{2}\frac{e}{m}(Rmv) = \frac{1}{2}\frac{e}{m}L$$

Our estimate of the proportionality factor relating magnetic dipole moment to angular momentum for a particle in an atom is this:

$$(\text{factor}) \approx \frac{1}{2}\frac{e}{m}$$

We will assume that this expression for the proportionality factor is valid for orbiting electrons, electrons spinning on their axes, and nucleons (protons and neutrons) spinning on their axes. How big is this factor?

For an electron: $\dfrac{1}{2}\dfrac{e}{m} = \dfrac{1}{2}\dfrac{(1.6 \times 10^{-19}\,\text{C})}{(9 \times 10^{-31}\,\text{kg})} = 8.9 \times 10^{10}\,\text{C/kg}$

For a nucleon: $\dfrac{1}{2}\dfrac{e}{m} = \dfrac{1}{2}\dfrac{(1.6 \times 10^{-19}\,\text{C})}{(1.7 \times 10^{-27}\,\text{kg})} = 4.7 \times 10^{7}\,\text{C/kg}$

This factor is nearly 2000 times smaller for a nucleon. Angular momentum is quantized, with similar values for electrons and nucleons, so the magnetic dipole moment of a single atom comes almost entirely from orbiting electrons or electrons spinning on their own axes.

Angular Momentum of an Orbiting or Spinning Electron

You may recall from Chapter 11 that both translational angular momentum (like the orbital angular momentum of an electron) and rotational angular momentum (like the angular momentum of an electron spinning on its axis) are quantized in atoms; only certain discrete values are observed. For an electron orbiting a nucleus, the angular momentum has to be an integer multiple of Planck's constant h divided by 2π, written as $\hbar$:

$$L = N\hbar$$

where $N = 0, 1, 2, 3, \ldots$, and $\hbar = 1.05 \times 10^{-34}\,\text{J}\cdot\text{s}$.

The electron's spin angular momentum is always one-half of $\hbar$, but there is a compensating factor of 2 in its relation to the magnetic dipole moment. (If you are not familiar with angular momentum quantization, you may wish to read Section 17.13, which uses more familiar classical reasoning about circular motion to estimate the orbital angular momentum of an electron. The answer is the same.)

Assuming one quantum of angular momentum, we have the following:

$$L = \hbar \quad \text{so} \quad \mu \approx \frac{1}{2}\frac{e}{m}\hbar$$

Evaluating this expression, we find that:

$$\mu \approx \left(\frac{1}{2}\right)\frac{(1.6\times 10^{-19}\,\text{C})}{(9\times 10^{-31}\,\text{kg})}(1.05\times 10^{-34}\,\text{J}\cdot\text{s})$$

$$\mu \approx 1\times 10^{-23}\,\text{A}\cdot\text{m}^2 \text{ per atom}$$

Comparing with Experiment

If you have a bar magnet, a compass, and a meter stick, you can determine the magnetic dipole moment of the bar magnet experimentally. The procedure for doing this is detailed in Problem P59.

In an earlier example we found that a particular magnet whose mass was 69.5 g has a magnetic dipole moment of 3.5 A·m². Assume that the magnetic dipole moment of this magnet is due to the fact that each atom has one unpaired electron contributing a magnetic dipole moment of $\mu \approx 1\times 10^{-23}$ A·m² (due either to orbital or spin angular momentum) and that the magnetic dipole moments of all the atoms are aligned in the same direction. What would be the predicted magnetic dipole moment of this bar magnet?

Estimated Magnetic Dipole Moment of a Bar Magnet

The bar magnet in an earlier example had a mass of 69.5 g. Assume that almost all of the atoms in this magnet are iron atoms. The mass of one mole of iron is 56 g.

$$N = \left(\frac{69.5\,\text{g}}{56\,\text{g/mol}}\right)(6.02\times 10^{23}\text{ atoms/mol}) = 7.5\times 10^{23}\text{ atoms}$$

$$\mu = N\mu_{\text{atom}} = (7.5\times 10^{23}\text{ atoms})\left(1\times 10^{-23}\frac{\text{A}\cdot\text{m}^2}{\text{atom}}\right) = 7.5\text{ A}\cdot\text{m}^2$$

QUESTION   The estimated value of the magnetic dipole moment of this magnet is 7.5 A·m², but the observed value is only 3.5 A·m². Is this good agreement or poor agreement?

This is excellent agreement! The values differ by a factor of only about 2. The fact that the prediction and measurement are this close is evidence that the basic assumptions of our very simple model—that each atom in a magnet is itself a small dipole, and that the field produced by the magnet is the sum of the fields of all the atoms—are reasonable assumptions. If the predicted and measured values had differed by a factor of a thousand or a million, we would have had to re-examine our model. A related problem is Problem P59.

Simplified Models

We made a number of simplifying assumptions in our model of an atomic dipole. You probably know from chemistry courses that more sophisticated atomic models do not assume that electrons orbit the nucleus in well-defined circular paths. In these models, there is only a probability for finding the electron at a particular place. Moreover, many electrons in atoms have spherically symmetric probability distributions (s orbitals), and such a symmetric distribution has a zero average angular momentum. However, electrons in nonspherically symmetric probability distributions (p, d, or f orbitals) have nonzero angular momentum, and can contribute a nonzero magnetic dipole moment. Additionally, we assumed that in every atom there is just one electron that contributed a magnetic field, but in some materials two or more unpaired electrons may contribute. Moreover, the "spinning ball" model of the electron isn't really adequate, because the most sensitive experiments are consistent with the notion that an electron is a true point charge and has zero radius! It is nevertheless an experimental fact that an electron does make a magnetic field as though it were a spinning ball of charge, and it can be useful to think of it literally as a spinning ball.

Nonetheless, the fact that our simplified model gave a prediction with an appropriate order of magnitude shows the power of reasoning with simple models.

The Modern Theory of Magnets

Modern theories of the nature of magnets take into account not only the contribution of individual atoms in a solid magnet, but also the effects of their interactions with each other. These more complex theoretical treatments suggest that the magnetic dipole moments due to the rotational angular momentum of spinning electrons contribute most of the magnetic field of a magnet, although the orbital angular momentum of electrons does make some contribution.

As we saw above, because the factor e/m is so much smaller for a proton than for an electron, we can ignore nuclear contributions to the magnetic field of your bar magnet. Nuclear magnetic dipole moments do play an important role in the phenomenon of nuclear magnetic resonance (NMR) and in the technology of magnetic resonance imaging (MRI) used in medicine, which is based on NMR.

Alignment of the Atomic Magnetic Dipole Moments

In many materials, an atom has no net orbital or spin magnetism. In most materials whose individual atoms do make a magnetic field, the orbital and spin motions in different atoms don't line up with each other, so the net field of the many atoms in a piece of the material averages out to zero. However, in a few materials, notably iron, nickel, cobalt, and some alloys of these elements, the orbital and spin motions in neighboring atoms line up with each other and therefore produce a sizable magnetic field. These unusual materials are called "ferromagnetic." The reason for the alignment can be adequately discussed only within the framework of quantum mechanics; basically the alignment is due to electric interactions between the atoms, not to the much weaker magnetic interactions.

Magnetic Domains

In an ordinary piece of iron that isn't a magnet, the iron is a patchwork of small regions, called "magnetic domains," within which the alignment of all the atomic magnetic dipole moments is nearly perfect. Normally, though, these domains are oriented in random directions, so the net effect is that this piece of iron doesn't produce a significant net magnetic field. Figure 17.48 shows a picture of the situation, in which the arrows indicate the magnetic dipole moments of the atoms.

If you use a coil to apply a large magnetic field to the iron, domains that happen to be nearly aligned with the applied field tend to grow at the expense of domains that have a different orientation (Figure 17.49). Also, with a sufficiently large applied magnetic field, the magnetic dipole moments of the atoms in a domain tend to rotate toward aligning with the applied field, just as a compass needle turns to align with an applied field. The result of both of these effects is to partially align the magnetic dipole moments of most of the

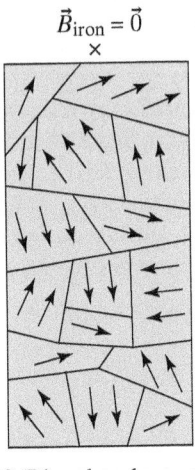

Figure 17.48 Disordered magnetic domains—the net magnetic field produced by the iron is nearly zero.

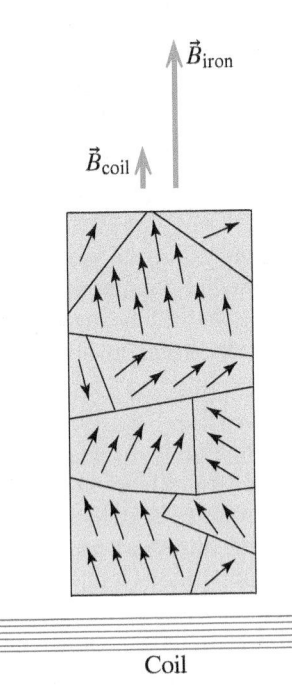

Figure 17.49 The magnetic field of a coil causes a partial ordering of the magnetic domains, which produces a significant nonzero magnetic field contributed by the iron.

atoms, which produces a magnetic field $\vec{B}_{iron}$ that may be much larger than the applied magnetic field. There is a kind of multiplier effect.

In the case of very pure iron, when you turn off the current in the coil, the piece of iron goes back nearly to its original disordered patchwork of domains and doesn't act like a magnet any more. In some other ferromagnetic materials, though, including some alloys of iron, when you remove the applied field the domains remain nearly aligned. In that case you end up with a permanent magnet, which is commonly made of the alloy alnico (alnico V contains 51% iron, 8% aluminum, 14% nickel, 24% cobalt, and 3% copper). Hitting a magnet with a hard blow may disorder the domains and make the metal no longer act like a magnet. Heating above a critical temperature also destroys the alignments.

Iron Inside a Coil

It is possible to observe this multiplier effect. A small magnetic field created by current-carrying loops of wire wrapped around a piece of iron can lead to a large observed magnetic field being contributed by the iron, due to the alignment of the magnetic dipole moments in the iron. Thanks to this multiplier effect, putting an iron core inside a current-carrying coil of wire makes a powerful "electromagnet" that can pick up iron objects. See EXP12 and EXP13.

Why Are There Multiple Domains?

Within one domain of a ferromagnetic material such as iron, strong electric forces between neighboring atoms make the atomic magnetic dipole moments line up with each other. Why don't all the atoms in a piece of iron spontaneously align with each other? Why does the piece divide into small magnetic domains with varying magnetic orientations in the absence of an applied magnetic field?

There is a simple way to get some insight into why this happens. Consider two possible patterns of magnetic domains in a bar of iron in the absence of an applied magnetic field, a single domain or two opposed domains (Figure 17.50).

You can simulate this situation by taking two permanently magnetized bar magnets and holding them in one of these two positions, as shown in Figure 17.51.

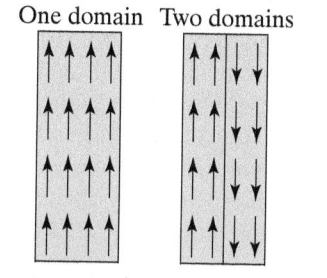

One domain   Two domains

Figure 17.50 Possible arrangements—one magnetic domain or two.

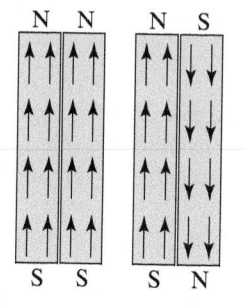

Figure 17.51 Arrange two bar magnets side by side, aligned parallel or antiparallel.

> QUESTION Which of these two magnet configurations do you find to be more stable, the parallel or the antiparallel alignment? By analogy, which is more likely for a bar of iron: that it have just one domain or that it split into two domains in the absence of an applied magnetic field?

You will find that if you start with the parallel alignment the magnets tend to flip into the antiparallel alignment. The domain structure of iron can be a compromise between a tendency for neighboring atoms to have their magnetic dipole moments line up with each other (a strong but short-range electric interaction), and a competing tendency for magnetic dipole moments to flip each other (a weaker but longer-range magnetic interaction).

However, the net effect depends critically on the details of the geometric arrangement. Although two long bar magnets that are side by side tend to line up with their magnetic dipole moments in opposite directions, two thin disk-shaped magnets tend to line up with their magnetic dipole moments in the same direction. What happens inside a magnetic material depends on the details of the geometrical arrangement of the atoms and on the details of the electric interactions between neighboring atoms.

17.13   *ESTIMATE OF ORBITAL ANGULAR MOMENTUM OF AN ELECTRON IN AN ATOM

We can estimate the orbital angular momentum of an electron in an atom by using the simple Bohr model of an atom, in which an electron orbits a nucleus at constant speed in a circular orbit, and applying the Momentum Principle. In this model (Figure 17.52), an outer electron is held in a circular orbit of radius R by the electric attraction to the rest of the atom (which has a net charge of $+e$).

We want to find the magnitude of the angular momentum of the electron:

$$|\vec{L}| = |\vec{r} \times \vec{p}|$$

Since in a circular orbit $\vec{r} \perp \vec{p}$, this reduces to:

$$L = Rp = Rmv \quad (v \ll c)$$

We know the radius of an atom (about 1×10^{-10} m) and the mass of an electron (9×10^{-31} kg), so we need to find the speed of the electron.

System: Electron, charge $-e$
Surroundings: Nucleus and outer electrons, with charge $+e$
Momentum Principle:

$$\frac{d\vec{p}}{dt} = \vec{F}_{\text{net}}$$

Parallel component (constant speed, so the parallel component of momentum is not changing):

$$\frac{dp}{dt}\hat{p} = \vec{0}$$

Perpendicular component (electric force on electron is toward the nucleus, perpendicular to momentum):

$$p\left|\frac{d\hat{p}}{dt}\right| = F\perp$$

$$(mv)\left(\frac{v}{R}\right) = e\left(\frac{1}{4\pi\varepsilon_0}\frac{e}{R^2}\right)$$

$$\frac{mv^2}{R} = \frac{1}{4\pi\varepsilon_0}\frac{e^2}{R^2}$$

$$v = \sqrt{\frac{1}{4\pi\varepsilon_0}\frac{e^2}{Rm}}$$

Plugging in the electron charge and mass, and assuming an approximate atomic radius of 1×10^{-10} m, we get:

$$v = \sqrt{\frac{(9 \times 10^9 \,\text{N·m}^2/\text{C}^2)(1.6 \times 10^{-19}\,\text{C})^2}{(1 \times 10^{-10}\,\text{m})(9 \times 10^{-31}\,\text{kg})}}$$

$$v \approx 1.6 \times 10^6 \ \text{m/s}$$

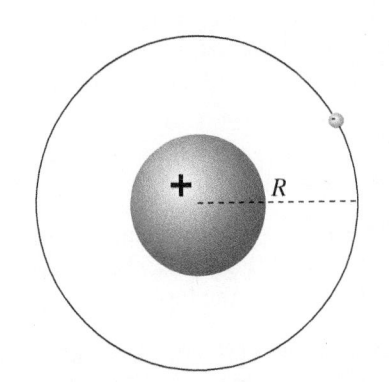

Figure 17.52 Simple model of an atom in which an outer electron orbits a positive inner core.

This gives a value for angular momentum of

$$L = Rmv = (1 \times 10^{-10} \text{ m})(9 \times 10^{-31} \text{ kg})(1.6 \times 10^6 \text{ m/s})$$

$$L \approx 1.4 \times 10^{-34} \text{kg} \cdot \text{m}^2/\text{s}^2$$

This is close to the quantum mechanical value of $\hbar = 1.05 \times 10^{-34}$ J · s. If we had reproduced Bohr's full calculation we would have obtained a slightly different value for R, and L would be exactly $\hbar$.

17.14 *MAGNETIC FIELD OF A SOLENOID

The analysis in this section offers one more example of how to apply the Biot–Savart law. It shows that the magnetic field is nearly uniform along the axis of a solenoid, far from the ends. A solenoid is a coil that is much longer than its radius.

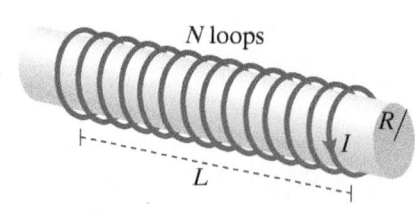

Figure 17.53 A solenoid of length L with N loops of radius R carrying current I.

Step 1: Cut Up the Distribution into Pieces and Draw $\Delta \vec{B}$

We consider a solenoid of length L that is made up of N circular loops wound tightly right next to each other, each of radius R (Figure 17.53). We consider each loop as one piece. The conventional current in the loops is I.

> QUESTION   Given what you know about individual current loops, what will be the direction of the magnetic field $\Delta \vec{B}$ contributed by each of the loops at *any* location along the axis of the solenoid in the figure above?

All of the contributions inside the solenoid will point to the right in the diagram.

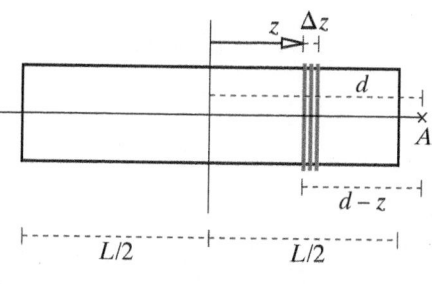

Figure 17.54 Defining an integration variable z for summing the contributions of the many loops.

Step 2: Write an Expression for the Magnetic Field Due to One Piece

Origin: Center of solenoid
Location of one piece: Given by z, so the integration variable is z; the distance from the loop to the observation location A is $d - z$ (Figure 17.54). Our result will be valid for any location along the axis, inside or outside the solenoid.

> QUESTION   How many closely packed loops are contained in a short length Δz of the solenoid?

There are N/L loops per meter, so the number of loops in a length Δz is $(N/L)\Delta z$. We know the magnetic field made by each loop along its axis.

> QUESTION   Show that the loops contained in the section Δz of the solenoid in Figure 17.54 contribute a magnetic field at the observation location in the z direction of this amount:

$$\Delta B_z = \frac{\mu_0}{4\pi} \frac{2\pi R^2 I}{[(d-z)^2 + R^2]^{3/2}} \frac{N}{L} \Delta z$$

Step 3: Add Up the Contributions of All the Pieces

The net magnetic field lies along the axis and is the summation of all the ΔB_z contributed by all the loops:

$$B_z = \sum \Delta B_z = \sum \frac{\mu_0}{4\pi} \frac{2\pi R^2 I N}{L} \frac{\Delta z}{[(d-z)^2 + R^2]^{3/2}}$$

Many of these quantities are the same for every piece and can be taken outside the summation as common multiplicative factors:

$$B_z = \frac{\mu_0}{4\pi} \frac{2\pi R^2 IN}{L} \sum \frac{\Delta z}{[(d-z)^2 + R^2]^{3/2}}$$

This can be turned into an integral, with z ranging from $-L/2$ to $+L/2$:

$$B_z = \frac{\mu_0}{4\pi} \frac{2\pi R^2 IN}{L} \int_{z=-L/2}^{z=+L/2} \frac{dz}{[(d-z)^2 + R^2]^{3/2}}$$

Let $u = d - z$, in which case $dz = -du$, and therefore the limits on the integration run from $u = d - (-L/2)$ to $u = d - (+L/2)$:

$$B_z = \frac{\mu_0}{4\pi} \frac{2\pi R^2 IN}{L} \int_{u=d+L/2}^{u=d-L/2} \frac{-du}{[u^2 + R^2]^{3/2}}$$

$$= \frac{\mu_0}{4\pi} \frac{2\pi R^2 IN}{L} \int_{u=d-L/2}^{u=d+L/2} \frac{du}{[u^2 + R^2]^{3/2}}$$

This integral can be found in standard tables of integrals:

$$B_z = \frac{\mu_0}{4\pi} \frac{2\pi R^2 IN}{L} \left[\frac{u}{R^2\sqrt{u^2 + R^2}} \right]_{u=d-L/2}^{u=d+L/2}$$

$$B_z = \frac{\mu_0}{4\pi} \frac{2\pi IN}{L} \left[\frac{d+L/2}{\sqrt{(d+L/2)^2 + R^2}} - \frac{d-L/2}{\sqrt{(d-L/2)^2 + R^2}} \right]$$

In this expression, d is the distance from the center of the solenoid.

Figure 17.55 is a plot of this expression for B_z for a particular solenoid, as a function of the distance from the center of the solenoid. Notice that the magnetic field is nearly uniform inside the solenoid, as long as you are not too near the ends. The field is even more uniform for solenoids that are longer and/or thinner than this one.

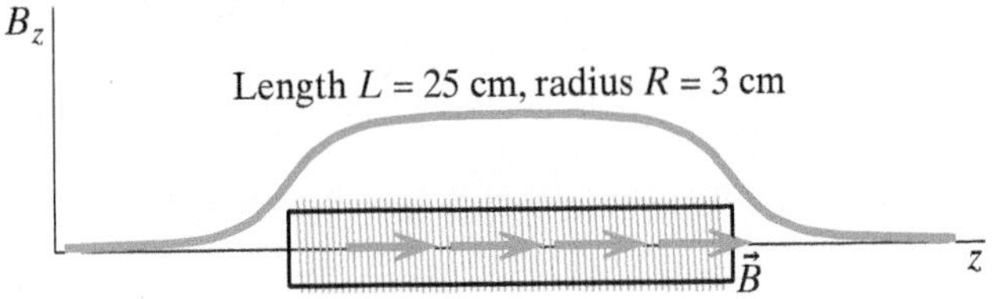

Figure 17.55

QUESTION If the radius R is small compared with the length L ($R \ll L$), show that at the center of the solenoid the magnetic field has a remarkably simple value:

$$B_z \approx \frac{\mu_0 NI}{L}$$

With $d = 0$ and $R \ll L$, the quantity in square brackets above reduces to

$$\frac{L/2}{\sqrt{(L/2)^2}} - \frac{-L/2}{\sqrt{(-L/2)^2}}$$

which is equal to 2, from which follows the simple form $B_z \approx \mu_0 NI/L$.

The function plotted above shows that this is also the approximate magnetic field inside the solenoid along much of the axis, not too close to the ends. Outside the solenoid, the magnetic field falls off rapidly. At a large distance the field falls off like $1/r^3$, since the solenoid then looks like a simple collection of current loops.

MAGNETIC FIELD INSIDE A LONG SOLENOID

$$B_z \approx \frac{\mu_0 NI}{L}$$

inside a long solenoid, with the radius of the solenoid $\ll L$.

The same result can be obtained with much less effort by using Ampere's law, which we will study in Chapter 21, where we will also show that the magnetic field is uniform across the cross section as well as along the centerline (not too near the ends).

Step 4: Check the Result

Does our final result make sense? In particular, do we have the right units? Comparing with the magnetic field for a single current element,

$$\frac{\mu_0}{4\pi} \frac{I\Delta\vec{l} \times \hat{r}}{r^2}$$

we easily verify that our answer does have the right units, since

$$\frac{NI}{L} \quad \text{has the same units as} \quad \frac{I\Delta l}{r^2}$$

(Remember that $\hat{r}$ is dimensionless.)

SUMMARY

The Biot–Savart Law

$$\vec{B} = \frac{\mu_0}{4\pi} \frac{q\vec{v} \times \hat{r}}{r^2} \quad \text{(single moving particle, Figure 17.56)}$$

$\hat{r}$ is a unit vector that points *from* the source charge *toward* the observation location.

$$\frac{\mu_0}{4\pi} = 1 \times 10^{-7} \frac{\text{T} \cdot \text{m}^2}{\text{C} \cdot \text{m/s}} \ exactly$$

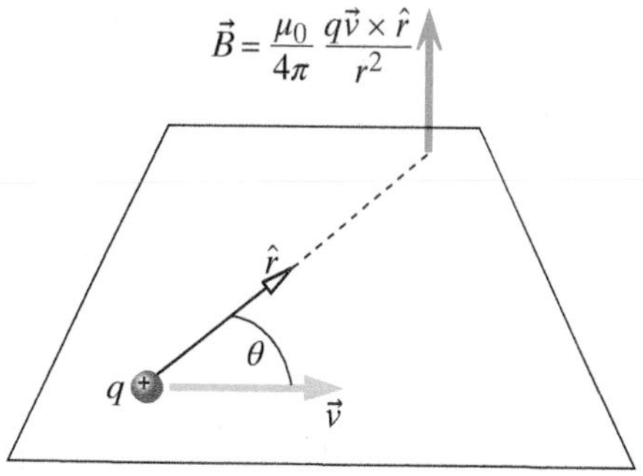

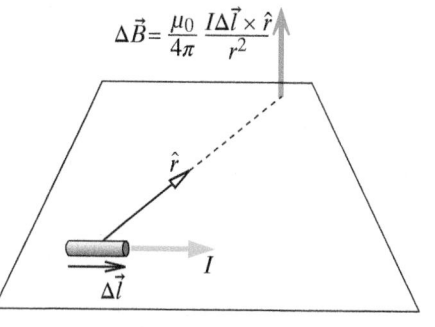

Figure 17.57

Figure 17.56

$$\Delta\vec{B} = \frac{\mu_0}{4\pi} \frac{I\Delta\vec{l} \times \hat{r}}{r^2} \quad \text{(current distribution, Figure 17.57)}$$

Electron current: $i = nA\bar{v}$
Conventional current: $I = |q|nA\bar{v}$
A magnetic field can be detected by a compass; the horizontal component of $B_{\text{Earth}} \approx 2 \times 10^{-5}$ T in much of the continental United States.

Magnetic Field of a Wire
For a straight wire of length L and conventional current I, at a perpendicular distance r from the center of the wire (Figure 17.58):

$$B_{\text{wire}} = \frac{\mu_0}{4\pi} \frac{LI}{r\sqrt{r^2 + (L/2)^2}} \approx \frac{\mu_0}{4\pi} \frac{2I}{r} \quad \text{for } r \ll L$$

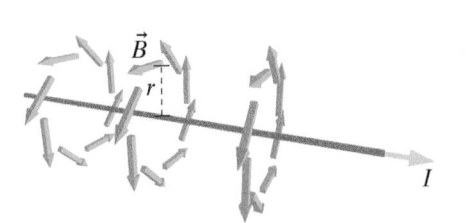

Figure 17.58

Magnetic Field of a Loop

For a circular loop of radius R and conventional current I, at a distance z from the center, along the axis (Figure 17.59):

$$B_{\text{loop}} = \frac{\mu_0}{4\pi} \frac{2\pi R^2 I}{(z^2 + R^2)^{3/2}} \approx \frac{\mu_0}{4\pi} \frac{2\mu}{z^3} \quad \text{for } z \gg R$$

Magnetic dipole moment of a loop: $\mu = (\pi R^2)I$

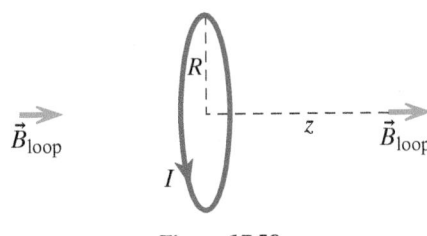

Figure 17.59

Magnetic Field inside a Long Solenoid

Inside a long solenoid (radius of solenoid $\ll L$)

$$B_z \approx \frac{\mu_0 N I}{L}$$

Note that a solenoid is very long compared to its radius—it is not the same thing as a coil composed of a few loops close together (Figure 17.60).

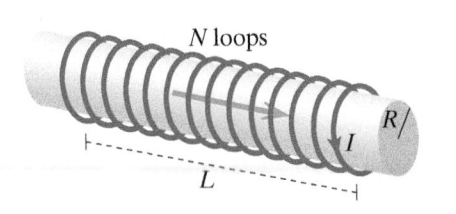

Figure 17.60

Atomic Model of Magnets

An atom can be a magnetic dipole, whose magnetic dipole moment is primarily due to the spin (rotational angular momentum) of unpaired electrons. The orbital angular momentum of electrons can also contribute.

The magnetic dipole moment of a bar magnet is the sum of the magnetic dipole moments of all its constituent atoms.

Ferromagnetic Material

It is organized into domains of aligned atomic magnetic dipole moments, and an applied field can orient these domains. Removal of the applied field may leave the material partially aligned, forming a permanent magnet.

EXPERIMENTS

Equipment

In the study of magnetic field you will need the following equipment:

- two flashlight batteries in a battery holder
- light bulbs of two kinds
- screw-in bulb sockets
- several short copper wires with clips on the ends ("clip leads" used as connecting wires)
- a long wire (about 2 m in length)
- a liquid-filled magnetic compass
- unmagnetized nails for Experiment EXP12

Simple Circuits

To observe the magnetic effects of electric currents, it is useful to construct simple circuits containing wires, light bulbs, and batteries. These are the simplest examples of systems in which we can observe the fundamental electric and magnetic properties of continuous electric currents.

The equipment needed for the experiments in this chapter is the following: two D cells (and it is useful to have a battery holder for them), flashlight bulbs (#48 and #14 if possible), sockets for these bulbs, insulated "hookup" wire, some clip leads (wires

with alligator clips on the end), a liquid-filled compass (air-filled compasses typically don't work well because the needle tends to get stuck on its pivot), and a bar magnet (a magnet with a north end and a south end).

Suitable equipment may be available in a laboratory, or you can purchase experiment kit EM-8675 from http://www.pasco .com.

Light Bulbs and Sockets

The filament of a light bulb (the very thin metal wire that glows) is made of tungsten, a metal that does not melt until reaching a very high temperature. A glowing tungsten wire would rapidly oxidize and burn up in air, so there is a vacuum or an inert gas such as argon inside the bulb.

The thin tungsten filament in the bulb strongly resists the passage of electrons. When the electron sea is forced to move through the tungsten, the mobile electrons collide with the positive cores (nuclei plus inner electrons), and this "friction" makes the metal get hot and glow.

EXP1 Experiments with simple circuits

(a) Using one battery, some connecting wires (insulated wires with clips on the ends, not bare Nichrome wire), and a round bulb (#14) *but no socket*, make the bulb light up. (See Figure 17.61.) If the bulb glows with a steady light, this is a "steady state." Make a diagram showing how you connected the circuit.

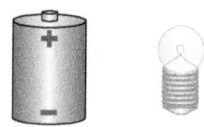

Figure 17.61

(b) Examine a light bulb carefully, and imagine slicing the bulb in half lengthwise. Figure 17.62 is a cutaway sketch:

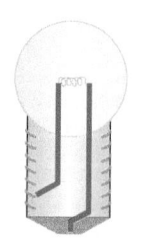

Figure 17.62

On a copy of the sketch, label the important parts and connections, indicating which parts of the bulb you think are metal conductors and which parts are insulators. Using a different color if available, show the conducting path that electric current follows through the bulb. What happens if you switch the connections to the battery? Does the light bulb still light?

(c) Closely examine the long #48 bulb and the round #14 bulb (Figure 17.63).

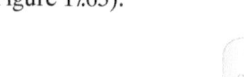

Figure 17.63 A long #48 bulb (left) and a round #14 bulb (right).

The tungsten filaments in both bulbs are about the same length, but perhaps you can see even with the naked eye that the filament in the long bulb is extremely thin—thinner than the filament in the round bulb. Would you guess it would be easier to push electric current through a thin filament (as in the long bulb) or through a thick filament (as in the round bulb)? Why? (At this point this is mostly just a guess, but we'll study this in detail later.)

(d) Examine a bulb *socket* (the receptacle into which a bulb is screwed), and imagine slicing the socket in half lengthwise. Make a cutaway sketch, and label the important parts and connections. Indicate which parts are metal conductors and which parts are insulators. On your sketch, trace the path (in a different color if available) along which electrons will move through the socket when there is a bulb in it. Screw a bulb into the socket and connect the socket to the battery with two connecting wires. Make sure the bulb lights.

(e) Connect a round bulb in a socket to two batteries in series (that is, one after the other) using connecting wires (Figure 17.64).

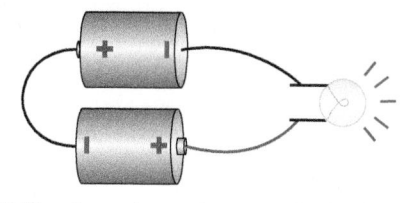

Figure 17.64 Two batteries and a round bulb in series. The socket is not shown.

To connect two batteries in series, put them in the battery holder in opposite directions, and connect them as shown (note that + is connected to −). Compare the brightness of the bulb with one battery and with two batteries in series. Because of this difference we'll usually use two batteries in series in our experiments.

EXP2 The magnetic effects of currents

Make a two-battery circuit with a round bulb in a socket. Place your magnetic compass on a flat surface under one of the wires as shown in Figure 17.65. Keep the compass away from steel objects, such as the steel-jacketed batteries and the alligator clips on the ends of your wires. If you are working on a steel table, you may need to put the compass on a thick book. (For flexibility in placement, you may find it useful to make a long wire by connecting two of your wires together.)

Connect the circuit so that the bulb glows, and do the following:

- Lift the wire up above the compass.
- Orient the wire to be horizontal and lined up with the compass needle. (Using a long wire may make it easier to do this.)
- Bring the aligned wire down onto the compass, as in Figure 17.65.

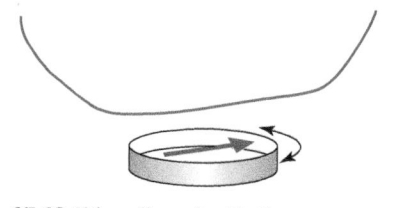

Figure 17.65 Wire aligned with the compass needle.

(a) What is the effect on the compass needle as you bring the wire down on top of the compass?

(b) What happens when the wire is initially aligned perpendicular instead of parallel to the needle as shown in Figure 17.66?

(c) Reverse the connections to the batteries, or reverse the direction of the wire over the compass, in order to force electrons through the circuit in the opposite direction. Again make the compass needle deflect. How is the deflection of the compass needle affected by changing the direction of the current?

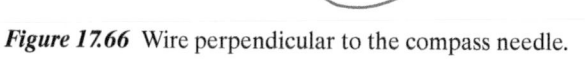

Figure 17.66 Wire perpendicular to the compass needle.

(d) Run the wire under the compass (Figure 17.67) instead of over the compass. What changes?

Figure 17.67 Wire under compass.

(e) To make sure that it is the current in the wire, and not the metal wire itself, that affects the compass, disconnect the batteries. When you bring the wire down on the compass, is there a deflection?

(f) Record the observed compass deflection in the following cases:

1. Two batteries and the (bright) round bulb
2. Two batteries and the (dim) long bulb
3. Two batteries, and just a long wire (no bulb). This is called a short circuit, and it puts a large drain on the batteries, so you should not leave this connected for many minutes.

The effect you have just observed was discovered by accident by the Danish scientist Oersted in 1820 while doing a lecture demonstration in a physics class. The phenomenon is often called "the Oersted effect." From your experiments, you should have drawn the following conclusions:

- The magnitude of the magnetic field produced by a current of moving electrons depends on the amount of current.
- A wire with no current running in it produces no magnetic field.
- The magnetic field due to the current appears to be perpendicular to the direction of the current.
- The direction of the magnetic field due to the current under the wire is opposite to the direction of the magnetic field due to the current above the wire.

EXP3 Electron current and battery
Use the right-hand rule to predict whether the compass needle should deflect to the left or right, assuming that electron current flows out of the negative end of the battery and into the positive end. Check to see that the needle does deflect in the predicted direction.

EXP4 The magnetic field of a long straight wire
Use clip leads to connect a wire about 2 m long to a battery, but don't make the final connection yet. Lay the wire down on the floor or a table, making a straight length as long as possible, heading north and south, as far from the steel-clad battery as possible.

- Hold the compass above the wire at a height where the compass deflection is $20°$ when you turn the current on and off. Mark on a vertically held piece of paper this height $r_{20°}$ of the compass needle above the wire, and record this height.
- Mark the paper at twice this height and use the second mark to place the compass at a height $2(r_{20°})$ above the paper, and record the compass deflection.

(a) Are your data consistent with the theoretical prediction that the magnitude of the magnetic field near a long straight wire is proportional to $1/r$, where r is the perpendicular distance to the wire?

(b) Using your value of r_{20}, determine the amount of current I in the wire.

Experimental Observations of a Coil of Wire
Next we will make measurements of a coil of wire consisting of multiple loops to make a larger magnetic field than a single loop would provide.

EXP5 The magnetic field of a coil of wire
Take an insulated wire about 2 m in length and wrap a coil of about $N = 20$ turns loosely around two fingers. Twist the ends

together to help hold the coil together, and remove the coil from your fingers. Use clip leads to connect the coil to a battery.

By appropriate placement and orientation of your compass relative to your N-turn current loop, measure the experimental magnetic field direction and relative magnitude at the locations indicated. Remember that you must always position the compass in such a way that the coil's magnetic field is perpendicular to the Earth's magnetic field.

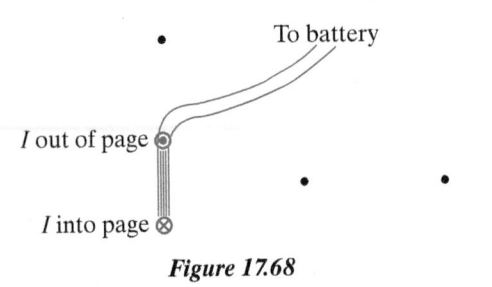

I out of page

I into page

Figure 17.68

On a diagram like that in Figure 17.68, draw the magnetic field vectors at all marked locations. Record the number of turns N, the approximate radius of the coil R, the distances to the measurement locations, and the compass deflections at those locations.

(a) Do your measurements agree with the predicted pattern of magnetic field around a coil of wire?

(b) Determine the amount of current I in the wire.

EXP6 A coil is a compass
For reasons that we will discuss in a later chapter, the magnetic dipole moment vector $\vec{\mu}$ acts just like a compass needle. In an applied magnetic field, a current-carrying loop rotates so as to align the magnetic dipole moment $\vec{\mu}$ with the field.

It is hard to observe this twisting with your own hanging coil, because the rather stiff wires prevent the coil from freely rotating, unless you suspend the entire apparatus from a thread, batteries and all. Or do this:

If you have access to kitchen equipment, you may be able to float your batteries and coil in a bowl or large glass or aluminum pan, or on a block of wood. (Avoid steel containers!) Then you may be able to observe the axis of the coil line up with the Earth's magnetic field, just as though the magnetic dipole moment of the coil were a compass needle.

EXP7 Directions of the magnetic field of a bar magnet
From its name, you might guess that a magnet makes a magnetic field. We say that a magnetic field is present if we see a compass needle twist, so let's see whether your bar magnet does make your compass needle deflect.

Place your compass to the right of your bar magnet, with the magnet oriented in such a way as to make the compass deflect to the east as shown in the accompanying figure. Make a similar diagram in your notebook.

We define the direction of a magnetic field $\vec{B}$ at a particular location as the direction that the "north" end of a compass points to when placed at that location.

(a) Assuming that the bar magnet makes a magnetic field along the axis of the magnet, draw a vector in your notebook to show the direction of the compass needle and another arrow to show the direction of the magnetic field $\vec{B}_m$ of your bar magnet at the present location of the compass. For future reference, write "N" for "north" on the end of the magnet that the magnetic field points out of (attach tape to write on if necessary).

(b) Move the compass to the left of your magnet, above your magnet, and below your magnet, as shown in Figure 17.69. For each location, record the direction of the compass needle and also record on your diagram the direction of the magnetic field due to the bar magnet at that location.

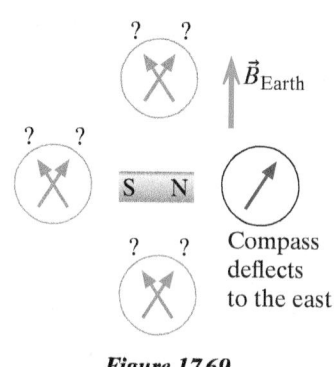

Figure 17.69

Place your compass at each of the locations shown above, relative to your bar magnet, and note the deflection of the compass needle. What is the direction of the magnet's magnetic field at these locations?

(c) Does this pattern of field look familiar? Where have we seen a similar pattern of field directions in space before?

(d) Suspend your magnet from a thread or a hair using a piece of tape, or float the magnet on a dish or plate in water. Note which end points toward the north. Evidently a compass consists simply of a magnet in the shape of a needle, mounted on a pivot.

EXP8 Magnets and matter
Obtain another magnet by working with a partner or finding a kitchen magnet. Make sure that you have labeled the north and south ends of both magnets.

(a) Describe briefly the interactions that the two magnets have with each other.

(b) You probably know that magnets don't interact strongly with anything but iron or steel objects (steel is mostly iron). Magnets also interact with nickel or cobalt objects, but these aren't so readily available. Check to see for yourself that your magnet doesn't interact with aluminum (aluminum foil), copper (a penny, or an electrical wire), or any nonmetal. Also notice that your magnet is strongly attracted to steel parts of your electricity kit.

(c) Check what happens with charged invisible tape, and see that although there is the usual attraction of a charged tape for any uncharged object, there doesn't seem to be any magnetic interaction, because either end of the magnet acts the same.

(d) Can you give any evidence that magnetic interactions pass through matter, as electric interactions do?

(e) Place two magnets in various positions near your compass to demonstrate to your satisfaction that the superposition principle holds for the magnetic fields of magnets; that is, the net field of the two magnets is the vector sum of the two fields. Give one example of your observation of the superposition principle for magnetic fields.

EXP9 Dependence on distance for a bar magnet
Again place the compass to the right of your bar magnet, oriented in such a way as to make the compass deflect 70° to the east. (See Figure 17.70.) *Do not work on a table that contains steel components*, which can interfere with the measurements. The

best place to make the measurements is outdoors, away from iron and steel objects.

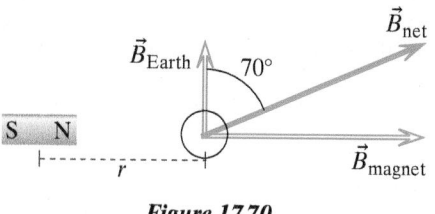

Figure 17.70

(a) At the location where you get a 70° deflection, measure the distance r from the center of the magnet to the center of the compass. A centimeter ruler is provided at the back of the book.

(b) Use vector analysis and the magnitude of the Earth's field to calculate the magnitude B_r of the magnetic field of the magnet at this distance r from the center of the magnet.

(c) Now move the magnet farther away by a factor of two; that is, place the magnet so that the distance from the center of the magnet to the center of the compass is $2r$. Record the new distance and the new compass deflection angle. Calculate the magnitude B_{2r} of the magnetic field of the magnet at this new distance from the magnet.

(d) By what ratio did the magnetic field of the magnet decrease when the distance from the magnet was doubled?

(e) The magnetic field of the magnet gets smaller with distance, and it is plausible to guess that the magnetic field of the magnet might vary as $1/r^n$, where n is initially unknown. According to your measurements, what is n? (You may wish to make measurements at some other distances to give further support to your analysis.)

(f) Extrapolating to a location near one end of your bar magnet, approximately how strong is the magnetic field in teslas near one end of your magnet? (Remember, measure r to the center of the magnet.)

(g) Place the compass north or south of the magnet, at the original distance r from the center of the magnet, and determine the magnitude of the magnetic field at this location. (See Figure 17.71.) How does this magnitude compare to your result in (b)?

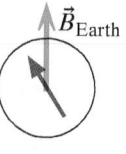

Figure 17.71

EXP10 Coils attract and repel like magnets
There is yet another way in which the behavior of a magnet and a current-carrying coil of wire is very similar. Hang your coil over the side of the table and bring your bar magnet near the hanging coil, along the axis of the coil. Reverse the magnet. Repeat from

the other side of the coil. Is this behavior similar to the interaction you have observed between two bar magnets?

EXP11 Cutting a magnet in two

Without actually cutting a magnet into two pieces, you can simulate such an operation by putting two bar magnets together north end to south end as in Figure 17.72, then taking them apart.

Figure 17.72

Do you in fact find that the put-together longer magnet acts just like a single magnet, and that after pulling them apart the two pieces still act like magnets?

If you have the equipment available, you can magnetize a soft iron nail by stroking it with a magnet; then use a hacksaw to cut it in half and investigate the properties of each half.

EXP12 The magnetic multiplier effect

Note: You need unmagnetized nails for this experiment. If the nails have ever been near a magnet, they may have become strongly magnetized. Check this by bringing a nail near the compass, first one end, then the other. If the nail is unmagnetized, both ends of the nail will affect the compass equally. However, if the nail is strongly magnetized, the two ends of the nail will affect the compass quite differently.

If a nail is strongly magnetized, bring your bar magnet near it (but not touching the nail) to magnetize the nail in the opposite direction. Check again with the compass and repeat as necessary, trying to reduce the magnetization to zero. Do not proceed until you have at least three unmagnetized nails.

Now you can study the magnetic multiplier effect. Take an insulated wire about 2 m in length and wrap a coil of about 20 turns loosely around two fingers. Twist the ends together to help hold the coil together and remove it from your fingers. Connect the coil to a battery and position your compass to the east or west of the coil so that the compass needle deflects away from north by about 5°. Turn off the current, so the compass deflection goes to zero.

(a) Insert an unmagnetized iron nail into the coil and record the compass deflection (if any). Now start the current again and observe what happens to the compass deflection. Add additional nails and observe and record what happens to the compass deflection.

(b) Explain briefly why each nail produces a sizable increase in the compass deflection when you turn on the current.

(c) To check whether the nails are permanently magnetized after being inserted into the coil, turn off the current. If the compass still deflects significantly without any current in the coil, the nails have become strongly magnetized. Are they?

(d) Bring one of these nails near the compass, first one end and then the other. Is the nail slightly magnetized?

EXP13 An electromagnet

Next, construct an "electromagnet." Take an insulated wire about 2 m in length and wrap it tightly around an iron nail, all along its length. You can make multiple layers of windings, back and forth along the nail, to increase the effects, but leave the ends of the nail sticking out so that you can touch the nail to other objects. Twist the ends of the wire together to help hold the coil together. Connect the coil to a battery and bring it near another nail.

Can you pick up the other nail? (If the battery is not fresh, you may not get enough current to lift the other nail, but you should be able to lift a paper clip!) Why does the other nail fall when you cut the current? Is there any evidence that the electromagnet nail retains some magnetization?

QUESTIONS

Q1 Given that $\vec{v} = \langle v_x, v_y, v_z \rangle$ and $\hat{r} = \dfrac{\langle x, y, z \rangle}{\sqrt{x^2 + y^2 + z^2}}$, write out $\vec{v} \times \hat{r}$ as a vector.

Q2 The vectors $\vec{a}$ and $\vec{b}$ both lie in the xz plane. The vector $\vec{c}$ is equal to $\vec{a} \times \vec{b}$. Which of these statements about $\vec{c}$ must be true? (1) The x and y components of $\vec{c}$ are zero. (2) The y and z components of $\vec{c}$ are zero. (3) The z and x components of $\vec{c}$ are zero.

Q3 How does the magnetic field of a moving point charge fall off with distance at a given angle: like $1/r$, $1/r^2$, or $1/r^3$?

Q4 Describe the magnetic field made by a charge that is not moving.

Q5 A current-carrying wire is laid on a table, oriented north–south. Electrons in the wire are flowing north. **(a)** What is the direction of the magnetic field at a location directly underneath the wire, due only to these moving electrons? **(b)** If a compass is placed under the wire, in which direction will the needle deflect?

Q6 A current carrying wire is laid on a table, oriented north–south. Electrons in the wire are flowing south. **(a)** What is the direction of the magnetic field at a location directly underneath the wire, due only to these moving electrons? **(b)** If a compass is placed under the wire, in which direction will the needle deflect?

Q7 Consider the portion of a circuit in Figure 17.73. The wire rests on top of the two compasses. When no current is running, both compasses point north (direction shown by the gray arrows). When current runs in the circuit, the needle of compass 1 deflects as shown. In what direction will the needle of compass 2 point? Draw a sketch indicating its deflection.

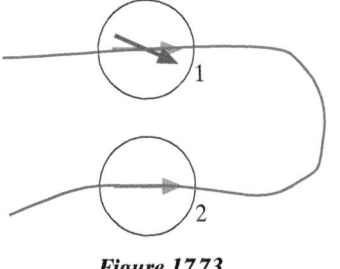

Figure 17.73

Q8 In a particular wire in a circuit, conventional current flows in the $+y$ direction. What is the direction of electron current in the wire?

Q9 In a particular wire in a circuit, electrons flow in the $+z$ direction. What is the direction of conventional current in the wire?

Q10 An iron bar magnet makes a pattern of magnetic field that looks just like the pattern of magnetic field outside a long current-carrying coil of wire. Are there currents in the iron? Explain briefly.

Q11 Suppose you have two alnico bar magnets, one with a mass of 100 g and one with a mass of a kg. At a distance of a meter from the center of either one, how would the magnetic field differ? Why?

Q12 Conventional current flows in a ring in the direction indicated in Figure 17.74. If you stand at location A, on the $+x$ axis, and look toward the ring, current flows clockwise. At each of the locations labeled by a letter, find the direction of the magnetic field at that location due to the current in the ring.

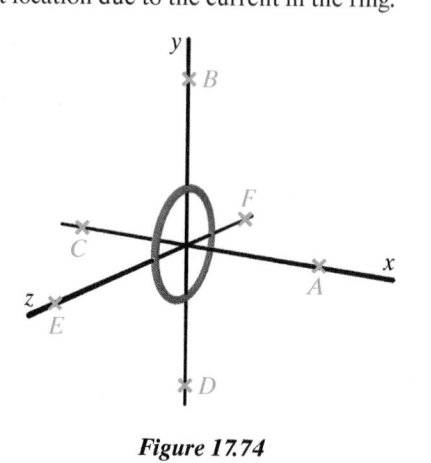

Figure 17.74

Q13 **(a)** How can you produce a magnetic field that is nearly uniform in a region? **(b)** How can you produce a magnetic field that falls off like $1/r$? **(c)** How can you produce a magnetic field that falls off like $1/r^2$? (Note that you *cannot* use just a short piece of current-carrying wire, because the other parts of the wire also contribute.) **(d)** How can you produce a magnetic field that falls off like $1/r^3$?

Q14 What is the direction of the magnetic field at the indicated locations inside and outside this current-carrying rectangular coil of wire in Figure 17.75? Explain briefly. (The direction of conventional current is shown.)

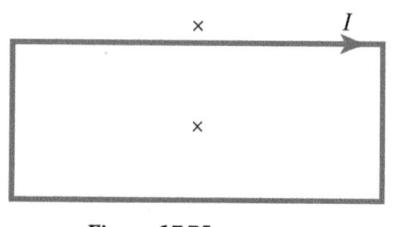

Figure 17.75

PROBLEMS

Section 17.3

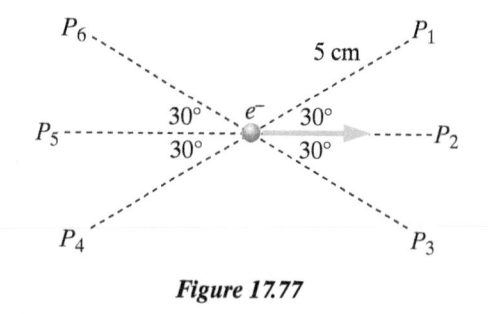

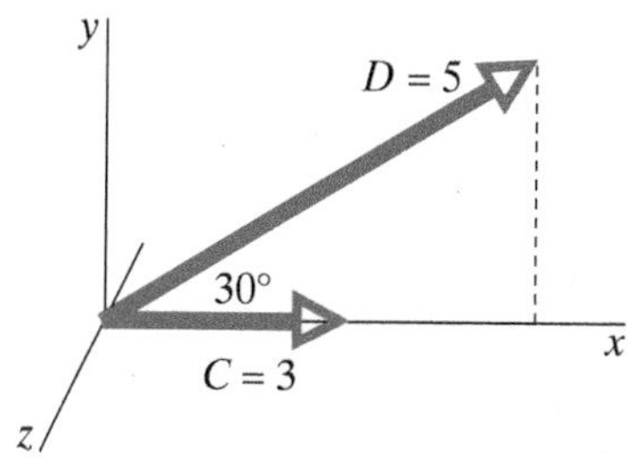

Figure 17.76

•P15 A vector $\vec{C}$ of magnitude 3 lies along the x axis, and a vector $\vec{D}$ of magnitude 5 lies in the xy plane, $30°$ from the x axis (Figure 17.76). What are the magnitude and direction of the cross product $\vec{C} \times \vec{D}$? What are the magnitude and direction of the cross product $\vec{D} \times \vec{C}$? Draw both vectors on a diagram.

•P16 If $\vec{v} = \langle 390, -480, 333 \rangle$ m/s and $\hat{r} = \langle 0.577, 0.577, -0.577 \rangle$ what is $\vec{v} \times \hat{r}$?

•P17 What is $\langle 2, 0, -5 \rangle \times \langle -5, 3, 5 \rangle$?

•P18 To get an idea of the size of magnetic fields at the atomic level, consider the magnitude of the magnetic field due to the electron in the simple Bohr model of the hydrogen atom. In the ground state the Bohr model predicts that the electron speed would be 2.2×10^6 m/s, and the distance from the proton would be 0.5×10^{-10} m. What is B at the location of the proton?

•P19 An electron is moving horizontally to the right with speed 4×10^6 m/s (Figure 17.77). Each of the indicated locations is 5 cm from the electron. What is the magnetic field due to this moving electron at the indicated locations? Give both magnitude and direction of the magnetic field at each location.

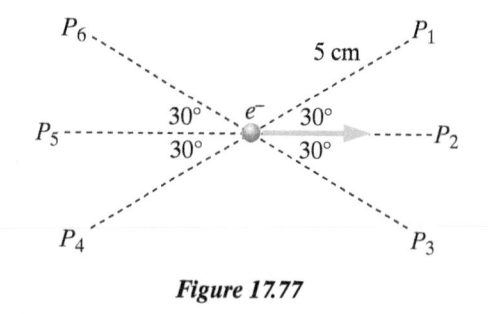

Figure 17.77

•P20 At a particular instant, a proton at the origin has velocity $\langle 2 \times 10^4, -2 \times 10^4, 0 \rangle$ m/s. You need to calculate the magnetic field at location $\langle 0.02, 0.06, 0 \rangle$ m due to the moving proton. **(a)** What is $\vec{r}$? **(b)** What is $\hat{r}$? **(c)** What is $\vec{v} \times \hat{r}$? **(d)** What is the magnetic field at the observation location due to the moving proton?

•**P21** In Figure 17.78 a proton is moving upward with speed 5×10^6 m/s. What is the magnetic field due to this moving proton at the indicated locations? Each location is a distance $r = 8$ cm from the proton. Give both magnitude and direction of the magnetic field at each location.

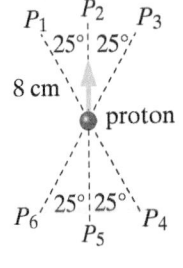

Figure 17.78

••**P22** At a particular instant a proton is at the origin, moving with velocity $\langle 3 \times 10^4, -2 \times 10^4, -7 \times 10^4 \rangle$ m/s. **(a)** At this instant, what is the electric field at location $\langle 4 \times 10^{-3}, 4 \times 10^{-3}, 3 \times 10^{-3} \rangle$ m, due to the proton? **(b)** At this instant, what is the magnetic field at the same location due to the proton?

••**P23** The electron in Figure 17.79 is traveling with a speed of 3×10^6 m/s.

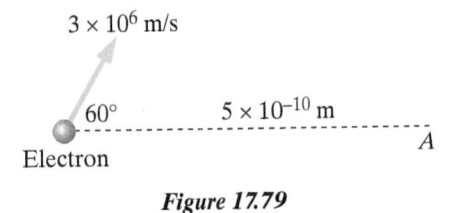

Figure 17.79

(a) At location A, what are the directions of the electric and magnetic fields contributed by the electron? **(b)** Calculate the magnitudes of the electric and magnetic fields at location A.

Section 17.6

•**P24** The electron current in a horizontal metal wire is 3.7×10^{18} electrons/s, and the electrons are moving to the right. What are the magnitude and direction of the conventional current?

•**P25** A conventional current of 12 A runs to the right in a horizontal metal wire. **(a)** How many electrons pass some point in the wire per second? **(b)** In what direction are the electrons moving? **(c)** The electron mobility in this metal is 2.1×10^{-4} (m/s)/(N/C) and the electric field in the wire is 0.15 N/C. What is the average drift speed of the electrons?

•**P26** You have used copper wires in your circuits. Let's calculate the mobile electron density n for copper. A mole of copper has a mass of 64 g (0.064 kg), and one mobile electron is released by each atom in metallic copper. The density of copper is about 9 g/cm³, or 9×10^3 kg/m³. Show that the number of mobile electrons per cubic meter in copper is 8.4×10^{28} m⁻³.

•**P27** Calculate the mobile electron density for nickel. A mole of nickel has a mass of 59 g (0.059 kg), and one mobile electron is released by each atom in metallic nickel. The density of nickel is about 8.8 g/cm³, or 8.8×10^3 kg/m³.

Section 17.7

•**P28** In a circuit consisting of a long bulb and two flashlight batteries in series the conventional current is about 0.1 A. What is the magnetic field 5 mm from the wire? (This is about how far away the compass needle is when you place the wire on top of the compass.) Is this a big or a small field?

•**P29** A current-carrying wire oriented north–south and laid over a compass deflects the compass 10° to the east. **(a)** What is the magnitude of the magnetic field made by the current? The horizontal component of Earth's magnetic field is about 2×10^{-5} T. **(b)** In what direction does the electron current flow in the wire?

•**P30** A straight wire of length 0.62 m carries a conventional current of 0.8 A. What is the magnitude of the magnetic field made by the current at a location that is a perpendicular distance 2.9 cm from the center of the wire? Use both the exact equation and the approximate equation to calculate the field.

•**P31** A long straight wire carries a conventional current of 0.9 A. What is the approximate magnitude of the magnetic field at a location a perpendicular distance of 0.035 m from the wire due to the current in the wire?

•**P32** A battery is connected to a Nichrome wire and a conventional current of 0.3 A runs through the wire. The wire is laid out in the form of a rectangle 50 cm by 3 cm. **(a)** What is the approximate magnetic field at the center of the rectangle? Give the direction as well as the magnitude. **(b)** What approximations did you make?

•**P33** A wire through which a current is flowing lies along the x axis in Figure 17.80. Connecting wires that are not shown in the diagram connect the ends of the wire to batteries (which are also not shown). Electron current flows through the wire in the $-x$ direction, as indicated in the figure. To calculate the magnetic field at location A due to the current in the wire, we divide the wire into pieces, approximate each piece as a point charge moving in the direction of conventional current, and calculate the magnetic field at the observation location due only to this piece; then add the contributions of all pieces to get the net magnetic field.

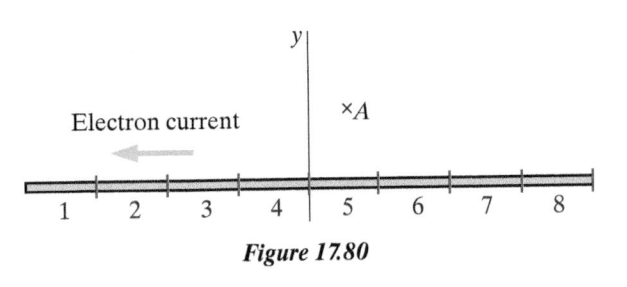

Figure 17.80

The wire is 1.3 m long, and is divided into eight pieces. The observation location A is located at $\langle 0.081, 0.178, 0 \rangle$ m. The conventional current running through the wire is 6.5 A. In this exercise you will calculate the magnetic field at the observation location due only to segment 4 of the wire. **(a)** What is the direction of conventional current in this wire? **(b)** How long is segment 4? **(c)** What is the magnitude of the vector $\Delta \vec{l}$ for segment 4? **(d)** What is the vector $\Delta \vec{l}$ for segment 4? **(e)** What is the location of the center of segment 4? **(f)** What is the vector $\vec{r}$ from source to observation location for segment 4? **(g)** What is the unit vector $\hat{r}$? **(h)** What is $\Delta \vec{l} \times \hat{r}$? **(i)** Calculate the magnetic field $\Delta \vec{B}$ at the observation location due only to the current in segment 4 of the wire.

••**P34** You place a long straight wire on top of your compass, and the wire is a height of 5 mm above the compass needle

(Figure 17.81). If the conventional current in the wire is $I = 0.2$ A and runs left to right as shown, calculate the approximate angle the needle deflects away from north and draw the position of the compass needle.

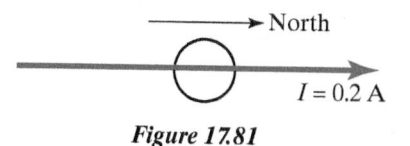

Figure 17.81

••P35 When you bring a current-carrying wire down onto the top of a compass, aligned with the original direction of the needle and 5 mm above the needle, the needle deflects by $10°$ (Figure 17.82).

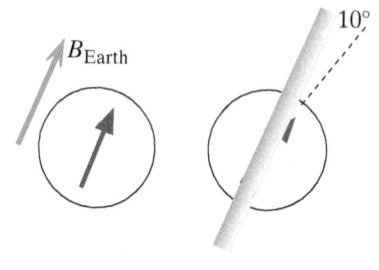

Figure 17.82

(a) Show on a diagram the direction of conventional current in the wire and the direction of the additional magnetic field made by the wire underneath the wire, where the compass needle is located. Explain briefly. **(b)** Calculate the amount of current flowing in the wire. The measurement was made at a location where the horizontal component of the Earth's magnetic field is $B_{Earth} \approx 2 \times 10^{-5}$ T.

••P36 A compass is placed inside a triangular coil of wire with three turns, as shown in Figure 17.83. Each side of the triangle has a length L. The compass is a perpendicular distance d from the center of each side of the triangle. The coil is in the xy plane; magnetic north (due to the Earth) is in the negative x direction. Conventional current runs in the coil as shown (clockwise, as viewed from a location on the $+z$ axis).

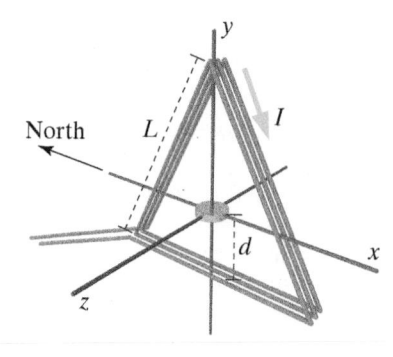

Figure 17.83

(a) While current is running through the coil, in what direction does the compass needle point? **(b)** Which of the following statements about how to model the coil are reasonable? List all that apply. (1) Since the coil is made from a single wire, we could model one turn of the coil as one long wire. (2) A circular loop is a poor model for one turn of the coil, because the distance from the wire to the compass varies so much. (3) One turn of the coil could be modeled as three short straight wires. **(c)** A current of

0.62 A runs through the wires. If $L = 8.3$ cm and $d = 2.4$ cm, what is the magnitude of the magnetic field at the location of the compass due to the coil? (Remember that the coil has three turns.) **(d)** What is the approximate magnitude of the compass deflection in degrees?

••P37 A thin wire is part of a complete electrical circuit that carries a current I. For this problem consider only the piece of wire of length d as shown in Figure 17.84. Answer the following questions based on this figure. **(a)** What are the magnitude and direction of the magnetic field due to the wire at Q (location $\langle -w, 0, 0 \rangle$? **(b)** Set up the integrals necessary to determine the x, y, and z components of the magnetic field at P (location $\langle -w, h, 0 \rangle$). The integrals must be in a form that can be evaluated (no cross products in the integrand), but you do not need to evaluate them. **(c)** What is the direction of the magnetic field at location P?

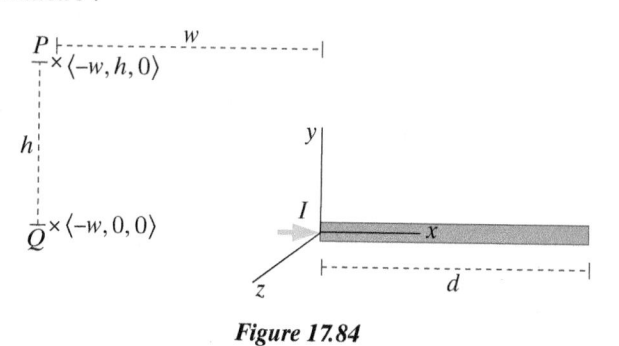

Figure 17.84

••P38 In Figure 17.85 a long current-carrying wire, oriented north–south, lies on a table (it is connected to batteries that are not shown). A compass lies on top of the wire, with the compass needle about 3 mm above the wire. With the current running, the compass deflects $11°$ to the west. At this location, the horizontal component of the Earth's magnetic field is about 2×10^{-5} T.

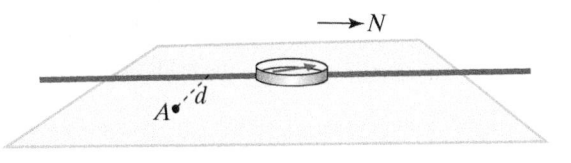

Figure 17.85

(a) What is the magnitude of the magnetic field at location A, on the tabletop, a distance 2.8 cm to the east of the wire, due only to the current in the wire? **(b)** What is the direction of the magnetic field at location A, due only to the current in the wire?

••P39 In Figure 17.86 a wire is connected to batteries (not shown) and current flows through the wire. The wire lies flat on a table, and you are looking down on it from above. The wire is laid on top of a compass, resting about 3 mm above the needle. The distance d between the straight wires is 4.5 cm. The compass needle deflects $17°$ from north, as shown. At this location the horizontal component of the Earth's magnetic field is 2×10^{-5} T.

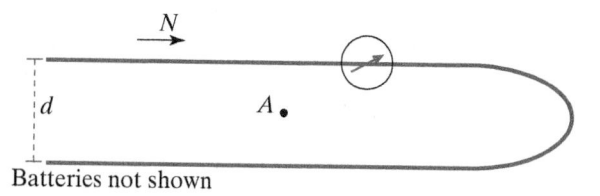

Batteries not shown

Figure 17.86

(a) What is the direction of the magnetic field at location A, midway between the straight wires, due to the current in the wire? **(b)** What is the approximate magnitude of the magnetic field at location A, due to the current in the wire? **(c)** What approximations or assumptions did you make in determining the magnetic field at location A?

••P40 At one time, concern was raised about the possible health effects of the small alternating (60-Hz) magnetic fields created by electric currents in houses and near power lines. In a house, most wires carry a maximum of 15 A (there are 15-A fuses that melt and break the circuit if this current is exceeded). The two wires in a home power cord are about 3 mm apart as shown in Figure 17.87, and at any instant they carry currents in opposite directions (both of which change direction 60 times per second).

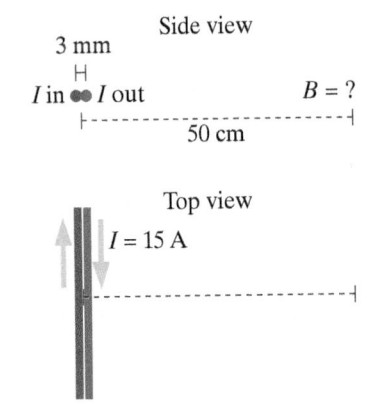

Figure 17.87

(a) Calculate the maximum magnitude of the alternating magnetic field, 50 cm away from the center of a long straight power cord that carries a current of 15 A. Both wires are at the same height as the observation location. **(b)** Explain briefly why twisting the pair of wires into a braid as shown in Figure 17.88 greatly decreases the magnetic field at the location discussed in (a), a generally useful technique.

Figure 17.88

(c) Optional: For two long parallel wires like those shown in part (a), a distance d apart and carrying conventional current I in opposite directions, find an *algebraic* equation for the magnetic field a perpendicular distance $r \gg d$ to the right of the wires.

(The magnitude of the field that you calculated in part (a) is very small compared to the Earth's magnetic field, but there were questions as to whether a very small alternating magnetic field might have health effects. After many detailed studies, the consensus of most scientists now seems to be that these small alternating magnetic fields are not a hazard after all.)

Section 17.8

•P41   (a) A loop of wire carries a conventional current of 0.8 A. The radius of the loop is 0.09 m. Calculate the magnitude of the magnetic field at a distance of 0.32 m from the center of the loop, along the axis of the loop. **(b)** What would be the magnitude of the magnetic field at the same location if there were 100 loops of wire in a coil instead of one loop?

•P42 A thin circular coil of wire of radius 5 cm consists of 100 turns of wire (Figure 17.89). If the conventional current in the wire is 4 A, what are the magnitude and direction of the magnetic field

at the center of the coil? (The direction of conventional current is shown.)

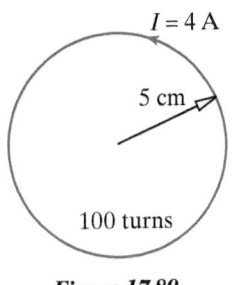

Figure 17.89

•P43 A conventional current of 6 A runs clockwise in a circular loop of wire in the xy plane, with center at the origin and with radius 0.08 m. Another circular loop of wire lies in the same plane, with its center at the origin and with radius 0.02 m. How much conventional current must run counterclockwise in this smaller loop in order for the magnetic field at the origin to be zero?

•P44 How much conventional current must you run in a solenoid with radius = 0.04 m and length = 0.35 m to produce a magnetic field inside the solenoid of 2×10^{-5} T, the approximate field of the Earth? The solenoid has 200 turns.

••P45 In Figure 17.90 two long wires lie very close together and carry a conventional current I as shown and each wire has a semicircular kink, one of radius R_1 and the other of radius R_2. Calculate the magnitude and direction of the magnetic field at the common center of the two semicircular arcs.

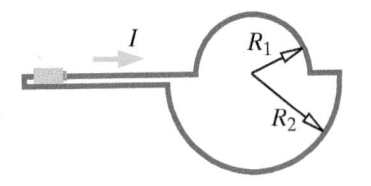

Figure 17.90

••P46 A very long wire carrying a conventional current of 3.5 A is straight except for a circular loop of radius 5.8 cm (Figure 17.91). Calculate the approximate magnitude and the direction of the magnetic field at the center of the loop.

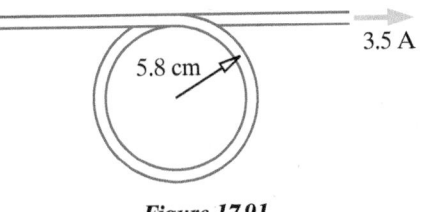

Figure 17.91

••P47 You can use measurements of the magnetic field of a coil to determine how much current your battery is supplying to the coil. Using your value of B (Experiment EXP5), determine the conventional current I through your coil. If this current is less than 3 A, you should replace the battery.

••P48 A thin circular coil of radius $r = 15$ cm contains $N = 3$ turns of Nichrome wire. A small compass is placed at the center of the coil, as shown in Figure 17.92. With the battery disconnected, the compass needle points to the right, in the plane of the coil. Assume that the horizontal component of the Earth's magnetic field is about $B_{Earth} \approx 2 \times 10^{-5}$ T.

When the battery is connected, a current of 0.25 A runs through the coil. Predict the deflection of the compass needle. If you have to make any approximations, state what they are. Is the deflection outward or inward as seen from above? What is the magnitude of the deflection?

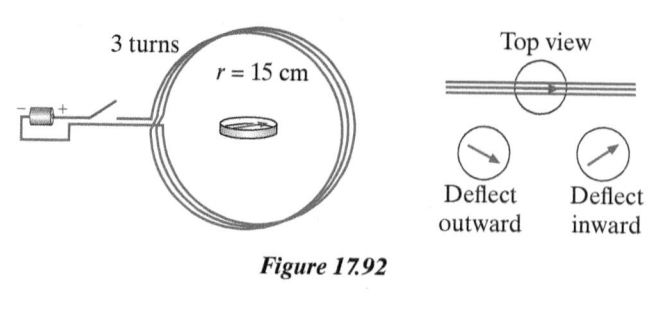

Figure 17.92

••P49 The circuit in Figure 17.93 consists of a battery and a Nichrome wire, through which runs a current I.

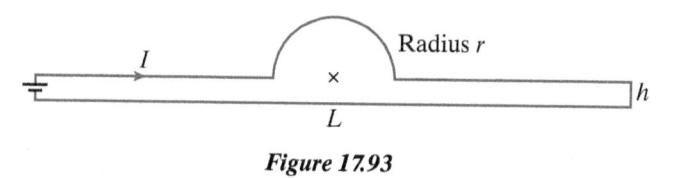

Figure 17.93

(a) At the location marked × (the center of the semicircle), what is the direction of the magnetic field? **(b)** At the location marked × (the center of the semicircle), what is the magnitude of the magnetic field? If you have to make any approximations, state what they are.

••P50 Two thin coils of radius 3 cm are 20 cm apart and concentric with a common axis (Figure 17.94). Both coils contain 10 turns of wire with a conventional current of 2 A that runs counterclockwise as viewed from the right side.

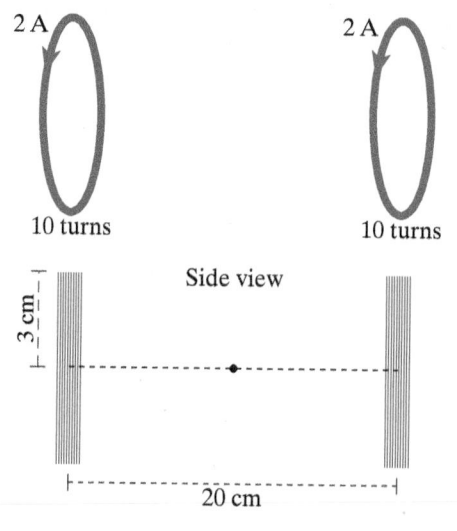

Figure 17.94

(a) What are the magnitude and direction of the magnetic field on the axis, halfway between the two loops, without making the approximation $z \gg r$? (For comparison, remember that the horizontal component of magnetic field in the United States is about 2×10^{-5} T). **(b)** In this situation, the observation location is not very far from either coil. How bad is it to make the $1/z^3$ approximation? That is, what percentage error results if you

calculate the magnetic field using the approximate equation for a current loop instead of the exact equation? **(c)** What are the magnitude and direction of the magnetic field midway between the two coils if the current in the right loop is reversed to run clockwise?

••P51 A conventional current I runs in the direction shown in Figure 17.95. Determine the magnitude and direction of the magnetic field at point C, the center of the circular arcs.

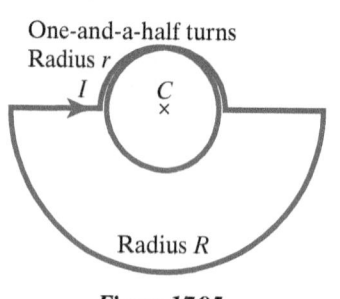

Figure 17.95

••P52 A square loop of wire with sides of length L and carrying a conventional current I lies in the yz plane with its center at the origin. **(a)** What is the magnitude of the magnetic field due to this loop at a location x on the x axis? **(b)** What is the approximate equation for the magnitude of the magnetic field at a location x that is much greater than L, in terms of the magnetic moment of the loop? **(c)** Is the result of part (b) consistent with our previously derived result for the magnetic field of a magnetic dipole?

Section 17.11

••P53 A compass originally points north; at this location the horizontal component of the Earth's magnetic field has a magnitude of 2×10^{-5} T. A bar magnet is aligned east–west, pointing at the center of the compass. When the center of the magnet is 0.25 m from the center of the compass, the compass deflects 70°. What is the magnetic dipole moment of the bar magnet?

••P54 In the region shown in Figure 17.96, the magnitude of the horizontal component of the Earth's magnetic field is about 2×10^{-5} T. Originally a compass placed at location A points north. Then a bar magnet is placed at the location shown in the diagram, with its center 16 cm from location A. With the magnet present, the compass needle points 60° west of north.

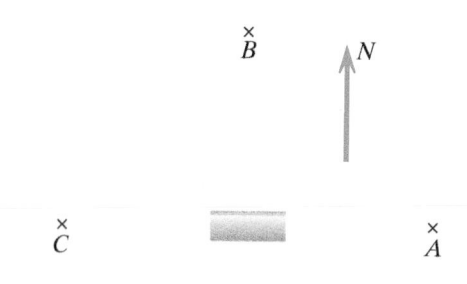

Figure 17.96

(a) Which pole of the magnet is nearer to location *A*? **(b)** What are the directions of the magnetic field at locations *B*, *C*, and *D*? **(c)** What is the magnetic dipole moment of the bar magnet?

••P55 A bar magnet with magnetic dipole moment $\langle 8,0,0 \rangle$ A·m² is located at the origin. A second bar magnet with magnetic dipole moment $\langle \mu,0,0 \rangle$ is located at $\langle 0.4,0.2,0 \rangle$ m. Calculate the value of μ that makes the magnetic field at $\langle 0.4,0,0 \rangle$ m be zero.

••P56 A bar magnet is aligned east–west, with its center 16 cm from the center of a compass (Figure 17.97). The compass is observed to deflect 50° away from north as shown, and the horizontal component of the Earth's magnetic field is known to be 2×10^{-5} T.

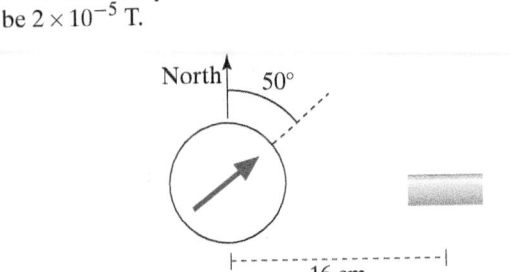

Figure 17.97

(a) Label the N and S poles of the bar magnet and explain your choice. **(b)** Determine the magnetic dipole moment of this bar magnet, including correct units.

••P57 A bar magnet with magnetic dipole moment 0.54 A·m² lies on the negative *x* axis, as shown in Figure 17.98. A compass is located at the origin. Magnetic north is in the negative *z* direction. Between the bar magnet and the compass is a coil of wire of radius 3.5 cm, connected to batteries not shown. The distance from the center of the coil to the center of the compass is 9.7 cm. The distance from the center of the bar magnet to the center of the compass is 22.5 cm. A steady current of 0.96 A runs through the coil. Conventional current runs clockwise in the coil when viewed from the location of the compass.

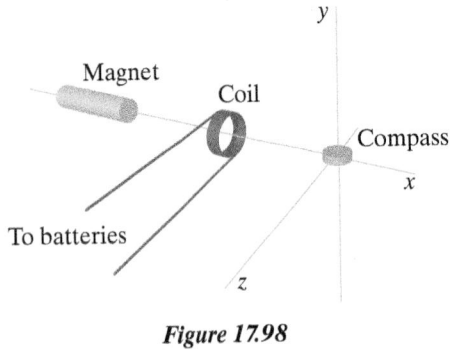

Figure 17.98

Despite the presence of the coil and the bar magnet, the compass points north. **(a)** Which end of the bar magnet is closer to the compass? **(b)** List all of the following statements that are correct: (1) At the location of the compass, the magnetic field due to the magnet is equal in magnitude to the magnetic field due to the coil. (2) It isn't necessary to take the magnetic field of the magnet into account, since the coil is in between the magnet and the compass. (3) Because not all parts of the magnet are the same distance from the compass, treating the magnet as a dipole located at the center of the magnet is an approximation. (4) The magnetic dipole moment of the coil is equal to the magnetic dipole moment of the compass. (5) At the location of the compass, the net magnetic field is equal to the magnetic field of the Earth. **(c)** How many turns of wire are in the coil?

Section 17.12

•P58 A particular alnico (aluminum, cobalt, nickel, and iron) bar magnet (magnet A) has a mass of 10 g. It produces a magnetic field of magnitude 6×10^{-5} T at a location 0.19 m from the center of the magnet, on the axis of the magnet. **(a)** Approximately what is the magnitude of the magnetic field of magnet A a distance of 0.38 m from the center of the magnet, along the same axis? **(b)** If you removed the original magnet and replaced it with a magnet made of the same material but with a mass of 50 g (magnet B), approximately what would be the magnetic field at a location 0.19 m from the center of the magnet, on the axis of the magnet?

••P59 The magnetic field along the axis of a bar magnet can be written like this:

$$B_{\text{axis}} \approx \frac{\mu_0}{4\pi} \frac{2\mu}{r^3}$$

(a) Use your own data from Experiment EXP9 to determine the magnetic dipole moment μ of your bar magnet. If you have mislaid those data, quickly repeat the measurements now. You will use the value of your magnetic dipole moment in later work. The magnetic dipole moment describes how strong a magnet is: the bigger the magnetic dipole moment, the bigger the magnetic field at some distance *r*. **(b)** Determine the mass of your magnet and thereby determine the number *N* of atoms in your magnet. Although your magnet is probably made of some alloy such as alnico V (51% iron, 8% aluminum, 14% nickel, 24% cobalt, and 3% copper), for simplicity assume it is made just of iron, which has a density of 8 g/cm³ and an atomic mass of 56 (that is, 6×10^{23} atoms weigh a total of 56 g). Assuming that each of the atoms has a magnetic dipole moment with a value estimated in Section 17.12, see how well the atomic model for a magnet fits the measured value of the magnetic dipole moment of your bar magnet.

COMPUTATIONAL PROBLEMS

More detailed and extended versions of some computational modeling problems may be found in the lab activities included in the *Matter & Interactions, 4th Edition*, resources for instructors.

••P60 A proton moves with a constant velocity of 4×10^4 m/s in the +*x* direction, along the *x* axis, starting at an initial location of $\langle -4 \times 10^{-10},0,0 \rangle$ m. **(a)** Write a program that calculates

and displays the magnetic field of the moving proton at four observation locations in the *yz* plane (two on the *y* axis and two on the *z* axis), each a distance of 8×10^{-11} m from the *x* axis. Use cyan arrows to represent the magnetic field. **(b)** Does the magnetic field at one of your fixed observation locations change or stay constant? **(c)** What do you observe if you change the speed of the proton? **(d)** What do you observe if you change the proton to an antiproton?

••P61 Start with the program you wrote in Problem P60. Modify the program so that there are many observation locations, positioned on a cylindrical surface aligned with the x axis.

••P62 Start with the program you wrote in either Problem P60 or P61. Add to your display the electric field made by the proton at your fixed observation locations. Make the electric field arrows orange. Note that you need to scale the arrows representing electric field differently from the arrows used to represent magnetic field, since these quantities have very different magnitudes and units.

•••P63 Calculate and display (as arrows) the magnetic field made by a square current-carrying loop whose center is at the origin and which lies in the yz plane, with axis in the $+x$ direction. Each side of the square loop is 0.1 m long and the current in the loop is 3 A. Represent the square loop as a curve object with $N = 40$ positions. Show the field at many locations on a circle of radius 1 m in the xy plane centered on the origin, so that you can get a feel for the magnetic field at locations both along the axes and away from the axes. Print the values of the magnetic field where the circle intersects the x and y axes and check that they agree with the analytical expression for the magnetic field of this square loop when far from the loop. When you have the program working properly, determine the approximate minimum value of N that gives good agreement with the analytical solution.

•••P64 This problem is similar to Problem P63 but with a circular loop instead of a square loop, where the radius of the loop is 0.05 m and the current is again 3 A. Represent the circular loop as a curve object with $N = 40$ positions. Show the field at many locations on a circle of radius 1 m in the xy plane centered on the origin, so that you can get a feel for the magnetic field at locations both along the axes and away from the axes. Print the values of the magnetic field where the circle intersects the x and y axes and check that they agree with the analytical expression for the magnetic field of this circular loop when far from the loop. When you have the program working properly, determine the approximate minimum value of N that gives good agreement with the analytical solution.

•••P65 A solenoid of length $L = 0.5$ m and radius $R = 3$ cm is wound with $N = 50$ turns of wire carrying a current $I = 1$ A. Its center line lies on the x axis, with the origin at the center of the solenoid (Figure 17.99). This is a loosely wound solenoid, so it is not clear how uniform will be the magnetic field in its interior. To explore this, do the following:

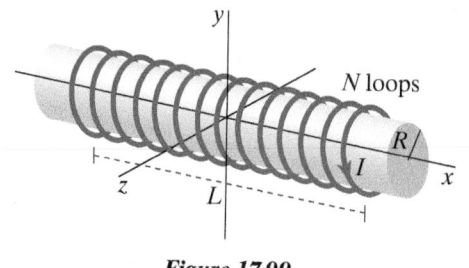

Figure 17.99

(a) Calculate and display magnetic field vectors at the locations in the xy plane, inside and outside of the solenoid, as shown in Figure 17.100. Do the pattern and direction of the magnetic field make sense? **(b)** Display the numerical value of the magnitude

of the magnetic field at one location, the center of the solenoid. Also display the theoretical numerical value of the magnetic field, in the approximation that this is a very long solenoid ($B \approx \mu_0 NI/L$). **(c)** What is the minimum number of steps around one turn that are necessary to obtain good agreement between the theoretical value and your numerical integration? What is your criterion for good agreement? **(d)** Vary the number of turns of the helix. What is the minimum number of turns necessary to get an approximately uniform field inside the solenoid? **(e)** How do the magnitude and direction of the magnetic field outside the solenoid compare to the magnitude and direction of the magnetic field inside the solenoid?

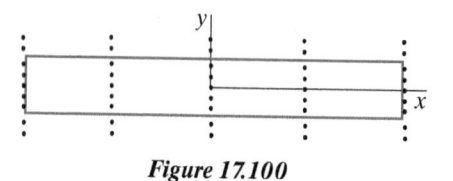

Figure 17.100

The code segment below creates a solenoid-shaped `curve`, and also creates a list of `arrows` at the specified observation locations. The `arrow` at the center of the array of observation locations is given the special name `center`.

```
from visual import *
scene.width = 800


L = 0.5            ## length of solenoid
R = 0.03           ## radius of solenoid
mzofp = 1e-7       ## mu0/4pi
I = 1.0            ## conventional current
Nturns = 75        ## number of turns
Nsteps = 20        ## number of line segments/turn

## a curve is a list of points
## start with an empty curve (no points)
solenoid = curve(color=color.yellow,
                 radius = 0.001)
dx = L/(Nturns*Nsteps)
omega = 2*pi*Nturns/L
x = -L/2
## add points to the curve
while x < L/2+dx:
    solenoid.append(pos=vector(x,
                    R*sin(x*omega),
                    R*cos(x*omega)))
    x = x + dx


## a list of arrows
## at observation locations with
## arbitrary directions initially
Barrows = []       ## empty list
dx = L/4
## skip y = 0.03 which is inside wire
ylist = [-0.05, -0.04, -0.02, -0.01,
                0., 0.01, 0.02, 0.04, 0.05]
x = -L/2
while x < L/2 + dx:
    i = 0
    while i < len(ylist):
        aa = arrow(pos=vector(x,ylist[i],0),
```

```
               axis=vector(0,0,0.005),          ## Barrows[i]
               color=color.cyan,                ## inner while loop: calculate field at
               shaftwidth=0.003)                ## Barrows[i].pos
        Barrows.append(aa)                       ## and scale axis of arrow appropriately
        i = i + 1                                ## ...
     x = x + dx

## outer while loop: choose observation location
```

A N S W E R S   T O   C H E C K P O I N T S

1 6×10^{18} electrons/s; 5.4×10^{21} electrons
2 5.4×10^{-6} T
3 (a) Figure 17.101; **(b)** Figure 17.102

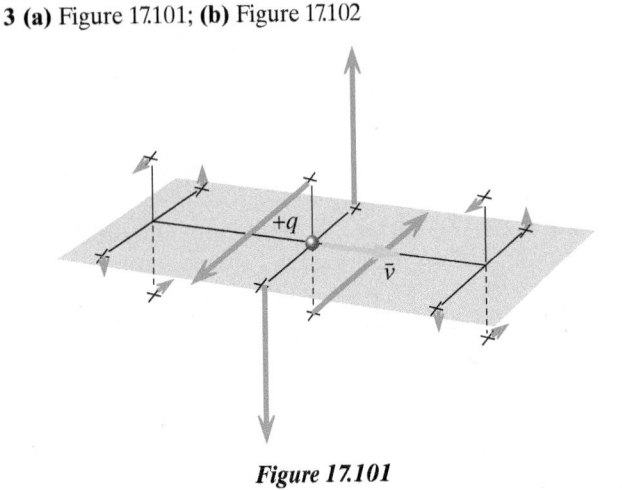

Figure 17.101

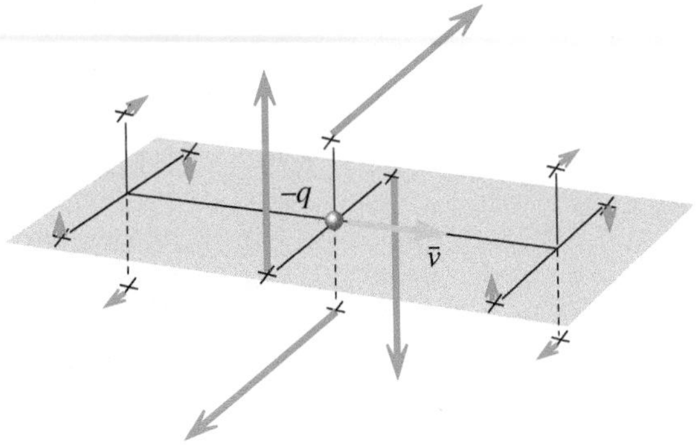

Figure 17.102

4 Electron current flows to the left, conventional current to the right.
5 2.4 A
6 9 A·m^2

Electric Field and Circuits

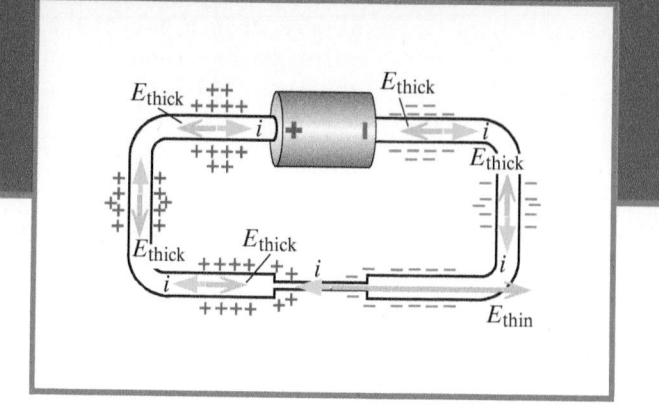

OBJECTIVES

After studying this chapter you should be able to

- Calculate the electric fields inside elements of a steady-state circuit, and depict these fields using arrows.
- Explain the role of charges on the surface of a circuit, and draw the approximate distribution of charges on the surface of the components of a simple steady-state circuit.
- Use the principles of conservation of energy and conservation of charge to analyze a circuit quantitatively, using microscopic quantities such as mobile charge density and charge mobility.

18.1 A CIRCUIT IS NOT IN EQUILIBRIUM

When an electric field is applied to a conductor, the mobile charges in the conductor experience forces, and begin to move in the direction of those forces. We have previously studied situations in which the flow of charges lasts for a very short time, until the buildup of charge at the surface of the conductor produces an electric field that is equal and opposite to the applied field, the net electric field inside the conductor becomes zero, and the system reaches equilibrium. In contrast, an electric circuit is a system that does not reach equilibrium. Despite the motion of charges, the net electric field inside the conductor does not go to zero, and charge flow continues for a long period in this nonequilibrium system.

Our goal in this chapter is to construct a microscopic model of what happens in an electric circuit, using familiar physics principles (the Energy Principle, the Momentum Principle), the concepts of electric field and potential, and the microscopic model of a metal that we developed in Chapter 14. This is a somewhat different view of electric circuits from the macroscopic view you may have seen in a previous physics course. We will be interested in questions such as:

- Are charges used up in a circuit?
- How is it possible to create and maintain a nonzero electric field inside a wire?
- What is the role of the battery in a circuit?

Looking ahead, once we understand the motion of charges in electric circuits, we will be able to explore magnetic forces. In the previous chapter we studied the magnetic fields made by moving charges. However, we did not discuss the effect of those magnetic fields on other charges. Just as moving charges make magnetic fields, only moving charges are affected by magnetic fields. One situation in which we can easily observe these effects is in electric circuits where currents are flowing. Before we can study magnetic forces in detail, we need to know more about the motion of charges in electric circuits.

18.2 CURRENT IN DIFFERENT PARTS OF A CIRCUIT

As a first step in constructing a microscopic model to explain current flow in a circuit, we need to address a crucial question, both experimentally and theoretically: What happens to the charges that flow through the circuit? Do they get used up? Is there a different amount of current in different parts of a series circuit, a circuit with a single path?

The Steady State

Once a circuit has been assembled, current keeps flowing at about the same rate for quite a long time. The brightness of a light bulb or the amount of compass deflection does not change noticeably over several minutes. In fact, since it takes quite a while for the batteries to run down under these circumstances, the current remains about the same for several hours. We say that these circuits are in a "steady state." The steady state is not the same as equilibrium.

EQUILIBRIUM VERSUS STEADY STATE

Equilibrium means that no current is flowing: $\overline{v} = 0$.

Steady state means that charges are moving ($\overline{v} \neq 0$), but their drift velocities at any location do not change with time, and there is no change in the deposits of excess charge anywhere.

Current in Different Parts of a Simple Circuit

One can measure the amount of current flowing in different parts of a circuit in the steady state by using compass deflections as an indicator of current, since the magnetic field made by a wire is proportional to the current in the wire.

Before making such a measurement, it is instructive to think about what we expect to see. Consider a circuit consisting of two batteries and a bulb as shown in Figure 18.1. Electron current flows from the negative terminal of the battery toward the bulb.

> QUESTION How would you expect the amount of electron current at location A to compare to the electron current at location B? Make a clear prediction—write it down. (It is ok to guess.)

Most people choose one of the following possibilities:

1. There should be no current at all at B, because all of the electron current coming from the negative end of the battery is used up in the bulb.
2. The current at B should be less than the current at A, because some of the current is used up to make the bulb give off light and heat.
3. The current should be the same at A and at B.

Before doing the experiment, let's consider these alternatives from a theoretical viewpoint, using what we know about electric interactions and the properties of metals.

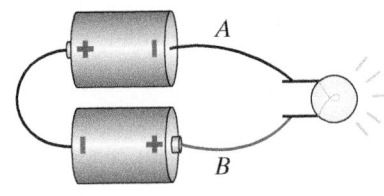

Figure 18.1 A circuit containing a bulb and two batteries in series.

Can Current Be Used Up in the Bulb?

It is reasonable to expect that something will be used up in the bulb in order to produce heat and light. Let's consider whether electric current can be the thing that is used up.

> QUESTION Current in a metal wire is simply the flow of electrons past a point. Can electrons be used up in the bulb?

Electrons alone cannot be destroyed, because this would violate the fundamental principle of conservation of charge. If negative particles were destroyed, the universe would become more and more positive!

We do know that electrons can react with positrons (positively charged anti-electrons), annihilating both particles and producing large amounts of energy. However, ordinary matter cannot coexist with antimatter for longer than a tiny fraction of a second, so there cannot be a supply of positrons in the circuit to annihilate electrons.

> QUESTION Could electrons just accumulate in the bulb, so that the current at B would be less than at A?

If electrons accumulated in the bulb, the bulb would become negatively charged. The negative charge of the bulb would become large enough to repel incoming electrons and stop the current. Since in the steady state current keeps flowing, this can't be happening. (Moreover, an increasingly negatively charged bulb would strongly attract nearby objects, which you don't observe.)

We have used proof by contradiction to rule out the possibility that electrons are annihilated or get stuck in the bulb, so we must conclude that current cannot be used up in a light bulb. Since we have ruled out both possibilities (1) and (2) by this argument, we expect that alternative (3) is correct: the current should be the same at A and at B.

Measuring Current All around a Circuit

We have predicted on theoretical grounds that current cannot be used up in a circuit. If you make careful measurements, you should indeed find the same current everywhere in a series circuit (a circuit with no parallel branches, such as the circuit in Figure 18.2). To make these measurements yourself using simple equipment, do Experiment EXP1 at the end of this chapter.

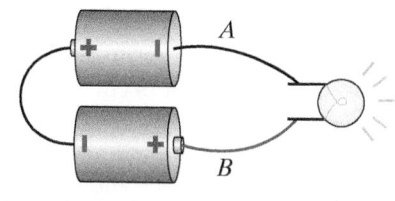

Figure 18.2 If we measure the current at A and at B, we should find these currents to be the same.

What Is Used Up in the Light Bulb?

As can be observed in Experiment EXP1, the amount of current entering and leaving a lighted bulb in a steady state circuit is the same. You may still feel uneasy about all this. You might ask, "If electrons don't get used up in the bulb, what makes the light that we see? You can't get something for nothing!"

> QUESTION Rub your hands together as hard as you can and as fast as you can for several seconds. What change do you observe in your hands?

Your hands get hot, but they are certainly not used up! However, the feeling that something must be used up to light the bulb is entirely reasonable.

> QUESTION What does get used up to warm your hands or to make a light bulb hot enough to emit visible light?

In both cases energy is being "used up," or rather transformed from a form of stored energy into thermal energy (and energy in the form of emitted light, in the case of the bulb). In order to move your hands you must use up some of the stored chemical energy in your body. Similarly, to force electrons through the filament, some of the chemical energy stored in the battery must be used up.

Forcing electrons through the filament heats the metal with a kind of friction, not unlike the friction that heats your hands when you rub them. In both cases chemical energy is converted into thermal energy, but without destroying the objects that rub against each other.

Since energy flows out of the system (the circuit) in the form of heat and light, we will need to take this into account when we study the energetics of circuits later in this chapter.

Current at a Node

We can extend these considerations to more complicated arrangements of wires. In general we have the following important principle, based on charge conservation in the steady state (a node is a junction of two or more conductors in a circuit):

THE CURRENT NODE RULE

In the steady state, the electron current entering a node in a circuit is equal to the electron current leaving that node.

This rule is not a fundamental principle, but rather a consequence of the fundamental principle of conservation of charge and the definition of steady state. It is also called "the Kirchhoff node rule."

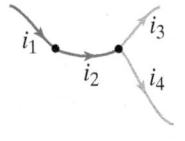

Figure 18.3 Electron current at different locations in part of a circuit.

EXAMPLE   **Current at a Node**

Figure 18.3 depicts a portion of a circuit. Suppose that in the steady state $i_1 = 3 \times 10^{19}$ electrons/s and $i_4 = 7 \times 10^{19}$ electrons/s. What is i_3?

Solution

$$i_1 = i_2$$
$$i_2 = i_3 + i_4$$
$$3 \times 10^{19} \text{ electrons/s} = i_3 + 7 \times 10^{19} \text{ electrons/s}$$
$$i_3 = -4 \times 10^{19} \text{ electrons/s}$$

What does the minus sign mean physically in this answer? It means that we guessed wrong about the direction of i_3. The arrow representing i_3 shown on the diagram is incorrect, and electrons flow in the opposite direction—into the node instead of out of it. In circuit diagrams we use arrows to indicate our guesses as to the direction of electron or conventional current, and symbols such as i_3, but if i_3 turns out to be negative, it just means that we guessed wrong as to which way the current was running.

QUESTION   Is the current always the same everywhere in a steady-state circuit?

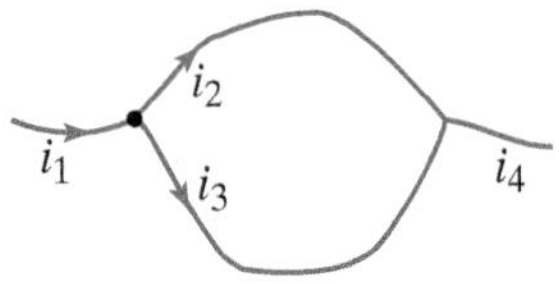

Figure 18.4 A portion of a circuit with two parallel branches.

No. Consider Figure 18.4. There are two parallel branches carrying currents i_2 and i_3. Assuming the direction of the arrows is correct, both i_2 and i_3 must be less than i_1; depending on the wires used, it may be the case that $i_2 \neq i_3$. We do know, however, that in this circuit $i_2 + i_3 = i_1$. One can of course express the node rule in terms of conventional current; we could write $I_2 + I_3 = I_1$.

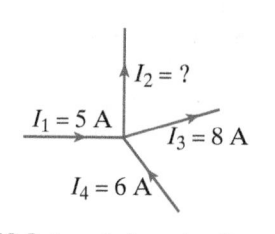

Figure 18.5 A node in a circuit.

Checkpoint 1 Figure 18.5 shows a portion of a steady-state circuit, with the magnitudes of the conventional currents given in amperes (A), and the directions given by arrows. Write the node equation for the circuit in Figure 18.5. What is the value of the outward-going current I_2? If I_4 were 1 A instead of 6 A, what would be the value of the outward-going current I_2? What is the meaning of the minus sign?

18.3 ELECTRIC FIELD AND CURRENT

In Chapter 14 we discussed the Drude model for the motion of mobile electrons in a metal and the relationship $\bar{v} = uE$: The average drift speed of the mobile electrons is proportional to the magnitude of the electric field. The proportionality constant u is called the electron mobility.

> QUESTION   According to the Momentum Principle, once an object is in motion, no force is required to keep it moving at constant speed. Given this, why is an electric field needed to keep a current flowing?

If the mobile electrons in a metal did not interact at all with the lattice of positively charged atomic cores, then once a current got started, it could flow forever. However, as we have observed, there is an interaction; the moving electrons do lose energy to the lattice, increasing the thermal motion of the atoms. We detect this by observing that the wire gets hot—in some cases hot enough to emit visible photons. Unless an electric field is present to increase the momentum of the mobile electrons repeatedly after collisions, their energy will quickly be dissipated (as energy can be dissipated by friction), and the current flow will stop.

Electrons Cannot Push Each Other through the Wire

> QUESTION   One might argue that in a circuit each mobile electron makes an electric field that affects the mobile electron to the left of it ("pushing" on its neighbor), thereby causing current to flow. Given our description of the mobile electron sea in a metal, what is wrong with this analysis?

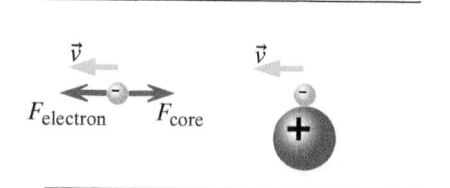

Figure 18.6 A mobile electron cannot push another mobile electron through a wire. The total force on a mobile electron due to the other mobile electrons and the positive cores is zero.

As we have stated previously, there can be no excess charge inside a conductor (and we'll see later that this also applies to the steady state as well as to equilibrium), so the number density of mobile electrons inside a metal wire must equal the number density of positive atomic cores—the inside of the wire is electrically neutral. In Figure 18.6 the repulsion of the mobile electron on the left by a mobile electron on the right is canceled by the attraction of a neighboring positive core, so the total force on a mobile electron due to the other mobile electrons and the positive cores is zero. The result is that the mobile electron sea behaves rather like an ideal gas, in which none of the particles interact with other particles. Electrons cannot continually push each other through the wire like peas pushed through a tube from one end of the tube. Rather, it must be other charges somewhere outside the wire that make an electric field throughout the wire that continually drives the electron current. At this point in our analysis it is not yet clear exactly where these charges are.

Magnitude of $\vec{E}$ in a Circuit

Because there are many mobile electrons in a metal wire, and because in most metals the electron mobility is quite high, the magnitude of the electric field needed to sustain a steady-state current in a wire is small.

EXAMPLE

Magnitude of $\vec{E}$ in a Circuit

In the previous chapter we calculated the drift speed in a copper wire to be 5×10^{-5} m/s for a typical electron current. The mobility of mobile electrons in copper is $u = 4.5 \times 10^{-3}$ (m/s)/(N/C). Calculate the magnitude of the electric field inside the wire.

Solution

$$\bar{v} = uE$$

$$E = \frac{5 \times 10^{-5} \text{ m/s}}{4.5 \times 10^{-3} \dfrac{\text{m/s}}{\text{N/C}}}$$

$$= 0.011 \text{ N/C}$$

Because copper is a very good conductor, only a small electric field is needed to sustain a current in a copper wire.

Electric Field and Drift Speed in Different Elements of a Circuit

We confirmed by experiment that in the steady state, the current was the same everywhere in a series circuit (a circuit without parallel branches). However, we know that the electron current i depends on the cross-sectional area A of a wire, since $i = nA\bar{v}$. The cross-sectional area of the tungsten filament of a light bulb is much less than the cross-sectional area of the copper connecting wires in the circuit.

In order to simplify thinking about this situation, let's first consider a circuit in which a wire leads into another, thinner wire of the same material (Figure 18.7).

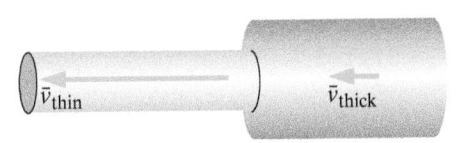

Figure 18.7 The drift speed of electrons in the thin and thick wires.

> QUESTION How must the drift speed of mobile electrons in the thin wire compare to the drift speed of mobile electrons in the thick wire?

Since the current must be the same in the thick and thin wires, the electron drift speed must be greater in the thin wire.

$$nA_{\text{thin}}\bar{v}_{\text{thin}} = nA_{\text{thick}}\bar{v}_{\text{thick}}$$

and therefore

$$\bar{v}_{\text{thin}} = \frac{A_{\text{thick}}}{A_{\text{thin}}}\bar{v}_{\text{thick}}$$

Note that this is very different from "compressible" flows such as highway traffic. On a highway the density of cars (how close they are to each other) can vary. For example, when a highway narrows down from two lanes to one, cars usually go slower in the narrow region but are correspondingly closer together (higher density). Because the electron sea in a metal is nearly incompressible (its density cannot change significantly), in the steady state the electrons must go faster in the narrow region. The density n of mobile electrons in copper is the same in both wires.

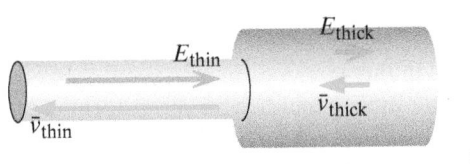

Figure 18.8 Electric field at locations inside the thin and thick wires.

> QUESTION If the electron drift speed is greater in the thin wire, how must the electric field in the thin wire compare to the field in the thick wire?

Since the drift speed must be greater in the thinner wire, the electric field in the thinner wire must also be larger, since $\bar{v} = uE$ (Figure 18.8).

Checkpoint 2 Suppose that a wire leads into another, thinner wire of the same material that has only half the cross-sectional area. In the "steady state," the number of electrons per second flowing through a cross section of the thick wire must be equal to the number of electrons per second flowing through a cross section of the thin wire. If the drift speed $\bar{v}_1$ in the thick wire is 4×10^{-5} m/s, what is the drift speed $\bar{v}_2$ in the thinner wire? If the electric field E_1 in the thick wire is 9×10^{-3} N/C, what is the electric field E_2 in the thinner wire?

Direction of Electric Field in a Wire

According to our measurements of the magnetic field made by the moving charges in our circuits, the current is the same in every part of a series circuit. Since conventional current $I = |q|nA\bar{v} = |q|nAuE$, and since conventional current flows in the direction of $\vec{E}$, this must mean that E is the same in every part of a wire with uniform properties ($|q|$, n, A, and u), and that $\vec{E}$ must be parallel to the wire at every location—even if the wire twists and turns, as indicated in Figure 18.9.

Proof That Field and Current Inside the Wire Are Uniform

Not only must the electric field follow the wire, but it also must be uniform across a cross section of the wire. Suppose for a moment that this isn't true—that instead, at different locations in the cross section of a straight wire, the drift speed and therefore the electric field are different (Figure 18.10). Consider the potential difference

$$\Delta V_{ABCDA} = -\int_A^A \vec{E} \cdot d\vec{l}$$

along the round-trip path $ABCDA$ in Figure 18.10 (from A to B to C to D and back to A).

QUESTION Calculate the contributions along each part of the path, $\Delta V_{AB}, \Delta V_{BC}, \Delta V_{CD}$, and ΔV_{DA}, and the sum ΔV_{ABCDA}. What can you conclude about E_1 and E_2?

The requirement that the round-trip potential difference be zero means that E_1 and E_2 have to be equal. Therefore the electric field must be uniform both along the length of the wire and also across the cross-sectional area of the wire (Figure 18.11). Since the drift speed is proportional to E, we find that the current is indeed uniformly distributed across the cross section. This result is true only for uniform cross section and uniform material, in the steady state. The current is not uniformly distributed across the cross section in the case of high-frequency (non-steady-state) alternating currents, because time-varying currents can create non-Coulomb forces, as we will see in a later chapter on Faraday's law.

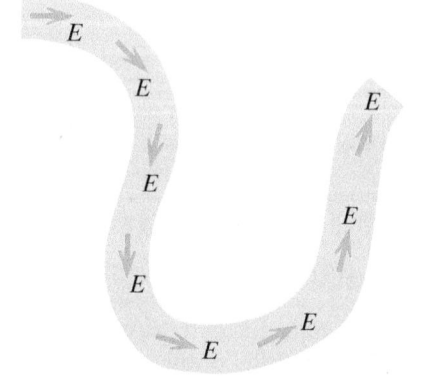

Figure 18.9 The electric field in a current-carrying wire must follow the wire.

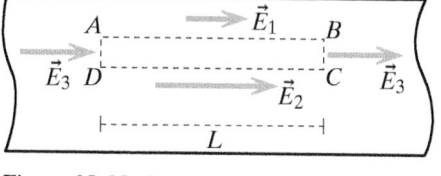

Figure 18.10 Could E vary at different locations in a uniform wire?

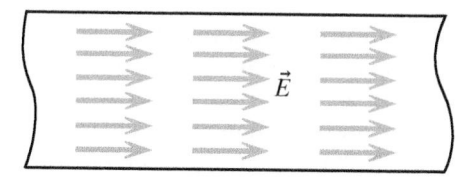

Figure 18.11 The electric field in a wire in a steady-state circuit must be uniform both along the wire and across the cross-sectional area of the wire.

18.4 WHAT CHARGES MAKE THE ELECTRIC FIELD INSIDE THE WIRES?

We have concluded that in a steady-state circuit:

- There must be a nonzero electric field in the wires.
- The magnitude of $\vec{E}$ must be the same throughout a wire of uniform cross section and material.
- The direction of the electric field at every location must be along the wire, since current flow follows the wire.

Figure 18.12 There is an electric field in the bulb filament. Is it charges in and on the battery that produce this field?

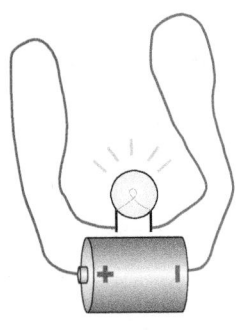

Figure 18.13 Bend the wires so the bulb is much closer to the battery. Presumably the electric field in the bulb should get much larger, but does it?

Figure 18.14 Rotate the bulb. Does the brightness change, due to a change in the parallel component of electric field?

Consider a very simple circuit consisting of a bulb connected by long wires to a battery (Figure 18.12). We have concluded that there must be an electric field inside the metal bulb filament, forcing electrons through the filament, heating it so much that it glows. Since electric fields are produced by charges, there must be excess charges somewhere to produce the electric field inside a wire in a circuit. Where might these charges be? What distribution of charges could produce a pattern of electric field following the wire?

Are the Excess Charges Inside the Conductors?

The bulb filament and the wires are conductors, so there can be no excess charge in the interior of the bulb filament and wires. In these metal objects, there are equal numbers of mobile electrons and positive atomic cores. The excess charges cannot be inside the wires, so they must be somewhere else.

Are the Excess Charges on the Battery?

One might reasonably assume that the charges that make the electric field are in and on the battery, since that's the active element in the circuit. Perhaps there are + charges on the + end of the battery, and − charges on the − end? In that case the field made by the battery will be something like the field of a dipole.

> QUESTION Suppose the bulb is 10 cm from the battery, and shining brightly (Figure 18.12). Move the bulb to a distance of about 1 cm from the battery, allowing the wires to bend (Figure 18.13). Very roughly, about how much larger is the electric field made by the battery at this closer location?

Although we're too close to the battery for the dipole-like electric field to be proportional to $1/r^3$, a rough estimate is that the electric field at 1 cm distance is about $1 \times 10^3 = 1000$ times as large as the field at 10 cm.

> QUESTION Now that the field made by the battery at the location of the bulb filament has increased by a factor of 1000 or so, what should happen to the electron current in the bulb and the brightness of the bulb?

For an ordinary metal, the electron current is proportional to the electric field: $i = nA\bar{v} = nAuE$. Therefore the electron current should increase by about a factor of 1000. The bulb should be enormously brighter than before! Yet this doesn't happen. As you have seen in working with a simple circuit, moving the bulb has no effect at all on the bulb brightness, or on the amount of current indicated by your compass.

Another test of the "charge on the battery" hypothesis also fails. To drive current through the bulb filament, the electric field must have a component parallel to the filament. Yet if you rotate the bulb, you don't see any change in the bulb brightness (Figure 18.14). If the charges responsible for making the electric field were solely in and on the battery, rotating the bulb ought to make a big difference in the brightness—the bulb might even go out.

The fact that the bulb doesn't get brighter when moved toward the battery, and does not go out when rotated, proves that charges in and on the battery cannot be the only contributors to the electric field in the bulb filament. There must be other charges somewhere else (and these charges can't be inside the wires or bulb filament, either, where there is no net charge, averaged over a few atomic diameters). This is deeply puzzling.

A "mechanical battery"

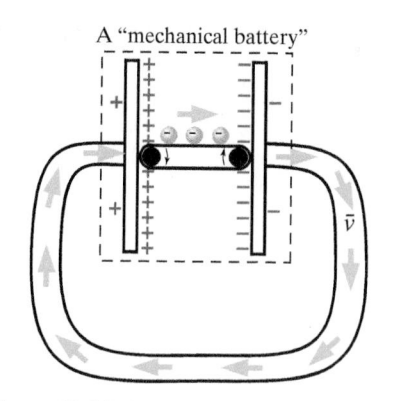

Figure 18.15 A "mechanical battery": a conveyor belt maintains a charge separation that drives a steady current in the wire.

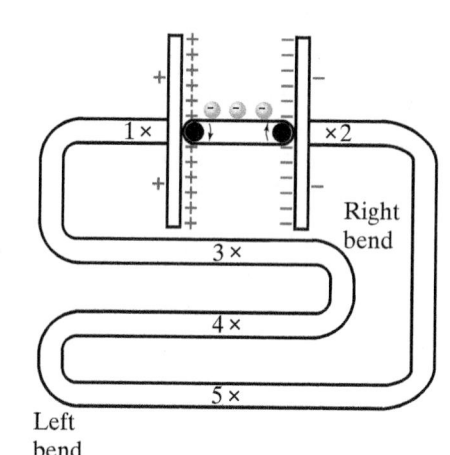

Figure 18.16 A circuit with a Nichrome wire and a mechanical battery.

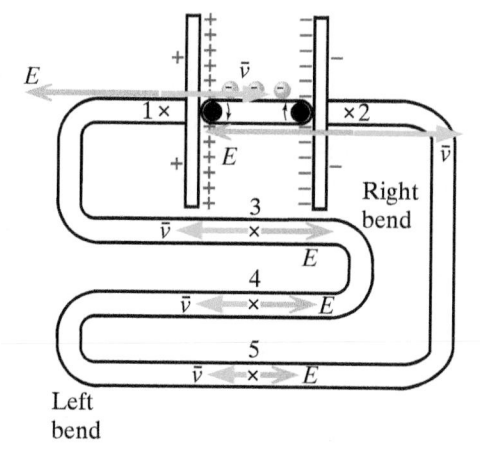

Figure 18.17 Electric field (orange) due solely to the mechanical battery, and electron drift velocities (green) that would result from this pattern of electric field. Some arrows are offset from the observation locations for clarity.

A Mechanical Battery

To simplify our further analysis of circuits, including in particular our search for the location of the charges responsible for producing the electric field in a bulb filament, we introduce a simplified model of a battery, a "mechanical battery" that is much easier to understand in detail than a chemical battery yet behaves in a circuit in ways that are similar to a chemical battery.

Our mechanical battery has a conveyor belt driven by a motor or a hand crank that pulls electrons out of one plate, making it positive, and pushes them onto another plate, making it negative (Figure 18.15). This action replenishes electrons that leave the negative plate and move through the wire with drift speed $\bar{v}$, and it removes electrons that enter the positive plate after traveling through the wire. You may have seen a Van de Graaff generator, which has a motor-driven conveyor belt like this that pumps lots of charge onto a metal sphere.

This combination of charged plates plus the motor-driven conveyor belt acts very much like a chemical battery. The main difference is that here we imagine moving the charges by mechanical means, and this mechanism is easier to understand than the mechanism for charge transfer in a chemical battery. As long as the motor is able to maintain the charge separation across the two plates, we can have a steady-state current running in the wire.

Field Due to the Battery

We will now consider a simple circuit consisting of a mechanical battery and a resistive Nichrome wire that has some twists and turns in it. (The bare wire in the experiment kit mentioned in the experiment section of Chapter 17 is Nichrome; the heating elements in toasters are made of such wire.) You know from your own experience with a simple circuit that twists and turns do not seem to affect the amount of steady-state current, which seems odd. It will seem much odder in a moment!

> QUESTION At the locations marked × on a copy of Figure 18.16 (locations 1 through 5), draw arrows representing the approximate electric field due solely to the charges on the metal plates of the mechanical battery. (The amount of charge on the belt is completely negligible compared to the charge on the plates.) Forget the presence of the wire for the moment.

> QUESTION Next, in a different color or with a different-looking arrow, indicate the drift velocity of the mobile-electron sea at those locations, assuming that the drift velocity is due solely to the electric field made by the charges on the plates of the mechanical battery.

Good grief! Figure 18.17 shows that we've got the electron current running *upstream* at location 4 in the wire! That can't be right in the steady state (the situation in which charge distributions and currents are not changing).

We are forced to conclude that in the steady state there must be some other charges somewhere that contribute to the net electric field in such a way that the electric field points upstream everywhere (giving an electron drift velocity downstream everywhere).

Charge Buildup on the Surface of the Wire

Remember that the electrons in the mobile-electron sea inside the metal effectively don't interact with each other, because their mutual repulsions are on the average canceled by attraction to the positive atomic cores so they can't push each other through the wire. Some other charges must contribute to the electric field that is responsible for pushing the electrons through the wire.

Take a look at the left bend in the wire in Figure 18.17. If the only charges were on the plates, electron current would flow toward the left bend from both neighboring sections. Here we have a section of wire that has electron current flowing into it but not out.

QUESTION   What effect will this have on this section of the wire? Be as specific as you can, bearing in mind that the wire is made of metal. Illustrate your thoughts on your copy of Figure 18.16.

If electrons are flowing into a section of wire from both ends, that section will gain a net negative charge. Since the wire is made of metal, electrons are free to move, and all excess charge will move to the surface of the wire. In Chapter 14 we asserted that excess charge goes to the surface of a metal. This assertion is plausible, because excess charges in the interior of a metal repel each other toward the surface.

More precisely, excess negative charge expands the mobile-electron sea so that some of the sea peeks out at the surface, giving a negative surface charge. Excess positive charge contracts the electron sea, allowing positive cores to peek out at the surface. A rigorous proof that all the excess charge goes to the surface of a metal requires Gauss's law, which we will study in Chapter 21.

Thus, negative charge accumulates on the surface of the left bend.

QUESTION   What happens at the right bend in the wire? Illustrate your thoughts on a copy of Figure 18.16.

At the right bend electrons flow out of both ends, leaving a net positive charge, also on the surface of the wire.

QUESTION   Focus your attention on location 4, a place where the electric field of the mechanical battery alone would drive current in the wrong direction. At location 4, what is the direction of the additional contribution to the electric field made by the excess + and − charges on the left and right bends?

The charges on the bends contribute an electric field to the left at location 4 that is opposite to the electric field of the mechanical battery. The "wrongly" directed net electric field is reduced in magnitude (Figure 18.18). However, as long as the net field at location 4 still points to the right (the "wrong" direction), more and more excess charge will pile up on the bends, and the electric field contributed by these charges will grow.

QUESTION   How far will this pileup process go? What stops it?

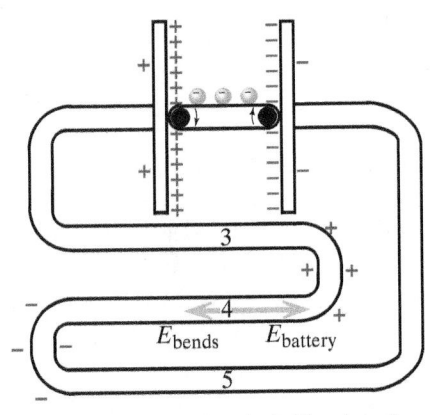

Figure 18.18 The electric field at location 4 is reduced by the contribution of the charges that have piled up on the bends.

The pileup will continue automatically until there is so much charge on the bends that the net field at 4 points to the left. Other charges must appear at various places on the surfaces of the wires in order that eventually the net field at locations 3, 4, and 5 all have the same magnitude. Only then will the electron current $i = nA\bar{v} = nAuE$ into a bend equal the electron current out of a bend, with no further change in the amount of charge on the surface of the bend.

We see examples of "feedback" in both the left bend and the right bend of the circuit. If the initial field is such that it drives current that is different in amount or direction from the steady-state current, surface charge automatically builds up in such a way as to alter the current to be more like the steady-state current. Moreover, once the steady state is established, there is "negative feedback": any deviation away from the steady state will produce a change in the surface charge that tends to restore the steady-state conditions.

We have deliberately emphasized the most dramatic aspect of this feedback, which is what happens at the bends in the wire. However, the resulting buildup of surface charge isn't limited to bends in the circuit. There must be positive and negative charge on the surface of the wires near the

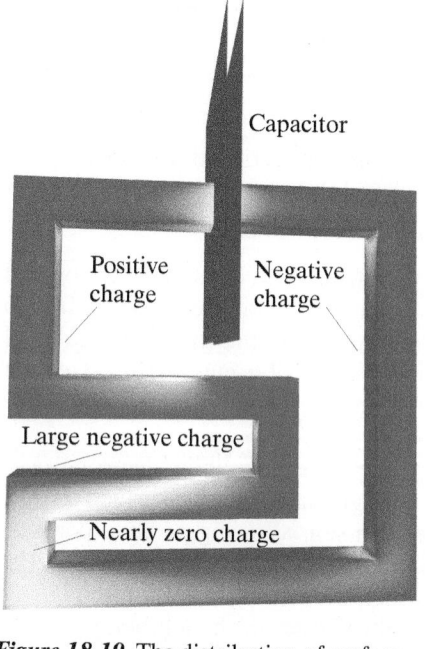

Figure 18.19 The distribution of surface charge on the twisty circuit discussed above, generated by a computational model. Red represents positive surface charge and blue represents negative surface charge. The intensity ("saturation") of the color is proportional to the amount of charge. Square wires were used to simplify the computation.

+ and − plates of the mechanical battery. The surface charge must arrange itself in such a way as to produce a pattern of electric field that follows the direction of the wire and has the same magnitude throughout the wire. Since the current is given by $i = nA\bar{v} = nAuE$, to have the same i everywhere in this series circuit we must have the same E everywhere inside the wires.

Figure 18.19 shows a detailed visualization of the distribution of surface charge, produced by a computational model to be described later, with a charged capacitor playing the role of the battery (we continually replenish the supply of charges on the capacitor plates). The calculations confirm the predictions we made using qualitative reasoning: there is an accumulation of positive charge on the right of the lower middle wire and of negative charge on the left. It is also evident that in this geometrically complicated circuit the distribution of surface charge is rather complex because some parts of the circuit polarize other parts. For example, note how the wire containing the location we've been considering is strongly polarized by the positive wire just above it and the negative wire just below it. There is however a general trend of the surface charge running from positive to negative around the circuit.

18.5 SURFACE CHARGE DISTRIBUTIONS

Despite the complex details of the pattern of charge on the twisty circuit discussed in the previous section, one thing stands out dramatically: the amount and sign of the charge on the surface of the wire varies. The general trend is clear: the wire to the left of the capacitor is strongly positive; the wire to the right is strongly negative, and there is only a small amount of charge in the middle of the circuit. This variation can be described as a gradient of surface charge. Mathematically, a gradient is a derivative describing the variation of some quantity in space. If we think of a circuit as a one-dimensional path, we can find the gradient of surface charge as we travel along the wire from one end to the other.

To see how a gradient of surface charge is necessary to produce a uniform field inside a wire in a steady-state circuit, let's start with a simple case. We'll consider a long straight wire in a circuit, and we'll look at a section of the middle of the wire, far from the battery. We'll model the surface of the wire as a series of rings of charge. In our simple model circuit each ring will be uniformly charged. In order to calculate the electric field inside the wire, we'll model each ring as a collection of point charges (Figure 18.20).

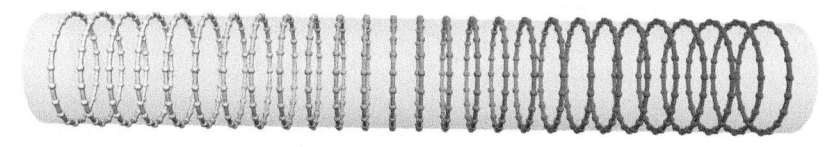

Figure 18.20 A model of the charge on the surface of a cylindrical wire as a series of charged rings, each approximated by a collection of point charges. Red is positive, blue is negative, and white is uncharged. The saturation of the color indicates the amount of charge. The surface charge on this wire changes linearly from slightly negative at the left end to strongly positive at the right end.

Figure 18.21 Two equally charged rings on the surface of a wire and their contributions to the net electric field (orange arrows) inside a wire in a circuit.

Two Equally Charged Rings

First, we'll consider two rings with equal charges (Figure 18.21). At a location on the axis, midway between the rings, the net electric field is zero. At locations not on the axis, the field is nonzero but very small, and does not point along the wire. No matter how large the charges on these rings are, they cannot create an electric field aligned with the wire.

QUESTION What is the gradient of charge along this two-ring wire?

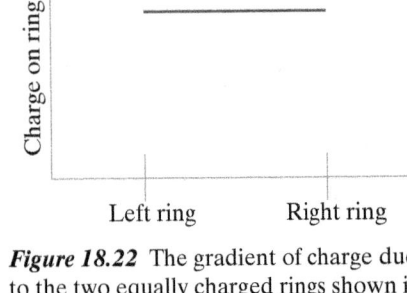

Figure 18.22 The gradient of charge due to the two equally charged rings shown in Figure 18.21 is zero (the gradient is the slope of the line).

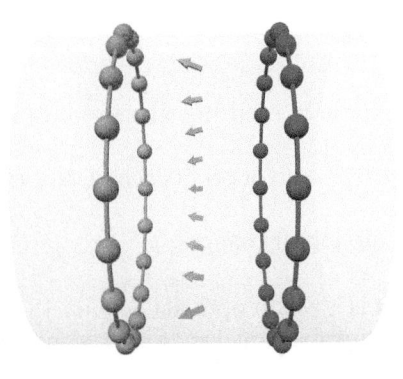

Figure 18.23 The electric field between two rings with different amounts of positive charge.

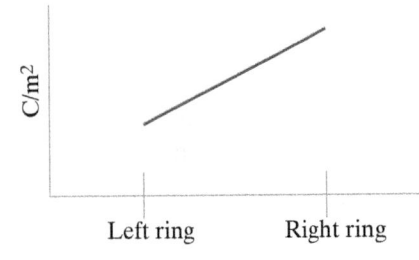

Figure 18.24 Between two rings with different charges, the charge gradient is nonzero.

If we plot a graph of ring charge vs. position, we see that the slope of this plot is zero—the charge does not vary from one ring to the next—so the gradient of surface charge is zero in this case (Figure 18.22). It is clear from this example that simply adding more equally charged rings will not produce a uniform electric field pointing along the wire. Let's try a different approach.

A Constant Gradient of Charge

The simplest configuration of charged rings that has a nonzero gradient of charge is two rings with different amounts of charge (Figure 18.23). This looks promising. Although both rings are positive, since one ring has more charge than the other the electric field between the rings is nonzero and points roughly along the axis. The charge gradient along the wire is not zero (Figure 18.24).

Let's extend our model of the charges on the surface of the wire, keeping the charge gradient constant, and extending it further along the wire in both directions (Figure 18.25).

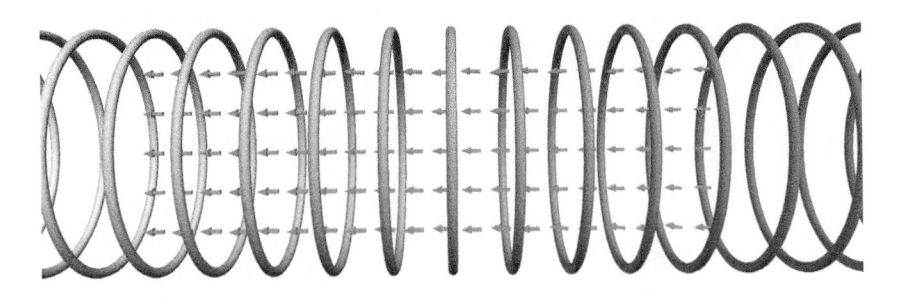

Figure 18.25 A closeup view of the electric field near the middle of the series of charged rings shown in Figure 18.20. To make it easier to see the orange arrows representing the electric field, the point charges making up the rings are not shown.

The result is surprisingly good—even though the charged rings are widely spaced, producing a "lumpy" distribution of charge instead of a smooth one, the electric field throughout the interior of the wire is remarkably uniform, both longitudinally and laterally, even near the surface of the wire. Evidently a constant gradient of charge density on the surface of a wire produces an electric field inside the wire that is uniform in magnitude along and across the wire and points along the wire.

Surface Charge Distribution on Circuits

Even the simplest circuit is more complex than a very long straight wire—it includes a battery, and the wire must bend around to connect to both ends of the battery. Therefore we should expect the distribution of surface charge on a circuit to be more complex than a simple constant gradient of charge. However, we might anticipate that:

- On long straight sections of wire that are far from other circuit elements there may be a nearly constant gradient of surface charge.
- The surface charge gradient will be largest in regions of the circuit in which the electric field is largest.
- In regions such as sharp bends in a wire, and locations where two different elements (wires of different diameter, or a wire and a battery) are connected, the distribution of surface charge may be quite complex.

In the following discussion, we show the results of a computational model that calculates the distribution of charge on the surface of several simple circuits.

QUESTION Why don't we just measure the distribution of charge on the surface of a simple circuit, instead of calculating it?

Even when there is a large gradient of charge on the surface of a wire in a circuit, the total amount of charge is so small that it is very difficult to measure. Section 18.11 describes a demonstration in which surface charge can indeed be detected on a very high voltage circuit, but this isn't possible with an ordinary circuit.

How the Model Calculates Surface Charge Distributions

Basically, the computational model used to calculate surface charge uses this reasoning:

- Start out with charge only on the battery (modeled as a capacitor whose charge is continually replenished).
- Use square wires to simplify the calculations. Divide the surface of the wires into many tiny square tiles (Figure 18.26).
- Now, repeatedly:
  - Calculate the electric field at the location of each tile due to all other charges on the circuit.
  - Use this field to calculate the flow of charge onto or off of each surface tile in the next time step Δt.
  - Repeat until the distribution of charge no longer changes.

A more detailed explanation of the model is given in optional Section 18.12. The computational techniques of implementing the model are complex and are beyond those taught in this book.

A Single Wire Circuit

Figure 18.27 shows the calculated distribution of surface charge on a very simple circuit: one resistive wire and a battery. Starting at the positive left plate of the capacitor and following the circuit counterclockwise around to the negative capacitor plate, we find that the surface charge starts out strongly positive, becoming less and less positive and reaching zero in the middle of the bottom wire, then becoming increasingly negative as we continue on around the circuit.

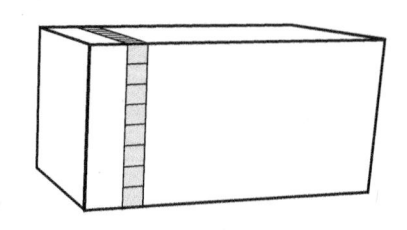

Figure 18.26 A (square) ring of charged tiles on the surface of a circuit wire.

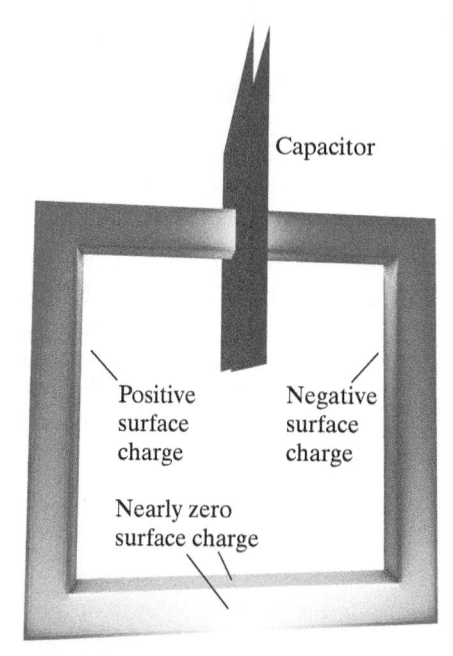

Figure 18.27 A model of a simple circuit. The resistive wire has a 6 mm by 6 mm square cross section. The circuit is 60 mm wide, and the capacitor gap is 2 mm, with a potential difference across the gap of 1.5 V.

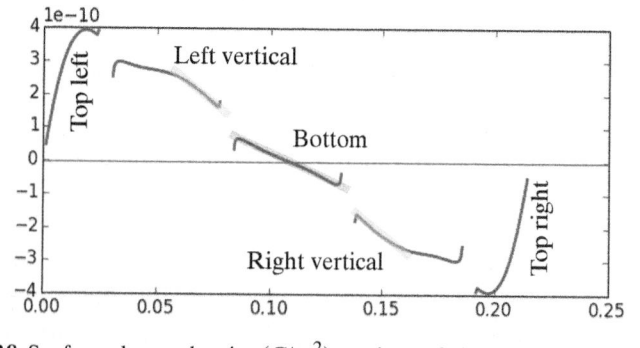

Figure 18.28 Surface charge density (C/m^2) on rings of charge along a path (meters) counterclockwise around the circuit of Figure 18.27, starting from the positive plate of the capacitor. We omit the charge situation at the square corners, where adjacent wires polarize each other. The charge gradient on the bottom wire, far from the battery, is constant (pink shading). The charge gradients on the bottom parts of the vertical wires (green shading) are also constant.

A graph of surface charge density vs. position in the circuit is shown in Figure 18.28. Surface charge density is calculated by adding up the charge on all the tiles in a single ring (Figure 18.26) and dividing by the total surface area of all those tiles. Figure 18.28 shows that there is a roughly constant gradient of surface charge density on those straight sections of wire that are farthest from the battery. The graph does not show the complications at the corners, which are in part due to the square geometry of the model circuit.

As can be seen from Figure 18.29, inside the wire the net electric field (shown in orange) produced by the capacitor and the surface charges is quite uniform, which is a check on the validity of the computational model. Figure 18.29 also shows the contributions to the net field from the battery and from charges on the surface of the wires. In the bottom wire, which is farthest from the battery, almost all of the electric field is due to the surface charge gradient on the wire.

Figure 18.29 The electric field at a number of locations around the circuit. The net electric field (orange), the field due to the capacitor (magenta), and the field due to all the surface charges (green) are shown. The fields are calculated along the centerline of the wire, but the arrows are displayed in front of the wire in order to make them visible. Note that to the left of the capacitor the field due to surface charge is actually opposite to the net field. At this location the large field due to the capacitor is opposed by the field due to the piled-up positive surface charge near the top left corner.

Electric Field Is Simple; Surface Charge Distributions Are Complicated

The main features of the surface charge distribution shown in Figure 18.27 are what we expected. We also expected some deviations from this pattern in special locations, and we do in fact see these. For example, in the inside corners you can see red puddles of positive surface charge inside blue negative regions, and blue puddles of negative surface charge inside red positive regions. These "wrong-sign" puddles are due to the polarizing effects of the capacitor field and the field of the neighboring wires. In these wrong-sign places one finds on the opposite side of the wire an extra amount of "correct-sign" charge; what counts is the total charge on a ring of tiles. Right next to the capacitor plates the charge density is reduced due to vertical components of the electric field contributed by the capacitor plates.

Although the graph of surface charge in Figure 18.28 shows that there is a roughly constant gradient of ring surface charge along the left, bottom, and right wires, Figure 18.29 shows that along the left and right wires the net field

Figure 18.30 In the lower parts of this taller version of the circuit the field contributed by the capacitor is very small, and there are constant surface charge gradients.

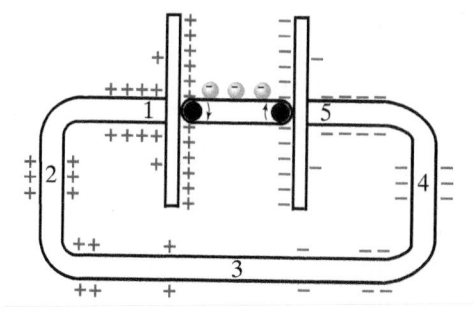

Figure 18.32 A geometrically simple circuit with approximate surface charge shown. The number of plus or minus signs indicates schematically the density of surface charge at that location. Charges are drawn only at sample locations, to make it possible to see differences in amount of charge.

has large contributions from the capacitor, not just from the nearby surface charge gradient. Note too that the horizontal component of the field due to the surface charge means that there is sizable lateral polarization of these wires. In contrast, in the bottom wire, far from the capacitor, almost all of the field is due to the gradient of surface charge on the bottom wire.

If we make the circuit taller (Figure 18.30), the field of the capacitor is very small in the lower parts of the circuit, where we see truly constant gradients of surface charge (Figure 18.31).

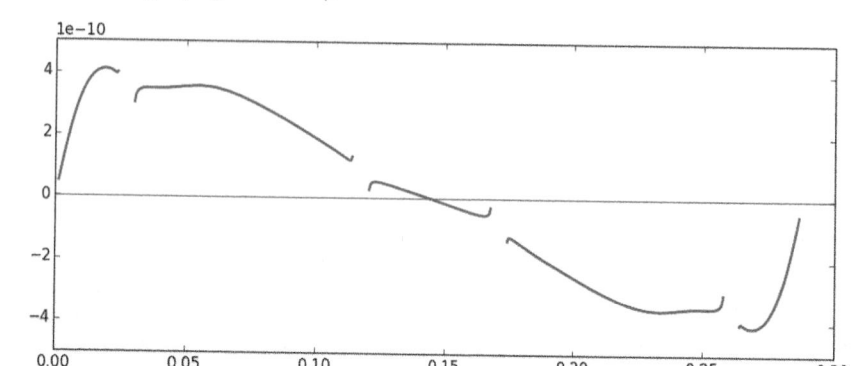

Figure 18.31 Surface charge along the circuit shown in Figure 18.30.

Although the surface charge distribution on a circuit is complex, and the details of the contributions of each charge to the net field are complex, the pattern of the net electric field inside the wires is very simple: the net field is uniform in magnitude and points along the wire (see the orange arrows in Figure 18.29), as we knew must be the case. We'll see that electric fields in circuits are easy to calculate, whereas it is very complicated to figure out where all the charges on the circuit are!

A Simplified Representation

In this chapter we will often draw approximate surface distributions like that in Figure 18.32. At selected locations we indicate high density of surface charge (C/m^2) with lots of +'s or −'s, and low density with few +'s or −'s. We emphasize that, as we've just seen, these charge distributions are not accurate in detail; our goal is to give a rough indication of the surface charge distribution, in a way that is easily written and read. Feedback in the circuit will tend to put some additional charge on the bends, the charges near location 2 will polarize the bottom wire, etc. We ignore the sideways polarization that occurs. The key point is that the surface charge near location 1 must be positive, and the surface charge near location 5 must be negative. When we connect a wire from one end of the battery to the other, there will be some kind of transition along the wire from + surface charge to − surface charge. There will even be some location at which there is zero surface charge (location 3 in Figure 18.32).

> **Checkpoint 3** At locations 1, 2, 3, 4, 5 on a diagram like Figure 18.32, draw the net electric field. Also draw the resulting electron drift velocity. Pay close attention to direction and relative magnitudes.

A Circuit with a Thin Resistor

A slightly more complex circuit, including a thin resistor and thick connecting wires, is shown in Figure 18.33. In Section 18.3 we reasoned that in order for the current to be the same everywhere in a single-loop circuit, the drift speed of electrons must be larger in wires whose cross-sectional area is smaller.

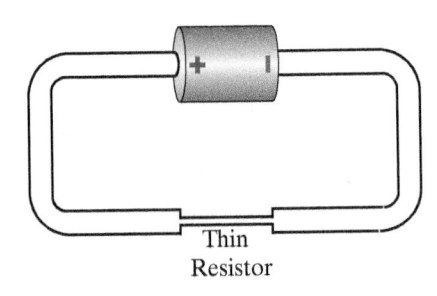

Figure 18.33 A circuit with a thin section acting as a resistor.

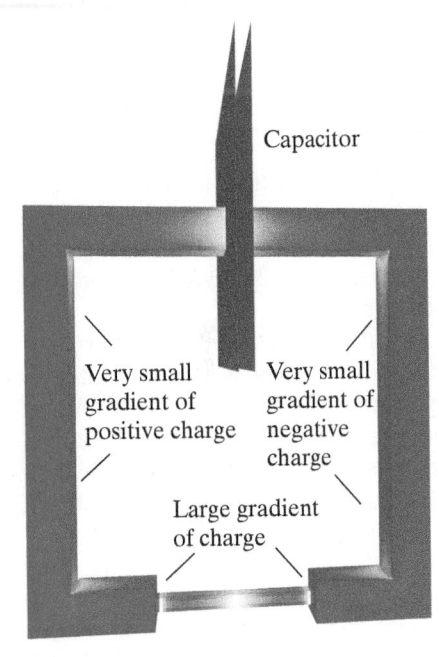

Figure 18.34 Computed surface charge on a circuit similar to that in Figure 18.33. Positive charge is red and negative charge is blue.

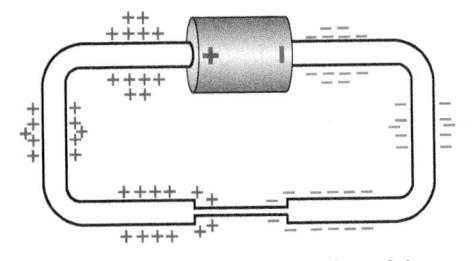

Figure 18.36 A schematic version of the surface charge distribution shown in Figure 18.34. The charge gradient across the thin resistor is large; the gradient on the thick wires is small.

QUESTION What can we conclude about the gradient of surface charge across the thin resistor in Figure 18.33, compared to the gradient of surface charge on the thick connecting wires?

Two steps in reasoning are needed to answer this question. First, since the drift speed must be larger in the thin resistor, we deduce that the electric field inside the resistor must be larger than that inside the thick wires, since $\bar{v} = uE$. Second, we infer that there must be a larger surface charge gradient across the resistor, in order to produce a larger electric field.

We can see both of these features in Figures 18.34 and 18.35. In Figure 18.34 the thick wires (6 mm × 6 mm) have little charge gradient—the amount of charge per m^2 varies very little on these wires. This small charge gradient is associated with a small electric field inside these wires. In contrast, the charge on the thin wire (2 mm × 2 mm) varies rapidly from positive to negative across the wire. This large gradient is associated with a large electric field inside the resistor.

In the graph shown in Figure 18.35 we can see that there is an extra pile-up of surface charge at the entrance and exit of the thin resistor and a large gradient along the resistor (pink shading), both of which contribute to a large electric field in the resistor. In contrast, the surface charge gradient on the left and right vertical wires is nearly zero (green shading), contributing to a small electric field inside these wires.

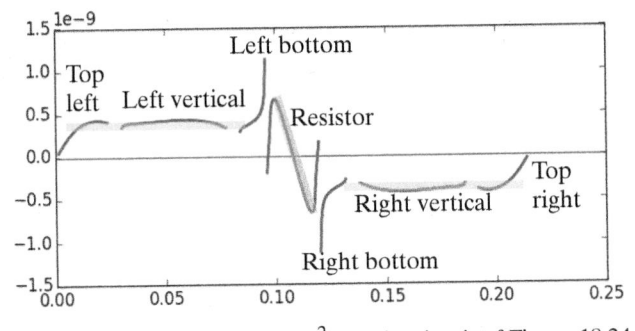

Figure 18.35 Surface charge density (C/m^2) on the circuit of Figure 18.34. The gradient of surface charge on the thick wires is very small (green shading), but the gradient of charge on the surface of the thin wire is large (pink shading).

In Figure 18.34 we again see complex charge patterns near junctions: there is wrong-sign charge density at each end of the resistor, due to polarization by the adjacent larger, strongly charged thick wires, which apply an electric field with large vertical components on the ends of the thin section.

Figure 18.36 shows a schematic version of the surface charge distribution shown in Figure 18.34. Gradients can be inferred by comparing the amount and sign of charge at neighboring locations. The large change in amount and sign of charge across the resistor indicates a large surface charge gradient.

Amount of Surface Charge

Despite the large number of electrons moving around in a circuit, you will find that the wires in one of your circuits will not repel a hanging charged tape. Try it! (The tape is of course attracted to neutral matter, which masks the tiny repulsion due to surface charges.) The large number of electrons moving inside the wires is balanced by an equally large number of positively-charged atomic cores, so the net charge inside the wire is zero and cannot repel charges by electric interaction. (The current does of course produce magnetic effects.)

The surface charges and the electric field they make are essential to driving and guiding the current through the interiors of the wires, but the amount of charge present on the wires in a circuit powered by two 1.5 V batteries is very small: in Figure 18.35 the maximum surface charge density is only about 1 nC per m². Little surface charge is needed in such a circuit; it is very easy to move electrons through metal, and very small electric fields are sufficient to drive sizable currents. We will see that in a typical circuit the magnitude of the electric field inside a copper wire is very small—typically much less than a volt per meter. Compare this with typical electrostatic fields, which can easily be large enough to create a spark (3×10^6 V/m). Evidently the amount of surface charge on a typical circuit is very small.

It is possible to run a circuit with a 10,000 V power supply and demonstrate how the surface charges can repel charged objects. The amount of surface charge is proportional to the voltage, and such circuits have much more surface charge than a 3 V circuit. A description of such a demonstration is given in Section 18.11.

18.6 CONNECTING A CIRCUIT: THE INITIAL TRANSIENT

When you complete a circuit by making the final connection, feedback forces a rapid rearrangement of surface charges leading to the steady state. This period of adjustment before establishing the steady state is called the initial transient. How does this work, and how long does the transient take? We'll look now in closer detail at how the steady state is established.

Consider the circuit in Figure 18.37, containing one chemical battery and two Nichrome wires with a gap between them. The system is in equilibrium, so $E = 0$ everywhere inside the wires. An approximate surface-charge distribution is indicated, with one wire charged positive and the other charged negative, ignoring the exact details of how these charges are distributed. (In actual fact, there are larger concentrations of surface charge near the gap, due to strong mutual polarization in the gap region, so this is a crude diagram.)

Let's look closely at the neighborhood of the gap. In Figure 18.38 we show the part of the electric field inside the wire that is contributed *solely* by the charges on the faces of the gap. You should convince yourself that these vectors are indeed drawn correctly.

> QUESTION On a similar diagram, sketch the part of the electric field that is due to all the *other* charges (on the ends of the battery and along the wires). Why is the field due to all the other charges exactly opposite to the field due to the gap charges?

The net field inside the wire must be zero, because at this time the system is in equilibrium. Therefore the field due to all other charges must be exactly equal in magnitude and opposite in direction to the field made by the charges on the faces of the gap (Figure 18.39).

Let's complete the circuit by connecting the Nichrome wires together. The charges on the facing ends of the wires neutralize each other, leaving a tube of surface charge on the outside of the wire, as shown in Figure 18.40 in a cross section of the region near the connection point. At this instant this surface charge distribution has a big, unstable discontinuity in it—closely neighboring areas on the outside of the wire have positive and negative charges.

> QUESTION In a diagram similar to Figure 18.40, use vectors to indicate the electric field at the locations marked × inside the wire just after closing the gap. (*Hint:* The gap charges are now gone.)

Immediately after closing the gap we've lost the contribution to the net field formerly made by the charges on the faces of the gap. Therefore the net field will look like the field of the "other" charges that we drew before.

Electrons in the mobile electron sea inside the metal will move due to these fields. Imagine that you could take a snapshot of the surface charge after the

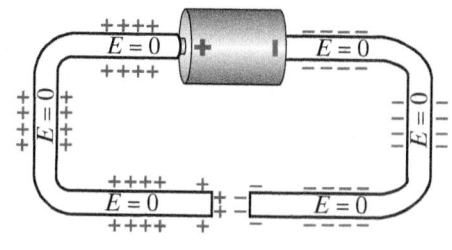

Figure 18.37 A circuit with a gap; approximate surface charge shown.

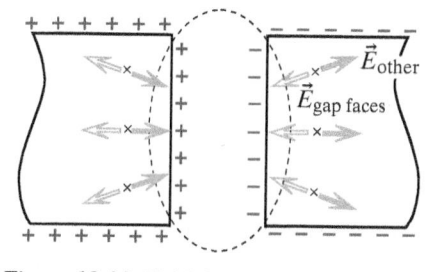

Figure 18.38 A close-up of the gap, showing the electric field contributed *solely* by the charges on the faces of the gap. At the locations marked ×, draw the electric field due to all *other* charges.

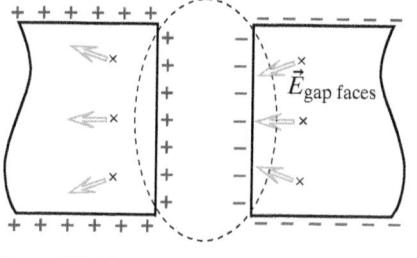

Figure 18.39 Field due to charges on the faces of the gap, and due to all others.

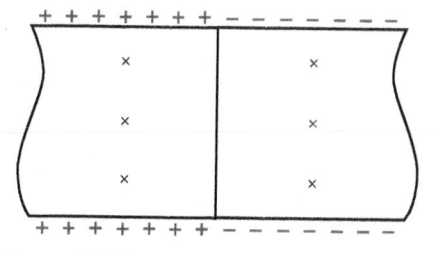

Figure 18.40 A cross section near the former gap, just after closing the gap.

electrons inside the wire have moved just a little bit (that is, after a very short time). Electrons moving toward a positive region of the surface will make the surface charge in that region less positive, while electrons moving away from a negative region of the surface will make the surface charge in that region less negative.

> QUESTION On your diagram corresponding to Figure 18.40, where there is a change in the surface charge, write what that change is (for example, increase of positive charge, decrease of negative charge, etc.). (Remember that a + sign on the surface does not represent some exotic positive charge. Rather, a + sign indicates a region where there is a deficiency of electrons, leaving partially exposed the metal's positive atomic cores.)

Figure 18.41 shows the new distribution of surface charge after a very short time (a fraction of a nanosecond!). The motion of electrons toward and away from the surface, driven by the new electric field, dilutes the surface charge near the location where the gap was, so that in the gap region the initial abrupt charge discontinuity turns into a more gradual change, indicated by placing fewer + or − symbols near the gap location. A nonzero electric field has come into being where the gap used to be, due to the + charges to the left and − charges to the right of that location.

In Figure 18.42 we see that the affected region, where there has been a rearrangement of the surface charge, grows rapidly outward from the gap location at approximately the speed of light, which is the speed with which the change in the distribution of charges is felt at places some distance away. At the instant sketched in Figure 18.42, the surface charges and the electric field at distant locations haven't been affected yet, because those regions haven't yet received the information that the gap has been closed! The electric field is still zero inside the wires except near the location of the gap.

It is the speed of light that determines the minimum time required to establish the steady state. Light travels 30 cm (about one foot) in a nanosecond (1×10^{-9} s). The time required to move the electrons toward or away from the surface is completely negligible, because the electrons only have to move an infinitesimal distance in order to establish a significant amount of surface charge, through slight expansion or contraction of the mobile-electron sea (compare with the molecular stretch associated with polarizing an insulator, which is less than the radius of a single proton!).

In just a few nanoseconds the rearrangement of the surface charges will extend all the way around the circuit. There is some wave-like sloshing around of the surface-charge distribution, which due to dissipation eventually settles down to the establishment of a "steady state," in which the surface charge and current no longer vary with time. In the steady state there is a variation of surface charge around the circuit, and the electric field has uniform magnitude throughout the inside of the wire (Figure 18.43).

In terms of the average drift speed $\bar{v}$ of the mobile charges:

- Equilibrium: $\bar{v} = 0$ (no current)
- Steady state: $\bar{v}$ = constant (constant, nonzero current)
- Initial transient: $\bar{v}$ changing (short process leading to the steady state)

Why a Light Comes on Right Away

As discussed earlier, the drift speed in one of your battery-and-bulb circuits is only about 5×10^{-5} m/s. However, a change in the circuit, such as making or breaking a connection, propagates its effects very rapidly through the circuit, at approximately the speed of light. It is not necessary for charges to move physically to a distant place to cause a change at that place: a rearrangement of

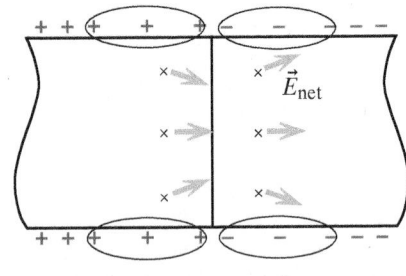

Figure 18.41 The dilution of the original surface charge. The circled regions at left become less positive, while the circled regions at right become less negative.

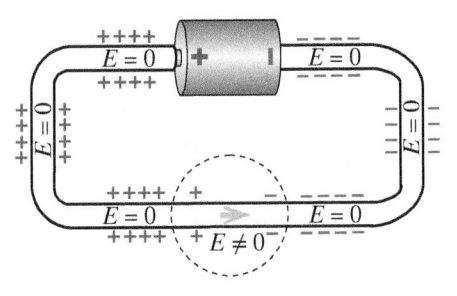

Figure 18.42 Parts of the circuit haven't yet been informed that the gap has been closed! The sphere enclosing the region of changed electric field expands at the speed of light.

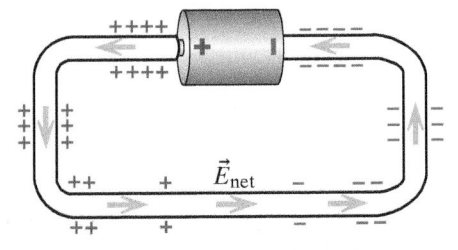

Figure 18.43 In the steady state the electric field has the same magnitude everywhere in the wire.

charges due to tiny movements produces a change at the speed of light in the electric field at a distance. It is important to keep in mind that the effects do take a nonzero amount of time to propagate (nanoseconds). The propagation is not instantaneous, but it is very fast.

When you turn on the light switch in your room (making the final connection in a circuit), it could take hours on the average for an electron in the switch to drift to the overhead light. However, the light comes on right away, because the rearrangement of surface charges in the circuit takes place at about the speed of light. The final steady state of the circuit is established in a few nanoseconds, after which the electron sea circulates slowly and majestically around and around, everywhere in the circuit, including through the switch and the wires and the light bulb. (Most lighting actually uses "alternating current," in which case the electron sea doesn't drift continuously but merely sloshes back and forth very short distances, everywhere in the circuit, 50 or 60 times per second.)

The electrons don't have to move from the switch to the light bulb; there are already plenty of electrons in the bulb filament. All that is required is to begin to establish the appropriate steady-state surface-charge distribution, very quickly, and then the electrons that are already in the filament start moving, propelled by the new nonzero electric field in the wire.

18.7 FEEDBACK

Feedback during the initial transient produces surface charge of the right amount to create the appropriate steady-state electric field. It also maintains these steady-state conditions, as we will show next.

Feedback Leads to Current Equalization

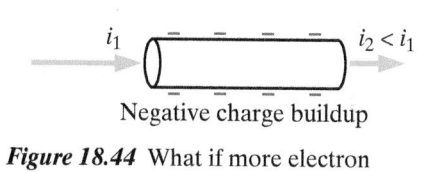

Negative charge buildup

Figure 18.44 What if more electron current enters than leaves a section of wire?

Consider a straight section of wire that at some moment has more incoming electron current than outgoing electron current, leading to a buildup of excess negative charge on the surface of this section of the wire (Figure 18.44).

This surface-charge buildup will tend to equalize the incoming and outgoing electron currents. The outer surface of this section of wire becomes more negative, and this negative surface charge tends to impede the incoming electrons and tends to speed up the outgoing electrons, thus making the incoming and outgoing electron currents $i = nA\bar{v}$ more nearly equal to each other. This feedback process will continue until the two currents are exactly equal to each other. At that point there will be no further change in the surface charge.

> QUESTION If the incoming electron current is smaller than the outgoing current, what will happen on the outer surface of this section of the wire?

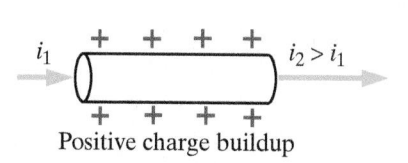

Positive charge buildup

Figure 18.45 What if less electron current enters than leaves a section of wire?

In this case (Figure 18.45) this section of wire becomes more positive, which speeds up the incoming electrons and slows down the outgoing electrons, so the incoming electron current increases and the outgoing electron current decreases. Again, the effect of this feedback mechanism is to equalize the current into and out of a section of wire.

Thus, whichever way there is an imbalance between incoming and outgoing current, the two currents will quickly become equal to each other due to the automatic feedback mechanism of changes in the surface charge. Earlier you saw that in the steady state the current everywhere in a series circuit is the same. Now you can see how the feedback mechanism creates and maintains this steady state.

Feedback Makes Current Follow the Wire

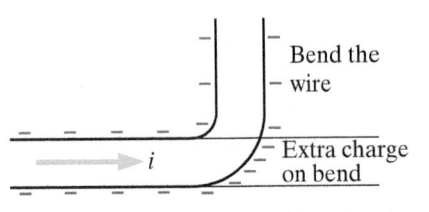

Figure 18.46 What happens if you bend a straight wire while current is running?

It is instructive to see what happens if you bend a straight wire while the current is running (Figure 18.46). The oncoming electrons don't immediately know what has happened and continue in their original direction, but as they run into the bend, electrons pile up on the outside of the bend, which tends to turn the oncoming electrons.

> QUESTION What determines how much charge piles up? (That is, what makes the pileup process come to a stop?)

The electrons pile up on the bend until there are enough there to repel oncoming electrons just enough to make them turn the corner (follow the wire) without running into the side of the wire. A calculation by Rosser indicates that a single extra electron on the bend is sufficient to turn a sizable current! (The required inward force is equal to mv^2/r, and the electron mass m and drift speed v are very small.) Not only does the feedback mechanism equalize incoming and outgoing current for any section of the wire, it also forces the electron current to follow the wire, no matter how the wire is bent or twisted.

For details of the calculation about the amount of charge needed to turn the current, see W. G. V. Rosser, "Magnitudes of surface charge distributions associated with electric current flow," *American Journal of Physics* **38** (1970), 265–266.

Summary of Feedback

We have learned that there must be some excess charge on the outside of the wire, all along the wire. Through the feedback mechanism, this surface charge will automatically arrange itself (or rearrange itself) in such a way as to ensure that the net electric field everywhere inside the wire points along the wire and has the appropriate magnitude to drive the appropriate amount of steady-state current. (The amount of this current is ultimately determined by the strength of the motor in our mechanical battery, and by the electron mobility u of the material that the wire is made of.)

This is very similar to the feedback we have seen in static electricity. If an external field is applied to an isolated metal block, charges automatically arrange themselves on the surface of the metal block in such a way as to make the electric field inside the metal block have a particular form: the electric field must be zero throughout the inside of the block in equilibrium.

To summarize:

> Feedback in a circuit leads to surface charges and steady-state current: $\vec{E}_{net} \neq \vec{0}$ inside a metal.

> Feedback in static electricity situations leads to equilibrium: $\vec{E}_{net} = \vec{0}$ inside a metal.

18.8 SURFACE CHARGE AND RESISTORS

We have used surface charges and electric fields to analyze a circuit made up of just a battery and a high-resistance Nichrome wire. In this section we will use the surface-charge model to analyze circuits involving "resistors"—sections of a circuit that resist the passage of electrons more than other parts of the circuit.

Narrow Resistors and Thick Wires

We will analyze a circuit similar to a circuit you assembled, consisting of a battery, copper wires, and a bulb. The very thin tungsten filament of the light bulb is an example of a "resistor"—a component that has a much higher

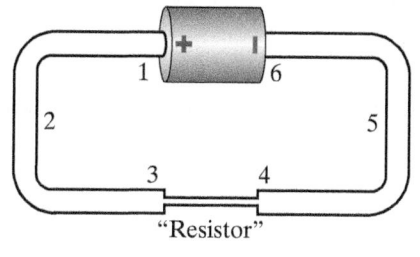

Figure 18.47 A simple circuit with a "resistor"—a thin section of the Nichrome wire.

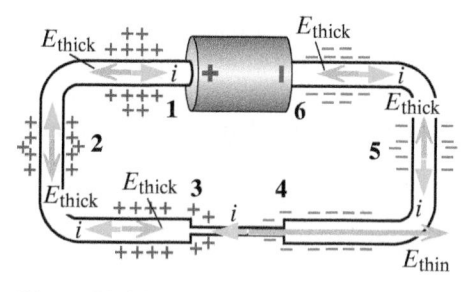

Figure 18.48 An approximate surface-charge distribution that is consistent with what we know about the electric field.

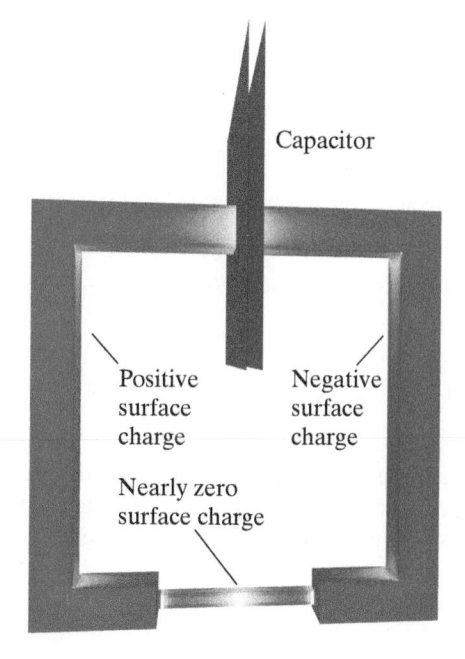

Figure 18.49 A detailed representation of the surface charge on a circuit with a thin resistor.

resistance to current flow than do the thick wires connecting the resistor to the battery.

For a simple example of a circuit containing a resistor, we'll again consider a battery and a high-resistance Nichrome wire, but we'll make part of the Nichrome wire very much thinner than the rest of the wire (Figure 18.47). We'll call the narrow section a "resistor," and there are two thick wires connecting the resistor to the battery.

Just after connecting the circuit but before the establishment of the steady-state surface charge, the electric field in the narrow resistor might temporarily be about the same as the electric field in the neighboring thick wires, leading to a comparable drift speed v in both sections. However, the number of electrons per second trying to enter the resistor is the large number $nA_{thick}\overline{v}$, whereas the number of electrons per second passing through the resistor is the small number $nA_{thin}\overline{v}$, so electrons pile up at location 3, the entrance to the resistor. (Remember that the mobile-electron density n is the same in the wires and the resistor, because they are both made of Nichrome in our example.)

Similarly, a deficiency of electrons, or + charge, accumulates at location 4, the exit from the resistor, because the number of electrons per second emerging from the resistor is small compared to the electron current in the thick wire that is removing electrons from the region.

> QUESTION Later, in the steady state, how does the electron current in the thick wires compare with the electron current in the resistor? Why?

Once the steady state has been established, the currents must have become the same. Reason by contradiction: if the currents weren't the same in the steady state, surface charge would be building up on the resistor, which would mean we hadn't reached the steady state yet. Charge piles up until the resulting increased electric field in the resistor gives a high enough drift speed to make an electron current that equals the electron current in the thick wires. Once these two currents are the same, there will be no more pileup.

> QUESTION In the steady state, how does the electric field E_{thin} in the resistor compare with the electric field in the thick wires? Why? On a diagram like Figure 18.47, draw the steady-state electric field in the resistor and in the neighboring thick wires, paying attention to relative magnitudes.

Since the steady-state current is the same in the thick and thin sections, we have the relation $nA_{thick}uE_{thick} = nA_{thin}uE_{thin}$. Therefore the electric field in the resistor must be quite a lot larger than the electric field in the thick sections. Quantitatively, $E_{thin} = (A_{thick}/A_{thin})E_{thick}$, since the mobile-electron density n and the mobility u are the same in the thick and thin wires (both made of Nichrome). The electric field has a uniform large magnitude throughout the resistor and a uniform small magnitude throughout the thick wires.

> QUESTION Now draw the approximate surface charge in the steady state. Be sure to make the distribution of surface charge consistent with the distribution of electric field that you drew already.

In the steady state the surface charge varies only a little along the thick wires, producing a small electric field E_{thick} and a small drift speed. Along the resistor there is a big variation from + at one end to − at the other, which makes a large field E_{thin} inside the thin resistor (Figure 18.48). This makes a large drift speed in the resistor (with the same electron current i as in the thick wires). We have an approximate charge distribution that is consistent with the distribution of electric field. Figure 18.49 again shows a detailed representation of the surface charge that you saw earlier, with a charged capacitor playing the role of the battery.

This resistor circuit offers another example of feedback in an electric circuit—an automatic adjustment leads to producing the electric fields that are needed in the steady state. Note that if the drift speed in the thick wires were momentarily to increase for any reason, extra charge would pile up at the resistor, which would tend to decrease the drift speed in the thick wires.

On the other hand, if the drift speed in the thick wires were to momentarily decrease, charge would drain through the resistor in such a way as to decrease the pileup of charge at the ends of the resistor, which would tend to increase the drift speed in the thick wires.

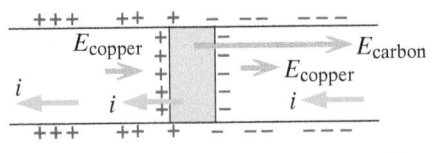

Figure 18.50 Make graphs of magnitude of electric field and of drift speed at locations around the circuit.

> **Checkpoint 4** Sketch a graph (Figure 18.50) of the magnitude of the electric field inside the wire in Figure 18.48 as a function of location around the circuit. In a different color, graph the drift speed of the electrons.

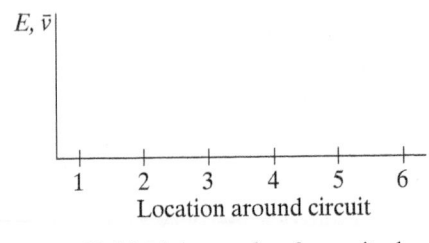

Figure 18.51 A portion of a circuit with a wide resistor.

A Good Analysis Strategy

The analysis in the preceding section is an example of a general strategy for understanding the relationship of charge and field in a circuit. Here is a summary of this strategy:

- Based on $i = nAuE$, draw the distribution of electric field in the wires.
- Next, draw an approximate charge distribution that is consistent with the distribution of electric field.

The reason for carrying out the analysis in this order is that the distribution of electric field is fairly simple and can be analyzed nearly exactly, whereas the details of the pattern of surface charge can be quite complicated.

What we know is that feedback will produce an arbitrarily complicated distribution of surface charge appropriate to the steady-state distribution of electric field. The pattern of electric field gives us quantitative information about the electron current; the pattern of surface charge gives us a clear sense of the mechanism that produces the pattern of electric field.

A Wide Resistor: Charges on the Interface

Another kind of resistor is wide rather than narrow but has low mobility. Consider copper wires leading up to a short piece of carbon having the same, wide cross section (Figure 18.51; we don't show the entire circuit). Carbon has a much smaller electron mobility u than copper, so the steady-state electric field has to be much larger in the carbon than in the copper in order to have the same current in both materials.

In the transient leading up to the steady state, electrons pile up not only on the outer surfaces of the copper and carbon, but also on the interfaces between the copper and the carbon, and these interface charges make a major contribution to the large electric field inside the carbon resistor.

Our assertion that "excess charge is always on the surface of a conductor" extends to this case, where the surface may be in contact with a different conducting material.

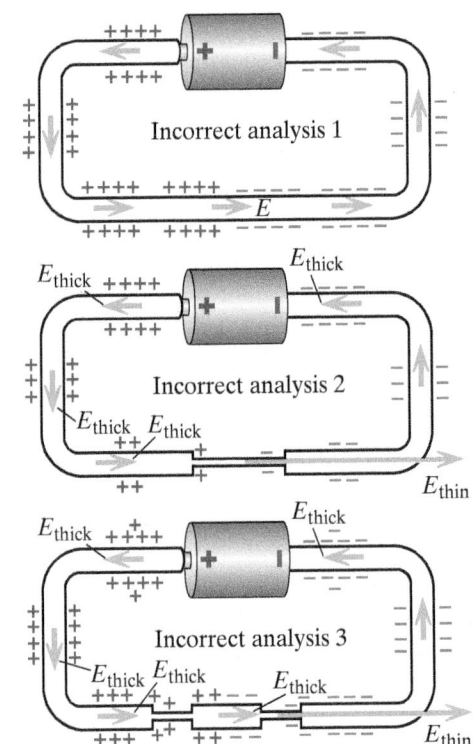

Figure 18.52 Circuits incorrectly analyzed: distribution of surface charge inconsistent with the pattern of electric field.

> **Checkpoint 5** The circuits in Figure 18.52 have been analyzed incorrectly; the distribution of surface charge is inconsistent with the distribution of electric field. Briefly identify what is inconsistent in each case.

18.9 ENERGY IN A CIRCUIT

Charge conservation plus the definition of a steady state led to the current node rule, which tells us that the current in a simple battery-and-bulb circuit must be the same everywhere in the circuit. But how many amperes of current flow? What determines the magnitude of the current? The Energy Principle provides the answer, plus the properties of batteries.

Consider a circuit containing various circuit elements such as resistors and batteries. We could discuss energy conservation in the circuit by considering one electron as it makes a complete trip around the circuit. The energy gained by the electron as it is transported across the mechanical battery is dissipated in collisions with atomic cores as it travels through the wire. Alternatively, we can analyze the energy per unit charge gained and lost in a trip around the circuit. We know that over any path the round-trip path integral of the electric field must be zero, as long as the current is steady or nearly steady. (This is not true in alternating current or AC circuits where there are time-varying currents, with time-varying magnetic fields, as we will see in Chapter 22 on Faraday's law.) If we follow a round-trip path through a circuit, through circuit element 1, circuit element 2, and so on, we find the following:

ENERGY CONSERVATION (THE LOOP RULE)

$$\Delta V_1 + \Delta V_2 + \cdots = 0 \quad \text{along any closed path in a circuit}$$

This is valid for circuits with steady or slowly time-varying currents.

This is essentially the Energy Principle, but on a per-coulomb basis. Remember that electric potential difference is defined as potential energy difference per unit charge: $\Delta V = \Delta U / q$. The loop rule says that you can't gain energy in a round trip. In circuits this energy principle is often called the "Kirchhoff loop rule," but its validity is not confined to circuits.

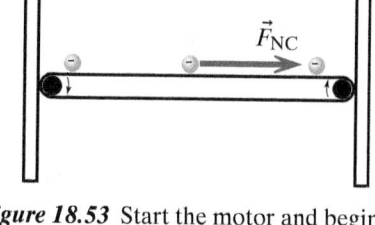

Figure 18.53 Start the motor and begin moving electrons from left to right.

Potential Difference Across a Battery

Along a wire or resistor of length L in which there is a uniform electric field of magnitude E, the magnitude of the potential difference is just EL. What determines how much potential difference the motor of a mechanical battery can maintain across the battery? Suppose you start with uncharged plates (Figure 18.53) and turn on the motor (with no wire connected to the plates), extracting electrons from the left-hand plate and transporting them to the right-hand plate. The conveyor belt, driven by the motor, exerts what we will call a "non-Coulomb" force $\vec{F}_{NC}$ on each electron. This being the only force, the belt accelerates electrons.

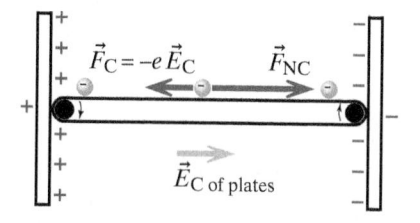

Figure 18.54 Charges build up on the plates and exert a "Coulomb" force on the electrons being transported.

As the motor continues to transport electrons to the right, charge builds up on the plates. The charges on the plates exert a "Coulomb" force $F_C = eE_C$ on the electrons being transported, in the opposite direction to F_{NC}, the non-Coulomb force that the conveyor belt exerts (Figure 18.54). There is less acceleration of the electrons.

By "Coulomb force" and "Coulomb field" we mean the force and field due to point charges, as given by Coulomb's law. We use the term "non-Coulomb" to refer to other kinds of interactions. The Coulomb and non-Coulomb forces oppose each other inside the battery.

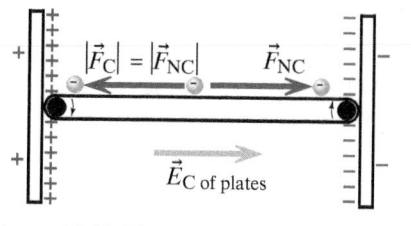

Figure 18.55 The Coulomb force grows to be as big as the non-Coulomb force.

Eventually there is enough charge on the plates to make $F_C = eE_C = F_{NC}$. At that point the motor cannot pump any more charge, and the plates are charged up as much as they can be (Figure 18.55). We see that the function of a mechanical battery (or of a chemical battery) is to produce and maintain a potential difference by pulling electrons out of the positive plate and pushing them onto the negative plate. The amount of potential difference is limited by and determined by the strength of the motor in a mechanical battery (or by the nature of the chemical reactions in a chemical battery).

If the distance between the (large) plates of a mechanical battery is s, and the electric field E_C of the charged plates is uniform between the plates, then the potential difference across the battery is

$$|\Delta V_{\text{batt}}| = E_C s = \frac{F_{\text{NC}} s}{e}$$

The quantity $F_{\text{NC}} s / e$, the energy input per unit charge, is a property of the battery and is called the "emf" of the battery (pronounced "e" "m" "f"). Historically, emf was an abbreviation for "electromotive force," which is a bad name, since emf is not a force at all but energy per unit charge. We will avoid this terminology by just using the abbreviation emf. In a chemical battery, the emf is a measure of the chemical energy per unit charge expended by the battery in moving charge through the battery. The emf of a flashlight battery is about 1.5 V, and the associated charge separation is due to chemical reactions in which electrons are reactants or products.

The emf of a chemical battery turns out to be nearly constant, independent of how much current is running through the battery (as long as the current is not too large), and approximately independent of how long the battery has been used (though it does decline with time). It is the approximate constancy of the emf that makes this a useful concept, and that makes it useful to analyze circuits in terms of emf and the potential differences in the circuit elements attached to the battery.

It is important to keep in mind that although the units of emf are volts, the emf is not a potential difference. Potential difference is a path integral of the electric field made by real charges. The emf is the energy input per unit charge and might in principle be gravitational or nuclear in nature. For example, our mechanical battery could be run by a falling weight.

ROLE OF A BATTERY

A battery maintains a potential difference across the battery. This potential difference is numerically equal to the battery's emf for batteries supplying a small amount of current. Later we will discuss how the potential difference decreases if the current is large.

Checkpoint 6 If the plates of a mechanical battery are very large compared to the distance s between the plates, and the area of a plate is A, show that the amount of charge Q on one of the plates of an isolated battery is determined by the battery's emf: $Q = A\varepsilon_0(\text{emf})/s$. For an ordinary flashlight battery, emf $= 1.5$ V. Make a rough estimate of the amount of charge in coulombs on the end plate of a battery, whose length $s = 5$ cm and radius $r = 1$ cm. (This is a very rough estimate because 5 cm isn't small compared to 1 cm.) How does this compare to the charge on a U tape? Should you expect to be able to repel a U tape with one end of a battery?

Internal Resistance

If we now connect a Nichrome wire from one plate of the mechanical battery to the other (Figure 18.56), we can achieve a steady state if the motor transports as many electrons per second from the + plate to the − plate as are flowing through the wire. The larger the current, the more electrons per second the motor must transport to keep the plates resupplied. If there is no resistance to the flow of charge inside the battery, we will have $F_C = eE_C = F_{\text{NC}}$.

In any real battery there is some resistance to the flow of charge through the battery, and this resistance is called the "internal resistance" of the battery.

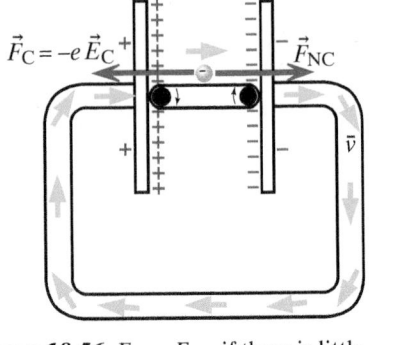

Figure 18.56 $F_C = F_{\text{NC}}$ if there is little internal resistance.

If the mobility of an ion of charge e through the battery is u, the drift velocity through the battery will be proportional to the net force per unit charge acting on the ion: $\bar{v} = u\left(F_{NC}/e - E_C\right)$. Since the motor force F_{NC} is fixed, the maximum drift speed is obtained when $E_C = 0$, which means that there is no charge on the ends of the battery: the motor isn't keeping up, and $|\Delta V_{batt}| = 0$. Much of the time we will make the simplifying assumption that the mobility u of ions inside the battery is very high (low internal resistance), so that we get a sizable drift speed even if F_C ($= eE_C$) is nearly as large as F_{NC}. Correspondingly, we assume that $|\Delta V_{batt}| \approx$ emf.

In Chapter 19 we will show how to model a battery when the internal resistance is not negligible. For now we will consider "ideal" batteries for which the internal resistance is negligible, and $|\Delta V_{batt}| =$ emf.

Field and Current in a Simple Circuit

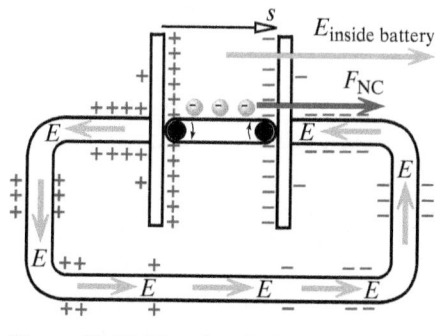

In Figure 18.57, note that the electric field inside the mechanical battery points from the $+$ end toward the $-$ end, in the opposite direction to the electric field in the neighboring wires. The $+$ end of a battery is at a higher potential than the $-$ end.

If we start at the negative plate of the battery and go counterclockwise in Figure 18.57, there is a potential increase of $+$ emf across the battery and a potential drop of $-EL$ along the wire of length L. The round-trip potential difference is zero:

$$\Delta V_{batt} + \Delta V_{wire} = 0$$
$$\text{emf} + (-EL) = 0$$
$$E = \frac{\text{emf}}{L}$$

We used the energy principle (round-trip potential difference is zero) to determine the magnitude of the electric field in the wire. If we know the properties of the wire (cross-sectional area A, electron density n, and electron mobility u), we can now determine the magnitude of the conventional current $I = enAuE$.

Figure 18.57 The electric field inside the mechanical battery is opposite to the electric field in the neighboring wires.

A Parallel Circuit: Two Different Paths

Figure 18.58 shows a parallel circuit with more than one path for current to follow. Start at the negative end of the battery and follow the dashed path clockwise through wires L_2 and L_3, and then through the battery to the starting location. There is a potential rise $+E_2 L_2$ along the part of the path of length L_2, a potential rise $+E_3 L_3$ along the rest of the path of length L_3, then a potential drop through the battery of amount $-$emf, yielding $E_2 L_2 + E_3 L_3 + (-\text{emf}) = 0$ for the round-trip potential difference. If we instead follow the path through L_1 and L_3, we find a round-trip potential difference $E_1 L_1 + E_3 L_3 + (-\text{emf}) = 0$.

These two equations imply (correctly) that $E_1 L_1 = E_2 L_2$, which makes sense: because the two wires have the same starting and ending points, they have the same potential differences. Also notice in Figure 18.58 that $i_3 = i_1 + i_2$, due to the current node rule (charge conservation in the steady state).

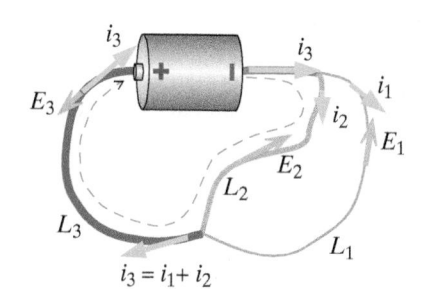

Figure 18.58 Follow the dashed path through wires L_2 and L_3.

Potential Difference Along Connecting Wires

In your own experiments you have probably noticed that in a circuit containing a light bulb or a Nichrome wire, the number or length of the connecting wires used has a negligible effect on the amount of current in the circuit. In a series circuit with thick copper wires and a light bulb, the electric field is large in the bulb and quite small in the thick copper wires, because the steady-state electron current $nA\bar{v} = nAuE$ must be the same in both. The electron mobility in cool copper is much higher than the electron mobility in hot tungsten, and typically the copper wires are much thicker than the tungsten bulb filament, so that not only is it the case that $E_{wires} \ll E_{bulb}$, but even for rather long wires $E_{wires}L_{wires} \ll E_{bulb}L_{bulb}$. Therefore, in the energy equation (emf $-E_{bulb}L_{bulb} - E_{wires}L_{wires} = 0$), the term $E_{wires}L_{wires}$ may be negligible; the work done by the battery goes mostly into energy dissipation in the bulb, with a small amount of energy dissipation in the thick copper wires.

Checkpoint 7 Explain briefly why the brightness of a bulb doesn't change noticeably when you use longer copper wires to connect it to the battery.

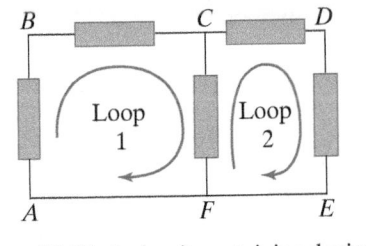

Figure 18.59 A circuit containing devices connected by thick copper wires.

General Use of the Loop Rule

The loop rule can be applied along any path through an arbitrarily complicated circuit. Consider a circuit made up of five circuit elements connected together by thick copper wires that have high electron mobility (Figure 18.59). The circuit elements could be resistors, batteries, capacitors, or even more complex devices such as semiconductor diodes or even refrigerators!

In Figure 18.60 we show the left portion of the complete circuit (loop 1). Start at location A and walk along the path labeled $ABCFA$. From A to B the potential difference is $\Delta V_1 = V_B - V_A$; from B to C the potential difference is $\Delta V_2 = V_C - V_B$; and from C to F the potential difference is $\Delta V_3 = V_F - V_C$. From F to A there is negligible potential difference $\Delta V_4 = V_A - V_F \approx 0$, because the thick wire has high electron mobility and large cross-sectional area, and so requires a negligibly small electric field to drive the current.

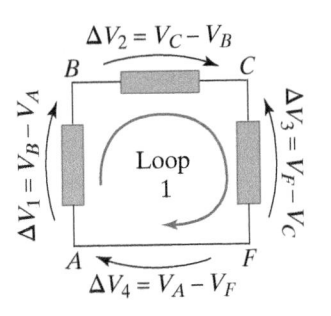

Figure 18.60 Walk around loop 1, following the path $ABCFA$.

Any round-trip potential difference must be zero, and the loop rule gives this:

$$\Delta V_1 + \Delta V_2 + \Delta V_3 + \Delta V_4 = 0$$
$$(V_B - V_A) + (V_C - V_B) + (V_F - V_C) + (V_A - V_F) = 0$$

We could write a similar equation corresponding to walking along loop 2 (path $FCDEF$). The idea is that you start anywhere you like on the circuit diagram and walk around a loop that brings you back to your starting location, adding up potential drops and rises as you go.

You can even walk along a round-trip path through the air, outside the circuit elements, but unless you know something about the electric field and potential differences outside the devices and connecting wires, this usually isn't very useful.

Checkpoint 8 In Figure 18.59, suppose that $V_C - V_F = 5$ V, and $V_D - V_E = 3.5$ V. What is the potential difference $V_C - V_D$? If the element between C and D is a battery, is the $+$ end of the battery at C or at D?

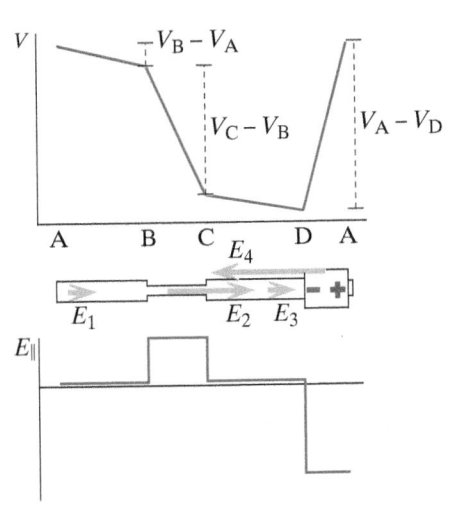

Figure 18.61 Potential rises and falls as you walk around a circuit. The electric field is the gradient of the potential. The connection from the $+$ end of the battery to the left wire is not shown.

Potential Around a Circuit

At the top of Figure 18.61 is a graph of potential around the circuit shown in the middle, along the path ABCDA. Below the circuit is a graph of $E_{\parallel}$, the component of the electric field in the direction that we walk, counterclockwise around the circuit. Inside the battery we are walking opposite to the direction of the electric field, so there is a potential rise equal in magnitude to the emf of the battery.

A big potential gradient (volts per meter) means a large field, so a steep slope on the potential graph corresponds to a large electric field (see bottom of Figure 18.61). Along the thick wires, the potential gradient is small and the field is small ($E = |\Delta V|/L$), yielding a small drift speed (wire AB, for instance).

Along the thin wire (BC) there is a large potential gradient corresponding to a large field and a high drift speed.

For this circuit with its two thick wires and thin resistor connected to a battery, all the potential drops (negative potential differences) plus the potential rise across the battery (numerically equal to the emf) add up to 0:

$$\Delta V_1 + \Delta V_2 + \Delta V_3 + \Delta V_{battery} = 0$$
$$(-E_1 L_1) + (-E_2 L_2) + (-E_3 L_3) + emf = 0$$

Checkpoint 9 What would be the potential difference ΔV_2 across the thin resistor in Figure 18.61 (between locations B and C) if the battery emf is 1.5 V? Do you have enough information to determine the current I in the circuit? Assume that the electric field in the thick wires is very small (so that the potential differences along the thick wires are negligible).

18.10 APPLICATIONS OF THE THEORY

We now have two principles on which to base the analysis of circuits. One has to do with charge conservation:

THE CURRENT NODE RULE

In the steady state, for the many electrons flowing into and out of a node,

electron current: net i_{in} = net i_{out}, where $i = nAuE$

conventional current: net I_{in} = net I_{out}, where $I = |q|nAuE$

The second principle has to do with energy:

ENERGY CONSERVATION (THE LOOP RULE)

In the steady state, for any round-trip path,

$$\Delta V_1 + \Delta V_2 + \cdots = 0$$

The current node rule refers to many electrons moving together, whereas the energy equation is on a per-unit-charge basis.

We will apply these two principles to predict the behavior of simple circuits. These predictions can be tested by measuring the currents in the circuits.

Application: Doubling the Length of a Wire

In each of the two circuits shown in Figure 18.62, a Nichrome wire (which has low electron mobility) is connected to a battery by thick copper connecting wires. The circuits differ only in the length of the Nichrome wire that forms part of the circuit. Suppose that you measure the compass deflection produced by the current in each circuit.

> QUESTION Based on our model for current in a circuit, if the current through a Nichrome wire of length L produces a compass deflection θ, what compass deflection would you expect from the current in the Nichrome wire of length $2L$?

Around the circuit loop emf $-EL = 0$, so if we double the length L, we'll get half as big an electric field (since the emf doesn't change). With half the field we get half the drift speed ($\bar{v} = uE$), half the electron current ($i = nAuE$), and half the conventional current ($I = |q|nAuE$). Therefore we expect half as large a magnetic field and half the compass deflection.

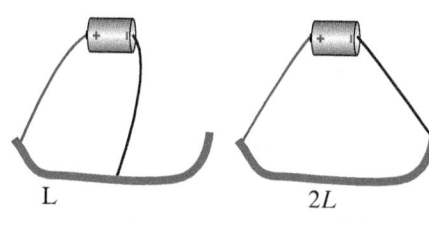

L   2L

Figure 18.62 How does the current change when we double the length of a Nichrome wire?

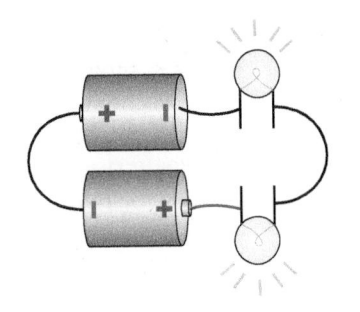

Figure 18.63 A circuit with two identical bulbs in series.

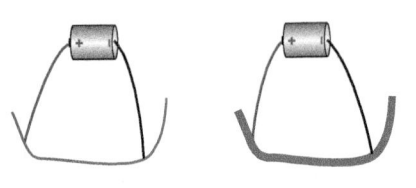

Figure 18.64 How does the current change when we double the cross-sectional area of a Nichrome wire?

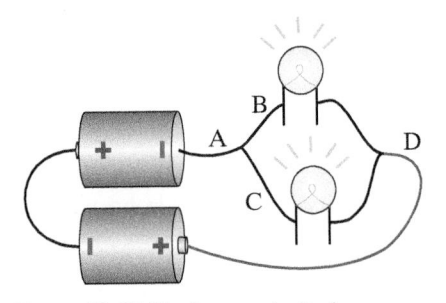

Figure 18.65 Placing two bulbs in parallel is analogous to doubling the cross-sectional area of a bulb filament.

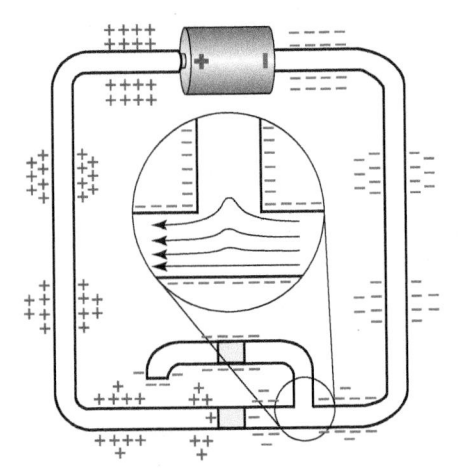

Figure 18.66 A circuit with a (dead-end) parallel branch.

Note that placing two identical bulbs in series is analogous to doubling the length of the bulb filament. However, if you compare a one-bulb circuit to a two-bulb series circuit (Figure 18.63), you will probably find more than half the current in the two-bulb circuit. The issue here is that the electron mobility in very hot tungsten (several thousand degrees!) is less than the electron mobility in cooler tungsten. The greater thermal agitation of the atomic cores leads to larger "friction." Note that the bulbs in the two-bulb circuit emit redder light than the single bulb, indicating that their filaments are not as hot.

Application: Doubling the Cross-Sectional Area of a Wire

Figure 18.64 shows two circuits in which a Nichrome wire is connected to a battery by thick copper connecting wires. The circuits differ only in the cross-sectional area of the Nichrome wire: the thicker wire has twice the cross-sectional area of the thinner wire.

QUESTION If the current in the circuit with the thin Nichrome wire is I, what compass deflection would you expect from the current in the circuit with the thick wire?

It seems plausible that doubling A should double the conventional current $I = |q|nAuE$. It is like having two wires connected to the battery, doubling the drain on the battery.

We can make this argument quantitative by considering energy. Since we have emf $- EL = 0$ around the loop, the electric field E in the wire does not change when we use a thicker wire, so the current in the second circuit should be $2I$. (This neglects internal resistance in the battery, which has the effect of limiting the amount of current that the battery can supply, so that for large cross-sectional areas and large currents, doubling a large cross-sectional area may not yield twice the current.)

Note that placing two bulbs in parallel, as shown in Figure 18.65, is analogous to doubling the cross-sectional area of a bulb filament.

Quantitative Measurements of Current with a Compass

QUESTION How should the current at location D in Figure 18.65 compare to the current in a circuit with a single bulb?

We should expect the current at D to be twice as large as the current in a circuit with only one bulb, for the reasons discussed above.

Experiments EXP1 through EXP5 are designed to let you explore the consequences of the node rule (charge conservation in the steady state) and the loop rule (energy conservation on a per-coulomb basis).

How Does Current Know How to Divide between Parallel Resistors?

When there are parallel branches in a circuit, as in Experiment EXP5, there is a question of how the current "knows" how to divide between the branches.

In Figure 18.66 we show very approximately the steady state of a circuit containing wide carbon resistors in two parallel branches. During the initial transient just after connecting the battery, electrons flowed into both branches, but the surface of the dead-end branch became charged so negatively that no more electrons could enter. In the inset, note the deflection of the electron stream lines due to the lack of charge immediately above the branch point, inside the dead-end branch. The wire is effectively a bit wider as a result of adding the dead-end branch.

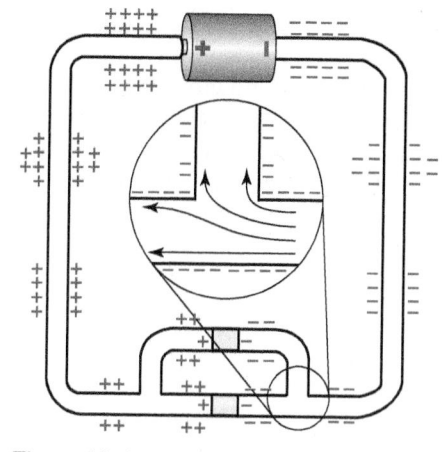

Figure 18.67 Complete the connection: the electron current splits.

Next we complete the parallel connection, which leads to a rearrangement of surface and interface charges (Figure 18.67). The inset shows how the new arrangement of surface charges guides and splits the oncoming electron current, with some of the electrons taking the upper branch and some the lower branch. Note that there is now a larger current through the battery and a larger gradient of surface charge along the wires to drive this larger current.

The steady-state current in each branch is determined by how well each branch conducts current. For example, if the mobility is low in one of the branches, surface charge will build up in such a way as to decrease the fraction of the current that goes down that branch. The extreme case is a dead end, where surface charge builds up until it prevents any current at all from going down the dead-end branch.

Application: Thin Filament Bulb and Thick Filament Bulb in Series

In the Experiments section of the previous chapter we described simple equipment that can be used to explore electric and magnetic phenomena. This equipment includes a thin-filament bulb (a #48 flashlight bulb) that has an elongated shape and a thick-filament bulb (a #14 flashlight bulb) that has a round shape. If you put a thin-filament bulb and a thick-filament bulb in series with each other (Figure 18.68), the thin-filament bulb lights up but the thick-filament bulb doesn't light up! What happened to the current that ran through the thin-filament bulb? We can explain this puzzling behavior. Consider that:

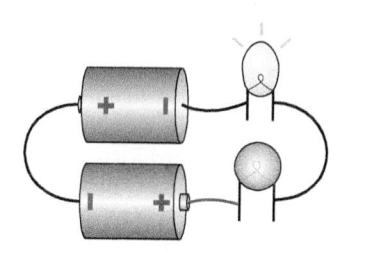

Figure 18.68 A thin-filament bulb and a thick-filament bulb in series—why doesn't the thick-filament bulb light up?

Both bulb filaments are made of tungsten.

The round, thick-filament bulb has large cross-sectional area A_r, length L_r, electric field E_r, and drift speed $\bar{v}_r$.

The long, thin-filament bulb has small cross-sectional area A_l, length L_l, electric field E_l, and drift speed $\bar{v}_l$.

Assume for the moment that the electron mobility is the same in both bulbs, even though in fact the glowing thin-filament bulb is very hot (which should mean low mobility) and the thick-filament bulb is apparently cool (since it isn't glowing, which should mean high mobility).

> QUESTION Using the current node rule, write an equation relating the drift speeds $\bar{v}_r$ and $\bar{v}_l$.

In the steady state, the number of electrons per second passing through each bulb must be the same:

$$nA_r\bar{v}_r = nA_l\bar{v}_l$$

> QUESTION Based on the equation you just wrote, write an equation relating the electric fields E_r and E_l, and state which is larger, E_r or E_l.

Assuming the same mobility in both filaments, since

$$\bar{v} = uE$$

we know that $nA_ruE_r = nA_luE_l$, so

$$E_l = \frac{A_rE_r}{A_l}$$

Since $A_r > A_l$, $E_l > E_r$.

> QUESTION Applying the loop rule, we find that $2\text{emf} = E_rL_r + E_lL_l$, which is approximately equal just to E_lL_l, since E_r is very small. Explain why the thick-filament bulb doesn't glow.

If the thick-filament bulb were alone in a circuit, it would be bright, and the electric field in the filament would be:

$$2\text{emf} = E_{r,\text{bright}}\, L$$

so

$$E_{r,\text{bright}} = \frac{2\text{emf}}{L}$$

In the two-bulb circuit (if the lengths of the filaments are the same) we have:

$$2\text{emf} = E_r L + E_l L = L\left(E_r + \frac{A_r E_r}{A_l}\right) \quad \text{(from previous exercise)}$$

$$= L E_r (1 + A_r/A_l)$$

so

$$E_r = \frac{2\text{emf}}{L(1 + A_r/A_l)} < \frac{2\text{emf}}{L}$$

Thus the electric field in the thick-filament bulb is significantly smaller than it is in the single-bulb circuit. We can estimate how much smaller by noting that one typically measures about 1/4 as much current in a circuit with a single thin-filament bulb as in a circuit with a single thick-filament bulb. This suggests that $A_r \approx 4 A_l$, so the electric field in the thick-filament bulb has been reduced by a factor of 5. Evidently this field does not produce sufficient current through the filament to make it glow.

Moreover, since the thick-filament bulb isn't glowing, its temperature is low and its mobility is much larger than usual. This further reduces the E_r required to drive the same current that runs through the thin-filament bulb.

What Determines Bulb Brightness?

For each electron that moves through a wire of length L due to an electric field E, the electric potential energy of the system (electron plus wire) decreases by an amount $e\Delta V = eEL$. This loss of electric potential energy is equal to the gain in kinetic energy of the electron, which is dissipated in collisions with the atoms in the wire, with consequent increase in the thermal energy and temperature of the wire. As the wire's temperature increases, the rate of energy transfer to the surroundings increases in the form of the emission of light. The temperature of the wire stops rising once the temperature is high enough that the rate of emitted energy (as light) has become equal to the rate of energy dissipation in the wire.

In a time Δt the number of electrons entering (and leaving) the wire is $nA\bar{v}\,\Delta t = nAuE\Delta t$, so the energy dissipation per second (power in joules per second, or watts) is $(eEL)(nAuE) = enAuLE^2$. We see that the rate of energy dissipation is proportional to the square of E, or alternatively to the square of the current $I = enAuE$. In the previous example the field E in the thick-filament bulb was reduced by a factor of 5, so the rate of energy dissipation in the wire was reduced by a factor of 25. This in turn means that the rate of energy emitted as light was reduced by a factor of 25. The bulb is so dim that you don't see it glow.

Application: Two Batteries in Series

It should now be clear why two batteries in series can drive more current through a wire than one battery can: the potential difference across two batteries in series is 2 emf, and this doubles the electric field everywhere in the circuit. Double the electric field, double the drift speed, double the current (if the mobility doesn't decrease due to a higher temperature).

In Experiment EXP6 you can test this prediction with a Nichrome wire rather than a light bulb. When you add a second battery, a bulb filament undergoes a sizable temperature increase that changes the mobility significantly (you can see the color become whiter, which is an indication of the higher filament temperature). With a length of Nichrome wire the temperature stays close to room temperature, so the mobility is nearly the same with one or two batteries.

EXAMPLE

A Circuit with a Wide Wire

The circuit in Figure 18.69 contains a low-resistance mechanical battery that exerts a non-Coulomb force F_{NC} to move electrons through the battery (with negligible internal resistance). The end plates of the battery are very large compared to the distance $0.1L$ between the plates (plates not drawn to scale). Two thin Nichrome wires of length L and cross-sectional area A connect the battery to a thick Nichrome wire of length $0.6L$ and cross-sectional area $4A$. The electron mobility of the Nichrome is u, and there are n mobile electrons per cubic meter in the Nichrome.

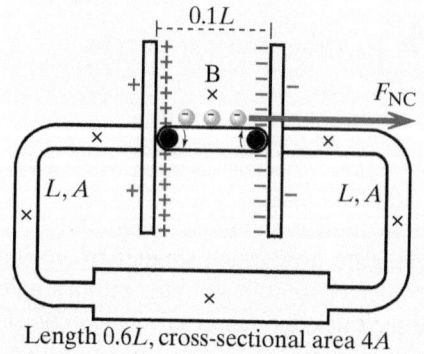

Length $0.6L$, cross-sectional area $4A$

Figure 18.69 Circuit with a wide wire.

(a) Show the electric field at the six locations marked with × (including location B between the plates). Pay attention to the relative magnitudes of the six vectors that you draw.

(b) Show the approximate distribution of charge on the surface of the Nichrome wires. Make sure that your distribution is compatible with the electric fields that you drew in part (a).

(c) Calculate the number of electrons that leave the battery every second, in terms of the given quantities L, A, n, u, and F_{NC} (and fundamental constants). Be sure to show all of your work.

(d) Calculate the magnitude of the electric field between the plates of the battery (location B).

Solution

(a) By the current node rule, current in the thin and thick wires will be equal. Since $i = nAuE$, E will be larger in wires with thinner cross-sectional area. This suggests the pattern of electric field shown in the wires in Figure 18.70.

In order for the round-trip potential difference to be zero, the electric field between the plates must point to the right and be large compared to the electric field in the wires, as shown in Figure 18.70.

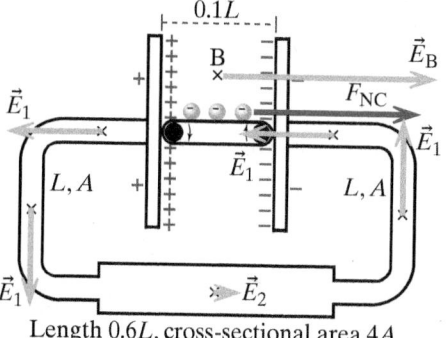

Length $0.6L$, cross-sectional area $4A$

Figure 18.70 Part (a), electric field at locations in the wire.

(b) A large gradient of surface charge is associated with a large electric field; a small gradient with a small field, as shown in Figure 18.71.

(c) Apply basic principles:

Current node rule: current in thin wire (1) = current in thick wire (2) gives

$$nAuE_1 = n(4A)uE_2$$

$\Delta V_{\text{round trip}} = 0$ (energy conservation), going counterclockwise from the negative plate of the mechanical battery:

$$\frac{F_{NC}(0.1L)}{e} - E_1L - E_2(0.6L) - E_1L = 0$$

From the current equation:

$$E_2 = \left(\frac{A}{4A}\right)E_1 = \frac{E_1}{4}$$

We now have two equations in two unknowns (E_1 and E_2). Substitute for E_2:

$$\frac{F_{NC}(0.1L)}{e} - 2(E_1L) - \left(\frac{E_1}{4}\right)0.6L = 0$$

$$E_1 = \frac{F_{NC}}{e(21.5)}$$

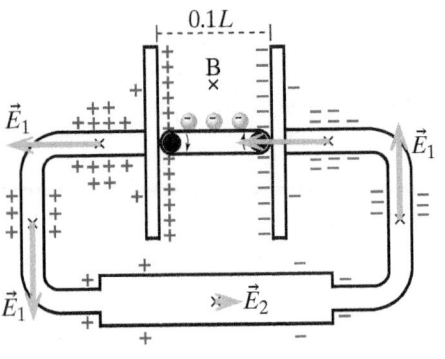

Figure 18.71 Part (b), an approximate surface-charge distribution.

Now we can solve for electron current (number of electrons/s leaving the battery):

$$i = nAuE_1 = nAu\left(\frac{F_{NC}}{e(21.5)}\right)$$

(d) The round-trip potential difference must be zero, so:

$$E_B(0.1L) - 2E_1L - \left(\frac{E_1}{4}\right)(0.6L) = 0$$

$$E_B = 21.5\left(\frac{F_{NC}}{e(21.5)}\right) = 21.5\,E_1$$

The electric field E_B inside the battery is 21.5 times as large as the electric field E_1.

18.11 DETECTING SURFACE CHARGE

The reason we can't detect the surface charge on the wires of the simple circuits we have been assembling is that the amount of charge on the wires is very small—very little charge is needed to make the small electric fields in the wires and bulb filaments of your circuits. However, it is possible to detect surface charge in a high-voltage circuit (around 10,000 V is required). The following interesting demonstration of surface charge was suggested by the science education group at the Weizmann Institute in Israel. Perhaps your instructor will be able to show this to you.

Warning: Working with high voltages is dangerous, so don't try this without an instructor present. Keep one hand behind your back when you are near a high-voltage circuit. Do not touch the bare wires!

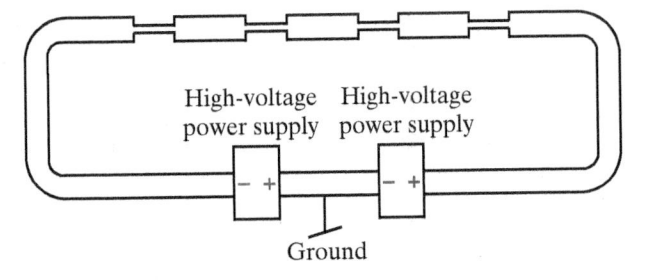

Figure 18.72 A high-voltage (low-current) circuit for observing surface charge.

In Figure 18.72 a chain of four identical high-resistance resistors is supported in air from Styrofoam columns, so that the resistor chain is far from other objects, including the table, to avoid unwanted polarization effects. The four resistors are represented in Figure 18.72 as thin wires. The resistors are connected by bare (uninsulated) thick wires to two high-voltage power supplies, which maintain one end of the resistor chain at a positive potential and the other end at a negative potential with respect to ground (the round hole in a three-prong socket). Note that you need to avoid touching the wires when the circuit is connected!

We model the resistors as very thin wires. The conventional current everywhere in this series circuit is $I = enA\bar{v} = enAuE$.

> QUESTION Given the expression for the current, sketch the pattern of electric field around the circuit.

Evidently the electric field must be very much larger inside the thin resistors than inside the thick connecting wires, and the electric field must have the same magnitude inside each resistor. As usual, the pattern of electric field is simple.

The feedback mechanism will lead to an arbitrarily complicated distribution of surface charge that produces this simple pattern of field.

QUESTION Sketch a plausible approximate distribution of surface charge, consistent with the pattern of electric field.

Very roughly, the electric field inside the leftmost resistor (AB) is in large part due to a large negative surface-charge density at A and a medium surface-charge density at B, with a large gradient along the resistor (Figure 18.73). This is an approximate analysis, because the field in the resistor is due not just to the nearby charges but to all charges. The electric field must have the same magnitude in the next resistor (BC) and is largely due to the medium surface-charge density at B and the nearly zero surface-charge density at C. Similar remarks apply to the other two resistors.

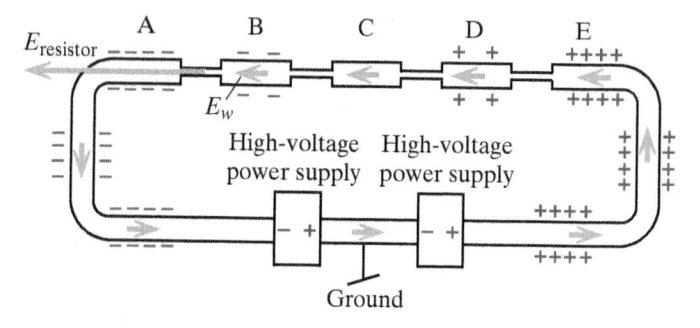

Figure 18.73 Electric field and surface charge for a high-voltage circuit.

Inside the thick connecting wires the electric field is very small, so we indicate a roughly uniform charge distribution along these wires (the details of the actual charge distribution, except for the sign of the charge, are presumably more complicated).

A neutral strip of aluminum foil is suspended by a hair from an insulating rod and brought close to the bare wire near location A.

QUESTION What do you expect will happen?

The strip is observed to be attracted to the wire (due to polarization of the strip by the surface charge on the wire) and then to jump away, due to being charged by contact with the bare wire and repelled by the surface charge at that location. The strip is found to be negatively charged, because it is repelled by a plastic pen rubbed through one's hair (a clear plastic pen is known to charge negatively). A similar effect is observed at E, but the strip is observed to be positively charged. A smaller effect is observed at B and D, where the surface-charge density is expected to be smaller. No effect is observed at point C, where by symmetry there is essentially no surface charge. These experimental observations confirm the validity of the approximate surface charge analysis.

The surface-charge density is proportional to the circuit voltage, and only at very high voltages is there enough charge to observe electrostatic repulsion in a mechanical system. It is difficult to observe repulsion by surface charge in a low-voltage circuit, because a charged object is of course attracted to neutral matter, and the attraction between a charged object and a circuit can mask the repulsion, if the surface charge on the circuit is small as it is in low-voltage circuits.

The object must be very lightweight so that one can observe the interaction at a sizable distance (a small piece of aluminum foil hanging from a thread works well). The problem is that at short distances there is competition between repulsion and the attraction that is due to polarization of neutral matter. The attractive part of the interaction falls off with distance much more rapidly than

$1/r^2$. As we saw in Chapter 14, the attraction between a point charge and a small neutral object is proportional to $1/r^5$, since the polarization of the neutral object is proportional to $1/r^2$, and the interaction between a point charge and a dipole is proportional to $1/r^3$.

18.12 *COMPUTATIONAL MODEL OF A CIRCUIT

The computational model used to calculate surface charge distributions can be applied to the polarization of a neutral metal block, as well as to a circuit. Consider the case of a positive point charge polarizing a neutral metal block (Figure 18.74). Imagine that at first there are no charges near the block, and then we suddenly bring the positive point charge into position to the left of the block.

As we saw in Chapter 14, at first there is a nonzero net field to the right inside the block, due solely to the point charge. This field drives mobile electrons to the left, and in a *very* short time Δt some excess electrons will appear to the left on the surface of the block. There will be a corresponding deficiency of electrons to the right. The time to create some nonzero surface charge is very short because there are plenty of mobile electrons everywhere inside the metal; the mobile electron sea has to shift only an extremely short distance in order for significant nonzero charges to build up on the surface.

Figure 18.74 A 4 mm × 4 mm × 8 mm metal block polarized by a positive point charge 12 mm to the left of the block. Shades of blue represent negative charge and shades of red represent positive charge.

These new surface charges, negative on the left and positive on the right, contribute an electric field inside the block that points to the left, so that the net field inside is reduced. In the next very short time interval, the (reduced) electric field, still to the right, will drive more electrons onto the left portion of the surface, with increased electron deficiency on the right. These charges will reduce the net field even more. Eventually there will be so much surface charge, negative on the left and positive on the right, that the sum of the field contributed by the point charge and the field contributed by the surface charge add up to zero.

In our computational model of this process we divide the surface into small square tiles, as though we were covering the surface with little postage stamps (in Figure 18.74 the tiles are 0.25 mm × 0.25 mm). In each step of an iterative procedure, we calculate the electric field just inside each tile based on the contributions of all the charged tiles and any external fixed charges. Knowing the field just inside a surface tile, we can determine how much charge will be added to or subtracted from that tile in the next short time interval.

The next iteration uses the new tile charges in the computation of the field near each tile. We graph the amount of charge on a representative tile and halt the iterations when we see that this tile's charge is no longer changing. We find that the electric field contributed by the final distribution of charged tiles, plus the field of the external charge in Figure 18.74, is as expected zero inside the metal block, which is a check on the procedure.

In order to get a more quantitative representation of the distribution of surface charge we consider the charge on (square) rings of tiles whose axes are the center line of the block as shown in Figure 18.75. Figure 18.76 is a graph of the computed surface charge density on such rings of charged tiles from left to right along the center line of the block. For each ring we plot the total charge of the ring divided by the surface area of the ring (the number of tiles times the area of each tile), which is the average surface charge on this ring. Note the approximately constant gradient of surface charge density except near the left and right faces of the block, where there are large buildups of charge.

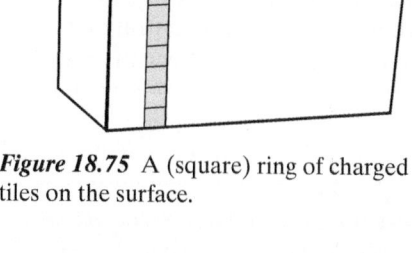

Figure 18.75 A (square) ring of charged tiles on the surface.

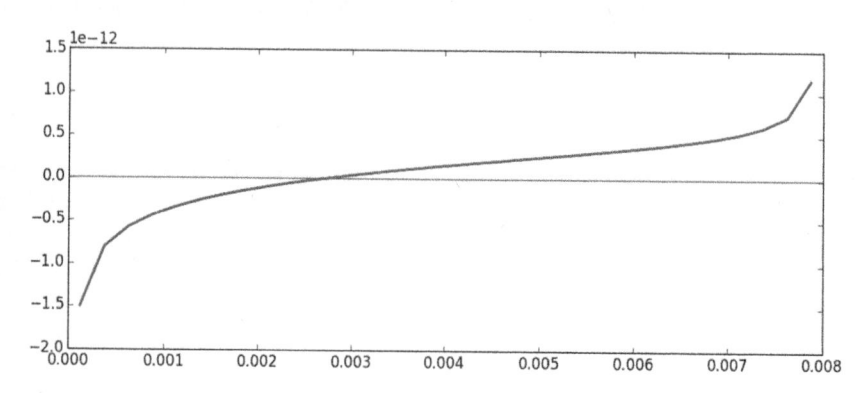

Figure 18.76 Surface charge density (C/m^2) of rings of charge surrounding a horizontal path (in meters) going left to right through the center of a polarized neutral metal block.

The computational model for determining the distribution of surface charge was implemented in VPython, exploiting the fast array processing provided by Python's NumPy module, which is imported by VPython. In preparation for the iterative loop that moves charge onto and off the surface tiles, we calculate for each tile the effect that every other tile would have if the other tile were a point charge of 1 C. To do this we calculate the field the other tile contributes just inside our tile, take the component E_{normal} along the outward-going unit vector that is normal to our tile, and multiply by $enAu\Delta t$ to get the change in our tile's charge in a time Δt.

After calculating this for every one of the N tiles, we have an N by N matrix of the effects of all tiles on all other tiles, on a per-coulomb basis. We add to these effects the effect of each tile's nearby field on itself, considering the tile to be a uniformly charged square plate. In a similar way, for each tile we compute what effect the fixed external charges will have.

Having prepared this information, each iteration of the iterative process consists of multiplying the current tile charges times their per-coulomb effects on all of the tile charges, and we add in the fixed effects of the external charges. After making a list of all the Δq's for all tiles, we update the charges on all the tiles. We graph the charge on a representative tile as a function of time and stop the iteration when we see that this charge is no longer changing. A conclusive test of the procedure is that when we calculate the electric field due to all of the computed tile charges and the external charges, the distribution of electric field is correct: zero inside a polarized metal block, nonzero and uniform inside circuit wires in the steady state.

Currents in the Interior

We need not concern ourselves with charge in the interior of the metal, thanks to a remarkable, special property of the $1/r^2$ electric force law that ensures that the interior of the metal will remain neutral at all times. Consider what

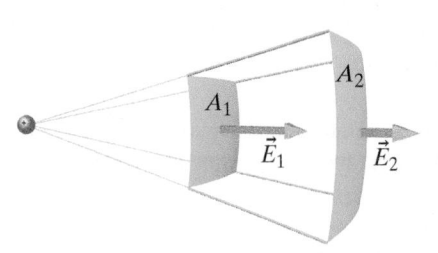

Figure 18.77 A volume similar to a truncated pyramid, located in the interior of the metal. The volume is bounded by two spherical faces centered on the point charge. A_1 is the area of the smaller face (nearer to the point charge), and A_2 is the area of the larger face (farther from point charge). On each face the electric field due to the point charge is uniform.

happens initially to a (neutral) region in the interior of our metal block, chosen to have the form of a truncated pyramid, with the apex of the pyramid at the location of the point charge (Figure 18.77).

The field of the point charge drives an electron current to the left, out of the near face of the truncated pyramid, of magnitude $i_1 = nAv = nA_1uE_1$, where A_1 is the area of the left face and E_1 is the field on the near face. The field of the point charge also drives an electron current into the far face of the region, of magnitude $i_2 = nAv = nA_2uE_2$, where E_2 is the field on the far face.

On the far side of the region the field is smaller but the area is bigger: the area A is proportional to r^2 (where r is the distance to the point charge) while the electric field is proportional to $1/r^2$, so that $A_1E_1 = A_2E_2$, which means that $i_1 = nA_1uE_1$ has the same magnitude as $i_2 = nA_2uE_2$. Also, no electrons enter or leave the sides of the truncated cone because there is no component of the field that is perpendicular to the surface there. Therefore, in the next short time Δt, the outflow of electrons is exactly the same as the inflow, so this initially neutral interior region of the metal stays neutral. In each time interval Δt, the same number of electrons flow out of the region as flow into it, so the interior of the metal remains neutral at all times. The only changes in charge are on the surface of the metal.

In Chapter 21 we will see, using a property of electric fields known as Gauss's law, that this result is general, that the net charge contained in any region in the interior of the metal, of any shape, will remain zero at all times. As charge accumulates on the surface of the metal block, the additional fields contributed by these surface charges also have the property of leaving all regions of the interior with a net charge of zero. As a result, the computational model only has to calculate changes in the amount of charge of the surface tiles and does not need to track what happens in the interior of the metal.

An Interactive Presentation

On the Wiley student website for this textbook is a program that lets you inspect these surface charge distributions in 3D and explore by dragging the mouse the electric fields contributed by the fixed charges and by the surface charges inside and outside the metal.

SUMMARY

New Concepts

- In a uniform wire, a uniform electric field drives a uniform steady-state current.
- A gradient of charge on the surface of the wires makes electric fields inside the wires.
- There is an initial transient, lasting a minimum of a few nanoseconds, in which feedback establishes the steady-state distribution of surface charges and currents.
- A battery maintains a charge separation and therefore a potential difference. The emf of the battery is the energy input per unit charge required to maintain the charge separation.

Analysis Techniques

In analyzing electric circuits we apply two rules:
The current node rule:

In the steady state:
electron current: net $i_{\text{in}} = $ net i_{out}, where

$$i = nAuE$$

conventional current: net $I_{\text{in}} = $ net I_{out}, where

$$I = |q|nAuE$$

q is the charge of a mobile charged particle; n is the density of mobile charges; A is the cross-sectional area of the conducting object, u is the mobility of the charged particles.

Energy conservation (the loop rule):
In the steady state, for any round-trip path,

$$\Delta V_1 + \Delta V_2 + \cdots = 0$$

Results

Our model of the microscopic behavior of circuits provides answers to several questions about circuit behavior:

1. *Where are the charges that are the sources of the electric field inside the wires?* These charges are on the surfaces of the wires, and on and in the battery. The necessary charge distribution is created and maintained by feedback.

2. *How can the light come on instantaneously despite the very slow drift speed of the mobile charges?* Electric fields propagate at the speed of light. The electrons inside the bulb are affected by the electric field which is quickly (in a few nanoseconds) established in the bulb filament.

3. *What does a battery actually do in a circuit?* A battery does a fixed amount of work per unit charge (emf), maintaining a charge separation and potential difference across itself, and consequently maintaining a distribution of surface charge all over the circuit.

EXPERIMENTS

EXP1 Current in different parts of a circuit

Assemble a circuit containing a thick-filament bulb and two batteries in series, as shown in Figure 18.78.

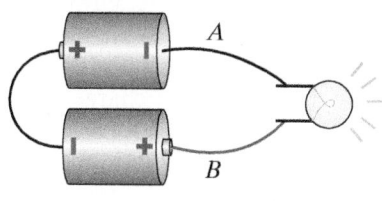

Figure 18.78

Use the following procedure to make careful measurements of the magnetic field produced by each wire. (Note that equal currents will produce equal magnetic fields, and hence equal compass deflections.)

- Disconnect one wire.
- Make sure the batteries and the steel alligator clips are not near the compass.
- Lay the wire in which you wish to measure current over the compass, carefully lining up the wire with the compass needle.
- Carefully reconnect the circuit without moving anything.
- Read the compass deflection to within 2° or better.

(a) Is the circuit in a steady state? Leave the compass under the wire at A, and observe it for 15–20 s. Does the deflection remain constant?

(b) Using the compass, measure the current at A and B, and record the observed compass deflections.

(c) Use this sensitive "null" test to compare the current in the two wires: run wires A and B side-by-side in opposite directions over the compass, then connect the circuit. The magnetic fields produced by the two wires are in opposite directions. If the currents are equal and the wires are very close together, the magnetic fields in the two wires will cancel each other and give zero deflection.

Making Careful Measurements of Currents

In the following experiments (Experiments EXP 16 through EXP5) we will use our model of circuits to predict the ratio of current in one circuit to current in another circuit, and then we will do experiments to test our predictions. In order to get good experimental results, we need to understand how to use a compass to make accurate measurements of current. There are two important issues to consider.

1. Is Magnetic Field Uniform All Along the Compass Needle?

The compass needle aligns with the net magnetic field at its location. As we have seen, the net magnetic field is the sum of the Earth's magnetic field and the magnetic field made by the current in the wire. In Chapter 17 we found that the magnitude of the magnetic field made by current in a long wire is inversely proportional to distance from the wire:

$$B = \frac{\mu_0}{4\pi} \frac{2I}{r}$$

If the distance from the wire to the center of the needle is significantly different from the distance from the wire to the ends of the needle, the magnetic field at the ends of the needle will be weaker than at the center. In this case the deflection of the needle will be smaller than it should be. If the wire lies on top of the compass, it is about 6 mm away from the center of the needle. If the needle deflects more than 15°, the ends of the needle are significantly farther away from the wire than the center is, as shown in Figure 18.79, and the deflection will not be proportional to the current.

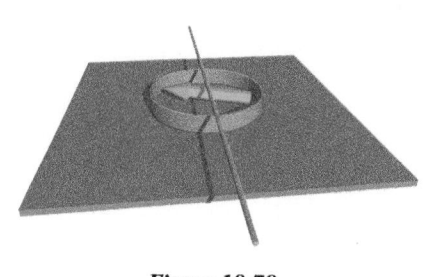

Figure 18.79

Elevating the Wire

By elevating the wire about 1 cm above the top of the compass, we minimize this effect. Now even if the needle deflects up to 40°, the ends and the center will still be approximately the same distance from the wire. An easy way to elevate the wire is to support it on two fingers, as shown in Figure 18.80.

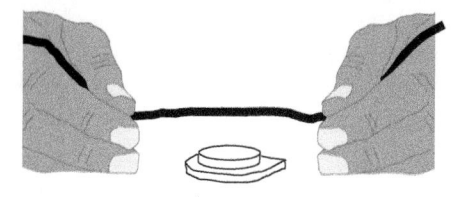

Figure 18.80

2. Is Current Proportional to Deflection Angle?

As can be seen from Figure 18.81, the tangent of the deflection angle is proportional to the magnetic field made by the current in the wire. For small deflection angles (measured in radians), $\theta \approx \tan\theta$. However, this approximation is not good for angles greater than about 0.3 rad ($17°$).

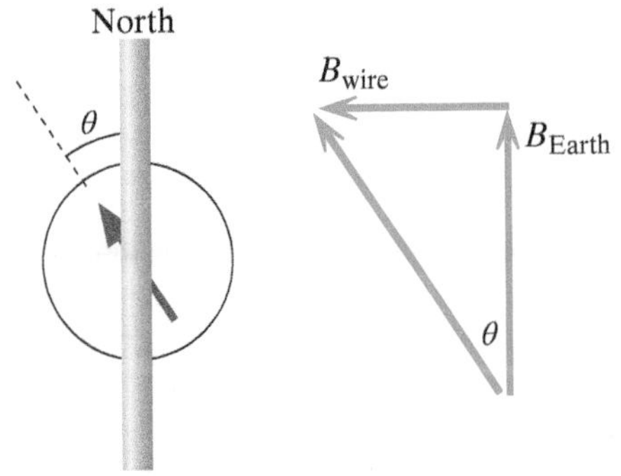

North

θ

B_{wire}

B_{Earth}

θ

Figure 18.81 $B_{\text{wire}}/B_{\text{Earth}} = \tan\theta$

Tangent of Deflection Angle

- For deflection angles less than $15°$, we can assume that the magnetic field, and hence the current in the wire, is proportional to the deflection angle.
- For deflection angles greater than $15°$, we need to use the tangent of the deflection angle to calculate magnetic field (and therefore current).

EXP2 Effect of twice the length

Here is an experiment in which we compare the electron current in a Nichrome wire of length L to the current in a Nichrome wire of length $2L$.

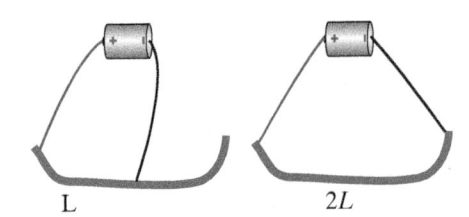

L   2L

Figure 18.82

> QUESTION What does our model for current in a circuit predict for the result of the experiment?

Around the circuit loop we have emf $- EL = 0$, so if we double the length L, we'll get half as big an electric field (since the emf doesn't change). With half the field we get half the drift speed ($\bar{v} = uE$), half the electron current ($i = nAuE$), and half the conventional current ($I = |q|nAuE$). Therefore we expect half as large a magnetic field and half the compass deflection.

Next make the measurement. Be careful when connecting Nichrome wire. If you run current through a short (< 10 cm) length of the wire it can get hot enough to burn you. (It may even get hot enough to glow.)

Following the directions given above, use the compass to compare the amount of current in the two different circuits shown in Figure 18.82:

(a) One battery (to keep the compass deflection small) and half the length of the thin Nichrome wire (about 20 cm). Don't cut the Nichrome wire, just connect clips to a portion half the length of the wire.

(b) One battery and the full length (about 40 cm) of the thin Nichrome wire.

Make careful, quantitative measurements of the angles, and record your measurements.

- Are you able to verify the prediction that the current is halved when you double the length of the Nichrome wire?
- Do you find that the current is unaffected if you bend the wire or add additional connecting wires?

EXP3 Current in a two-bulb series circuit

Two identical light bulbs in series (Figure 18.83) are essentially the same as one light bulb with twice as long a filament, since the copper connecting wires contribute little energy dissipation. From the preceding experiment with doubling the length of a Nichrome wire, we might expect to get half the current compared with just one bulb.

(a) Following the directions given above, use a compass to measure the current in the circuit with two thick filament bulbs in series shown in Figure 18.83, and record the observed compass deflections. Compare your observations to your prediction.

(b) Compare the compass deflection you observe in the two-bulb circuit with the compass deflection you observe in a one-bulb circuit. Record your measurements (by removing one bulb from the circuit).

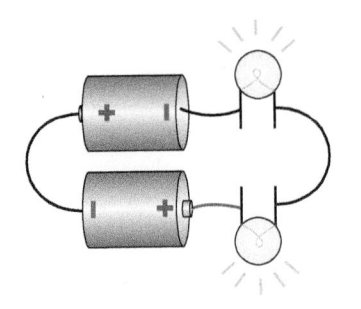

Figure 18.83

You presumably found that the two-bulb circuit had more than half the current of the one-bulb circuit (if necessary, repeat this measurement carefully). The issue here is that the electron mobility in very hot tungsten (several thousand degrees!) is less than the electron mobility in cooler tungsten. The greater thermal agitation of the atomic cores leads to larger "friction." Note that the bulbs in the two-bulb circuit emit redder light than the single bulb, indicating that their filaments are not as hot.

Explain why the current in the two-bulb series circuit is more than half the current in the one-bulb circuit. How is the electric field affected? How is the drift speed affected?

Why didn't we see this effect when we experimented with doubling the length of the Nichrome wire?

EXP4 Effect of doubling the cross-sectional area

Figure 18.84 illustrates an experiment in which we see what happens to the current in a Nichrome wire when we double its cross-sectional area.

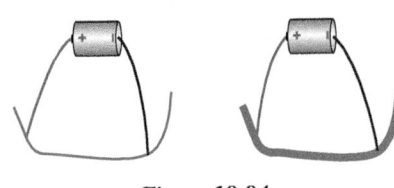

Figure 18.84

QUESTION Explain why the electron current in the wire should increase by a factor of two if the cross-sectional area of the wire doubles.

In Figure 18.84 it seems plausible that doubling A should double the conventional current $I = |q|nAuE$. It is like having two wires in parallel, doubling the drain on the battery.

We can make this argument quantitative by considering energy. Since we have emf $- EL = 0$ around the loop, the electric field E in the wire does not change when we use a thicker wire, so the current doubles. (This neglects internal resistance in the battery, which has the effect of limiting the amount of current that the battery can supply, so that for large cross-sectional areas and large currents, doubling a large cross-sectional area may not yield twice the current.)

Use the compass to compare the amount of current in two different circuits (Figure 18.84): one battery (to keep the compass deflection small) and the same lengths of thin and thick Nichrome wires.

Use only one battery. Connect the connecting wires about 30 cm apart on each Nichrome wire (don't cut the Nichrome wire). Following the directions given above, make careful, quantitative measurements of the angles.

The thick Nichrome wire has a cross section about twice the cross section of the thin Nichrome wire. Your instructor may give you a more precise ratio of the areas. What do you measure for the ratio of the current in the thick wire to the current in the thin wire? Does this agree with predictions?

Your own data are probably consistent with what more precise experiments confirm. These measurements imply that the drift speed of individual electrons is the same in the thick wire and in the thin wire (because the electric field is the same in the two circuits if the length of the wires is the same), and that the many-electron current is distributed uniformly over the cross section.

If the drift speed were not the same everywhere across the cross section of the wire, we couldn't simply write $nA\bar{v}$ for the electron current, using the same $\bar{v}$ for all the electrons passing that point in the circuit. In Section 18.3 we proved that the current is indeed uniform across the cross section.

EXP5 A parallel circuit
Connect two thick filament bulbs in "parallel" (that is, one beside the other), as shown in Figure 18.85.

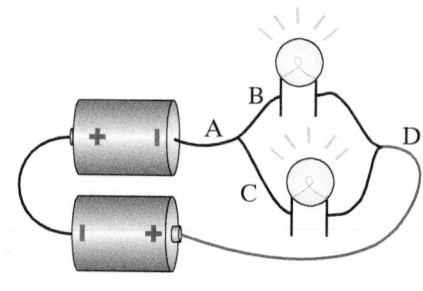

Figure 18.85

(a) How does the brightness of the thick-filament bulbs in this parallel circuit compare to the brightness of two thick-filament bulbs in series?

(b) Unscrew one of the bulbs in the parallel circuit. What effect does this have on the brightness of the remaining bulb?

(c) Based on these observations, what relationship would you expect to see between the compass deflections at locations A, B, C, and D, in this circuit? Record your prediction.

(d) Measure the compass deflections at locations A, B, C, and D, and record your measurements. Do your observations agree with your predictions in (c)?

In the preceding experiment you should again have seen that the current entering a location is equal to the current leaving the same location in the steady state. In particular, the current at A should be equal to the sum of the currents at B and C. If you did not observe this, you should repeat the experiment. We can also think of the two bulbs in parallel as equivalent to increasing the cross-sectional area of one of the bulb filaments.

We can analyze the situation quantitatively. If we follow the path through one of the bulbs, we have $2\,\text{emf} - EL = 0$, ($2\,\text{emf}$ for two batteries), and L is the length of the bulb filament (neglecting dissipation in the connecting wires). If we follow the path through the other thick filament bulb, we also have $2\,\text{emf} - EL = 0$. Thus the electric field $E = 2\,\text{emf}/L$ is the same in both bulb filaments, and it is the same as there would be with only one bulb present.

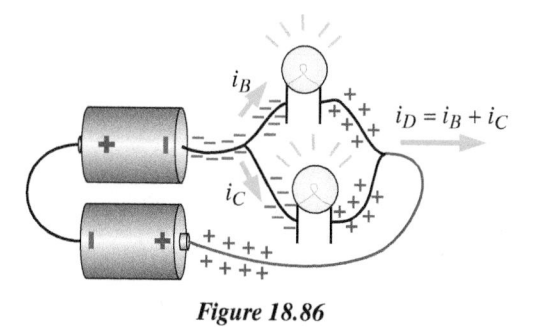

Figure 18.86

Figure 18.86 shows an approximate surface-charge distribution on the circuit. Each bulb is flanked by positive and negative charges, which make a large electric field in the bulb filament. Removing one of the bulbs doesn't change the surface-charge distribution very much, and the brightness of the remaining bulb hardly changes. The batteries of course have to deliver more current ($i_B + i_C$) to the two parallel bulbs than they do to one bulb.

EXP6 Effect of two batteries
Use the compass to compare the amount of current in two different circuits: one battery and a length of thin Nichrome wire, and two batteries and the same length of thin Nichrome wire. Following the directions given with Experiment EXP21, make careful, quantitative measurements of the angles.

Are you able to verify the prediction that the current doubles when you use two batteries? (If you wish, try three or more batteries with the same length of thin Nichrome wire—but don't use more than two batteries with your light bulbs, because it may burn them out.)

QUESTIONS

Q1 What is the most important general difference between a system in steady state and a system in equilibrium?

Q2 Describe the following attributes of a *metal wire* in steady state vs. equilibrium:

Metal Wire	Steady-State	Equilibrium
Location of excess charge Motion of mobile electrons E inside the metal wire		

Q3 How can there be a nonzero electric field inside a wire in a circuit? Isn't the electric field inside a metal always zero?

Q4 State your own theoretical *and* experimental objections to the following statement: In a circuit with two thick-filament bulbs in series, the bulb farther from the negative terminal of the battery will be dimmer, because some of the electron current is used up in the first bulb. Cite relevant experiments.

Q5 Since there is an electric field inside a wire in a circuit, why don't the mobile electrons in the wire accelerate continuously?

Q6 Why don't all mobile electrons in a metal have exactly the same speed?

Q7 There are very roughly the same number of iron atoms per m^3 as there are copper atoms per m^3, but copper is a much better conductor than iron. How does u_{iron} compare with u_{copper}?

Q8 In the few nanoseconds *before* the steady state is established in a circuit consisting of a battery, copper wires, and a single bulb, is the current the same everywhere in the circuit? Explain.

Q9 During the initial transient leading to the steady state, the electron current going into a bulb may be greater than the electron current leaving the bulb. Explain why and how these two currents come to be equal in the steady state.

Q10 At a typical drift speed of 5×10^{-5} m/s, an electron traveling at that speed would take about 100 min to travel through one of your connecting wires. Why, then, does the bulb light immediately when the connecting wire is attached to the battery?

Q11 Criticize the statement below on theoretical and experimental grounds. Be specific and precise. Refer to your own experiments, or describe any new experiments you perform: "A flashlight battery always puts out the same amount of current, no matter what is connected to it."

Q12 What is the difference between emf and electric potential difference?

Q13 Compare the direction of the average electric field inside a battery to the direction of the electric field in the wires and resistors of a circuit.

Q14 In the circuit shown in Figure 18.87, bulbs 1 and 2 are identical in mechanical construction (the filaments have the same length and the same cross-sectional area), but the filaments are made of different metals. The electron mobility in the metal used in bulb 2 is three times as large as the electron mobility in the metal used in bulb 1, but both metals have the same number of mobile electrons per cubic meter. The two bulbs are connected in series to two batteries with thick copper wires (like your connecting wires).

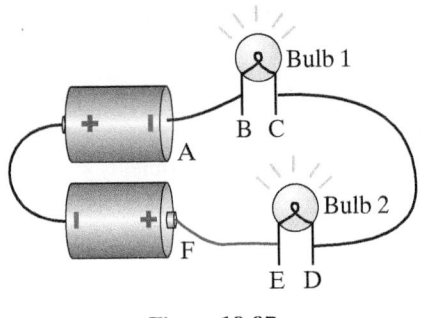

Figure 18.87

(a) In bulb 1, the electron current is i_1 and the electric field is E_1. In terms of these quantities, determine the corresponding quantities i_2 and E_2 for bulb 2, and explain your reasoning. **(b)** When bulb 2 is replaced by a wire, the electron current through bulb 1 is i_0 and the electric field in bulb 1 is E_0. How big is i_1 in terms of i_0? Explain your answer, including explicit mention of any approximations you must make. Do not use ohms or series-resistance equations in your explanation, unless you can show in detail how these concepts follow from the microscopic analysis introduced in this chapter. **(c)** Explain why the electric field inside the thick copper wires is very small. Also explain why this very small electric field is the same in all of the copper wires, if they all have the same cross-sectional area. **(d)** Figure 18.88 is a graph of the magnitude of the electric field at each location around the circuit when bulb 2 is replaced by a wire. Copy this graph and add to it, on the same scale, a graph of the magnitude of the electric field at each location around the circuit when both bulbs are in the circuit. The very small field in the copper wires has been shown much larger than it really is in order to give you room to show how that small field differs in the two circuits.

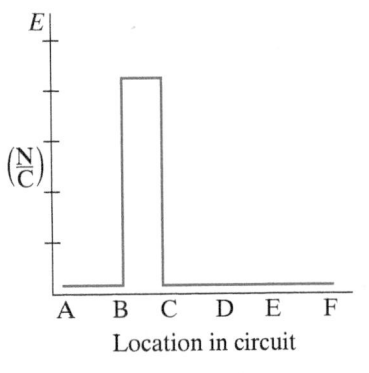

Figure 18.88

Q15 Some students intended to run a light bulb off two batteries in series in the usual way, but they accidentally hooked up one of the batteries backwards, as shown in Figure 18.89 (the bulb is shown as a thin filament).

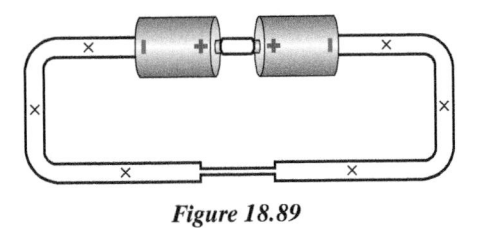

Figure 18.89

(a) Use +'s and −'s to show the approximate steady-state charge distribution along the wires and bulb. **(b)** Draw vectors for the electric field at the indicated locations inside the connecting wires and bulb. **(c)** Compare the brightness of the bulb in this circuit with the brightness the bulb would have had if one of the batteries hadn't been put in backwards. **(d)** Try the experiment to check your analysis. Does the bulb glow about as you predicted?

Q16 For each of the following experiments, state what effect you observed (how the current in the circuit was affected) and why, in terms of the relationships:

$$\text{electron current} = nA\bar{v} \text{ and } \bar{v} = uE$$

Experiment	Effect on Current (Be Quantitative)	Circle the Parameter(s) That Changed and Explain Briefly
Double the length of a Nichrome wire		$n\ A\ u\ E$
Double the cross-sectional area of a Nichrome wire		$n\ A\ u\ E$
Two identical bulbs in series compared to a single bulb		$n\ A\ u\ E$
Two batteries in series compared to a single battery		$n\ A\ u\ E$

Q17 In a table like the one shown, write an inequality comparing each quantity in the steady state for a narrow resistor and thick connecting wires, which are made of the same material as the resistor.

Electron Current in Resistor	<, =, or >	Electron Current in Thick Wires
n_R		n_W
A_R		A_W
u_R		u_W
E_R		E_W
v_R		v_W

Q18 A steady-state current flows through the Nichrome wire in the circuit shown in Figure 18.90.

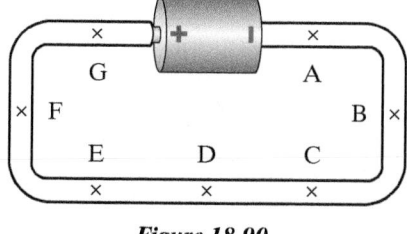

Figure 18.90

Before attempting to answer the following questions, draw a copy of this diagram. All of the locations indicated by letters are inside the wire. **(a)** On your diagram, show the electric field at the locations indicated, paying attention to relative magnitude. **(b)** Carefully draw pluses and minuses on your diagram to show the approximate surface charge distribution that produces the electric field you drew. Make your drawing show clearly the differences between regions of high surface charge density and regions of low surface-charge density. Use your diagram to determine which of the following statements about this circuit are true. (1) There is some excess negative charge on the surface of the wire near location B. (2) Inside the metal wire the magnitude of the electric field is zero. (3) The magnitude of the electric field is the same at locations G and C. (4) The electric field points to the left at location G. (5) There is no excess charge on the surface of the wire. (6) There is excess charge on the surface of the wire near the batteries but nowhere else. (7) The magnitude of the electric field inside the wire is larger at location G than at location C. (8) The electric field at location D points to the left. (9) Because the current is not changing, the circuit is in static equilibrium.

Q19 In the circuit shown in Figure 18.91, all of the wire is made of Nichrome, but one segment has a much smaller cross-sectional area. On a copy of this diagram, using the same scale for magnitude that you used in the previous question for Figure 18.90, show the steady-state electric field at the locations indicated, including in the thinner segment. Before attempting to answer these questions, draw a copy of this diagram. All of the locations indicated by letters are inside the wire.

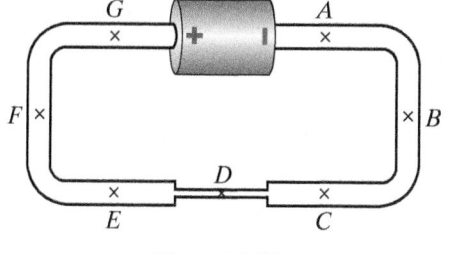

Figure 18.91

(a) On your diagram, show the electric field at the locations indicated, paying attention to relative magnitude. Use the same scale for magnitude as you did in the previous question. **(b)** Carefully draw pluses and minuses on your diagram to show the approximate surface charge distribution that produces the electric field you drew. Make your drawing show clearly the differences between regions of high surface charge density and regions of low surface-charge density. Use your diagram to determine which of the following statements about this circuit are true. (1) There is a large gradient of surface charge on the wire between locations C and E. (2) The electron current is the same at every location in this circuit. (3) Fewer electrons per second pass location E than location C. (4) The magnitude of the electric field at location G is smaller in this circuit than it was in the previous circuit (Figure 18.90). (5) The magnitude of the electric field is the same at every location in this circuit. (6) The magnitude of the electric field at location D is larger than the magnitude of the electric field at location G. (7) There is no surface charge at all on the wire near location G. (8) The electron current in this circuit is less than the electron current in the previous circuit (Figure 18.90).

PROBLEMS

Section 18.2

•**P20** Which of the following statements about a metal wire *in equilibrium* are true? List all that apply. (1) There cannot be excess charges on the surface of the wire. (2) There is no net flow of mobile electrons inside the wire. (3) There are no excess charges in the interior of the wire. (4) There may be excess charges in the interior of the wire. (5) The net electric field everywhere inside the wire is zero. (6) There may be excess charges on the surface of the wire. (7) The interior of the metal wire is neutral. (8) There may be a constant flow of mobile electrons inside the wire. (9) The electric field inside the wire may be nonzero but uniform.

•**P21** Which of the following statements about a metal wire *in the steady state* are true? List all that apply. (1) There is a constant flow of mobile electrons inside the wire. (2) The net electric field everywhere inside the wire is zero. (3) There are no excess charges in the interior of the wire. (4) There cannot be excess charges on the surface of the wire. (5) There is no net flow of mobile electrons inside the wire. (6) There may be excess charges on the surface of the wire. (7) There may be excess charges in the interior of the wire. (8) There may be a nonzero, uniform electric field inside the wire. (9) The interior of the metal wire is neutral.

•**P22** Electron current $i = nA\bar{v} = nA\,uE$: **(a)** What are the units of electron current? **(b)** What is n? What are its units? **(c)** What is A? What are its units? **(d)** What is $\bar{v}$? What are its units? **(e)** What is u? What are its units?

•**P23** In the circuit shown in Figure 18.92, what are the relationships among i_A, i_B, i_C, and i_D? How much current flows through the lower battery?

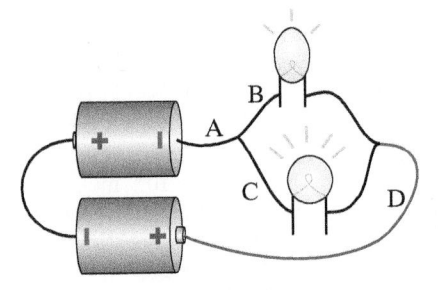

Figure 18.92

•**P24** A steady-state current runs in the circuit shown in Figure 18.93. The narrow resistor and thick connecting wires are made of the same material.

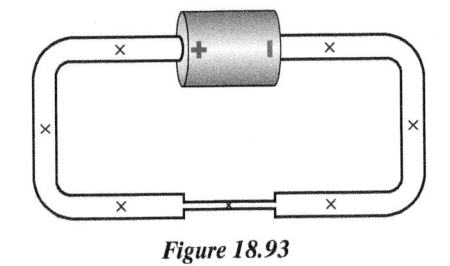

Figure 18.93

Which of the following quantities are greater in the thin resistor than in the thick wire? List all that apply: u, n, E, v, A, i, or none of these.

•**P25** Consider the circuit fragment in Figure 18.94. **(a)** What is the absolute value of the outward-going conventional current I_2?

(b) In this case, did we make the right guess about the direction of the conventional current I_2? **(c)** Suppose I_3 is 20 A; what is the absolute value of the outward-going conventional current I_2? **(d)** In this case, did we make the right guess about the direction of the conventional current I_2?

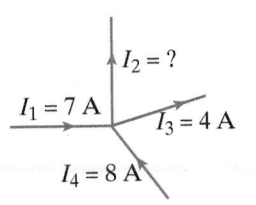

Figure 18.94

•**P26** In the circuit in Figure 18.95 the narrow resistor is made of the same material as the thick connecting wires. In the steady state, which graph in Figure 18.96 correctly shows the magnitude of the electric field at locations around the circuit?

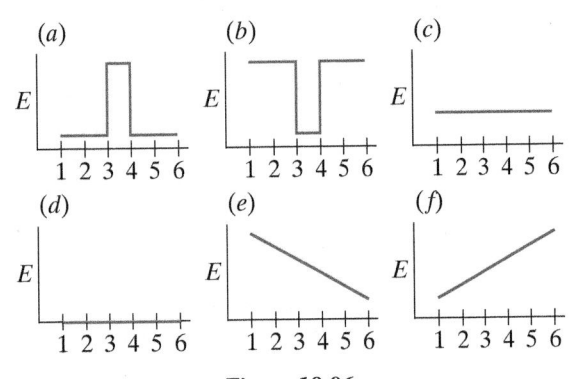

Figure 18.95

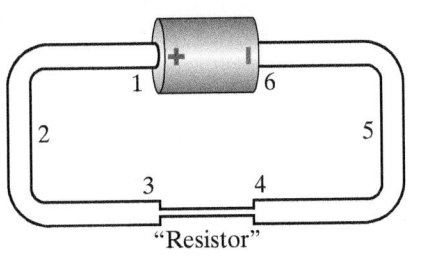

Figure 18.96

In the steady state, which graph in Figure 18.97 correctly shows the drift speed of the electrons at locations around the circuit?

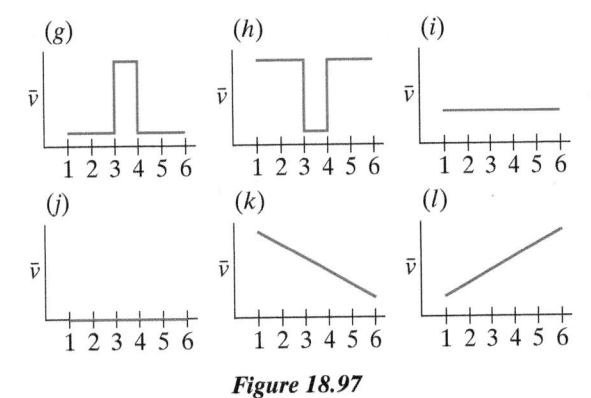

Figure 18.97

Section 18.3

•**P27** The drift speed in a copper wire is 7×10^{-5} m/s for a typical electron current. Calculate the magnitude of the electric field E inside the copper wire. The mobility of mobile electrons in copper is 4.5×10^{-3} (m/s)/(N/C). (Note that though the electric field in the wire is very small, it is adequate to push a sizable electron current through the copper wire.)

•**P28** Suppose that a wire leads into another, thinner wire of the same material that has only a third the cross-sectional area. In the steady state, the number of electrons per second flowing through the thick wire must be equal to the number of electrons per second flowing through the thin wire. If the drift speed $\bar{v}_1$ in the thick wire is 4×10^{-5} m/s, what is the drift speed $\bar{v}_2$ in the thinner wire?

•**P29** In the circuit in Figure 18.98 a mechanical battery keeps a steady-state current running in a wire that has rather low electron mobility.

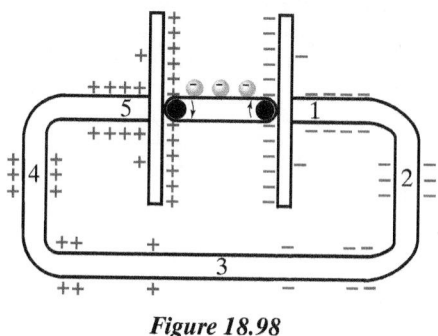

Figure 18.98

Which of the following statements about the circuit are true? List all that apply. (1) The electric field inside the wire varies in magnitude, depending on location. (2) At location 3 inside the wire the electric field points to the right. (3) The electric field is zero at all locations inside the metal wire. (4) At location 3 the electric field points to the left. (5) The magnitude of the electric field inside the wire is the same at all locations inside the wire. (6) Mobile electrons inside the wire push each other through the wire. (7) The nonzero electric field inside the wire is created by the moving electrons in the wire. (8) At every location inside the wire the direction of the electric field is parallel to the wire. (9) The nonzero electric field inside the wire is created by excess charges on the surface of the wire and in and on the mechanical battery.

•**P30** Suppose that a wire leads into another, thinner wire of the same material that has only half the cross-sectional area. In the steady state, the number of electrons per second flowing through the thick wire must be equal to the number of electrons per second flowing through the thin wire. If the electric field E_1 in the thick wire is 1×10^{-2} N/C, what is the electric field E_2 in the thinner wire?

•**P31** Suppose that wire A and wire B are made of different metals and are subjected to the same electric field in two *different* circuits. Wire B has 6 times the cross-sectional area, 1.3 times as many mobile electrons per cubic centimeter, and 4 times the mobility of wire A. In the steady state, 2×10^{18} electrons enter wire A every second. How many electrons enter wire B every second?

••**P32** Inside a chemical battery it is not actually individual electrons that are transported from the + end to the − end. At the + end of the battery an "acceptor" molecule picks up an electron

entering the battery, and at the − end a different "donor" molecule gives up an electron, which leaves the battery. Ions rather than electrons move between the two ends to transport the charge inside the battery.

When the supplies of acceptor and donor molecules are used up in a chemical battery, the battery is dead, because it can no longer accept or release electrons. The electron current in electrons per second, times the number of seconds of battery life, is equal to the number of donor (or acceptor) molecules in the battery.

A flashlight battery contains approximately half a mole of donor molecules. The electron current through a thick-filament bulb powered by two flashlight batteries in series is about 0.3 A. About how many hours will the batteries keep this bulb lit?

••**P33** All of the wires in the circuit shown in Figure 18.99 are made of the same material, but one wire has a smaller radius than the other wires.

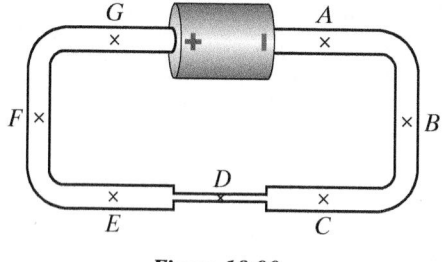

Figure 18.99

Which of the following statements are true of this circuit in the steady state? List all that apply. (1) The drift speed of electrons passing location D is greater than the drift speed of electrons passing location G. (2) The magnitude of the electric field is the same at each location labeled by a letter. (3) The electric field at location F points up. (4) The electric field at G is larger in magnitude than the electric field at C. (5) The number of electrons passing location B each second is the same as the number of electrons passing location D each second.

The radius of the thin wire is 0.22 mm, and the radius of the thick wire is 0.55 mm. There are 4×10^{28} mobile electrons per cubic meter of this material, and the electron mobility is 6×10^{-4} (m/s)/(V/m). If 6×10^{18} electrons pass location D each second, how many electrons pass location B each second? What is the magnitude of the electric field at location B?

••**P34** Figure 18.100 is a top view of a portion of a circuit containing three identical light bulbs (the rest of the circuit including the batteries is not shown). The connecting wires are made of copper.

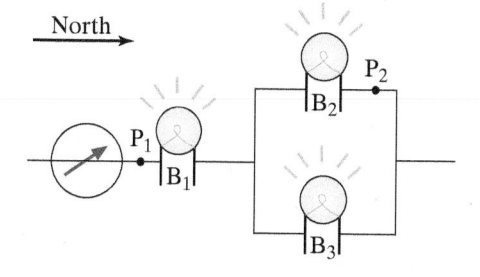

Figure 18.100

(a) The compass is placed on top of the wire, and it deflects 20° away from north as shown (the wire is underneath the compass). In what direction are the electrons moving at location P_1? How do you know? **(b)** In the steady state, 3×10^{18} electrons pass location P_1 every second. How many electrons pass location P_2 every second? Explain briefly. **(c)** Give the relative brightnesses of bulbs B_1, B_2, and B_3. Explain briefly. **(d)** There are 6.3×10^{28} mobile electrons per cubic meter in tungsten. The cross-sectional area of the tungsten filament in bulb B_1 is $0.01 \, \text{mm}^2$ (which is $1 \times 10^{-8} \, \text{m}^2$). The electron mobility in hot tungsten is $1.2 \times 10^{-4} \, \text{(m/s)/(N/C)}$. Calculate the electric field inside the tungsten filament in bulb B_1. Give both the direction and the magnitude of the electric field.

••P35 Consider the circuit containing three identical light bulbs shown in Figure 18.101. North is indicated in the diagram. Compasses are placed under the wires at locations A and B.
(a) The magnitude of the deflection of compass A is $13°$ away from north. In what direction does the needle point? Draw a sketch. **(b)** What is the magnitude of the deflection of the needle of compass B? In what direction does the needle point? Draw a sketch. Explain your reasoning. **(c)** In the steady state 1.5×10^{18} electrons per second enter bulb 1. There are 6.3×10^{28} mobile electrons per cubic meter in tungsten. The cross-sectional area of the tungsten filament in bulb 1 is $1 \times 10^{-8} \, \text{m}^2$. The electron mobility in hot tungsten is $1.2 \times 10^{-4} \, \text{(m/s)/(N/C)}$. Calculate the electric field inside the tungsten filament in bulb 3. Give both the direction and the magnitude of the electric field.

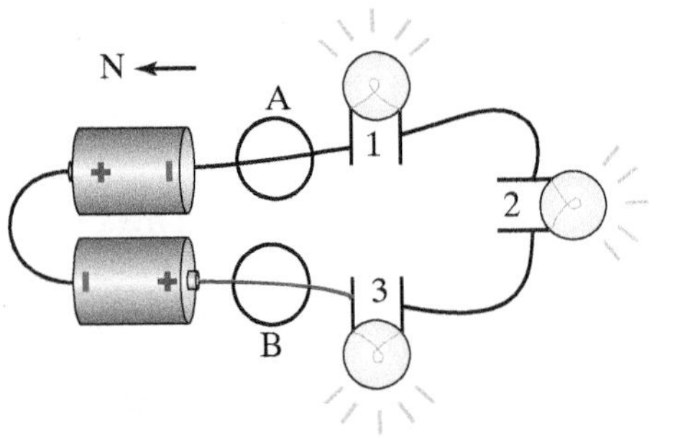

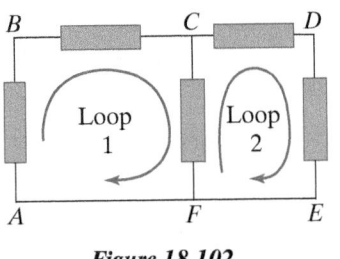

Figure 18.101

Section 18.10
Base your analyses in the following problems on the principles discussed in this chapter. If you wish to use equations derived elsewhere, you must justify them in terms of the microscopic analysis in terms of field that was introduced in this chapter.

•P36 **(a)** The emf of a particular flashlight battery is 1.7 V. If the battery is 4.5 cm long and the radius of the cylindrical battery is 1 cm, estimate roughly the amount of charge on the positive end plate of the battery. **(b)** Is this amount of charge sufficient to repel noticeably a positively charged piece of invisible tape?

•P37 A Nichrome wire 30 cm long and 0.25 mm in diameter is connected to a 1.5 V flashlight battery. What is the electric field inside the wire? Why don't you have to know how the wire is bent? How would your answer change if the wire diameter were 0.35 mm? (Note that the electric field in the wire is quite small compared to the electric field near a charged tape.)

•P38 Why does the brightness of a bulb not change noticeably when you use longer copper wires to connect it to the battery? (1) Very little energy is dissipated in the thick connecting wires. (2) The electric field in connecting wires is very small, so emf $\approx$ $E_{\text{bulb}} L_{\text{bulb}}$. (3) Electric field in the connecting wires is zero, so

emf $= E_{\text{bulb}} L_{\text{bulb}}$. (4) Current in the connecting wires is smaller than current in the bulb. (5) All the current is used up in the bulb, so the connecting wires don't matter.

•P39 A Nichrome wire 48 cm long and 0.25 mm in diameter is connected to a 1.6 V flashlight battery. What is the electric field inside the wire? Next, the wire is replaced by a different Nichrome wire with the same length, but diameter 0.20 mm. Now what is the electric field inside the wire?

•P40 In a circuit with one battery, connecting wires, and a 12 cm length of Nichrome wire, a compass deflection of $6°$ is observed. What compass deflection would you expect in a circuit containing two batteries in series, connecting wires, and a 36 cm length of thicker Nichrome wire (double the cross-sectional area of the thin piece)? Explain.

•P41 In Figure 18.102, suppose that $V_C - V_F = 8$ V, and $V_D - V_E = 4.5$ V.

Figure 18.102

(a) What is the potential difference $V_C - V_D$? **(b)** If the element between C and D is a battery, is the + end of the battery at C or at D?

•P42 What would be the potential difference $V_C - V_B$ across the thin resistor in Figure 18.103 if the battery emf is 3.5 V? Assume that the electric field in the thick wires is very small (so that the potential differences along the thick wires are negligible). Do you have enough information to determine the current I in the circuit?

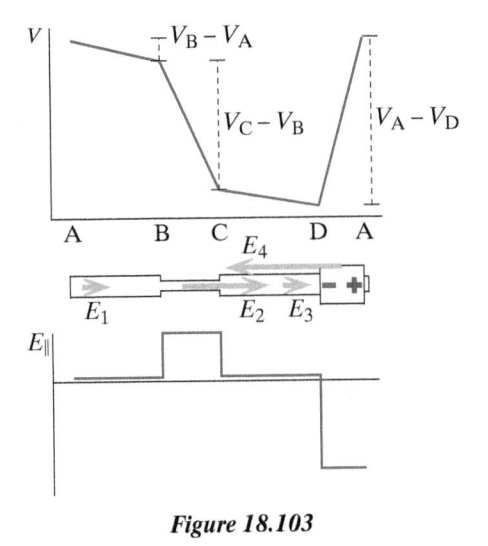

Figure 18.103

••P43 A circuit is constructed from two batteries and two wires, as shown in Figure 18.104. Each battery has an emf of 1.3 V. Each

wire is 26 cm long and has a diameter of 7×10^{-4} m. The wires are made of a metal that has 7×10^{28} mobile electrons per cubic meter; the electron mobility is 5×10^{-5} (m/s)/(V/m). A steady current runs through the circuit. The locations marked by × and labeled by a letter are in the interior of the wire.
(a) Which of these statements about the electric field in the interior of the wires, at the locations marked by ×'s, are true? List all that apply. (1) The magnitude of the electric field at location G is larger than the magnitude of the electric field at location F.

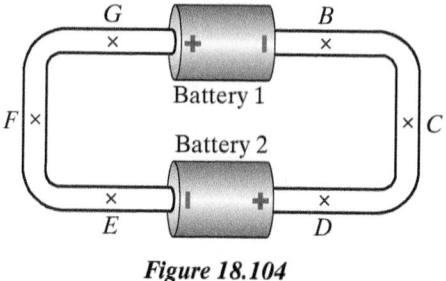

Figure 18.104

(2) At every marked location the magnitude of the electric field is the same. (3) At location B the electric field points to the left. **(b)** Write a correct energy conservation (round-trip potential difference) equation for this circuit, along a round-trip path starting at the negative end of battery 1 and traveling counterclockwise through the circuit (that is, traveling to the left through the battery, and continuing on around the circuit in the same direction). **(c)** What is the magnitude of the electric field at location B? **(d)** How many electrons per second enter the positive end of battery 2? **(e)** If the cross-sectional area of both wires were increased by a factor of 2, what would be the magnitude of the electric field at location B? **(f)** Which of the diagrams in Figure 18.105 best shows the approximate distribution of excess charge on the surface of the circuit?

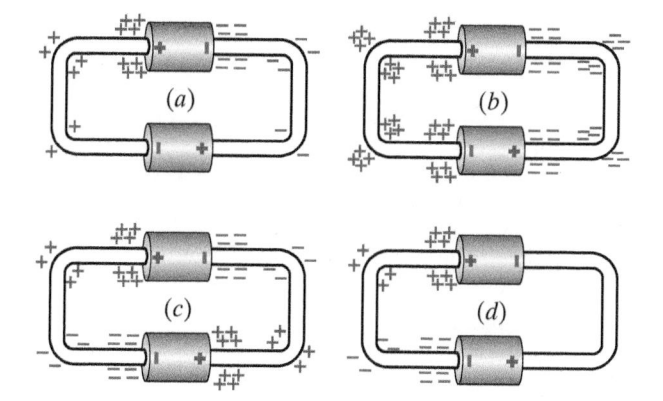

Figure 18.105

••P44 Three identical light bulbs are connected to two batteries as shown in Figure 18.106.
(a) To start the analysis of this circuit you must write energy conservation (loop) equations. Each equation must involve a round-trip path that begins and ends at the same location. Each segment of the path should go through a wire, a bulb, or a battery (not through the air). How many valid energy conservation

(loop) equations is it possible to write for this circuit? **(b)** Which of the following equations are valid energy conservation (loop) equations for this circuit? E_1 refers to the electric field in bulb 1; L refers to the length of a bulb filament. Assume that the electric field in the connecting wires is small enough to neglect.

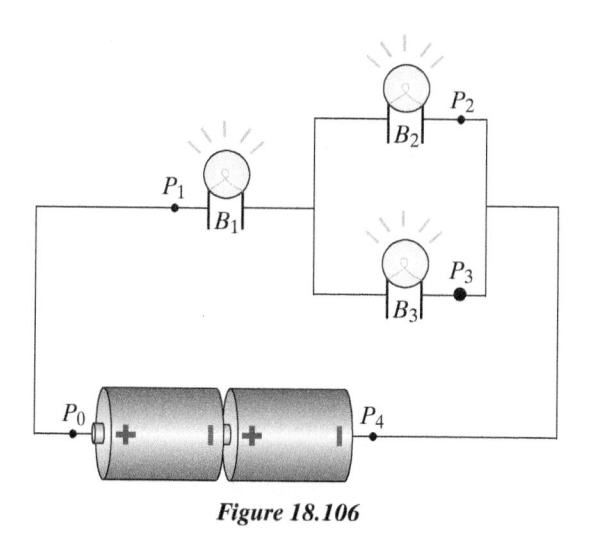

Figure 18.106

(1) $+E_2 L - E_3 L = 0$, (2) $E_1 L - E_3 L = 0$, (3) $+2\,\text{emf} - E_2 L - E_3 L = 0$, (4) $E_1 L - E_2 L = 0$, (5) $+2\,\text{emf} - E_1 L - E_2 L = 0$, (6) $+2\,\text{emf} - E_1 L - E_3 L = 0$, (7) $+2\,\text{emf} - E_1 L - E_2 L - E_3 L = 0$.
(c) It is also necessary to write charge conservation equations (node) equations. Each such equation must relate electron current flowing into a node to electron current flowing out of a node. Which of the following are valid charge conservation equations for this circuit? (1) $i_1 = i_3$, (2) $i_1 = i_2$, (3) $i_1 = i_2 + i_3$.

Each battery has an emf of 1.5 V. The length of the tungsten filament in each bulb is 0.008 m. The radius of the filament is 5×10^{-6} m (it is very thin!). The electron mobility of tungsten is 1.8×10^{-3} (m/s)/(V/m). Tungsten has 6×10^{28} mobile electrons per cubic meter. Since there are three unknown quantities, we need three equations relating these quantities. Use any two valid energy conservation equations and one valid charge conservation equation to solve for E_1, E_2, i_1, and i_2.

••P45 The circuit shown in Figure 18.107 consists of a single battery, whose emf is 1.8 V, and three wires made of the same material but having different cross-sectional areas. Each thick wire has cross-sectional area 1.4×10^{-6} m^2 and is 25 cm long. The thin wire has cross-sectional area 5.9×10^{-8} m^2 and is 6.1 cm long.

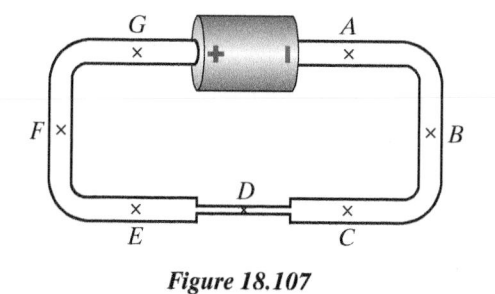

Figure 18.107

In this metal, the electron mobility is 5×10^{-4} (m/s)/(V/m), and there are 4×10^{28} mobile electrons/m^3.

(a) Which of the following statements about the circuit in the steady state are true? (1) At location B the electric field points toward the top of the page. (2) The magnitude of the electric field at locations F and C is the same. (3) The magnitude of the electric field at locations D and F is the same. (4) The electron current at location D is the same as the electron current at location F. **(b)** Write a correct energy conservation (loop) equation for this circuit, following a path that starts at the negative end of the battery and goes counterclockwise. **(c)** Write a correct charge conservation (node) equation for this circuit. **(d)** Use the appropriate equation(s), plus the equation relating electron current to electric field, to solve for the magnitudes E_D and E_F of the electric field at locations D and F. **(e)** Use the appropriate equation(s) to calculate the electron current at location D in the steady state.

••P46 In the circuit shown in Figure 18.108, two thick copper wires connect a 1.5 V battery to a Nichrome wire. Each thick connecting wire is 17 cm long and has a radius of 9 mm. Copper has 8.4×10^{28} mobile electrons per cubic meter and an electron mobility of 4.4×10^{-3} (m/s)/(V/m). The Nichrome wire is 8 cm long, and has a radius of 3 mm. Nichrome has 9×10^{28} mobile electrons per cubic meter and an electron mobility of 7×10^{-5} (m/s)/(V/m).

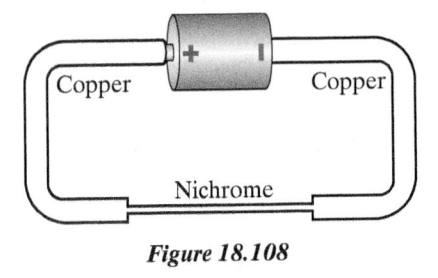

Figure 18.108

(a) What is the magnitude of the electric field in the thick copper wire? **(b)** What is the magnitude of the electric field in the thin Nichrome wire?

••P47 When a single thick-filament bulb of a particular kind and two batteries are connected in series, 3×10^{18} electrons pass through the bulb every second. When two batteries connected in series are connected to a single thin-filament bulb, with a filament made of the same material and the same length as that of the thick-filament bulb but smaller cross section, only 1.5×10^{18} electrons pass through the bulb every second.

(a) In the circuit shown in Figure 18.109, how many electrons per second flow through the thin-filament bulb? **(b)** What approximations or simplifying assumptions did you make? **(c)** Show approximately the surface charge on a diagram of the circuit.

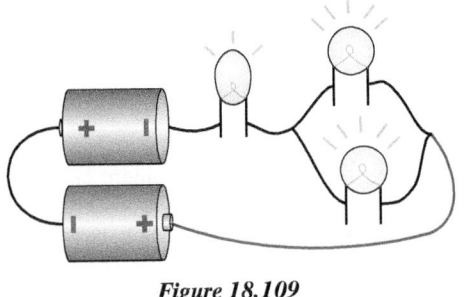

Figure 18.109

••P48 In the circuit shown in Figure 18.110, the two thick wires and the thin wire are made of Nichrome.

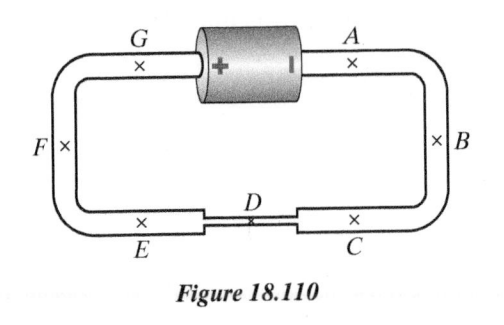

Figure 18.110

(a) Show the steady-state electric field at the locations indicated, including in the thin wire. **(b)** Carefully draw pluses and minuses on your own diagram to show the approximate surface-charge distribution in the steady state. Make your drawing show clearly the differences between regions of high surface-charge density and regions of low surface-charge density. **(c)** The emf of the battery is 1.5 V. In Nichrome there are $n = 9 \times 10^{28}$ mobile electrons per m^3, and the mobility of mobile electrons is $u = 7 \times 10^{-5}$ (m/s)/(N/C). Each thick wire has length $L_1 = 20\,cm = 0.2\,m$ and cross-sectional area $A_1 = 9 \times 10^{-8}\,m^2$. The thin wire has length $L_2 = 5\,cm = 0.05\,m$ and cross-sectional area $A_2 = 1.5 \times 10^{-8}\,m^2$. (The total length of the three wires is 45 cm.)   In the steady state, calculate the number of electrons entering the thin wire every second. Do not make any approximations, and do not use Ohm's law or series-resistance equations. State briefly where each of your equations comes from.

••P49 Three identical thick-filament bulbs are in series as shown in Figure 18.111. Thick copper wires connect the bulbs. In the steady state, 3×10^{17} electrons leave the battery at location A every second.

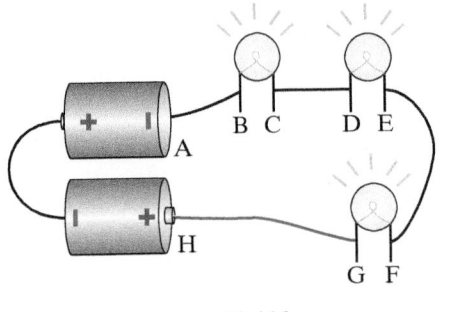

Figure 18.111

(a) How many electrons enter the second bulb at location D every second? If there is insufficient information to give a numerical answer, state how it compares with 3×10^{17}. Justify your answer carefully. **(b)** Next, the middle bulb (at DE) is replaced by a wire, as shown in Figure 18.112. Now how many electrons leave the batteries at location A every second? Explain clearly! If you have to make an approximation, state what it is. Do not use ohms or series-resistance equations in your explanation, unless you can show in detail how these concepts follow from the microscopic analysis introduced in this chapter.

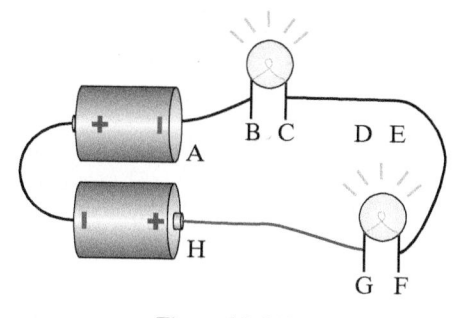

Figure 18.112

(c) Check your analysis by trying the experiment with a partner. Try to find three thick-filament bulbs that glow equally brightly when in series with each other, because bulb construction varies slightly in manufacturing. Remember to arrange the circuit so that the largest compass deflection is no more than 15°. Report the deflections that you observe. Does the experiment agree with your prediction? (If not, can you explain the discrepancy? Be specific. For example, if the current is larger than predicted, explain why it is larger than predicted.) **(d)** Finally, the last bulb (at FG) is replaced by a bulb identical in every way except that its filament has twice as large a cross-sectional area, as shown in Figure 18.113. Now how many electrons leave the batteries at location A every second? Explain clearly! If you have to make an approximation, state what it is.

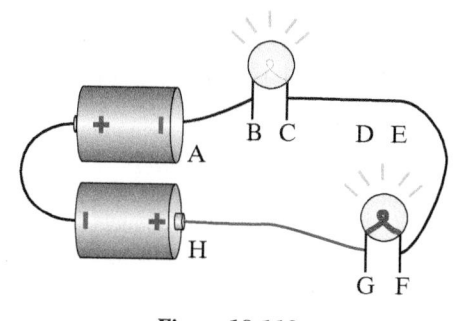

Figure 18.113

••P50 The following questions refer to the circuit shown in Figure 18.114, consisting of two flashlight batteries and two Nichrome wires of different lengths and different thicknesses as shown (corresponding roughly to your own thick and thin Nichrome wires).

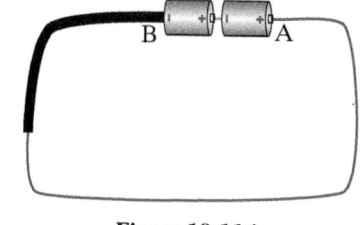

Figure 18.114

The thin wire is 50 cm long, and its diameter is 0.25 mm. The thick wire is 15 cm long, and its diameter is 0.35 mm. **(a)** The emf of each flashlight battery is 1.5 V. Determine the steady-state electric field inside each Nichrome wire. Remember that in the steady state you must satisfy both the current node rule and energy conservation. These two principles give you two equations for the two unknown fields. **(b)** The electron mobility in room-temperature Nichrome is about 7×10^{-5} (m/s)/(N/C). Show that it takes an electron 36 min to drift through the two

Nichrome wires from location B to location A. **(c)** On the other hand, about how long did it take to establish the steady state when the circuit was first assembled? Give a very approximate numerical answer, not a precise one. **(d)** There are about 9×10^{28} mobile electrons per cubic meter in Nichrome. How many electrons cross the junction between the two wires every second?

•P51 A Nichrome wire 75 cm long and 0.25 mm in diameter is connected to a 1.7 V flashlight battery. **(a)** What is the electric field inside the wire? **(b)** Next, the Nichrome wire is replaced by a wire of the same length and diameter, and same mobile electron density but with electron mobility 4 times as large as that of Nichrome. Now what is the electric field inside the wire? **(c)** The electron current in the first circuit (Nichrome) is i_1. The electron current in the second circuit (wire with higher mobility) is i_2. Which of the following statements is true? (1) $i_2 = i_1$, (2) $i_2 < i_1$, (3) $i_2 > i_1$, (4) Not enough information is given to compare the two currents.

••P52 Two circuits are assembled using 1.5 V batteries, thick copper connecting wires, and thin-filament bulbs (Figure 18.115). Bulbs A, B, and C are identical thin-filament bulbs. Compasses are placed under the wires at the indicated locations. **(a)** On sketches of the circuits, draw the directions the compass needles will point, and indicate the approximate magnitude of the compass deflections in degrees. Note that the deflection is given at one location. If you do not

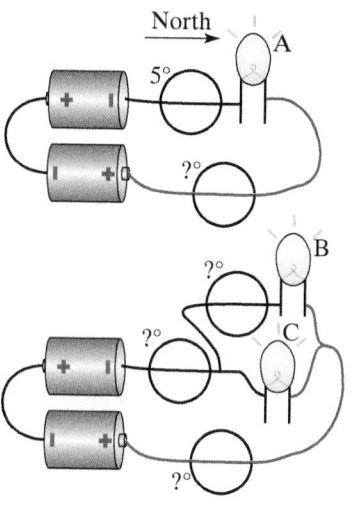

Figure 18.115

have enough information to give a number, then indicate whether it will be greater than, less than, or equal to 5°. (Assume that the compasses are adequately far away from the steel-jacketed batteries.) **(b)** Briefly explain your reasoning about the magnitudes of the compass deflections. **(c)** On the same sketches of the circuits, show very approximately the distribution of surface charge. **(d)** The tungsten filament in each of the bulbs is 4 mm long with a radius of 6×10^{-6} m. Calculate the electric field inside each of the three bulbs, E_A, E_B, and E_C.

••P53 A circuit is assembled that contains a thin-filament bulb and a thick-filament bulb as shown in Figure 18.116, with four compasses placed underneath the wires (we're looking down on the circuit). When the thin-filament bulb is unscrewed from its socket, the compass on the left (next to the battery) deflects 15°. When the thin-filament bulb is screwed back in and the thick-filament bulb is unscrewed, the compass on the left deflects 4°.

North

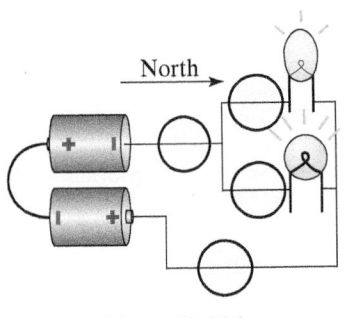

Figure 18.116

With both bulbs screwed into their sockets, draw the orientations of the needle on each compass, and write the compass deflection in degrees beside the compass. Explain briefly.

•••P54 A solid metal sphere of radius R carries a uniform charge of $+Q$. Another solid metal sphere of radius r carries a uniform charge $-q$. The amount of charge is not enough to cause breakdown in air. The two spheres are very far apart (distance $\gg R$ and distance $\gg r$). At $t = 0$ a very thin wire of length L is connected to the two spheres (Figure 18.117). The mobility u of mobile electrons in this wire is very small, and the wire conducts electrons so poorly that it takes about an hour for the system to reach equilibrium. In a short time Δt (a few seconds) how many electrons leave the sphere of radius r? There are n mobile electrons per cubic meter in the wire, and the wire has a constant cross-sectional area A. Explain your work and any approximations you need to make.

+Q
R
L
−q
r

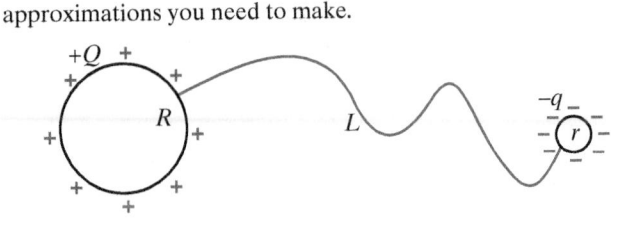

Figure 18.117

ANSWERS TO CHECKPOINTS

1 $I_1 + I_4 = I_2 + I_3$; $I_2 = 3$ A; $I_2 = -2$ A, where the minus sign means that the current is entering the node rather than leaving. It is not true that current must be the same in every part of a circuit, only that the current entering a node must equal the current leaving a node; here $i_1 = i_2 + i_3$.
2 $\overline{v_2} = 8 \times 10^{-5}$ m/s; $E_2 = 1.8 \times 10^{-2}$ N/C
3 See Figure 18.118.

$\bar{v}$
E

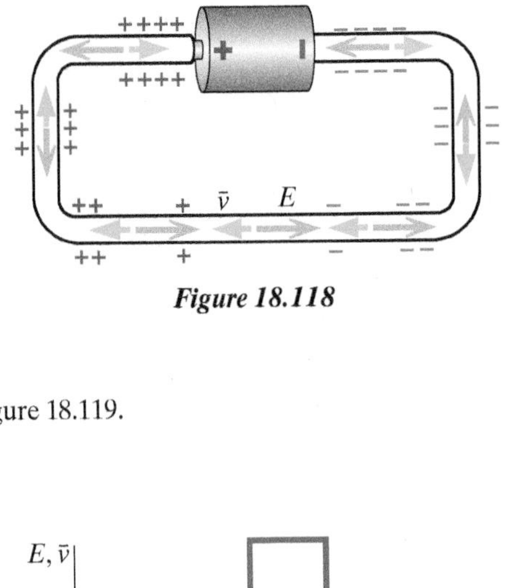

Figure 18.118

4 See Figure 18.119.

$E, \bar{v}$
E
$\bar{v}$
1
2
3
4
5
6
Location around circuit

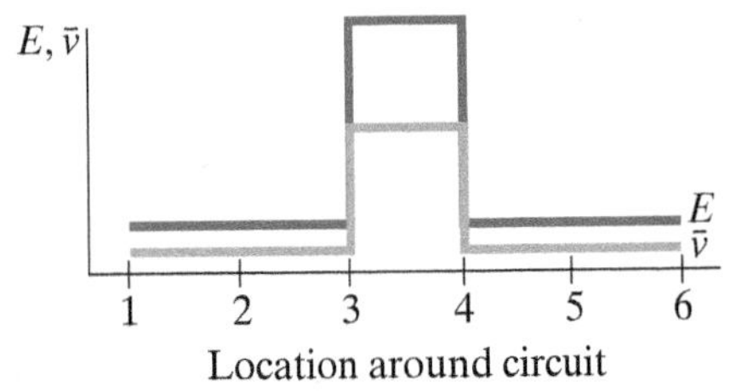

Figure 18.119

5 See Figure 18.120.

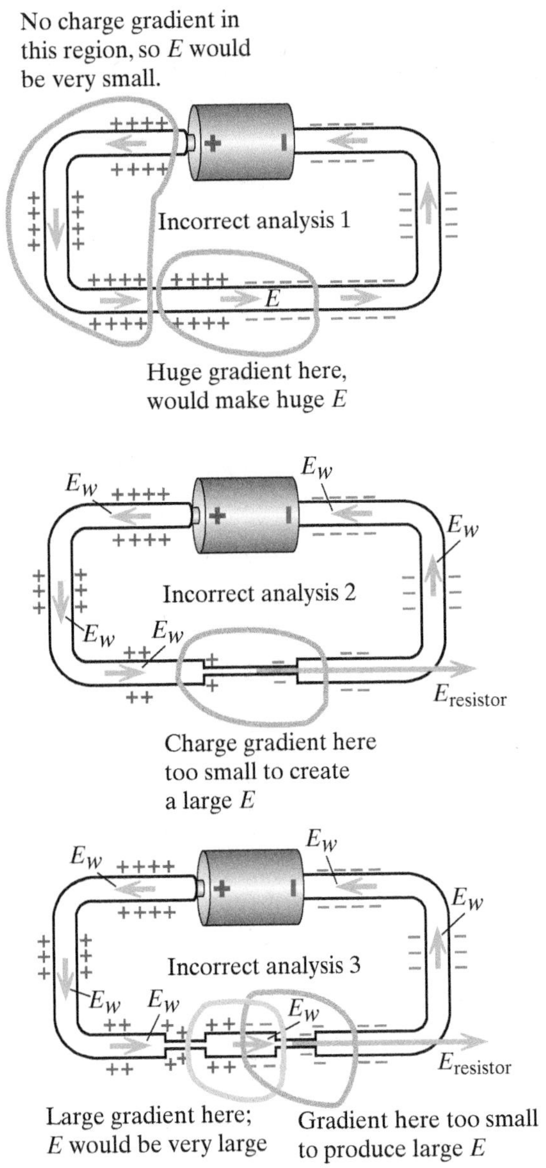

No charge gradient in this region, so E would be very small.
Incorrect analysis 1
E
Huge gradient here, would make huge E
Incorrect analysis 2
E_W
$E_{resistor}$
Charge gradient here too small to create a large E
Incorrect analysis 3
E_W
$E_{resistor}$
Large gradient here; E would be very large
Gradient here too small to produce large E

Figure 18.120

6 8.3×10^{-14} C. This is much less than the charge on a tape (about 1×10^{-8} C). One would not expect to observe repulsion, since the repulsive force would probably be smaller than the force attracting the tape to the battery (which had been polarized by the charged tape).

7 The electric field in the thick copper connecting wires is very small, so emf $\approx E_{bulb} L_{bulb}$ and the electric field in the bulb filament depends almost entirely on the length of the filament itself.

8 $+1.5$ V; $+$ end of battery is at C.

9 Approximately 1.5 V; not enough information to determine I.

19

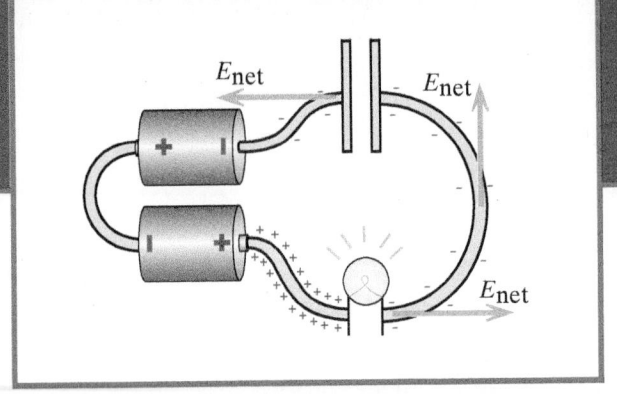

Circuit Elements

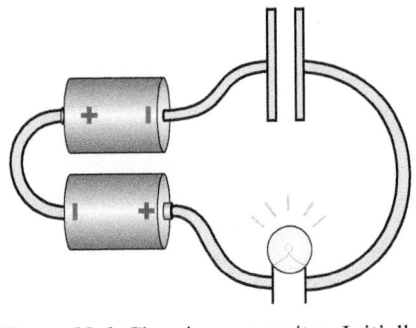

Figure 19.1 Charging a capacitor. Initially the bulb glows brightly, but with time the bulb dims and finally goes out.

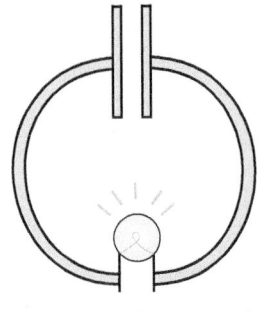

Figure 19.2 Discharging a capacitor. Initially the bulb glows brightly, but with time the bulb dims and finally goes out.

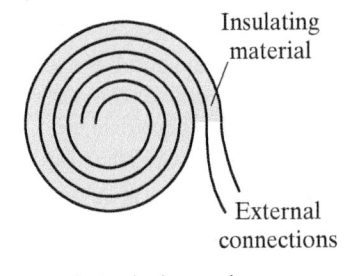

Insulating material

External connections

Figure 19.3 Typical capacitor construction—a coiled-up sandwich of metal foils separated by insulating material.

OBJECTIVES

After studying this chapter you should be able to

- Qualitatively and quantitatively explain the behavior of a circuit containing a capacitor.
- Analyze a circuit using macroscopic quantities such as resistance and capacitance.

19.1 CAPACITORS

Experiment EXP1 at the end of this chapter shows an interesting phenomenon. In the experiment you connect batteries in series with a certain kind of capacitor (1 farad, F) and a flashlight bulb (Figure 19.1), and what you observe in this "charging" process is that initially the bulb glows brightly, as brightly as if there were no capacitor in the circuit, despite the fact that no current can flow across the gap between the capacitor plates. As time goes on, however, the bulb gets dimmer and dimmer and eventually goes out after many seconds.

After the bulb has stopped glowing, if you remove the batteries and attach the bulb to the capacitor ("discharging," Figure 19.2), the bulb again glows brightly but then gets dimmer and dimmer and eventually stops glowing after many seconds.

There are two different time scales for events in this circuit. The bulb lights instantly when a connection is made (surface charge rearrangement takes only nanoseconds), but the current in the circuit declines slowly over a period of many seconds—the two time scales differ by a factor of 1×10^{10}. We call the slow process a "quasi-steady state," in which the currents change slowly with time as the circuit slowly comes into equilibrium, with the current coming to a stop. (It should be mentioned that in high-frequency circuits with capacitors measured in picofarads, 1×10^{-12} F, there may be little difference in these two time scales. Also, surface charge rearrangement in a circuit as big as a national power grid will take far longer than a nanosecond.)

Capacitor Construction

The capacitors we've discussed in previous chapters consisted of two very large flat plates separated by a gap. However, it wouldn't be very practical to try to use such a large device in a circuit. A capacitor of the kind used in circuits often consists of a sandwich of two strips of metal foil separated by insulating material and coiled up into a small package (Figure 19.3). Note that there is no conducting path through the capacitor, even though you are able to light a bulb in series with the capacitor. Charges cannot jump from one plate to the other.

Discharging a Capacitor

Using what we already know, we can explain the charging and discharging of a capacitor in terms of the fundamental properties of charge and field. First we'll discuss the discharging process, which is easier to understand. We'll consider a simple circuit like that shown in Figure 19.2.

Our reasoning will focus on the net electric field at a location inside the wire, and the effect it has on the electrons at that location in the wire. We'll think about it step by step, iteratively, choosing a time step that is long compared to the time required to discharge the capacitor (several seconds), but short compared to the time it takes surface charge to arrange itself on the wire and battery (nanoseconds). Figure 19.4 shows the pattern of electric field in the connecting wires 0.01 s after connecting the bulb to the charged capacitor, while the bulb is glowing brightly.

Just outside the capacitor, we know that the electric field at a location inside the wire points away from the positive plate and toward the negative plate, due in part to the fringe field of the capacitor. After the quick transient (a matter of nanoseconds), the electron current everywhere in the connecting wires must be the same, and surface charge will build up to make a uniform electric field everywhere in the wires. There is a big field inside the bulb filament (not shown), across which is a big surface charge gradient, going from + to −.

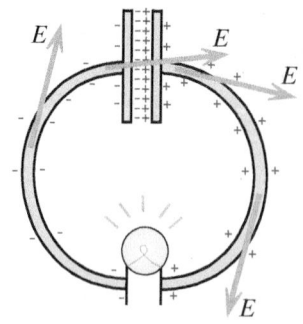

Figure 19.4 Charge and electric field 0.01 s after connecting the bulb to the charged capacitor. The bulb is bright.

> QUESTION In Figure 19.4, electron current will move opposite to the electric field. In the next short time interval, given the direction of electron current, what will happen to the amount of charge on the negative plate of the capacitor? On the positive plate?

Since electron current moves away from the negative plate, the negative plate will lose some of its negative charge. Since electron current moves toward the positive plate, that plate will become less positively charged. Since the plates now have less charge on them, they contribute a smaller fringe field in the connecting wires, and the electric field everywhere in the circuit gets smaller, as seen in Figure 19.5. Smaller electric field means smaller electron current, so the bulb glows less brightly.

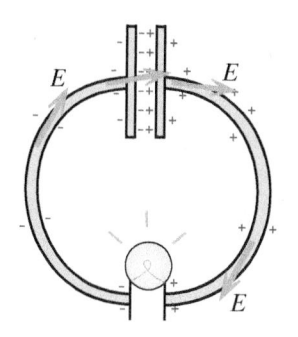

Figure 19.5 Charge and electric field 1 s after connecting the bulb to the charged capacitor. The bulb is less bright than before.

> QUESTION As the process continues, what will be the final state?

With time, the capacitor plates will lose more and more of their charge, so the electric field will get smaller and smaller, and the electron current will get smaller and smaller. The bulb will get dimmer and dimmer and finally stop glowing visibly when the current is so small that the bulb filament isn't hot enough to glow. Eventually the capacitor plates lose all their charge and the current stops completely (Figure 19.6). The circuit is now in equilibrium.

The dimming of the light bulb is gradual because the less charge there is on the capacitor plates, the smaller the fringe field and the smaller the current draining charge off the plates. The amount of charge that leaves the plates in the first second is larger than the amount that leaves in the second second, which in turn is larger than the amount that leaves in the third second. The bulb glows brightly for a short time but dimly for a long time.

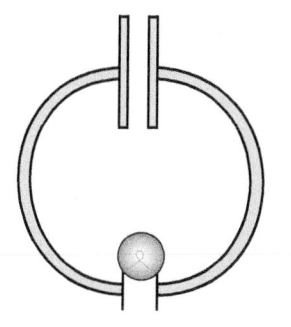

Figure 19.6 After many seconds there is no more charge on the capacitor plates, and the current is zero (equilibrium).

> QUESTION Since the positive and negative charges on the plates of the capacitor are attracted to each other, why does charge leave the capacitor and go through the light bulb?

In the reasoning above we focused on the electric field in the wire close to but outside the plates. Initially, right next to the negative plate, the electric field in the wire is not zero and points toward the negative plate, as shown in Figure 19.4. Electrons in the wire at this location feel a force and move away from the negative plate; this will create a positively charged region inside the wire unless electrons flow off the negative plate into the wire. Electrons in the wire near the positive plate likewise move toward the positive plate, under the influence of the fringe electric field in the wire.

Charging a Capacitor

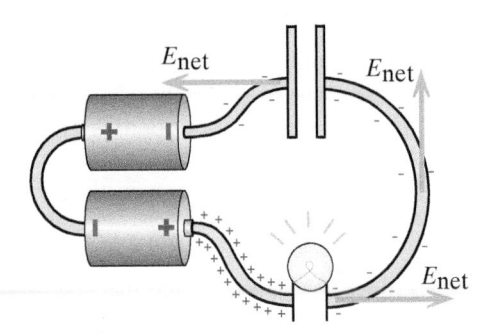

Figure 19.7 Charge and electric field 0.01 s after connecting the circuit. The bulb is as bright as in a noncapacitor circuit.

We can use similar reasoning to explain the initial charging of the capacitor in a circuit containing a battery. As we did above, we will consider the electric field at a location inside the wire, and the effect this field has on electrons in the wire at that location. Again we will reason step by step, considering this as an iterative process.

We'll choose a time step of about 0.01 s, which is short compared to the total time it takes to charge the capacitor (a few seconds), but long compared to the time it takes the surface charges on the battery and wire to rearrange themselves (a few nanoseconds). Because the capacitor is initially uncharged, during the first time step the capacitor's contribution to the net field inside the wire is nearly zero (Figure 19.7).

> QUESTION Since the capacitor is initially uncharged, how bright will the bulb be initially, compared to a circuit containing just batteries and the bulb?

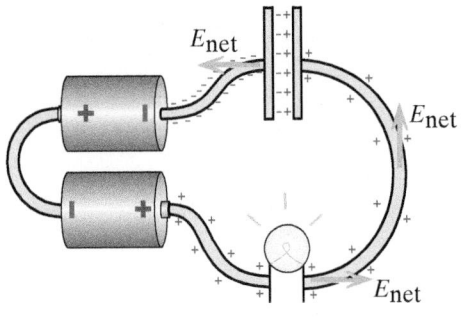

Figure 19.8 Charge and electric field 1 s after connecting the circuit. The bulb is less bright than before.

Evidently the initial current will be the same as it would be in a simple bulb circuit, and the bulb just as bright as usual (Figure 19.7). However, after several time steps, when current has been running in the wires for a second or more, a significant amount of + and − charge will have built up on the capacitor plates.

> QUESTION One second after the circuit is connected, will the fringe field of the capacitor affect the net electric field in the neighboring wire—will it increase or decrease the net field? What will happen to the current?

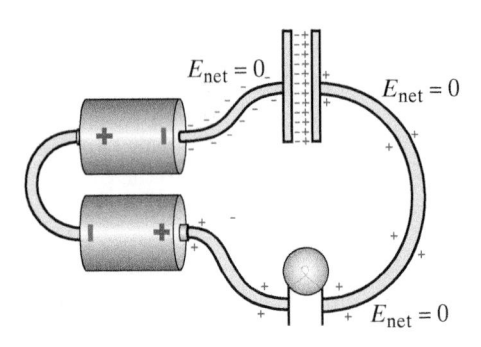

Figure 19.9 After many seconds the current is zero (equilibrium).

The fringe field of the capacitor acts in a direction opposite to the conventional current, thus decreasing the net field. As a result, the current will decrease (Figure 19.8). Eventually the current stops completely when there is enough charge on the capacitor plates to create a large enough fringe field to counteract the field made by all the other charges. The light stops glowing (Figure 19.9). We see that for both charging and discharging, a key to explaining the qualitative behavior of the circuit is the varying fringe field of the capacitor.

The initial establishment of surface charges takes a few nanoseconds, whereas it takes several seconds to reach equilibrium. During the several seconds while the bulb is glowing, the current is not constant, but it changes so slowly that it makes sense to describe the situation as a "quasi-steady state."

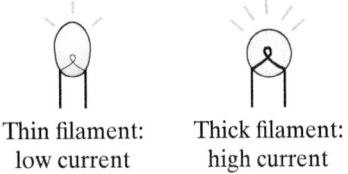

Thin filament: Thick filament:
low current   high current

Figure 19.10 When a thin-filament bulb is connected to a battery, the current is small and the bulb is dim. With a thick-filament bulb the current is high and the bulb is bright.

The Effect of the Resistor

You might wonder what effect the light bulb in the circuit has on the charging process and on the final state of the capacitor. Suppose that you have two light bulbs, one with a thick filament and the other with a thin filament (Figure 19.10). If you charge a capacitor through a thin-filament bulb, the bulb is initially dim (low current) and takes a long time to go out. In contrast, if you charge the capacitor through a thick-filament bulb, the bulb is initially bright (high current) but glows for only a short time.

QUESTION Should the final amount of charge on the capacitor plates depend on what kind of bulb is used during the charging process? Consider the following two arguments.

- Since charging the capacitor through the thin-filament bulb goes on much longer than charging through the thick-filament bulb, one might argue that the capacitor gets charged up a lot more when the thin-filament bulb is used.
- Alternatively, one might argue that since the thick-filament bulb glows much more brightly, the capacitor would get charged up a lot more even though the bulb goes out much quicker.

In the final state there is no current, so the only potential differences are across the battery and the capacitor. Therefore the potential difference across the capacitor is equal to that across the battery, no matter which bulb is in the circuit, and the charge on the plates is the same no matter which bulb we use. With the thin-filament bulb a small current runs for a long time, and with the thick-filament bulb a large current runs for a short time, but the total amount of charge is the same in either case. Experiment EXP2 is a test of the effect of the kind of bulb on the final amount of charge on the capacitor plates.

In Section 19.7 we will see how to predict the current and the amount of charge on the capacitor quantitatively at any time during the charging or discharging process.

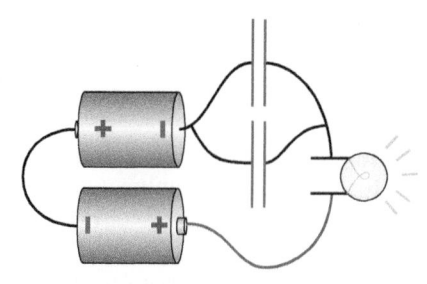

Figure 19.11 Two capacitors in parallel; Experiment EXP3.

Capacitors in Circuits

Capacitors are used in various ways in circuits, some of which are indicated in Figures 19.11–19.13.

Parallel capacitors: Two capacitors in parallel as shown in Figure 19.11 can be thought of as one capacitor with plates that have twice the area (see Checkpoint 1). Experiment EXP3 shows that the bulb glows twice as long with this arrangement.

An isolated bulb: In Experiment EXP4, shown in Figure 19.12, you can light a bulb despite the fact that there is no conducting path between the bulb and the battery! In the isolated section of the circuit, electrons flow from the plate on the right of the capacitor on the top through the bulb onto the plate on the right of the capacitor on the bottom, at the same time that there is current in the other wires.

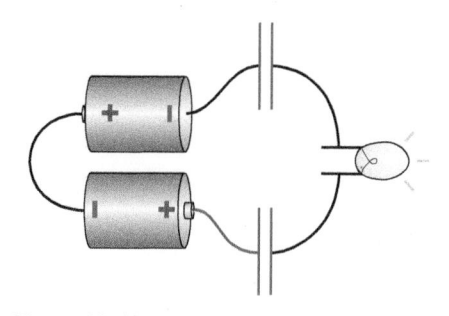

Figure 19.12 A glowing bulb completely isolated from the batteries; Experiment EXP4.

Reversing a capacitor: In Experiment EXP5 you can see that reversing the capacitor in a circuit after charging it leads to an interesting effect. Temporarily, it's like putting an extra battery in the circuit.

Compensating for brief power interruptions: Experiment EXP6 (shown in Figure 19.13) helps understand how these capacitors can be useful. What you find if you intermittently break the connection to the batteries for fractions of a second is that the bulb stays lit despite being disconnected from the battery. The current is supplied from the capacitor, which had been charged by the batteries. We say that the capacitor is connected "in parallel" with the batteries, rather than in series.

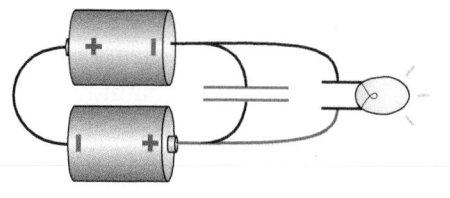

Figure 19.13 What capacitors are used for; Experiment EXP6.

The large-capacity capacitor used in these experiments is designed to be used in this way, connected in parallel to a power supply in a computer, to supply the computer circuits with current during momentary power outages. For a given voltage, these capacitors hold almost a million times more charge than ordinary capacitors!

More generally, capacitors placed in parallel are used in all kinds of electric circuits to even out rapid changes in voltage. For example, in a power supply, parallel capacitors are used to filter out high-frequency (AC) voltage changes that are riding on top of a constant (DC) voltage. You might say that a capacitor makes a circuit behave in a sluggish manner, being unable to change conditions rapidly.

Energy Stored in a Capacitor

Let's consider briefly the energy aspects of charging and discharging a capacitor through a light bulb.

> QUESTION In the charging phase, from a time just before the circuit is completed to the time when there is essentially no more current, state whether there is energy gain or loss in the battery, in the bulb, in the capacitor, and in the surroundings. (*Hint:* Think especially carefully about the initial and final states of the bulb.)

The battery loses energy, the bulb neither loses nor gains (it wasn't glowing originally, and it isn't glowing afterward either), the capacitor gains energy (enough to power a bulb later), and the surroundings gain energy (light and heat transfer from the bulb).

The Current Node Rule in a Capacitor Circuit

The current node rule in circuits results from charge conservation in the steady state. Figure 19.14 shows a simple application of the node rule to a node in a circuit where the incoming current I_1 supplies the outgoing currents I_2 and I_3, so that $\sum (I_{\text{entering}}) = I_1 - I_2 - I_3 = 0$.

Figure 19.15 shows a more complex situation. Consider locations 1 and 2 as nodes. While the capacitor is being charged, there is current running onto one plate of the capacitor without any current coming off that plate (region 1 in Figure 19.15), and as a result the charge on the plate is changing. This is clearly not a steady state. Similarly, current comes off of the other plate (region 2 in Figure 19.15), which changes the charge on that plate.

However, if you look at the capacitor as a whole (region 3 in Figure 19.15), just as much current leaves one plate as enters the other plate. Otherwise the two plates together would acquire a nonzero charge that would quickly repel or attract electrons in such a way as to restore the net charge of the capacitor to be zero. So during charging or discharging of a capacitor the current node rule applies to the capacitor as a whole, even though it doesn't apply to just one plate of the capacitor.

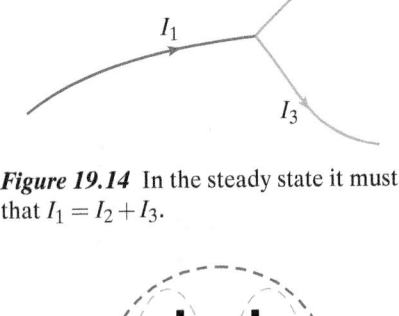

Figure 19.14 In the steady state it must be that $I_1 = I_2 + I_3$.

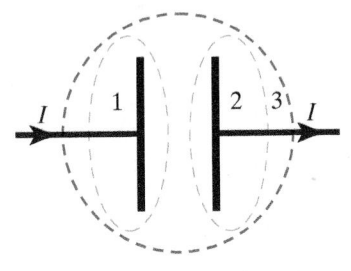

Figure 19.15 In a non-steady-state situation, the current node rule does not apply.

Capacitance

The capacitors you used in your circuit experiments may have had something like "0.47 F" or "1 F" marked on them, which would mean that they have a "capacitance" of about 0.5 or 1 farad (a unit honoring Michael Faraday). Let's see what is meant by this "capacitance" rating.

The more charge on the capacitor plates ($+Q$ and $-Q$), the bigger the electric field in the gap, and the greater the potential difference ΔV across the gap (Figure 19.16). Thus the absolute magnitude of the potential difference $|\Delta V|$ is proportional to the amount of positive charge Q on the positive plate, and conversely Q is proportional to $|\Delta V|$. We define capacitance C as this proportionality constant:

DEFINITION OF CAPACITANCE

$$Q = C|\Delta V|$$

The units of capacitance are coulombs per volt, or farads.

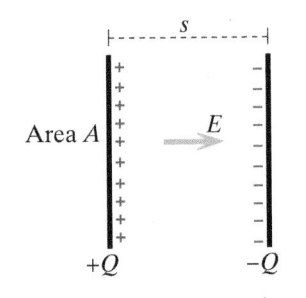

Figure 19.16 The more charge there is on the plates, the bigger the electric field and the bigger the potential difference.

QUESTION By using what you know about E and ΔV, show that $C = (\varepsilon_0 A)/s$ in a parallel-plate capacitor with plate area A and plate separation s.

Since the magnitude of the electric field inside the gap is

$$E = (Q/A)/\varepsilon_0$$

we can calculate the potential difference as

$$|\Delta V| = [(Q/A)/\varepsilon_0]s$$

which shows that the potential difference is proportional to the charge Q on one of the plates, as we expect. The capacitance is therefore

$$C = Q/|\Delta V| = (\varepsilon_0 A)/s \quad \text{(parallel-plate capacitor)}$$

Note that the capacitance depends on the geometry of the capacitor: the area of the plates and the separation between them.

The concept of capacitance is useful in the detailed analysis of circuits, and we will use it later to study capacitor circuits quantitatively. Bear in mind, however, that capacitance is just a derived quantity relating more fundamental quantities: charge and potential difference (and ultimately charge and field).

Reflection: Capacitors in Circuits

Circuits containing capacitors display a behavior that may be called a "quasi-steady state." The current in the capacitor circuit is approximately steady over a short time, but only approximately: the current slowly decreases and finally goes to zero—equilibrium is attained.

Actually, even an ordinary battery-and-bulb circuit is not in a true steady state, because when the chemicals in the battery are used up, the system comes into equilibrium (no current). There is an important difference, however. In a capacitor circuit the current continually decreases. In a battery-powered circuit, however, the current may stay nearly constant for many hours, then drop to zero in a much shorter time.

We were able to understand the behavior of a capacitor circuit in terms of the fringe field of the capacitor competing with the electric field of the other charges (battery charges and surface charges). The more charge on the capacitor, the larger the fringe field. When charging the capacitor, equilibrium is finally reached when there is so much charge on the capacitor that the fringe field is large enough to cancel the other contributions to the electric field.

We emphasized a qualitative analysis of capacitor circuits to get at the fundamental issues, and to illustrate the power of reasoning with charge and field. To analyze such circuits quantitatively we need to apply the energy principle in our analysis. Later in this chapter we do this, and we will find that we can quantitatively predict the rate of decrease of the current.

The design of a capacitor has an effect on the behavior of a capacitor circuit. By thinking about what happens in the first fraction of a second when the current decreases slightly, and how the state of the system at the end of this time interval will affect what happens next, we can predict qualitatively how capacitor circuits will behave. In Section 19.7 we will see how to make these predictions quantitative.

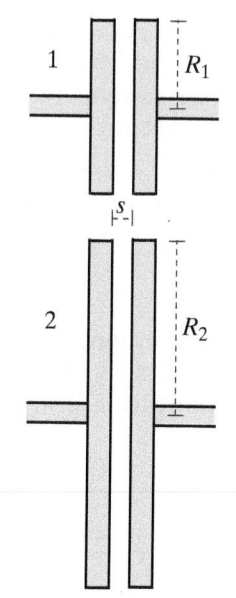

Figure 19.17 What is the effect of increasing the size of the capacitor plates?

Checkpoint 1 Consider two capacitors whose only difference is that the plates of capacitor number 2 are larger than those of capacitor number 1 (Figure 19.17). Neither capacitor has an insulating layer between the plates. They are placed in two different circuits having similar batteries and bulbs in series with the capacitor.

Show that in the first fraction of a second the current stays more nearly constant (decreases less rapidly) in the circuit with capacitor number 2. Explain your reasoning in detail. *Hint:* Show charges on the metal plates, and consider the electric fields they produce in the nearby wires. Remember that the fringe field near a plate outside a circular capacitor is approximately

$$\frac{Q/A}{\varepsilon_0} \left(\frac{s}{2R} \right)$$

A more extensive analysis shows that this trend holds true for the entire charging process: the larger capacitor ends up with more charge on its plates.

19.2 RESISTORS

Our study of electric circuits has been based on a microscopic picture of the motion of electrons inside the wires, due to electric fields produced by charges on and in the battery and on the surfaces of the wires. This microscopic picture gives us insight into the fundamental physical mechanisms of circuit behavior. However, it is not easy to measure electric field, surface charge, electron drift speed, or mobility directly.

On the other hand, it is easy to use relatively inexpensive commercially available meters to measure conventional current (rather than electron current), potential difference (rather than electric field), and "resistance" (rather than mobility). For practical purposes, it is useful to describe and analyze circuits in terms of these macroscopic quantities. We will link what we already know about circuits at the atomic level to a macroscopic description in terms of quantities that can be measured easily with standard meters.

To analyze the behavior of circuits in detail at the microscopic level, we have found it necessary to take into account both properties of particular materials and the geometry of particular circuit elements, such as thin and thick wires. It is sometimes useful to use a single number to summarize the properties of a particular object.

Conductivity: Combining the Properties of a Material

Conventional current is $I = |q|nA\bar{v} = |q|nAuE$, where $|q|$ is the absolute value of the charge of one of the mobile "charge carriers" (electrons or holes), n is the number density of the charge carriers, A is the cross-sectional area of the wire, u is the mobility of the charge carriers, and E is the electric field inside the wire that drives the mobile charges. This equation for conventional current mixes together three quite different kinds of quantities: $|q|$, n, and u are properties of the material, A describes the geometry of the material, and E is the electric field due to charges outside the material (and/or on its surface).

It is helpful to group the material properties together: $I = (|q|nu)AE$. Then if we divide both sides of the equation by the area A, we arrive at the following way of describing the situation:

$$J = \frac{I}{A} = (|q|nu)E = \sigma E$$

where J = current density (A/m^2) and σ = conductivity = $(|q|nu)$.

By dividing the current by the cross-sectional area, we obtain a "current density" J in amperes per square meter, which is independent of the geometry of the situation. The current density is proportional to the applied electric field, and the proportionality constant σ is called the "conductivity," which lumps together all the relevant properties of the material, $|q|nu$. (σ is a lowercase Greek sigma.)

We can consider $\vec{J}$ to be a vector pointing in the direction of the conventional current flow, which is in the direction of the applied electric field $\vec{E}$ (Figure 19.18):

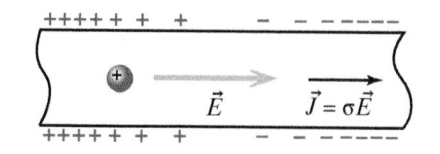

Figure 19.18 $\vec{J}$ is in the direction of the electric field, and in the direction of motion of a fictitious positive charge.

CURRENT DENSITY AND CONDUCTIVITY

$$\vec{J} = \sigma\vec{E}$$

The current density $J = (I/A)$; units are A/m^2.
The conductivity $\sigma = |q|nu$ depends on

- $|q|$, the absolute value of the charge on each carrier;
- n, the number of charge carriers per m^3; and
- u, the mobility of the charge carriers.

A higher conductivity means that a given applied electric field $\vec{E}$ gives larger current densities. The conductivity is a property of the material and has nothing to do with what shape the material has or with how large an electric field is applied to the material.

Conductivity with Two Kinds of Charge Carrier

When electric forces are applied to salt water, both the sodium ions (Na$^+$) and the chloride ions (Cl$^-$) move and constitute an electric current. If the electric field is to the right in the water, the force on the Na$^+$ ions is to the right, but the force on the Cl$^-$ ions is to the left (Figure 19.19).

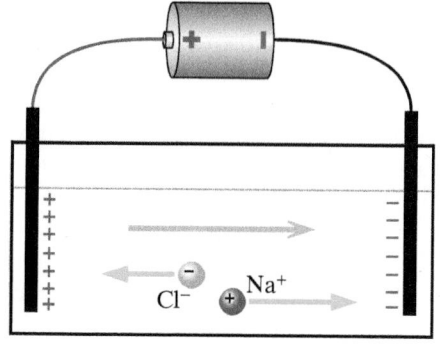

Figure 19.19 Both positive and negative charges move in salt water.

The flow of Na$^+$ ions to the right constitutes a conventional current to the right, and the flow of Cl$^-$ ions to the left also constitutes a conventional current to the right, because the motion of these negative ions has the effect of tending to make the region on the right less negative and hence more positive, with an effect similar to the flow of Na$^+$ ions to the right.

If the charge on the positive ion is q_1 (e in the case of Na$^+$ ions), the positive-ion contribution to the conventional current in coulombs per second is $q_1 n_1 A v_1$. If the charge on the negative ion is q_2 (the charge on a Cl$^-$ ion is $q_2 = -e$), the negative-ion contribution to the conventional current in coulombs per second is $|q_2|n_2 A v_2$. We have to take the absolute value of the charge because the negative ions contribute conventional current in the same direction as do the positive ions.

Note that the mobilities (and therefore the drift speeds) may differ for the two kinds of ions. Larger ions usually have lower mobilities because it is more difficult to move them through the water.

The conventional current in amperes (coulombs per second) is the sum of the conventional current due to the positive charges and the conventional current in the same direction due to the negative charges:

$$I = |q_1|n_1 A v_1 + |q_2|n_2 A v_2$$

Note that both ionic currents contribute to the total current, and that the ionic charge q_1 or q_2 may be a multiple of e. For example, the charge of Zn^{++} is $2e$.

The current density $J = I/A = |q_1|n_1u_1E + |q_2|n_2u_2E = \sigma E$, so the conductivity in this situation where there are two kinds of charge carrier is

CONDUCTIVITY WITH TWO KINDS OF CHARGE CARRIERS

$$\sigma = |q_1|n_1u_1 + |q_2|n_2u_2$$

Resistance Combines Conductivity and Geometry

Often when we have compared two different electric circuits, we have said such things as, "The cross-sectional area goes up, the length decreases, and the mobility increases." It is practical to have a single quantity to describe a resistor that encapsulates both the properties of the material (mobility, mobile-carrier number density) and the geometrical properties of the particular object (cross-sectional area, length). This quantity is called "resistance."

Consider the potential difference ΔV from one end to the other of a wire of constant cross-sectional area A, length L, and uniform composition (Figure 19.20). In the steady state, the electric field inside the wire has the same magnitude everywhere (otherwise the drift speed and current would vary) and is everywhere parallel to the wire (otherwise additional charge would be driven to the surface). Since potential difference is defined as

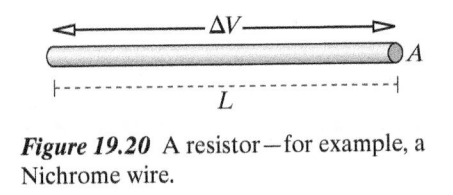

Figure 19.20 A resistor—for example, a Nichrome wire.

$$\Delta V = -\int_i^f \vec{E} \bullet d\vec{l}$$

we have $|\Delta V| = EL$. The conventional current density $J = I/A$ is the same everywhere on the cross-sectional area A of the wire, as was proven in the previous chapter (if E varied across the cross section, one could construct a path with a nonzero round-trip potential difference, and uniform E implies uniform v and therefore uniform J).

> QUESTION For some materials $J = \sigma E$. For such materials, write an equation for I in terms of σ and ΔV (consider I and ΔV to both have positive values).

Combining the two equations for ΔV and J, we find a relation between two macroscopic quantities, conventional current I and potential difference ΔV:

$$J = \sigma E$$
$$\frac{I}{A} = \sigma \frac{\Delta V}{L}$$

It is useful to rearrange this relation between current and potential difference in the following way:

$$I = \frac{\Delta V}{L/(\sigma A)} = \frac{\Delta V}{R}$$

where the resistance R is measured in volts per ampere or ohms (abbreviated Ω and named for Georg Ohm, the German physicist who first studied resistance quantitatively in the early 1800s). We have this definition:

DEFINITION OF RESISTANCE

$$R = \frac{L}{\sigma A}$$

Note that resistance depends on

- the properties of the material: $\sigma = |q|nu$
- the geometry of the resistor: L, A

These aspects are lumped into one quantity—resistance. We can make a large-resistance resistor in a variety of ways. The resistor could have a large length L, or a small cross-sectional area A, or a low conductivity σ. Sometimes resistance is expressed in terms of the "resistivity" $\rho = 1/\sigma$. In that case, $R = \rho L/A$.

We can of course rearrange $I = \Delta V/R$ and write $\Delta V = RI$. Usually, however, it is appropriate to think of the potential difference as causing a current, not of the current causing a potential difference, and the form $I = \Delta V/R$ reminds us that the current is the result of applying an electric field, with an associated potential difference.

Note that we haven't developed any new physical principles. We have simply rewritten $I = |q|nA\bar{v} = |q|nA(uE)$ in terms of potential difference, so that we have $I = \Delta V/R$.

Reflection: Connecting Macroscopic and Microscopic Viewpoints

We can relate the analysis we've just done to our earlier microscopic analysis of a current in a material with a single charge carrier, such as electrons in a metal:

$$\bar{v} = uE$$

$$I = |q|nA\bar{v} = |q|nA(uE)$$

Taking ΔV to be positive, $\Delta V = EL$ for a straight wire, so $E = \Delta V/L$.

$$I = |q|nA\left(u\frac{\Delta V}{L}\right) = \frac{\Delta V}{\left(\dfrac{L}{|q|nAu}\right)}$$

$I = \dfrac{\Delta V}{R}$, where $R = \dfrac{L}{\sigma A}$, with $\sigma = |q|nu$. You should derive this relationship yourself without referring to notes.

For some kinds of scientific work the microscopic picture in terms of electric field, mobility, and drift speed ($\bar{v} = uE$) is more useful. However, when you need to make contact with the technological world of voltmeters, ammeters, and ohmmeters, you also need to be able to describe a situation in terms of macroscopic quantities such as potential difference (voltage), resistance, and conventional current.

Here is a side-by-side comparison of the two views:

Microscopic view	Macroscopic view
$\bar{v} = uE$	$J = \sigma E$
$i = nA\bar{v} = nAuE$	$I = |q|nA\bar{v} = \left(\dfrac{1}{R}\right)\Delta V$

In other books you may see this equation written as $V = IR$. This can lead to confusion between potential at one location and the potential difference between two locations, so we always write $\Delta V = IR$.

Ohmic and Non-Ohmic Materials

As you may have seen in experiments with light bulbs, the mobility u of the tungsten in a light bulb filament varies with the temperature. For example, the current through one bulb is less than twice the current through two bulbs in series, because the mobility is lower at higher temperatures. Since conductivity and resistance both depend on mobility ($\sigma = |q|nu$ and $R = L/(\sigma A)$), both these quantities may change as the current through an object varies.

In some materials the conductivity σ is nearly constant, independent of the amount of current flowing through the resistor. We call such constant conductivity materials "ohmic." An ohmic material has the property that $I = \Delta V/R$, where the resistance $R = L/(\sigma A)$ is constant, independent of the amount of current. No material is truly ohmic, because conductivity depends somewhat on temperature, and higher currents tend to raise the temperature of the material. However, many materials can be considered to be approximately ohmic as long as the temperature doesn't change very much.

Resistors made of ohmic materials are called "ohmic" resistors. Doubling the length L of an ohmic resistor doubles the resistance R and cuts the current in half if the potential difference ΔV across the resistor is kept constant. Doubling the cross-sectional area A cuts the resistance in half and doubles the current if the potential difference ΔV across the resistor is kept constant.

Nichrome wires, and the carbon resistors used in electronic circuits, are nearly ohmic resistors, as long as not enough current runs through them to raise the temperature significantly. In metals the conductivity decreases somewhat at higher temperatures, and metals are only approximately ohmic. You can think of the electrons as having more frequent collisions with atoms that have increased random motion at higher temperatures. You may have seen experimental evidence for non-ohmic behavior in the previous chapter, where the current through two bulbs in series was more than half the current through one bulb.

Checkpoint 2 By actual measurement, the current through a thin-filament bulb when connected to two batteries in series (3 V) is about 100 mA; connected to one battery (1.5 V) the current is about 80 mA; and connected to a small voltage of only 50 mV the current is about 6 mA. (Different thin-filament bulbs may differ from these values somewhat.) Using the equation $I = \Delta V / R$, what is R for each of these cases? Is R a constant? Is a thin-filament bulb an ohmic resistor over this whole range of currents?

After calculating R, notice that from 0 to 3 V a thin-filament bulb is not ohmic, although in the range from 1.5 to 3 V the resistance changes by only about 50% and the bulb can be said to be very roughly ohmic.

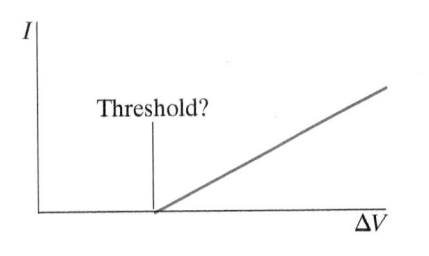

Figure 19.21 Mobile electrons, not bound to atoms—ohmic behavior.

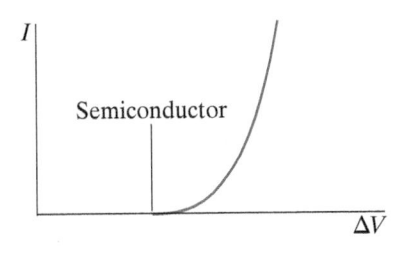

Figure 19.22 If electrons in a metal were bound to the atoms, the current might behave like this.

Semiconductors

In a metal the current is zero when the potential difference is zero, but the slightest potential difference produces some current, as shown in Figure 19.21. Such behavior implies that at least some of the electrons in a metal are truly free, not bound to the atoms. If they were all bound, we would have to apply some field just to free some up, and then more field to increase the current, as shown in Figure 19.22.

Pure silicon and germanium are non-ohmic in a spectacular and extremely useful way. In these "semiconductor" materials, the density of mobile electrons n and therefore the conductivity $\sigma = enu$ (where u is the mobile electron mobility) depend exponentially on the voltage. Applying a slight voltage produces almost no current. After applying enough voltage to get a small current, doubling this voltage may make the current go up by a factor of a hundred rather than by a factor of two! (See Figure 19.23.)

The key difference between a metal and a semiconductor lies in n, the number of available charge carriers per unit volume. In a metal, n is a fixed number (one mobile electron or hole per atomic core). In a semiconductor, at very low temperatures there are no mobile electrons or holes and the material acts nearly like an insulator for small applied fields. However, the electrons are not tightly bound to the atomic cores and it is possible to free up some electrons by applying a large enough electric field, so n is variable and increases with increasing potential difference.

Raising the temperature can also free up some electrons (or holes), and the effect of increasing n more than compensates for the higher rate of collisions with the atomic cores, so that for a semiconductor the resistance decreases with increasing temperature. This is also true for carbon: the resistance of a carbon resistor decreases somewhat with increasing temperature.

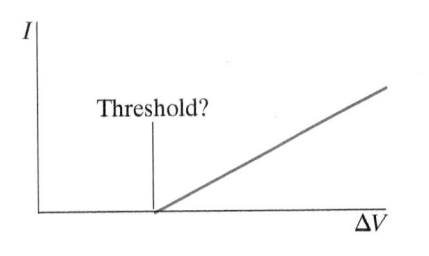

Figure 19.23 Behavior of a pure semiconductor such as silicon or germanium.

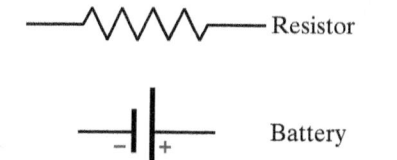

Capacitor

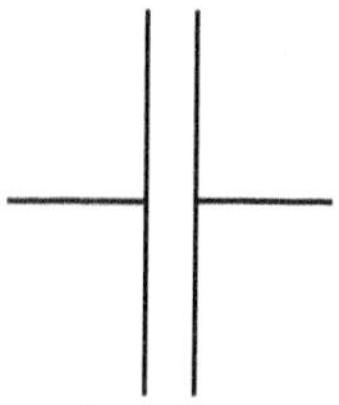

Figure 19.24 The conventional symbol for a capacitor.

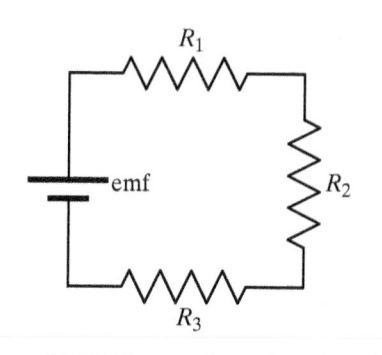

Resistor

Battery

Figure 19.25 Conventional symbols for resistors and batteries.

Other Non-Ohmic Circuit Elements

Not all resistors are ohmic, and we have seen that even those called ohmic are only approximately so if the temperature changes with current. Moreover, circuit elements other than resistors are definitely not ohmic.

Capacitors aren't ohmic. If you double the current flowing onto and off of the capacitor plates, the capacitor gets charged faster, but it is not the case that ΔV is proportional to I for a capacitor. The voltage is proportional to the charge on one of the plates rather than being proportional to the current: $|\Delta V| = Q/C$, where as usual Q is the charge on the positively charged plate of the capacitor. Therefore capacitors are not ohmic devices.

Batteries aren't ohmic. If you double the current through a battery (by changing the circuit elements attached to the battery), the battery voltage hardly changes at all. The battery voltage certainly doesn't double and will actually decrease slightly. It definitely is not the case that ΔV is proportional to I for a battery.

It is extremely important to apply $I = \Delta V/R$ *only* to resistors. This equation is not a fundamental physics principle with broad validity. It is just an expression of the fact that some materials are (approximately) ohmic.

19.3 CONVENTIONAL SYMBOLS AND TERMS

In conventional circuit diagrams wires are represented as straight lines, and stylized symbols are used to represent circuit elements. A capacitor is shown as though we had unrolled the coiled-up capacitor and made external connections to the center of the metal plates (Figure 19.24). The insulating material between the plates is usually not indicated.

Resistors are represented by jagged lines. The symbol for a battery is two parallel lines of unequal length. The longer line represents the positive end of the battery and the short line represents the negative end. Figure 19.25 shows the symbols for a resistor and a battery.

Series Resistors

When circuit elements are connected along a single path with no branches, they are said to be in series. Figure 19.26 shows three ohmic resistors in series with an ideal battery (one with negligible internal resistance).

We can show that several ohmic resistors in series are equivalent to one ohmic resistor with a resistance that is the sum of the individual resistances. Consider the circuit of Figure 19.26.

The current I is the same everywhere in this series circuit.

> QUESTION Write the energy-conservation loop equation for this circuit, using the fact that $\Delta V = RI$ for an ohmic resistor.

We can rearrange the loop equation, $\text{emf} - R_1 I - R_2 I - R_3 I = 0$, in this form:

$$\text{emf} = (R_1 + R_2 + R_3)I = R_{\text{equivalent}} I$$

Figure 19.26 Three resistors in series with a battery.

Evidently a series of ohmic resistors acts like a single ohmic resistor, with a resistance equal to the sum of all the resistances:

$$R_{\text{equivalent}} = R_1 + R_2 + R_3$$

This is not an entirely obvious result! The proof depends on the following:

- the definition of resistance $R = L/(\sigma A)$;
- the assumption that the conductivity σ is independent of the current for the material in these resistors (ohmic resistors);

- the fact that the magnitude of the sum of all the resistor potential drops is equal to the potential rise across the battery, which is numerically equal to the emf of the battery; and
- the fact that in a series circuit the steady-state current I is the same in every element, due to charge conservation, and an unchanging charge distribution in the steady state.

Because $R = L/(\sigma A) = \rho L/A$, if the various resistors are all made of the same material and have the same cross-sectional area, note that the series-resistance equation is equivalent to just adding the various lengths: $L_{equivalent} = L_1 + L_2 + L_3$.

Parallel Resistors

Instead of connecting circuit elements in series, it is possible to connect them in parallel, creating several branches in a circuit. Figure 19.27 shows three resistors connected in parallel.

We can show that several ohmic resistors in parallel are equivalent to one ohmic resistor with a resistance that is the reciprocal of the sum of reciprocals of the individual resistances. Consider the circuit of Figure 19.27, with three different ohmic resistors in parallel.

The potential difference is the same across each resistor in this parallel circuit and is numerically equal to the emf of the battery.

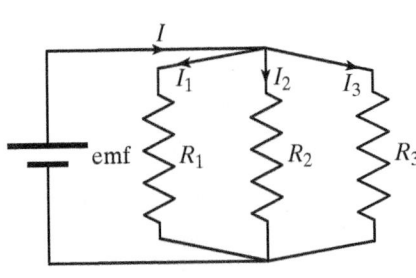

Figure 19.27 Three resistors in parallel.

QUESTION Write the current node equation for this circuit, using the fact that $I = \Delta V/R$ for an ohmic resistor:

We can rearrange the current equation, $I - \dfrac{emf}{R_1} - \dfrac{emf}{R_2} - \dfrac{emf}{R_3} = 0$, in this form:

$$I = \left(\frac{1}{R_1} + \frac{1}{R_2} + \frac{1}{R_3}\right)(emf) = \frac{emf}{R_{equivalent}}$$

Evidently ohmic resistors in parallel act like a single ohmic resistor, with a resistance whose reciprocal is equal to the sum of the reciprocals of the individual resistances:

$$\frac{1}{R_{equivalent}} = \frac{1}{R_1} + \frac{1}{R_2} + \frac{1}{R_3}$$

Because $1/R = \sigma A/L$, if the various resistors are all made of the same material and have the same length, note that the parallel-resistance equation is equivalent to just adding the various cross-sectional areas: $A_{equivalent} = A_1 + A_2 + A_3$. This corresponds to what you see if you connect two bulbs in parallel to a battery: there is twice as much current through the battery, as though you had one bulb with twice the cross-sectional area.

Checkpoint 3 When glowing, a thin-filament bulb has a resistance of about 30 Ω and a thick-filament bulb has a resistance of about 10 Ω. If they are in parallel, what is their equivalent resistance? How much current goes through two 1.5 V flashlight batteries in series if a thin-filament bulb and a thick-filament bulb are connected in parallel to the batteries?

19.4 WORK AND POWER IN A CIRCUIT

If you move a small amount of charge Δq from one location to another, the small amount of electric potential energy change ΔU_e is equal to the charge Δq times the potential difference ΔV, $\Delta U_e = (\Delta q)(\Delta V)$, since electric potential is electric potential energy per unit charge (joules per coulomb, or volts). If this work is done in a short amount of time Δt, we can calculate the power, which is energy per unit time:

$$\text{Power} = \frac{\Delta U_e}{\Delta t} = \frac{(\Delta q)(\Delta V)}{\Delta t}$$

However, $\Delta q/\Delta t$ is the amount of charge moved per unit time, which is the current I. Since $\Delta q/\Delta t = I$, we have the following important equation:

POWER FOR ANY KIND OF CIRCUIT COMPONENT

$$\text{Power} = I\Delta V$$

You should be able to repeat this argument without referring to your notes. This is a very general result, applying to any kind of circuit component. Note that the power dissipated in the resistor is $I(RI) = RI^2$, which can also be written as $(\Delta V/R)\Delta V = (\Delta V)^2/R$.

> **Checkpoint 4** Power in a battery: In a certain circuit, a battery with an emf of 1.5 V generates a current of 3 A. What is the output power of the battery? Include appropriate units. (If on the other hand you were charging a rechargeable battery with a charging current of 3 A, this would be the input power to the battery.)

EXAMPLE

Two Different Bulbs in Series

The circuit in Figure 19.28 consists of two flashlight batteries, a thick-filament bulb, a thin-filament bulb, and very low-resistance copper wires.

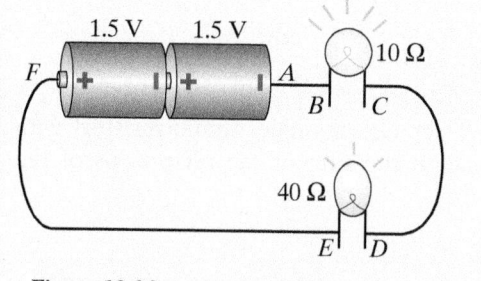

(a) Calculate the number of electrons that flow out of the battery every second at location A, in the steady state. Assume that in these steady-state conditions the resistance of the thick-filament bulb is 10 Ω and the resistance of the thin-filament bulb is 40 Ω.

(b) Show the approximate surface charge on the steady-state circuit.

(c) Draw an accurate graph of potential around the circuit, with the x axis running from A to F to A. Label the y axis with numerical values of the potential.

Figure 19.28 A circuit with a thin-filament bulb and a thick-filament bulb.

(d) The tungsten filament in the thin-filament bulb is 8 mm long and has a cross-sectional area of 2×10^{-10} m². How big is the electric field inside this metal filament?

(e) What is the power output of the battery?

Solution

(a) Loop rule: $2\,\text{emf} - IR_{10} - IR_{40} = 0$

$$I = \frac{2\,\text{emf}}{(R_{10} + R_{40})} = \frac{2(1.5\,\text{V})}{(10\,\Omega + 40\,\Omega)} = 6 \times 10^{-2}\,\text{A}$$

$$i = \frac{I}{|q|} = \frac{6 \times 10^{-2}\,\text{A}}{1.6 \times 10^{-19}\,\text{C}} = 3.8 \times 10^{17}\,\text{electrons/s}$$

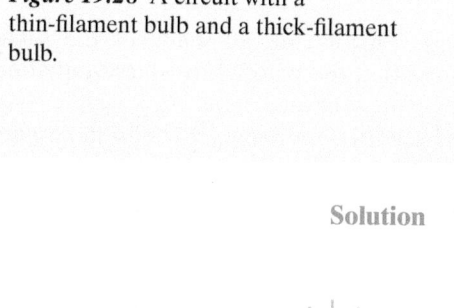

Figure 19.29 Approximate surface charge distribution on a circuit in steady state.

(b) See Figure 19.29. A bigger electric field is needed in the thin-filament bulb (because it has higher resistance), so the surface charge gradient across the thin-filament bulb is much bigger than the surface charge gradient across the thick-filament bulb. The surface charge gradient across the copper wires is extremely small, since only a very small electric field is needed in the wires. This gradient is too small to show up on this diagram, so the surface charge is shown as approximately uniform on the wires.

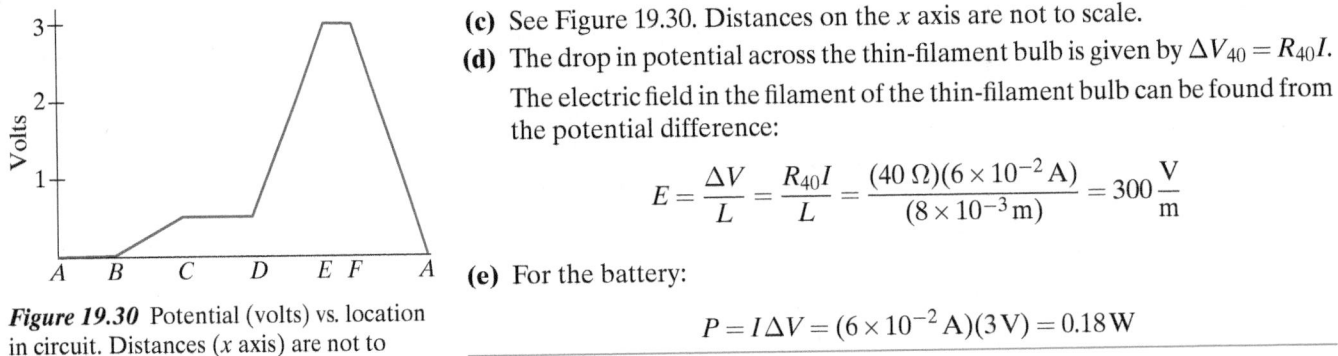

Figure 19.30 Potential (volts) vs. location in circuit. Distances (x axis) are not to scale.

(c) See Figure 19.30. Distances on the x axis are not to scale.

(d) The drop in potential across the thin-filament bulb is given by $\Delta V_{40} = R_{40}I$.

The electric field in the filament of the thin-filament bulb can be found from the potential difference:

$$E = \frac{\Delta V}{L} = \frac{R_{40}I}{L} = \frac{(40\,\Omega)(6 \times 10^{-2}\,\text{A})}{(8 \times 10^{-3}\,\text{m})} = 300\,\frac{\text{V}}{\text{m}}$$

(e) For the battery:

$$P = I\Delta V = (6 \times 10^{-2}\,\text{A})(3\,\text{V}) = 0.18\,\text{W}$$

19.5 BATTERIES

We have been treating batteries as if they were ideal devices: that is, as if they always managed to maintain the same potential difference across themselves regardless of what they are connected to. Real batteries, however, don't quite succeed in maintaining a potential difference equal to the battery emf, due to "internal resistance"—the resistance of the battery itself to the passage of current. We will show that a real battery can be modeled as an ideal, resistance-less battery in series with the internal resistance of the battery.

Consider again our mechanical "battery" consisting of two large charged plates whose charge is continually replenished by transporting electrons on a conveyor belt from the $+$ plate to the $-$ plate. The electrons on the motor-driven belt are acted on both by the non-Coulomb force exerted by the motor, $\vec{F}_{\text{NC}}$, and by the Coulomb force exerted by the charges on the plates, $\vec{F}_{\text{C}} = -e\vec{E}_{\text{C}}$. These two forces oppose each other (Figure 19.31).

In general, the non-Coulomb force F_{NC} has to be somewhat larger than the Coulomb force $F_{\text{C}} = eE_{\text{C}}$ because of finite mobility of the charges moving through the battery. In a chemical battery, this takes the familiar form of collisions, leading to an average drift speed. The drift speed of ions through the battery is proportional to the net force $(F_{\text{NC}} - eE_{\text{C}})$. If σ is the conductivity for moving ions through the battery and A is the cross-sectional area of the battery, the current density $J = I/A$ is given not by the usual $J = \sigma E$ but by

$$J = \frac{I}{A} = \sigma \times (\text{force per unit charge}) = \sigma \left(\frac{F_{\text{NC}}}{e} - E_{\text{C}} \right)$$

Rearranging this equation, we have

$$E_{\text{C}} = \frac{F_{\text{NC}}}{e} - \frac{I/A}{\sigma}$$

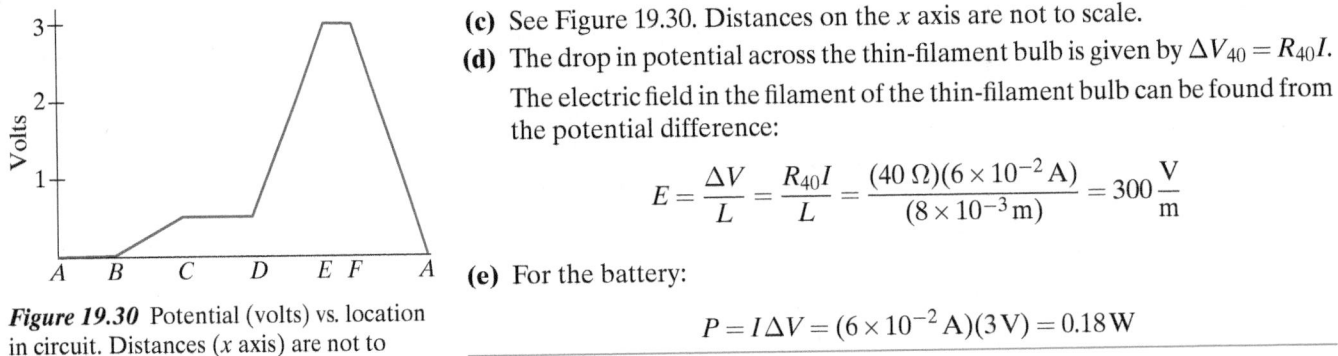

Figure 19.31 A mechanical battery with some friction in the mechanism.

If the (short) length of our mechanical battery is s, the electric field is nearly uniform, and the potential difference across the battery is $\Delta V = E_{\text{C}}s$, so

$$\Delta V = \frac{F_{\text{NC}}s}{e} - \frac{s}{\sigma A}I$$

Note, however, that $F_{\text{NC}}s/e$ is the work per unit charge or emf of the battery, and $s/(\sigma A)$ has the units of resistance. We call $s/(\sigma A)$ the "internal resistance" r_{int} of the battery, and finally we have

$$\Delta V_{\text{battery}} = \text{emf} - r_{\text{int}}I$$

This says: If there is very little internal resistance r_{int} to movement of ions through the battery (or very little current I), the potential difference ΔV across the battery is numerically nearly equal to the emf of the battery. Up until now we have made this assumption to simplify the discussion. However, if the current I through the battery is large enough, the term $r_{\text{int}}I$ may be sizable, and the potential difference across the battery (and therefore across the rest of the circuit) may be significantly less than the battery emf.

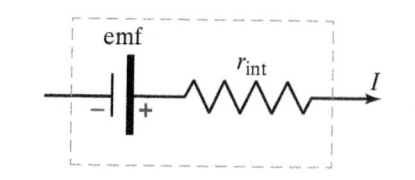

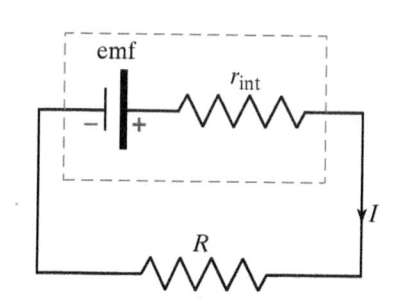

Figure 19.32 A real battery can be modeled as an ideal battery in series with an internal resistance.

Figure 19.33 A nonideal battery connected to a resistor.

Since $\Delta V_{\text{battery}} = \text{emf} - r_{\text{int}}I$, a real (nonideal) battery can be modeled as an ideal battery with the internal resistance in series, so that the potential difference across the battery is less than the emf by an amount $r_{\text{int}}I$, the potential drop across the internal resistance (Figure 19.32).

If a resistor R is connected to a real, nonideal battery (Figure 19.33), energy conservation (round-trip potential difference = 0) gives

$$(\text{emf} - r_{\text{int}}I) - RI = 0$$

Solving for the current I, we have

$$I = \frac{\text{emf}}{R + r_{\text{int}}}$$

Let's see how this works out in practice. A fresh 1.5 V alkaline flashlight battery has an internal resistance r_{int} of about 0.25 Ω.

QUESTION   What current would we get when we connect various resistors to the battery? Calculate the various values to get an idea of the effect:

Ideal Battery $r_{\text{int}} = 0$		Real Battery $r_{\text{int}} = 0.25$ Ω
100 Ω:	0.015 A	100 Ω:
10 Ω:	0.15 A	10 Ω:
1 Ω:	1.5 A	1 Ω:
0 Ω (short circuit):	infinite!	0 Ω (short circuit):

The currents with the real battery are 0.01496 A, 0.146 A, 1.2 A, and 6 A. There are several things to notice in this table: With a real battery, with nonzero internal resistance, you don't get 10 times the current through a 1 Ω resistor that you get through a 10 Ω resistor. Also, the internal resistance determines the maximum current that you can get out of a battery—this battery cannot deliver more than 6 A, no matter how small a resistor you attach to it. However, if the current is small, the behavior is nearly ideal.

An example of the role of a battery's internal resistance is seen in the requirements for a car battery, which must deliver a very large current to the starting motor. The battery must be built in such a way as to have a particularly low internal resistance. There are very large metal electrodes inside a car battery (large cross-sectional area A), which helps reduce the internal resistance to the transport of ions.

We should point out that the internal resistance r_{int} of a chemical battery is only approximately constant. As the battery is used more and more, the internal resistance increases. However, at any particular moment, the equation $\Delta V_{\text{battery}} = \text{emf} - r_{\text{int}}I$ does relate the potential difference across the battery to the current through the battery (using the values of emf and internal resistance valid at that moment). Experiments EXP7 to EXP9 let you explore internal resistance.

Checkpoint 5 What is the potential difference ΔV across a 1.5 V battery when it is short-circuited, if the internal resistance is 0.25 Ω and the resistance of the connecting wire is much less than 0.25 Ω? What is the average net electric field E inside the battery? For a mechanical battery, how much charge is on the battery plates?

This again shows the difference between the non-Coulomb emf of the battery, which does not change no matter what you connect to the battery, and the potential difference ΔV across the battery, which depends on how much charge the battery is able to maintain on its ends. Only for an ideal battery with no internal resistance is ΔV exactly equal to the emf. For real batteries that are discharging, ΔV is always somewhat less than the emf (and if you are charging a real rechargeable battery, you have to apply a ΔV that is somewhat greater than the emf).

19.6 AMMETERS, VOLTMETERS, AND OHMMETERS

An important application of circuit theory is to explain the operation of ammeters, voltmeters, and ohmmeters used for measuring conventional currents, potential differences, and resistances. Modern digital "multimeters" combine these functions into one compact instrument, with a switch that lets you choose between measuring current, voltage, or resistance, and with a convenient digital readout, complete with sign. Here are the most important things to know about ammeters and voltmeters:

- An ammeter must be inserted into the circuit in series with the circuit element whose current you want to measure. An ammeter must have a very small resistance, so as not to alter significantly the circuit into which it has been inserted (Figure 19.34).
- A voltmeter must be placed in parallel with the circuit element along which you want to measure the potential difference. A voltmeter must have a very high resistance, so as not to create a significant alternative pathway for current, which would alter the circuit (Figure 19.35).

Ammeter
(low resistance)

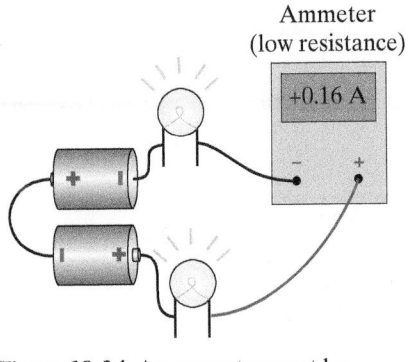

Figure 19.34 An ammeter must be inserted in series with the circuit element whose current you want to measure. The ammeter must have low resistance so as not to alter the circuit significantly.

Using an Ammeter

An ammeter must be inserted into a circuit, and the current to be measured is brought into the + socket and out the − socket. If conventional current flows into the + socket, the ammeter displays a positive number. When conventional current flows into the − socket, the ammeter indicates a negative current reading.

AMMETER SIGN CONVENTION

If conventional current flows into the socket marked "+" an ammeter indicates a positive current.

(This is the opposite of the convention for batteries, since conventional current flows out of the + terminal of a battery.)

Voltmeter
(high resistance)

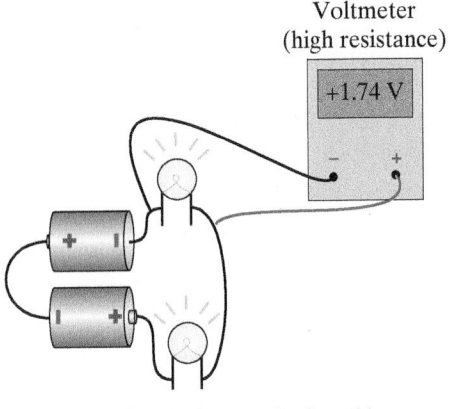

Figure 19.35 A voltmeter is placed in parallel with the circuit element along which you want to measure the potential difference. The voltmeter must have high resistance so as not to alter the circuit significantly by providing an alternative current pathway.

A Simple Ammeter

If you have been doing experiments with currents and compasses, you have been using a simple ammeter, in the form of a compass whose needle is deflected by a passing current. If you could run known amounts of current over the compass, you could calibrate this ammeter by determining the relationship between the known current and the observed deflection angle. This is a "noninvasive" ammeter, because you don't have to alter the circuit in order to be able to detect and measure the current in a wire.

A serious practical problem in using standard ammeters is that if you want to measure the current somewhere in an existing circuit, you must break the circuit at the place of interest and insert your ammeter. In order that this make as little change as possible in the current to be measured, ammeters are designed to have very low resistance, so that if you insert the ammeter it's as though you're just lengthening the wire a bit.

AN AMMETER SHOULD HAVE LOW RESISTANCE

An ammeter should have as small a resistance as possible, so as not to change the current that you are trying to measure.

QUESTION Suppose that you are trying to measure the current through the bulb in the circuit shown in Figure 19.36. What is wrong with the placement of the ammeter?

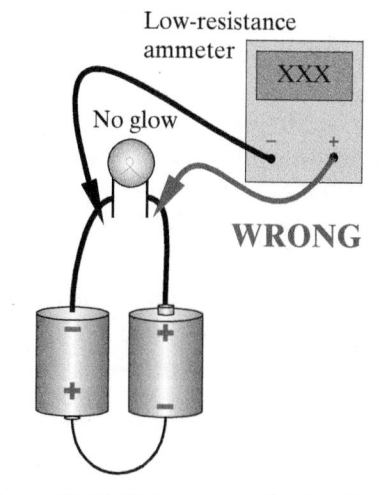

Figure 19.36 This ammeter is placed incorrectly in the circuit.

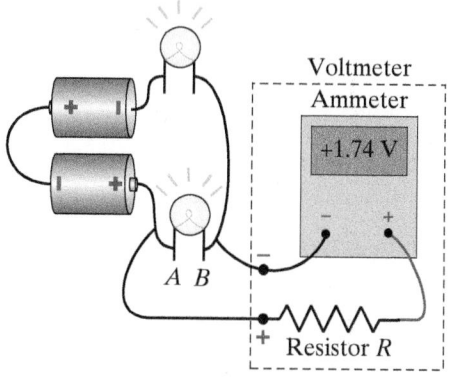

Figure 19.37 An ammeter with a resistor in series forms a voltmeter.

Connecting the ammeter in parallel to the bulb means that there is a very low-resistance path for the current (a "short circuit"). Most current will go through the ammeter, with very little current going through the bulb. The bulb no longer glows, and the large current measured by the ammeter has nothing to do with the tiny current that goes through the bulb but rather is a measure of the maximum current these batteries can deliver.

The fact that an ammeter must be inserted into a circuit may not seem like a problem with simple circuits, but it can be a big problem if the circuit of interest is built in such a way that you can't break in to insert your ammeter without damaging the circuit. Moreover, a commercial ammeter may blow a fuse or even be permanently damaged if used incorrectly as shown in Figure 19.36, because a current of several amperes will run through the very low-resistance ammeter, and many commercial ammeters cannot handle such large currents.

Voltmeters Measure Potential Difference

If we add a series resistor to an ammeter, in principle we can make a voltmeter to measure potential differences. Suppose that we attach an ammeter across a circuit element, through a resistor whose resistance R is very large (Figure 19.37).

> QUESTION If the measured current through the ammeter is I, what is the potential difference $V_A - V_B$ in this circuit, in terms of R and I?

The potential difference along the resistor is $V_A - V_B = RI$. A small current I running through the resistor is measured by the ammeter, and the potential difference is RI. If the ammeter is relabeled in volts (RI), the device reads potential differences directly. A voltmeter can be thought of as an ammeter with a series resistance, all in one box (shown inside dashed lines in Figure 19.37).

Notice that a voltmeter has two leads and measures potential difference, not potential! Commercial voltmeters have their + and − sockets arranged as follows:

VOLTMETER SIGN CONVENTION

If the potential is higher at the socket marked "+"
a voltmeter indicates a positive potential difference.

If the − socket is at the higher potential, the voltmeter displays a *negative* potential difference. The meter in Figure 19.37 will read positive.

An ammeter should have very small resistance so that inserting the ammeter into a circuit should disturb the circuit as little as possible. A voltmeter, however, should have as large a resistance as possible, so that placing the voltmeter in parallel with a circuit element will disturb the circuit as little as possible, with only a tiny current being deflected through the voltmeter.

A VOLTMETER SHOULD HAVE HIGH RESISTANCE

A voltmeter (used in parallel) should have very large resistance.
An ammeter (used in series) should have very small resistance.

There is a practical limit to the voltmeter resistance. The larger the resistance of the voltmeter, the more sensitive must be the ammeter part of the voltmeter. Commercial voltmeters often have resistances as high as 10 MΩ (10 megohms or 10×10^6 ohms), and the ammeter portion of the voltmeter is correspondingly sensitive. Even a commercial voltmeter with a resistance of 10 MΩ is useless for measuring potential differences in a circuit consisting of 100 MΩ resistors, because placing a 10 MΩ voltmeter in parallel with a 100 MΩ resistor drastically alters the circuit.

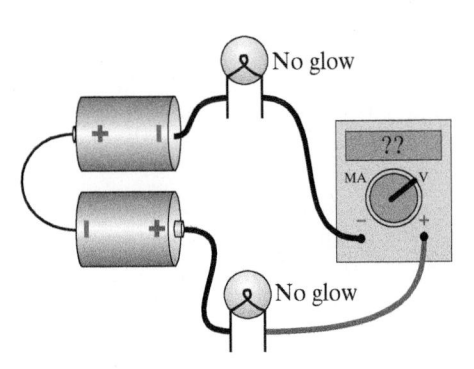

Figure 19.38 This voltmeter is placed incorrectly in the circuit.

Checkpoint 6 Two lab partners measured the current in a series circuit by placing a digital multimeter (set to read current in milliamperes) in series between the two thick-filament bulbs (Figure 19.38). Then they wanted to measure the voltage across a bulb, so they simply switched the multimeter to the V setting (volts). They were surprised that the bulbs stopped glowing. What did they do wrong? What did their voltmeter read?

Ohmmeters

Commercial multimeters can act not only as ammeters and voltmeters but also as ohmmeters to measure the resistance of an element that has been removed from a circuit. The ohmmeter section of the multimeter consists of an ammeter in series with a small voltage source (for example, 50 mV). When you connect an unknown resistor to the ohmmeter, the small voltage drives a small current through the resistor and ammeter. The ohmmeter is calibrated to display the applied voltage divided by the observed current, which is the resistance (assuming that the resistance of the ammeter is small compared to the resistance you are trying to measure). You can see that you have to remove the resistor of interest from its circuit before you can use the ohmmeter, because an ohmmeter is an active device with its own voltage source.

How Commercial Meters Are Really Constructed

We have seen that one could construct a voltmeter from a sensitive ammeter with a large resistor in series. Modern digital voltmeters read voltage directly, by timing how long it takes a known current to discharge a capacitor! A digital ammeter is basically a voltmeter with a very small resistance in parallel, and the voltmeter reads the very small potential difference across the small resistance. A digital ohmmeter drives a known current through the unknown resistor and measures the resulting potential difference. Independent of the construction details, it is important to understand that an ammeter has very low resistance, and a voltmeter has very high resistance.

19.7 QUANTITATIVE ANALYSIS OF AN RC CIRCUIT

We are now in a position to analyze quantitatively a series circuit containing an ideal battery with known emf, a resistor R, and a capacitor C—a so-called "RC" circuit. Recall that the potential difference across a capacitor is $\Delta V = Q/C$, where Q is the charge on the positive plate and C is the capacitance (Section 19.1; for a parallel-plate capacitor, $C = K\varepsilon_0 A/s$, where A is the area of one of the plates, s is the gap distance between the plates, and K is the dielectric constant of the material filling the gap, as discussed in Chapter 16). The energy equation for the RC circuit in Figure 19.39 is

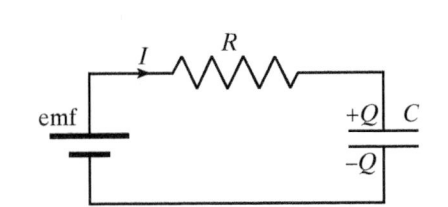

Figure 19.39 An RC circuit.

$$\Delta V_{\text{round trip}} = \text{emf} - RI - Q/C = 0$$

The Final State

The simplest situation to analyze is the final equilibrium state of the circuit, in which the current has dropped to zero, and the capacitor is fully charged. Then the term RI is zero, and the loop equation gives this:

$$\Delta V_{\text{round trip}} = \text{emf} - Q/C = 0$$
$$Q = \text{emf} \cdot C$$

We see that the final charge on the capacitor depends only on the emf and the capacitance. The resistor in the circuit determines how long it takes to reach equilibrium, but it has no effect on the final charge on the capacitor.

The Initial Situation

Recall what happens when a battery and thick-filament bulb are connected in series to a (discharged) capacitor. Immediately after making the final connection the light bulb is very bright (and then it gets dimmer). The initial brightness looks about the same as you get in a circuit containing only a thick-filament bulb without a capacitor.

Let's see how we can understand this quantitatively. Rearrange the energy equation to solve for I in terms of the other quantities:

$$I = \frac{\text{emf} - Q/C}{R}$$

QUESTION Use this equation for I to explain why the initial brightness is the same as though the capacitor weren't there. (*Hint:* What is Q initially?)

The capacitor is initially uncharged, so initially $Q = 0$, and plugging this value for Q into the equation for I gives us $I = \text{emf}/R$. This is the same current we would have in a simple circuit without the capacitor. Having understood the initial situation, let's see what happens next.

The rate at which charge Q accumulates on the positive capacitor plate is equal to $I = dQ/dt$. To put it another way, in a time dt the plate receives an additional amount of charge $dQ = Idt$. We rewrite the energy equation:

$$I = \frac{dQ}{dt} = \frac{\text{emf} - Q/C}{R}$$

$$dQ = \left(\frac{\text{emf} - Q/C}{R}\right) dt$$

This is in a form suitable for doing a numerical integration. At each instant, we know Q (initially zero), so we can calculate the change dQ from this equation. Then $Q + dQ$ is the new value of Q, and we can use this new Q to calculate a new value of dQ, and so on.

Numerical Integration

Let's take a couple of steps in this numerical integration to see how things go. In the first time interval dt, $dQ = (\text{emf}/R)dt$, since the initial value, Q_0, is zero. The new value, Q_1, is

$$Q_1 = Q_0 + dQ = (\text{emf}/R)dt$$

Now the capacitor has a small nonzero charge Q_1. In the second time interval, we have

$$dQ = \left(\frac{\text{emf} - Q_1/C}{R}\right) dt$$

$$Q_2 = Q_1 + dQ$$

Because Q_1 is nonzero, the increase in charge dQ is smaller this time. As time goes on, and Q gets larger and larger, dQ for each additional time step will be smaller and smaller. Therefore a graph of Q vs. time will look like the upper graph of Figure 19.40. The graph is the result of a numerical integration using

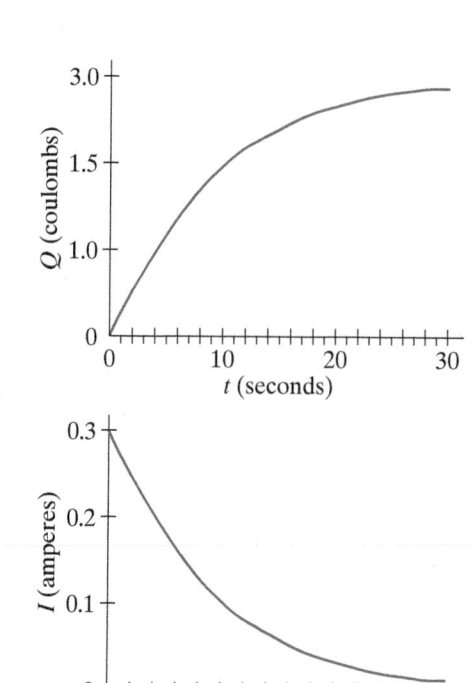

Figure 19.40 Charge and current as a function of time.

$R = 10\ \Omega$ (approximately the resistance of a glowing thick-filament bulb) and $C = 1.0$ F, with two 1.5 V batteries in series.

The lower graph of Figure 19.40 is the current flowing into the resistor, calculated as $I = dQ/dt$. As Q increases, I decreases, because the fringe field of the capacitor increasingly opposes the current flow. In terms of potential, as Q increases, the potential difference across the capacitor increases, which means that the potential difference across the resistor must decrease, and therefore $I = \Delta V_{\text{resistor}}/R$ decreases.

Analytical Solution

The graph in Figure 19.40 looks like some kind of exponential function. Let's make a guess that the current is in fact given by

$$I = I_0 e^{-at}$$

where I_0 is the initial current (which we found earlier to be $I_0 = \text{emf}/R$), and a is an unknown constant. Consider this line of reasoning:

$$I = \frac{dQ}{dt} = \frac{\text{emf} - Q/C}{R} \quad \text{(energy equation in terms of } I \text{ and } Q\text{)}$$

$$\frac{dI}{dt} = -\frac{1}{RC}\frac{dQ}{dt} = -\frac{1}{RC}I \quad \text{(since emf}/R \text{ is constant)}$$

$$\frac{dI}{dt} = \frac{d}{dt}(I_0 e^{-at}) = -aI_0 e^{-at} = -aI = -\frac{1}{RC}I \quad \text{(using our guess for } I\text{)}$$

This last equation is true for all time if $a = 1/RC$, so we have a solution:

CURRENT IN A SERIES RC CIRCUIT

$$I = \left(\frac{\text{emf}}{R}\right) e^{-t/RC}$$

This solution has all the right properties. When $t = 0$, the exponential is equal to 1, and $I = \text{emf}/R$, so the initial brightness is indeed the same as if there were no capacitor. As t becomes very large, the exponential goes to 0, and the current goes to zero, which is what is observed in an RC circuit.

We used a very common and powerful method for finding an analytical solution for a differential equation (an equation involving derivatives): Guess the form of the solution (containing unknown constants such as a), plug this form and its derivatives into the differential equation, and determine what values the constants must have to satisfy the equation ($a = 1/RC$).

Since $I = dQ/dt$, $Q = \int_0^t I\,dt$, and it is easy to show this:

CHARGE IN A SERIES RC CIRCUIT

$$Q = C(\text{emf})\left[1 - e^{-t/RC}\right]$$

This solution has all the right properties. When $t = 0$, the exponential is equal to 1, and $Q = 0$ (no charge yet on the positive plate of the capacitor). As t becomes very large, the exponential goes to 0, and Q approaches $C(\text{emf})$. This is correct: $\Delta V_{\text{capacitor}} = Q/C = C(\text{emf})/C = \text{emf}$. When the current stops, the potential difference across the capacitor becomes equal to the potential difference across the battery.

> **QUESTION** What is the numerical value of the final charge on one plate of the capacitor in this circuit, in terms of coulombs?

We find that $Q_F = C \times \text{emf} = (1.0\ \text{farad})(3\ \text{V}) = 3.0$ C. That's a huge amount of charge! Compare, for example, with the amount of charge on a strip of invisible tape, which is of the order of 1×10^{-8} C.

Actually, our analytical solution of the differential equation is physically only an approximation, because as the current changes the resistor changes temperature and changes resistance. This is particularly significant when the resistor is a light bulb, since its resistance drops a great deal as the current decreases and it cools off.

The *RC* Time Constant

A rough measure of how long it takes to reach final equilibrium is the "time constant" RC. When the time $t = RC$, the factor $e^{-t/RC}$ has fallen from the value $e^0 = 1$ to the value

$$e^{-t/RC} = e^{-1} = \frac{1}{e} = \frac{1}{2.718} = 0.37$$

QUESTION Calculate the time constant for a circuit with $R = 10 \, \Omega$ (approximately the resistance of a glowing thick-filament bulb) and $C = 1.0 \, \text{F}$. This is the time when the current has decreased from the original 0.30 A to 0.11 A (0.37 times 0.30 A).

The time constant $RC = (10 \, \Omega)(1.0 \, \text{F}) = 10 \, \text{s}$. The bulb will be very dim with this current. This is roughly consistent with what is observed with a thick-filament bulb and a 1 F capacitor, which loses most of its brightness in this amount of time. Figure 19.41 shows graphs of Q vs. t and I vs. t for the RC circuit with the thick-filament bulb, with the 10 s RC time constant indicated.

QUESTION Show that the power dissipated in the bulb at $t = RC$ is only 14% of the original power.

The power dissipated in the bulb is $I\Delta V = RI^2$, so a reduction in the current by a factor of 0.37 gives a reduction in the power by a factor of $(0.37)^2 = 0.136$.

QUESTION Calculate the time constant for a circuit in which you replace the thick-filament bulb with a thin-filament bulb (whose resistance R when glowing is typically about 30 Ω, which may vary from one bulb to another).

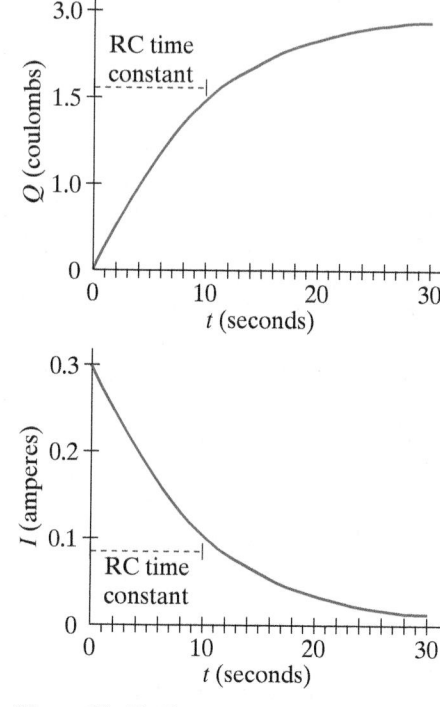

Figure 19.41 Charge and current as functions of time.

The time constant $RC = (30 \, \Omega)(1 \, \text{F}) = 30 \, \text{s}$, which agrees roughly with observations with a 1 F capacitor.

We now have a complete quantitative description of the behavior of an RC circuit. In experiments where one times how long it takes for the current to decrease to a point where the bulb no longer glows, this is not the same as timing how long it takes for the current to get smaller by a factor of e^{-1}.

19.8   REFLECTION: THE MACRO–MICRO CONNECTION

In this chapter we made connections between a microscopic and macroscopic view of resistance. In our original microscopic view, individual electrons acquire a drift speed proportional to the electric field, with $\bar{v} = uE$. This electric field is made by all the charges in and on the battery and on the surface of the conducting circuit elements. Electron current i is measured in number of electrons per second.

In the macroscopic view, the electric field made by all the charges produces a conventional current density $J = \sigma E$ and a conventional current $I = JA$ measured in coulombs per second (amperes). The integral of the electric field is potential difference, and any round-trip integral of the electric field due to point charges must equal zero, which represents energy conservation per unit charge. We also used a conventional-current version of the steady-state current node rule.

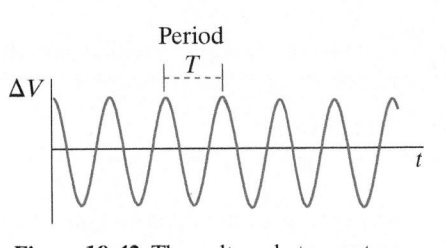

Figure 19.42 The voltage between two slots in a wall socket varies sinusoidally.

19.9 *WHAT ARE AC AND DC?

The steady constant current in a circuit consisting of a battery and a resistor is called "direct current" or DC. Home appliances such as a floor lamp use "alternating current" or AC. Two slots of the wall socket supply a sinusoidally varying voltage ("AC voltage") that drives a sinusoidally varying current in the lamp (Figure 19.42). (A third wall socket connection goes to ground for safety reasons.) In many countries including the United States the frequency is 60 Hz (60 hertz means 60 cycles of a sine wave every second, with a period $T = 1/60$ s). Many other countries, including those in Europe, use 50 Hz AC. For reasons discussed below, what is called "110 volts AC" is actually a voltage that varies sinusoidally between −155 volts and +155 volts.

Although current in an incandescent floor lamp rises and falls sinusoidally, the light doesn't seem to flicker for two reasons. First, humans have difficulty perceiving flickering that occurs 50 or 60 times per second. For example, movies are presented in the form of 24 still pictures per second, but you perceive a continuous image. Second, the temperature of an incandescent bulb filament cannot change rapidly; there isn't time in a fraction of a 60 Hz cycle for the filament to cool down. However, with a fluorescent lamp, where the light is produced nonthermally in sync with the applied voltage, you may see strobe-like images of your fingers if you wave your fingers rapidly back and forth while looking through your fingers at the lamp.

Very high voltages (100,000 V or more) are used in power transmission lines to reduce energy losses between a generating station and a remote city. If R is the resistance of the transmission line, you want the voltage drop RI along the line to be small, so you want the current I to be small. The power delivered to the line is $I\Delta V$, so for I to be small you need ΔV to be large. However, very high voltages would be extremely dangerous in the home. It is relatively cheap and efficient to convert high AC voltages to low AC voltages using "transformers" whose operation is based on Faraday's law, which we will study in a later chapter. Until quite recently, converting high DC voltages to low DC voltages was difficult and inefficient (though this is changing with modern semiconductor devices) so historically it made sense to use AC in distributing electrical power.

Electronic devices such as computers need low DC voltages. The small box between a wall socket and a laptop uses semiconductor elements to convert the AC voltage to 12 volts DC; similar AC to DC converters are built into other devices such as desktop computers and radios.

AC Power

The reason a sinusoid varying between −155 V and +155 V is called "110 volts AC" is that it delivers the same energy in one second as would a constant 110 V. Consider a circuit containing an AC power supply and a resistor, such as a floor lamp. With $\Delta V = \Delta V_{max} \cos(\omega t)$, we'll integrate the instantaneous power over one cycle (with period T and $\omega = 2\pi/T$) and see how much energy is delivered to a resistor R in one cycle:

$$\text{energy} = \int_0^T \frac{\Delta V^2}{R} dt$$

$$= \frac{\Delta V_{max}^2}{R} \int_0^T \cos(\omega t)^2 dt$$

$$= \frac{\Delta V_{max}^2}{R} \int_0^T \frac{1}{2}[1 + \cos(2\omega t)]dt \quad \text{(using a trig identity)}$$

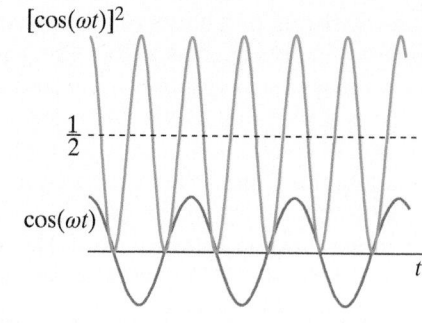

Figure 19.43 The average value of the square of a cosine is 1/2.

$$\text{energy} = \frac{\Delta V_{max}^2}{R}\frac{1}{2}T$$

The average power is the energy delivered in time T divided by the time T:

$$\text{average power} = \frac{1}{2}\frac{\Delta V_{max}^2}{R}$$

The factor of 1/2 comes ultimately from the fact that the average value of the square of a cosine is 1/2 (Figure 19.43).

The "root mean square" (rms) voltage is defined as $\Delta V_{rms} = \sqrt{\frac{1}{2}}\Delta V_{max}$, in which case the average power can be written in a form that looks like the equation for DC power:

$$\text{average power} = \frac{\Delta V_{rms}^2}{R}$$

The "110 volts AC" available at the wall socket means $V_{rms} = 110$ V, which corresponds to $\Delta V_{max} = \sqrt{2}\Delta V_{rms} = 155$ V.

Phase Shift

An unusual aspect of AC circuits is that the current need not be in phase with the voltage. For example, if an AC voltage is maintained across an RC circuit, this applied voltage peaks later than the current (Figure 19.44). We say that the current in the circuit "leads" the applied voltage. The energy equation (loop rule) is

$$\Delta V_{max}\cos(\omega t) - RI - \frac{Q}{C} = 0$$

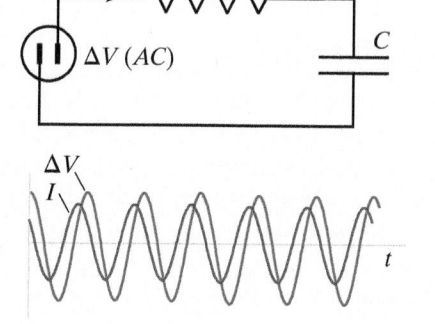

Figure 19.44 An AC voltage is applied to an RC circuit. The current in the circuit peaks earlier than does the applied ΔV.

where $Q = \int I dt$, or $dQ/dt = I$; the rate at which charge increases on a plate is equal to the charge per second entering the plate, which is the current I. Differentiating the energy equation with respect to t we have

$$-\Delta V_{max}\omega \sin(\omega t) - R\frac{dI}{dt} - \frac{I}{C} = 0$$

By substituting into the energy equation it is possible to show that the current is

$$I = I_{max}\cos(\omega t + \phi)$$

where ϕ is called the "phase angle" and expresses the result that the current I peaks at a different time than the applied voltage. The values for ϕ and I_{max} can be obtained from the following relations, which emerge from the solution process:

$$\tan\phi = \frac{1}{RC\omega}$$

$$I_{max} = \frac{\Delta V_{max}}{\sqrt{R^2 + 1/(C\omega)^2}}$$

The frequency-dependent factor $\sqrt{R^2 + 1/(C\omega)^2}$ is called the "impedance" and plays the role of resistance in this circuit. If the capacitance C or the angular frequency ω is very large, the impedance is approximately equal to R, and $I = \Delta V_{max}/R$, just as in a simple resistor circuit. Also, in this case $\tan\phi \approx 0$ and $\phi \approx 0$, so the current is in phase with the applied voltage as it is for a DC circuit.

19.10 *ELECTRONS IN METALS

It is possible to show experimentally that the mobile charged particles in a metal are indeed electrons. For example, magnetic forces can lead to a transverse polarization of a current-carrying wire, and the sign of this polarization is different for positive and negative moving charges (this is called the "Hall effect," and we will study this in Chapter 20).

A different kind of experiment was carried out in 1916–1917 by Tolman and Stewart, which showed that the mobile charge carriers in a metal are indeed electrons, which was not known for certain before their experiment. We will describe a simpler experiment that is conceptually equivalent to the experiment they actually carried out.

Suppose that we accelerate a metal bar, which means that we accelerate the atomic cores, which are bound to each other (Figure 19.45).

On the average, the mobile electrons normally feel no force; the mutual repulsion of the mobile electrons is effectively canceled by the attraction to the atomic cores. Therefore as the bar and atomic cores accelerate, the mobile electrons are left behind. As a result, they pile up at the trailing end of the bar (they can't easily leave the bar). This polarizes the bar, with negative charge on the trailing surface and positive charge (deficiency of electrons) on the leading surface, as shown in Figure 19.45.

The surface charges produce an electric field E to the left, which means that there is a force eE to the right on electrons inside the metal (Figure 19.45). As the polarization increases, E increases until the electric force eE is equal to the mass of the electron times the acceleration of the metal bar, at which point the acceleration of the mobile electrons matches the acceleration of the bar, and then the polarization no longer increases.

The observed direction of the polarization (negative at the left end) led the experimenters to conclude that the mobile particles are indeed negatively charged. You should run through the argument to see that if the movable charges had been positive, the trailing end of the bar would have been positive instead of negative.

Although this experiment does not give separate values for the charge and the mass of the mobile particles, it does provide a measurement of the ratio q/m. There had already been measurements of the ratio e/m for electrons moving in a vacuum, in a device similar to an old television cathode-ray tube. The modern value for this ratio for the electron is about

$$\frac{e}{m} = \frac{(1.6 \times 10^{-19}\,\text{C})}{(9 \times 10^{-31}\,\text{kg})} = 1.7 \times 10^{11}\,\text{C/kg}$$

Tolman and Stewart measured the acceleration a, and they used a voltmeter to measure the potential difference ΔV from one end of the metal bar to the other. The potential was lower at the left end of the bar, which shows that the mobile charge carriers are negative. Moreover, the experimental value of q/m had the same value as that obtained for electrons in other experiments. This was very strong evidence that currents in metals consist of moving electrons.

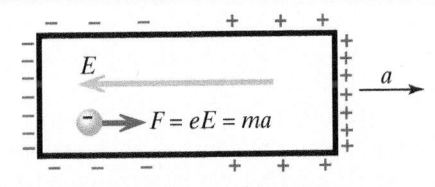

Figure 19.45 In an accelerated metal bar the mobile electrons are initially left behind, polarizing the bar.

> **Checkpoint 7** If the length of the bar is L, express q/m in terms of the quantities measured by Tolman and Stewart (a and ΔV).

19.11 *A COMPLICATED RESISTIVE CIRCUIT

We can use energy conservation (loop equations) and the current node rule to determine how much current would flow in each element of a complicated multiloop circuit made up of batteries and ohmic resistors (resistors for which the current is proportional to the potential difference). Analyzing complex

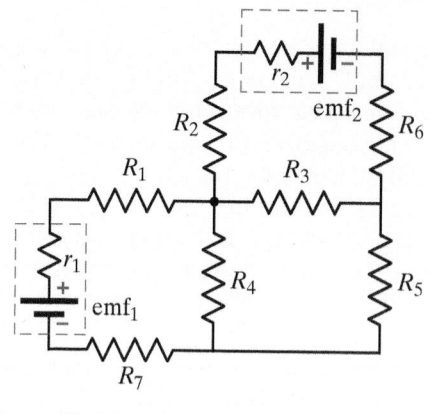

Figure 19.46 A circuit with unknown currents, which we will determine.

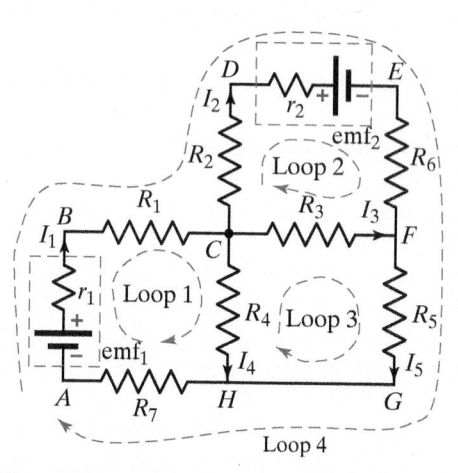

Figure 19.47 Name the unknown currents, draw paths along which to apply energy conservation.

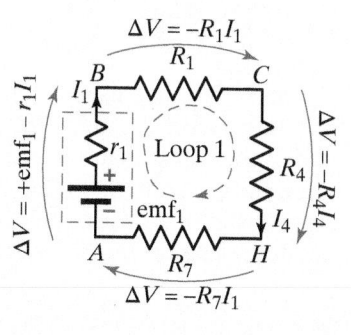

Figure 19.48 Apply energy conservation for a round-trip path, loop 1 (*ABCHA*).

circuits like the one discussed in this section is more a matter of technology than fundamental physics, so we won't analyze many such circuits. However, it is useful to analyze at least one such circuit as another illustration of the application of energy conservation and the current node rule. We will use the loop and node rules to generate as many independent equations as we have unknown currents, and then solve the system of equations. Assume that the emf's, battery internal resistances, and R's are known for the circuit shown in Figure 19.46, but the currents are unknown. We will determine the current in each part of this circuit. The first step in the solution is to assign names and directions to the unknown currents (I_1 through resistor R_1, I_2 through resistor R_2, etc.). We don't know what the actual directions of these unknown conventional currents are, so we guess, and we record our guesses on a diagram, as shown in Figure 19.47. We also draw possible paths to follow for writing loop equations for energy conservation.

We don't have to assign a different current variable to every branch. Consider conventional current I_1. On the circuit diagram we have assumed that it flows from A toward B. This also means that we have assumed the same amount of current I_1 goes from B to C through resistor R_1, and from H to A through resistor R_7. Similar comments apply to the other currents. These simplifications are trivial applications of the current node rule.

First let's use energy conservation for loop 1 (the path *ABCHA*). We can start anywhere around the loop; let's start at location A and walk around the loop, adding up the potential differences we encounter on our walk (Figure 19.48). The potential rises across the battery (the electric field inside the battery is opposite to the direction we're walking) and the potential falls along the resistors, since we are walking in the assumed direction of the conventional current, which runs in the direction of the electric field from higher potential to lower potential.

When we get back to our starting location at A, the net potential change along our walk should be zero, since potential difference in a round trip is zero (and $V_A - V_A = 0$):

$$\text{Loop 1:} \quad (\text{emf}_1 - r_1 I_1) + (-R_1 I_1) + (-R_4 I_4) + (-R_7 I_1) = 0$$

Note that in writing this loop rule we implicitly used the current node rule to show that the current is the same I_1 through the battery, r_1, R_1, and R_7. We also used the property of ohmic resistors to be able to relate the current through R_1 to the potential difference along R_1 (as $-R_1 I_1$).

If I_1 turns out to be -1.3 A when we solve for the unknown currents, that will merely mean that we guessed the wrong direction and that I_1 actually runs from C to B (and that C is at a higher potential than B). Thus there is no real penalty for guessing the wrong current directions.

Next consider loop 2 (path *CDEFC*), starting at location C (Figure 19.49). When we get back to our starting location at C, the net potential change along our walk should be zero:

$$\text{Loop 2:} \quad (-R_2 I_2) + (-\text{emf}_2 - r_2 I_2) + (-R_6 I_2) + (R_3 I_3) = 0$$

Comments on loop 2: From D to E there is a potential drop (not a rise) across the battery numerically equal to $(-\text{emf}_2 - r_2 I_2)$, because we are traversing the battery from its high-potential + end to its low-potential − end, in the direction of the electric field inside the battery. (If we've guessed right about the direction of I_2, this battery is being charged rather than discharged.) From F to C there is a potential *rise* $+R_3 I_3$ (not a drop) because we are heading upstream against the current I_3. That is, the assumed direction of I_3 (from C to F) implies that the potential at C is higher than the potential at F. Again,

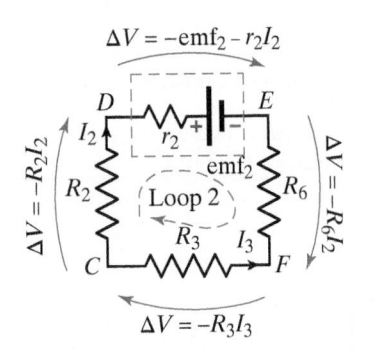

$\Delta V = -\text{emf}_2 - r_2 I_2$

$\Delta V = -R_2 I_2$

$\Delta V = -R_6 I_2$

$\Delta V = -R_3 I_3$

Figure 19.49 Apply energy conservation for a round-trip path, loop 2 (*CDEFC*).

if we have made the wrong assumption, all that will happen is that the final numerical answer for I_3 will have a negative sign.

> **QUESTION** In the same manner, use energy conservation for loop 3 (path *HCFGH*) and for loop 4 (path *ABCDEFGHA*).

The equation for loop 4 around the outside of the circuit is the algebraic sum of the equations for the other three loops and is not an independent equation. Only three of these four loop equations are actually significant in solving for the unknown currents, although it doesn't matter which three we choose to use in the solution.

Loop 3: $(R_4 I_4) + (-R_3 I_3) + (-R_5 I_5) = 0$

Loop 4:

$$(\text{emf}_1 - r_1 I_1) + (-R_1 I_1) + (-R_2 I_2) + (-\text{emf}_2 - r_2 I_2) + (-R_6 I_2) + \\ (-R_5 I_5) + (-R_7 I_1) = 0$$

> **QUESTION** Now write current node rule equations for nodes *C*, *H*, and *F*, paying attention to signs (+ for incoming current, − for outgoing current).

Adding the equations for node *C* and node *F* gives an equation that is equivalent to the equation for node *H*. Only two of these three node equations are actually independent, and we can use any two of them in solving for the unknown currents.

Node *C*: $I_1 - I_2 - I_3 - I_4 = 0$
Node *F*: $I_2 + I_3 - I_5 = 0$
Node *H*: $I_4 + I_5 - I_1 = 0$

We have a total of five independent equations (three independent loop equations and two independent node equations).

It can be shown that in any resistive circuit there are as many independent loop equations as there are minimal-sized loops (that is, ignoring combined loops such as loop 4). The number of independent node equations can be determined by subtracting the number of independent loop equations from the number of independent currents.

Our five equations are sufficient to be able to solve for the five unknown currents in terms of the known emf's and known resistances. This solution step can involve extremely tedious algebraic manipulation and should usually be turned over to a symbolic manipulation package such as Maple or Mathematica. Of course in some very simple cases it may be easy to solve the equations, but in general the algebra can be quite forbidding.

Once the unknown currents have been determined (typically by using a computer program to solve the equations), we can know essentially everything there is to know about the circuit. For example, we can calculate the potential drop $\Delta V = -RI$ across each resistor, since we know R and I for each resistor. We can therefore also calculate the potential difference between any two locations in the circuit.

> **Checkpoint 8** Refer to Figure 19.47, and suppose that $\text{emf}_1 = 12\,\text{V}$ and $R_1 = 20\,\Omega$, and that after we solved the equations we found that I_1 was 0.4 A. What is the potential difference $V_C - V_B$? How much power is dissipated in R_1? What is the power expenditure of battery 1 (some of which is dissipated in the battery itself due to internal resistance r_1)?

S U M M A R Y

New Concepts

In a capacitor circuit there is a slow transient leading to a final equilibrium state.

Current can run (for awhile) even if there is a gap in a circuit. The fringe field of the capacitor in the neighboring wires is critical for understanding the behavior of a capacitor circuit in terms of charge and electric field.

Conventional current I:

- is in the direction of motion of positive carriers
- is measured in coulombs per second

Conductivity $\sigma = |q|nu$ is a property of a material;

$$J \equiv \frac{I}{A} = \sigma E$$

Ohms and nonohmic resistors

Results

Charging and discharging times are larger with larger plates, smaller gap, or with an insulator in the gap.

Resistance $R = \dfrac{L}{\sigma A}$ incorporates the geometry of a resistor.

$I = \dfrac{\Delta V}{R}$ is the current through a resistor.

$\Delta V = RI$ is the potential difference across a resistor.

Power $= I\Delta V$ for any circuit element.

Potential difference across a real, nonideal battery (Figure 19.50):

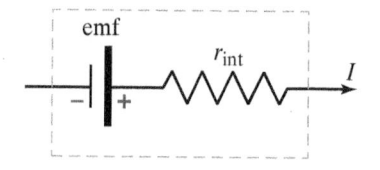

emf

r_{int}

I

Figure 19.50 Equivalent nonideal battery.

Capacitors in circuits can be analyzed in terms of $\Delta V = Q/C$, where C is the capacitance, and the charge Q and current I vary exponentially with time, with a time constant RC.

Capacitance $C = Q/|\Delta V|$ depends on the geometry of the capacitor (Section 19.1; for a parallel-plate capacitor, $C = K\varepsilon_0 A/s$, where A is the area of one of the plates, s is the gap distance between the plates, and K is the dielectric constant of the material filling the gap).

E X P E R I M E N T S

Capacitors in Circuits

EXP1 Charging and discharging a capacitor
Find the capacitor in your experiment kit or lab (your instructor will tell you what it looks like).

> **Caution: Do not exceed the voltage rating marked on the capacitor—typically a few volts. (If the rating is 2.5 V, it is normally okay to use two ordinary 1.5 V batteries.)**

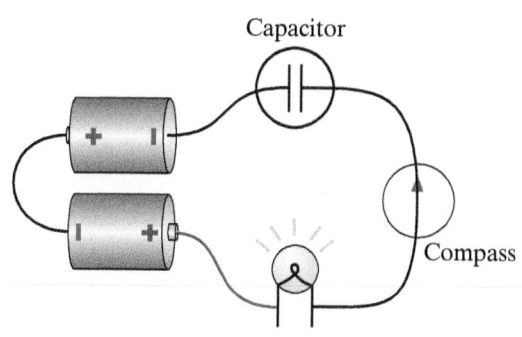

Capacitor

Compass

Figure 19.51 Charging a capacitor.

- To make sure that the capacitor is initially uncharged, start by connecting a wire across the capacitor for a couple of seconds in order to discharge the capacitor.
- Figure 19.51: Connect the discharged capacitor in series with a thick-filament bulb (#14), and run the wire over a compass. The process is called "charging the capacitor." What do you observe?

- After the bulb stopped glowing, was there still current flowing? (You may need to place the wire under the compass in order to see small deflections easily.)
- Figure 19.52: Remove the batteries and connect the bulb directly to the capacitor. The process is called "discharging the capacitor." What do you observe?

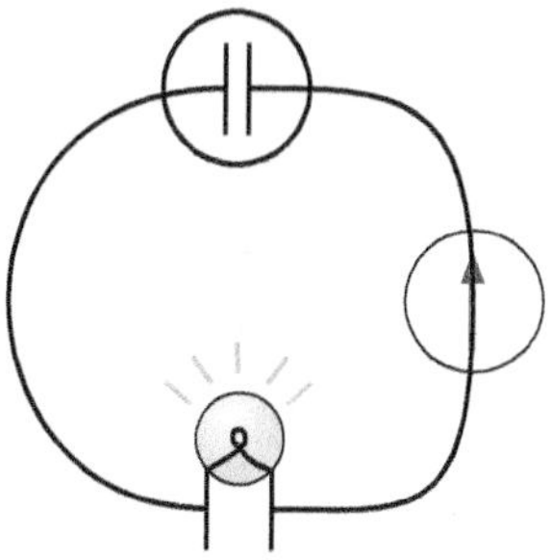

Figure 19.52 Discharging a capacitor.

- Repeat both stages of the experiment (charging and discharging), but this time use a thin-filament bulb (#48) with the two batteries and the capacitor. What difference does the choice of bulb make?
- Sketch graphs of current through the bulb vs. time for both kinds of bulbs. On your graphs label the time axis, and note which curve refers to which bulb. Pay special attention to the magnitude of the initial current (you may need to repeat the experiment to observe this).

You see that the time can be varied by varying the resistance. If you like, you can experiment with different lengths or thicknesses of your Nichrome wires, which provide a wide range of charging and discharging times.

EXP2 The effect of different bulbs on final charge
Design an experiment that will show how the final amount of charge on each plate of the capacitor depends on which bulb is involved in charging the capacitor. You might discuss your ideas with other students before carrying out your experiments. (If you wish, you can conduct your study using different lengths or thicknesses of Nichrome wire rather than two kinds of light bulbs.)

Be careful in your experiments. Before charging a capacitor be sure that it doesn't already have some charge on its plates: discharge it fully by connecting a wire across it for a couple seconds. Repeat any numerical measurements several times to make sure you know how reproducible the measurements are: 15.43 s and 15.67 s may not be significantly different!

Describe your experiments, results, and conclusions.

EXP3 Two capacitors in parallel
Work with another group so that you can try lighting a thick-filament bulb with two capacitors in parallel as shown in Figure 19.53. What do you predict for the charging time compared to a circuit with one capacitor? (Again, consider what happens in the first fraction of a second, and what effect this will have on the process in the next time interval.) What do you observe?

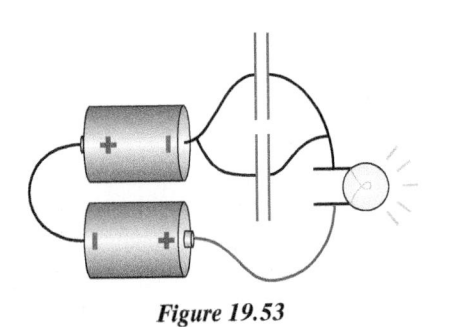

Figure 19.53

EXP4 Lighting an isolated bulb
Work with another group so that you can make a circuit with two capacitors as shown in Figure 19.54. Note that the thin-filament bulb is in a conducting "island" completely isolated from the batteries.

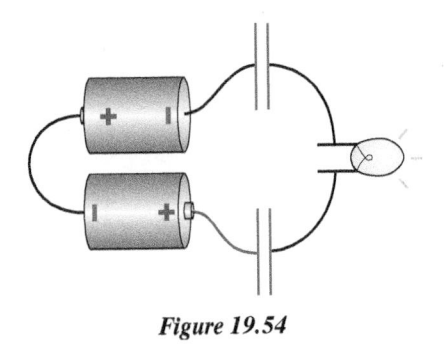

Figure 19.54

Before running the circuit, predict in detail what you expect will happen, and explain your reasoning. Try to work out agreement among the students in the two groups. Remember that during the few-nanosecond transient, the uncharged capacitors have no effect on the electric field.

Now construct the two-capacitor circuit and observe what happens. Discuss any discrepancies between the prediction you made and what you see. You might use compasses to verify your theory about the currents in the various parts of the circuit.

EXP5 Reversing the capacitor
Follow the following instructions carefully—otherwise you might damage the capacitor or the bulb.

Charge the capacitor using one battery and a thick-filament bulb (use only one battery!). Now that the capacitor is charged, what do you predict would happen if you reversed the connections to the capacitor, leaving all other connections intact, including the battery? Explain your reasoning, including an appropriate diagram.

Now try it. What do you observe? If the observation is different from your prediction, explain what was wrong with the prediction. Why is it important to use only one battery?

EXP6 The usefulness of a capacitor
Connect the capacitor and a thin-filament bulb in parallel rather than in series (Figure 19.55). Intermittently break the connection to the batteries for very brief fractions of a second. What happens to the brightness of the bulb during the very brief time that the batteries are disconnected, compared to what happens without a capacitor in the circuit? Explain.

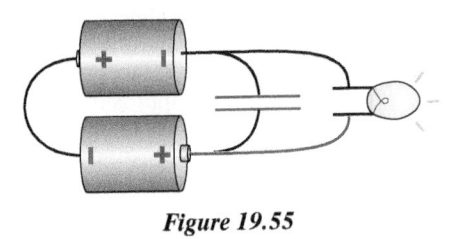

Figure 19.55

Internal Resistance of a Battery

EXP7 Two bulbs in parallel
Connect two thick-filament bulbs in parallel to two batteries in series. Unscrew one of the two bulbs. When you screw it back in, you may be able to see a slight change in the brightness of the other bulb. Why is this?

EXP8 Two batteries in parallel
Connect two batteries in parallel to a thin-filament bulb. You will see that the bulb has about the same brightness (dimness) as it has with just one battery, because the potential difference is essentially unchanged. What would you expect concerning how long the batteries would last, compared with just one battery? Why?

EXP9 The short-circuit current
Hold a copper connecting wire high enough above a compass that when the wire is connected by other connecting wires to short-circuit a single battery, you get about a 20° compass deflection. Try a different battery, trying not to move the wire. What compass deflections do you get for each battery? For the two batteries in series? For the two batteries in parallel? Are your measurements consistent with your predictions in the exercises above for the maximum current from batteries in series and in parallel?

QUESTIONS

Q1 How is the initial current through a bulb affected by putting a capacitor in series in the circuit? Explain briefly.

Q2 How is the charging time for a capacitor correlated with the initial current? That is, if the initial current is bigger, is the charging time longer, shorter, or the same? Explain briefly.

Q3 Consider two capacitors whose only difference is that the plates of capacitor number 2 are closer together than those of capacitor number 1 (Figure 19.56). Neither capacitor has an insulating layer between the plates. They are placed in two different circuits having similar batteries and bulbs in series with the capacitor.

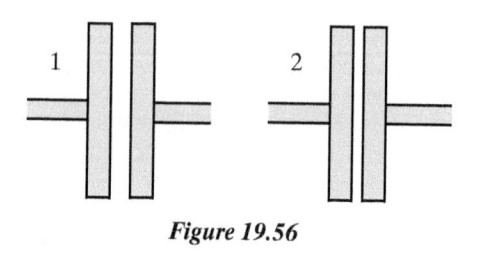

Figure 19.56

Show that in the first fraction of a second the current stays more nearly constant (decreases less rapidly) in the circuit with capacitor number 2. Explain your reasoning in detail. *Hint:* Show charges on the metal plates, and consider the electric fields they produce in the nearby wires. Remember that the fringe field near a plate outside a circular capacitor is approximately

$$\frac{Q/A}{\varepsilon_0}\left(\frac{s}{2R}\right)$$

A more extensive analysis shows that this trend holds true for the entire charging process: the capacitor with the narrower gap ends up with more charge on its plates.

Q4 The insulating layer between the plates of a capacitor not only holds the plates apart to prevent conducting contact but also has a big effect on charging. Consider two capacitors whose only difference is that capacitor number 1 has nothing between the plates, while capacitor number 2 has a layer of plastic in the gap (Figure 19.57). They are placed in two different circuits having similar batteries and bulbs in series with the capacitor.

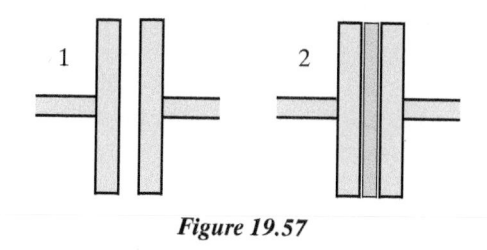

Figure 19.57

Show that in the first fraction of a second the current stays more nearly constant (decreases less rapidly) in the circuit with capacitor number 2. Explain your reasoning in detail. *Hint:* Consider the electric fields produced in the nearby wires by this plastic-filled capacitor.

Suppose that the plastic is replaced by a different plastic that polarizes more easily. In the same circuit, would this capacitor keep the current more nearly constant or less so than capacitor 2?

A more extensive analysis shows that this trend holds true for the entire charging process: the capacitor containing an easily polarized insulator ends up with more charge on its plates. The capacitor you have been using is filled with an insulator that polarizes extremely easily.

Q5 Suppose that instead of placing an insulating layer between the plates of the capacitor shown in Figure 19.57, you inserted a metal slab of the same thickness, just barely not touching the plates. In the same circuit, would this capacitor keep the current more nearly constant or less so than capacitor 2 in Question Q4? Explain why this is essentially equivalent to making a capacitor with a shorter distance between the plates.

Q6 How does the final (equilibrium) charge on the capacitor plates depend on the particular resistor (for example, the kind of bulb or the length of Nichrome wire) in the circuit during charging? Explain briefly.

Q7 How does the final (equilibrium) charge on the capacitor plates depend on the size of the capacitor plates? On the spacing between the capacitor plates? On the presence of a plastic slab between the plates?

Q8 Researchers have developed an experimental capacitor that takes about eight hours (!) to discharge through a thin-filament bulb. They propose using this in electric cars to provide a burst of energy for emergency situations (it would be charged at a slow rate during normal driving). Describe in general terms the key design elements of this extraordinary capacitor.

Q9 Give a complete but brief explanation for the behavior of the current during the discharging of a capacitor in a circuit consisting of a capacitor and light bulb. Include detailed diagrams. Explain, don't just describe.

Q10 Give a complete but brief explanation for the behavior of the current during the charging of a capacitor in a circuit consisting of batteries, bulb, and capacitor in series. Include detailed diagrams. Explain, don't just describe.

Q11 Figure 19.58 shows three circuits labeled A, B, and C. All the thin-filament bulbs, capacitors, and batteries are identical and are like the equipment you used in class. The capacitors are initially uncharged. In each circuit the batteries are connected for a short time T and then disconnected. The time T is only 10% of the total charging time through a single thin-filament bulb, so that the bulb brightness doesn't change much during the time T.

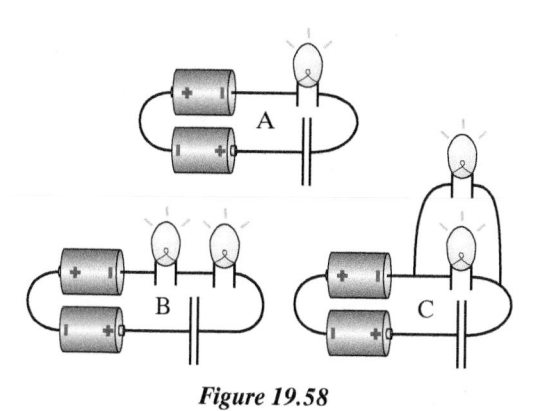

Figure 19.58

(a) In which circuit (A, B, or C) does the capacitor now have the most charge? Explain. **(b)** In which circuit (A, B, or C) does the capacitor now have the least charge? Explain. **(c)** Design and carry out experiments to check your answers. Describe your experiments and the numerical results. Before each experiment, connect a wire across the capacitor for a few seconds to fully discharge the capacitor. One way to compare the amount of charge stored in the capacitor during the time T is to finish charging it through a single thin-filament bulb, and see how much less time is required than when you start with a discharged capacitor.

Q12 The two circuits shown in Figure 19.59 have different capacitors but the same batteries and thin-filament bulbs. The capacitors in circuit 1 and circuit 2 are identical except that the capacitor in circuit 2 was constructed with its plates closer together. Both capacitors have air between their plates. The capacitors are initially uncharged. In each circuit the batteries are connected for a time that is short compared to the time required to reach equilibrium, and then they are disconnected. In which circuit (1 or 2) does the capacitor now have more charge? Explain your reasoning in detail.

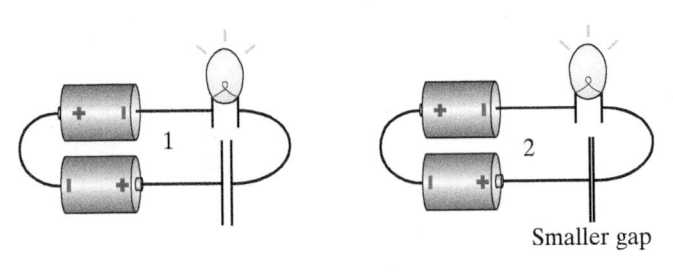

Figure 19.59

Q13 The circuit shown in Figure 19.60 is composed of two batteries, two identical thin-filament bulbs, and an initially uncharged capacitor. Three compasses are placed on the desktop underneath the wires, all pointing north before the circuit is closed. Then the gap in the wire is closed, and the compass on the left immediately shows about a $20°$ deflection.

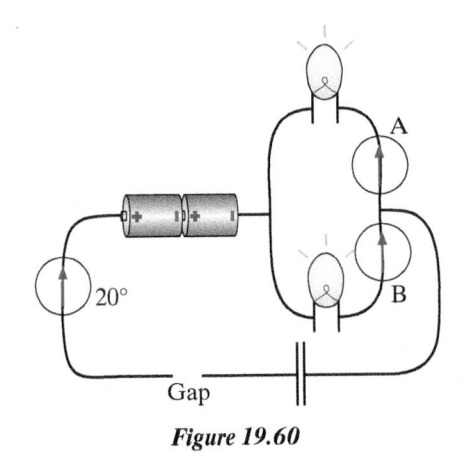

Figure 19.60

(a) On a diagram, draw the new needle positions on all three compasses at this time, and write the approximate angle of deflection beside the compasses labeled A and B. Explain carefully. **(b)** When this circuit was connected with one of the bulbs removed from its socket, the single bulb glowed for T seconds. In the two-bulb circuit shown in Figure 19.60, predict how long the two bulbs would glow (in terms of T). Explain

carefully. **(c)** If possible, carry out the experiment, observing initial compass deflections and length of time of glow with one or two thin-filament bulbs. State your numerical results. Do your observations agree with your predictions?

Q14 The circuit shown in Figure 19.61 consists of two flashlight batteries, a large air-gap capacitor, and Nichrome wire. The circuit is allowed to run long enough that the capacitor is fully charged with $+Q$ and $-Q$ on the plates.

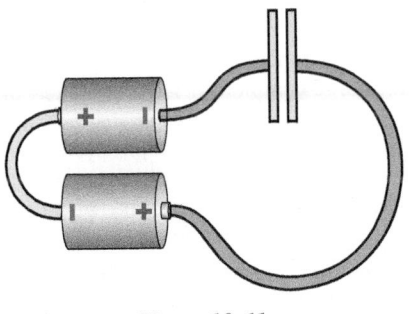

Figure 19.61

Next you push the two plates closer together (but the plates don't touch each other). Describe what happens, and explain why in terms of the fundamental concepts of charge and field. Include diagrams showing charge and field at several times.

Q15 The charge on an isolated capacitor does not change when a sheet of glass is inserted between the capacitor plates, and we find that the potential difference decreases (because the electric field inside the insulator is reduced by a factor of $1/K$). Suppose instead that the capacitor is connected to a battery, so that the battery tries to maintain a fixed potential difference across the capacitor. **(a)** A light bulb and an air-gap capacitor of capacitance C are connected in series to a battery with known emf. What is the final charge Q on the positive plate of the capacitor? **(b)** After fully charging the capacitor, a sheet of plastic whose dielectric constant is K is inserted into the capacitor and fills the gap. Does any current run through the light bulb? Why? What is the final charge on the positive plate of the capacitor?

Q16 An isolated capacitor consisting of two large metal disks separated by a very thin sheet of plastic initially has equal and opposite charges $+Q$ and $-Q$ on the two disks. Then we connect a long thin Nichrome wire to the capacitor as shown in Figure 19.62.

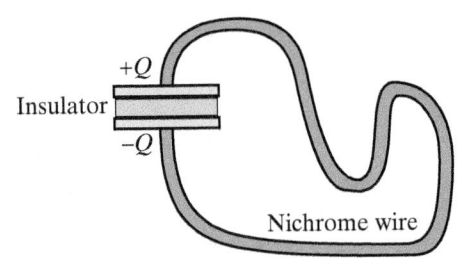

Figure 19.62

Describe and explain in detail what happens, including the first few nanoseconds, the approach to the final state, and the final state. Base your explanation on the fundamental concepts of charge and field; do not use the concept of potential in your explanation. To allow room for drawing charges and fields, make your diagram like Figure 19.62, with the thickness of the capacitor and the wire greatly exaggerated.

Q17 A capacitor with a slab of glass between the plates is connected to a battery by Nichrome wires and allowed to charge completely. Then the slab of glass is removed. Describe and explain what happens. Include diagrams. If you give a direction for a current, state whether you are describing electron current or conventional current.

Q18 A capacitor is connected to batteries by Nichrome wires and allowed to charge completely. Then the plates are suddenly moved farther apart. Describe what happens and explain in detail why it happens, based on fundamental physical principles. If you give a direction for a current, state whether you are describing electron current or conventional current. Include appropriate diagrams to support your explanation.

Q19 Two 1.5 V batteries in series were connected to a capacitor consisting of two very large metal disks placed very close to each other, so that the disks became charged (left disk positive). The batteries were then removed.

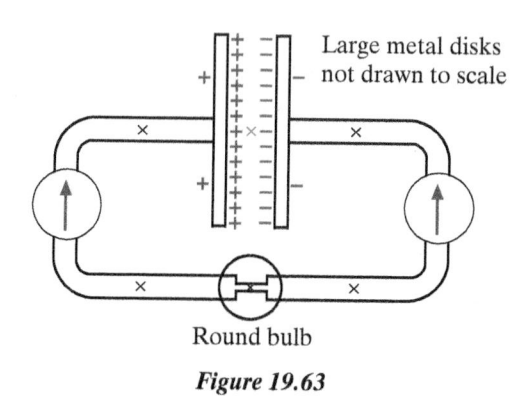

Large metal disks not drawn to scale

Round bulb

Figure 19.63

Next, thick copper wires and a thick-filament bulb were attached to the charged capacitor, as indicated in Figure 19.63. Two magnetic compasses lying on top of the copper wires initially pointed north. You find experimentally that when this thick-filament bulb is connected directly to the two batteries in series you observe a compass deflection of 15°. **(a)** Briefly describe in detail what you would observe about the bulb and both compasses during the next minute. You do not need to explain yet why this happens. Just describe in detail what you would see. **(b)** At a time that is 0.01 s after assembling this circuit, draw and label the electric field at the six locations marked × on the diagram. Pay attention to the relative magnitudes of your vectors. Make sure you have clearly labeled your electric field vectors (for example, E_{bulb}, etc.). **(c)** Sketch roughly a possible charge distribution on the diagram. If necessary for clarity, add comments on the diagram about the charge distribution. **(d)** The length of the left copper wire is L, the length of the right copper

wire is also L, the length of the bulb filament is L_{bulb}, and the gap between the capacitor plates is s. The diameter of the copper wires is d, and the diameter of the bulb filament is d_{bulb}. The mobility in the copper is u, and the mobility in the hot tungsten is u_{bulb}. The number of mobile electrons per unit volume is approximately the same in the copper and the tungsten. At a time that is 0.01 s after assembling this circuit, write two valid physics equations involving the magnitudes of the electric fields, using the labels that you gave for the electric fields in part (b). What are the principles underlying these physics equations? **(e)** In part (a) you described what you would see happening with the bulb and compasses over the course of a minute. Now explain qualitatively but in detail why this happens. Be brief but be complete in your explanation.

Q20 Someone said, "Current takes the path of least resistance." What's wrong with this statement?

Q21 What are the units of conductivity σ, resistivity ρ, resistance R, and current density J?

Q22 Which of the following are ohmic resistors? For those that aren't, briefly state why they aren't. **(a)** Nichrome wire, **(b)** a thin filament light bulb, **(c)** a plastic rod, **(d)** salt water, **(e)** silicon (a semiconductor)

Q23 A desk lamp that plugs into a 120 V wall socket can use a 60 W or a 100 W light bulb. Which bulb has the larger resistance? Explain briefly.

Q24 **(a)** If the current through a battery is doubled, by what factor is the battery power increased? **(b)** If the current through a resistor is doubled, by what factor is the power dissipation increased? **(c)** Explain why these factors are the same or different (depending on what you find).

Q25 State whether the following statement is true or false, and briefly explain why: "In the two circuits shown in Figure 19.64, the battery output power is greater in circuit 2 because there is an additional resistor R_2 dissipating power."

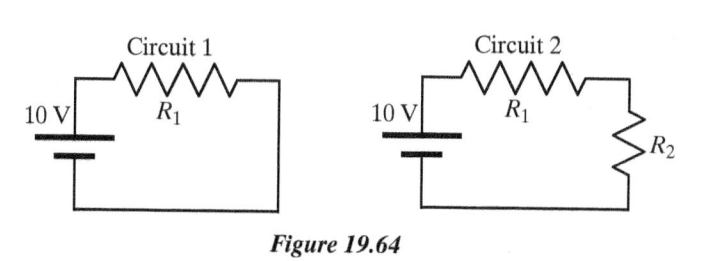

Figure 19.64

Q26 Why can birds perch on the bare wire of a 100,000 V power line without being electrocuted?

Q27 Should an ammeter have a low or high resistance? Why? Should a voltmeter have a low or high resistance? Why?

Q28 You are marooned on a desert island full of all kinds of standard electrical apparatus including a sensitive voltmeter, but you don't have an ammeter. Explain how you could use the voltmeter to measure currents.

PROBLEMS

Section 19.1

•**P29** The capacitor in Figure 19.65 is initially charged, then the circuit is connected. Which graph in Figure 19.66 best describes the current through the bulb as a function of time?

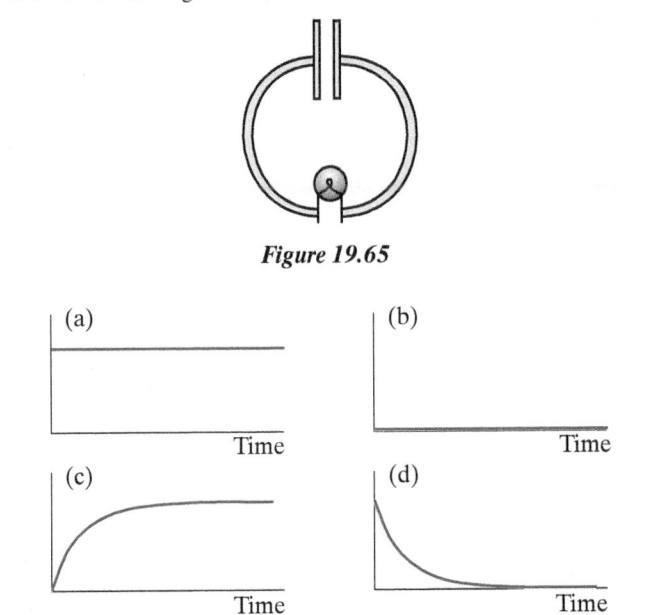

Figure 19.65

(a)

Time

(b)

Time

(c)

Time

(d)

Time

Figure 19.66

•**P30** The capacitor in Figure 19.67 is initially uncharged, then the circuit is connected. Which graph in Figure 19.66 best describes the current through the bulb as a function of time?

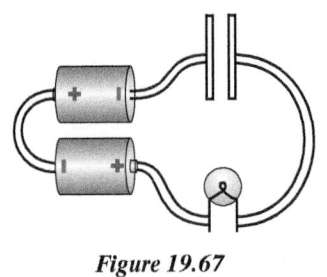

Figure 19.67

•**P31** The capacitor in Figure 19.67 is initially uncharged, then the circuit is connected. Which graph in Figure 19.66 best describes the absolute value of the charge on the left plate as a function of time?

•**P32** The capacitor in Figure 19.68 is initially uncharged, then the circuit is connected. Which graph in Figure 19.66 best describes the magnitude of the fringe field of the capacitor at location A (inside the connecting wire) as a function of time?

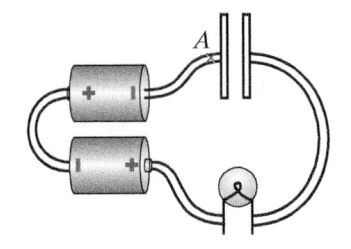

Figure 19.68

•**P33** The capacitor in Figure 19.68 is initially uncharged, then the circuit is connected. Which graph in Figure 19.66 best describes the magnitude of the net electric field at location A (inside the connecting wire) as a function of time?

•**P34** A particular capacitor is initially charged. Then a high-resistance Nichrome wire is connected between the plates of the capacitor, as shown in Figure 19.69. The needle of a compass placed under the wire deflects 20° to the east as soon as the connection is made. After 60 s the compass needle no longer deflects. **(a)** Which of the diagrams in Figure 19.69 best indicates the electron current at three locations in this circuit? (1) 0.01 s after the circuit is connected, (2) 15 s after the circuit is connected, (3) 120 s after the circuit is connected. **(b)** Which of the diagrams in Figure 19.70 best indicates the net electric field inside the wire at three locations in this circuit? (1) 0.01 s after the circuit is connected, (2) 15 s after the circuit is connected, (3) 120 s after the circuit is connected.

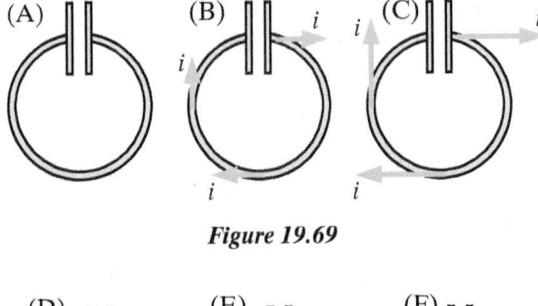

Figure 19.69

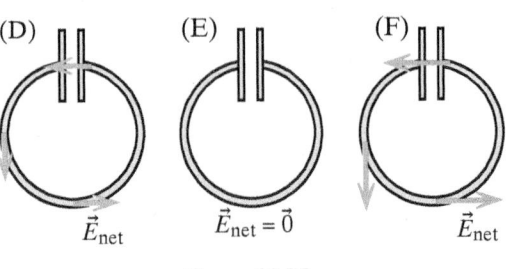

Figure 19.70

•**P35** Which of the following statements about the discharging of a capacitor through a light bulb are correct? Choose all that are true. (1) The fringe field of the capacitor decreases as the charge on the capacitor plates decreases. (2) Electrons flow across the gap between the plates of the capacitor, thus reducing the charge on the capacitor. (3) The electric field at a location inside the wire is due to charge on the surface of the wires and charge on the plates of the capacitor. (4) Electrons in the wires flow away from the negative plate toward the positive plate, reducing charge on the plates.

•**P36** When a particular capacitor, which is initially uncharged, is connected to a battery and a small light bulb, the light bulb is initially bright but gradually gets dimmer, and after 45 s it goes out. The diagrams in Figure 19.71 show the electric field in the circuit and the surface charge distribution on the wires at three different times (0.01 s, 2 s, and 240 s) after the connection to the bulb is made. The diagrams in Figure 19.71 show the pattern of electric field in the wires and charge on the surface of the wires at three different times (0.01 s, 8 s, and 240 s) after the connection to the battery is made. Which of the diagrams best represents the state of the circuit at each time specified? **(a)** 0.01 s after the connection is made, **(b)** 8 s after the connection is made, **(c)** 240 s after the connection is made

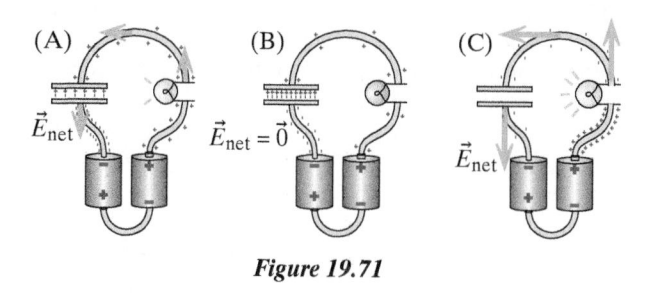

Figure 19.71

•**P37** In circuit 1 (Figure 19.72), an uncharged capacitor is connected in series with two batteries and one light bulb. Circuit 2 (Figure 19.72) contains two light bulbs identical to the bulb in circuit 1; in all other respects it is identical to circuit 1. In circuit 1, the light bulb stays lit for 25 s. The following questions refer to these circuits. You should draw diagrams representing the fields and charges in each circuit at the times mentioned, in order to answer the questions.

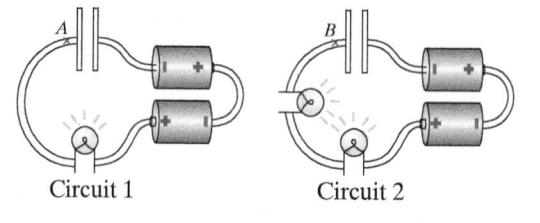

Circuit 1   Circuit 2

Figure 19.72

(a) One microsecond (1×10^{-6} s) after connecting both circuits, which of the following are true? Choose all that apply: (1) The net electric field at location A in circuit 1 is larger than the net electric field at location B in circuit 2. (2) At location A in circuit 1, electrons flow to the left. (3) At location A in circuit 1, the electric field due to charges on the surface of the wires and batteries points to the right. (4) In circuit 1 the potential difference across the capacitor plates is equal to the emf of the batteries. (5) The current in circuit 1 is larger than the current in circuit 2. **(b)** Two seconds after connecting both circuits, which of the following are true? Choose all that apply: (1) There is more charge on the plates of capacitor 1 than there is on the plates of capacitor 2. (2) There is negative charge on the right plate of the capacitor in circuit 1. (3) At location B in circuit 2 the net electric field points to the right. (4) At location B in circuit 2 the fringe field of the capacitor points to the right. (5) At location A in circuit 1 the fringe field of the capacitor points to the left. **(c)** Which of the graphs in Figure 19.73 represents the amount of charge on the positive plate of the capacitor in circuit 1 as a function of time? **(d)** Which of the graphs in Figure 19.73 represents the current in circuit 1 as a function of time?

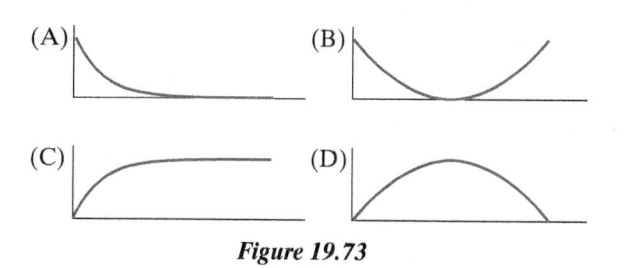

Figure 19.73

•**P38** A certain capacitor has rectangular plates 56 cm by 24 cm, and the gap width is 0.20 mm. What is its capacitance? We see that typical capacitances are very small when measured in farads. A 1 F capacitor is quite extraordinary! Apparently it has a very large area A (all wrapped up in a small package), and a very small gap s.

•**P39** A certain capacitor has rectangular plates 59 cm by 33 cm, and the gap width is 0.27 mm. If the gap is filled with a material whose dielectric constant is 2.9, what is the capacitance of this capacitor?

•**P40** Suppose that you charged a 2.5 F capacitor with two 1.5 V batteries. How much charge would be on each plate in the final state? How many excess electrons would be on the negative plate?

•**P41** You connect a 9 V battery to a capacitor consisting of two circular plates of radius 0.08 m separated by an air gap of 2 mm. What is the charge on the positive plate?

•**P42** Two circular plates of radius 0.12 m are separated by an air gap of 1.5 mm. The plates carry charge $+Q$ and $-Q$ where $Q = 3.6 \times 10^{-8}$ C. **(a)** What is the magnitude of the electric field in the gap? **(b)** What is the potential difference across the gap? **(c)** What is the capacitance of this capacitor?

•**P43** You may have noticed that while discharging a capacitor through a light bulb, the light bulb glows just about as brightly, and for just about as long, as it does while charging the same capacitor through the same bulb. Let E stand for the energy emitted by the light bulb (as light and heat) in the discharging phase, from just before the bulb is connected to the capacitor until the time when there is essentially no more current. In terms of $+E$ or $-E$, what was the energy change of the battery, capacitor, bulb, and surroundings during the charging phase, and during the discharging phase? One answer is already given in the following table:

	Charging	**Discharging**
Battery:		
Bulb:		
Capacitor:		
Surroundings:		$+E$

It is somewhat surprising that we can get this much information out of one simple observation.

••**P44** As shown in Figure 19.74, a spherical metal shell of radius r_1 has a charge Q (on its outer surface) and is surrounded by a concentric spherical metal shell of radius r_2 which has a charge $-Q$ (on its inner surface).

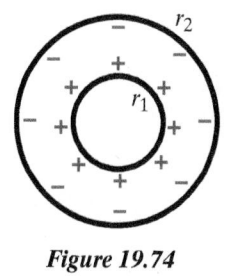

Figure 19.74

(a) Use the definition of capacitance: $Q = C|\Delta V|$ to find the capacitance of this spherical capacitor. **(b)** If the radii of the spherical shells r_1 and r_2 are large and nearly equal to each other, show that C can be written as $\varepsilon_0 A/s$ (which is also the equation for the capacitance of a parallel-plate capacitor) where $A = 4\pi r^2$ is the surface area of one of the spheres, and s is the small gap distance between them ($r_2 = r_1 + s$).

Section 19.2

•P45 In gold at room temperature, the mobility of mobile electrons is about 4.3×10^{-3} (m/s)/(V/m), and there are about 5.9×10^{28} mobile electrons per cubic meter. Calculate the conductivity of gold, including correct units.

•P46 Consider a silver wire with a cross-sectional area of 1 mm^2 carrying 0.3 A of current. The conductivity of silver is 6.3×10^7 (A/m^2)/(V/m). Calculate the magnitude of the electric field required to drive this current through the wire.

•P47 When a thin-filament light bulb is connected to two 1.5 V batteries in series, the current is 0.075 A. What is the resistance of the glowing thin-filament bulb?

•P48 In copper at room temperature, the mobility of mobile electrons is about 4.5×10^{-3} (m/s)/(V/m), and there are about 8×10^{28} mobile electrons per m^3. Calculate the conductivity σ and include the correct units. In actual practice, it is usually easier to measure the conductivity σ and deduce the mobility u from this measurement.

•P49 (a) A carbon resistor is 5 mm long and has a constant cross section of 0.2 mm^2. The conductivity of carbon at room temperature is $\sigma = 3 \times 10^4$ per ohm·m. In a circuit its potential at one end of the resistor is 12 V relative to ground, and at the other end the potential is 15 V. Calculate the resistance R and the current I. **(b)** A thin copper wire in this circuit is 5 mm long and has a constant cross section of 0.2 mm^2. The conductivity of copper at room temperature is $\sigma = 6 \times 10^7$ ohm^{-1}m^{-1}. The copper wire is in series with the carbon resistor, with one end connected to the 15 V end of the carbon resistor, and the current you calculated in part (a) runs through the carbon resistor wire. Calculate the resistance R of the copper wire and the potential $V_{\text{at end}}$ at the other end of the wire.

You can see that for most purposes a thick copper wire in a circuit would have practically a uniform potential. This is because the small drift speed in a thick, high-conductivity copper wire requires only a very small electric field, and the integral of this very small field creates a very small potential difference along the wire.

•P50 When a single thin-filament bulb is connected to a 1.5 V battery, the current through the battery is about 80 mA. If you add another thin-filament bulb in parallel, the battery current of course increases to 160 mA. Is the battery ohmic? That is, is the current through the battery proportional to the potential difference across the battery?

•P51 A certain ohmic resistor has a resistance of 40 Ω. A second resistor is made of the same material but is three times as long and has half the cross-sectional area. What is the resistance of the second resistor? What is the effective resistance of the two resistors in series?

•P52 In the circuit shown in Figure 19.75, the emf of the battery is 7.9 V. Resistor R_1 has a resistance of 23 Ω, and resistor R_2 has a resistance of 44 Ω. A steady current flows through the circuit. **(a)** What is the absolute value of the potential difference across R_1? **(b)** What is the conventional current through R_2?

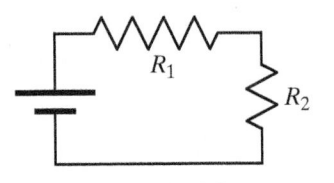

Figure 19.75

•P53 When a thick-filament bulb is connected to one flashlight battery, the current is 0.20 A. When you use two batteries in series, the current is not 0.40 A but only 0.33 A. Briefly explain this behavior.

•P54 In Figure 19.76 the resistance R_1 is 10 Ω, R_2 is 5 Ω, and R_3 is 20 Ω. If this combination of resistors were to be replaced by a single resistor with an equivalent resistance, what should that resistance be?

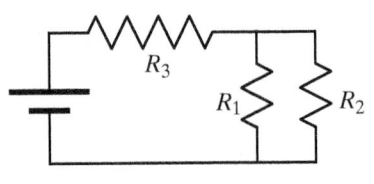

Figure 19.76

•P55 Suppose that you charge a 1 F capacitor in a circuit containing two 1.5 V batteries, so the final potential difference across the plates is 3 V. How much charge is on each plate? How many excess electrons are on the negative plate?

•P56 In the circuit shown in Figure 19.77 the emf of the battery is 7.4 V. Resistor R_1 has a resistance of 31 Ω, resistor R_2 has a resistance of 47 Ω, and resistor R_3 has a resistance of 52 Ω. A steady current flows through the circuit.

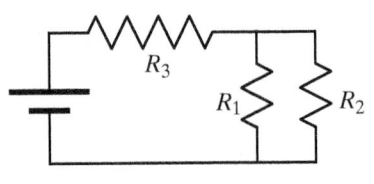

Figure 19.77

(a) What is the equivalent resistance of R_1 and R_2? **(b)** What is the equivalent resistance of all three resistors? **(c)** What is the conventional current through R_3?

•P57 Consider a copper wire with a cross-sectional area of 1 mm^2 (similar to your connecting wires) and carrying 0.3 A of current, which is about what you get in a circuit with a thick-filament bulb and two batteries in series. Calculate the strength of the very small electric field required to drive this current through the wire.

••P58 The conductivity of tungsten at room temperature, 1.8×10^7 (A/m^2)/(V/m), is significantly smaller than that of copper. At the very high temperature of a glowing light-bulb filament (nearly 3000 kelvins), the conductivity of tungsten is 18 times smaller than it is at room temperature. The tungsten filament of a thick-filament bulb has a radius of about 0.015 mm. Calculate the electric field required to drive 0.20 A of current through the glowing bulb and show that it is very large compared to the field in the connecting copper wires.

••P59 A circuit consists of a battery, whose emf is K, and five Nichrome wires, three thick and two thin as shown in

Figure 19.78. The thicknesses of the wires have been exaggerated in order to give you room to draw inside the wires. The internal resistance of the battery is negligible compared to the resistance of the wires. The voltmeter is not attached until part (e) of the problem.

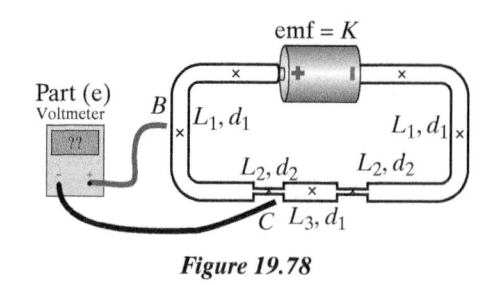

Figure 19.78

(a) Draw and label appropriately the electric field at the locations marked × inside the wires, paying attention to appropriate relative magnitudes of the vectors that you draw. **(b)** Show the approximate distribution of charges for this circuit. Make the important aspects of the charge distribution very clear in your drawing, supplementing your diagram if necessary with very brief written descriptions on the diagram. Make sure that parts (a) and (b) of this problem are consistent with each other. **(c)** Assume that you know the mobile-electron density n and the electron mobility u at room temperature for Nichrome. The lengths (L_1, L_2, L_3) and diameters (d_1, d_2) of the wires are given on the diagram. Calculate accurately the number of electrons that leave the negative end of the battery every second. Assume that no part of the circuit gets very hot. Express your result in terms of the given quantities (K, L_1, L_2, L_3, d_1, d_2, n, and u). Explain your work and identify the principles you are using. **(d)** In the case that $d_2 \ll d_1$, what is the *approximate* number of electrons that leave the negative end of the battery every second? **(e)** A voltmeter is attached to the circuit with its + lead connected to location B (halfway along the leftmost thick wire) and its − lead connected to location C (halfway along the leftmost thin wire). In the case that $d_2 \ll d_1$, what is the approximate voltage shown on the voltmeter, including sign? Express your result in terms of the given quantities (K, L_1, L_2, L_3, d_1, d_2, n, and u).

••P60 Using thick connecting wires that are very good conductors, a Nichrome wire ("wire 1") of length L_1 and cross-sectional area A_1 is connected in series with a battery and an ammeter (this is circuit 1). The reading on the ammeter is I_1. Now the Nichrome wire is removed and replaced with a different wire ("wire 2"), which is 2.5 times as long and has 5.5 times the cross-sectional area of the original wire (this is circuit 2).

In the following questions, a subscript 1 refers to circuit 1, and a subscript 2 refers to circuit 2. It will be helpful to write out your solutions to the following questions algebraically before doing numerical calculations. (*Hint:* Think about what is the same in these two circuits.) **(a)** What is the value of I_2/I_1, the ratio of the conventional currents in the two circuits? **(b)** What is the value of R_2/R_1, the ratio of the resistances of the wires? **(c)** What is the value of E_2/E_1, the ratio of the electric fields inside the wires in the steady state?

•••P61 A long iron slab of width w and height h emerges from a furnace, as shown in Figure 19.79. Because the end of the slab near the furnace is hot and the other end is cold, the electron mobility increases significantly with the distance x.

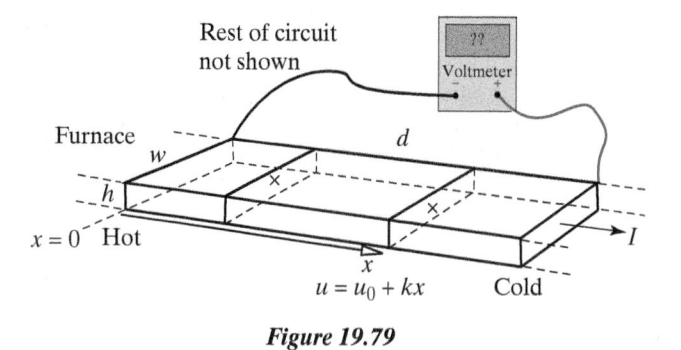

Figure 19.79

The electron mobility is $u = u_0 + kx$, where u_0 is the mobility of the iron at the hot end of the slab. There are n iron atoms per cubic meter, and each atom contributes one electron to the sea of mobile electrons (we can neglect the small thermal expansion of the iron). A steady-state conventional current runs through the slab from the hot end toward the cold end, and an ammeter (not shown) measures the current to have a magnitude I in amperes. A voltmeter is connected to two locations a distance d apart, as shown. **(a)** Show the electric field inside the slab at the two locations marked with ×. Pay attention to the relative magnitudes of the two vectors that you draw. **(b)** Explain why the magnitude of the electric field is different at these two locations. **(c)** At a distance x from the left voltmeter connection, what is the magnitude of the electric field in terms of x and the given quantities w, h, d, u_0, k, n, and I (and fundamental constants)? **(d)** What is the sign of the potential difference displayed on the voltmeter? Explain briefly. **(e)** In terms of the given quantities w, h, d, u_0, k, n, and I (and fundamental constants), what is the magnitude of the voltmeter reading? Check your work. **(f)** What is the resistance of this length of the iron slab?

Section 19.4

••P62 A battery with negligible internal resistance is connected to a resistor. The power produced in the battery and the power dissipated in the resistor are both P_1. Another resistor of the same kind is added, so the circuit consists of a battery and two resistors in series. **(a)** In terms of P_1, now how much power is dissipated in the first resistor? **(b)** In terms of P_1, now how much power is produced in the battery? **(c)** The circuit is rearranged so that the two resistors are in parallel rather than in series. In terms of P_1, now how much power is produced in the battery?

••P63 Two flashlight batteries in series power a circuit consisting of four ohmic resistors as shown in Figure 19.80, connected by copper wires with negligible resistance. The internal resistance of the batteries is negligible. Two magnetic compasses are placed underneath the wires, initially pointing northward before closing the circuit. A very-high-resistance voltmeter is connected as shown. The resistance R_1 is 20 Ω, R_2 is 20 Ω, R_3 is 30 Ω, and R_4 is 15 Ω.

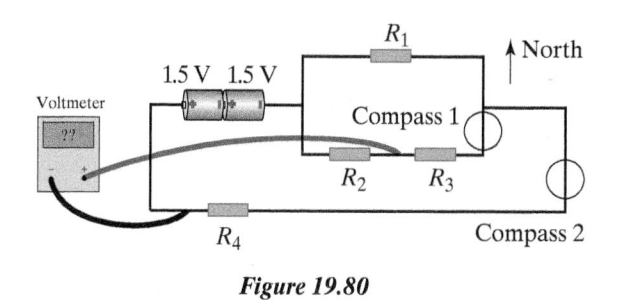

Figure 19.80

(a) Starting from fundamental principles, write equations that could be solved to determine the conventional current through each resistor, and show the directions of these currents on the diagram. Explain where each of your equations comes from. Do not use equations for parallel and series resistances (but you are free to use such equations to check your work if you like). If these equations are solved, we find that the current through R_1 is 0.073 A and the current through R_4 is 0.102 A. **(b)** Sketch the positions of the compass needles. Compass 1 deflects by 3 degrees. What is the deflection angle of compass 2? **(c)** The resistors are made of a material that has 8×10^{28} free electrons per cubic meter and a mobility of 3×10^{-5} (m/s)/(N/C). Find E_2, the magnitude of the electric field in resistor R_2, which is in the form of a short wire with a constant cross-sectional area of 6×10^{-10} m^2. **(d)** Is the field E_1 in resistor R_1 larger, smaller, or the same as E_2? **(e)** What is the length of resistor R_2? **(f)** What does the voltmeter read, including sign? **(g)** How much energy in joules is expended by the batteries in moving a singly charged ion from one end of one of the batteries to the other end of that same battery? **(h)** How much power output is there from one of the batteries?

••P64 Consider the two circuits depicted in Figure 19.81. In circuit 1, ohmic resistor R_1 dissipates 5 W; in circuit 2, ohmic resistor R_2 dissipates 20 W. The wires and batteries have negligible resistance. The circuits contain 10 V batteries.

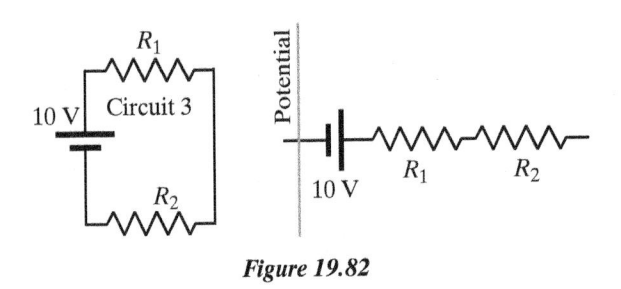

Figure 19.81

(a) What is the resistance of R_1 and of R_2? **(b)** Resistor R_1 is made of a very thin metal wire that is 3 mm long, with a diameter of 0.1 mm. What is the electric field inside this metal resistor? **(c)** The same resistors are used to construct circuit 3 (Figure 19.82), using the same 10 V battery as before. Make a complete accurate graph of potential versus location for circuit 3. Label the y axis numerically. Be sure to explain the significant features of your graph.

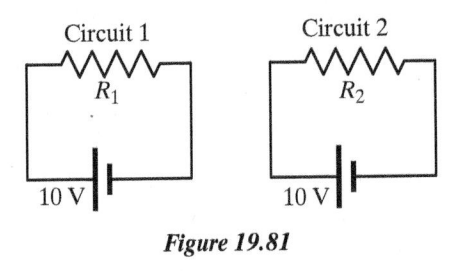

Figure 19.82

(d) On a copy of the diagram of circuit 3, place +'s and −'s to indicate the distribution of surface charge everywhere on the circuit. Assume that both resistors are far from the battery. **(e)** In circuit 3, calculate the number of electrons entering R_1 every second and the number of electrons entering R_2 every second. **(f)** What is the power output of the battery in circuit 3?

••P65 Work and energy with a capacitor: A capacitor with capacitance C has an amount of charge q on one of its plates, in which case the potential difference across the plates is $\Delta V = q/C$ (definition of capacitance). The work done to add a small amount of charge dq when charging the capacitor is $dq(\Delta V) = dq(q/C)$. Show by integration that the amount of work required to charge up the capacitor from no charge to a final charge Q is $\frac{1}{2}(Q^2/C)$. Since this is the amount of work required to charge the capacitor, it is also the amount of energy stored in the capacitor. Substituting $Q = C\Delta V$, we can also express the energy as $\frac{1}{2}C(\Delta V)^2$.

••P66 A circuit is made of two 1.5 V batteries and three light bulbs as shown in Figure 19.83.

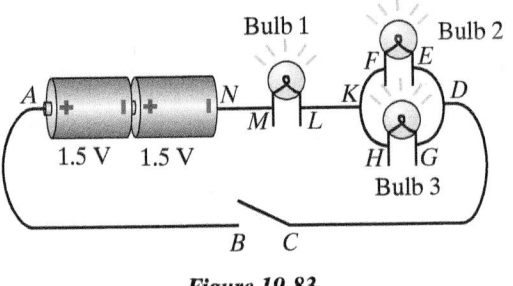

Figure 19.83

When the switch is closed and the bulbs are glowing, bulb 1 has a resistance of 10 Ω, bulb 2 has a resistance of 40 Ω, bulb 3 has a resistance of 30 Ω, and the copper connecting wires have negligible resistance. You can also neglect the internal resistance of the batteries. **(a)** Make a copy of the circuit diagram. With the switch open, indicate the approximate surface charge with +'s and −'s on your diagram. **(b)** With the switch open, find the potential differences $V_B - V_C$ and $V_D - V_K$. **(c)** After the switch is closed and the steady state is established, the currents through bulbs 1, 2, and 3 are I_1, I_2, and I_3, respectively. Write loop and node equations that could be solved to determine these three unknown currents, but do not solve the equations. Label on the diagram what current directions, loops, and nodes you are using, and explain which equation is which. In order to learn about the general approach, do not use equations for series and parallel resistance in this problem. (You can use these equations to check your work.) **(d)** In terms of the unknown currents I_1, I_2, and I_3, what is the potential difference $V_C - V_F$ (with the switch closed)? **(e)** In terms of the unknown currents I_1, I_2, and I_3, how much power is delivered by the batteries (with the switch closed)? **(f)** Solve your equations and give values for I_1, I_2, and I_3. **(g)** How many electrons leave the battery at location N every second? **(h)** What is the numerical value of $V_C - V_F$? **(i)** What is the numerical value of the power delivered by the batteries? **(j)** The tungsten filament in the 40 Ω bulb is 8 mm long and has a cross-sectional area of 2×10^{-10} m^2. How big is the electric field inside this metal filament?

Section 19.5

•P67 A certain 6 V battery delivers 12 A when short-circuited. How much current does the battery deliver when a 1 Ω resistor is connected to it?

••P68 **(a)** You short-circuit a 9 V battery by connecting a short wire from one end of the battery to the other end. If the current in the short circuit is measured to be 18 A, what is the internal resistance of the battery? **(b)** What is the power generated by the battery? **(c)** How much energy is dissipated in the internal resistance every second? (Remember that one watt is one joule per second.) **(d)** This same battery is now connected to a 10 Ω resistor. How much current flows through this resistor? **(e)** How much power is dissipated in the 10 Ω resistor? **(f)** The leads to a voltmeter are placed at the two ends of the battery of this circuit containing the 10 Ω resistor. What does the meter read?

••**P69** A 9 V battery is connected to a 100 Ω resistor, and a voltmeter shows the potential difference across the battery to be 6.7 V. **(a)** What is the internal resistance of the battery? **(b)** If the resistor is replaced by a very low-resistance wire, what does the voltmeter read? **(c)** What is the current through this wire?

••**P70** **(a)** Suppose that you connect two identical flashlight batteries in series to get a 3 V emf. Write a loop equation to show that the maximum (short-circuit) current these batteries can deliver is exactly the same as the maximum current from one battery. **(b)** Alternatively, suppose that you connect two identical flashlight batteries in parallel, which gives you just a 1.5 V emf. Write loop and node equations to show that the maximum (short-circuit) current these batteries can deliver is double that of one battery.

Section 19.6

•**P71** You connect an ammeter to a 1.5 V battery whose internal resistance is 0.1 ohm. **(a)** What does the ammeter read? **(b)** Replace the ammeter with a voltmeter. What does the voltmeter read?

••**P72** A 40-cm-long high-resistance wire with rectangular cross section 7 mm by 3 mm is connected to a 12 V battery through an ammeter, as shown in Figure 19.84. The resistance of the wire is 50 Ω. The resistance of the ammeter and the internal resistance of the battery can be considered to be negligibly small compared to the resistance of the wire.

Leads to a high-resistance voltmeter are connected as shown, with the − lead connected to the inner edge of the wire, at the top (location A), and the + lead connected to the outer edge of the wire, at the bottom (location C). The distance d along the wire between voltmeter connections is 5 cm.

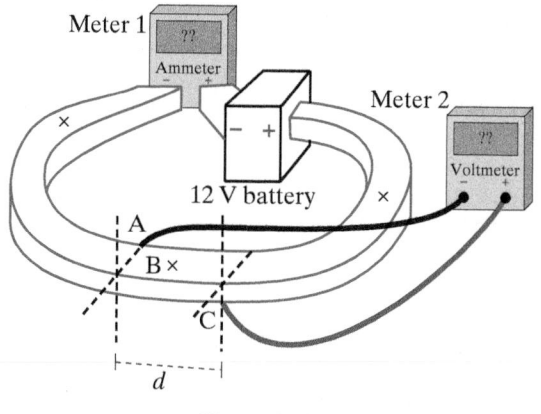

Figure 19.84

(a) On a copy of the diagram, show the approximate distribution of charge. **(b)** On a copy of the diagram, draw the electric field inside the wire at the 3 locations marked ×. **(c)** What is the magnitude of the electric field at location B? **(d)** What does the voltmeter read, both magnitude and sign? **(e)** What does the ammeter read, both magnitude and sign? **(f)** In a 60 s period,

how many electrons are released from the − end of the battery? **(g)** There are 1.5×10^{26} free electrons per cubic meter in the wire. What is the drift speed v of the electrons in the wire? **(h)** What is the mobility u of the material that the wire is made of? **(i)** Switch meter 1 from being an ammeter to being a voltmeter. Now what do the two meters read? **(j)** The 12 V battery is removed from the circuit and both the ammeter and voltmeter are connected in parallel to the battery. The voltmeter reads 1.8 V, and the ammeter reads 20.4 A. What is the internal resistance of the battery?

••**P73** Two resistors each with resistance of 4×10^6 Ω are connected in series to a 60 V power supply whose internal resistance is negligible. You connect a voltmeter across one of these resistors, and this voltmeter has an internal resistance of 1×10^6 ohms. What is the reading on the voltmeter?

Section 19.7

•**P74** A 20 Ω resistor and a 2 F capacitor are in series with a 9 V battery. What is the initial current when the circuit is first assembled? What is the current after 50 s?

••**P75** A resistor with resistance R and an air-gap capacitor of capacitance C are connected in series to a battery (whose strength is "emf"). **(a)** What is the final charge on the positive plate of the capacitor? **(b)** After fully charging the capacitor (so there is no current), a sheet of plastic whose dielectric constant is K is inserted into the capacitor and fills the gap. Explain why a current starts running in the circuit. You can base your explanation either on electric field or on electric potential, whichever you prefer. **(c)** What is the initial current through the resistor just after inserting the sheet of plastic? **(d)** What is the final charge on the positive plate of the capacitor after inserting the plastic?

••**P76** The deflection plates in an oscilloscope are 10 cm by 2 cm with a gap distance of 1 mm. A 100 V potential difference is suddenly applied to the initially uncharged plates through a 1000 Ω resistor in series with the deflection plates. How long does it take for the potential difference between the deflection plates to reach 95 V?

••**P77** A capacitor consists of two rectangular metal plates 3 m by 4 m, placed a distance 2.5 mm apart in air (Figure 19.85). The capacitor is connected to a 9 V power supply long enough to charge the capacitor fully, and then the battery is removed.

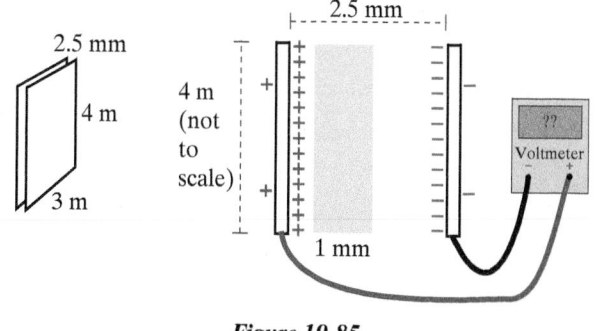

Figure 19.85

(a) Show that there will not be a spark in the air between the plates. **(b)** How much charge is on the positive plate of the capacitor?

With the battery still disconnected, you insert a slab of plastic 3 m by 4 m by 1 mm between the plates, next to the positive plate, as shown in Figure 19.85. This plastic has a dielectric constant of 5. **(c)** After inserting the plastic, you connect a voltmeter to the capacitor. What is the initial reading of the voltmeter? **(d)** The voltmeter has a resistance of 100 MΩ (1×10^8 Ω). What does the voltmeter read 3 s after being connected?

Section 19.11

•**P78** For the circuit shown in Figure 19.86, which consists of batteries with known emf and ohmic resistors with known resistance, write the correct number of energy-conservation and current node rule equations that would be adequate to solve for the unknown currents, but do not solve the equations. Label nodes and currents on the diagram, and identify each equation (energy or current, and for which loop or node).

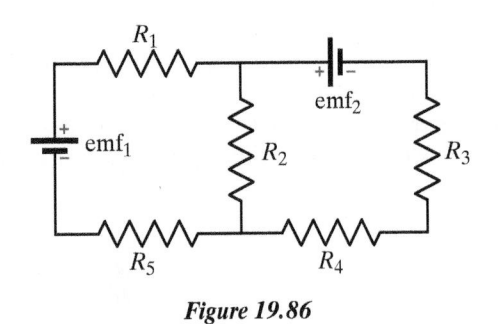

Figure 19.86

••**P79** A circuit consists of two batteries (with negligible internal resistance), six ohmic resistors, and connecting wires that have negligible resistance. The resistance R_1 is 10 Ω, R_2 is 20 Ω, R_3 is 30 Ω, R_4 is 12 Ω, R_5 is 15 Ω, and R_6 is 20 Ω. Unknown currents I_1, I_2, I_3, I_4, I_5, and I_6 have their directions marked on the circuit diagram in Figure 19.87.

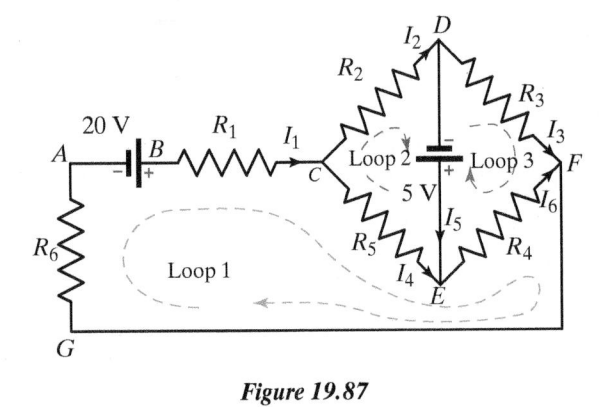

Figure 19.87

(a) Write down a set of equations that could be solved for the six unknown currents. Make sure you can explain how you got these equations. **(b)** When a correct set of equations is solved, the currents are as follows (to the nearest milliampere): $I_1 = 0.4394$ A, $I_2 = 0.3312$ A, $I_3 = 0.0065$ A, $I_4 = 0.1082$ A, $I_5 = 0.3247$ A, $I_6 = 0.4329$ A. Check your equations by substituting in these numbers. **(c)** Suppose that you connect the negative lead of a voltmeter to location G and the positive lead of the voltmeter to location C. What does the voltmeter read, including both magnitude and sign? **(d)** What is the power output of the 5 V battery? **(e)** Resistor R_4 is made of a very thin metal wire that is 3 mm long, with a diameter of 0.1 mm. What is the electric field inside this metal resistor?

••**P80** A circuit contains two batteries (with negligible internal resistance) and five ohmic resistors (Figure 19.88). The connecting wires have negligible resistance. The letters A through H are shown to make it possible to refer to specific parts of the circuit.

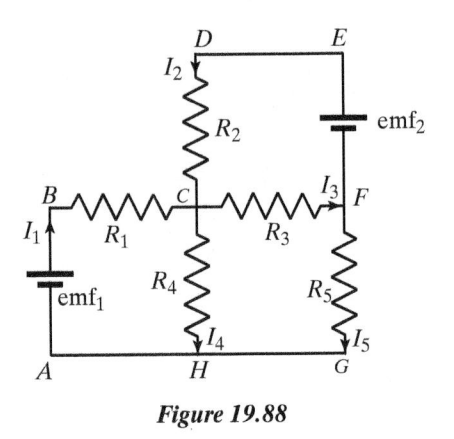

Figure 19.88

(a) Write all the equations necessary to solve for the unknown currents I_1, I_2, I_3, I_4, and I_5, whose directions are indicated on the circuit diagram. Do not solve the equations, but do explain very clearly what your equations are based on, and to what they refer.

Assume that a computer program has solved your equations in terms of the known battery voltages and known resistances, so that the currents I_1, I_2, I_3, I_4, and I_5 are known. **(b)** In terms of the known quantities, calculate $V_D - V_A$, and check that your sign makes sense. **(c)** In terms of the known quantities, calculate the power produced in battery number 2.

ANSWERS TO CHECKPOINTS

1 Consider the very first short time interval—say, 0.01 s. At the beginning of this interval both capacitors have very little charge. The electric field in the wires is due almost entirely to the battery and surface charge on the wires, and is the same in both cases. Thus the number of electrons flowing onto the left plate and off

of the right plate is approximately the same in the first 0.01 s. At the end of this time interval, both capacitors have nearly the same small amount of charge, q. This is the situation illustrated in Figure 19.89.

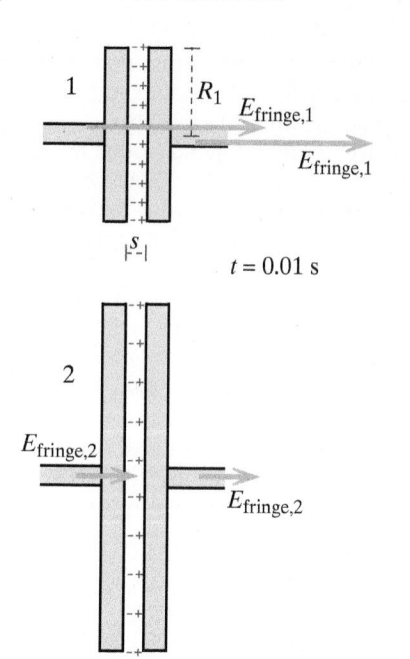

$$E_{\text{fringe},1}$$

$t = 0.01$ s

$$E_{\text{fringe},2}$$

Figure 19.89

The fringe field of the smaller capacitor is this:

$$E_{\text{fringe},1} \approx \frac{q/A}{\varepsilon_0} \left(\frac{s}{2R_1} \right) = \frac{1}{\varepsilon_0} \frac{q}{\pi R_1^2} \left(\frac{s}{2R_1} \right)$$

The fringe field of the larger capacitor is smaller, since $R_2 > R_1$:

$$E_{\text{fringe},2} = \frac{1}{\varepsilon_0} \frac{q}{\pi R_2^2} \left(\frac{s}{2R_2} \right) < E_{\text{fringe},1}$$

The smaller fringe field of the large capacitor means a larger *net* field and larger drift speed, so the current flowing onto the large capacitor has decreased less than the current flowing onto the small capacitor. See further discussion in Section 19.7.

2 30 Ω; 18.8 Ω; 8.3 Ω; R is not constant, and the bulb filament is not ohmic for this range of currents

3 7.5 Ω; 0.4 A

4 4.5 W

5 nearly 0; nearly 0; nearly 0

6 A voltmeter has very high resistance, so the current is extremely small; the voltmeter reads 3 V ($+3 - 3 = 0$ round-trip potential difference).

7 $q/m = aL/\Delta V$

8 -8.0 V; 3.2 W; 4.8 W

20

Magnetic Force

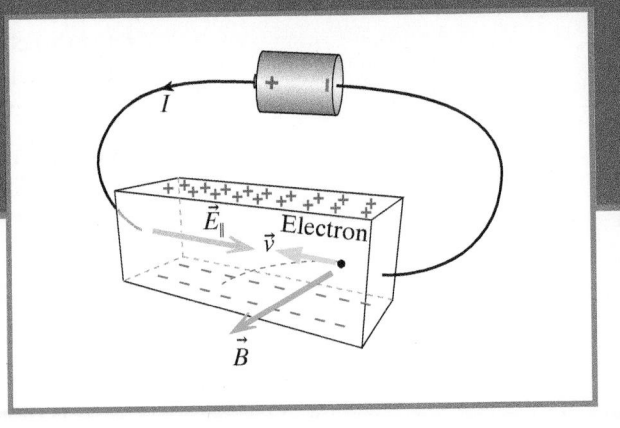

OBJECTIVES

After studying this chapter you should be able to

- Calculate the magnetic force on a moving charged particle, and on a current-carrying wire.
- Determine the sign of the mobile charges in a conductor using the Hall effect.
- Explain motional emf and calculate its magnitude.

Review: Magnetic Field of Moving Charge

Recall that the magnetic field associated with a moving charge can be calculated from the Biot–Savart law:

$$\vec{B} = \frac{\mu_0}{4\pi} \frac{q\vec{v} \times \hat{r}}{r^2}$$

From this law and the definition of current we derived an expression for the magnetic field contributed by a small current element:

$$\Delta \vec{B} = \frac{\mu_0}{4\pi} \frac{I\Delta\vec{l} \times \hat{r}}{r^2}$$

and we were able to integrate this form of the Biot–Savart law to predict the magnetic field of various configurations of currents, such as straight wires, loops, and solenoids.

In this chapter we will study the effects of magnetic fields. Often our analysis will require two steps: first, find the field at a location made by a moving charge or a current, and second, find the effect of this field on a different moving charge. Just as the electric field made by a charge does not affect the source charge itself, so the magnetic field made by a moving charge affects other moving charges, not the source charge.

20.1 MAGNETIC FORCE ON A MOVING CHARGE

In the cathode ray tube (CRT) displays that were formerly used in home television sets and computer monitors (not the flat-panel displays used today), electrons were accelerated to fairly high speeds by electric fields. The moving electrons were deflected vertically by a magnetic field. This magnetic field was horizontal and was produced by currents in coils positioned as shown in Figure 20.1 (not shown are other coils whose magnetic field deflected the electrons from side to side).

There is a 3D geometrical relationship between the direction of the magnetic force on a moving charge and

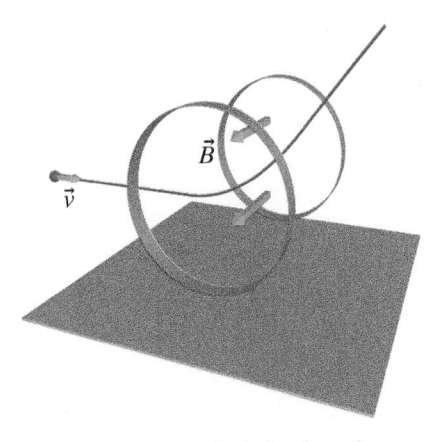

Figure 20.1 Magnetic deflection of a moving electron. Current-carrying coils are the source of the magnetic field.

- the direction of the applied magnetic field,
- the direction of the velocity of the moving charge, and
- the sign of the moving charge.

Experiments in which these parameters are systematically varied show that the magnetic force on a moving charge is given by the following equation:

MAGNETIC FORCE

$$\vec{F}_{\text{magnetic}} = q\vec{v} \times \vec{B}$$

q is the charge (including sign) of the moving charge.
$\vec{v}$ is the velocity of the moving charge.
$\vec{B}$ is the applied magnetic field (in tesla).

Since $\vec{F}_{\text{magnetic}} = q\vec{v} \times \vec{B}$, we can see that a tesla has the units of F/qv:

$$T = \frac{\text{N}}{\text{C} \cdot \text{m/s}} = \frac{\text{N/C}}{\text{m/s}}$$

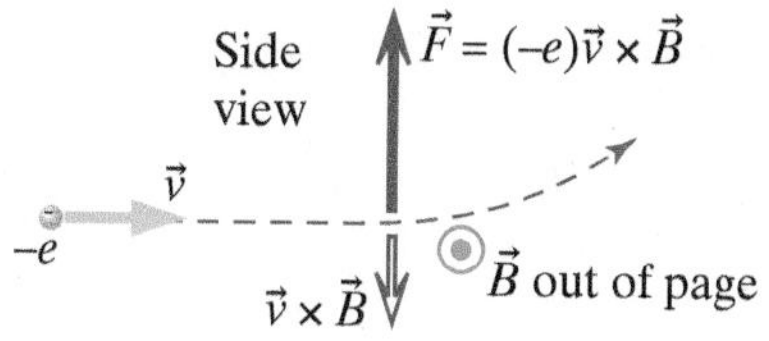

Side view $\quad \vec{F} = (-e)\vec{v} \times \vec{B}$

$\vec{v}$

$-e$

$\vec{v} \times \vec{B}$ $\qquad \vec{B}$ out of page

Figure 20.2 Side view of the magnetic deflection of the moving electron.

Use the right-hand rule for cross products to verify that the equation for magnetic force gives the observed direction of magnetic force in Figure 20.1 and in the side view shown in Figure 20.2. Remember that the electron charge $q = -e$, which means that the magnetic force on a moving electron is in the opposite direction to the direction of the cross product $\vec{v} \times \vec{B}$.

The equation $\vec{F} = q\vec{v} \times \vec{B}$ summarizes the experimental observations of the effect of a magnetic field on a moving charge.

> QUESTION What do these observations say about the effect that a magnetic field has on a charge that is not moving? How does this compare with the effect that an electric field has on a charge that is not moving?

If the charge is not moving, the magnetic field has no effect on it, whereas electric fields affect charges even if they are at rest.

Magnitude of the Magnetic Force

According to the equation $\vec{F}_{\text{magnetic}} = q\vec{v} \times \vec{B}$, there should be a magnetic force on a charged piece of invisible tape when you move it near a magnet. However, the magnitude of this magnetic force is extremely small compared to the familiar electric force acting between the tape and the magnet (due to the charged tape polarizing the metal).

The magnetic force on a single moving electron in a CRT is significant because the particle's acceleration $\vec{a} = d\vec{v}/dt$ is proportional to the ratio q/m:

$$\left| \frac{d\vec{p}}{dt} \right| = qvB\sin\theta$$

$$\left| \frac{d\vec{v}}{dt} \right| = \frac{q}{m}vB\sin\theta \quad (\text{for } v \ll c)$$

The q/m ratio is enormous for a single electron:

$$(1.6 \times 10^{-19}\ \text{C})/(9 \times 10^{-31}\ \text{kg}) \approx 1 \times 10^{11}\ \text{C/kg}$$

but q/m is very small for a charged tape:

$$(1 \times 10^{-8}\ \text{C})/(1 \times 10^{-4}\ \text{kg}) \approx 1 \times 10^{-4}\ \text{C/kg}$$

Magnetic forces on moving objects that have been charged by rubbing will always be tiny, because only an extremely small fraction of the atoms in the object have lost or gained an electronic charge. Moreover, the magnetic force is proportional to the speed of the moving particle, and the speed of electrons in a CRT is very high (it may be nearly 1×10^8 m/s).

> **Checkpoint 1** In Chapter 14 we found that a charged invisible tape has a net charge of about 1×10^{-8} C. You could move it past a magnet with a speed of about 10 m/sec. In Chapter 17, we saw that the magnetic field near the end of a small bar magnet could be around 0.1 T. Compare the approximate magnetic force on a moving invisible tape near a magnet with the approximate electric force that two charged tapes exert on each other when they are 10 cm apart.

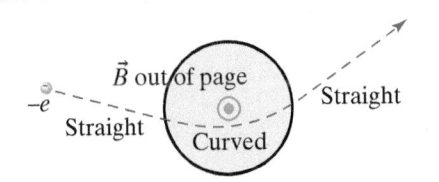

Figure 20.3 Electron trajectory through a limited region of magnetic field.

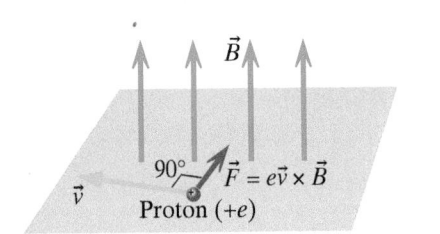

Figure 20.4 Circular motion of a proton in a region of uniform magnetic field.

Circular (or Helical) Motion in a Magnetic Field

In a CRT the region of magnetic field that deflects the electrons is mainly confined to the region of the deflection coils, and the trajectory of an electron is nearly straight before entering the magnetic field region and after leaving it (Figure 20.3).

However, if there is a large enough region of uniform magnetic field, a charge can go around and around in a circle, and the curved portion of the electron trajectory in a CRT is a circular arc (if the magnetic field is fairly uniform). There are many other examples of circular orbits of charged particles in magnetic fields, including particle trajectories in particle accelerators, and the orbits in the magnetic field of the Earth of high-altitude electrons that produce the Northern Lights.

In Figure 20.4 a proton moves horizontally with speed v in a large region of uniform vertically upward magnetic field (neglect the much smaller effects of gravity). The magnetic force $\vec{F} = q\vec{v} \times \vec{B}$ is perpendicular to the velocity $\vec{v}$, which means that the magnetic force will change the direction of the velocity without changing the magnitude of the velocity (the speed v), and no work is done by the magnetic force in this situation. If the motion is not horizontal, the component of velocity parallel to the magnetic field will not be changed, so the particle's path will be a helix (spiral) instead of a circle.

Circular Motion at Any Speed

This situation is similar to that of a rock going around in a circle at the end of a rope, or the Moon orbiting the Earth in a circular orbit. Consider the general case of a particle traveling at any speed (including near the speed of light). Recall that for a particle traveling in a circular orbit of radius R at constant speed, the rate of change of the particle's momentum points toward the center of the circle, and has this magnitude:

$$\left| \frac{d\vec{p}}{dt} \right| = p\left(\frac{v}{R}\right) \quad \text{where } p = \gamma m v = \frac{1}{\sqrt{1 - (v/c)^2}} m v$$

The quantity v/R is the angular speed ω (omega), which has units of radians/s, so

$$p\left(\frac{v}{R}\right) = \omega p = \omega \gamma m v$$

The Momentum Principle tells us that the rate of change of the particle's momentum is equal to the net force acting on the particle, which in this case is the magnetic force $\vec{F} = q\vec{v} \times \vec{B}$, so

$$\omega \gamma m v = |q\vec{v} \times \vec{B}| = |q| v B \sin 90°$$

$$\omega = \frac{|q| B}{\gamma m} \quad \text{valid at any speed}$$

Circular Motion at Low Speeds

An important special case is circular motion of a charged particle at speeds small compared to the speed of light c, in which case $\gamma \approx 1$ and the angular speed ω is (approximately) independent of the speed v:

$$\omega \approx \frac{|q|B}{m} \quad \text{if } v \ll c$$

An alternative derivation of this result starts from the nonrelativistic form of the Momentum Principle. If $v \ll c$, $\vec{p} \approx m\vec{v}$, so

$$\left|\frac{d\vec{p}}{dt}\right| = p\left(\frac{v}{R}\right) \approx \frac{v(mv)}{R} = \frac{mv^2}{R}$$

$$\frac{d\vec{p}}{dt} \approx m\frac{d\vec{v}}{dt} = \vec{F}_{\text{net}}$$

$$\frac{mv^2}{R} = |q|vB\sin 90°$$

$$m\omega = |q|B \quad \text{or} \quad \omega = \frac{|q|B}{m}$$

Time Required for a Circular Orbit

Let $T =$ the time required for a charged particle to go around once in a complete circular orbit.

> QUESTION Show that $T = 2\pi\dfrac{m}{|q|B}$ for nonrelativistic speeds ($v \ll c$).

Since by definition the angular speed ω is equal to 2π rad per the amount of time T required to go around once, $\omega = 2\pi/T$, and $\omega \approx |q|B/m$ if $v \ll c$, the result follows. This is an unusual result, because it is independent of the speed v and radius R. It says that the particle will take the same amount of time to go around in a circle in a uniform magnetic field, no matter how fast it is going (as long as $v \ll c$). The faster it goes, the bigger the radius of the circle, but the longer distance around the bigger circle is exactly compensated by the higher speed. This independence of the orbit time is the basis for the cyclotron, to be discussed in a moment.

Determining the Momentum of a Particle

In a region of known magnetic field, we can find the momentum of a particle of known charge (even one traveling at a speed close to the speed of light) by measuring the radius of curvature of its motion in a magnetic field.

$$\left|\frac{d\vec{p}}{dt}\right| = p\left(\frac{v}{R}\right) = |q|vB \quad \text{at any speed, so} \quad p = |q|BR$$

This result is used to measure momentum in high-energy particle experiments.

> **Checkpoint 2** In Figure 20.5 there is a magnetic field going into the page, produced by a large current-carrying coil (not shown). Would an electron go in a clockwise or counterclockwise circle? Explain briefly. (*Hint:* Try both directions, and see which direction is consistent with the magnetic force.) Note that we're not asking what magnetic field is made by the moving electron, but what effect the coil's magnetic field has on the moving electron.

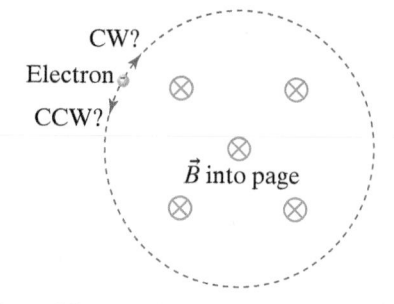

Figure 20.5 In which direction will the electron go in this uniform magnetic field (Checkpoint 2)?

Application: A Cyclotron

One of the principal methods for studying the nucleus of an atom is to shoot high-velocity protons at the nucleus and observe reactions that take place. Because they repel each other electrically, in order to drive a proton into a nucleus we have to accelerate the proton through a potential difference of a million volts or more. For example, there is a million-volt potential difference from very far away (infinity) to the surface of a carbon nucleus ($q = 6e$), whose radius is about 1×10^{-14} m. A compact device for creating the equivalent of such large potential differences is the cyclotron. Ernest Lawrence invented the idea for a cyclotron in 1929 [see Richard Rhodes, *The Making of the Atomic Bomb*, Simon & Schuster (1986), pp. 145–148; this book contains an excellent history of early 20th-century physics].

A cyclotron consists of two D-shaped metal boxes (called "dees") placed in a strong magnetic field (Figure 20.6 and Figure 20.7). Protons can enter the dees through slits on the flat faces. Near the center of the cyclotron is a source of low-velocity ionized hydrogen (protons). Inside the metal dees the electric field is zero (because polarization of the metal box contributes an electric field inside the box that cancels any electric fields of external charges—see Chapter 21), and a proton follows a circular path in the magnetic field. As a proton goes from the left dee to the right dee, it receives a kick to the right due to a potential difference across the two dees. This increases the proton's momentum, so its path in the right dee has a larger radius of curvature.

While the proton is inside the right dee (where the electric field is zero), we reverse the potential difference across the dees so that when the proton goes from the right dee to the left dee at the bottom of the drawing, it again receives a kick, this time to the left, resulting in a higher momentum and a larger radius of curvature in the left dee. By repeatedly kicking the proton through a modest potential difference, we can accelerate the proton up to very high momentum and energy that would otherwise require a practically unattainable potential difference of millions of volts. Problems P32 and P82 deal with the operation of a cyclotron.

As we found earlier in this section, if $v \ll c$, the period of a proton's orbit does not change as its speed increases, so it is possible to apply a sinusoidally oscillating potential difference $\Delta V = \Delta V_0 \cos(\omega t)$ across the dees to accelerate the proton.

However, if the equivalent accelerating potential is more than about 100 million volts, as it is in modern accelerators, the speed v of a proton becomes a sizable fraction of the speed of light c, and relativistic effects are significant, so that

$$\omega = \frac{2\pi}{T} = \frac{|q|B}{\gamma m} \quad \text{depends on } v$$

At high speeds γ is no longer approximately equal to 1, so the period of the motion is no longer independent of the speed. Either the frequency of the accelerating voltage or the strength of the magnetic field in a cyclotron must be adjusted during the acceleration to compensate.

The largest modern circular accelerator is the Large Hadron Collider (LHC) at CERN (European Organization for Nuclear Research) in Geneva, Switzerland. In the final stage of acceleration, protons travel in a circle having a circumference of 27 km (17 mi)! The high-vacuum ring is buried underground and passes through parts of Switzerland and France. A smaller accelerator injects protons into this ring with a speed very close to the speed of light, $v = 0.999998c$, and in the large ring their speed gets even closer as they are accelerated to a speed of $v = 0.9999999c$. Although the speed hardly changes, the energy change is enormous, with $E = \gamma mc^2$ going from $480mc^2$ to $7500mc^2$. The magnetic field is increased during the acceleration to keep the protons at the same radius despite the large change in γ. At the maximum

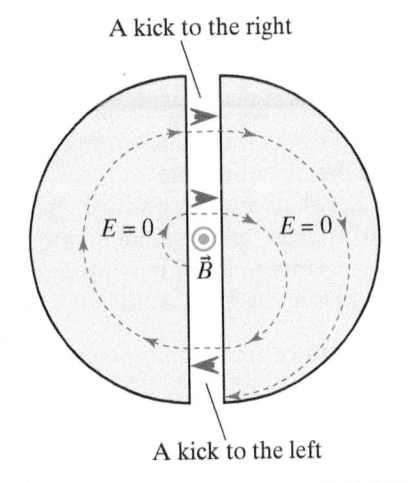

A kick to the right

$E = 0$ $\vec{B}$ $E = 0$

A kick to the left

Figure 20.6 A cyclotron. Magnetic field is uniform in the $+y$ direction; the proton orbits in the x-z plane.

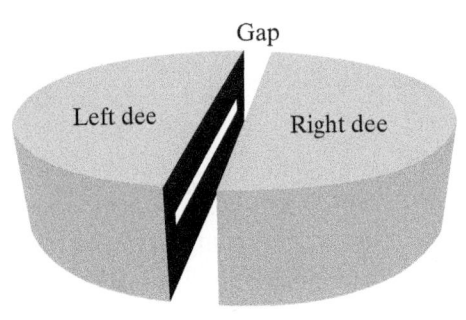

Gap

Left dee   Right dee

Figure 20.7 3D view of a cyclotron.

momentum, $B = \gamma mv/eR$ is 5.6 T, which is provided by electromagnets containing superconducting coils. (Actually, these magnets don't cover the entire circumference, and there are straight sections between the bending regions; the magnets reach 8.4 T.) The time to go around once is nearly constant as in a cyclotron, but for a completely different reason: the nearly constant speed is close to c. For more information, search online for LHC, or the campy and informative video, "Large Hadron Rap."

20.2 MAGNETIC FORCE ON A CURRENT-CARRYING WIRE

Often we are interested in the magnetic force exerted on a large number of charges moving through a wire in a circuit. Because the net force on a system (the wire) is the sum of all the forces acting on its individual components, adding up these forces will give the net force on the entire wire.

Let's calculate the magnetic force on a bunch of moving positive charges contained in a small volume with length Δl and cross-sectional area A (Figure 20.8). Their average drift velocity is $\vec{v}$. If there are n moving charges per unit volume, there are $nA\,\Delta l$ moving charges in this small volume.

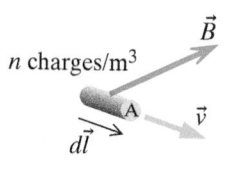

n charges/m^3

Figure 20.8 Calculate the magnetic force on a short current-carrying wire.

QUESTION Show that the net magnetic force on all the moving charges in this short length of wire is

$$\Delta \vec{F}_{\text{magnetic}} = I\,\Delta \vec{l} \times \vec{B}$$

where $\Delta \vec{l}$ is a vector with magnitude Δl, pointing in the direction of the conventional current I.

Since there are $nA\,\Delta l$ moving charges affected by the magnetic field, the total force on this piece of wire is

$$(nA\Delta l)(q\vec{v} \times \vec{B}) = (qnA\vec{v})(\Delta \vec{l} \times \vec{B})$$

Note that $\vec{v} = |\vec{v}|\hat{v}$, so $\vec{v}(\Delta l) = |\vec{v}|\hat{v}(\Delta l) = |\vec{v}|(\hat{v}\Delta l) = |\vec{v}|\Delta \vec{l}$, and the conventional current is $I = qnA\vec{v}$, so we get

$$\Delta \vec{F}_{\text{magnetic}} = I\Delta \vec{l} \times \vec{B}$$

QUESTION In most metal wires the moving charges are negative electrons traveling in the opposite direction. Explain briefly why the equation is valid even if the actual charge carriers are negative.

The electrons drift in a direction opposite to the conventional current, and their charge is negative. Therefore the quantity $q\vec{v}$ has two canceling minus signs, and the equation in terms of conventional current works fine, even if the actual charge carriers are negative electrons.

FORCE ON A SHORT LENGTH OF CURRENT-CARRYING WIRE

$$\Delta \vec{F}_{\text{magnetic}} = I\Delta \vec{l} \times \vec{B}$$

I is the conventional current; $|\Delta \vec{l}|$ is the length of the wire segment; $\Delta \vec{l}$ points in the direction of the conventional-current flow; $\vec{B}$ is the applied magnetic field.

This is not a new fundamental equation. It is merely the magnetic force added up for all the individual moving charges in a piece of wire. One way to keep this straight is to memorize the derivation. The key point is that there are $nA\,\Delta l$ electrons in a short length of wire, each moving with average speed $\bar{v}$, so that the sum of all the $|q|\bar{v}$'s is $(nA\,\Delta l)|q|\bar{v} = (|q|nA\bar{v})\Delta l = I\Delta l$.

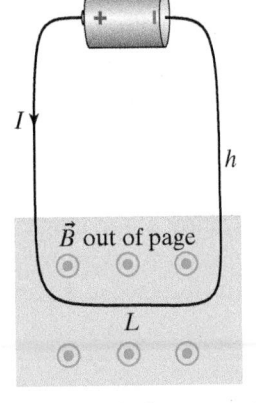

Figure 20.9 Magnetic forces on a loop of wire carrying conventional current I.

QUESTION Figure 20.9 shows a circuit consisting of a battery and a long wire, part of which is in a uniform magnetic field pointing out of the page. What is the direction of the force on those parts of the wire that are in the magnetic field?

Along each wire, the magnetic force points outward from the loop (toward the left along the left wire, down along the bottom wire, and toward the right along the right wire). Note that you get this result whether you consider positive charges drifting in the direction of the conventional current or electrons drifting in the opposite direction.

QUESTION Given those force directions, what is the net force on the wire in terms of the magnetic field B, current I, height h, and length L? (*Note:* Add up the contributions of lots of $\Delta\vec{l}$'s.)

The outward forces on the left and right wires cancel (though they tend to stretch the loop), and the net magnetic force is down. The magnitude is $(IL)B\sin 90° = IBL$.

QUESTION To get an idea of the numerical order of magnitude, suppose $B = 1$ T, $I = 10$ A, $h = 50$ cm, and $L = 10$ cm. What is the magnitude of the force?

The magnetic force depends only on L, not h, and $IBL = 1$ N, a rather small force (1 N is approximately the gravitational force on a small apple). Related experiments: EXP1–EXP3.

Keep in mind that there is really only one magnetic force expression, not two. Always remember that $\Delta\vec{F}_{\text{magnetic}} = I\Delta\vec{l} \times \vec{B}$ is simply the result of adding up the effects on many moving charges in a short length of wire, each of which experiences a magnetic force $\vec{F} = q\vec{v} \times \vec{B}$.

Application: Forces between Parallel Wires

An important application of the theory is the calculation of the magnetic force that parallel wires exert on each other, because the ampere is defined as the amount of conventional current in two very long parallel wires, a meter apart, which gives a force of one wire on the other of 2×10^{-7} N per meter of length, from which it follows that $\mu_0/4\pi$ is *exactly* equal to 1×10^{-7} T · m/A.

Figure 20.10 shows two very long parallel straight wires of length L, a distance d apart. The upper wire carries a conventional current I_1 and the lower wire carries a conventional current I_2 in the same direction. There are no magnets present.

Step 1, magnetic field due to upper wire: In order to calculate the force that the upper wire exerts on the lower wire, first we need to determine the magnitude and direction of the magnetic field $\vec{B}_1$ produced by the upper wire (carrying I_1) at the location of the lower wire. In Chapter 17 we calculated the magnetic field near a long straight wire. Using that result we have $B_1 \approx [\mu_0/(4\pi)][2I_1/d]$. The direction is into the page, using the right-hand rule.

Step 2, magnetic force on lower wire: Now we can calculate the magnitude of the magnetic force $\vec{F}_{21}$ exerted on the lower wire by the field produced by the upper wire:

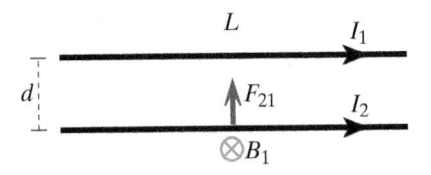

Figure 20.10 Parallel current-carrying wires exert magnetic forces on each other.

$$F_{21} = I_2 L B_1 \sin 90° = I_2 L \left(\frac{\mu_0}{4\pi} \frac{2I_1}{d} \right)$$

Using the right-hand rule with $I\Delta\vec{l} \times \vec{B}$, the direction of the force is up, so that the lower wire is attracted to the upper wire.

Force on upper wire: Find the magnitude and direction of the magnetic force $\vec{F}_{12}$ exerted on the upper wire. This requires two steps:

QUESTION First, what are the magnitude and direction of the magnetic field produced by the lower wire (carrying I_2) at the location of the upper wire?

We have $B_2 = [\mu_0/(4\pi)][2I_2/d]$. The direction is out of the page.

QUESTION What are the magnitude and direction of the magnetic force $\vec{F}_{12}$ exerted on the upper wire by the field produced by the lower wire? Is Newton's third law obeyed (reciprocity of forces)? That is, is it the case that $\vec{F}_{12} = -\vec{F}_{21}$?

The magnitude of the force on the upper wire is this:

$$F_{12} = I_1 L B_2 \sin 90° = I_1 L \left(\frac{\mu_0}{4\pi} \frac{2I_2}{d} \right)$$

Using the right-hand rule with $I\Delta\vec{l} \times \vec{B}$, the direction of the force is down, so that the upper wire is attracted to the lower wire. The forces on the two wires have the same magnitude and opposite directions, so in this case the reciprocity of the magnetic forces holds. (However, this need not always be the case: see Question Q3 and Problem P30.)

Currents in Opposite Directions

QUESTION If the two currents run in the same direction, we see that the magnetic interaction is an attraction. If the two currents run in opposite directions, is the magnetic interaction an attraction or a repulsion? Prove your assertion by showing the direction of a magnetic field and the force it exerts.

In this case you should find that two currents running in opposite directions repel each other. For electric forces we say that "likes repel, unlikes attract," but for the magnetic forces between two parallel current-carrying wires, we can say "likes attract, unlikes repel."

Checkpoint 3 To get a feel for the size of the effects, calculate the magnitude of the force per meter if $I_1 = I_2 = 10$ A and the distance between the wires is $d = 1$ cm.

20.3 COMBINING ELECTRIC AND MAGNETIC FORCES

If a charged particle is moving in a magnetic field, it is subject to a magnetic force $\vec{F}_m = q\vec{v} \times \vec{B}$. If there is an electric field in the region, the particle is also subject to an electric force $\vec{F}_e = q\vec{E}$.

$\vec{F} = q\vec{E} + q\vec{v} \times \vec{B}$ net force due to electric and magnetic fields

This net electric plus magnetic force is called the "Lorentz force." In one sense, the terms in this equation are a kind of definition of electric and magnetic fields in terms of the observable effects that these fields have on charges.

Checkpoint 4 (a) Consider the situation in Figure 20.11, in which there is a uniform electric field in the x direction and a uniform magnetic field in the y direction. For each example of a proton at rest or moving in the x, y, or z direction, what is the direction of the net electric and magnetic force on the proton at this instant? (b) In each example in Figure 20.12, what is the direction of the net electric and magnetic force on an electron at this instant?

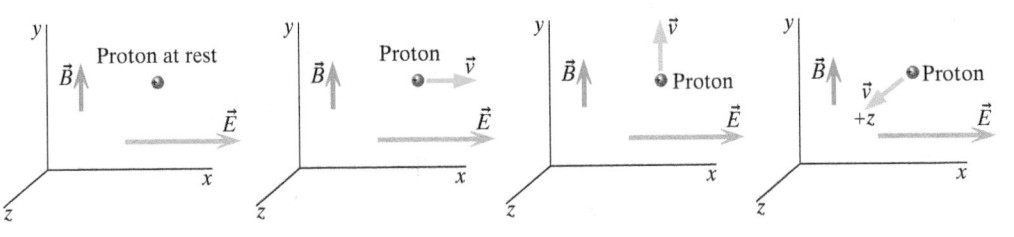

Figure 20.11 A proton in electric and magnetic fields.

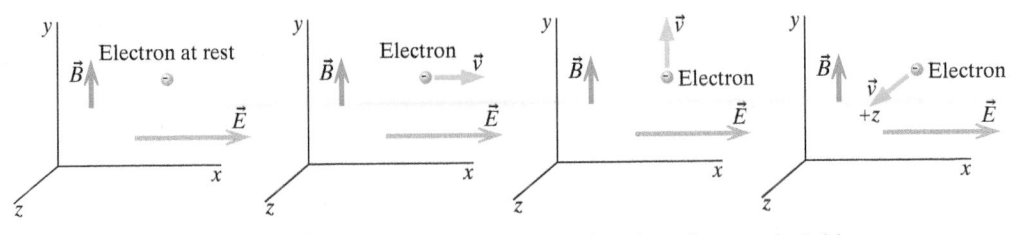

Figure 20.12 An electron in electric and magnetic fields.

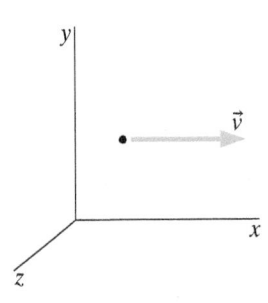

Figure 20.13 Sketch electric and magnetic fields in three-dimensional space so that the net electric and magnetic force on the particle is zero.

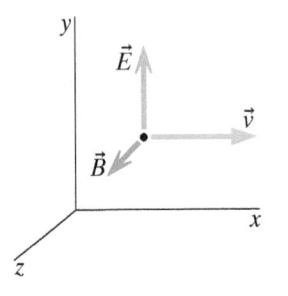

Figure 20.14 One configuration of fields that can allow a charged particle to travel in a straight line.

Application: A Velocity Selector

It is possible to arrange electric and magnetic fields so that the combined electric and magnetic force $\vec{F} = q\vec{E} + q\vec{v} \times \vec{B}$ acting on a positive charge $+q$ moving with velocity $\vec{v}$ is zero.

QUESTION   With the velocity $\vec{v}$ in the x direction (Figure 20.13), try to sketch a three-dimensional diagram of an arrangement of $\vec{E}$ and $\vec{B}$ that could lead to a net electric and magnetic force of zero, in which case the particle will move with constant velocity through the region.

Several arrangements are possible, one of which is shown in Figure 20.14.

QUESTION   If the magnetic field in this situation has a magnitude B, what must be the magnitude E of the electric field?

Since the magnetic and electric forces must be equal and opposite to produce a zero net force,

$$eE = evB\sin 90°$$
$$E = vB$$

QUESTION   Now suppose that a particle with the same charge q comes through this region, traveling in the same direction but at a slower speed. Explain briefly why it will *not* move in a straight line.

Now $qE > qvB$, so the net force isn't zero.

QUESTION   If a particle with the correct velocity but with a *negative* charge $-q$ comes through, will it pass straight through or not? (Make a diagram.)

Yes. Both the electric force and the magnetic force are flipped 180°, and the net force is still zero.

Thus a region containing crossed electric and magnetic fields (that is, perpendicular to each other) forms a kind of velocity selector, allowing particles to pass straight through only if they have the chosen speed and direction. You might notice that the cross product $\vec{E} \times \vec{B}$ points in the direction of the velocity that can pass straight through.

EXAMPLE

Electron and Current Loop

A two-turn circular loop of wire of radius R is fed a large conventional current I from long straight wires (Figure 20.15). A charge of magnitude $+Q$ is a distance $3R$ to the right of the center of the loop. An electron traveling upward with speed v passes through the center of the loop, and the net force on this electron at this instant is zero. The Earth's magnetic field is negligible in this region. What are the magnitude and direction of the current I? Assume that all other quantities are known.

Solution

The key idea here is that $\vec{F}_{\text{magnetic}} + \vec{F}_{\text{electric}} = 0$ at this instant.

Direction of current

At the location of the electron, the electric field due to the positive charge is in the $-x$ direction, so the electric force on the electron is in the $+x$ direction, as shown in Figure 20.16. Since at the instant specified the net force on the electron is zero, the magnetic force on the electron must be equal and opposite to the electric force. $\vec{F}_{\text{magnetic}}$ is in the $-x$ direction, but the electron's charge is negative, so $\vec{v} \times \vec{B}$ must be in the $+x$ direction, as shown in Figure 20.17. By trial and error with the right-hand rule (thumb in direction of $\vec{v} \times \vec{B}$, fingers initially pointing up), we find that $\vec{B}$ must be in the $+z$ direction. Applying the right-hand rule for a current loop, we find that conventional current must flow counterclockwise around the loops to make a magnetic field in this direction.

Magnitude of current

Since the net force on the electron is zero, we know that:

$$|\vec{F}_{\text{magnetic}}| = |\vec{F}_{\text{electric}}|$$

$$\frac{1}{4\pi\varepsilon_0} \frac{eQ}{(3R)^2} = evB$$

$$B = \frac{\frac{1}{4\pi\varepsilon_0} \frac{Q}{9R^2}}{v}$$

We found in an earlier chapter that the magnetic field of a loop is this:

$$B_{\text{loop}} = \frac{\mu_0}{4\pi} \frac{2\pi R^2 I}{(z^2 + R^2)^{3/2}}$$

At the center of the loops $z = 0$, so $B_{\text{loop}} = \frac{\mu_0}{4\pi} \frac{2\pi I}{R}$, and for two loops:

$$\frac{\frac{1}{4\pi\varepsilon_0} \frac{Q}{9R^2}}{v} = 2\frac{\mu_0}{4\pi} \frac{2\pi I}{R}$$

$$I = \frac{1}{\varepsilon_0 \mu_0} \frac{Q}{36\pi R v}$$

This quantity can be shown to have units of C/s, or amperes.

Two turns, radius R

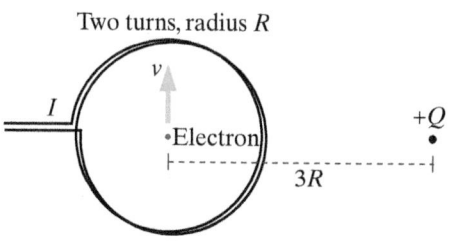

Figure 20.15 An electron inside two turns of current-carrying wire.

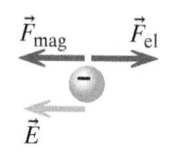

Figure 20.16 The directions of the electric and magnetic forces on the electron.

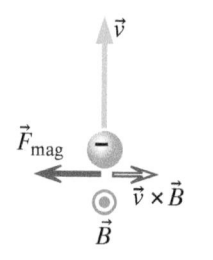

Figure 20.17 Finding the direction of the magnetic field.

Checkpoint 5 If an electron traveling at a velocity of $\langle 0, 0.9c, 0 \rangle$ passes through a region where the magnetic field is $\langle 1, 0, 0 \rangle$ T, what electric field would be necessary to keep the electron traveling in a straight line? Give $\vec{E}$ as a vector.

20.4 THE HALL EFFECT

Magnetic forces on moving charged particles inside a conductor can cause the conductor to polarize. This is the basis of the Hall effect.

The Initial Transient: Polarization Develops

Mobile electrons moving inside a metal wire can be deflected to the surface of the wire by a magnetic field that is perpendicular to the wire. In Figure 20.18, large coils (not shown) apply a uniform magnetic field of magnitude B throughout the region of the circuit. As usual, mobile electrons move through the wire in a direction opposite to the electric field driving the current. We will call this electric field $\vec{E}_{\parallel}$, because it is parallel to the wire, as we saw in Chapter 18.

Verify for yourself that the moving electrons will be deflected toward the bottom surface of the wire (shown as a rectangular bar for simplicity). This will make the bottom surface negatively charged and the top surface positively charged (due to a deficiency of electrons). These charges are in addition to the usual surface charges, which for clarity we don't show on the diagram.

The Steady State: Magnetic Force Balanced by Electric Force

As time goes on, more and more electrons will pile up on the bottom of the wire. However, this piling up won't continue forever. When enough electrons have piled up on the bottom (with a corresponding depletion of electrons on the top surface), these extra surface charges will produce a new electric field that applies a vertical electric force to the moving electrons inside the wire.

QUESTION What is the direction of this new electric field?

Inside the metal wire, the new electric field points down, pointing away from the positive charge on the top and toward the negative charge on the bottom. We'll call this the "transverse" electric field, $\vec{E}_{\perp}$.

QUESTION When no more electrons pile up on the bottom, what is the magnitude $E_{\perp}$ of the transverse (perpendicular) electric field, in terms of the drift speed $\bar{v}$ of the free electrons and the magnetic field B? (*Hint:* If no more electrons are piling up, what do you know about the electric and magnetic forces on the moving electrons?)

Pileup will cease when the electric force grows to be as large as the magnetic force, at which time the net vertical force becomes zero, and the electrons move horizontally through the wire, with no vertical deflection. When this happens we know the vertical force on an electron is zero, so the magnetic force must equal the perpendicular electric force:

$$|\vec{F}_{\text{electric}\perp}| = |\vec{F}_{\text{magnetic}}|$$
$$eE_{\perp} = e\bar{v}B\sin 90°, \quad \text{so} \quad E_{\perp} = \bar{v}B$$

Note the similarity between this analysis and the analysis of the velocity selector in Section 20.3.

QUESTION Suppose that you attach the leads of a digital voltmeter to the top and bottom surfaces as shown in Figure 20.19. Determine the magnitude and sign of the voltage displayed on the voltmeter (the wire has height h, depth d, and length L; there is a uniform magnetic field of magnitude B).

Since the electric field points down inside the wire, the electric potential is higher on top. The electric field has the uniform value $E_{\perp} = \bar{v}B$ throughout the wire, so the voltmeter will read a positive value $+E_{\perp}h = +\bar{v}Bh$.

QUESTION How big is this voltage if there is no magnetic field?

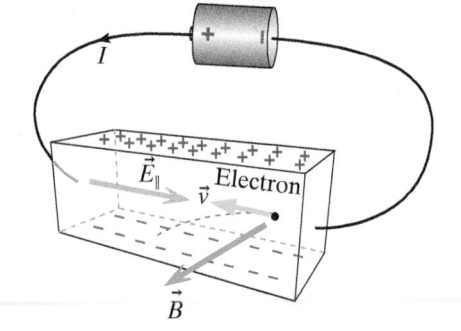

Figure 20.18 Magnetic deflection of electrons in a conducting bar. $\vec{E}_{\parallel}$ is the electric field due to the battery and surface charge on the bar (not shown), which drives the current. Polarization shown is due to magnetic deflection of electrons during the transient phase.

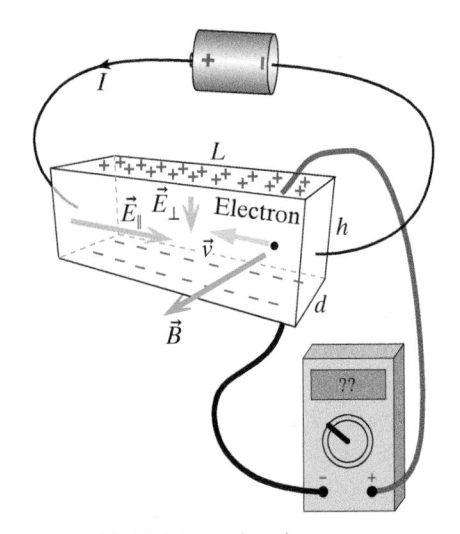

Figure 20.19 Measuring the transverse potential difference. The voltmeter leads are directly across from each other on the top and bottom surfaces.

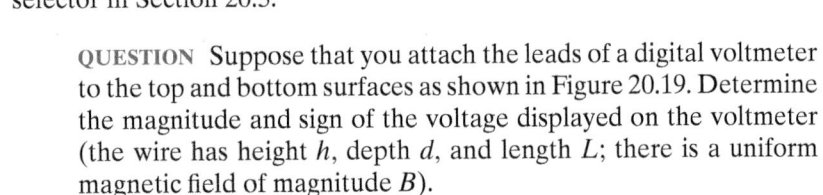

Since the potential difference is $\bar{v}Bh$, the voltage will be zero if $B = 0$.

The appearance of a sideways or transverse potential difference across a current-carrying wire in the presence of a magnetic field is called the "Hall effect," named for its discoverer.

Let's see approximately how big the Hall voltage is in a metal. Consider a bar of iron of length 5 cm, height 2 cm, and depth 0.5 mm, carrying a conventional current of 15 A to the right in the presence of a perpendicular magnetic field of magnitude 2 T, as shown in Figure 20.19. There are about 8.4×10^{28} mobile electrons per cubic meter in iron.

The magnetic force involves the average drift speed of the moving charge:

$$\bar{v} = \frac{I}{|q|nA} = \frac{15 \text{ A}}{(1.6 \times 10^{-19} \text{ C})(8.4 \times 10^{28} \text{ m}^{-3})(0.02 \text{ m})(5 \times 10^{-4} \text{ m})}$$
$$= 1.11 \times 10^{-4} \text{ m/s}$$

In the steady state the vertical force (the force perpendicular to current flow) on a moving charge is zero:

$$F_{\text{magnetic}} = F_{\text{electric}\perp} \quad \text{and therefore} \quad evB = eE_\perp$$
$$E_\perp = \bar{v}B = (1.11 \times 10^{-4} \text{ m/s})(2 \text{ T}) = 2.22 \times 10^{-4} \text{ V/m}$$
$$\Delta V_{\text{Hall}} = E_\perp h = (2.22 \times 10^{-4} \text{ V/m})(0.02 \text{ m}) = 4.5 \times 10^{-6} \text{ V}$$

This is a very small voltage, but it can be measured with a sensitive voltmeter whose leads are carefully aligned vertically across the bar.

Determining the Density of Mobile Charges

We can use measurements of the Hall voltage and the conventional current to determine the density of mobile charges in a metal.

$$\Delta V_{\text{Hall}} = E_\perp h = \bar{v}Bh = \left(\frac{I}{|q|nA}\right)(Bh)$$
$$n = \frac{I}{|q|\Delta V_{\text{Hall}}A}(Bh)$$

For some monovalent metals such as sodium, Hall effect measurements show n to be the same as the number of atoms per cubic meter, indicating that each atom gives up one electron into the free-electron sea. In other metals there may be more than one electron contributed per atom.

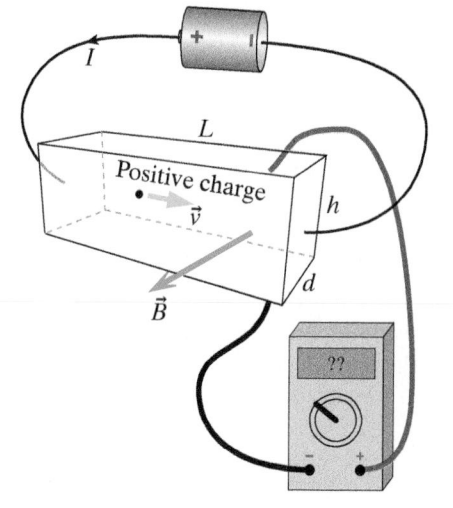

Figure 20.20 What happens if the charge carriers are positive? The voltmeter leads are directly across from each other on the top and bottom surfaces.

Determining the Sign of the Mobile Charges

The Hall effect offers a very special insight into the nature of the "charge carriers" (mobile charges) responsible for a current. Most electrical measurements come out the same whether electrons are moving inside the wire or positive particles (holes in a metal) are moving in the opposite direction. However, the Hall effect differs in the two cases.

Consider what would happen if *positive* particles (holes in the case of aluminum or zinc) were moving to the *right* in the wire as in Figure 20.20, instead of electrons moving to the left.

QUESTION On a diagram like Figure 20.20, show the charge pileup on the top and bottom surfaces and the electric field due to that pileup. What would be the sign of the voltage displayed on the voltmeter?

Now it is the bottom of the wire that has additional positive surface charge, rather than the top being positive as it was with negative charge carriers. The transverse electric field now points up, and the voltmeter reads a negative voltage.

HALL EFFECT AND THE SIGN OF CHARGE CARRIERS

By measuring the Hall effect for a particular material, we can determine the sign of the moving particles that make up the current.

In the steady state, the Hall effect is an example of the straight-line motion of charged particles through crossed electric and magnetic fields. Remember that the surface charges that accumulate due to the Hall effect are not the only charges on the wires; there are also the usual surface charges responsible for the longitudinal electric field in the wires. The distinction is that the ordinary surface charges produce electric fields inside the wire in the direction of the conventional current, whereas the Hall effect surface charges produce a transverse electric field $\vec{E}_\perp$, perpendicular to the current.

> **Checkpoint 6** Calculate the Hall voltage ΔV_{Hall} for the case of a ribbon of copper 5 mm high and 0.1 mm deep, carrying a current of 20 A in a magnetic field of 1 T. For copper, $q = -e$, and $n = 8.4 \times 10^{28}$ free electrons per cubic meter (one per atom).
>
> This is a very small voltage, but it can be measured with a sensitive voltmeter. In semiconductors the density of charge carriers n may be quite small compared to a metal, in which case the Hall effect voltage can be quite large, since it is inversely proportional to n.
>
> It is instructive to compare this transverse Hall voltage in copper with the voltage you would measure *along* the copper ribbon. Calculate the potential difference you would measure along 5 mm of the ribbon; the conductivity of copper is
>
> $$\sigma = 6 \times 10^7 \, \frac{\text{A/m}^2}{\text{V/m}}$$
>
> Since this voltage is much larger than the Hall voltage, experimenters must take special precautions to place their voltmeter leads exactly across from each other, to avoid a spurious reading due to the potential difference along the bar. One way to check for this problem is to remove the magnetic field, in which case the Hall voltage should go to zero.

Practical Applications of the Hall Effect

Measuring the Hall effect potential difference for a material can provide information about the sign of the mobile charges and about the number of them per unit volume. For example, Hall effect measurements have confirmed that in some metals and semiconductors, it is positive holes in the electron sea rather than the electrons themselves that are the dominant charge carriers. Hall effect measurements of mobile electron density for some monovalent metals such as sodium and copper show n to be the same as the number of atoms per cubic meter, which shows that each atom gives up one electron into the free-electron sea. In other metals there may be more than one electron contributed per atom.

Another practical use of the Hall effect turns the relationship around. There are commercial devices that measure the magnitude and direction of an unknown magnetic field by observing the Hall voltage in a material in which the sign and number density of the mobile charges are known.

Why Does a Current-Carrying Wire Move?

There is a subtle point about the magnetic force that a magnetic field exerts on a current-carrying wire. The magnetic force only acts on moving particles, which in the case of a copper wire are the drifting electrons. The stationary positive atomic cores do not experience a magnetic force—so why does the entire wire move?

The answer to this puzzle is closely related to the Hall effect. In Figure 20.21, consider again a current-carrying metal wire in a magnetic field, with a growing number of electrons piling up on the bottom surface (and a deficiency of electrons accumulating on the top surface). Again, for clarity we don't show the usual surface charges along the wire; we just show the extra surface charges due to the magnetic deflection of the electrons.

The extra surface charges exert an electric force $eE_\perp$ upward on the moving electrons, which in the steady state just balances the magnetic force evB. The extra surface charges also exert an electric force $eE_\perp$ on the positive atomic cores, and this is an unbalanced force, because the atomic cores initially are not moving and hence are not subject to a magnetic force.

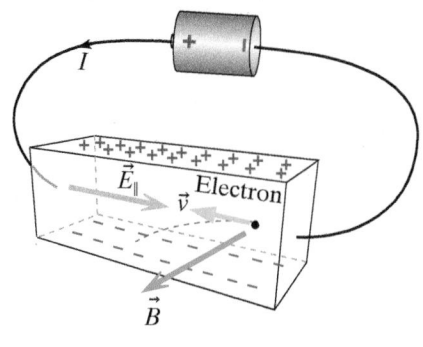

Figure 20.21 Deflected electrons drag the wire along too.

QUESTION What is the direction of this electric force exerted by the surface charges on the atomic cores?

Evidently the positively charged atomic cores are pulled downward. The electric force due to the Hall effect surface charges forces the wire as a whole to move in the direction of $\Delta\vec{F}_{\text{magnetic}} = I\Delta\vec{l} \times \vec{B}$, and the magnitude of the electric force is equal to this force. The overall effect is that in Figure 20.21 the atomic cores are dragged downward as the electrons are dragged downward. The motion of the wire is an electric side effect of the magnetic force on the moving electrons.

Another way of thinking about this is to say that the electrons are bound electrically to the metal wire—they can't simply fall out of the wire. Thus when the electrons are pushed down by the magnetic force, they drag the wire along with them. Note that ultimately it is the battery that supplies energy to the wire.

EXAMPLE

Determining Electrical Properties

A bar made out of a new conducting material is 9 cm long with a rectangular cross section 4 cm wide and 2 cm deep. The bar is inserted into a circuit with a 1.5 V battery and carries a constant current of 0.7 A (Figure 20.22). The resistance of the copper connecting wires and the ammeter, and the internal resistance of the battery, are all negligible compared to the resistance of the bar.

Using large coils not shown in the diagrams, a uniform magnetic field of 1.7 T is applied perpendicular to the bar (to the left, as shown in Figure 20.22). A voltmeter is connected across the bar, with the connections across the bar carefully placed directly across from each other as shown. The voltmeter reads −0.29 mV. Remember that a voltmeter reads positive if the positive lead is connected to a higher potential location than the negative lead. There is only one kind of mobile charge in the bar material. **(a)** What is the sign of the mobile charges in the bar, and which way do they move? Explain carefully, using diagrams to support your explanation. **(b)** How long does it take for a mobile charge to go from one end of the bar to the other? **(c)** What is the mobility u of the mobile charges? **(d)** If each mobile charge is singly charged ($|q| = e$), how many mobile charges are there in 1 m^3 of this material?

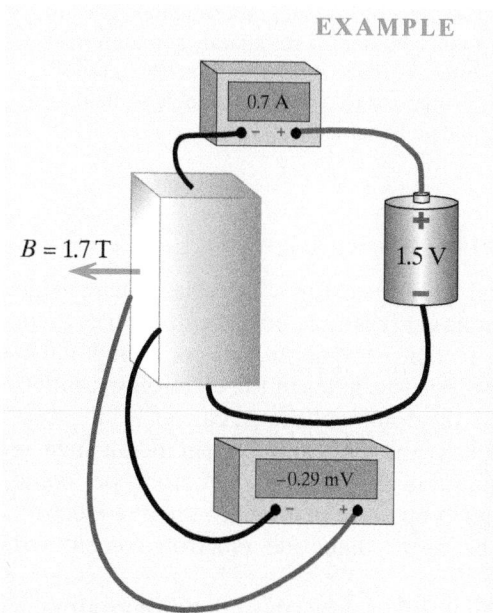

Figure 20.22 Measuring electrical properties of a new material.

Solution **(a)** This requires trial and error—we assume that the mobile charges have a particular sign, and see whether this would lead to the observed polarization. First we need to draw $\vec{E}_{\parallel}$, the field established by the batteries and surface charge on the bar, which drives the current in the bar (see Figure 20.23).

If the mobile charges are positive, they will move in the direction of $\vec{E}_{\parallel}$. As is shown in Figure 20.23, this would lead to positive charge buildup on the back of the bar (magnetic force into the page). The voltmeter would give a positive reading, because $\vec{E}_{\perp}$ would point out of the page. This is not what is observed, so the mobile charges must not be positive.

If the mobile charges are negative, they will move opposite to $\vec{E}_{\parallel}$, as shown in Figure 20.24. The magnetic force would again be into the page, making the back of the bar negative, and creating $\vec{E}_{\perp}$ pointing into the page. The voltmeter reading would be negative, as is observed. We can conclude that the mobile charges are negative.

(b) We need to find the average drift speed of the mobile charges, which we can find from the magnetic force on an electron. In the steady state, the transverse magnetic and electric forces balance:

$$eE_{\perp} = e\bar{v}B$$

$$\frac{\Delta V_{\perp}}{d} = \bar{v}B \quad \text{because} \quad E_{\perp} = \frac{\Delta V_{\perp}}{d}$$

$$\bar{v} = \frac{\Delta V_{\perp}}{Bd} = \frac{0.29 \times 10^{-3} \text{ V}}{(1.7 \text{ T})(0.02 \text{ m})} = 8.5 \times 10^{-3} \text{ m/s}$$

$$\Delta t = \frac{0.09 \text{ m}}{8.5 \times 10^{-3} \text{ m/s}} = 10.6 \text{ s}$$

(c) The drift speed of the mobile charges depends on the longitudinal electric field and on the charge mobility.

$$\bar{v} = uE_{\parallel}$$

$$E_{\parallel} = (1.5 \text{ V})/(0.09 \text{ m}) = 16.7 \text{ V/m}$$

$$u = \frac{\bar{v}}{E_{\parallel}} = \frac{(8.5 \times 10^{-3} \text{ m/s})}{(16.7 \text{ V/m})} = 5 \times 10^{-4} (\text{m/s})/(\text{V/m})$$

(d) $I = enA\bar{v}$, so $n = I/(eA\bar{v})$.

$$n = \frac{(0.7 \text{ A})}{(1.6 \times 10^{-19} \text{ C})(0.02 \text{ m})(0.04 \text{ m})(8.5 \times 10^{-3} \text{ m/s})} = 6.4 \times 10^{23} \text{ m}^{-3}$$

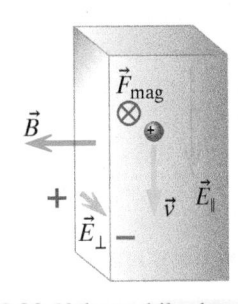

Figure 20.23 If the mobile charges are positive their velocity will be in the direction of $\vec{E}_{\parallel}$. The magnetic force will be into the page, so the back of the bar will become positive.

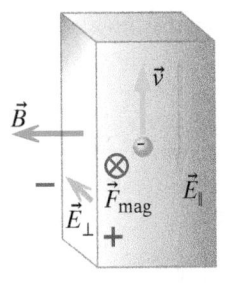

Figure 20.24 If the mobile charges are negative, their velocity will be opposite to $\vec{E}_{\parallel}$. The magnetic force will again be into the page, so the back of the bar will become negative.

20.5 MOTIONAL EMF

A magnetic field exerts a force on a current-carrying wire. This effect can be reversed: moving a wire through a magnetic field makes a current run in the wire, which provides a way to generate electricity from mechanical work.

The Initial Transient: Polarization Develops

Consider a metal bar of length L that is moving through a region of uniform magnetic field, out of the page, with magnitude B (Figure 20.25). Because a mobile electron inside the moving metal bar is moving to the right, it experiences a magnetic force.

> QUESTION If the bar is moving at a speed v, what are the magnitude and direction of the magnetic force on this electron?

The magnetic force $\vec{F} = (-e)\vec{v} \times \vec{B}$ points upward, and the magnitude of the force is $evB\sin(90°) = evB$. As a result of this magnetic force, the mobile-electron sea will shift upward.

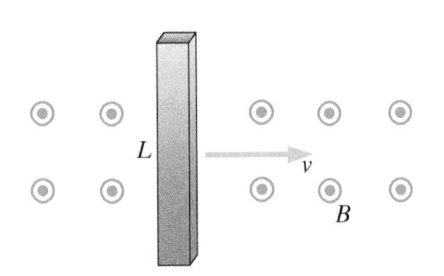

Figure 20.25 A metal bar moving at constant velocity through a region of uniform magnetic field into the page.

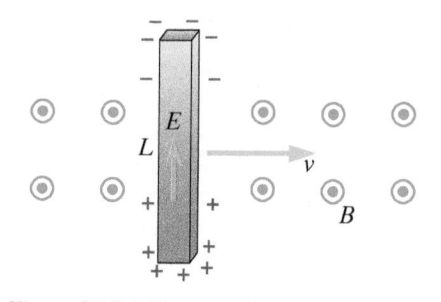

Figure 20.26 The metal bar is polarized as shown, with a downward Coulomb electric field inside the metal.

QUESTION What will be the resulting approximate charge distribution on the metal bar? What is the direction of the electric field $\vec{E}$ produced by this charge distribution inside the metal?

The magnetic force on the moving bar polarizes the bar so that it becomes negative at the top and positive at the bottom. As a result, there is an electric field $\vec{E}$ inside the metal that points upward (Figure 20.26). A mobile electron inside the metal is subject to an electric force due to this electric field as well as to the magnetic force.

The Steady State: Magnetic Force Balanced by Electric Force

QUESTION If the bar is kept moving at a constant speed v, how big will E be compared to B, in the steady state? Why? (*Hint:* Why doesn't the polarization keep increasing as more and more electrons pile up at one end of the bar?)

At first the only force on a mobile electron is the upward magnetic force, but the more the mobile-electron sea shifts upward, the larger the downward electric force $-e\vec{E}$ on an electron. Polarization continues until the electric field grows so large that the net upward force $(evB - eE)$ is zero. Therefore in the steady state we have $E = vB$.

This situation is a bit odd. Here we have a nonzero electric field inside a metal in equilibrium (bar moving at constant speed, no accelerations). There is no current, because the electric force is balanced by a magnetic force: the net force on a mobile electron inside the metal is zero.

QUESTION What is the potential difference ΔV from one end of the bar to the other? Which end is at the higher potential?

Since the electric field $E = vB$, and v and B are the same throughout the bar, the electric field E is uniform, and the potential difference is simply $\Delta V = EL = vBL$, with the lower (positive) end of the bar at the higher potential (moving downward in the bar, we go against the direction of the electric field, which means we're moving toward higher potential).

QUESTION Once the bar is completely polarized, how much force must we apply to the bar to keep it moving at a constant speed v?

In the steady state there is no current running in the bar, so there is no net force on the bar due to the magnetic field, and the polarized bar will coast along with no force required to keep it going.

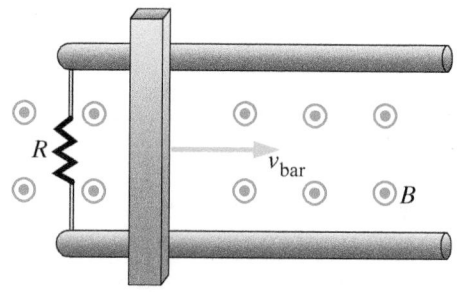

Figure 20.27 Pull the metal bar at constant speed along frictionless metal rails. A resistor is connected between the rails.

Motional Emf Can Drive Current

QUESTION Does this polarized bar remind you of any object we've studied previously? What object or device have we seen before that is most similar to the polarized bar?

This looks like a battery, because a charge separation is maintained from one end to the other by a non-Coulomb force, in this case a magnetic force. We can exploit this effect to generate an electric current.

Suppose that you pull the metal bar at a constant speed v_{bar} along frictionless metal rails through the uniform magnetic field (Figure 20.27). The moving metal bar and the metal rails have negligible resistance, but a resistor with resistance R is connected between the rails. A current will run in this circuit, driven by a "battery" whose role is played by the moving bar, and the phenomenon is called "motional emf." We show an approximate surface-charge distribution on the circuit (Figure 20.28). Electrons are driven

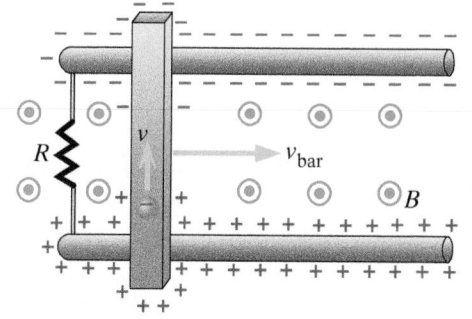

Figure 20.28 Approximate surface-charge distribution on the circuit.

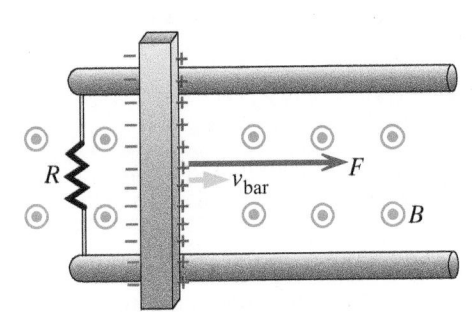

Figure 20.29 When the bar starts to move, initially the mobile electrons are left behind, so the bar polarizes left–right.

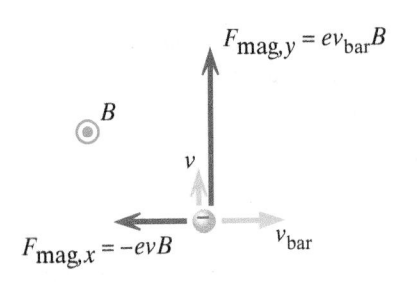

Figure 20.30 The x and y components of the magnetic force on a mobile electron in the bar.

the "wrong" way through the moving bar, moving toward the negative end of the bar.

To understand motional emf in detail, we'll start with the bar at rest, so there is no emf and no current, and we will reason step by step. To begin, you apply a force of magnitude F to the right to start the bar moving. Because the net electric force on a mobile electron in a metal is effectively zero, the first thing that happens when the bar starts moving to the right is that the mobile electrons are left behind (see the optional section "Electrons in Metals" near the end of Chapter 19 for another example of this). The bar almost instantaneously becomes polarized as shown in Figure 20.29. Now there is a nonzero electric force on a mobile electron, due to the bar having become polarized, so the electrons begin to move to the right. Another way of describing the situation is to say that you pull the lattice of positive ions to the right and the mobile electrons in turn are pulled to the right by electric interactions with the positive charges.

QUESTION In the presence of the magnetic field out of the page ($+z$), as shown in Figure 20.27, what is the direction of the magnetic force on the mobile electrons that are starting to move to the right?

There will be a small upward magnetic force on a mobile electron, in the $+y$ direction.

As a result of this magnetic force, a mobile electron will acquire a small $+y$ component of velocity in addition to its component of velocity in the $+x$ direction. We'll write the vector velocity of the electron as $\vec{v}_{electron} = \langle v_{bar}, v, 0 \rangle$, where v_{bar} is the speed of the bar and v is the vertical component, as shown in Figures 20.28 and 20.30.

QUESTION At this instant, what are the components of the magnetic force on a mobile electron? The magnitude of the magnetic field is B and the electron charge is $-e$.

$\vec{F}_{mag} = -e\vec{v}_{electron} \times \vec{B} = \langle -evB, ev_{bar}B, 0 \rangle$. The magnetic force on a mobile electron now has a component in the $-x$ direction as well as in the $+y$ direction (Figure 20.30). Both components are very small just after applying the force F to the bar, because the velocity of the mobile electron is small. Considering the bar as the system of interest, the x component of the net force on the bar is $F_{net,x} = F - NevB$, where N is the number of mobile electrons in the bar.

As long as $F_{net,x}$ is nonzero, the bar will continue to accelerate to the right. However, higher speed means that the vertical component $ev_{bar}B$ of the magnetic force is bigger, which increases the vertical component v of the mobile electron's velocity. The faster the bar moves, the bigger is v and the larger the force $NevB$ to the left on the bar, due to all the magnetic forces on the mobile electrons.

As time goes on, the net force acting on the bar gets smaller and smaller. Eventually the net force becomes zero, and the bar moves at constant speed with you pulling to the right and the $-x$ portion of the magnetic force pulling to the left. As you keep pulling, you do work to maintain a charge separation as shown in Figure 20.28, and a current runs through the resistor.

QUESTION You can see that there is an electric field inside the bar, due to the charges maintained above and below the center of the bar. Why do the electrons move the "wrong" way through the bar, vertically upward?

The electron current through the resistor is continually depleting the charges on the ends of the moving bar, with the result that the electric field inside the bar is always slightly less than is needed to balance the magnetic force. The downward electric force eE is slightly less than the upward magnetic force evB, so the electrons move toward the negative end of the bar.

The magnetic force evB is a non-Coulomb force (F_{NC} in our earlier discussions of batteries). The work done by this force in moving an electron from one end of the bar to the other is

$$F_{NC}L = ev_{bar}BL$$

QUESTION The emf is the work done per unit charge in moving a charge from one end of a "battery" to the other, so what is the emf of this "battery"?

The non-Coulomb work per unit charge is $(ev_{bar}BL)/e = v_{bar}BL$. If the resistance of the bar is negligible, $\Delta V = \text{emf} = v_{bar}BL$. If the bar has some resistance r_{int}, this resistance is like the internal resistance of any other kind of battery, and $\Delta V = \text{emf} - r_{int}I$ (see the discussion of batteries with internal resistance in the section "Batteries" in Chapter 19).

It is interesting to see that although the vertical component of the magnetic force does positive work on a mobile electron, the horizontal component does negative work, and the total work is zero, which is consistent with the fact that the magnetic force is always perpendicular to the electron's velocity. The power input to the electron due to the magnetic force is $\vec{F}_{mag} \bullet \vec{v}_{electron}$, which is zero, as you can verify by taking the dot product of $\vec{F}_{mag} = \langle -evB, ev_{bar}B, 0 \rangle$ and $\vec{v}_{electron} = \langle v_{bar}, v, 0 \rangle$. We'll see in a moment that the work that drives the current is due to the force that you apply to the bar to keep it going.

The emf that is created by moving a wire in a magnetic field is called "motional emf." This effect is reminiscent of the Hall effect, because when you move a wire through a magnetic field, the magnetic force creates a polarization (and electric field) in a direction perpendicular to the motion.

QUESTION Does it look like the potential difference $\Delta V = -\oint \vec{E} \bullet d\vec{l}$ for a round trip around the circuit in Figure 20.31 could be equal to zero, as should be the case for the path integral of electric fields due to point charges? (In Figure 20.31, pay attention to the directions of the electric field in the moving bar, in the rails, and in the resistor.)

Going around the loop clockwise from the bottom end of the moving bar, we get a small potential drop in the lower rail, a large potential drop in the resistor, a small drop in the upper rail, and a large potential increase in the moving bar. Therefore it does look plausible that the net ΔV for the round trip could be zero, which we know should be true.

This situation is quite similar to a circuit with a mechanical battery in it, where charges move through the mechanical battery in the "wrong" direction, due to non-Coulomb forces, while the round-trip potential difference is zero. Here the non-Coulomb force is the magnetic force.

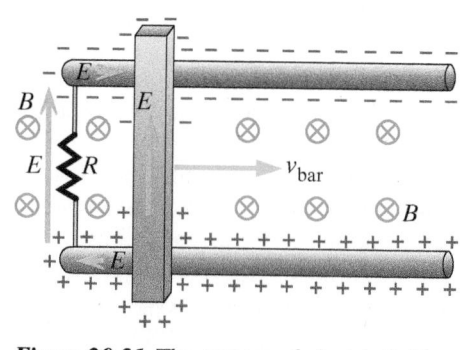

Figure 20.31 The pattern of electric field in the circuit driven by the moving bar.

Checkpoint 7 Can you light a flashlight bulb by moving a connecting wire rapidly past your bar magnet? Let's put in the numbers. You found that near your bar magnet the magnetic field was about 0.1 T, and the length of wire affected by the magnetic field might be about a centimeter. You might be able to move the wire at a speed of 10 m/s past the magnet. What is the emf you can produce? How does this compare with the emf of a flashlight battery?

Small Size of Motional Emf

From the preceding exercise dealing with your flashlight bulb you can see that it is hard to see the effects of motional emf with batteries and bulbs. To get a sizable emf you need larger magnetic field B extending over larger regions (L) with very fast motion (v_{bar}) of the wires through the magnetic field and many turns of wire, each contributing to the net emf.

Also, if you move a wire rapidly past your magnet, the emf doesn't last very long. Finally, a light bulb or compass isn't a very sensitive detector of small currents. Consequently, to study this motional emf you need somewhat sensitive lab equipment. Your instructor may arrange a demonstration of these effects or a formal laboratory exercise where you can observe the phenomena for yourself.

Forces on the Moving Bar and Energy Conservation

If the metal bar were to slide effortlessly along the rails at a constant speed v_{bar}, and the resistor were to continually warm the surroundings, we would be getting something for nothing. Thus there must be an energy input into the system consisting of the moving bar, the resistor circuit, and the magnet that provides the magnetic field. The magnetic force is internal to this system; the only external force is the force you continuously apply to the bar. You must exert a force on the bar to keep it moving at constant speed (Figure 20.32), because there is a horizontal component of the magnetic force on the moving bar, due to the conventional current I in the bar.

> QUESTION What is the direction of this magnetic force? What are the direction and magnitude of the force that we have to exert on the bar to keep it going at a constant speed?

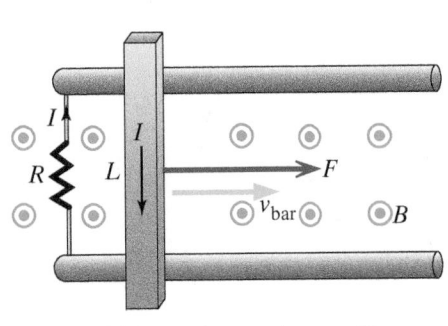

Figure 20.32 You have to keep pulling on the bar in order to keep current running in this circuit.

Taking just the bar as the system for a moment, we've seen that there is a $-x$ component of magnetic force of magnitude $NevB$ opposing the force we apply. N is the number of mobile electrons in the bar, which is equal to the electron density n (electrons per unit volume) times the volume of the bar AL, where A is the cross-sectional area of the bar. Therefore the magnitude of the magnetic force opposing us is $NevB = (nAL)evB = (enAv)LB$, which is ILB, the usual expression for the force on a current-carrying wire with a magnetic field perpendicular to the wire.

In the steady state, in order for the bar to keep moving at constant speed, the x component of the net force on the bar must be zero, so we have to exert a force of magnitude ILB to the right.

In a short time Δt we move the bar a distance Δx. We do an amount of work $F\Delta x$ in a time Δt, so we have to supply energy at a rate of $F\Delta x/\Delta t = Fv_{bar}$.

> QUESTION Is this power input Fv_{bar} numerically equal to the power dissipated in the resistor?

We have power $= (ILB)v_{bar} = I(LBv_{bar}) = I(\text{emf})$, but $I(\text{emf})$ is equal to the power output of the battery that is input to the resistor. Moreover, $I = \text{emf}/R$, so the power dissipated in the resistor is $\text{emf}^2/R = RI^2$, as expected.

This is the basic principle of electric generators: mechanical power is converted into electric power. Commercially, the mechanical energy is supplied by falling water at power dams, or by expanding steam in a turbine (where the steam is heated by chemical or nuclear reactions—burning coal or oil or natural gas, or fissioning uranium). A practical generator has a rotating coil rather than a sliding bar, and we'll discuss practical generators later in this chapter.

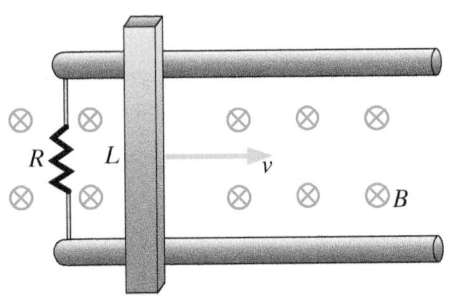

Figure 20.33 Pull the metal bar at constant speed along frictionless metal rails. A resistor R is connected between the rails.

Checkpoint 8 Here is a review of the basic issues: A metal rod of length L is dragged horizontally at constant speed v_{bar} on frictionless conducting rails through a region of uniform magnetic field into the page of magnitude B (Figure 20.33). There is a resistance R. What is the magnitude of the induced emf? What is the direction of the conventional current I? What is the magnitude of the current I? What are the magnitude and direction of the force you have to apply to keep the rod moving at a constant speed v_{bar}?

EXAMPLE   **A Metal Bar Moving Downward**

In Figure 20.34 a horizontal metal bar is moving downward in the plane of the page with electrical contact along two metal vertical bars that are a distance w apart. Throughout this region there is a uniform horizontal magnetic field with magnitude B out of the plane of the page, made by large coils that are not shown. At the instant that the moving bar has a speed v, what is the reading on the voltmeter? Give magnitude and sign, and explain briefly.

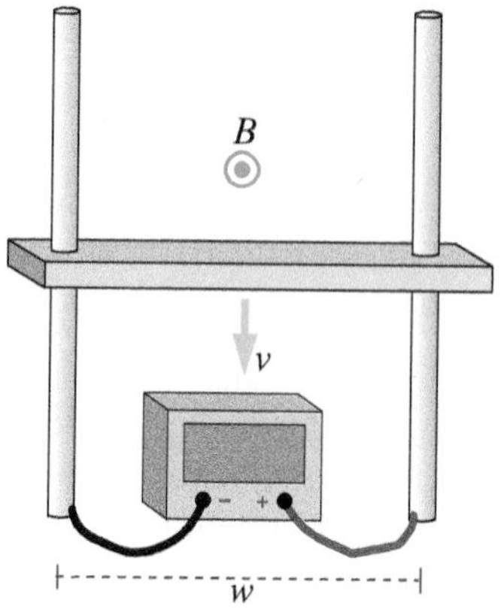

Solution   The charged particles in the bar are moving through a magnetic field, and therefore experience a magnetic force. It does not matter what sign we assume for the mobile charges, so we will assume they are positive because it simplifies keeping track of signs.

$\vec{v} \times \vec{B}$ is to the left, so the bar polarizes as shown in Figure 20.35.

The polarization takes an extremely short time (less than nanoseconds), after which for every mobile charge in the bar we have the following (note that the voltmeter has very high resistance, so there is almost no current):

$$\vec{F}_{\text{magnetic}} + \vec{F}_{\text{electric}} = 0$$
$$qvB \sin(90°) = qE$$
$$E = vB$$

Since E is uniform throughout the bar,

$$|\Delta V| = \left| -\int \vec{E} \cdot d\vec{l} \right| = vBw$$

The left end of the bar is at a higher potential than the right end. Since the high potential end is connected to the negative terminal of the voltmeter, the reading on the voltmeter will be negative. The reading on the voltmeter is therefore $-vBw$.

Figure 20.34 As the metal bar moves downward, what does the voltmeter read?

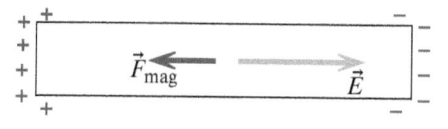

Figure 20.35 Magnetic force on a positive charge, resulting polarization of the bar, and electric field due to polarization.

20.6 MAGNETIC FORCE IN A MOVING REFERENCE FRAME

The magnetic field produced by a moving charge depends on the velocity of the charge, and this leads to the curious result that a stationary Bob may observe a different magnetic field than a moving Alice sees. Suppose that Bob holds a stationary charge, which of course produces a pure electric field and no magnetic field. If Alice runs past Bob, she sees a mixture of electric and magnetic fields produced by what is for her a moving charge.

The magnetic force experienced by a moving charge also depends on the velocity of the charge, which is another reason why observations of magnetic effects may differ in different moving reference frames. If Bob holds two charged tapes that repel each other, he describes this by saying that one tape produces a purely electric field, and the other tape is affected by that electric field. Alice, however, as she moves past Bob says that one (moving) tape produces a mixture of electric and magnetic fields, and the other (moving) tape feels both electric and magnetic forces.

Two Moving Protons

We are now in a position to do a detailed calculation of these odd effects. We'll consider two protons initially traveling parallel to each other with the same speed v, a distance r apart, and we will calculate the magnitudes and directions of the electric and magnetic forces that proton 1 exerts on proton 2 at this moment (Figure 20.36). The electric force is easy. Proton 1 makes an electric field $\vec{E}_1$ at the location of proton 2, and proton 2 experiences an electric force $\vec{F}_{21,e} = q_2 \vec{E}_1$:

$$F_{21,e} = \frac{1}{4\pi\varepsilon_0} \frac{e^2}{r^2}, \quad \text{downward}$$

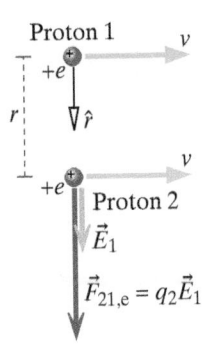

Figure 20.36 Proton 1 exerts an electric force on proton 2.

We're making an approximation when we use Coulomb's law to calculate the electric force. Due to relativistic effects, the electric field $\vec{E}_1$ made by the moving charge (proton 1) is somewhat different from the electric field made by a stationary charge. We'll say a bit more about this approximation later.

The magnetic force involves cross products both in the field and in the force. To determine the magnetic force $\vec{F}_{21,m}$ acting on proton 2, first we need to calculate the magnetic field $\vec{B}_1$ made by proton 1 at the location of proton 2. It helps enormously to draw the unit vector on the diagram (from source to observation point; that is, from proton 1 to proton 2).

> QUESTION Give the direction and the magnitude of $\vec{B}_1 = \frac{\mu_0}{4\pi} \frac{q_1 \vec{v}_1 \times \hat{r}}{r^2}$.

The magnetic field $\vec{B}_1$ is into the page at the location of proton 2, and its magnitude is $B_1 = [\mu_0/(4\pi)](ev/r^2)\sin 90°$.

> QUESTION Now that you know the direction and magnitude of the magnetic field $\vec{B}_1$, determine the magnitude and direction of the magnetic force exerted on proton 2, $\vec{F}_{21,m} = q_2 \vec{v}_2 \times \vec{B}_1$.

The direction is upward, and the magnitude is $F_{21,m} = \frac{\mu_0}{4\pi} \frac{e^2 v^2}{r^2}$.

The Ratio of the Magnetic and Electric Forces

Figure 20.37 summarizes the situation. The magnetic force on proton 2 is opposite to the electric force, and the net force on proton 2 is reduced from what it would be for a stationary proton.

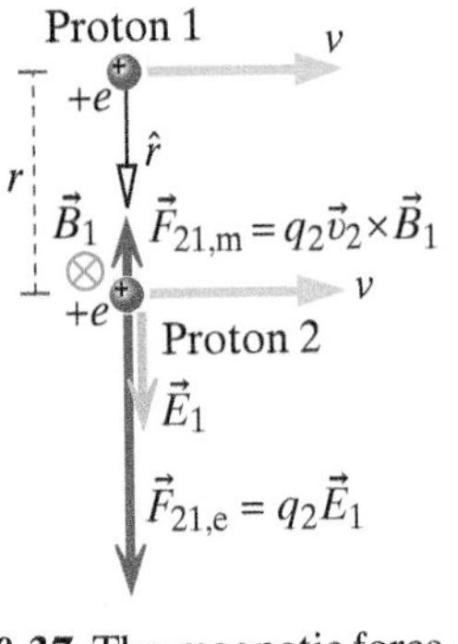

> QUESTION What is the ratio of the magnetic force to the electric force on proton 2 due to proton 1?

$$\frac{F_{21,m}}{F_{21,e}} = \frac{\left(\dfrac{\mu_0}{4\pi} \dfrac{e^2 v^2}{r^2}\right)}{\left(\dfrac{1}{4\pi\varepsilon_0} \dfrac{e^2}{r^2}\right)}$$

$$= \frac{v^2}{\left(\dfrac{1}{4\pi\varepsilon_0} \Big/ \dfrac{\mu_0}{4\pi}\right)} = \frac{v^2}{(9 \times 10^9 / 1 \times 10^{-7})} = \frac{v^2}{(3 \times 10^8)^2} = \frac{v^2}{c^2}$$

Figure 20.37 The magnetic force on proton 2 is opposite to the electric force.

This is a surprise! Why should the speed of light show up in a comparison between electric and magnetic forces? What does the speed of light have to do with this? There are some deep issues here. (We did not show units in the calculation above, but they do work out correctly.)

This result indicates that at ordinary speeds ($v \ll c$) the magnetic interaction between two charged particles is much weaker than the electric interaction. However, if v approaches the speed of light c, the magnetic force may be comparable to the electric force. Even though at low speeds magnetic forces are inherently weak compared to electric forces, magnetic forces are more important in many industrial applications, because the magnetic force on a current-carrying wire is typically much larger than an electric force on the wire, since the net charge of the wire is nearly zero (neglecting the small amount of surface charge).

QUESTION Now determine the full electric and magnetic force, $\vec{F} = q\vec{E} + q\vec{v} \times \vec{B}$, on proton 2 due to proton 1 (magnitude and direction).

The electric force is downward and the magnetic force is upward, so the net force on proton 2 is

$$\frac{1}{4\pi\varepsilon_0}\frac{e^2}{r^2} - \frac{\mu_0}{4\pi}\frac{e^2v^2}{r^2} = \frac{1}{4\pi\varepsilon_0}\frac{e^2}{r^2}\left(1 - \frac{v^2}{c^2}\right), \quad \text{downward}$$

QUESTION Describe the subsequent trajectories of the protons.

The protons repel each other, so their trajectories are curving lines moving apart (Figure 20.38).

Checkpoint 9 As a numerical example, suppose that $v = 1500$ m/s, a very high speed typical of a molecule of hydrogen gas at room temperature. What is the numerical ratio of the magnetic force to the electric force for protons moving at this speed?

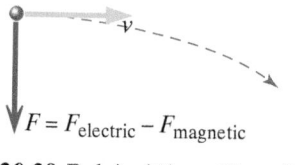

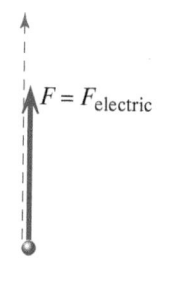

$F = F_{\text{electric}} - F_{\text{magnetic}}$

$F = F_{\text{electric}} - F_{\text{magnetic}}$

Figure 20.38 Bob is sitting still, and he sees the protons move along curving paths.

Alice and Bob and Einstein

The fact that the net force is smaller on the moving protons may seem innocuous, but it has startling consequences. Suppose that Bob is sitting still and sees the protons go by. The protons repel each other along curving paths, with a force somewhat weakened by the magnetic effects (Figure 20.38).

Suppose that Alice is running along beside the protons, with the same speed v that the protons have initially. She sees the two protons repel each other straight away from each other (Figure 20.39). The repulsion is purely electric, with no weakening of this electric repulsion (the magnetic field of one proton is zero at the location of the other proton, because $\vec{v} \times \hat{r} = 0$ along the line of motion of the proton).

Let both Alice and Bob use accurate timers to measure how long it takes for the protons to hit the floor and ceiling of the room, starting from halfway between floor and ceiling.

QUESTION Given the difference in the force that accelerates the protons vertically, which observer will measure a shorter time for the protons to reach the floor and ceiling?

Alice sees a pure electric repulsion, while Bob sees the electric repulsion partially counteracted by an attractive magnetic force. To take a concrete example, suppose Bob's timer advances by 20 ns during this process. Alice's timer might advance by only 15 ns (if v is a large fraction of the speed of light), corresponding to her observation that the protons get to the floor and ceiling in less time due to the greater force of repulsion.

But that's crazy! Surely Alice and Bob wouldn't measure different amounts of time for the same process, would they? It was Einstein who thought very

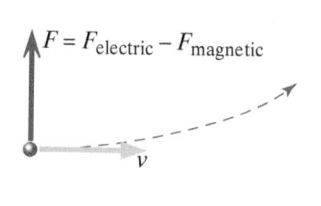

$F = F_{\text{electric}}$

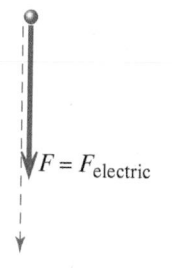

$F = F_{\text{electric}}$

Figure 20.39 Alice runs along beside the protons, at speed v; she sees the protons move along straight paths.

deeply about this and related paradoxes and concluded that, much as it violates common sense, it has to be the case that Alice's timer seems to run slower for Bob than his own timer!

This sounds utterly absurd, yet a great variety of measurements of this kind all agree with Einstein's radical proposal that time runs at different rates for different observers. One of the reasons that Einstein was driven to these strange notions was that by the time that he proposed his special theory of relativity (1905), it was clear that the existing theories of electricity and magnetism seemed to lead to paradoxes, as we have seen in the observations of Alice and Bob. In fact, the title of Einstein's 1905 paper on relativity was "On the electrodynamics of moving bodies."

The net force (electric minus magnetic) observed by Bob is always smaller than the purely electric force observed by Alice, and Alice and Bob disagree on how long it takes the protons to move apart, and on whether their timers are marking time correctly. Yet both Alice and Bob can correctly predict how much their own timers will advance during the time it takes for the protons to hit the floor and ceiling.

The details of our calculations are not correct, because relativistic aspects of the fields have not been taken into account. Both the electric field and the magnetic field of a fast-moving charge are different from the fields given by Coulomb's law and the Biot–Savart law for a slow-moving charge. However, both fields are modified by the same factor $1/\sqrt{1 - v^2/c^2}$, so the ratio of the magnetic force to the electric force is equal to v^2/c^2 for all speeds. Alice and Bob will indeed observe different times for the same process. See Section 20.11 at the end of this chapter for a more detailed discussion of the nature of electric and magnetic fields in moving reference frames.

An Asymmetry in the Measurements

You may wonder about the asymmetry in the measurements that Alice and Bob make. If Bob thinks that Alice's timer runs slow, why doesn't Alice think that Bob's timer runs slow? After all, Alice should be able to take the point of view that she is standing still while Bob moves past her.

In reality, Bob cannot make his measurements unaided. Rather, Bob and his friend Fred must station themselves an appropriate distance apart in the room. Bob records the time (on Bob's timer) when the protons pass by him, and he also records the time that he sees on Alice's timer at that moment. Fred records the later time (on Fred's timer) when the protons reach him and hit the floor and ceiling, and he also records the time that he sees on Alice's timer at that moment. Two people, Bob and Fred, are needed to make these measurements. Otherwise Bob would have to apply corrections to his measurements, due to the time it takes for light from Alice's timer to reach him from the distant location.

After Fred has made his measurements, Bob and Fred compare notes and find that the times that they wrote down differ by 20 ns (they had previously synchronized their timers). Their two views of Alice's timer indicate that her timer advanced only 15 ns. Bob and Fred describe Alice's timer as running slow. The situation is different for Alice, because she looks at two *different* timers, Bob's and Fred's. The asymmetry lies in the fact that there are two observers sitting in the room (Bob and Fred), but only one observer moving through the room (Alice).

On the other hand, suppose that Alice sits at the front of a space ship moving past Bob, and her friend Sue sits further back in the space ship. Alice notes the time on her own timer and on Bob's timer when she passes Bob, and Sue does the same when she passes Bob. When Alice and Sue compare notes, they will describe Bob's timer as seeming to run slow.

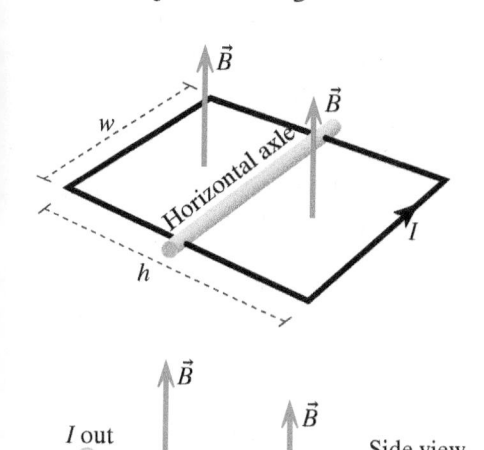

Figure 20.40 A current-carrying loop connected to an insulating horizontal axle is free to rotate.

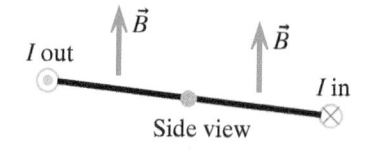

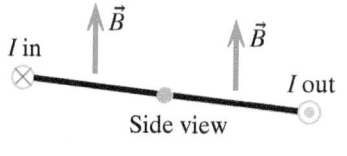

Figure 20.41 Which of these orientations is stable?

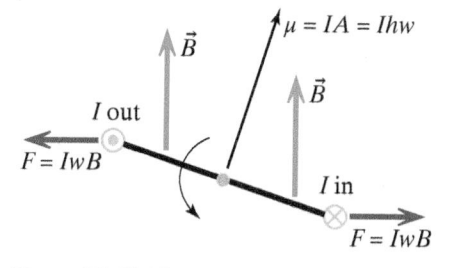

Figure 20.42 The magnetic dipole moment of the loop is a vector that tends to line up with the applied magnetic field.

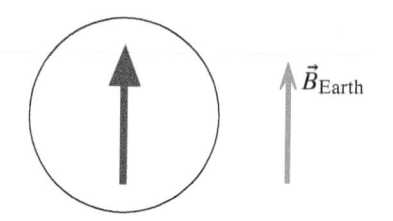

Figure 20.43 What is the direction of the atomic magnetic dipole moments inside this compass needle, which points north?

20.7 MAGNETIC TORQUE

A magnetic field can cause a current-carrying coil of wire to twist. You may have seen this for yourself, if you were able to suspend a coil of wire and see it twist in the magnetic field of a bar magnet (or you may have floated a coil and seen it twist in the magnetic field of the Earth). Moreover, we have seen that a compass needle can be thought of as a collection of atomic current loops, and you have certainly seen a compass needle twist in a magnetic field.

QUESTION In Figure 20.40, what is the direction of the magnetic force on each of the four sides of the loop?

Consider a rectangular current-carrying loop of wire in a uniform magnetic field (that is, the magnetic field has the same direction and magnitude throughout this region). The loop is free to rotate on an insulating horizontal axle (Figure 20.40).

On the sides of length h the magnetic force is horizontal and points outward, tending to stretch the loop. On the other sides, of length w, the magnetic forces are horizontal and tend to make the loop twist on the axle. This may be easiest to see in the side view shown in Figure 20.40.

QUESTION In the side view, will the loop rotate clockwise (to the right) or counterclockwise (to the left)?

The magnetic forces act to twist the loop counterclockwise.

QUESTION When the plane of the loop is perpendicular to the magnetic field, the magnetic forces don't exert any twist. However, one of these orientations is unstable: if you nudge it slightly, the loop will flip over. Which of the two orientations shown in side view in Figure 20.41 is stable, and which is unstable?

The situation on the top is stable: the magnetic forces twist the loop back toward the horizontal plane. The situation on the bottom is unstable: a small displacement away from the horizontal plane leads to magnetic forces that rotate it even farther out of the plane.

It is often easier to talk about the twist on a current-carrying loop of wire in terms of the "magnetic dipole moment" (see Chapter 17). The magnetic dipole moment $\vec{\mu}$ of a current-carrying loop of wire is defined as a vector pointing in the direction of the magnetic field that the loop makes along its axis, which is given by a right-hand rule (this magnetic field made by the loop is of course different from the applied magnetic field that makes the loop twist). The magnitude of the magnetic dipole moment is $\mu = IA$.

Our rectangular loop has an area $A = wh$, so its magnetic dipole moment is $\mu = IA = Ihw$. In Figure 20.42, note that the coil tends to twist in a direction to make $\vec{\mu}$ line up with the applied magnetic field $\vec{B}$.

Checkpoint 10 Inside the magnetized compass needle shown in Figure 20.43, which points north, what is the direction of a typical atomic magnetic dipole moment $\vec{\mu}$? If the needle is moved away from north and then released, what will happen? Explain briefly.

At last we have an explanation in terms of the fundamental principles of magnetism for the behavior of a compass needle, and why it always lines up with the applied magnetic field. (A practical compass has some friction to damp out the oscillations: without friction the needle would swing back and forth forever, like a frictionless pendulum. Many compasses are filled with liquid to damp out the oscillations.)

Quantitative Torque

The fact that a magnetic dipole twists in a magnetic field is very important. The torque provided by each of the magnetic forces around the axle is equal to the distance from the axle (the "lever arm") times the component of the force perpendicular to the lever arm. The twist applied to this rectangular loop is due to the magnetic forces on the w-lengths of the loop, as is seen in Figure 20.44, where the magnetic torque is out of the page (there is a clockwise twist).

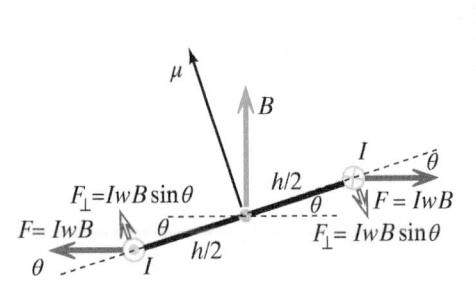

Figure 20.44 The magnetic forces on this rectangular current-carrying loop tend to twist it clockwise (there is a torque $\vec{\mu} \times \vec{B}$ into the page).

The perpendicular component of each force is $F_\perp = IwB\sin\theta$, and the lever arm is $h/2$. (Note that the other component of F, and the forces on the h lengths of the loop, just tend to stretch the loop without causing a twist.) Each of the forces exerts a torque of $(F_\perp)(h/2)$, so the total torque is $2IwB\sin\theta(h/2) = (Iwh)B\sin\theta$.

Since $\mu = Iwh$, we can express the torque in the form $\vec{\mu} \times \vec{B}$, because this cross product has a magnitude $|\vec{\mu}||\vec{B}|\sin\theta = \mu B\sin\theta$. The direction of this torque vector is along the axle around which the loop rotates. If the thumb of your right hand points in the direction of $\vec{\mu} \times \vec{B}$, your fingers curl around in the direction that the coil will twist. In Figure 20.44, the torque vector is into the page, and the coil rotates clockwise.

We showed that the torque is $\mu B\sin\theta$ for a rectangular loop, but a loop of any shape can be approximated by a set of current-carrying rectangles along whose adjacent sides the current cancels (Figure 20.45). This shows that the torque is $\mu B\sin\theta$ for any kind of loop.

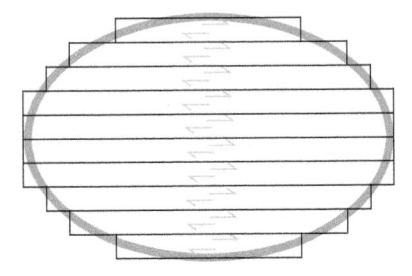

Figure 20.45 A loop of any shape can be approximated by a set of rectangular loops.

TORQUE ON A MAGNETIC DIPOLE

$$\vec{\tau} = \vec{\mu} \times \vec{B}$$

20.8 POTENTIAL ENERGY FOR A MAGNETIC DIPOLE

If allowed to pivot freely, a magnetic dipole turns toward alignment with the applied magnetic field. That means that an aligned magnetic dipole is associated with lower potential energy in the magnetic field, and the system tries to go to the lower potential-energy configuration. If we calculate how much work it takes to move the magnetic dipole out of alignment, we will have a measure of the increased potential energy associated with being out of alignment.

We consider the rectangular loop shown in Figure 20.46, and we calculate how much work we have to do to move it from an initial angle $\theta = \theta_i$ to a final angle $\theta = \theta_f$. We move slowly, without changing the kinetic energy of the loop. The loop is always very nearly in equilibrium.

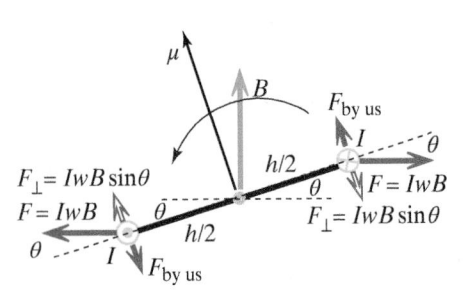

Figure 20.46 Calculating how much work is required to turn a current loop in a magnetic field.

As we rotate the loop through a small angle $d\theta$, both of the forces that we exert act through a distance $(h/2)d\theta$, because arc length is radius $(h/2)$ times angle ($d\theta$ measured in radians). If we force the loop to rotate from the initial angle $\theta = \theta_i$ to a final angle $\theta = \theta_f$, the amount of work we do by exerting our two forces must be calculated by an integral, because F isn't constant. The work that we do goes into changing the magnetic potential energy U_m of the system consisting of the magnetic dipole in the magnetic field:

$$\text{work} = \Delta U_m = \int_{\theta_i}^{\theta_f} 2IwB\sin\theta \left(\frac{h}{2}d\theta\right) = IwhB\int_{\theta_i}^{\theta_f} \sin\theta\, d\theta$$

$$\Delta U_m = IwhB[-\cos\theta]_{\theta_i}^{\theta_f} = -IwhB[\cos\theta_f - \cos\theta_i]$$

$$\Delta U_m = \Delta(-\mu B\cos\theta) \text{ since } \mu = IA = Iwh$$

It is customary to define the potential energy of a magnetic dipole in a magnetic field by choosing the zero of potential energy in such a way that one writes this:

$$U_m = -\mu B\cos\theta$$

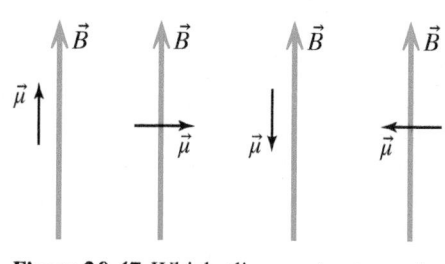

Figure 20.47 Which alignment represents the lowest potential energy? The highest?

This can also be written as a dot product:

THE POTENTIAL ENERGY FOR A MAGNETIC DIPOLE

$$U_m = -\vec{\mu} \bullet \vec{B}$$

This convention corresponds to calling the potential energy zero when $\theta = 90°$, since $\cos(90°) = 0$.

QUESTION Given that $U_m = -\vec{\mu} \bullet \vec{B}$, which of the alignments in Figure 20.47 corresponds to the lowest possible potential energy? To the highest possible potential energy?

The lowest energy corresponds to $\vec{\mu}$ and $\vec{B}$ pointing in the same direction, so that $U_m = -\mu B \cos 0 = -\mu B$. This is also the most stable configuration; if you twist the magnetic dipole away from this position, it will be twisted back by the magnetic field. The highest energy corresponds to $\vec{\mu}$ and $\vec{B}$ pointing in opposite directions, so that $U_m = -\mu B \cos 180° = +\mu B$. This is an unstable configuration; a slight displacement will lead to a large twist. The other two configurations correspond to medium energy: $U_m = 0$ (because $\cos 90° = 0$).

This picture of the potential energy of a magnetic dipole in a magnetic field is very important in atomic and nuclear physics. In a magnetic field, atoms or nuclei can make transitions between higher and lower energy states that correspond to different orientations of the magnetic dipole moments of the atoms or nuclei in the magnetic field.

For example, in the magnetic imaging equipment now used in hospitals, the patient lies inside a large solenoid. In the presence of the magnetic field of the solenoid, nuclei in the patient's body have slightly different energies depending on whether the nuclear magnetic dipole moment of the nucleus is in the direction of or opposite to the applied field. Sensitive equipment detects transitions between these two slightly different energy levels.

Checkpoint 11 What is the energy difference between the highest and lowest states?

Force on a Magnetic Dipole

We have successfully explained the twist of a magnetic dipole in a magnetic field. There remains the question of how to explain the net force on a magnetic dipole in a magnetic field. For example, how can we explain the attraction or repulsion between two magnets?

We can show that if a magnetic dipole is placed in a nonuniform magnetic field, it experiences a force. Figure 20.48 shows a circular current loop placed near the north pole of a bar magnet, along the axis of the magnet. The magnetic field of the bar magnet diverges with an angle θ as shown and so is nonuniform.

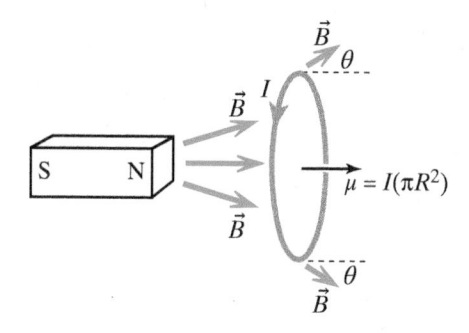

Figure 20.48 A bar magnet exerts a force on a current-carrying loop.

QUESTION Show that the net force on the current loop is to the left, with magnitude $F = I(2\pi R)B \sin \theta = (2\mu B/R) \sin \theta$, where R is the radius of the circular loop.

As shown in Figure 20.49, each segment of the ring experiences a magnetic force $d\vec{F} = I\Delta \vec{l} \times \vec{B}$. The horizontal component of the force is $dF \cos(90° - \theta) = dF \sin \theta$. The vertical components of the forces on two opposite pieces of the ring cancel, so the net force is in the $-x$ direction (toward the bar magnet).

$$F_{\text{net}} = IB \sin \theta \int dl = IB(2\pi R) \sin \theta$$

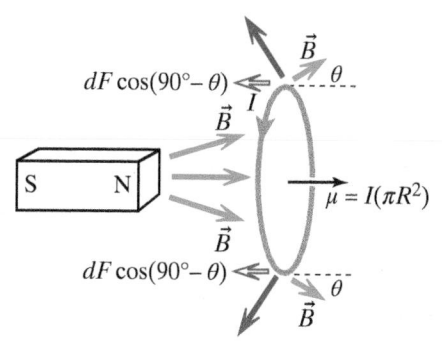

Figure 20.49 Force on a current loop by a bar magnet.

Since $\mu = IA = I(\pi R^2)$,

$$F_{\text{net}} = \frac{\mu(2B \sin \theta)}{R}$$

If we knew how the magnetic field spread out in angle (that is, how big θ is at a distance R off-axis), we could calculate the net force F. Without that information, we can nevertheless determine F by using an argument based on potential energy, which we will do in the next section.

> **Checkpoint 12** In what direction would the net force be if the bar magnet were turned around, or the current in the loop reversed? What can you say about the net force if the field were uniform ($\theta = 0$)?

Calculating the Force on a Magnetic Dipole

Using an energy argument, we can calculate the force on a magnetic dipole due to a nonuniform magnetic field. For example, we can calculate the force on one magnet due to another magnet.

If we move a magnetic dipole from a region of low magnetic field to a region of high magnetic field, the magnetic potential energy $U_m = -\vec{\mu} \bullet \vec{B}$ will change. If released from rest, the object will spontaneously move to the place where the potential energy is lower, picking up kinetic energy as it goes. If we prevent the kinetic energy from changing, by moving the object at a slow constant speed, we have to exert a force and do some work. By calculating how much work we have to do, we can determine how big a force we have to apply, and this is numerically equal to the magnetic force on the object. This will let us calculate how big a force one magnet exerts on another magnet.

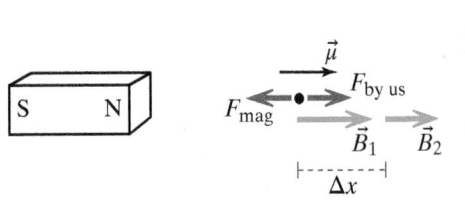

Figure 20.50 A bar magnet acts on a magnetic dipole.

Consider the situation shown in Figure 20.50. A magnetic dipole is aligned with the magnetic field made by a bar magnet, and the magnet exerts a force F_{magnetic} to the left on the magnetic dipole moment, as we saw in the previous section. We exert a force $F_{\text{by us}}$ to the right, infinitesimally larger than F_{magnetic}, and we move the magnetic dipole moment a short distance Δx to the right.

The magnet makes a nonuniform magnetic field (proportional to $1/x^3$ if we are far from the magnet), and we are moving the magnetic dipole from a larger magnetic field B_1 to a smaller magnetic field B_2. The magnetic potential energy increases from $-\mu B_1$ to the somewhat less negative $-\mu B_2$. Therefore we have to do an amount of work

$$F_{\text{by us}} \Delta x = \Delta U_m = (-\mu B_2) - (-\mu B_1) = -\mu(B_2 - B_1) = -\mu \Delta B$$

This is greater than zero because $B_2 < B_1$. The magnitude of the force we have to exert is this:

$$F_{\text{by us}} = \frac{\Delta U_m}{\Delta x} = -\mu \frac{\Delta B}{\Delta x} \rightarrow -\mu \frac{dB}{dx} \quad \text{(a positive value; } dB/dx < 0\text{)}$$

The force that the magnet exerts to the left on the magnetic dipole moment is numerically equal to the force that we exert. This magnitude is also correct in the case of repulsion. If the magnetic dipole moment is at some angle to the magnetic field, there is both a net force and a twist.

This is a special case of the general result that the force associated with a potential energy is the negative gradient of that potential energy:

$$F_x = -\frac{dU}{dx} \quad \text{(general result)}$$

$$F_x = -\frac{d(-\vec{\mu} \bullet \vec{B})}{dx} = \mu \frac{dB}{dx} \quad \text{(our specific case)}$$

In this specific case, the quantity $\vec{\mu} \bullet \vec{B}$ is decreasing in the $+x$ direction, so the equation predicts that the x component of the force on the magnetic dipole moment is in the $-x$ direction, which is correct.

Note that there is no force if the field is uniform ($dB/dx = 0$), in agreement with our previous results, though there is a twist. This is the situation with a compass needle affected by the magnetic field of the Earth, which hardly varies over short distances and therefore applies an extremely weak net force, but there is a twist due to equal and opposite forces that align the needle to point north.

We can now calculate the force that one bar magnet exerts on another. The magnetic field along the axis of one magnet, far from that magnet, is

$$B_1 \approx \frac{\mu_0}{4\pi} \frac{2\mu_1}{x^3}$$

where μ_1 is the total magnetic dipole moment of all the atomic magnetic dipole moments in the magnet.

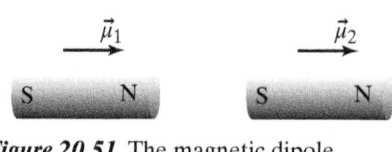

Figure 20.51 The magnetic dipole moment μ_2 of the second magnet lies along the axis of μ_1 of the first magnet.

QUESTION Show that the attractive force that this magnet would exert on a magnetic dipole moment μ_2 lying along the axis of μ_1 (Figure 20.51) is

$$F \approx 3\mu_2 \left(\frac{\mu_0}{4\pi} \frac{2\mu_1}{x^4} \right)$$

Just take the negative gradient of the magnetic potential energy to obtain this result. This says that two magnets should attract or repel each other with a force proportional to $1/x^4$.

Moreover, using our measurements of the magnetic dipole moments of our bar magnets in Chapter 17, we should be able to calculate how close one of our magnets must be to another magnet in order to pick it up. See Experiment EXP4.

Magnetic Force and Work

Because the basic magnetic force involves a cross product, the magnetic force $q\vec{v} \times \vec{B}$ on an individual moving charged particle acts perpendicular to the velocity of the particle (and its momentum). Therefore the magnetic force can change the direction of the particle's momentum but not its magnitude or its kinetic energy. Another way of seeing this is to note that the quantity $\vec{F} \cdot d\vec{r} = 0$, so no work is done on the particle.

However, applying a magnetic field can cause a compass needle to begin to rotate; the needle gains rotational kinetic energy. The directions of the magnetic torque and the needle's rotation are the same, so the work done is nonzero. When a bar magnet picks up another bar magnet, the second magnet accelerates upward, gaining kinetic energy. Again, the magnetic force and the displacement are in the same direction, so work is nonzero. Note that these are not cases of magnetic forces exerted on point particles.

Additionally, if a charged particle is bound to other particles by electric forces, the magnetic force may indirectly result in changes in kinetic energy. Recall that magnetic forces acting on a current of mobile electrons in a wire can indirectly move the stationary positive cores, because the electrons are bound to the positive cores by electric forces, and the battery that drives the current can supply energy.

The Stern–Gerlach Experiment

An important milestone in our understanding of atoms was the Stern–Gerlach experiment (1922). Silver was vaporized in an oven, and a beam of neutral silver atoms was defined by a series of slits (Figure 20.52). On theoretical grounds, each silver atom was expected to have angular momentum, with a magnetic dipole moment proportional to the angular momentum. The beam of neutral atoms passed through a strongly nonuniform magnetic field, and then struck a cold glass plate onto which the atoms condensed, leaving a visible silvery trace on the glass. (The entire apparatus was in a vacuum to avoid collisions with air molecules.)

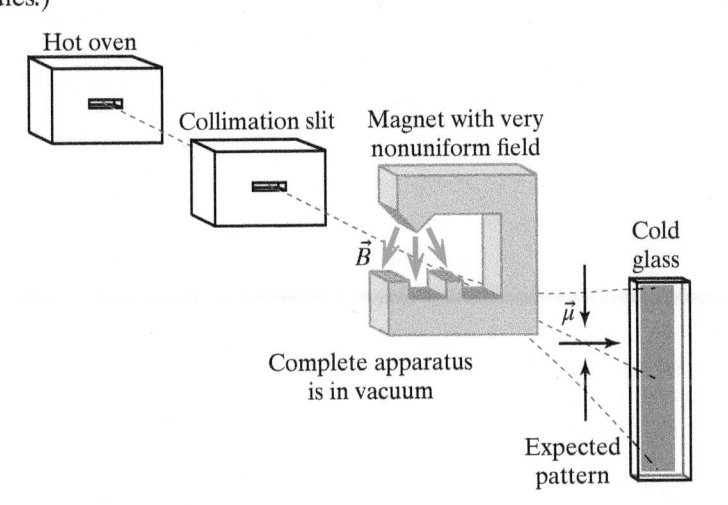

Figure 20.52 Classically, one expects a continuous pattern of silver deposited on the glass.

If the atoms have magnetic dipole moments, they should experience a vertical force, up or down in varying degrees depending on the orientation of their magnetic dipole moments. One would expect a random orientation of the magnetic dipole moments, so that there should be a vertical band of silver deposited on the glass. For example, if the magnetic dipole moment happens to be horizontal, the atom goes straight through without any deflection. (The motion is actually more complicated because the magnetic dipole moment oscillates back and forth in the magnetic field, like the motion around north of an undamped compass needle, unless it happens to point in the direction of the magnetic field. This does not affect the general conclusions, however.)

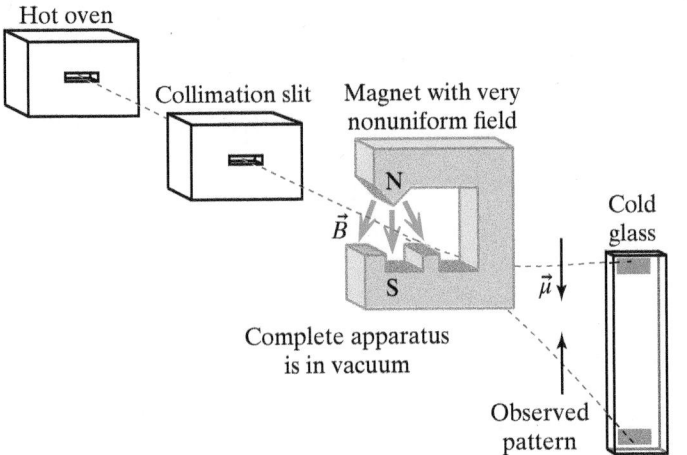

Figure 20.53 What one actually observes is a quantized pattern of silver deposited on the glass.

Instead of being spread out uniformly as one would expect, what Stern and Gerlach actually observed was that the atoms always hit either at the top or at the bottom of the expected region (Figure 20.53). This was interpreted as indicating that the orientation of the angular momentum of the atom is "quantized"—that whenever the atom is observed in this way we find that a component of angular momentum can have only certain quantized values. This distinctive quantization effect is typical of the behavior of atoms and subatomic particles.

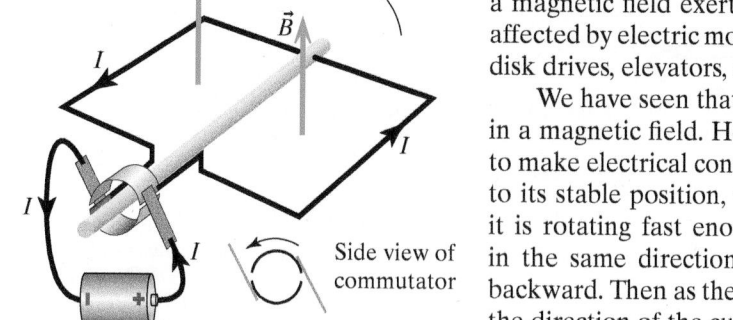

Figure 20.54 A split-ring commutator rotates with the coil, and as the coil reaches its most stable position the current in the coil reverses direction.

20.9 MOTORS AND GENERATORS

Electric motors are the most important everyday application of the torque that a magnetic field exerts on a current-carrying coil. Our daily lives are greatly affected by electric motors in refrigerator compressors, water pumps, computer disk drives, elevators, automobile starter motors, and so on.

We have seen that a current-carrying coil twists to its most stable position in a magnetic field. How can we get the coil to turn continuously? The key is to make electrical connections to the coil in such a way that just as it is coming to its stable position, we reverse the direction of the current. Assuming that it is rotating fast enough at that moment, the coil will continue on around in the same direction to the new stable position rather than simply going backward. Then as the coil approaches the new stable position we again switch the direction of the current.

A simple way to achieve continuous rotational motion of the coil is with a "split-ring commutator" that automatically changes the direction of the current through the coil at just the right moment (Figure 20.54). Springy metal tabs or brushes make contact between the battery and the commutator. Whereas actual motors have a wide variety of designs, this simple single-loop motor illustrates the basic principle involved in motors driven by direct current (DC, as opposed to alternating current, AC).

Electric Generators

It would be inconvenient to build a generator in the form of a bar riding on rails, such as we considered earlier in this chapter. It is mechanically awkward to move a bar or loop back and forth in a linear motion, and it is difficult to avoid having a lot of friction between a bar and supporting rails. Commercial generators rotate a loop of wire in a magnetic field rather than moving a bar back and forth. This has the same effect of achieving a charge separation along part of the rotating loop, but rotary motion is much easier to arrange mechanically, and with relatively low friction on the axle.

Rotating Loop in an External Magnetic Field

Suppose that we turn a crank to rotate a loop of dimensions $w \times h$ at a constant angular speed $\omega = d\theta/dt$ (in radians per second) inside a uniform magnetic field made by some large magnet or coils (Figure 20.55). The tangential speed of the left or right wire is $v = \omega(h/2)$, since the wire is a distance $h/2$ from the axle.

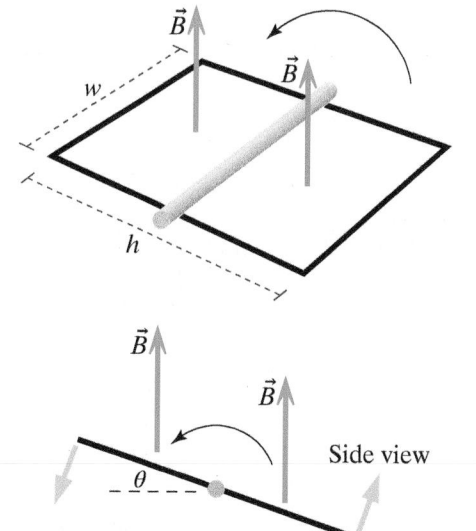

Figure 20.55 Rotating a loop in a magnetic field produces a current.

QUESTION Look at the magnetic forces on conventional positive charges inside the wires. At the moment shown in Figure 20.55, in which direction will conventional current run in this loop—clockwise or counterclockwise?

The magnetic forces drive current out of the page on the right wire and into the page on the left wire. Therefore there will be a clockwise conventional current, as seen from above (Figure 20.56). (As far as the front and back wires are concerned, the magnetic forces merely polarize them, perpendicular to the wires.)

QUESTION At a moment when the loop is at an angle θ to the horizontal, what is the magnitude of the magnetic force that acts on a charge carrier q in the right wire?

The magnetic force on a charge carrier q in the wire is $\vec{F} = q\vec{v} \times \vec{B}$, where $\vec{v}$ is the velocity of the wire. The magnitude of this force is $F = qvB\sin\theta$.

QUESTION   At a moment when the loop is at an angle θ to the horizontal, what is the motional emf created along the right wire in Figure 20.55?

The magnetic force is uniform throughout the right wire at this instant, so the motional emf, non-Coulomb work per unit charge, is given by the magnetic force times the length w, divided by q:

$$\text{emf along left wire} = \frac{(qvB\sin\theta)w}{q} = vBw\sin\theta$$

The left wire contributes the same emf, acting in the same direction around the loop, so the emf around the loop is this:

$$\text{emf around loop} = 2vBw\sin\theta$$

Since $v = \omega(h/2)$ and $\sin\theta = \sin(\omega t)$, the emf can be written as

$$\text{emf around loop} = \omega B(hw)\sin(\omega t)$$

where hw is the area of the loop.

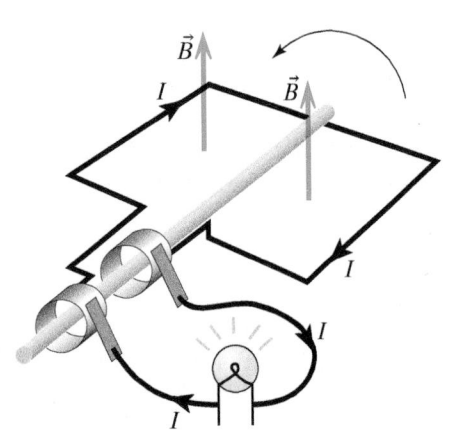

Figure 20.56 If you rotate the loop you can light the bulb, whose resistance is R.

We can take current out of the loop through metal tabs pressed against rotating rings connected to the loop, as shown in Figure 20.56.

QUESTION   If a light bulb with resistance R is connected as shown in Figure 20.56, what is the magnitude of the current I in the loop at this instant, assuming that the loop itself has negligible resistance?

The current will be $I = \omega B(hw)\sin(\omega t)/R$. We see that if we rotate a loop at constant speed inside a magnetic field, we generate an emf and can drive current through a light bulb. This is the basic principle of the electric generators used to power our cities, farms, and factories. The only significant difference from this simple generator is that a commercial generator has many turns in the loop to get a bigger emf. Having N turns is equivalent to having N single-loop generators in series, so we get N times the single-loop emf.

The emf produced in this rotating loop is proportional to $\sin(\omega t)$, so it will drive current through a bulb first one way, then the other way, with a sinusoidal variation. Such currents are called alternating currents or AC, and this is what is supplied at wall sockets. This is in contrast to the steady one-directional current maintained by a battery (or a bar sliding on rails through a uniform magnetic field), which is called direct current or DC.

If the frequency of the alternating current is high enough, a bulb filament doesn't have time to cool off very much between peaks in the cycle, so that the light seems steady to the eye (and there is persistence of vision in the eye, as well). This is the situation with incandescent bulbs in reading lamps, where the frequency f is 60 Hz in some countries and 50 Hz in others (with the angular frequency $\omega = 2\pi f$ in radians per second).

Checkpoint 13 For what value of the angle $\theta = \omega t$ is the instantaneous emf a maximum? A minimum? Rotate a 40-turn rectangular loop that is 0.1 m by 0.2 m in a uniform magnetic field of 2 T, turning at a rate of 30 revolutions (2π rad) per second. What is the maximum emf?

Power Required to Turn a Generator

Presumably we have to exert ourselves to rotate the loop in the magnetic field—we're not going to light the bulb for nothing! Let's see how much power

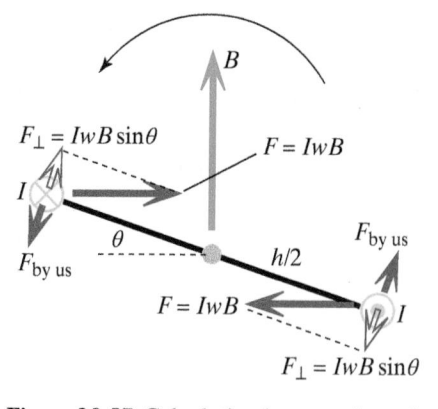

Figure 20.57 Calculating how much work is required to turn a current loop in a magnetic field.

we have to input to make the loop rotate. In the previous section we calculated the current I through an attached light bulb (as a function of the angle $\theta = \omega t$ of the loop to the horizontal). In order to turn the loop in Figure 20.57, we have to exert a varying force $F_{\text{by us}} = IwB\sin\theta$ at both ends of the loop, perpendicular to the loop, to balance the perpendicular component $F_\perp = IwB\sin\theta$ of the magnetic force on the loop.

When we rotate the loop through a small angle $\Delta\theta$, we move each end a short distance $(h/2)\Delta\theta$ and do a small amount of work

$$\Delta W = 2[IwB\sin\theta]\left[\frac{h}{2}\Delta\theta\right]$$

Divide this result by Δt and take the limit as Δt approaches zero to calculate the mechanical power dW/dt we supply:

$$\frac{dW}{dt} = 2[IwB\sin\theta]\left[\frac{h}{2}\frac{d\theta}{dt}\right] = I[Bwh\omega\sin\theta]$$

Given your calculation in the preceding section that $I = Bwh\omega\sin(\theta)/R$ when a bulb of resistance R is attached (where $\theta = \omega t$), we can write this:

$$\frac{dW}{dt} = I[RI] = RI^2$$

This is the power dissipated in the resistor at this instant. Thus we don't get something for nothing; the output power of the resistor that heats the room is equal to the input power that we supply. An electric "generator" might better be called a "converter," because it doesn't create electric power for free; it merely converts mechanical power into electric power.

20.10 *CASE STUDY: SPARKS IN AIR

We now have all the tools and concepts we need to discuss in detail a very complex electric phenomenon: the occurrence of sparks in air. You may occasionally see a spark when you flip a switch, unplug a power cord, or when you separate clothing that has stuck together in the dryer or pull a blanket away from a sheet in the winter. During a spark, charge is transferred from one object to another. How can such sparks occur in air?

Like other gases, air is an excellent insulator, consisting mostly of neutral nitrogen and oxygen molecules. (If air were a good conductor, a charged tape would very quickly become discharged.) There are no movable charged particles—all the electrons are bound to neutral gas molecules. During a spark, however, electrons are ripped out of molecules (ionization), leaving free electrons and positive N_2^+ and O_2^+ ions, all of which can move and transfer significant amounts of charge from one place to another through the air.

An ionized gas is called a plasma. Plasma physics is a domain of physics concerned with the behavior of plasmas, which are found in stars, fusion reactors, and sparks.

The goal of the following sections is to figure out a physically reasonable mechanism for the creation of sparks in air. You will be asked to estimate various physical quantities, in order to judge whether alternative models of sparks are physically reasonable. We will see that sometimes simple calculations make it possible to completely rule out particular models of complex phenomena!

To explain a spark, we must explain these aspects of the phenomenon:

How can electric charge move through air?
Why does a spark last only a short time?
Why is light given off?
And finally, how can the air become ionized?

A Long Chain of Reasoning

There is considerable interest in understanding sparks, since they are one of the most dramatic kinds of electric phenomena. In addition, the kind of explanation we will find for sparks provides a good example of an important kind of scientific reasoning, in which there is a long chain of reasoning. Each of the steps in the explanation is relatively simple; the complexity lies in linking many simple steps into one long, rigorous, logical chain.

How Can Electric Charge Move through Air? Two Models

We will explore two different possible models for how charge might travel through the air, which is normally not a conductor. We will first examine a seemingly plausible model for the situation but then show that this first model for conduction doesn't work. One of the powerful capabilities of science is the ability to rule out possible models or explanations, and we will illustrate this kind of reasoning here.

Suppose that we have two metal balls, one negatively charged and one positively charged. When the two balls are brought close together (but not touching), there is a spark between them. How and why does this happen?

Model 1: Electrons Jump between Balls

It is tempting to guess that during a spark electrons could simply jump from the negatively charged ball to the positively charged ball. Let's try to determine whether this is a reasonable model or not. One important issue is how far, on average, an electron could travel through air before colliding with a gas molecule and losing most of its energy. Since molecules in air are relatively far apart, compared to those in a solid or liquid, perhaps an electron could travel all the way from the negative ball to the positive ball without colliding with anything. We can determine this by estimating what is called the "mean free path" of a particle in air. You may recall a similar argument from Supplement S1 on gases; we review it here.

Mean Free Path

The mean free path is the distance that a particle can travel, on average, before colliding with another particle.

> QUESTION Qualitatively, what factors should the mean free path of a particle in a gas depend on?

You may have guessed that the mean free path d of a particle should be inversely proportional to the density of the gas surrounding it—the higher the gas density, the smaller the mean free path. You may also have guessed that d should be inversely proportional to the size of the gas molecules, or more precisely, since we are interested in collisions, the cross-sectional area of a gas molecule—the larger the molecule, the smaller the mean free path. Both of these factors will indeed prove to be important. If you guessed that the speed of the particle was a factor, think again; because we are not interested in the time between collisions, but only in the distance a particle can travel between collisions, speed does not matter.

It is not difficult to estimate the mean free path d of an electron in air, based on a simple geometrical insight. Draw a cylinder along the direction of motion of the electron, with length d and cross-sectional area A equal to the cross-sectional area of an air molecule (Figure 20.58). The geometrical significance of this cylinder is that if the path of the electron comes within one molecular radius of a molecule, the electron will hit the molecule, so any molecule whose center is inside the cylinder will be hit.

By definition, d is the average distance the electron will travel before colliding with a molecule, so the cylinder drawn in Figure 20.58 should on average contain about one molecule.

> QUESTION What is the volume of air that contains on average one molecule?

At standard temperature and pressure (one atmosphere at 0°C), one mole of gas occupies 22.4 L. One liter is one cubic decimeter—that is, one liter would fill a cube whose sides were 0.1 m long, and whose volume is $(0.1 \text{ m})^3$.

$$\left(\frac{22.4 \text{ L}}{1 \text{ mol}}\right)\left(\frac{1 \text{ mol}}{6.02 \times 10^{23} \text{ molecules}}\right)\left(\frac{(0.1 \text{ m})^3}{\text{L}}\right) = 3.72 \times 10^{-26} \text{ m}^3$$

> QUESTION To guarantee that an electron traveling through a cylinder of volume 3.72×10^{-26} m^3 will collide with a molecule, what should be the radius of the cylinder?

The radius of the cylinder should be the radius of a molecule. We can use the approximate value of 1.5×10^{-10} m for this radius.

> QUESTION Given the radius and the volume of our hypothetical cylinder containing one molecule, what is the length d of the cylinder?

$$\text{volume} = \pi r^2 d$$

$$d = \frac{\text{volume}}{\pi r^2} = \frac{3.72 \times 10^{-26} \text{ m}^3}{\pi (1.5 \times 10^{-10} \text{ m})^2} \approx 5 \times 10^{-7} \text{ m}$$

Thus we see that a free electron in air can travel approximately 5×10^{-7} m before colliding with an air molecule or ion.

> QUESTION During a spark in air, the positive ball becomes less positive and the negative ball becomes less negative. Do electrons jump through the air from one ball to the other?

Apparently not! Since a free electron can travel only about 5×10^{-7} m before colliding with a gas molecule and losing much of its energy, this is not a possible mechanism. On the basis of this estimate, we can rule out Model 1 for charge transfer through air.

Model 2: Positive Ions and Electrons Move in Ionized Air

If many of the oxygen and nitrogen molecules in air became ionized, we would have a gas composed of charged particles, all of which are free to move. In other words, the air would now contain both free electrons and positive N_2^+ and O_2^+ ions, and would therefore be a conductor. Leaving aside for the moment the question of how air could become ionized, let's consider the mechanism of charge transfer in ionized air. Just as in a metal wire, there is now a "sea" of mobile charged particles.

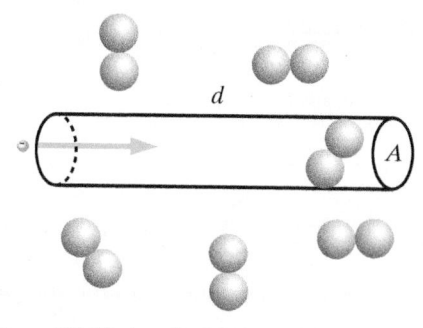

Figure 20.58 A cylindrical volume containing on average one air molecule.

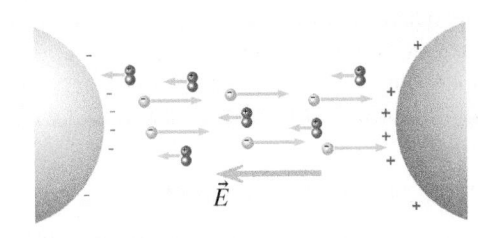

Figure 20.59 Free electrons and positively charged ions (O_2^+ or N_2^+) move through the ionized air between the balls.

QUESTION In which direction will the electrons in Figure 20.59 move? In which direction will the positive ions move?

All of the free electrons in the ionized air will move opposite to the electric field, toward the positively charged ball. The positive ion sea will be driven by the electric field toward the negatively charged ball.

QUESTION How would the negative ball become less negative? How would the positive ball become less positive?

As nearby positive ions come in contact with the negative ball, they can acquire electrons, making neutral molecules once again. The negative ball therefore becomes less negative. Free electrons near the positive ball can come in contact with the ball and join the free electron sea in the metal, making the ball less positive.

QUESTION Does any particle jump from one ball to the other?

No, since there are a very large number of charged particles near each ball, the seas of free electrons and positive ions need move only a very small distance to supply adequate charge to neutralize both balls.

Model 2 (ionized air) appears to be significantly better than Model 1 (jumping electrons), because in the ionized air model no particle travels farther than around one mean free path. We still need to answer the question of how the air might become ionized, though; we will return to this shortly.

> **Checkpoint 14** At high altitudes the air is less dense. If the density of air were half that at sea level, how would the mean free path of an electron in air change? If the radius of an air molecule were twice as large, how would the mean free path change?

Why Does a Spark Last Only a Short Time?

Assuming that the air can somehow become ionized (we haven't yet explained how), with plenty of mobile electrons and ions, what happens next? Electrons drift toward the positive ball, and the positive ions drift (much more slowly) toward the negative ball. Since the low-mass electrons move so much more quickly, the ionized air conducts in a manner that is rather similar to electric conduction in a metal. There is a "sea" of free electrons that drifts through the air. There is a kind of "start-stop" motion of an individual electron, in that it is accelerated until it collides with a molecule and comes nearly to a stop, then accelerates again. This accelerated "start-stop" motion has some average speed, called the "drift speed." Note the similarity to the Drude model of mobile electrons in a metal.

QUESTION What happens when an electron collides with the positive metal ball? Does the electric field in the air between the balls change?

The electron fills an electron deficiency in the positively charged metal ball, making the ball less positively charged. As a result, the magnitude of the electric field at all locations in the air between the spheres decreases slightly.

QUESTION What happens when an ion collides with the negative metal ball? What happens to the ion? Does the electric field in the air between the balls change?

The ion can pick up an excess electron from the negatively charged metal ball, and the ion becomes an ordinary neutral air molecule. The negatively charged ball is now less negatively charged. Again, as a result, the magnitude of the electric field at all locations in the air between the spheres decreases slightly.

We'll see later why it is that as long as the electric field is larger than about 3×10^6 V/m throughout a region of air, the air remains heavily ionized, but if the electric field becomes too small, the ions and electrons recombine with each other, and the gas is no longer a conductor.

QUESTION Why does the spark last only a very short time?

The excess charge on both metal balls is quickly reduced to an amount that is not sufficient to make a large enough electric field to keep the air heavily ionized, and the spark goes out. However, if you continually resupply the metal balls with charge by connecting the balls to a battery or power supply, you may be able to maintain a steady spark. A dramatic example is the very bright spark maintained between two carbon electrodes in a searchlight or commercial movie projector (a "carbon arc" light). A gentler example is the continuous glow seen in neon lights, which contain neon at low pressure.

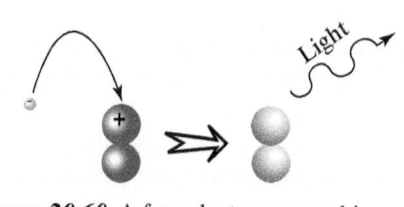

Figure 20.60 A free electron recombines with a positive ion to produce a neutral molecule. The excess energy is given off as visible light.

Why Is Light Given Off?

What causes the light that we see emitted by a spark (why are photons emitted)? Actually, the emission of light is a sort of side effect of the ionization of the air. Occasionally a free electron comes near enough to a positive ion (not necessarily the original ion) to be attracted and recombine with the ion to form a neutral molecule. When the free electron and the ion recombine (Figure 20.60), there is a transition from the high-energy unbound state (free electron and ion) to a low-energy bound state (Figure 20.61), with the emission of a photon.

It is important to understand that the emission of light goes on simultaneously with charge conduction through the ionized air. We will see later that as long as the applied electric field is big enough, neutral molecules are continually being ionized, at the same time that some free electrons and ions are recombining to make neutral molecules (with the emission of light). Although the light that we see is the most obvious aspect of a spark, the light can be considered to be essentially a mere side effect of conduction in a gas.

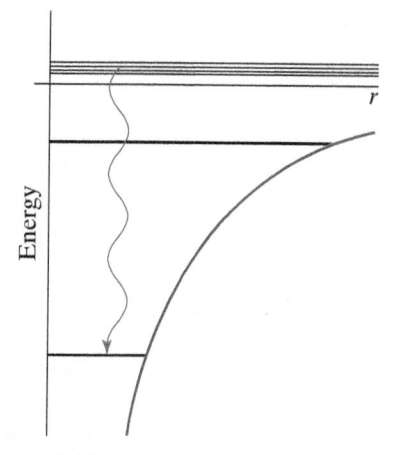

Figure 20.61 The transition of electron plus ion from an unbound (high-energy) state to a bound (lower-energy) state is accompanied by emission of a photon.

We have now seen the main aspects of sparks in ionized gas. Sparks are often referred to as "gas discharges," and a heavily ionized gas whose net charge is zero is called a "plasma." Plasma physics is an important area of contemporary research.

There remains the big question, how does the gas get ionized in the first place? That is the subject of the remaining sections.

How Does the Air Become Ionized? Two Models

It is observed experimentally that an electric field of about 3×10^6 V/m is sufficient to ionize air and make it a conductor. We will try to come up with a model of this process. To determine whether our model is a good one, we will compare the predictions of the model to experimental observations—in particular, the observation that the critical field value is $E_{\text{crit}} = 3 \times 10^6$ V/m.

As we did with models for conduction in the air, we will explore two different possible models for how the air can become ionized. First we'll examine a seemingly plausible model but then show that this model for ionization doesn't work. Again, it is an important and powerful aspect of science that it is often possible to rule out proposed explanations even if you don't know what the correct explanation is. After discarding the first model for ionization we'll discuss another model that explains many aspects of sparks.

Model 2A: A Strong Electric Field Pulls an Electron out of a Molecule

Since an atom or molecule is polarized by an electric field, we might suppose that an electric field of magnitude 3×10^6 V/m is large enough to pull an electron completely out of an air molecule, creating a positive ion and a free electron. Let's check this model.

> QUESTION Knowing what you know about atoms, estimate about how big an electric field would be needed to rip an outer electron out of a neutral atom.

If you could not do this calculation, think again for a moment before reading farther. You really do know enough to make this calculation. In particular, remember that the radius of an atom is about 1×10^{-10} m. For simplicity you may want to think about a hydrogen atom.

You know that a spherical charged object acts like a point charge (outside of the sphere). Think of an outer electron as being attracted by the rest of the atom, which has a charge $+e$ (Figure 20.62). The electric field made by this $+e$ charge that acts on the outer electron is

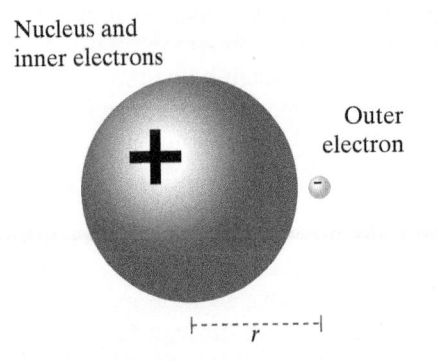

Figure 20.62 An outer electron interacts with the nucleus and inner electrons of the atom.

$$E = \frac{1}{4\pi\varepsilon_0}\frac{e}{r^2} = (9\times10^9\ \text{N}\cdot\text{m}^2/\text{C}^2)\frac{(1.6\times10^{-19}\,\text{C})}{(1\times10^{-10}\,\text{m})^2} = 1.4\times10^{11}\ \text{V/m}$$

where $r = 1 \times 10^{-10}$ m is the approximate radius of an atom. We would have to apply at least that big a field in order to pull an outer electron out of an atom.

The enormous difference between the experimentally observed field of 3×10^6 V/m and the value predicted by our model for the electric field necessary to ionize an atom means that we can rule out the possibility of direct atomic ionization by the applied electric field. This is an excellent example of one of the strengths of science: it is often possible to rule out a proposed model for a process, even if we can't figure out a better explanation.

Having ruled out direct atomic ionization by the applied electric field, what other process can we think of?

Model 2B: Fast-Moving Charged Particles Knock Electrons out of Atoms

It takes a dramatic event to ionize air, because it requires a very large force to rip electrons out of air molecules. A fast-moving charged particle that collides with an atom or molecule can knock out an electron, leaving a singly ionized ion behind. Where would such energetic charged particles come from? As it happens, there are fast-moving charged particles passing through your body at this very moment, ionizing some atoms and molecules in your body! Some of these charged particles are "muons" produced by cosmic rays in nuclear reactions at the top of the atmosphere. Others are electrons, positrons, or alpha particles (helium nuclei) emitted by radioactive isotopes present in trace quantities in materials in and around you.

This process is the cause of some potentially dangerous events. A neutral atom in your DNA that suddenly turns into an ion can cause biochemical havoc and genetic mutations. The passage of fast-moving charged particles through the tiny and very sensitive components in a computer chip can change a memory bit with potentially disastrous consequences. To guard against such disasters, both DNA and computer circuits have checks built into them to try to compensate for damage caused by the unavoidable passage of high-speed charged particles.

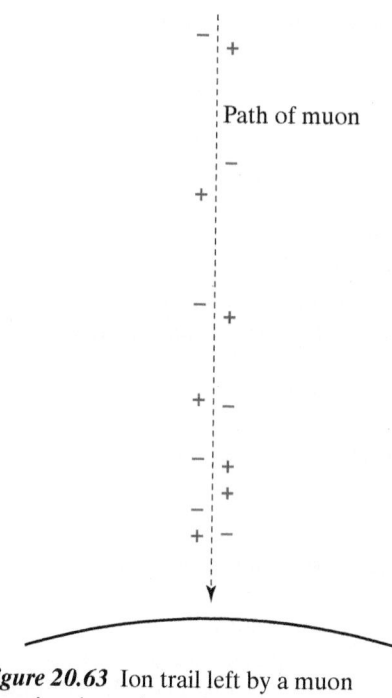

Path of muon

Figure 20.63 Ion trail left by a muon plunging down through the Earth's atmosphere.

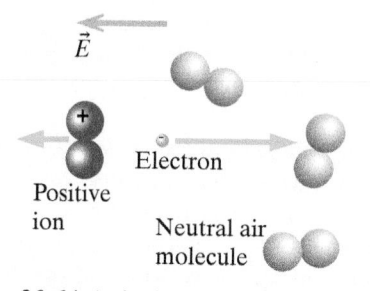

$\vec{E}$

Positive ion

Electron

Neutral air molecule

Figure 20.64 A single free electron is accelerated by the electric field, initiating a chain reaction.

Invaders from Outer Space

"Cosmic rays," consisting mostly of very high energy protons coming from outside our Solar System, strike the nuclei of molecules in the Earth's upper atmosphere. The resulting nuclear reactions produce a spray of other particles. Many of the products of these reactions are particles such as protons, neutrons, and pions that interact very strongly with other nuclei in air molecules. For that reason these particles typically don't go very far through the upper atmosphere. It is mainly positive and negative muons (μ^+ and μ^-) and neutral neutrinos that are able to penetrate the atmosphere and reach the surface of the Earth.

Electrons, muons, and neutrinos do not interact with nuclei through the strong interaction. Electrons and muons, being charged, do interact electrically with atoms and nuclei. Electrons have so little mass that they undergo large deflections and don't travel very far through the air. Muons are essentially just massive electrons, behaving in almost every way just like electrons except for having a mass about 200 times the electron mass. Because of their large mass, the muons undergo small deflections and can travel long distances through the air.

The neutral neutrinos, lacking electric charge, don't have any electric interactions with matter, and like electrons and muons they have extremely weak nonelectric interactions with matter. The interactions of neutrinos with matter are so unimaginably weak that most of them go right through the entire Earth, coming out the other side with no change!

As the rapidly moving charged muons plunge downward through the air, they have enough energy to occasionally knock electrons out of air molecules, producing free electrons and positive ions (Figure 20.63). Each muon leaves a trail of slightly ionized air. Muons produced by cosmic rays are not the only cause of ionization in the air. Materials around you contain trace quantities of radioactive nuclei that can emit high-speed electrons, positrons, or alpha particles (helium nuclei, He^{2+}). These rapidly moving charged particles can also ionize the air.

If the ionization were extensive, air would be a good conductor at all times, but the number of free electrons and ions produced by cosmic-ray muons and natural radioactivity is much too small to make the air be a good conductor. However, in the next section we will see that the small amount of ionization produced in this way can act as a trigger for large-scale ionization.

It is surprising that the muons actually manage to reach the Earth's surface, because a positive or negative muon at rest has an average lifetime of about 2 μs before splitting up (decaying) into a positron (if μ^+) or electron (if μ^-), a neutrino, and an antineutrino. Even if moving at nearly the speed of light, a muon would be expected to travel only about $(3 \times 10^8 \text{ m/s})(2 \times 10^{-6} \text{ s}) = 600$ m before decaying, which is far too short a distance to reach the ground from the top of the atmosphere where the muon was produced.

However, as we saw in Section 20.6, Einstein's special theory of relativity predicts that time elapses more slowly for a fast-moving object, so that in the reference frame of the muon it takes less than 2 μs to reach the ground, and many muons plunge through your room every minute.

A Chain Reaction

If there happens to be a single free electron in the air (due to muons or natural radioactivity), it can be accelerated by an applied electric field due to nearby charged objects such as the charged metal balls shown earlier (Figure 20.64).

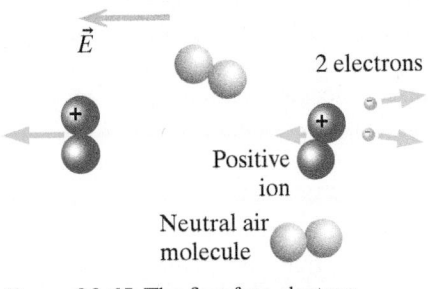

$\vec{E}$

2 electrons

Positive
ion

Neutral air
molecule

Figure 20.65 The first free electron ionizes an air molecule, producing a second free electron.

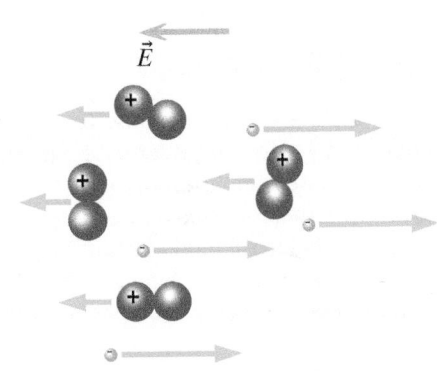

$\vec{E}$

Figure 20.66 As the chain reaction progresses, more and more charged particles are freed up, and the air becomes a conductor.

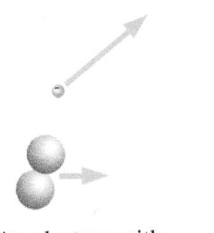

Figure 20.67 An electron with insufficient energy does not remove an electron from the air molecule.

If the electron gets going fast enough before colliding with an air molecule, it can knock another electron out of that molecule, so now there are two free electrons and two ions (Figure 20.65).

These two free electrons can knock out two more electrons, so the number of free electrons grows rapidly: 2, 4, 8, 16, 32, 64, and so on. This is an avalanche, or "chain reaction." The air becomes significantly ionized with a large number of mobile charged particles and is now a rather good conductor (Figure 20.66). Note that the air is still neutral overall, just as the interior of a metal wire is neutral overall.

Of course, the positive ions created when molecules are ionized are also affected by the applied electric field, and also begin to move under the influence of the field. However, because their mass is much larger than the mass of an electron, these ions move much more slowly. Therefore it makes sense to focus on free electrons as the particles whose collisions ionize a molecule when they collide with it.

Electric Field Required to Trigger a Chain Reaction

The few free electrons due to cosmic rays and radioactivity frequently bump into neutral air molecules, but these collisions won't start a chain reaction unless the electrons are traveling fast enough to rip electrons out of the molecules. Let's see whether we can predict from fundamental principles and the structure of matter the magnitude of the critical electric field E_{crit} that would be required to give a free electron a high enough kinetic energy to trigger a chain reaction.

> QUESTION What energy input is required to remove an outer electron from an oxygen or nitrogen molecule?

When a molecule is ionized, an outer electron is moved from its average location a distance R from a nucleus to a location far enough away that the field due to the positive ion is very small. The change in potential energy of the electron plus positive ion system is therefore:

$$\Delta U = (-e)\Delta V = (-e)(V_\infty - V_R)$$

As we saw in the previous discussion of Model 2A, near the ion the electric field due to the ion is very large compared to any applied field, so we only need to consider the potential difference due to the positive ion, which has a charge $+e$. Estimating the radius R of the ion to be about 1×10^{-10} m, we find that

$$(V_\infty - V_R) = 0 - \frac{1}{4\pi\varepsilon_0}\frac{e}{r}$$

$$= -\frac{1}{4\pi\varepsilon_0}\frac{e}{(1 \times 10^{-10}\text{ m})}$$

$$= -14.4\text{ V}$$

so $\Delta U = -1.6 \times 10^{-19}\text{C} \cdot (-14.4\text{V}) = 2.3 \times 10^{-18}\text{ J}$

This increase in the potential energy of the system must come from the kinetic energy of the fast-moving electron that collides with the molecule. (For simplicity we're ignoring the kinetic energy of the outer electron bound in the molecule, though it happens that 2.3×10^{-18} J is in fact close to the ionization energy for molecular nitrogen and molecular oxygen.) If the incoming electron's kinetic energy K is less than 2.3×10^{-18} J, it just bounces off or excites the air molecule to a higher quantum state without removing an electron (Figure 20.67). However, if K is equal to or greater than the ionization energy of an air molecule, after the collision there can be two electrons where there had been one (Figure 20.68).

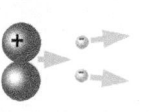

Figure 20.68 An electron with sufficient energy knocks another electron out of an air molecule during a collision.

Both of these electrons in Figure 20.68 can now be accelerated on average through a distance d, and both may ionize molecules, leading to the existence of 4 free electrons, then 8, 16, 32, 64, and so on—a chain reaction. (After a typical ionizing collision both electrons are moving rather slowly, so each time they start nearly from rest, and the collision energy is always about the same.)

QUESTION   Could a free electron that starts from rest gain adequate kinetic energy in one mean free path to ionize a molecule?

Considering the system of free electron plus the sources of the applied field:

$$\Delta K + \Delta U = 0$$
$$(K_f - 0) + (-e)\Delta V = 0$$
$$K_f = -(-e)\Delta V$$
$$\Delta V = \frac{K_f}{e} = \frac{2.3 \times 10^{-18}\,\text{J}}{e} = +14.4\,\text{V}$$

On average a free electron will travel only a distance of one mean free path before colliding with a molecule (Figure 20.69). To gain adequate kinetic energy, the free electron must travel through a potential difference of $+14.4\,\text{V}$ while traveling this distance. Unlike the electron that is bound to the molecule, the free electron is affected only by the applied electric field. Assuming that the applied field is uniform over this distance, and assuming that the electron travels a distance of one mean free path, which we estimated earlier to be 5×10^{-7} m:

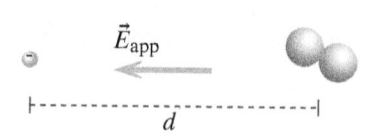

Figure 20.69 Mean free path d.

$$\Delta V = E_{app} \cdot d$$
$$+14.4\,\text{V} \approx E_{app} \cdot 5 \times 10^{-7}\,\text{m}$$
$$E_{app} \approx 2.9 \times 10^7 \frac{\text{V}}{\text{m}}$$

Model 2 therefore predicts that the applied electric field required to initiate a spark must be approximately 3×10^7 V/m.

QUESTION   Should we consider this good or poor agreement with the experimental value of 3×10^6 V/m?

For the previous model (direct ionization by the applied field), the prediction of the model was off by 5 orders of magnitude (1×10^5) from the experimental value. It is satisfying that with the chain reaction model we predict the right order of magnitude (we're off by a factor of 10 instead of being off by a factor of 1×10^5). This tells us that we now have a pretty good intuitive feeling for the issues.

To make an accurate prediction would involve difficult calculations because of the statistical nature of the process. In our simple calculations we assumed that every electron goes one mean free path and picks up the same amount of kinetic energy. However, occasionally an electron happens to go much farther than d, and a smaller electric field is sufficient to accelerate such an electron to a high enough energy to ionize the molecule that it finally hits. It also isn't necessary that every electron ionize an atom in order to create a chain reaction. That is why our simple calculation has overestimated the magnitude of electric field required to cause a spark.

Additional Tests of the Models

It is possible to make additional tests of the two models by comparing their predictions with other experimental observations, such as the effect on the critical field of changing the density of the gas. If a physical model explains a variety of different kinds of experimental results, this increases our confidence in the validity of the model.

> QUESTION If the density of air were half as large as it is at sea level, according to each model how would this affect the electric field required to ionize air, which is normally about 3×10^6 V/m?

Halving the density of the air would double the mean free path d. According to model 1 (electric field large enough to detach an electron) this would make no difference. According to model 2 (accelerated free electron collides with molecule), since the kinetic energy of the free electron (starting from rest) is $\Delta K = K_{\text{final}} - 0 = eEd$, doubling d would decrease E_{critical} by a factor of two. What is observed experimentally is that the required electric field is indeed proportional to the air density, supporting model 2.

An ordinary long fluorescent light tube or a neon sign exemplifies this. If the gas density inside the tube were not very low, it would be difficult to create a sufficiently high electric field throughout the tube to sustain a discharge (a continuous spark). Because the gas density in the tube is low, the electric field required is also low, making such lights practical. On the other end of the scale, gases at high pressures (hence high densities) are sometimes used as insulators.

> **Checkpoint 15** If you observed that under certain conditions an electric field three times as great as usual were required to ionize air, what might you conclude about the mean free path under these conditions? Approximately how large an electric field would be required to cause a spark in a gas at standard temperature and pressure if the ionization energy for the molecules of this gas is 1.5 times the ionization energy for air molecules, and the cross-sectional area of the molecules is 0.8 times the cross-sectional area of air molecules?

Drift Speed of Free Electrons in a Spark

We can now estimate the average "drift speed" of free electrons in a spark. Knowing that the kinetic energy acquired in one mean free path is 2.4×10^{-18} J, we can determine the typical final speed v_f of an electron when it hits a molecule:

$$\frac{1}{2}mv_f^2 \approx 2.3 \times 10^{-18} \text{ J}$$

$$v_f \approx \left(\frac{2(2.3 \times 10^{-18} \text{ J})}{9 \times 10^{-31} \text{ kg}}\right)^{1/2} = 2 \times 10^6 \text{ m/s}$$

$$\bar{v} \approx \frac{1}{2}v_f = 1 \times 10^6 \text{ m/s}$$

The statistical average speed is difficult to compute precisely, since there is a distribution of free paths. Evidently, however, the average drift speed is of the order of 1×10^6 m/s.

Propagation of Ionization

The magnitude of the electric field is not uniform in the region between two oppositely charged metal balls, and it is largest near the balls. (Do you see why?) Suppose that the field near the surface of the negative ball is bigger than 3×10^6 V/m, so that the air near the surface of this ball becomes strongly ionized. How can the ionization extend into other regions where E is smaller?

Positive ions are attracted toward the negative ball, and electrons are repelled into regions of the air where the electric field was less than 3×10^6 V/m. These electrons increase the electric field in the un-ionized region, and their large number facilitates a chain reaction, so the spark penetrates into the air farther and farther from the ball. It is possible to form an unbroken column of heavily ionized air leading from one ball to the other, even though initially the field was stronger than 3×10^6 V/m only near the negatively charged ball. Once the air is heavily ionized, an electric field significantly less than 3×10^6 V/m is sufficient to keep the spark going for a while. Recombination of electrons and ions along this column makes the column glow, and you see a spark connecting the two balls.

Something very similar happens in a lightning storm. A column of ionized air propagates from the negatively charged bottom of a cloud downward toward the ground. The speed of propagation from cloud to ground is somewhat slower than the average drift speed you calculated (about 1×10^6 m/s), because the propagation occurs in steps, with pauses between steps due to the statistical nature of triggering a chain reaction in the neighboring air. Once the column of ionized air reaches all the way to the ground, a large current runs through this ionized column. The visible lightning flash is of course due to the recombination of ions and electrons in this column. The large current heats the column explosively, and the rapidly expanding gas makes the noise heard as thunder.

20.11 *RELATIVISTIC FIELD TRANSFORMATIONS

Alice and Bob observed different electric and magnetic fields in their different reference frames, though they both could correctly predict the motion of the protons in their own reference frames, in terms of the fields they observe and the electric and magnetic forces.

The theory of relativity predicts how measurements of electric and magnetic fields transform when you change from one reference frame to another. We state these transform equations without proof. In the following, the "unprimed" quantities (E_x, etc.) are the fields that are observed in the lab frame (fixed to our laboratory). The "primed" quantities (E_x', etc.) are the fields that are observed in a reference frame that is moving with speed v in the $+x$ direction relative to the lab frame (Figure 20.70).

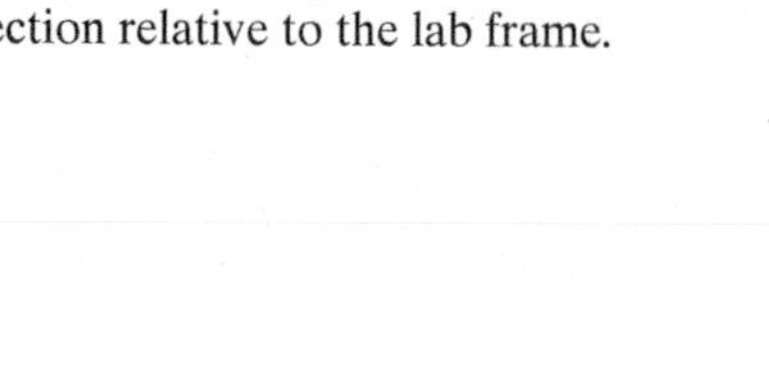

Figure 20.70 The moving reference frame moves with speed v in the $+x$ direction relative to the lab frame.

$$E_x' = E_x \qquad E_y' = \frac{(E_y - vB_z)}{\sqrt{1 - v^2/c^2}} \qquad E_z' = \frac{(E_z + vB_y)}{\sqrt{1 - v^2/c^2}}$$

$$B_x' = B_x \qquad B_y' = \frac{\left(B_y + \dfrac{v}{c^2}E_z\right)}{\sqrt{1 - v^2/c^2}} \qquad B_z' = \frac{\left(B_z - \dfrac{v}{c^2}E_y\right)}{\sqrt{1 - v^2/c^2}}$$

Magnetic Field of a Moving Charged Particle

Let's apply these transforms to a positively charged particle that is at rest in the lab frame. It of course makes only an electric field, yet if you run past the particle, you observe it in motion, and you observe not only an electric field but also a magnetic field.

In the lab frame the electric field of the stationary charge is given by:

$$\vec{E} = \frac{1}{4\pi\varepsilon_0} \frac{q}{r^2} \hat{r}$$

We would like to determine the fields made by this charge in the moving reference frame. In the moving reference frame the particle is seen to be moving with speed v in the $-x$ direction (Figure 20.71).

In the lab frame there is no magnetic field. However, straight above the moving particle, E_y is positive, so we have this in the moving reference frame:

$$B'_z = \frac{\left(B_z - \frac{v}{c^2}E_y\right)}{\sqrt{1 - v^2/c^2}} = \frac{-\frac{v}{c^2}E_y}{\sqrt{1 - v^2/c^2}}$$

At low speeds, this is approximately

$$B'_z = -\frac{v}{c^2}\left(\frac{1}{4\pi\varepsilon_0}\frac{q}{r^2}\right) \quad (v \ll c)$$

The magnetic field above the positively charged particle is in the $-z$ direction (Figure 20.71). This is consistent with the prediction of the Biot–Savart law for the direction of magnetic field above the charge. In an exercise at the end of this section you are asked to show that the transforms predict the usual curly magnetic field all the way around the moving charge.

Let's evaluate the magnitude of the magnetic field in the moving frame in terms of $\mu_0/4\pi$. In analyzing the electric and magnetic forces between two moving protons (the Alice and Bob episode), we found that

$$\frac{1}{\mu_0\varepsilon_0} = c^2 \quad \text{so} \quad \frac{1}{4\pi\varepsilon_0} = c^2\frac{\mu_0}{4\pi}$$

Therefore we can write

$$B'_z = -\frac{v}{c^2}\left(\frac{1}{4\pi\varepsilon_0}\frac{q}{r^2}\right) = -\frac{\mu_0}{4\pi}\frac{qv}{r^2}$$

This is exactly the magnitude of the magnetic field of the moving charge that is predicted by the Biot–Savart law, which is valid for $v \ll c$. This is partial verification that the field transforms are consistent with what you already know about electric and magnetic fields.

ELECTRIC AND MAGNETIC FIELDS ARE INTERRELATED

Electric fields and magnetic fields are closely related to each other, since a mere change of reference frame mixes the two together.

In many situations one can consider magnetic fields to be a relativistic consequence of electric fields. For example, if you run past a stationary charge you will observe a magnetic field, but if you stand still you will not observe a magnetic field. However, it is not always the case that magnetic fields are merely the result of changing the reference frame. For example, a stationary electron or proton or neutron acts like a magnetic dipole and produces a magnetic field, and there is no inertial reference frame in which that magnetic field is zero.

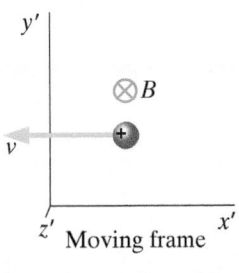

y'

$\otimes B$

v

z' Moving frame x'

Figure 20.71 In the moving reference frame the particle moves with speed v in the $-x$ direction. A magnetic field appears in this frame.

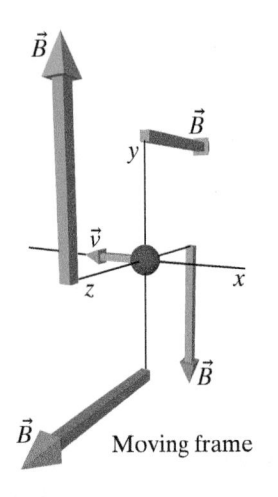

Figure 20.72 Use the field transforms to show that in the moving frame there is a curly magnetic field all the way around the moving positive charge.

Checkpoint 16 Use the field transforms to find the direction of the magnetic field in front of, behind, and below the positively charged particle in the moving reference frame, and therefore show that there is a curly magnetic field all the way around the moving charge, as expected (Figure 20.72).

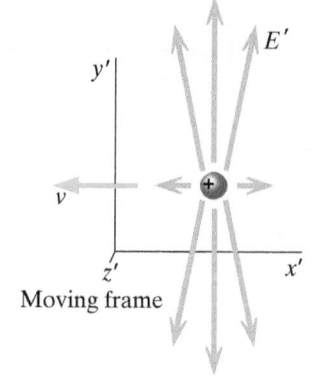

Figure 20.73 The electric field of the particle in the moving frame, where the particle is moving rapidly.

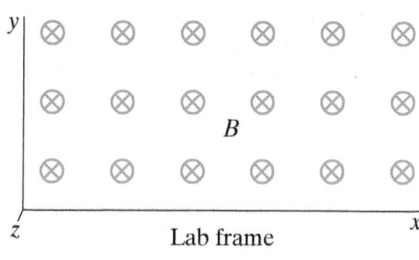

Figure 20.74 A region in the lab frame of uniform magnetic field in the $-z$ direction.

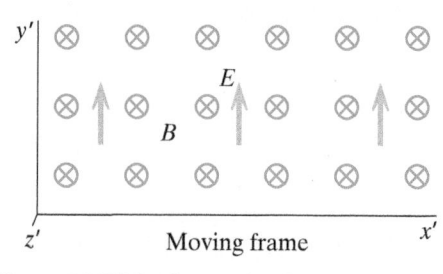

Figure 20.75 In the moving frame there is an electric field in the $+y$ direction of magnitude $E = vB$.

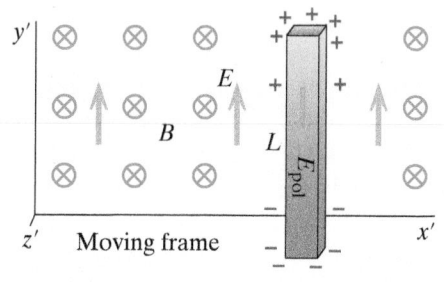

Figure 20.76 A metal bar at rest in the moving frame is polarized by the upward applied electric field. The polarization charges produce a downward electric field $E_{pol} = vB$.

Electric Field of a Rapidly Moving Charged Particle

If the speed of the moving reference frame is very large (very near c), the electric field of the particle is altered significantly in the moving frame. The magnitude of the electric field in front of and behind the particle in the moving frame is unchanged: $E'_x = E_x$. However, straight off to the side of the particle the electric field is much larger at high speeds, since we have

$$E'_y = \frac{(E_y + vB_z)}{\sqrt{1 - v^2/c^2}} = \frac{E_y}{\sqrt{1 - v^2/c^2}}$$

which is much greater than E_y if v approaches c. Figure 20.73 shows the general pattern of electric field in the moving frame, near the charge that is rapidly moving in this reference frame. The region of high field strength becomes a thinner and thinner pancake as the speed of the particle approaches the speed of light.

Moving through a Region of Uniform Magnetic Field

The previous examples involved a situation in which there was no magnetic field in the rest frame of the particle but there were both electric and magnetic fields in a frame where the particle was moving. Next we'll consider a region of uniform magnetic field in the lab frame, but no electric field, and see what fields are present in the moving reference frame. In the lab frame the magnetic field is $\langle 0, 0, -B \rangle$, so $B_z = -B$ (Figure 20.74).

> QUESTION Suppose that you move with speed v in the $+x$ direction through this region. Using the field transforms given above, what electric field would you observe?

The transforms show that in the moving reference frame, you will observe an upward electric field of magnitude $E = vB$ ($+y$ direction; see Figure 20.75), though there was no electric field in the lab frame:

$$E'_y = \frac{0 - vB_z}{\sqrt{1 - v^2/c^2}} = \frac{0 - v(-B)}{\sqrt{1 - v^2/c^2}} \approx vB \quad \text{if } v \ll c$$

Suppose that we are in this moving frame, and we observe a metal bar that is at rest in this frame. The metal bar polarizes because there is an upward electric field of magnitude $E = vB$ everywhere in this region (Figure 20.76). The bar polarizes until there is sufficient charge buildup to produce a downward electric field of magnitude $E_{pol} = vB$, at which point the net electric field inside the metal is zero, and there is equilibrium.

In the lab frame, the metal bar moves at constant speed v in the $+x$ direction through a region of uniform magnetic field and zero electric field (Figure 20.77). We discussed this situation earlier in the chapter in connection with motional emf. In the lab frame the moving metal bar polarizes due to the magnetic force qvB on the moving mobile charges. The polarization charges produce a downward electric field E_{pol} ($-y$ direction). At equilibrium it must be that $qE_{pol} = qvB$, so $E_{pol} = vB$.

The amount of charge buildup is the same in either reference frame, but the explanation that observers in the two frames give for the cause of the charge buildup is different. In the moving reference frame, where the bar is at rest, there is an upward electric field of magnitude $E = vB$ that polarizes the bar until this electric field becomes balanced by the downward electric field $E_{pol} = vB$ of the polarization charges. In the lab frame, where the bar is

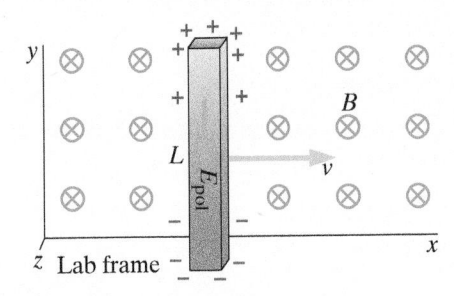

Figure 20.77 In the lab frame the moving metal bar is polarized by the magnetic force on the moving charges. The polarization charges produce a downward electric field $E_{\text{pol}} = vB$.

in motion, there is a magnetic force that polarizes the bar until the magnetic force becomes balanced by the electric force of the polarization charges; the downward electric field has magnitude $E_{\text{pol}} = vB$.

THE PRINCIPLE OF RELATIVITY

There may be different mechanisms for different observers in different reference frames, but all observers can correctly predict what will happen in their own frames, using the same relativistically correct physical laws.

Moving through a Velocity Selector

As another example, consider the velocity selector discussed earlier and shown in Figure 20.78. In the lab frame a charged particle traveling with speed v moves in the x direction through a region where there is an electric field of magnitude E in the $+y$ direction and a magnetic field of magnitude B in the $+z$ direction. If $E = vB$, the charged particle passes through with no deflection, because the electric force of magnitude qE is equal and opposite to the magnetic force of magnitude qvB.

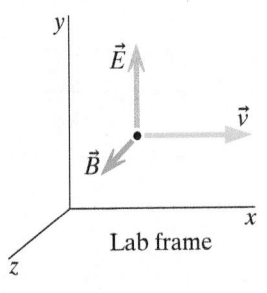

Figure 20.78 A velocity selector. If $E = vB$, the charged particle passes through with no deflection.

> QUESTION Suppose you move with speed v in the $+x$ direction through this region, without the charged particle present. Using the field transforms given above, what electric and magnetic fields would you observe?

The only fields you would observe in the moving frame are these:

$$E'_y = \frac{(E_y - vB_z)}{\sqrt{1 - v^2/c^2}} = \frac{E - vB}{\sqrt{1 - v^2/c^2}} = \frac{vB - vB}{\sqrt{1 - v^2/c^2}} = 0$$

$$B'_z = \frac{\left(B_z - \frac{v}{c^2}E_y\right)}{\sqrt{1 - v^2/c^2}} = \frac{B - \frac{v}{c^2}E}{\sqrt{1 - v^2/c^2}} = \frac{B - \frac{v}{c^2}(vB)}{\sqrt{1 - v^2/c^2}} = \frac{(1 - v^2/c^2)}{\sqrt{1 - v^2/c^2}}B$$

> QUESTION Given that you observe these fields, what will happen to a charged particle initially at rest in your reference frame?

A charge at rest is unaffected by the magnetic field, and there is no electric field, so there is no force on the charged particle. This makes sense, because a particle at rest in your reference frame moves with constant velocity in the lab frame, straight through the velocity selector.

Note again that observers in the two reference frames differ in their analyses of the situation, but both correctly predict what happens to the charged particle. You predict that a charged particle at rest will remain at rest, because there is no electric field, and the net electric and magnetic force is zero. An observer in the lab frame calculates an electric force and a magnetic force, determines that they are equal and opposite to each other, and predicts that the particle will have a uniform velocity since the net force on the particle is zero.

An Alternative Form of the Transforms

You may have noticed terms that look like a cross product in the transforms for the y and z components of the fields—the components perpendicular to the motion. We can write the transforms for the fields parallel to the motion and perpendicular to the motion in the following way:

$$E'_{\parallel} = E_{\parallel} \qquad\qquad B'_{\parallel} = B_{\parallel}$$

$$E'_{\perp} = \frac{(\vec{E} + \vec{v} \times \vec{B})_{\perp}}{\sqrt{1 - v^2/c^2}} \qquad B'_{\perp} = \frac{\left(\vec{B} - \dfrac{\vec{v} \times \vec{E}}{c^2}\right)_{\perp}}{\sqrt{1 - v^2/c^2}}$$

Checkpoint 17 Try using these versions of the transforms to reanalyze one or more of the phenomena treated earlier: fields of a moving charge, the metal bar moving through a magnetic field, or the velocity selector.

If you would like to know more about relativity, there are many fine books on the subject, or you might like to take a course on relativity. Here are two excellent introductory books:

R. Resnick, *Introduction to Special Relativity* (Wiley, 1968)
E. Taylor and J. Wheeler, *Spacetime Physics* (Freeman, 1992)

SUMMARY

Net Electric and Magnetic Force

$$\vec{F} = q\vec{E} + q\vec{v} \times \vec{B}$$
(called the "Lorentz force")

Force on a Short Length of Current-Carrying Wire

$$\Delta \vec{F}_{\text{magnetic}} = I\Delta\vec{l} \times \vec{B}$$

Circular Motion in a Uniform Magnetic Field

$$\omega = \frac{|q||\vec{B}|}{\gamma m} \quad \text{at any speed}$$

$$\omega \approx \frac{|q||\vec{B}|}{m} \text{ if } v \ll c$$

Momentum $|\vec{p}| = |q||\vec{B}|R$, valid even for relativistic speeds.

The *Hall effect:* A sideways electric field in a current-carrying wire in a magnetic field due to the magnetic force on moving particles. The Hall effect allows us to determine the sign and number density of the mobile charges in a particular material.

Motional emf is the integral of the magnetic force that acts on mobile charges in a wire moving through a magnetic field.

A current-carrying coil or magnetic dipole moment experiences a torque in a magnetic field, and twists to align with the applied magnetic field.

Magnetic torque on a magnetic dipole is

$$\vec{\tau} = \vec{\mu} \times \vec{B}$$

Potential energy associated with a magnetic dipole in an applied magnetic field is $U_m = -\vec{\mu} \bullet \vec{B}$, and the force on a magnetic dipole is

$$F_x = -\frac{dU_m}{dx} = \frac{d(\vec{\mu} \bullet \vec{B})}{dx}$$

Electric field required to ionize air:

$$E_{\text{critical}} = 3 \times 10^6 \,\text{V/m}$$

When the applied electric field is sufficiently large to accelerate free electrons to the energy required to ionize a nitrogen or oxygen molecule, a chain reaction can occur.

The pattern of electric and magnetic fields is different in stationary and moving reference frames. A phenomenon such as the polarization of a metal bar may have different explanations in different reference frames. The following transforms relate the fields $\vec{E}'$ and $\vec{B}'$ in a reference frame moving at speed v to the fields $\vec{E}$ and $\vec{B}$ in a stationary reference frame.

$$E'_x = E_x \qquad E'_y = \frac{(E_y - vB_z)}{\sqrt{1 - v^2/c^2}} \qquad E'_z = \frac{(E_z + vB_y)}{\sqrt{1 - v^2/c^2}}$$

$$B'_x = B_x \qquad B'_y = \frac{\left(B_y + \dfrac{v}{c^2}E_z\right)}{\sqrt{1 - v^2/c^2}} \qquad B'_z = \frac{\left(B_z - \dfrac{v}{c^2}E_y\right)}{\sqrt{1 - v^2/c^2}}$$

EXPERIMENTS

EXP1 Observing magnetic force
Observe for yourself the force that a magnetic field exerts on a current-carrying wire. Short-circuit one or more batteries and observe how a current-carrying wire deflects near your bar magnet (make and break the circuit to see the direction of the deflection). Use your 2-m-long wire and let a portion of it hang as freely as possible over the edge of the desk with a weight on the top part, to keep the lower part of the wire from moving when you connect and disconnect the circuit (Figure 20.79). When you connect the wire you can see the wire jerk sideways slightly. The force is small, so observe closely.

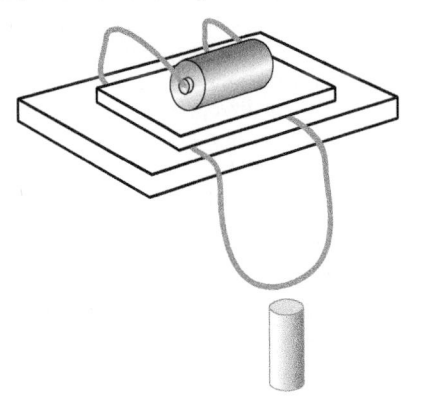

Figure 20.79

Try different directions of current and magnetic field (using the N and S markings on your magnet obtained previously, or by checking with your compass), and take turns with a partner in predicting the direction of the deflection. Make sketches of the various situations and use the equation to explain what you observe.

EXP2 Bar magnet and coil
Wrap your long wire into a coil, then hang the coil by its connecting wires and bring your bar magnet near the hanging coil.

(a) Under what conditions does the bar magnet repel the coil? Under what conditions does the bar magnet attract the coil?

(b) Sketch a situation in which the coil twists under the influence of the bar magnet. Explain briefly in terms of what forces act on what parts of the coil.

EXP3 Series and parallel batteries
(a) If you assume that all of your batteries have the same internal resistance, what would you predict would happen to the magnetic force if you had more batteries in series? In parallel? An ordinary flashlight battery has a resistance of about a quarter or a half an ohm; the resistance of your wire is much smaller than that.

(b) Using various battery combinations (including combining batteries with your partner), observe the deflection of a wire or a coil by your bar magnet. What do you observe, and how do your observations compare with your predictions? (*Hint on technique:* Compare the effects of different numbers of batteries in parallel by connecting some of them in and out of the circuit while observing the deflection of the wire.)

EXP4 Magnet picks up magnet
Prediction: Determine the magnetic dipole moments μ_1 and μ_2 of two magnets, and weigh the magnets. Predict the numerical value of the distance x at which the magnetic attraction between these two magnets just equals the weight Mg of one magnet.
Measurement: Place a magnet upright on the desk and try to lift it with the other magnet. Measure the distance x between their centers at the point where the lower magnet is lifted, and compare with your theoretical prediction. There is a ruler on the inside back cover of this textbook. In using the center-to-center distance we are averaging over distance, for a force that varies like $1/x^4$. This works out better than one might expect.

QUESTIONS

Q1 Suppose that a proton has a component of velocity parallel to the magnetic field as well as perpendicular to it (Figure 20.80). What is the effect of the magnetic field on this parallel component of the velocity? What will the trajectory of the proton look like?

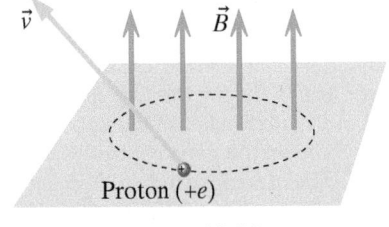

Figure 20.80

Q2 An electron is moving with a speed v in the plane of the page (Figure 20.81), and there is a uniform magnetic field B into the page throughout this region; the magnetic field is produced by some large coils that are not shown. Draw the trajectory of the electron, and explain qualitatively.

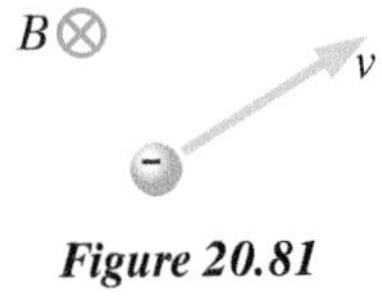

$B \otimes$ v

Figure 20.81

Q3 A proton and an electron are moving at a particular instant as shown in Figure 20.82.

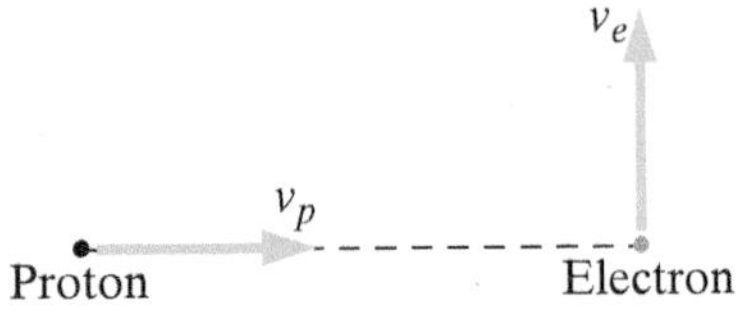

v_e

v_p

Proton   Electron

Figure 20.82

(a) On a similar diagram, show the directions of the electric field and magnetic field that the proton produces at the location of the electron. **(b)** Show the directions of the electric force and the magnetic force that the proton exerts on the electron. **(c)** Similarly, show the directions of the electric field and magnetic field that the electron produces at the location of the proton. **(d)** Show the directions of the electric force and the magnetic force that the electron exerts on the proton. **(e)** Consider carefully your results from parts (b) and (d). Are the magnetic forces on electron and proton in opposite directions? Does reciprocity hold for magnetic forces? **(f)** Will the total momentum of the two particles remain constant? Is this a violation of conservation of momentum for an isolated system?

There is momentum associated with the electric and magnetic fields of these particles, as we will see in a later chapter. Although the momentum of the proton plus the momentum of the electron does change, the sum of the particle momenta plus the field momentum does not change.

Q4 A copper wire with square cross section carries a conventional current I to the left (as in Figure 20.83). There is a magnetic field B perpendicular to the wire. Describe the direction of $E_\perp$, the transverse electric field inside the wire due to the Hall effect, and explain briefly.

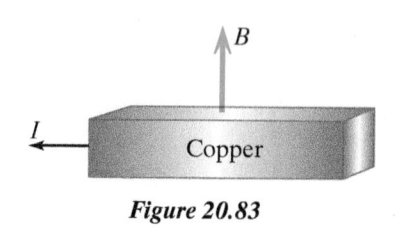

Figure 20.83

Q5 You hold your magnet perpendicular to a bar of copper that is connected to a battery (Figure 20.84). Describe the directions of the components of the electric field at the center of the copper bar, both parallel to the bar and perpendicular to the bar. Explain carefully.

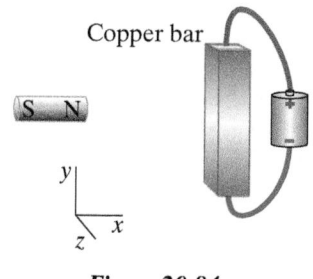

Figure 20.84

Q6 The Hall effect can be used to determine the sign of the mobile charges in a particular conducting material. A bar of a new kind of conducting material is connected to a battery as shown in Figure 20.85. In this diagram, the x axis runs to the right, the y axis runs up, and the z axis runs out of the page, toward you. A voltmeter is connected across the bar as shown, with the leads placed directly opposite each other along a vertical line. In order to answer the following questions, you should draw a careful diagram of the situation, including all relevant charges, electric fields, magnetic fields, and velocities.

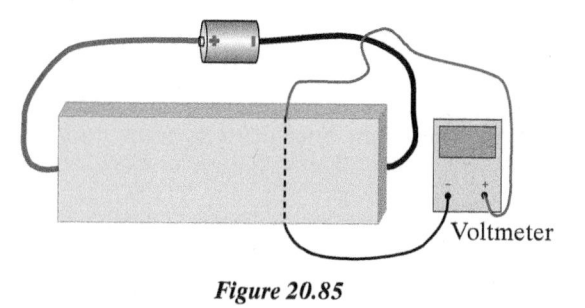

Figure 20.85

Initially, there is no magnetic field in the region of the bar. **(a)** Inside the bar, what is the direction of the electric field $\vec{E}$ due to the charges on the batteries and surface of the wires and the bar? This is the electric field that drives the current in the bar. **(b)** If the mobile charges in the bar are positive, in what direction do they move when current runs? **(c)** If the mobile charges in the bar are negative, in what direction do they move when current runs? **(d)** In this situation (zero magnetic field), what is the sign of the reading on the voltmeter?

Next, large coils (not shown) are moved near the bar, and current runs through the coils, making a magnetic field in the $-z$ direction (into the page). **(e)** If the mobile charges in the bar are negative, what is the direction of the magnetic force on the mobile charges? **(f)** If the mobile charges in the bar are negative, which of the following things will happen? (1) Positive charge will accumulate on the top of the bar. (2) The bar will not become polarized. (3) Negative charge will accumulate on the left end of the bar. (4) Negative charge will accumulate on the top of the bar. **(g)** If the mobile charges in the bar are positive, what is the direction of the magnetic force on the mobile charges? **(h)** If the mobile charges in the bar are positive, which of these things will happen? (1) Positive charge will accumulate on the top of the bar. (2) The bar will not become polarized. (3) Positive charge will accumulate on the right end of the bar. (4) Negative charge will accumulate on the top of the bar.

You look at the voltmeter and find that the reading on the meter is -5×10^{-4} volts. **(i)** What can you conclude from this observation? (Remember that a voltmeter gives a positive reading if the positive lead is attached to the higher potential location.) (1) There is not enough information to figure out the sign of the mobile charges. (2) The mobile charges are negative. (3) The mobile charges are positive.

Q7 A metal wire with mobile electrons and square cross section carries a conventional current I in the $-y$ direction. There is a magnetic field B in the $+x$ direction. Draw a diagram illustrating the situation, to help you answer the following questions. **(a)** What is the direction of E_{parallel}, the electric field that causes the current to flow? **(b)** What is the direction of the drift velocity of the mobile electrons? **(c)** What is the direction of the magnetic force on the moving electrons? **(d)** What is the direction of $E_{\text{transverse}}$, the electric field due to the Hall effect, inside the wire? **(e)** If the mobile charges had been positive (holes) instead of negative, what would have been the direction of the magnetic force on the moving positive charges? **(f)** If the mobile charges had been positive instead of negative, what would have been the direction of the transverse electric field?

Q8 You have a battery with known emf $=$ K, a sensitive voltmeter, and a slab of high-resistance metal in which the mobile charges are electrons with mobility u; there are n electrons per cubic meter. The metal slab has length L, height h, and width w. You also have plenty of very low-resistance wires you can use to connect the components in Figure 20.86 together. There is a uniform magnetic field pointing up throughout the region where the slab is located; this magnetic field is produced by a large magnet that is not shown. The goal is to figure out the magnitude B of this magnetic field, using the available equipment.

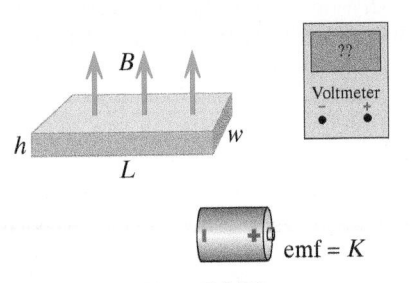

Figure 20.86

(a) Draw wires on the diagram to show how you would connect the equipment. **(b)** Explain in detail, including appropriate diagrams, how you will determine B, and give an algebraic expression for B involving the reading ΔV on the voltmeter and the known quantities (K, u, n, L, h, w).

Q9 In which direction will conventional current flow through the resistor in Figure 20.87? What will be the direction of the magnetic force on the moving bar?

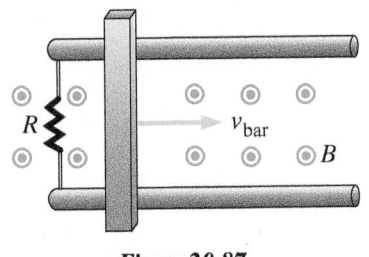

Figure 20.87

Q10 Two metal wires of length w are attached to two metal wires of length h to form a rectangle of dimensions w by h. The rectangle rotates counterclockwise on a frictionless axle (due to external forces) with constant angular velocity ω in a uniform upward magnetic field B (Figure 20.88).

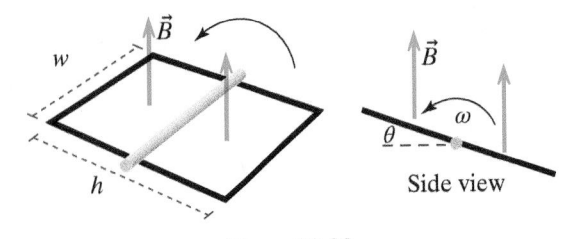

Figure 20.88

(a) At the moment when the rectangle is at an angle θ from the horizontal, what is the magnitude of the emf around the circuit and the direction of the current I? **(b)** Why are external forces required to rotate the loop, even though the rectangle rotates at a constant angular speed on a frictionless axle? Show the direction of the external forces on a diagram.

Q11 See Figure 20.89. If you travel at speed v along with the moving bar, in your reference frame the charged particles in the bar are not moving. Explain why, in your moving reference frame, the bar still polarizes.

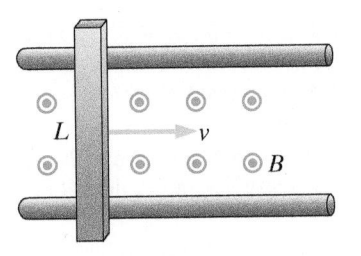

Figure 20.89

Q12 A current-carrying rectangular loop of wire is mounted on a pivot, held at an angle to a uniform magnetic field. If the loop is released, will it turn clockwise or counterclockwise? Explain briefly, and use Figure 20.90 in your explanation.

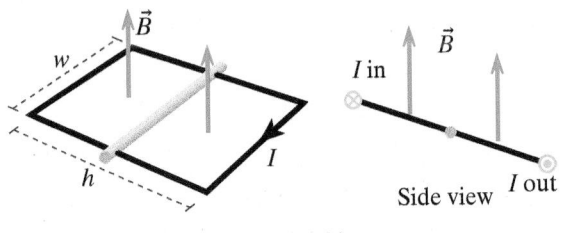

Figure 20.90

Q13 A bar magnet is mounted on a pivot and released in the presence of a magnetic field $\vec{B}$ made by coils that are not shown in Figure 20.91. Does the magnet initially turn clockwise or counterclockwise? Explain qualitatively in detail why it turns, and why it turns in the predicted direction.

Figure 20.91

Q14 Figure 20.92 shows a region of nonuniform magnetic field. The field is stronger to the right. What is the initial direction of the magnetic force on a magnetic dipole in each of the four orientations shown in the diagram? (*Hint:* A magnetic dipole if free to move goes to a region of lower potential energy. You can also check with what you know about the interaction of two magnets.)

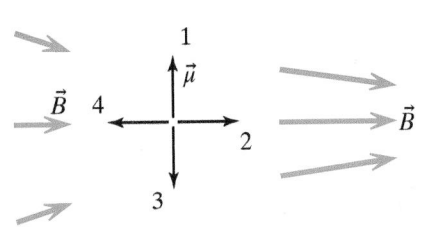

Figure 20.92

Q15 Put together what you know about ferromagnetic materials and about the forces on magnetic moments to explain qualitatively why a magnet can pick up a paper clip.

Q16 A student said, "During a spark, protons are pulled out of the positive ball and jump across to the negative ball." Give at least two reasons why this analysis cannot be correct.

Q17 "When you connect a metal wire between two oppositely charged metal blocks, electrons on the negatively charged block jump to the positively charged block." Explain briefly what is wrong with this statement.

Q18 What would happen if an electron with energy less than the ionization energy of a nitrogen molecule collided with a nitrogen molecule?

Q19 At high altitudes the air is less dense. If the density of air were half that at sea level, how would d, the mean free path of an electron in air, change? (1) d would be the same. (2) d would be twice as long. (3) d would be half as long. (4) d would be 1/4 as long. (5) d would be 4 times as long.

Q20 If the radius of an air molecule were twice as large, how would d, the mean free path of an electron in air, change? (1) d would be the same. (2) d would be twice as long. (3) d would be half as long. (4) d would be 1/4 as long. (5) d would be 4 times as long.

Q21 If you observed that under certain conditions an electric field six times as great as usual were required to ionize air, what might you conclude about the mean free path under these conditions? Give your answer as a ratio of the mean free path under these conditions to the mean free path at STP.

Q22 Metal sphere 1 is charged negatively, and metal sphere 2 is not charged in Figure 20.93.

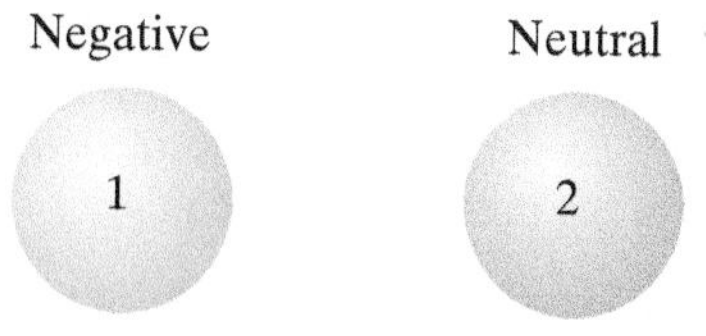

Negative   Neutral

(1)   (2)

Figure 20.93

(a) Using the conventions for diagrams described and used in this textbook, show the charge distribution for both spheres. **(b)** The spheres are moved closer to each other (using insulating supports so as not to change their charge). When they are a certain short distance apart, a spark is seen for a brief instant. Explain qualitatively what determines this distance for the spark to be produced. (Why doesn't the spark occur when the spheres are farther apart? What's special about this distance?) **(c)** After the spark stops, show the charge distribution for both spheres, and explain briefly but completely how this new charge distribution came about.

Q23 In very dry weather, if you shuffle across the carpet wearing rubber-soled shoes and then bring your finger near a metal object such as a doorknob, you will probably get a shock and see a spark. How this can occur is puzzling, since rubber is an insulator, so charge can't move through the soles of your shoes. Explain this process in detail, with appropriate diagrams. Make sure that you answer the following questions: **(a)** Carry out an experiment to determine the sign of the charge on the sole of your shoe after rubbing it on carpet, wool, or other cloth. Explain the test you did. (If you are unable to obtain results, choose a sign to use in the rest of the discussion.) **(b)** Draw a diagram that includes both the shoe and the rest of your body. Include all relevant charges and fields. **(c)** Suppose that a spark occurred when your finger was 1 cm from the doorknob. Draw two diagrams showing your hand, the doorknob, all relevant charges, and all contributions to the electric field at relevant locations: when your finger is 1.5 cm from the doorknob (no spark yet), and when your finger is 1 cm from the doorknob (spark starts). On the basis of these diagrams, explain why the spark starts only when your finger is close enough (1 cm) to the doorknob and not farther away. **(d)** At what location do you think the spark starts, and why (why not somewhere else)? **(e)** Why does the spark stop?

PROBLEMS

Section 20.1

•**P24** An electron is moving in the y direction at a speed $v = 2 \times 10^7$ m/s at a point where there is a magnetic field $B = 3.5$ T in the z direction (Figure 20.94). What are the magnitude and direction of the magnetic force on the moving electron? Draw the force vector on a diagram of the situation.

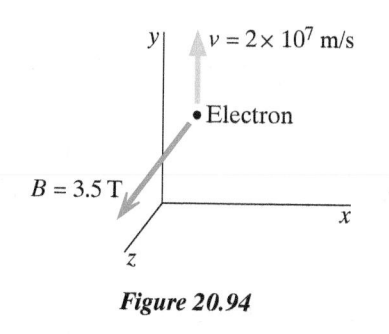

Figure 20.94

•**P25** At a particular instant a proton is traveling in the $-x$ direction, with speed 5×10^5 m/s. At the location of the proton there is a magnetic field of magnitude 0.24 T in the $-y$ direction, due to current running in a nearby coil. What are the direction and magnitude of the magnetic force on the proton?

•**P26** At a particular instant an electron is traveling in the $-x$ direction, with speed 4×10^5 m/s. At the location of the electron there is a magnetic field of magnitude 0.27 T in the $+z$ direction, due to a large bar magnet. What are the direction and magnitude of the magnetic force on the electron?

•**P27** At a particular instant an electron is traveling in the $+z$ direction, with speed 8×10^5 m/s. At the location of the electron there is a magnetic field of magnitude 0.32 T in the $-z$ direction, due to a large bar magnet. What are the direction and magnitude of the magnetic force on the electron?

•**P28** To become familiar with the order of magnitude of magnetic effects, consider the situation in a television cathode ray tube (not a flat-panel TV). In order to bend the electron trajectory so that the electron hits the top of the screen rather than going straight through to the center of the screen, you need a radius of curvature in the magnetic field of about 20 cm (Figure 20.95). If the electrons are accelerated through a 15,000 V potential difference, they have a speed of 0.7×10^8 m/s. Calculate the magnitude of the magnetic field required to make the electrons hit the top of the screen. Is the magnetic field into or out of the page?

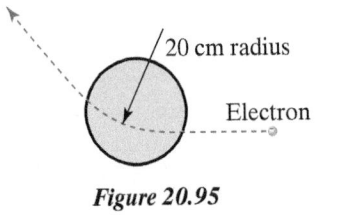

Figure 20.95

•P29 An alpha particle (consisting of two protons and two neutrons) is moving at constant speed in a circle, perpendicular to a uniform magnetic field applied by some current-carrying coils, making one clockwise revolution every 80 ns (Figure 20.96). If the speed is small compared to the speed of light, what is the numerical magnitude B of the magnetic field made by the coils? What is the direction of this magnetic field?

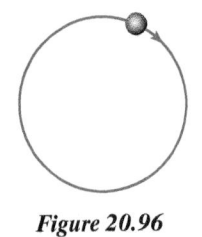

Figure 20.96

••P30 An electron and a proton are both in motion near each other as shown in Figure 20.97; at this instant they are a distance r apart. The proton is moving with speed v_p at an angle θ to the horizontal, and the electron is moving straight up with speed v_e.

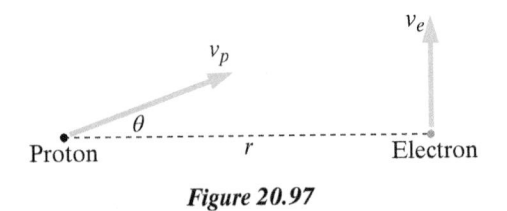

Figure 20.97

(a) Calculate the x and y components of the force that the proton exerts on the electron. First calculate the electric field and magnetic field that the proton produces at the location of the electron, then calculate the forces that these fields exert on the electron. (b) Calculate the x and y components of the force that the electron exerts on the proton. First calculate the electric field and magnetic field that the electron produces at the location of the proton, then calculate the forces that these fields exert on the proton. (In a completely correct quantitative calculation we would have to use the relativistically correct fields.) (c) Consider carefully your results. Are the magnetic forces on electron and proton equal and opposite? Does reciprocity hold for magnetic forces? (d) Will the total momentum of the two particles remain constant? Is this a violation of conservation of momentum for an isolated system?

See the comments at the end of Question Q3.

••P31 A mass spectrometer is a tool used to determine accurately the mass of individual ionized atoms or molecules, or to separate atoms or molecules that have similar but slightly different masses. For example, you can deduce the age of a small sample of cloth from an ancient tomb by using a mass spectrometer to determine the relative abundances of carbon-14 (whose nucleus contains 6 protons and 8 neutrons) and carbon-12

(the most common isotope, whose nucleus contains 6 protons and 6 neutrons).

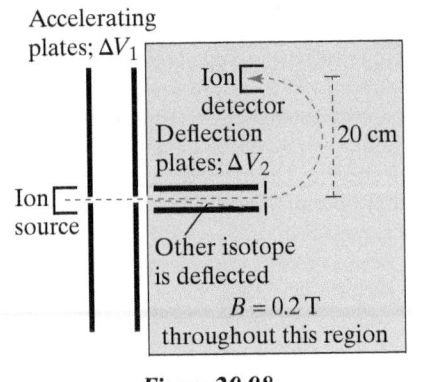

Figure 20.98

A particular kind of mass spectrometer is shown in Figure 20.98. Carbon from the sample is ionized in the ion source at the left. The resulting singly ionized $^{12}C^+$ and $^{14}C^+$ ions have negligibly small initial velocities (and can be considered to be at rest). They are accelerated through the potential difference ΔV_1. They then enter a region where the magnetic field has a fixed magnitude $B = 0.2$ T. The ions pass through electric deflection plates that are 1 cm apart and have a potential difference ΔV_2 that is adjusted so that the electric deflection and the magnetic deflection cancel each other for a particular isotope: one isotope goes straight through, and the other isotope is deflected and misses the entrance to the next section of the spectrometer. The distance from the entrance to the fixed ion detector is a distance of 20 cm.

There are controls that let you vary the accelerating potential ΔV_1 and the deflection potential ΔV_2 in order that only $^{12}C^+$ or $^{14}C^+$ ions go all the way through the system and reach the detector. You count each kind of ion for fixed times and thus determine the relative abundances. The various deflections ensure that you count only the desired type of ion for a particular setting of the two voltages.

Carry out the following calculations, and give brief explanations of your work: (a) Determine which accelerating plate is positive (left or right), which deflection plate is positive (upper or lower), and the direction of the magnetic field. (b) Determine the appropriate numerical values of ΔV_1 and ΔV_2 for ^{12}C. Carry out your intermediate calculations algebraically, so that you can use the algebraic results in part (c). (c) Determine the appropriate numerical values of ΔV_1 and ΔV_2 for ^{14}C.

Background: In organic material the ratio of ^{14}C to ^{12}C depends on how old the material is, which is the basis for carbon-14 dating. ^{14}C is continually produced in the upper atmosphere by nuclear reactions caused by cosmic rays (high-energy charged particles from outer space, mainly protons), and ^{14}C is radioactive with a half-life of 5700 y. When a cotton plant is growing, some of the CO_2 it extracts from the air to build tissue contains ^{14}C that has diffused down from the upper atmosphere. After the cotton has been harvested, however, there is no further intake of ^{14}C from the air, and the cosmic rays that create ^{14}C in the upper atmosphere can't penetrate the atmosphere and reach the cloth. Thus the amount of ^{14}C in cotton cloth continually decreases with time, while the amount of nonradioactive ^{12}C remains constant.

••P32 The design and operation of a cyclotron (Figure 20.99) was discussed in Section 20.1. (a) Show that the period of the motion, the time between one kick to the right and the next kick in the same direction, does not depend on the current speed of the proton at speeds small compared to the speed of light. As a result, we can place across the dees a simple sinusoidal potential difference having this period and achieve continual acceleration out to the maximum radius of the cyclotron. **(b)** One of Ernest Lawrence's first cyclotrons, built in 1932, had a diameter of only about 30 cm and was placed in a magnetic field of about 1 T. What was the frequency (= 1/period, in hertz = cycles per second) of the sinusoidal potential difference placed across the dees to accelerate the protons?

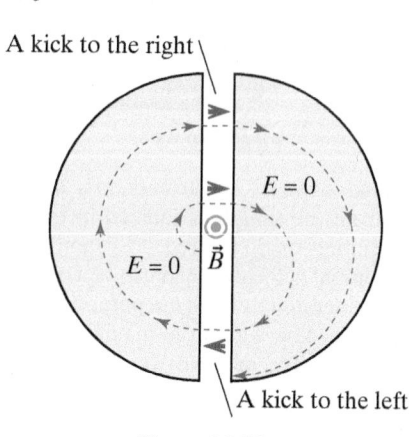

Figure 20.99

(c) Show that the equivalent accelerating potential of this little cyclotron was about a million volts! That is, the kinetic energy gain from the center to the outermost radius was $\Delta K = e\Delta V_{eq}$, with $\Delta V_{eq} = 1 \times 10^6$ V. (*Hint:* Calculate the final speed at the outermost radius.) **(d)** If the sinusoidal potential difference applied to the dees had an amplitude of 500 V (that is, it varied between +500 and −500 V), show that it took about 65 μs for a proton to move from the center to the outer radius.

••P33 The thin circular coil in Figure 20.100 has radius $r = 15$ cm and contains $N = 3$ turns of Nichrome wire. The coil has a resistance of 6 Ω. A small compass is placed at the center of the coil. With the battery disconnected, the compass needle points to the right, in the plane of the coil. The apparatus is located near the equator, where the Earth's magnetic field B_{Earth} is horizontal, with a magnitude of about 4×10^{-5} T.

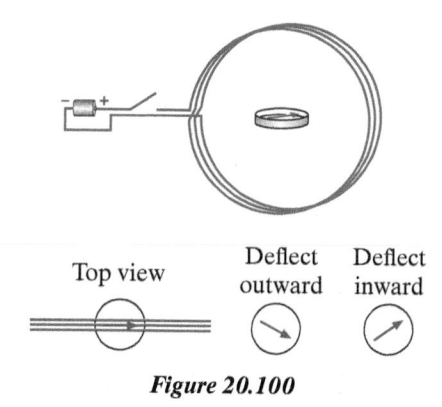

Figure 20.100

(a) When the 1.5 V battery is connected, predict the deflection of the compass needle. If you have to make any approximations, state what they are. Is the deflection outward or inward as seen from above? What is the magnitude of the deflection? **(b)** The

compass is removed. The current in the coil continues to run. An electron is at the center of the coil and is moving with speed $v = 5 \times 10^6$ m/s into the page (perpendicular to the plane of the coil). In addition to the magnetic fields in the region due to the Earth and the coil, there is an electric field at the center of the coil (due to charges not shown in Figure 20.100), and this electric field points upward and has a magnitude $E = 250$ V/m. What are the magnitude and direction of the force on the electron at this moment? (You can neglect the gravitational force, which easily can be shown to be negligible.)

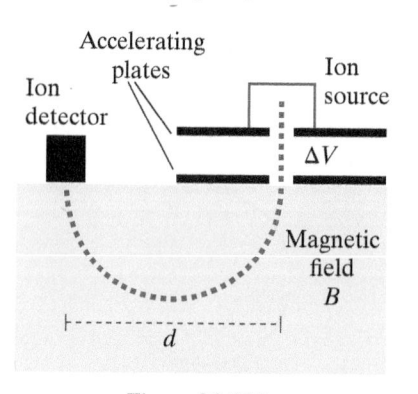

Figure 20.101

••P34 In the simple mass spectrometer shown in Figure 20.101, positive ions are generated in the ion source. They are released, traveling at very low speed, into the region between two accelerating plates between which there is a potential difference ΔV. In the shaded region there is a uniform magnetic field $\vec{B}$; outside this region there is negligible magnetic field. The semicircle traces the path of one singly charged positive ion of mass M, which travels through the accelerating plates into the magnetic field region, and hits the ion detector as shown. Determine the appropriate magnitude and direction of the magnetic field $\vec{B}$, in terms of the known quantities shown in Figure 20.101. Explain all steps in your reasoning.

••P35 A long solenoid with diameter 4 cm is in a vacuum, and a lithium nucleus (4 neutrons and 3 protons) is in a clockwise circular orbit inside the solenoid (Figure 20.102). It takes 50 ns $(50 \times 10^{-9}$ s) for the lithium nucleus to complete one orbit.

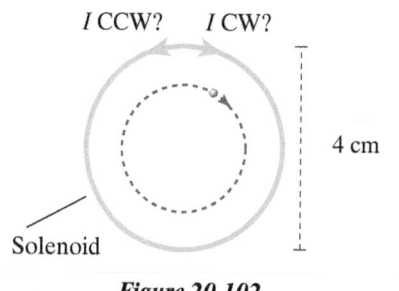

Figure 20.102

(a) Does the current in the solenoid run clockwise or counterclockwise? Explain, including physics diagrams. **(b)** What is the magnitude of the magnetic field made by the solenoid?

••P36 A long straight wire suspended in the air carries a conventional current of 8.2 A in the $-x$ direction as shown in Figure 20.103 (the wire runs along the x axis). At a particular instant an electron at location $\langle 0, -0.003, 0 \rangle$ m has velocity $\langle -1.5 \times 10^5, -1.8 \times 10^5, 0 \rangle$ m/s.

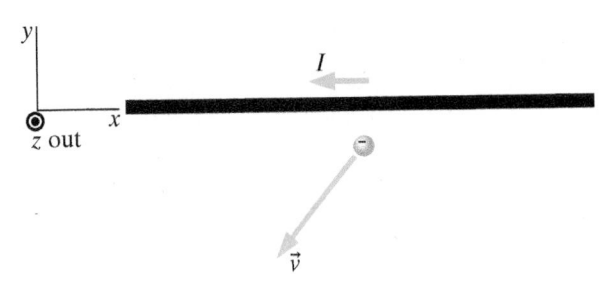

Figure 20.103

(a) What is the (vector) magnetic field due to the wire at the location of the electron? **(b)** What is the (vector) magnetic force on the electron due to the current in the wire? **(c)** If the moving particle were a proton instead of an electron, what would be the (vector) magnetic force on the proton?

Section 20.2

•P37 A wire is oriented along the x axis. It is connected to two batteries, and a conventional current of 1.8 A runs through the wire in the $+x$ direction. Along 0.25 m of the length of the wire there is a magnetic field of 0.54 T in the $+y$ direction, due to a large magnet nearby. At other locations in the circuit, the magnetic field due to external sources is negligible. What are the direction and magnitude of the magnetic force on the wire?

•P38 A 1.5 V battery is connected to a resistive wire whose resistance is 5 Ω. The wire is laid out in the form of a rectangle 50 cm by 3 cm (Figure 20.104). What are the approximate magnitude and direction of the magnetic force that the upper part of the wire exerts on the lower part of the wire in the diagram? Explain briefly.

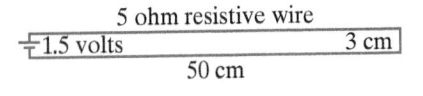

Figure 20.104

•P39 A current-carrying wire is oriented along the y axis. It passes through a region 0.6 m long in which there is a magnetic field of 4.5 T in the $+z$ direction. The wire experiences a force of 14.9 N in the $-x$ direction. **(a)** What is the magnitude of the conventional current in the wire? **(b)** What is the direction of the conventional current in the wire?

•P40 If your flashlight battery has an internal resistance of about 1/4 Ω, and your bar magnet produces a magnetic field of about 0.1 T near one end of the magnet, what is the approximate magnitude of the magnetic force on 5 cm of a wire that short-circuits the battery (Figure 20.105)? Indicate the direction of this force on a diagram. Check your direction experimentally.

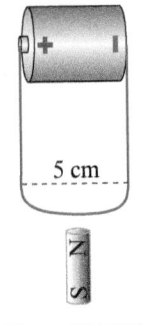

Figure 20.105

••P41 A Nichrome wire has the shape of two quarter-circles of radius a and b, connected by straight sections as shown in

Figure 20.106. The wire carries conventional current I in the direction shown.

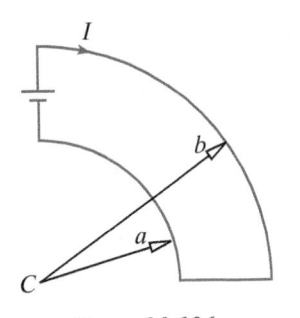

Figure 20.106

(a) Calculate the direction and magnitude of the magnetic field at point C, the center of the quarter-circles. Briefly explain your work, including a diagram. **(b)** At a particular instant, an electron is at point C, traveling to the right with speed v. Calculate the direction and magnitude of the magnetic force on the electron. Briefly explain your work, including a diagram.

••P42 A long wire carries a current I_1 upward, and a rectangular loop of height h and width w carries a current I_2 clockwise (Figure 20.107). The loop is a distance d away from the long wire. The long wire and the rectangular loop are in the same plane. Find the magnitude and direction of the net magnetic force exerted by the long wire on the rectangular loop. Briefly explain your work, including a diagram.

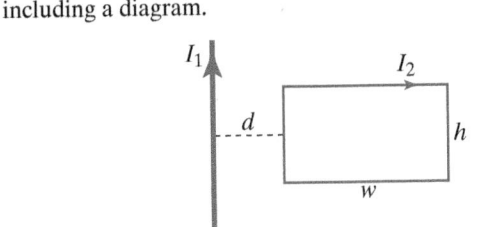

Figure 20.107

••P43 A metal rod of length $L = 12$ cm and mass $m = 70$ g has metal loops at both ends, which go around two metal poles (Figure 20.108). The rod is in good electrical contact with the poles but can slide freely up and down. The metal poles are connected by wires to a battery, and a current of $I = 5$ A flows through the rod. A magnet supplies a large uniform magnetic field B in the region of the rod, large enough that you can neglect the magnetic fields due to the 5 A current. The magnetic field is oriented so as to have the maximum effect on the rod. The rod sits at rest a distance $d = 4$ cm above the table. What are the magnitude and direction of the magnetic field in the region of the rod? Briefly explain your work, including a diagram.

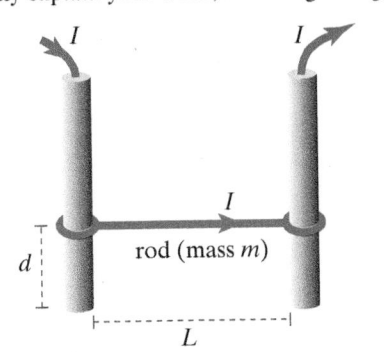

Figure 20.108

Section 20.3

•**P44** An object of charge q is moving in the x direction with speed v. Throughout the region there is a uniform electric field in the y direction of magnitude E (Figure 20.109). Determine a direction and magnitude B for a uniform magnetic field such that the net electric and magnetic force on the moving charge is zero. Draw $\vec{B}$ on a diagram of the situation.

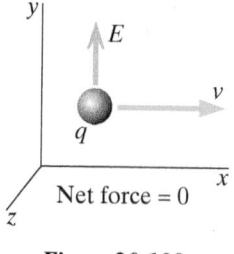

Figure 20.109

•**P45** In Figure 20.110 a particle with a charge of $+9$ nC travels to the left in a straight line at constant speed through a region where the electric field is 3800 V/m in the $-y$ direction and the magnetic field is 0.4 T in the $+z$ direction. What is the speed of the particle?

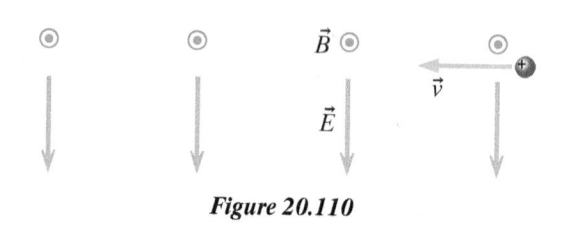

Figure 20.110

•**P46** An electron travels with velocity $\langle 5 \times 10^5, 0, 0 \rangle$ m/s. It enters a region in which there is a uniform magnetic field of $\langle 0, 0.9, 0 \rangle$ T. **(a)** What is the (vector) magnetic force on the electron? **(b)** Despite the magnetic force, the electron continues to travel in a straight line at constant speed. You conclude that there must be another force acting on the electron. Since you know there is also an electric field in this region, you decide that the other force must be an electric force. What is this (vector) electric force? **(c)** What is the (vector) electric field in this region that is responsible for the electric force?

•**P47** A proton traveling with speed 3105 m/s in the $+x$ direction passes through a region in which there is a uniform magnetic field of magnitude 0.6 T in the $+y$ direction. You want to keep the proton traveling in a straight line at constant speed. To do this, you can turn on an apparatus that can create a uniform electric field throughout the region. What (vector) electric field should you apply?

•**P48** A proton moves at constant velocity in the $+y$ direction, through a region in which there is an electric field and a magnetic field. The electric field is in the $+x$ direction and has magnitude 200 V/m. The magnetic field is in the $-z$ direction and has magnitude 0.95 T. **(a)** What is the magnitude of the net force on the proton? **(b)** What is the speed of the proton?

••**P49** Two long straight wires carrying 8 A of conventional current are connected by a three-quarter-circular arc of radius 0.035 m (Figure 20.111). (The rest of the circuit is far enough away

that it contributes little magnetic field in this region.) Electric fields are also present in this region, due to charges not shown on the drawing. An electron is moving to the right with a speed of 5.7×10^5 m/s as it passes through the center C of the arc, and at that instant the net electric and magnetic force on the electron is zero.

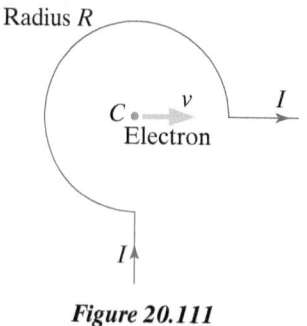

Figure 20.111

(a) What is the direction of the magnetic field at the center C due to the current? **(b)** What is the direction of the electric field at the center C? **(c)** What is the magnitude of the magnetic field at the center C due to the current? **(d)** What is the magnitude of the electric field at the center C?

••**P50** An electron travels with velocity $\langle 0, 0, -2.9 \times 10^6 \rangle$ m/s in a region between a pair of very large charged plates, separated by a distance of 4.5 mm. In this region there is also a uniform magnetic field. The velocity of the electron remains constant throughout this region. The magnetic field, due to large coils outside the region, is measured and is found to be $\langle 0, -0.27, 0 \rangle$ T.

On paper, draw a diagram showing the electron's velocity, the direction of the magnetic field in the region, the direction of the magnetic force on the electron, the direction of the electric force on the electron, and the direction of the electric field in the region. Use your diagram to answer the following questions. **(a)** What is the direction of the magnetic force on the electron? **(b)** What is the direction of the electric force on the electron? **(c)** What is the direction of the electric field in the region? **(d)** What is the magnitude of the magnetic force on the electron? **(e)** What is the absolute value of the potential difference between the plates?

••**P51** A cathode ray tube (CRT), as shown in Figures 20.112 and 20.113, is an evacuated glass tube. A current runs through a filament at the right end of the tube.

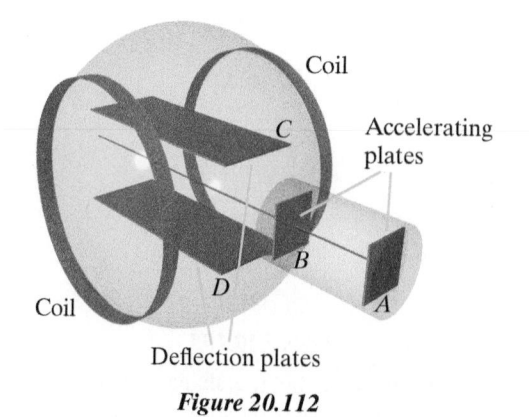

Figure 20.112

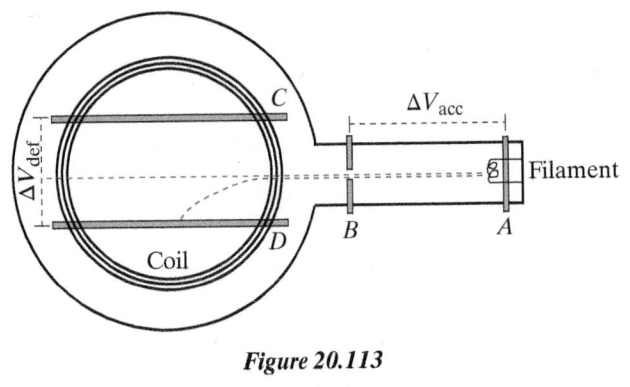

Figure 20.113

When the metal filament gets very hot, electrons occasionally escape from it. These electrons can be accelerated away from the filament by applying a potential difference ΔV_{acc} across the metal plates labeled A and B in the diagram.

The electrons pass through a hole in plate B and enter the glass sphere. There they pass between the two horizontal metal "deflection" plates labeled C and D. A potential difference ΔV_{def} can be applied across these plates, to deflect the beam of electrons.

In front of and in back of the glass sphere are two coils, through which current can be run to produce a magnetic field in the region between the deflection plates. The coils are oriented so that they both produce magnetic fields into the page in this region.
(a) Which of the accelerating plates, A or B, has a positive charge? **(b)** With no current in the coils, if the potential difference across the deflection plates, V_{def}, is zero, the electrons in the beam travel in a straight line, as indicated in Figure 20.113 by the dashed blue path. However, if V_{def} is not zero, the electron beam is deflected downward, following the path indicated on the diagram by the dashed red line. If the beam follows the dashed red line, what is the direction of the electric field between the deflection plates? **(c)** If a current runs through the coils, there will be a magnetic field in the region between the deflection plates. If the magnetic field made by the coils points into the page in the region between the plates, what is the direction of the magnetic force on an electron in the beam? (You can neglect the effect of the Earth's magnetic field, which is small.) **(d)** The accelerating potential difference ΔV_{acc} is measured to be 3.1 kV. What is the speed of an electron after it passes through the hole in plate B? Each of the two coils has 320 turns. The average radius of the coil is 6 cm. The distance from the center of one coil to the electron beam is 3 cm. If a current of 0.5 A runs through the coils, what is the magnitude of the magnetic field at a location on the axis of the coils, midway between the coils? (The electron beam passes through this location.) When deciding whether to use an exact or an approximate equation here, consider the relative magnitudes of the distances involved. **(e)** In a particular experiment, the accelerating potential difference ΔV_{acc} is set to 3.1 kV. The distance between the deflection plates is 8 mm. A current of 0.5 A runs through the coils, and the potential difference ΔV_{def} is adjusted until the electron beam again follows the straight line path indicated by the dashed blue line. In this situation (electron beam traveling in a straight line), which of the following statements are true? (1) The magnetic force on an electron in the beam is in the direction of its motion. (2) The electric and magnetic forces on an electron in the beam are in opposite directions. (3) The magnitude of the electric field in the region is equal to the magnitude of the magnetic field in the region. (4) The angle between the electric field and the magnetic field in

the region is 180°. (5) The net force on an electron in the beam is zero. (6) The electric force on an electron in the beam is equal in magnitude to the magnetic force on the electron. **(f)** What is the value of ΔV_{def} required to make the electron beam travel in a straight line?

••P52 In Figure 20.114 two straight wires carrying conventional current I are connected by a three-quarter-circular arc of radius R_1 and a one-quarter-circular arc of radius R_2. Electric fields are also present in this region, due to charges not shown on the drawing.

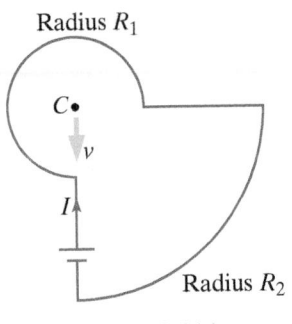

Figure 20.114

An electron is moving down with speed v as it passes through the center C of the arc, and at that instant the net electric and magnetic force on the electron is zero. (The Earth's magnetic field is negligible here.) **(a)** What is the direction of the magnetic field at the center C due to the current I? Explain briefly. **(b)** What is the direction of the electric field at the center C? Explain your reasoning clearly. **(c)** What is the magnitude of the magnetic field at the center C due to the current I? **(d)** What is the magnitude of the electric field at the center C? Explain briefly.

••P53 In Figure 20.115 two long straight wires carrying a large conventional current I are connected by one-and-a-quarter turns of wire of radius R. An electron is moving to the right with speed v at the instant that it passes through the center of the arc. You apply an electric field $\vec{E}$ at the center of the arc in such a way that the net force on the electron at this instant is zero. (You can neglect the gravitational force on the electron, which is easily shown to be negligible, and the magnetic field of the coil is much larger than the magnetic field of the Earth.)

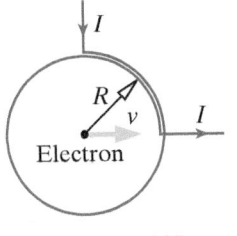

Figure 20.115

Determine the direction and magnitude of the electric field $\vec{E}$. Be sure to explain your work fully; draw and label any vectors you use.

••P54 In Figure 20.116 a battery with known emf $= K$ is connected to two large parallel metal plates. Each plate has a length L and width W, and the plates are a very short distance s apart. The plates are surrounded by a vertical thin circular coil of radius R containing N turns through which runs a steady conventional current I. The center of the coil is at the center of the gap between the plates. At a certain instant, a proton (charge

+e, mass M) travels through the center of the coil to the right with speed v, and the net force on the proton at this instant is zero (neglecting the very weak gravitational force). What are the magnitude and direction of conventional current in the coil? Explain clearly.

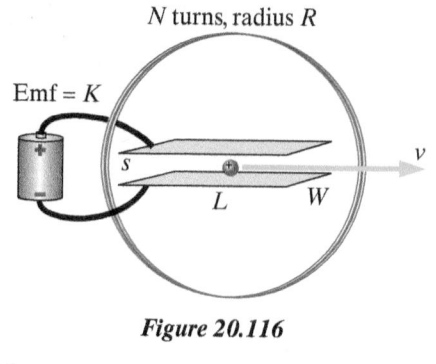

Figure 20.116

Section 20.4

••P55 A slab made of unknown material is connected to a power supply as shown in Figure 20.117. There is a uniform magnetic field of 0.7 T pointing upward throughout this region (perpendicular to the horizontal slab). Two voltmeters are connected to the slab and read steady voltages as shown. The connections across the slab are carefully placed directly across from each other. Assume that there is only one kind of mobile charges in this material, but we don't know whether they are positive or negative.

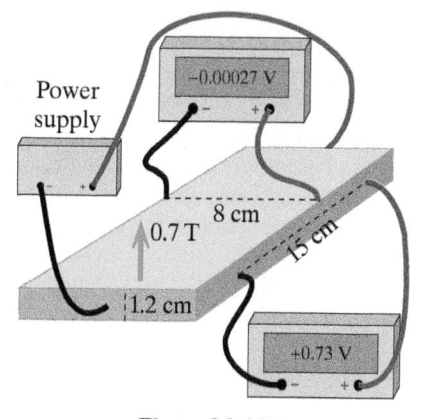

Figure 20.117

(a) Determine the (previously unknown) sign of the mobile charges, and state which way these charges move inside the slab. Explain carefully, using diagrams to support your explanation. **(b)** What is the drift speed $\bar{v}$ of the mobile charges? **(c)** What is the mobility u of the mobile charges? **(d)** The current running through the slab was measured to be 0.3 A. If each mobile charge is singly charged ($|q| = e$), how many mobile charges are there in 1 m³ of this material? **(e)** What is the resistance in ohms of a 15 cm length of this slab?

••P56 An experiment was carried out to determine the electrical properties of a new conducting material. A bar was made out of the material, 18 cm long with a rectangular cross section 3 cm high and 0.8 cm deep. The bar was part of a circuit and carried a steady current (Figure 20.118 shows only part of the circuit). A uniform magnetic field of 1.5 T was applied perpendicular to the bar, coming out of the page (using some coils that are not shown).

   Two voltmeters were connected along and across the bar and read steady voltages as shown (mV = millivolt = 1×10^{-3} V). The

connections across the bar were carefully placed directly across from each other to eliminate false readings corresponding to the much larger voltage along the bar. Assume that there is only one kind of mobile charge in this material.

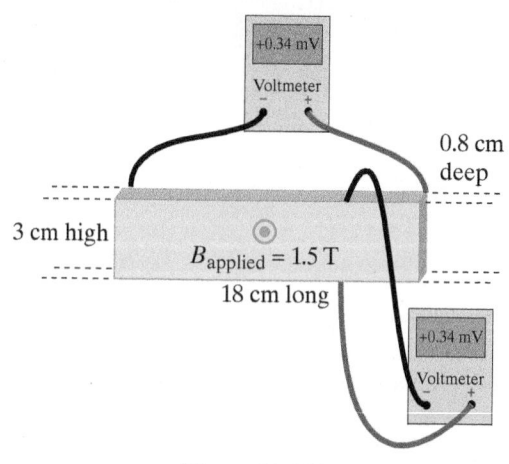

Figure 20.118

(a) What is the sign of the mobile charges, and which way do they move? Explain carefully, using diagrams to support your explanation. **(b)** What is the drift speed v of the mobile charges? Explain your reasoning. **(c)** What is the mobility u of the mobile charges? **(d)** The current running through the bar was measured to be 0.6 A. If each mobile charge is singly charged ($|q| = e$), how many mobile charges are there in 1 m³ of this material? **(e)** What is the resistance in ohms of this length of bar?

••P57 In Figure 20.119 a bar of length L, width w, and thickness d is positioned a distance h underneath a long straight wire that carries a large but unknown steady current I_{wire} (w is small compared to h, and h is small compared to the length of the long straight wire). The material that the bar is made of is known to have n positive mobile charge carriers ("holes") per cubic meter, each of charge e, and these are the only mobile charge carriers in the bar. An ammeter measures the small current I passing through the bar.

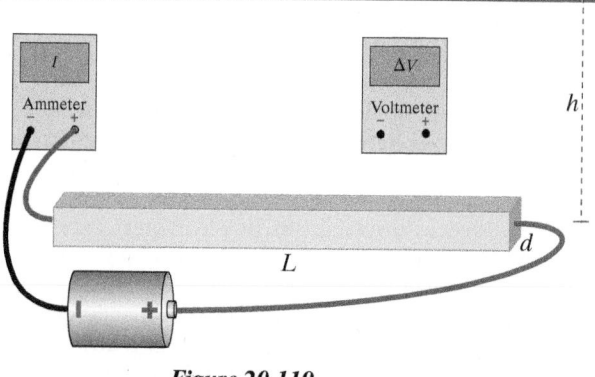

Figure 20.119

(a) Draw connecting wires from the voltmeter to the bar, showing clearly how the ends of the connecting wires must be positioned on the bar in order to permit determining the amount of current in the long straight wire I_{wire}. If the voltmeter reading ΔV is positive, does the current in the long straight wire run to the left or to the right? Explain briefly.

(b) If the voltmeter reading is ΔV, what is the current in the long straight wire, I_{wire}? Express your answer for the large current I_{wire} only in terms of the known quantities: h, L, w, d, n, $I_{ammeter}$, ΔV, and known physical constants such as e.

••P58 In Figure 20.120 the center of a large bar magnet is 20 cm from a thin plate of high-resistance material 12 cm long, 2.5 cm high, and 0.1 cm thick that is connected to a 240 V power supply whose internal resistance is negligible. The bar magnet is perpendicular to the plate.

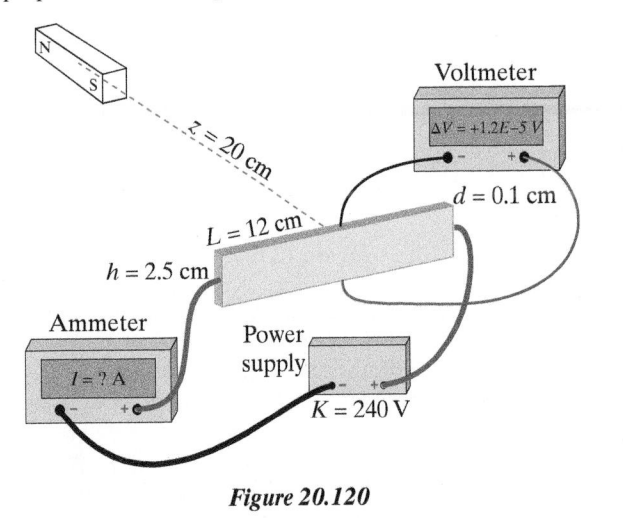

Figure 20.120

The mobility of charge carriers in the thin plate is 3.5×10^{-4} (m/s)/(V/m), and the number density of the singly charged charge carriers is 4×10^{23} m^{-3}. A voltmeter is connected precisely vertically across the thin plate and reads $+1.2 \times 10^{-5}$ V. A low-resistance ammeter is in series with the rest of the circuit. **(a)** Are the charge carriers electrons or holes? Explain, including a physics diagram. **(b)** What is the magnetic dipole moment of the bar magnet? Include units. **(c)** What does the ammeter read, including sign?

••P59 In Figure 20.121 a bar 11 cm long with a rectangular cross section 3 cm high and 2 cm deep is connected to a 1.2 V battery and an ammeter. The resistance of the copper connecting wires and the ammeter, and the internal resistance of the battery, are all negligible compared to the resistance of the bar.

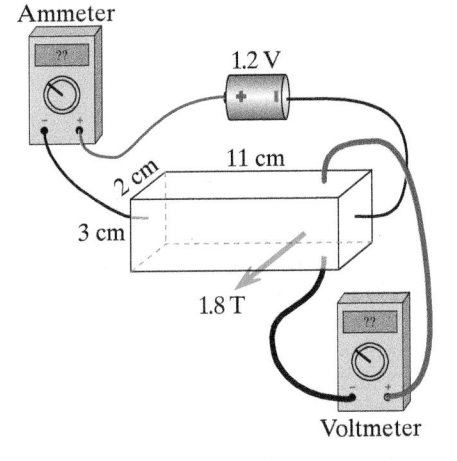

Figure 20.121

Using large coils not shown on the diagram, a uniform magnetic field of 1.8 T was applied perpendicular to the bar (out

of the page, as shown). A voltmeter was connected across the bar, with the connections across the bar carefully placed directly across from each other.

The mobile charges in the bar have charge $+e$, their density is 7×10^{23}/m^3, and their mobility is 3×10^{-5} (m/s)/(V/m).

Predict the readings of the voltmeter and ammeter, including signs. Explain carefully, using diagrams to support your explanation. Remember that a voltmeter reads positive if the $+$ terminal is connected to higher potential, and that an ammeter reads positive if conventional current enters the $+$ terminal.

Section 20.5

•P60 A neutral iron bar is dragged to the left at speed v through a region with a magnetic field B that points out of the page (Figure 20.122). Which diagram (1–5) best shows the state of the bar?

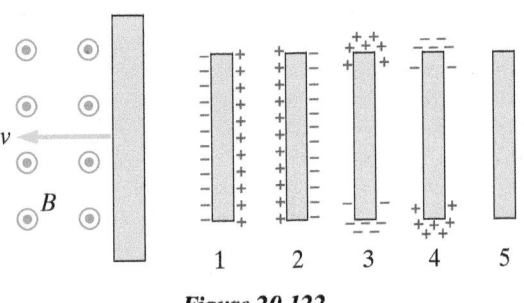

Figure 20.122

•P61 A neutral copper bar oriented horizontally moves upward through a region where there is a magnetic field out of the page. Which diagram (1–5) in Figure 20.123 correctly shows the distribution of charge on the bar?

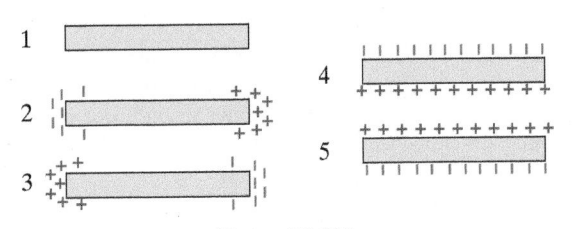

Figure 20.123

•P62 A metal rod of length L slides horizontally at constant speed v on frictionless insulating rails through a region of uniform upward magnetic field of magnitude B (Figure 20.124).

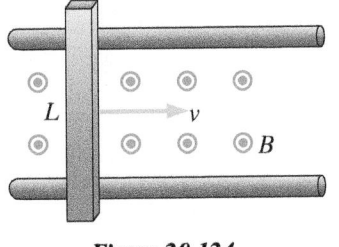

Figure 20.124

On a diagram, show the polarization of the rod and the direction of the Coulomb electric field inside the rod. Explain briefly. What is the magnitude E_C of the Coulomb electric field inside the rod? What is the potential difference across the rod? What is the emf across the rod? What are the magnitude and direction of the force you have to apply to keep the rod moving at a constant speed v?

•P63 A neutral metal rod of length 0.45 m slides horizontally to the left at a constant speed of 7 m/s on frictionless conducting rails through a region of uniform magnetic field of magnitude 0.4 T, directed out of the page as shown in Figure 20.125.

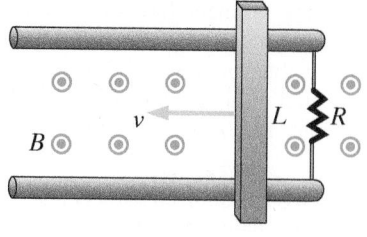

Figure 20.125

Before answering the following questions, draw a diagram showing the polarization of the rod, and the direction of the Coulomb electric field inside the rod.
(a) Which of the following statements is true? (1) The top of the moving rod is positive. (2) The top of the moving rod is negative. (3) The right side of the moving rod is positive. (4) The right side of the moving rod is negative. (5) The moving rod is not polarized. **(b)** After the initial transient, what is the magnitude of the net force on a mobile electron inside the rod? **(c)** What is the magnitude of the electric force on a mobile electron inside the rod? **(d)** What is the magnitude of the magnetic force on a mobile electron inside the rod? **(e)** What is the magnitude of the potential difference across the rod? **(f)** In what direction must you exert a force to keep the rod moving at constant speed?

••P64 A metal bar of mass M and length L slides with negligible friction but with good electrical contact down between two vertical metal posts (Figure 20.126). The bar falls at a constant speed v. The falling bar and the vertical metal posts have negligible electrical resistance, but the bottom rod is a resistor with resistance R. Throughout the entire region there is a uniform magnetic field of magnitude B coming straight out of the page. Explain every step clearly and in detail.

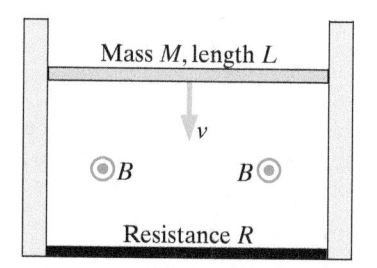

Figure 20.126

(a) Calculate the amount of current I running through the resistor. **(b)** On a diagram, clearly show the surface-charge distribution all the way around the circuit and the direction of the conventional current I. Explain. **(c)** Calculate the constant speed v of the falling bar. **(d)** Show that the rate of change of the gravitational energy of the Universe is equal to the rate of energy dissipated in the resistor.

••P65 In Figure 20.127 a rectangular loop of wire with width w, height h, and resistance R is dragged at constant speed v to the right through a region where there is a uniform magnetic field of magnitude B out of the page. The magnetic field is negligibly small outside that region.

For each stage of the motion, determine the current I in the loop and the force $\vec{F}$ required to maintain the constant speed. There are five stages. **(a)** Before entering the magnetic field region. **(b)** While the loop is part way into the magnetic field region (this is the stage that is shown). **(c)** While the loop is entirely inside the magnetic field region. **(d)** While the loop is part way out of the magnetic field region. **(e)** After leaving the magnetic field region.

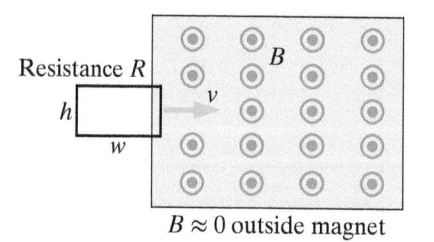

$B \approx 0$ outside magnet

Figure 20.127

Explain briefly in each case. In addition to determining the magnitude of $\vec{F}$ and I for each stage, draw a diagram for each of the five stages and show the direction of $\vec{F}$ and I on the diagram. Also show the approximate surface-charge distribution on the loop.

••P66 In Figure 20.128 on the left is a region of uniform magnetic field B_1 into the page, and adjacent on the right is a region of uniform magnetic field B_2 also into the page. The magnetic field B_2 is smaller than B_1 ($B_2 < B_1$). You pull a rectangular loop of wire of length w, height h, and resistance R from the first region into the second region, on a frictionless surface. While you do this you apply a constant force F to the right, and you notice that the loop doesn't accelerate but moves with a constant speed.

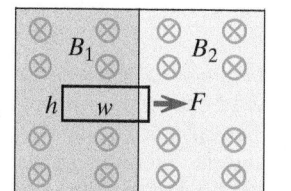

Figure 20.128

Calculate this constant speed v in terms of the known quantities B_1, B_2, w, h, R, and F, and explain your calculation carefully. Also show the approximate surface-charge distribution on the loop.

••P67 A neutral metal rod of length 60 cm falls toward the Earth. The rod is horizontal and oriented east west. **(1)** Which end of the rod, east or west, has excess electrons? Explain, using physics diagrams. **(2)** At a moment when the rod's speed is 4 m/s, approximately how many excess electrons are at the negative end of the rod?

••P68 A "unipolar" generator consists of a copper disk of radius R rotating in a uniform, steady magnetic field B perpendicular to the disk, out of the page (Figure 20.129). (The magnetic field is produced by large coils carrying constant current, not shown in the diagram.) Sliding contacts are made at the center (on the axle) and at the rim of the disk, and the wires are connected to a voltmeter. If the outer rim travels counterclockwise at a speed v, what does the voltmeter read (magnitude and sign)?

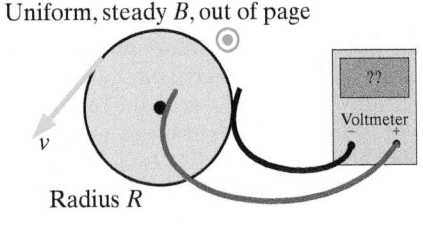

Uniform, steady B, out of page

??

Voltmeter

v

Radius R

Figure 20.129

••P69 In Figure 20.130 a rectangular loop of wire $L = 15$ cm long by $h = 3$ cm high, with a resistance of $R = 0.3$ Ω, moves with constant speed $v = 8$ m/s to the right, emerging from a rectangular region where there is a uniform magnetic field into a region where the magnetic field is negligibly small. In the region of magnetic field, the magnetic field points into the page, and its magnitude is $B = 1.2$ T.

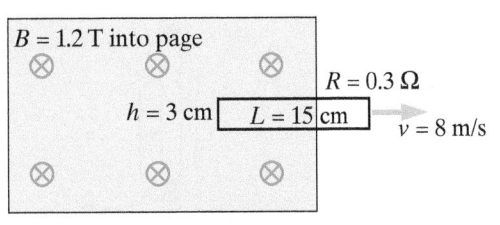

$B = 1.2$ T into page

⊗ ⊗ ⊗

$R = 0.3$ Ω

$h = 3$ cm $L = 15$ cm

⊗ ⊗ ⊗

$v = 8$ m/s

Figure 20.130

(a) On a diagram, show the approximate charge distribution on the moving loop. **(b)** What are the direction and magnitude of conventional current in the loop? Explain briefly. **(c)** Which of the following are true? (1) The magnetic force only stretches the loop; the net magnetic force on the loop is zero. (2) Because a current flows in the loop, there is a magnetic force on the loop. (3) The magnetic force on the loop is in the same direction as the velocity of the loop. **(d)** What are the direction and magnitude of the force that you have to apply in order to make the loop move at its constant speed?

Section 20.6

•P70 A proton moves with speed v in the $+x$ direction. In the lab (stationary) reference frame, what electric and magnetic fields would you observe a distance r above the proton (in the $+y$ direction)? In a reference frame moving with the proton, what electric and magnetic fields would you observe at the same location? Be quantitative.

Section 20.7

•P71 A bar magnet whose magnetic dipole moment is $\langle 4,0,1.5 \rangle$ A·m^2 is suspended from a thread in a region where external coils apply a magnetic field of $\langle 0.8,0,0 \rangle$ T. What is the vector torque that acts on the bar magnet?

•••P72 A bar magnet whose mass is 0.02 kg and whose magnetic dipole moment is $\langle 6,0,0 \rangle$ A·m^2 is suspended on a low-friction pivot in a region where external coils apply a magnetic field of $\langle 1.5,0,0 \rangle$ T. You rotate the bar magnet slightly in the horizontal plane and release it. **(a)** What is the angular frequency of the oscillating magnet? **(b)** What would be the angular frequency if the applied magnetic field were $\langle 3,0,0 \rangle$ T?

Section 20.8

•P73 A bar magnet whose magnetic dipole moment is 15 A·m^2 is aligned with an applied magnetic field of 4 T. How much work must you do to rotate the bar magnet 180° to point in the direction opposite to the magnetic field?

•P74 The center of a bar magnet whose magnetic dipole moment is $\langle 8,0,0 \rangle$ A·m^2 is located at the origin. A second bar magnet whose magnetic dipole moment is $\langle 3,0,0 \rangle$ A·m^2 is located at $x = 0.12$ m. What is the vector force on the second magnet due to the first magnet?

Section 20.9

•P75 A thin rectangular coil 8 cm by 5 cm has 50 turns of copper wire. It is made to rotate with angular frequency 100 rad/s in a magnetic field of 1.2 T. **(a)** What is the maximum emf produced in the coil? **(b)** What is the maximum power delivered to a 50 ohm resistor?

Section 20.10

•P76 In order for a spark to occur, it is necessary to ionize the air, which is not usually a conductor. One possible model for the process by which air becomes ionized is this: if a sufficiently strong electric field were applied to an atom, it would be possible to pull an outer electron out of the atom, leaving a positively charged ion and a free electron. We will estimate the strength of the electric field required to pull an electron out of an atom. **(a)** Consider the interaction of a single outer electron in a nitrogen atom with the "atomic core" (all the other charged particles in the atom—the protons in the nucleus and all the other electrons). What is the net charge of the atomic core? **(b)** If the radius of the atom is approximately 1×10^{-10} m, what is the magnitude of the electric field due to the atomic core at the location of the outer electron? **(c)** What is the magnitude of the electric field you would have to apply in order to pull the outer electron out of the atom? **(d)** It is observed experimentally that an applied electric field of 3×10^6 V/m is sufficient to cause a spark in air. What is the ratio of the electric field you calculated to the observed field needed to start a spark? **(e)** What should we conclude about this model? (1) Using more significant figures would not improve the agreement. (2) We need to think of a different physical explanation of how air gets ionized. (3) If our calculation used more accurate values the numbers would probably agree. (4) Since the discrepancy is so large, this model must be wrong.

•P77 A different model for how air becomes ionized in a spark focuses on energy. Suppose that a very energetic particle collided with a nitrogen or oxygen atom and knocked out an electron, ionizing the molecule. Estimate the kinetic energy required for this process. **(a)** Consider an outer electron in a nitrogen atom, interacting with the atomic core (nucleus plus all other electrons). What is the net charge of the atomic core? **(b)** Assuming the radius of the nitrogen atom is about 1×10^{-10} m, what is the electric potential at the location of this outer electron? **(c)** Ionizing the atom means moving the outer electron a very large distance away from the atomic core. What is the electric potential a very large distance from the atomic core? **(d)** What is the potential difference between these two locations? **(e)** What is the change in electric potential energy of the system of the electron plus the atomic core in this process? **(f)** How much kinetic energy would an incoming particle need in order to ionize the atom in a collision between the particle and the atom?

•P78 We will consider the possibility that a free electron acted on by an electric field could gain enough energy to ionize an air molecule in a collision. **(a)** Consider an electron that starts from rest in a region where there is an electric field (due to some charged objects nearby) whose magnitude is nearly constant. If the electron travels a distance d and the magnitude of the electric field is E, what is the potential difference through which the electron travels? (Pay attention to signs: Is the electron traveling with the electric field or opposite to the electric field?) **(b)** What is the change in potential energy of the system in this process? **(c)** What is the change in the kinetic energy of the electron in this process? **(d)** We found the mean free path of an electron in air to be about 5×10^{-7} m, and in the previous question you calculated the energy required to knock an electron out of an atom. What is the magnitude of the electric field that would be required in order for an electron to gain sufficient kinetic energy to ionize a nitrogen molecule? **(e)** The electric field required to cause a spark in air is observed to be about 3×10^6 V/m at STP. What is the ratio of the magnitude of the field you calculated in the previous part to the observed value at STP? **(f)** What is it reasonable to conclude about this model of how air becomes ionized? (1) Since we used accurate numbers, this is a huge discrepancy, and the model is wrong. (2) Considering the approximations we made, this is pretty good agreement, and the model may be correct.

•**P79** Explain briefly why there is a limit to how much charge can be placed on a metal sphere in the classroom. If the radius of the sphere is 15 cm, what is the maximum amount of charge you can place on the sphere? (Remember that a uniform sphere of charge makes an electric field outside the sphere as though all the charge were concentrated at the center of the sphere.)

••**P80** A Van de Graaff generator pulls electrons out of the Earth and transports them on a conveyor belt onto a nearly spherical metal shell (Figure 20.131). The diameter of this generator's metal shell is 24 cm.

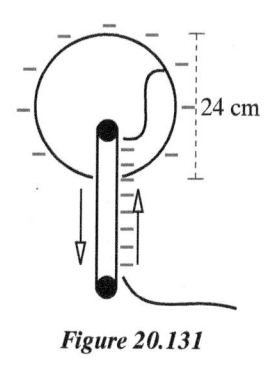

Figure 20.131

(a) Suppose that the conveyor belt in the Van de Graaff generator is running so fast that the generator succeeds in building up and maintaining just enough charge on the metal shell to cause the air to steadily glow bluish near the surface of the shell. Under these conditions, how much net charge $|Q|$ is on the metal shell? Calculate a numerical value for $|Q|$ and explain briefly. (Remember that the electric field outside a uniformly charged sphere is like that of a point charge located at the center of the sphere.) **(b)** Under these conditions (with the air steadily glowing), the Van de Graaff generator continually delivers additional electrons to the metal shell and yet the net charge Q on the shell does not change. Explain briefly but in detail why the charge on the shell does not change.

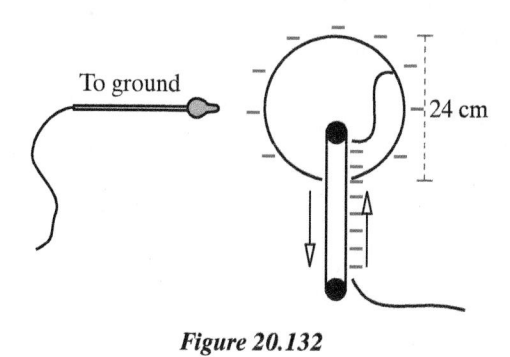

To ground

24 cm

Figure 20.132

(c) Now assume that the Van de Graaff generator is run more slowly, and the buildup of charge on the metal shell is limited by the inability of the motor to force any more electrons onto the negatively charged shell. The air no longer glows. We have a pear-shaped piece of metal that is attached by a metal wire to the earth (grounded). When we bring the grounded piece of metal to a location near the Van de Graaff metal shell, as shown in Figure 20.132, we observe a big spark. Explain why a spark occurs now but doesn't occur without the additional piece of metal nearby. **(d)** This spark lasts only a very short time. Why? **(e)** If we keep holding the grounded piece of metal 5 cm from the metal shell, we observe that there are big, brief sparks every 2 s. If we reduce the distance to 2 cm, the sparks that we observe occur more frequently, and they are less intense. Explain briefly.

COMPUTATIONAL PROBLEMS

••**P81** Starting with the skeleton program given below, write a program to predict the motion of a proton in a constant magnetic field. The skeleton program draws a "floor" and displays a uniform magnetic field, which initially has the value $\langle 0, 0.2, 0 \rangle$ T. Before beginning it may be helpful to review a previous program in which you predicted the motion of an object under the influence of one or more forces. Recall that you must:

Repeat

- Calculate the net force $\vec{F}_{net}$ acting on the system.
- Update the momentum of the system: $\vec{p}_f = \vec{p}_i + \vec{F}_{net}\,\Delta t$.
- Update the position: $\vec{r}_f = \vec{r}_i + \vec{v}_{avg}\,\Delta t$.

Use $\vec{p}_f/m$ to approximate the $\vec{v}_{avg}$ in each step.

Once your program is working, do the following: **(a)** Using an initial velocity of $\langle 2 \times 10^6, 0, 0 \rangle$ m/s, change the `while` statement to `while t < 3.34e-7:` and run the program. The proton should just make one complete orbit; if it does not, review your code. **(b)** What is the approximate radius of the proton's path? **(c)** What happens to the radius of the path if you double the initial speed? **(d)** How does doubling the speed change the time to complete one orbit? **(e)** What happens to the radius of the path if you halve the initial speed of 2×10^6 m/s? **(f)** How does halving the speed change the time to complete one orbit? **(g)** Restore the proton's initial velocity to its original value, and double the magnetic field. What changes? **(h)** Restore the magnetic field to its original value. Change the proton's initial velocity so that it is parallel to the magnetic field. Describe the path followed by the proton, and explain. **(i)** Change the initial velocity back to its original value, then add a significant $+y$ component to the velocity. Increase the time allowed in the loop sufficiently to allow the proton to make five complete orbits. Describe the proton's path, and explain. **(j)** Change the proton to an antiproton, whose mass is the same as that of the proton, but whose charge is $-e$. What changes?

```
from visual import *
scene.width = 800
scene.height = 800
## CONSTANTS ##
mzofp = 1e-7  ## mu-zero-over-four-pi
qe = 1.6e-19
mproton = 1.7e-27
B0 = vector(0,0.2,0)
bscale = 1
#### THIS CODE DRAWS A GRID ##
#### AND DISPLAYS MAGNETIC FIELD ##
xmax = 0.4
dx = 0.1
yg = -0.1
x = -xmax
while x < xmax+dx:
    curve(pos=[(x,yg,-xmax),(x,yg,xmax)],
          color=(.7,.7,.7))
    x = x + dx
z = -xmax
while z < xmax+dx:
    curve(pos=[(-xmax,yg,z),(xmax,yg,z)],
          color=(.7,.7,.7))
    z = z + dx
x = -xmax
dx = 0.2
while x < xmax+dx:
    z = -xmax
    while z < xmax+dx:
        arrow(pos=(x,yg,z),
              axis=B0*bscale,
              color=(0,.8,.8))
        z = z + dx
    x = x + dx
#### OBJECTS AND INITIAL CONDITIONS ##
particle = sphere(pos=vector(0,0.15,0.3),
                  radius=1e-2,
                  color=color.yellow,
                  make_trail=True)
## make trail easier to see (thicker) ##
particle.trail_object.radius = particle.
radius/3
vparticle = vector(-2e6,0,0)
p = mproton*vparticle
qparticle = qe
deltat = 5e-11
t = 0
###########################################
while True:
    rate(500)
    ## YOUR CODE GOES HERE ##
```

•••P82 Write a computer program to compute and display the motion of a proton inside a cyclotron (see Section 20.1). The first cyclotrons were small desktop devices; we will model a small device in order to understand the basic physical phenomena. Consider a cyclotron whose radius is 5 cm, with a gap of 0.5 cm between the dees (Figure 20.133).

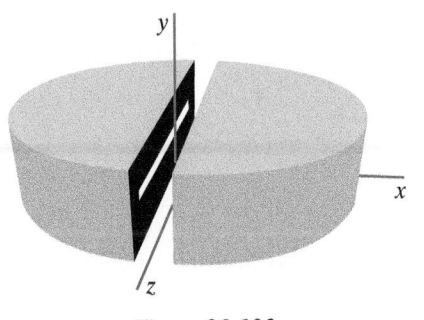

Figure 20.133

External magnets create a uniform magnetic field of 1.0 T in the $+y$ direction throughout the device. A proton starts essentially from rest at the center of the cyclotron. The maximum potential difference between the dees is 5000 V. Note that the electric field inside the dees is 0 at all times. Figure 20.133 shows that the metal dees form a nearly closed conducting enclosure. **(a)** What is the angular frequency ω of the sinusoidally oscillating potential difference you will need to apply? **(b)** Compute and display the trajectory of a proton in the cyclotron; display the trail of the proton. You may assume that the proton does not reach a speed sufficiently close to the speed of light to require a relativistically correct calculation, but do check this assumption. To get your program going, you can use a large value of dt (such as 1E-10 s), but you will need to decrease this value later. **(c)** Plot the proton's kinetic energy K (in eV) vs. time. Explain the shape of the graph you observe. **(d)** What happens if you use a different angular frequency for the potential difference? For example, try an angular frequency of 0.6ω. **(e)** If the proton gains a significant amount of kinetic energy while inside the dees, your dt is too large. Adjust your dt appropriately. What value of dt gives reasonable results? **(f)** How much time does the proton take to reach the outer edge of the cyclotron? **(g)** What is K (in eV) when the proton reaches the outer edge of the cyclotron? **(h)** Given your answer to (g), how many orbits must the proton have made before reaching the outer edge of the cyclotron? **(i)** What is the final speed of the proton? Were you justified in assuming that $v \ll c$?

ANSWERS TO CHECKPOINTS

1 $F_{\text{magnetic}} \approx 1 \times 10^{-8}$ N; $F_{\text{electric}} \approx 9 \times 10^{-5}$ N, much larger.

2 Clockwise; $q\vec{v} \times \vec{B} = -e\vec{v} \times \vec{B}$ points inward and turns the electron's momentum.

3 $F = 2 \times 10^{-3}$ N per meter of wire.

4 (a) See Figure 20.134 below. **(b)** See Figure 20.135 below.

5 $\vec{E} = \langle 0, 0, 2.7 \times 10^8 \rangle$ V/m. Note that a particle traveling that fast would presumably be in an evacuated chamber, so an electric field of this magnitude would not cause a spark.

6 $\Delta V_{\text{Hall}} = 1.5 \times 10^{-5}$ V; $\Delta V_{\text{wire}} = 3.3 \times 10^{-3}$ V $\gg \Delta V_{\text{Hall}}$

7 0.01 V ($\ll 1.5$ V)

8 emf $= v_{\text{bar}} BL$; I counterclockwise; $I = v_{\text{bar}} BL/R$; $F = v_{\text{bar}} B^2 L^2/R$ to the right

9 $F_{\text{magnetic}}/F_{\text{electric}} = 2.5 \times 10^{-11}$

10 Atomic magnetic dipole moments point north; swings toward north; magnetic dipoles twist to align with the magnetic field.

11 $2\mu B$

12 Net force would reverse and be repulsive; net force would be zero.

13 Emf is maximum when $\theta = 90°$ (velocity at right angle to magnetic field), minimum when $\theta = 0$ (velocity in the direction of magnetic field, so cross product is zero); 302 V

14 The mean free path would double; the mean free path would be one-quarter as long.

15 The mean free path must be one-third as long; 3.6×10^6 V/m.

16 Below the charge in Figure 20.72, $E_y < 0$ and $E_z = 0$, so

$$B'_y = \frac{\left(B_y + \frac{v}{c^2} E_z\right)}{\sqrt{1 - v^2/c^2}} = 0$$

$$B'_z = \frac{\left(B_z - \frac{v}{c^2} E_y\right)}{\sqrt{1 - v^2/c^2}} = \frac{-\frac{v}{c^2} E_y}{\sqrt{1 - v^2/c^2}} > 0$$

Therefore $B'_y = 0$ and $B'_z > 0$, as expected.

In front of the charge in Figure 20.72, $E_y = 0$ and $E_z > 0$, so

$$B'_y = \frac{\left(B_y + \frac{v}{c^2} E_z\right)}{\sqrt{1 - v^2/c^2}} = \frac{\frac{v}{c^2} E_z}{\sqrt{1 - v^2/c^2}} > 0$$

$$B'_z = \frac{\left(B_z - \frac{v}{c^2} E_y\right)}{\sqrt{1 - v^2/c^2}} = 0$$

Behind the charge in Figure 20.72, $E_y = 0$ and $E_z < 0$, so

$$B'_y = \frac{\left(B_y + \frac{v}{c^2} E_z\right)}{\sqrt{1 - v^2/c^2}} = \frac{\frac{v}{c^2} E_z}{\sqrt{1 - v^2/c^2}} < 0$$

$$B'_z = \frac{\left(B_z - \frac{v}{c^2} E_y\right)}{\sqrt{1 - v^2/c^2}} = 0$$

Therefore $B'_y < 0$ and $B'_z = 0$, as expected.

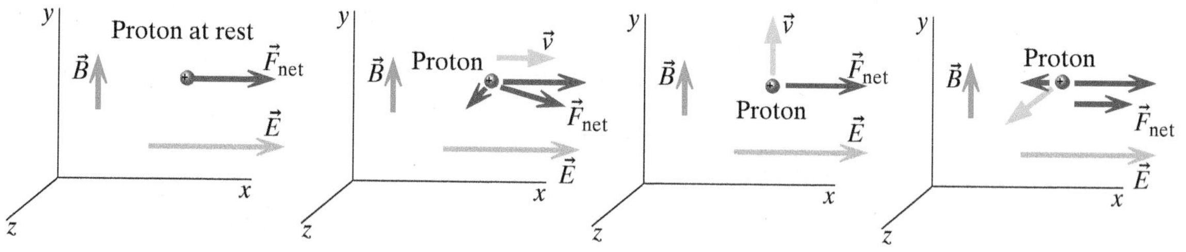

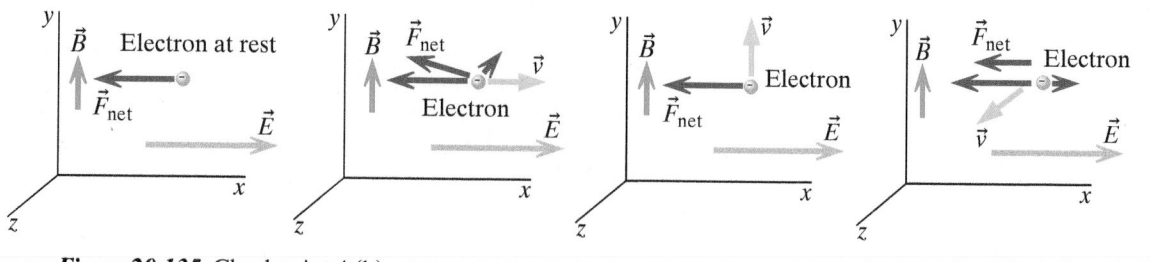

Figure 20.134 Checkpoint 4 (a).

Figure 20.135 Checkpoint 4 (b).

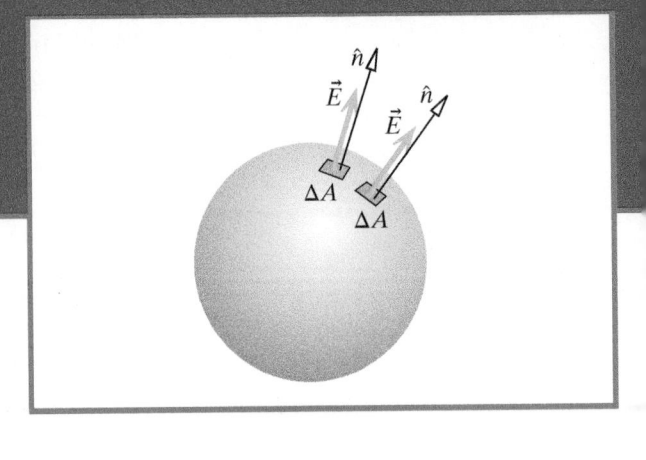

Patterns of Field in Space

OBJECTIVES

After studying this chapter you should be able to

- Mathematically relate the net electric field over a closed surface to the net charge enclosed.
- Mathematically relate the net magnetic field around a closed loop to the net current enclosed.

Electric charges produce electric and magnetic fields, and we know how to calculate or estimate the magnitude and direction of the field due to a particular distribution of charges (including moving charges). Sometimes, however, it is useful to reason in the other direction: from an observed three-dimensional pattern of electric or magnetic field it may be possible to infer what charges are responsible for this pattern. Gauss's law relates patterns of electric field over a closed surface to the amount of charge inside that closed surface. Ampere's law relates patterns of magnetic field around a closed path to the amount of current passing through that closed path.

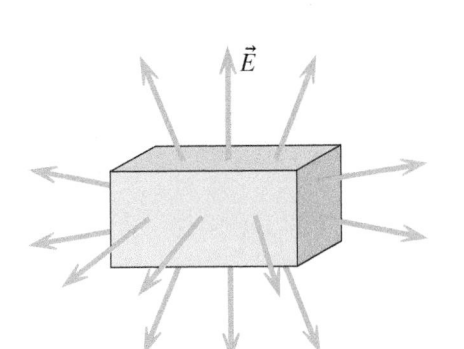

Figure 21.1 What's in this box?

21.1 PATTERNS OF ELECTRIC FIELD: GAUSS'S LAW

You are already familiar with some spatial patterns of electric field. Figure 21.1 shows an imaginary but opaque box that contains some charge. Suppose that you can't look inside the box, but you can measure the electric field all over the surface of the box.

> QUESTION Can you conclude that there is some charge in the box? Is it positive or negative?

The fact that the electric field points outward all over the box implies that there must be some positive charge inside the box.

Figure 21.2 shows another box (for clarity, the electric field on the surface is displayed with the head rather than the tail of the vector at the observation location). Again, suppose that you can't look inside the box, but you can measure the electric field all over the surface of the box.

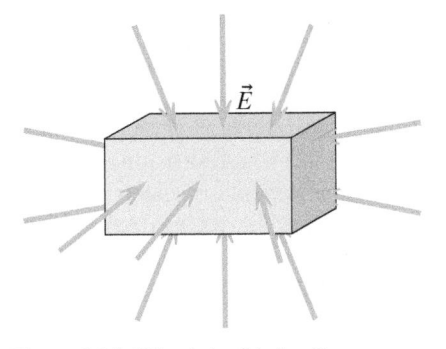

Figure 21.2 What's in this box?

> QUESTION Can you conclude that there is some charge in the box? Is it positive or negative?

Since the electric field points into the box, all over its surface, there must be some negative charge inside the box.

Suppose that you find a box on whose surface the electric field has a pattern like that in Figure 21.3.

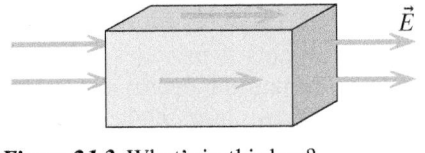

Figure 21.3 What's in this box?

> QUESTION Can you conclude that there is some charge in the box? Is it positive or negative? Can you think of a charge distribution that could produce this pattern of electric field?

The electric field points out of the box on part of the surface, and into the box on other parts of the surface. A possible charge distribution that would produce such a pattern of electric field is a very large positively charged plate to the left of the box, or a very large negatively charged plate to the right of the box, or plates of both kinds, in which case the box lies in the gap between the plates of a capacitor. (Another possibility would be that the box is a mathematical surface drawn inside a circuit wire, with a uniform electric field in the direction of the wire.)

It appears that if we find this kind of distribution of electric field over the surface of the box, we can conclude that there is no net charge inside the box. Although at first glance it might appear that there could be an electric dipole inside the box, a dipole would produce the observed pattern of electric field on the ends of the box ($\vec{E}$ to the right) but not on the sides of the box ($\vec{E}$ to the left), so this is not a possibility.

Suppose that you find a pattern of electric field on a box as shown in Figure 21.4.

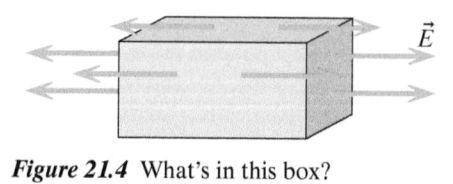

Figure 21.4 What's in this box?

> QUESTION Can you think of a distribution of charge that could produce such a pattern of electric field on the surface of the box? (*Hint:* There may be some charge outside the box in addition to the charge inside the box.)

A charge distribution that could produce this pattern of field over the surface of the box is a very large vertical positively charged plate that cuts through the middle of the box, making electric field to the left and right of the plate. In this case there is positive charge inside the box (and outside, too).

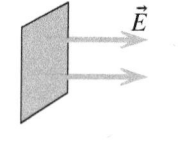

Figure 21.5 What can you conclude?

Closed and Open Surfaces

On the other hand, if your measurements of the electric field don't surround a 3D region, you may not be able to conclude much. Figure 21.5 shows a pattern of electric field on just one face of a box.

> QUESTION Give three different charge configurations that could be responsible for the field pattern observed in Figure 21.5.

Possibilities include a very large positively charge plate to the left of the surface, a very large negatively charged plate to the right of the surface, or both. We see an important difference between an "open" surface such as this one and a "closed" surface such as a box or a spherical surface, which surrounds a region in space. Qualitatively, it seems clear that measurements of electric field all over a closed surface can tell us quite a bit about the charge inside that closed surface. However, measurements of electric field on an open surface can't even tell us whether nearby charges are positive or negative.

Gauss's law is

- a quantitative relationship between
- measurements of electric field on a closed surface and
- the amount and sign of the charge inside that closed surface.

We will study Gauss's law for the additional insights it provides about the relationship between charge and field, and for certain important results that can be obtained using Gauss's law that are almost impossible to obtain using Coulomb's law alone, even though Gauss's law and Coulomb's law are essentially equivalent to each other (if the charges aren't moving).

Moreover, the concept of "flux" that arises in the context of Gauss's law plays an important role in our study of the effects of a time-varying magnetic field, Faraday's law, in the next chapter.

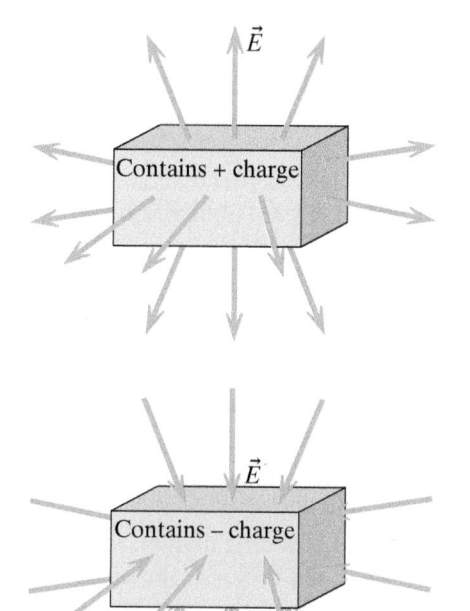

Figure 21.6 Compare patterns of electric field on these three boxes.

21.2 DEFINITION OF "ELECTRIC FLUX"

It is clear that there is some connection between the pattern of electric field on a closed surface and the amount of charge inside that surface, but how can we turn this into a quantitative relationship? We need to define a quantitative measure of the amount and direction of electric field over an entire surface. This measure is called "electric flux." We will develop a list of properties that a quantitative definition of electric flux should have.

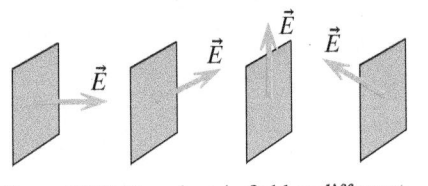

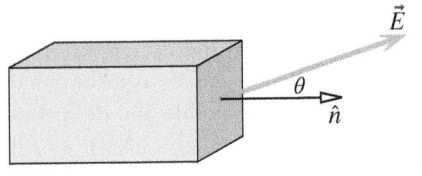

Figure 21.7 The electric field at different angles to a surface.

Property 1: Direction of Electric Field

Compare the electric field patterns on the surfaces of the three boxes shown in Figure 21.6, which contain positive, negative, and zero charge.

Evidently one property of "electric flux" is that it should be positive where the electric field points out of the box, negative where the electric field points into the box, and zero where the electric field does not pierce the surface. In that way, the total flux for the third box in Figure 21.6 would consist of +1 unit on the right face, −1 unit on the left face, and 0 on the other faces, for a total of zero, which corresponds to the zero charge that is inside the box.

We want the sign of electric flux to be related to the angle the electric field makes with the surface. This angle factor varies between +1 (outward and perpendicular to the surface) and −1 (inward and perpendicular to the surface), and is zero if the electric field is parallel to the surface. These values of +1, −1, and 0 are the extremes, and we need some way to express varying degrees of "outward-goingness" in the range from −1 to +1.

Figure 21.8 A unit vector pointing outward.

Figure 21.7 shows a series of open surfaces where this angle factor varies smoothly from +1 down to 0, and on to negative values. This suggests parameterizing the angle factor as the cosine of an angle between the electric field vector and the "outward-going normal," a unit vector $\hat{n}$ perpendicular to the surface and pointing outward from the closed surface (Figure 21.8).

> QUESTION Show that the cosine of this angle θ has the appropriate properties we're looking for: +1 if $\vec{E}$ points straight outward, −1 if $\vec{E}$ points straight inward, 0 if $\vec{E}$ is parallel to the surface, and an intermediate value between +1 and −1 in other cases.

If $\vec{E}$ and $\hat{n}$ are in the same direction, $\theta = 0°$ and $\cos\theta = +1$. If $\vec{E} \perp \hat{n}$, $\theta = 90°$ and $\cos\theta = 0$. If $\vec{E}$ is opposite to $\hat{n}$, $\theta = 180°$ and $\cos\theta = -1$.

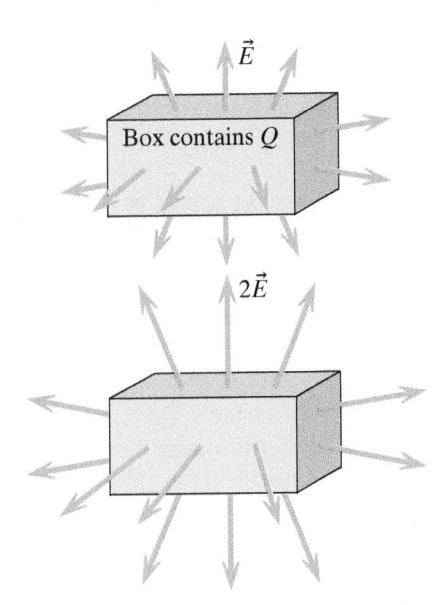

Property 2: Magnitude of Electric Field

How should the electric flux depend on the magnitude of the electric field? Consider the electric field patterns on the surfaces of the two boxes shown in Figure 21.9.

> QUESTION The upper box contains a charge Q. If the electric field vectors are twice as long on the surface of the lower box, how much charge is inside the lower box?

Figure 21.9 The magnitude of the field on the surface of the lower box is twice as large as the field on the upper box.

Doubling an amount of charge doubles the electric field made by that charge. Evidently the lower box contains an amount of charge $2Q$, so the definition of electric flux should have the property that twice the field gives twice the flux. Therefore electric flux ought to be proportional to $|\vec{E}|$ as well as being proportional to $\cos\theta$. Apparently the definition of electric flux should contain the product $E\cos\theta$.

We can write $E\cos\theta$ as $\vec{E}\bullet\hat{n} = E|\hat{n}|\cos\theta$, because the dot product involves the cosine of the angle between the two vectors, and the magnitude of a (dimensionless) unit vector $|\hat{n}|$ is 1 (Figure 21.10).

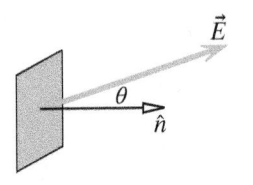

Figure 21.10 $E\cos\theta = \vec{E}\bullet\hat{n} = E|\hat{n}|\cos\theta$

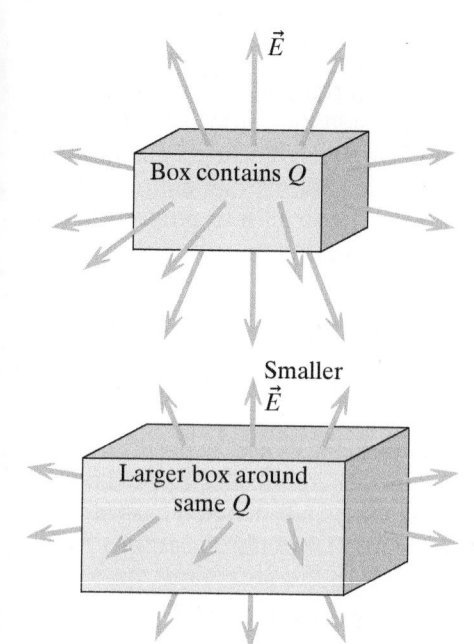

Box contains Q

Smaller $\vec{E}$

Larger box around same Q

Figure 21.11 A larger box containing the same charge has smaller electric field on its surface.

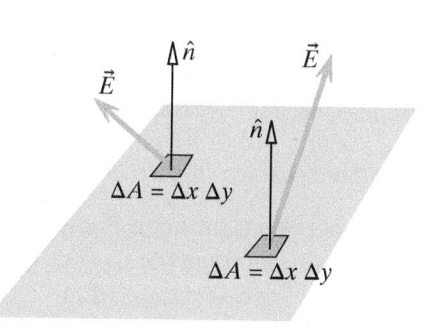

$\hat{n}$

$\vec{E}$

$\vec{E}$

$\hat{n}$

$\Delta A = \Delta x \, \Delta y$

$\Delta A = \Delta x \, \Delta y$

Figure 21.12 Electric flux on a surface is calculated by adding up the contributions $\vec{E} \cdot \hat{n} \Delta A$ on each little piece of area $\Delta A = \Delta x \, \Delta y$.

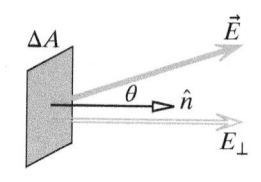

ΔA

$\vec{E}$

θ $\hat{n}$

$E_\perp$

Figure 21.13 $\vec{E} \cdot \hat{n} \Delta A = E_\perp \Delta A$

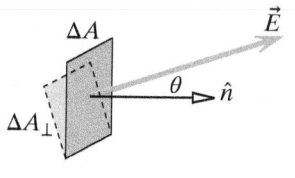

ΔA

$\vec{E}$

θ $\hat{n}$

$\Delta A_\perp$

Figure 21.14 $\vec{E} \cdot \hat{n} \Delta A = E \Delta A_\perp$

QUESTION In Figure 21.10, show that the dot product $\vec{E} \cdot \hat{n}$ is $+E$ if the electric field is parallel to $\hat{n}$, $-E$ if the electric field is opposite to $\hat{n}$, and 0 if $\vec{E}$ is parallel to the surface.

Outward: $\vec{E} \cdot \hat{n} = \langle +E, 0, 0 \rangle \cdot \langle 1, 0, 0 \rangle = +E + 0 + 0 = +E$

Inward: $\vec{E} \cdot \hat{n} = \langle -E, 0, 0 \rangle \cdot \langle 1, 0, 0 \rangle = -E + 0 + 0 = -E$

Parallel: $\vec{E} \cdot \hat{n} = \langle 0, E, 0 \rangle \cdot \langle 1, 0, 0 \rangle = 0 + 0 + 0 = 0$

Property 3: Surface Area

Finally, consider the two boxes shown in Figure 21.11. The small box contains a charge Q, and there are no charges outside the small box. If we enclose the same charge Q in a larger box, the electric field vectors on the surface of the larger box are smaller, because the surfaces are farther from the enclosed charge.

We need to define the electric flux in such a way that the flux is the same for the large electric fields on the surface of the small box or for the small electric fields on the surface of the large box. This suggests that the definition of electric flux ought to involve the area A of a surface, not just $\vec{E} \cdot \hat{n}$, in order to compensate for how big we draw the enclosing surface. Maybe electric flux should be defined as $\vec{E} \cdot \hat{n} A$. In fact, if you double all the distances, the surface area increases by a factor of four, while the magnitude of the electric field decreases by a factor of four. Thus including A in the definition ought to compensate exactly for changing the size of the box.

The magnitude and direction of $\vec{E}$ may vary from location to location on a large surface, so we should define electric flux on a small area ΔA over which there is little variation in the electric field.

Taking all these factors into account, perhaps a useful definition of electric flux over a small area $\Delta A = \Delta x \, \Delta y$ would be $\vec{E} \cdot \hat{n} \Delta A$, and we would add up all such contributions over an extended surface to get the electric flux over that entire surface (Figure 21.12), called Φ_{el} (Φ is uppercase Greek phi).

DEFINITION OF ELECTRIC FLUX ON A SURFACE

$$\Phi_{el} = \sum_{surface} \vec{E} \cdot \hat{n} \Delta A$$

Perpendicular Field or Perpendicular Area

We define electric flux as the sum of $\vec{E} \cdot \hat{n} \Delta A$ over a surface. This definition takes into account the direction of the electric field, the magnitude of the electric field, and the surface area. One of the ways to calculate the flux is to think of $E \cos \theta$ as the component of electric field perpendicular to the surface, $E_\perp = E \cos \theta$, so that the flux on a small area is $E_\perp \Delta A$ (Figure 21.13).

Alternatively, we can calculate the flux on a small area as the product of the electric field and the projection of the surface area ΔA perpendicular to the direction of the electric field (Figure 21.14). If we call this perpendicular projection of the area $\Delta A_\perp$, the flux is $E \Delta A_\perp$.

Thus we have three equivalent ways of writing the flux: $\vec{E} \cdot \hat{n} \Delta A$, $E_\perp \Delta A$, or $E \Delta A_\perp$. In many books you will find yet another notation for electric flux: $\vec{E} \cdot \Delta \vec{A}$. In this notation $\Delta \vec{A} = \hat{n} \Delta A$. $\Delta \vec{A}$ is considered to be a vector pointing in the direction of the outward-going normal $\hat{n}$, with a magnitude ΔA.

Checkpoint 1 In Figure 21.15 the magnitude of the electric field is 1000 V/m, and the field is at an angle of $30°$ to the outward-going normal. What is the flux on the small rectangle whose dimensions are 1 mm by 2 mm?

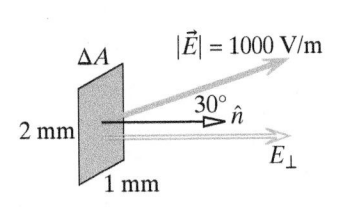

Figure 21.15 Find the electric flux on the rectangle.

Adding Up the Flux

The electric flux on a small area ΔA is $\vec{E} \bullet \hat{n} \, \Delta A$. To calculate the electric flux over an extended surface we need to divide the surface into little areas, calculate the flux on each little area, and then add up all these contributions to the flux. Symbolically, we have

$$\text{electric flux on a surface} = \sum \vec{E} \bullet \hat{n} \, \Delta A$$

If we divide the surface into an infinite number of infinitesimal areas, this summation becomes an integral. This is similar to what we did in Chapter 15 to calculate the electric field due to charges distributed over a ring or a disk:

$$\text{electric flux on a surface} = \Phi_{el} = \int \vec{E} \bullet \hat{n} \, dA$$

This "surface integral" can also be written in the following equivalent forms:

$$\Phi_{el} = \int E_{\perp} \, dA = \int E \, dA_{\perp} = \int \vec{E} \bullet d\vec{A}$$

We have seen that it is the electric flux on closed surfaces, such as a box, that tells us something about what charge is inside that closed surface. Whenever we need to add up the electric flux all over a closed surface, we can remind ourselves of this with the following notation:

$$\text{electric flux on a closed surface} = \oint \vec{E} \bullet \hat{n} \, dA = \sum_{\text{closed surface}} \vec{E} \bullet \hat{n} \, \Delta A$$

A circle superimposed on an integral sign is a standard notation for "completeness," in this case, that the flux should be integrated over an entire closed surface—that is, a surface that completely encloses a volume in space.

QUESTION What is the net flux Φ_{el} on the closed surface shown in Figure 21.16?

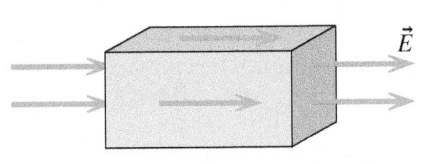

Figure 21.16 The electric field is uniform on each face of this closed surface.

The net flux on this surface is zero. Because the electric field on each face is uniform, it is sufficient in this case to take each ΔA to be the entire area of a face. The flux on the right end is $+EA_{\text{end}}$, the flux on the left end is $-EA_{\text{end}}$, and the flux on all other sides is 0 because on those sides $\vec{E} \perp \hat{n}$.

21.3 GAUSS'S LAW

We can now state Gauss's law, which we will then prove:

GAUSS'S LAW

$$\sum_{\text{closed surface}} \vec{E} \bullet \hat{n} \, \Delta A = \frac{\sum q_{\text{inside}}}{\varepsilon_0}$$

or

$$\oint \vec{E} \bullet \hat{n} \, dA = \frac{\sum q_{\text{inside}}}{\varepsilon_0}$$

When we add up the electric flux on the entire area of a closed surface, this quantitative measure of the pattern of electric field does turn out to be proportional to the amount of charge inside that closed surface, as we had guessed from the patterns we saw at the beginning of this chapter.

Proving Gauss's Law

What do we need to do in order to prove Gauss's law?

- Show that the proportionality constant is indeed $1/\varepsilon_0$.
- Show that the size and shape of the surface don't matter.
- Show that it is true for any number of point charges inside the surface.
- Show that charges outside the surface contribute zero net flux.

The following sections give a formal proof of the correctness of Gauss's law, by addressing each of these issues. The proof is a geometry-based proof rather than an algebra-based proof. By now you should be convinced that there is indeed some kind of relationship between the charge inside a closed surface and the total electric flux on that closed surface. The formal proof nails this down, quantitatively.

Determining the Proportionality Constant

We have postulated that the net electric flux on a closed surface is proportional to the charge inside:

$$\sum_{\text{closed surface}} \vec{E} \bullet \hat{n}\Delta A = k \sum q_{\text{inside}}$$

and that the proportionality constant k is equal to $1/\varepsilon_0$.

To show this, we will calculate the net electric flux for a surface for which we can make the calculation easily: a closed spherical surface surrounding a positive point charge $+Q$ at the center of the sphere (Figure 21.17). It is extremely important to understand that this spherical surface is of our own choosing and is purely mathematical in nature: it has nothing to do with any actual spherical shell made of some material.

Note that because it is hard to draw and interpret three-dimensional diagrams, we will usually draw two-dimensional diagrams that are cross-sectional slices through the 3D surface. You must, however, keep in mind that the spherical surface we have drawn around the charge is a closed 3D surface that encloses the charge Q.

> QUESTION We need to divide the spherical surface into small pieces ΔA and evaluate $\vec{E} \bullet \hat{n}\,\Delta A$ for each piece. Because the closed spherical surface is centered on the point charge, what can you say about the angle θ in $\vec{E} \bullet \hat{n} = E\cos\theta$, at all locations on the surface? (See Figure 21.18.)

Since both the electric field $\vec{E}$ and the outward-going normal $\hat{n}$ are perpendicular to our chosen spherical surface at all locations on the surface, the angle between the two vectors is zero, and $\cos\theta = +1$.

> QUESTION We need to add up $E\cos\theta\,\Delta A$ over the entire surface. What can you say about the magnitude of E at all locations on the surface, if the radius of the surface is r?

Because we chose our mathematical surface to be centered on the charge Q, the magnitude of the electric field at all locations on the sphere is simply

$$E = \frac{1}{4\pi\varepsilon_0}\frac{Q}{r^2}$$

> QUESTION The total surface area of a sphere is $4\pi r^2$. If you add up all the pieces $E\cos\theta\Delta A$ over the entire surface, what do you get, in terms of the charge Q at the center of the sphere?

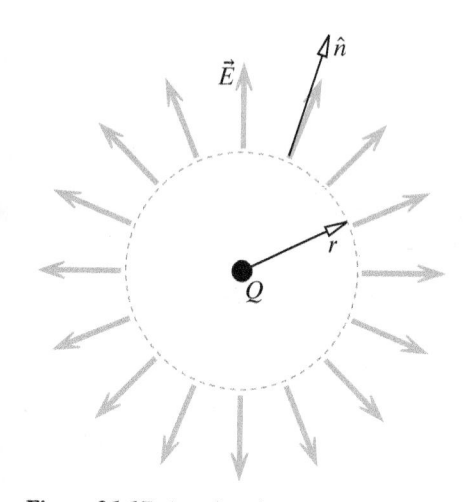

Figure 21.17 A point charge Q surrounded by a mathematical spherical surface of our choosing, of radius r.

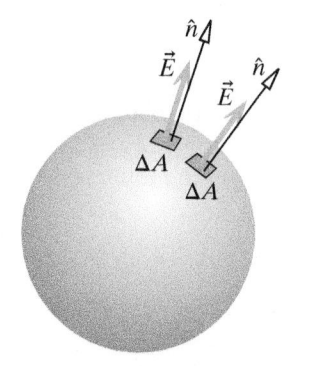

Figure 21.18 We need to calculate the flux on every little piece ΔA of the surface.

$E \cos\theta \, \Delta A$ is simply equal to $E \, \Delta A$ (since $\cos\theta = +1$), and E is constant, so the net flux over the entire surface is given by

$$\sum_{\text{closed surface}} \vec{E} \bullet \hat{n} \, \Delta A = \left(\frac{1}{4\pi\varepsilon_0} \frac{Q}{r^2} \right) (+1)(4\pi r^2) = \frac{Q}{\varepsilon_0}$$

We expect the total sum of electric flux on our chosen closed surface to be proportional to the charge inside the closed surface:

$$\sum_{\text{closed surface}} \vec{E} \bullet \hat{n} \, \Delta A = k q_{\text{inside}}$$

QUESTION What is the proportionality factor k?

Evidently for the spherical surface we get $k = 1/\varepsilon_0$, since the charge inside is simply equal to Q. This looks promising. We've got a quantitative measure of the electric field pattern on a closed surface (the total electric flux), and at least in this one case of a spherical surface centered on a point charge the total electric flux is equal to $(1/\varepsilon_0)$ times the charge inside the surface.

QUESTION What if the charge at the center of our chosen sphere were negative $(-Q)$? Would the relationship still hold? Check that it does, by considering what $\cos\theta$ is when the inside charge is negative.

In this case the flux would be negative (due to the cosine factor being -1), and proportional to the negative charge inside $(-Q)$. This is additional evidence that the $\cos\theta$ term is an appropriate part of the definition of flux.

The Size of the Surface Doesn't Matter

Note that since r does not appear in the final result, apparently this relationship doesn't depend on the radius r we chose for the spherical surface. Why is this? Suppose that we consider another spherical surface with twice the radius. We draw cross sections of the two spherical surfaces S_1 and S_2 (Figure 21.19).

QUESTION How does E_2 on surface S_2 differ from E_1 on surface S_1? How does the total surface area differ (the surface area of a sphere is $4\pi r^2$)? How does the total electric flux differ? Is $\sum_{\text{closed surface}} \vec{E} \bullet \hat{n} \, \Delta A = q_{\text{inside}}/\varepsilon_0$ valid for both surfaces?

The magnitude of the electric field drops by a factor of 4, since it is proportional to $1/r^2$. The surface area increases by the same factor of 4, since the surface area is proportional to r^2. Therefore the electric flux on the outer surface is exactly the same as on the inner surface, and the same proportionality factor $k = 1/\varepsilon_0$ holds. This example shows why we needed to include the area in the definition of electric flux. Note that this result depends critically on the fact that the electric field of a point charge is proportional to $1/r^2$. If it were proportional to $1/r^{2.001}$, say, the electric flux would be smaller on the outer surface than on the inner surface, and the concept of electric flux wouldn't be very useful.

We have shown that

$$\sum_{\text{closed surface}} \vec{E} \bullet \hat{n} \, \Delta A = \frac{q_{\text{inside}}}{\varepsilon_0}$$

is true no matter what the sign of q_{inside} is, or how large we draw the enclosing spherical surface.

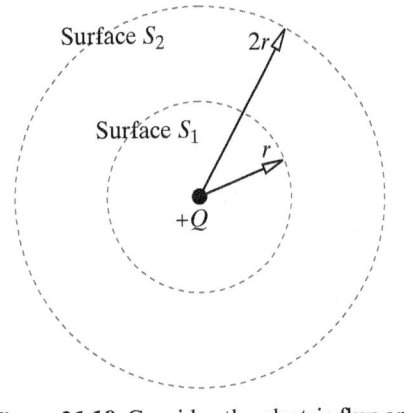

Figure 21.19 Consider the electric flux on two different spherical surfaces, chosen by us (and shown in cross section).

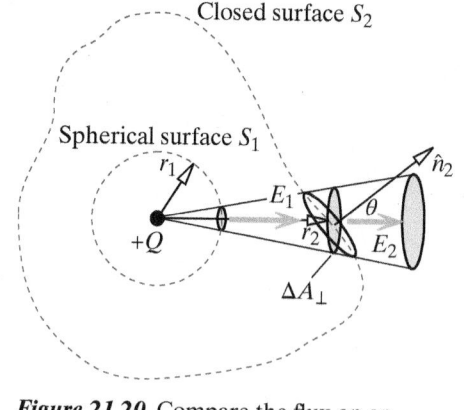

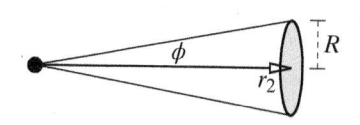

Figure 21.20 Compare the flux on an arbitrary closed surface with the flux on a small spherical surface.

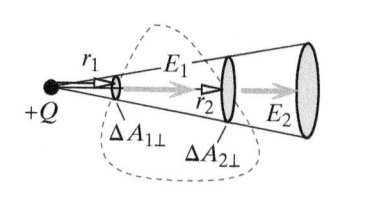

Figure 21.21 The perpendicular area $\Delta A_\perp = \pi R^2$, which is proportional to r_2^2.

The Shape of the Surface Doesn't Matter

Next consider an arbitrary closed surface surrounding the point charge, with a small spherical surface of radius r_1 for comparison (Figure 21.20). Remember that only cross sections of the surfaces are shown; the two dashed curves really represent closed surfaces that enclose the charge Q. We draw a cone that includes a small area of both surfaces.

The quantity $\Delta A_\perp = \Delta A \cos\theta$ in Figure 21.20 is the projection of the actual area ΔA onto a spherical surface drawn through that location, centered on the point charge (Figure 21.21). The radius R of this circular projected area is proportional to the distance r_2 from the charge Q; $R = r_2 \tan\phi$. Therefore the area $\Delta A_\perp$ is proportional to r_2^2.

> QUESTION Explain why the electric flux on the portion of the outer surface enclosed by the cone in Figure 21.20 is equal to the electric flux on the corresponding portion of the inner spherical surface. That is, why is $E_2 \Delta A_{2\perp}$ equal to $E_1 \Delta A_{1\perp}$?

We have $E_2/E_1 = r_1^2/r_2^2$, since the magnitude of the electric field is proportional to $1/r^2$. We also have $\Delta A_{2\perp}/\Delta A_{1\perp} = r_2^2/r_1^2$, since the circular projected area is proportional to r^2. Therefore $(E_2 \Delta A_{2\perp})/(E_1 \Delta A_{1\perp})$ is equal to 1.

> QUESTION If in every direction we find that the electric flux on any portion of the outer surface is the same as the electric flux on the projection of that area onto the small sphere, how does the total flux on the outer surface compare to the total flux on the inner sphere?

Evidently the total electric flux on the arbitrarily shaped outer surface is exactly equal to the total electric flux on the inner sphere. This is really encouraging! We've shown that

$$\sum_{\substack{\text{closed surface}}} \vec{E} \bullet \hat{n}\, \Delta A = \frac{q_{\text{inside}}}{\varepsilon_0}$$

is valid not only for an enclosing spherical surface but also that it holds true for any enclosing surface whatsoever. Again, the key reason why this works is the $1/r^2$ property of Coulomb's law: the electric field on a larger enclosing surface is exactly enough smaller to make the product of field and area come out to be exactly the same. If electric field didn't fall off exactly like $1/r^2$, electric flux wouldn't be a useful concept, because the total flux on differently shaped enclosing surfaces would differ.

Charges Outside the Surface Contribute Zero Net Flux

For the equation to be valid in all cases, it should be true that the total flux is zero if there is no charge inside the closed surface, even if there are other charges *outside* the closed surface. Consider the situation in Figure 21.22, where a point charge $+Q$ is outside a closed surface (shown in cross section).

> QUESTION Draw a cone from the charge, cutting through the near and far sides of the closed surface. Again note that $\Delta A_\perp = \Delta A \cos\theta$ is the projection of the actual area ΔA at the two locations where the cone pierces the surface. Show that the magnitude of the electric flux on the inner surface and on the outer surface is the same but that they differ in sign.

Figure 21.22 A point charge outside the surface (shown dashed) contributes zero net flux.

We can use the same argument we made before about the flux on two surfaces enclosed within a cone. However, in this case the flux on the first surface is negative (electric field pointing into the volume), whereas on the second surface the flux is positive (electric field pointing out of the volume). Therefore the fluxes add up to zero inside the cone.

> QUESTION Since this is true for any pair of inner and outer sections of the closed surface, what is the total electric flux over the entire surface due to the point charge outside the closed surface?

This cancellation holds true for any cone we draw, so the net flux added up over the entire surface is zero.

Point charges outside a closed surface certainly do produce electric field and electric flux on the surface, but when we add up all the contributions we find the total electric flux due to these outside charges is zero:

$$\sum_{\text{closed surface}} \vec{E} \bullet \hat{n}\,\Delta A = \frac{q_{\text{inside}}}{\varepsilon_0}$$

Superposition and Gauss's Law

Consider three point charges, two inside and one outside of a mathematical closed surface (Figure 21.23). For each individual charge we have

$$\sum_{\text{closed surface}} \vec{E}_1 \bullet \hat{n}\,\Delta A = Q_1/\varepsilon_0$$

$$\sum_{\text{closed surface}} \vec{E}_2 \bullet \hat{n}\,\Delta A = Q_2/\varepsilon_0$$

$$\sum_{\text{closed surface}} \vec{E}_3 \bullet \hat{n}\,\Delta A = 0 \quad \text{(since } Q_3 \text{ is outside the surface)}$$

If we add these up, we have

$$\sum_{\text{closed surface}} (\vec{E}_1 + \vec{E}_2 + \vec{E}_3) \bullet \hat{n}\,\Delta A = \frac{Q_1 + Q_2}{\varepsilon_0}$$

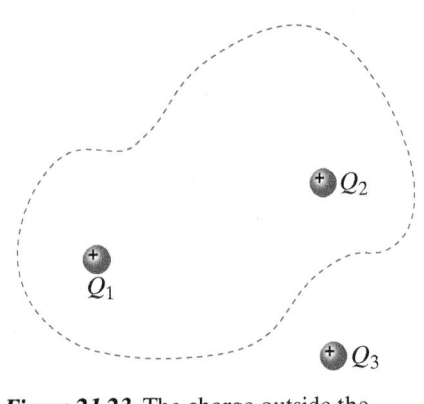

Figure 21.23 The charge outside the surface contributes zero to the net electric flux.

However, $\vec{E}_1 + \vec{E}_2 + \vec{E}_3$ is the net electric field due to all the charges, both inside and outside the surface, and it is this net field that we would measure on the surface. Generalizing to any number of point charges inside and outside a closed surface, we have proved what is called Gauss's law:

GAUSS'S LAW

$$\sum_{\text{closed surface}} \vec{E} \bullet \hat{n}\,\Delta A = \frac{\sum q_{\text{inside}}}{\varepsilon_0}$$

or

$$\oint \vec{E} \bullet \hat{n}\,\Delta A = \frac{\sum q_{\text{inside}}}{\varepsilon_0}$$

This is Gauss's law, the quantitative form of what we saw qualitatively: the pattern of electric field on a closed surface tells us something about the charges that are inside that closed surface. Now we see that by expressing the pattern of electric field on a surface quantitatively in terms of electric flux, we know exactly how much charge is inside our chosen surface, even though the electric field we measure may be partly due to charges outside the surface.

Gauss's Law and Coulomb's Law

Since we derived Gauss's law from Coulomb's law, it might seem that we should refer to "Gauss's theorem." However, it is possible in a similar way to derive

Coulomb's law from Gauss's law, so the two laws seem completely equivalent, despite their very different appearance.

However, Gauss's law and Coulomb's law are not equivalent if the charges are moving, and the two laws differ drastically in their predictions if the charges are moving at speeds that are comparable to the speed of light. The issue is retardation. The electric field at a location in space due to a fast-moving charge moving in a general way cannot be calculated from Coulomb's law but involves a complicated calculation that takes into account the past history of the charge's motion: the electric field at the observation location *now* depends on the location of the source charge *then*.

Unlike Coulomb's law, it turns out that Gauss's law is correct in all cases. Suppose that a large number of observers with synchronized watches are deployed all over a closed surface. By prior agreement, at 9:35 AM they all measure the electric flux on their own small piece of the surface and then report their measurements.

The sum of all these electric flux measurements is equal to the sum of all the charges (divided by ε_0) that were inside the surface at 9:35 AM, even if some of the charges were moving at speeds comparable to the speed of light. Thus Gauss's law is consistent with the special theory of relativity, but Coulomb's law isn't. (There is a practical limitation to this measurement: since the observers can't transmit their reports faster to us than the speed of light, by the time we add up all the flux contributions and thereby determine how much charge was inside the surface at 9:35 AM, the time is already quite a bit later!)

EXAMPLE   **A Cylindrical Surface**

The electric field is measured over a cylindrical Gaussian surface of radius 5 cm and height 15 cm, as shown in Figure 21.24. Everywhere on the curved part of the surface, the electric field $\vec{E}_1$ is found to be constant in magnitude (550 N/C), and to point radially outward. On the flat end caps, the field $\vec{E}_2$ varies in magnitude, but also points radially outward. **(a)** Find the amount of charge enclosed by the surface. **(b)** If the source of the field is a long uniformly charged thin rod, part of which is inside the surface, what is the linear charge density (charge per unit length, Q/L) on the long rod?

Solution   **(a)** On the curved portion of the surface, $\vec{E}_1 \| \hat{n}$ everywhere, so $\vec{E}_1 \cdot \hat{n} = \left| \vec{E}_1 \right|$.

$$\Phi_{\text{side}} = \left| \vec{E}_1 \right| (2\pi rh) = (550\,\text{N/C})(2\pi)(0.05\,\text{m})(0.15\,\text{m}) = 25.9\,\text{V} \cdot \text{m}$$

Everywhere on the flat top of the surface, $\vec{E}_2 \perp \hat{n}$, so $\vec{E}_2 \cdot \hat{n} = 0$ and $\Phi_{\text{top}} = 0$. Everywhere on the flat bottom of the surface, $\vec{E}_2 \perp \hat{n}$, so $\vec{E}_2 \cdot \hat{n} = 0$ and $\Phi_{\text{bottom}} = 0$.

$$\oint \vec{E} \cdot \hat{n}\, dA = (25.9 + 0 + 0)\,\text{V} \cdot \text{m} = 25.9\,\text{V} \cdot \text{m}$$

Applying Gauss's law:

$$\oint \vec{E} \cdot \hat{n}\, dA = \frac{q_{\text{inside}}}{\varepsilon_0}$$

$$q_{\text{inside}} = (25.9\,\text{V} \cdot \text{m})\left(8.85 \times 10^{-12} \frac{\text{C}^2}{\text{N} \cdot \text{m}^2} \right)$$

$$= 2.29 \times 10^{-10}\,\text{C}$$

(b) If the Gaussian cylinder encloses a section of a long uniformly charged rod, the charge density on the rod is

$$\frac{Q}{L} = \frac{2.29 \times 10^{-10}\,\text{C}}{0.15\,\text{m}} = 1.53 \times 10^{-9} \frac{\text{C}}{\text{m}}$$

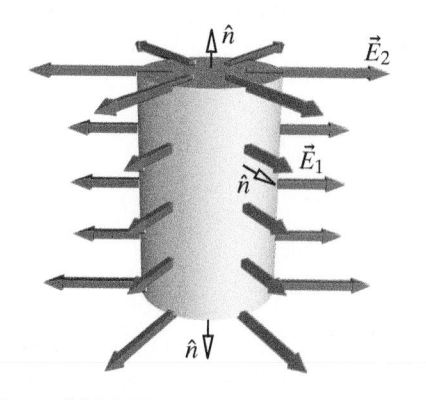

Figure 21.24 The electric field on a cylindrical Gaussian surface.

21.4 REASONING FROM GAUSS'S LAW

We've used the total electric flux on a closed surface to find out how much charge is inside that closed surface. This means that we are using knowledge of the electric field to find out something about charges. We can in some cases use Gauss's law in the other direction, to derive the magnitude of the electric field due to a charge distribution. We'll study several such examples. To apply Gauss's law you

- choose a (mathematical) closed surface that passes through locations where you know $\vec{E}$ or want to find $\vec{E}$, and
- that encloses a region in which you know what charges there are or want to find out what charges there are.

The Electric Field of a Large Plate

Suppose we don't know or don't remember the equation for the electric field near the center of a large uniformly charged plate. All we know is that near the center of the plate the electric field must be perpendicular to the plate, as shown in Figure 21.25.

EXAMPLE **The Electric Field of a Large Charged Plate**

Use Gauss's law to determine the equation for the electric field near the center of a very large uniformly charged plate (Figure 21.25). The area of one surface of the plate is A, and the total amount of charge on the plate is Q.

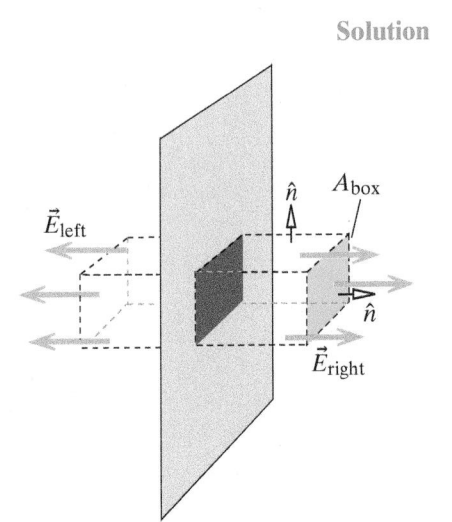

Figure 21.25 A portion of a large uniformly charged plate, with a Gaussian surface in the form of a box whose end area (shaded green) is A_{box}. The portion of the plate inside the box is shown darker than the rest of the plate, for clarity.

Solution We choose a box-shaped Gaussian surface, centered on the plate, as shown in Figure 21.25. Let the area of the end face of the Gaussian surface be A_{box}. Assume that we don't know the magnitude of either E_{right} or E_{left}, but by symmetry we can assume that they have the same magnitude $E_{\text{right}} = E_{\text{left}} = E$.

First, we find the total flux in terms of the unknown field E. If we draw the unit normal vector $\hat{n}$ on each face of the box, we see that there is nonzero electric flux only on the ends of the box, so the total electric flux over the entire box is the flux on the two ends, $\Phi_{\text{net}} = \oint \vec{E} \cdot \hat{n} \, dA = 2E(\cos 0)A_{\text{box}}$.

Second, we find an expression for the amount of charge inside the box. The fraction of the total area of the plate that is inside the box is (A_{box}/A), so the amount of charge inside the box is $Q(A_{\text{box}}/A)$.

Now we can apply Gauss's Law:

$$2EA_{\text{box}} = \frac{(Q/A)A_{\text{box}}}{\varepsilon_0}$$

$$E = \frac{(Q/A)}{2\varepsilon_0}$$

Contrast this simple calculation of the electric field near a large plate with the lengthy calculation for a disk that you did in Chapter 15!

Here you see the power of Gauss's law. In situations where we have some partial knowledge about the electric field (here it was the direction and symmetry of the field), it may be possible to determine the electric field due to a complex distribution of charges without carrying out a complicated integral.

Note that the area A_{box} of the end face of our chosen mathematical Gaussian surface drops out of the calculation. This must happen, because the physical result for the electric field cannot depend on the size of some mathematical surface that we happen to choose for our own convenience.

The Electric Field of a Uniform Spherical Shell

As a second example of using partial knowledge about the electric field pattern to deduce the magnitude of the electric field, consider a uniformly charged spherical shell. Recall from Chapter 15 that integrating Coulomb's law to get the electric field of this charge distribution is quite difficult.

EXAMPLE

The Electric Field of a Uniformly Charged Spherical Shell

Use Gauss's law to find equations for the electric field due to a uniformly charged hollow sphere of charge Q, at locations both outside and inside the sphere.

Solution

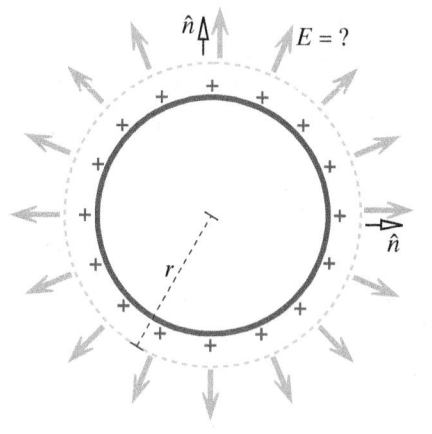

Figure 21.26 A spherical Gaussian surface (green) is drawn outside a uniformly charged spherical shell.

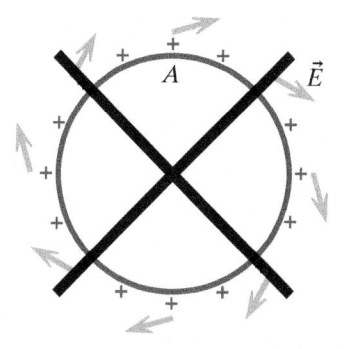

Figure 21.27 If the electric field of the charged sphere were not radial, it would have to look like this, which is impossible.

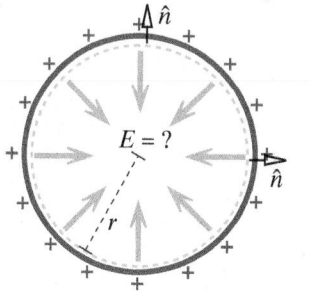

Figure 21.28 A spherical Gaussian surface (green) is drawn inside a uniformly charged spherical shell.

To find the field outside the shell, we draw a spherical Gaussian surface of radius r outside the sphere, centered on the sphere (Figure 21.26). The magnitude of the field on this surface is unknown, but because of the symmetry of the sphere and the Gaussian surface we reason that the field must be radial and uniform in magnitude. If the electric field weren't radial, by symmetry the pattern would have to look something like the one shown in Figure 21.27, but in that case the round-trip integral of electric field, $V_A - V_A$, wouldn't be zero; we can therefore rule out this field pattern.

Everywhere on the spherical Gaussian surface $\vec{E} \| \hat{n}$, so

$$\oint \vec{E} \cdot \hat{n} \, dA = EA = E(4\pi r^2)$$

All of the charge Q on the sphere is inside the surface. By Gauss's law,

$$\oint \vec{E} \cdot \hat{n} \, dA = \frac{q_{\text{inside}}}{\varepsilon_0}$$

$$E(4\pi r^2) = \frac{Q}{\varepsilon_0}$$

$$E = \frac{1}{4\pi\varepsilon_0} \frac{Q}{r^2}$$

Since the radius r could have any value, this result is valid for any location outside the sphere.

To find the field inside the shell, we draw a spherical Gaussian surface of radius r inside the charged sphere, as shown in Figure 21.28. Again, we conclude that $\vec{E}$ must be radial, because of the symmetry of the situation. The electric flux on the inner surface is now $-E(\cos 0)(4\pi r^2)$ (because the $\vec{E}$ we have drawn is opposite to $\hat{n}$), but the charge inside the inner surface is zero. By Gauss's law,

$$\oint \vec{E} \cdot \hat{n} \, dA = \frac{q_{\text{inside}}}{\varepsilon_0}$$

$$E(4\pi r^2) = \frac{0}{\varepsilon_0}$$

$$E = 0$$

We see that E must be zero at any location inside the sphere.

Note that we placed the Gaussian surface at the locations at which we wanted to determine the magnitude of the electric field—this is a key part of this kind of reasoning. In this example, Gauss's law allowed us to determine the magnitude of the electric field inside and outside of a uniformly charged spherical shell with very simple calculations. In contrast, determining the electric field of a uniformly charged spherical shell by integrating Coulomb's law over the charge distribution involves rather difficult trigonometry and integral calculus (see Chapter 15). The calculation using potential in Chapter 16 was also much more difficult than using Gauss's law.

The Electric Field of a Uniform Cube

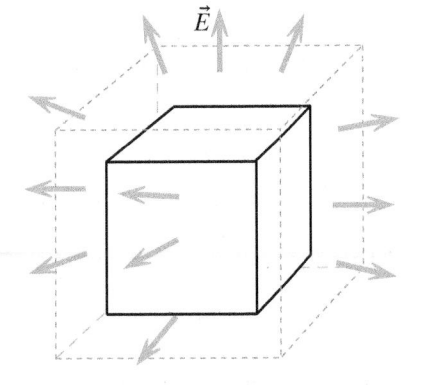

Figure 21.29 Trying to use Gauss's law to determine the electric field of a charged cube.

Next let's try to use Gauss's law to determine the electric field near a cube that is uniformly charged positive. This cube must be made of an insulator, because on a metal cube the charge would not be distributed uniformly. It seems plausible to choose a cubical box as the Gaussian surface surrounding the charged cube (Figure 21.29).

However, the electric field on a face of the cubical Gaussian surface is not uniform in direction or magnitude, as it was in the case of the large plate or the sphere.

QUESTION Is Gauss's law valid for this cubical Gaussian surface?

Gauss's law is always valid, no matter what the situation. It is definitely the case that the total electric flux over the entire Gaussian surface is numerically equal to the charge on the cube, divided by ε_0.

QUESTION Explain why we can't use Gauss's law to determine the electric field near a cube.

Although Gauss's law is valid, it doesn't help. The magnitude of the electric field varies all over the surface, and so does the angle between the electric field and the outward-going normal. There's no symmetry to simplify the integral, and neither E nor $\cos\theta$ can be factored out of the flux calculation. Therefore we can't carry out the flux calculation in terms of a single unknown quantity E.

Proving Some Important Properties of Metals

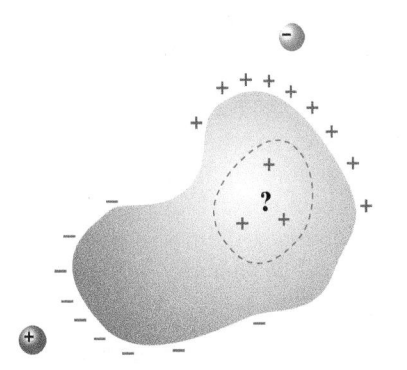

Figure 21.30 Can there be excess charge in the interior of a metal?

We have repeatedly stated that there cannot be any excess charge in the interior of a metal: any excess charge must be on a surface. Using Gauss's law, we can now prove this for the case of a metal in equilibrium. We will use "proof by contradiction," by assuming that there is some excess charge inside the metal and using Gauss's law to demonstrate a contradiction.

Consider a piece of metal that has been polarized by external charges and is in equilibrium. We assume (incorrectly) that there is some excess charge in the interior, as shown, and we draw a Gaussian surface (closed mathematical surface) that encloses that excess charge (Figure 21.30). The Gaussian surface is drawn in such a way that all elements of the surface are inside the metal. (Remember that this surface is just a mathematical construct, so it's fine to put it inside a solid object.)

QUESTION Since the metal is in equilibrium, what is the magnitude of the electric field everywhere on the Gaussian surface? What is the total electric flux on that surface? Using Gauss's law, what can you conclude about the charge inside the Gaussian surface?

Everywhere inside a metal in equilibrium, the electric field is zero, because if it were nonzero, electrons would flow and increase the polarization of the metal, until eventually in equilibrium the net field in the interior would become zero. Since the field is zero on our mathematical surface, which is inside the metal, the total electric flux on that surface must be zero. Gauss's law tells us that the charge inside the surface must be zero.

Thus we reach a contradiction, and our initial assumption must be wrong: in equilibrium there cannot be any excess charge in the interior of a metal—any excess charge therefore must be on the surface. Note that in this analysis we knew something about the electric field ($E = 0$), and from that we could deduce something about the charges.

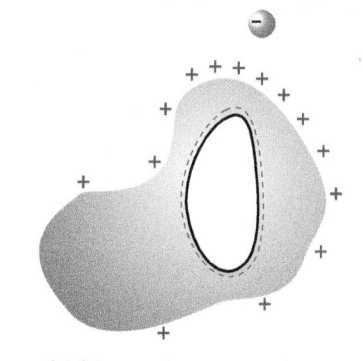

Figure 21.31 A hole inside a charged, polarized piece of metal.

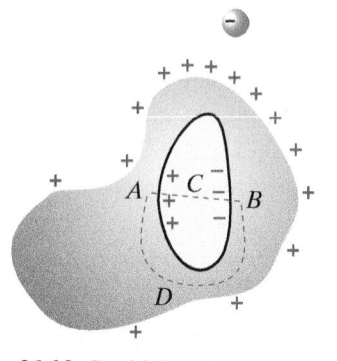

Figure 21.32 Could there be a nonzero electric field inside the hole, as long as the net charge on the surface of the hole is equal to zero?

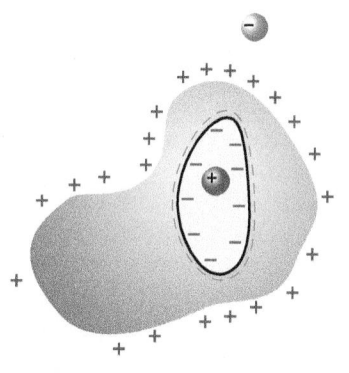

Figure 21.33 Now the hole is not empty but contains a +5 nC charge.

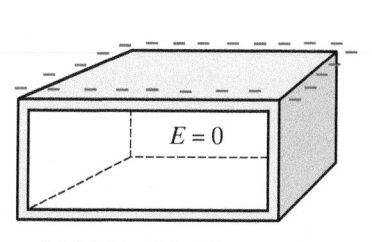

Figure 21.34 $E = 0$ inside an empty metal box, even though the box has a net charge and is polarized by an external charge.

QUESTION Suppose that there is a hole inside the metal as shown in Figure 21.31. Can there be charge on the inner surface? Using a Gaussian surface drawn in the metal but close to the boundary of the hole (Figure 21.31), what can you conclude about the net charge on the surface of the hole?

Again, all over a Gaussian surface inside a metal in equilibrium the electric field is zero, so the total electric flux is zero, and charge enclosed inside the Gaussian surface must be zero.

Is the electric field inside the hole zero or nonzero? It might conceivably be nonzero if there were + and − charges on this interior surface, as long as they add up to zero. However, we can show that E must be zero inside the hole. Consider the potential difference calculated along two different paths: A to C to B, and A to D to B (Figure 21.32).

QUESTION Along the path A to D to B, through the metal, what is the potential difference ΔV? Since potential difference is independent of path, what must be the potential difference along the path A to C to B?

All locations along the path A to D to B pass through the metal, where the electric field is zero, so the potential difference (path integral of the electric field) must be zero. There would be a potential difference along the path A to C to B if there were an electric field inside the hole made by charges on the surface of the hole, but the potential difference $V_B - V_A$ must be independent of path. Therefore we have reached a contradiction, and our assumption that there are charges on the surface of the hole must be false.

We have proved that in equilibrium

- there are no charges at all on the surface of an empty hole in a metal;
- the electric field is zero everywhere inside an empty hole in a metal.

Having gone through formal reasoning to show that an empty hole in a metal in equilibrium contains no charges or electric field, it may be helpful to think about this important result in a less formal way. If the metal were solid, the electric field in equilibrium would be zero throughout, and there would be no excess charge anywhere in the interior, as we concluded earlier. Imagine somehow removing lots of neutral, unpolarized atoms from the interior, as though they could be eaten away with acid. These neutral, unpolarized atoms contributed nothing to the electric field, so removing them will not affect the distribution of charge or field. Hence we conclude that the hole that we've made inside the metal will contain neither field nor charge.

QUESTION On the other hand, suppose that there is a charge of +5 nC floating inside the hole (Figure 21.33). Using Gauss's law with the indicated Gaussian surface, what can you conclude about the charge on the inner surface of the hole?

Since the flux on the chosen surface must be zero (because the electric field is zero all over this surface in the metal), the net charge inside the surface must be zero. Therefore there must be −5 nC of charge distributed on the inner surface of the hole. Thus in this case, where we have deliberately placed extra charge in the hole, there are charges and fields inside the hole.

"Screening"

We used Gauss's law (and potential theory) to show that in equilibrium the electric field is zero not only in the interior of a metal but even in an empty hole inside the metal. This has important practical consequences. A metal box "screens" its contents from the effects of charges outside the box (Figure 21.34).

For many purposes the box doesn't even have to have solid metal sides. A box with some holes in the sides, or a box with sides made of wire mesh or screen, can be quite effective in reducing the field inside the box.

> QUESTION How does this "screening" really work? Are the electric fields produced by the outside charges blocked from getting inside the box? What about the principle that electric fields pass right through matter? Can you explain the seeming paradox?

Electric field cannot be blocked; according to the superposition principle, the field contributed by the outside charges is definitely present in the box. However, the electric field contributed by the surface charges on the polarized metal is equal and opposite to the field due to the outside charges, making a net field that is zero.

Gauss's Law Applied to Circuits

All of the examples so far have dealt with equilibrium. However, Gauss's law expresses a completely general relationship between charges and electric field, so it applies to electric circuits, too. Figure 21.35 shows a portion of a current-carrying wire of constant cross section A and uniform composition.

> QUESTION In the steady state, excess charges arrange themselves on the surface of such a wire so that the electric field E has the same magnitude at different locations along the wire, and is everywhere parallel to the wire. In Figure 21.35 the dashed tube-shaped Gaussian surface nearly fills the wire. Use Gauss's law to prove that there is no excess charge in the interior of the wire.

Along the sides of the tube the electric field is parallel to the surface and contributes no flux. The right end of the tube contributes an amount of flux EA, but the left end contributes $-EA$, so the total electric flux on the Gaussian surface is zero, which means that there must not be any net charge inside the Gaussian surface; all excess charge is on the outer surface of the wire.

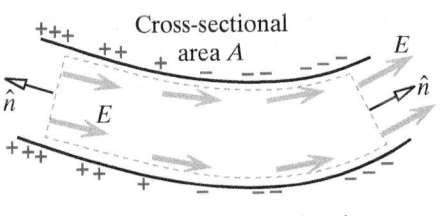

Figure 21.35 A Gaussian surface just inside a current-carrying wire in a circuit.

EXAMPLE **Charge at an Interface**

Figure 21.36 shows a portion of a circuit in which a steady-state electron current i is running as shown. A wire made of metal 1 joins a wire made of metal 2; both wires have uniform circular cross section of radius r. The electron mobility u_2 in metal 2 is much smaller than the electron mobility u_1 in metal 1. There are n_1 mobile electrons per unit volume in metal 1 and n_2 in metal 2; $n_1 > n_2$. Determine the amount of charge on the interface between metal 1 and metal 2, including sign. Explain carefully. Explicitly state all assumptions, and show how all quantities were evaluated.

Solution Start from fundamental principles:
By the current node rule, $i_1 = i_2$, so $n_1 A u_1 E_1 = n_2 A u_2 E_2$ and therefore

$$E_2 = \frac{n_1 u_1}{n_2 u_2} E_1$$

This makes sense, because E_2 must be greater than E_1. In order to create a bigger electric field in metal 2, it is likely that there is some excess charge at the interface.

Since we don't want to include charge on the outer surface of the wire, we choose a Gaussian surface inside the wire. We consider a tube-shaped Gaussian surface just inside the surface of the wire, shown by the dashed line in

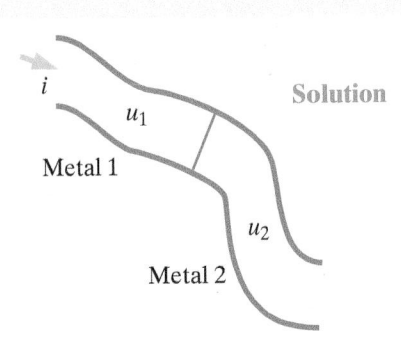

Figure 21.36 Current runs through a part of a circuit containing two different kinds of wire.

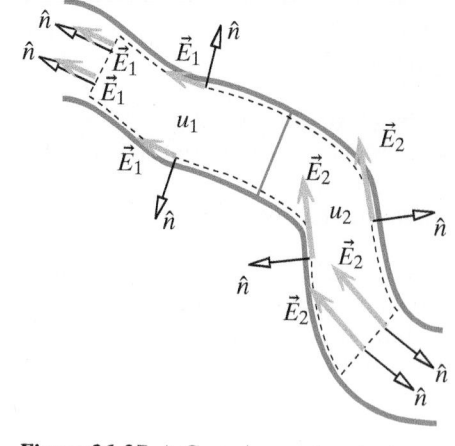

Figure 21.37 A Gaussian surface just inside the wire, with electric field and normal vector shown at several locations.

Figure 21.37. Any charge inside the Gaussian surface must be on the interface between the wires, since we just showed that in a uniform section of wire there could be no charge inside the wire.

By Gauss's law,

$$\oint \vec{E} \bullet d\vec{A} = \frac{\sum q_{inside}}{\varepsilon_0}$$

Assumptions: The electric field inside the wire is everywhere parallel to the wire, and is uniform in magnitude inside a wire of uniform composition. (We proved these assertions in Chapter 18.)

Given the direction of the electron current, the electric fields in the wire must be as shown in Figure 21.37. Evaluating the flux on different parts of the surface:

Outer surface of wire: flux $= 0$ because $\vec{E} \perp \hat{n}$
Top end (metal 1): flux $= E_1(\pi r^2)\cos 0° = +E_1(\pi r^2)$
Bottom end (metal 2): flux $= E_2(\pi r^2)\cos 180° = -E_2(\pi r^2)$

$$\oint \vec{E} \bullet d\vec{A} = 0 + E_1(\pi r^2) - E_2(\pi r^2)$$

Therefore the amount of charge on the interface is this, since $E = i/(nA\,u)$:

$$\sum q_{inside} = \varepsilon_0 (E_1(\pi r^2) - E_2(\pi r^2))$$
$$\sum q_{inside} = \varepsilon_0 i \left(\frac{1}{n_1 u_1} - \frac{1}{n_2 u_2} \right)$$

This quantity is negative, which makes sense, because the charge on the interface should be negative in order to increase E_2 and decrease E_1.

21.5 GAUSS'S LAW FOR MAGNETISM

There is a dramatic difference between electric and magnetic dipoles. The individual positive and negative electric charges ("monopoles") making up an electric dipole can be separated from each other, and these point charges make outward-going or inward-going electric fields (which vary like $1/r^2$). One might expect a magnetic dipole such as a bar magnet to be made of positive and negative magnetic monopoles, but although scientists have searched carefully, no one has ever found an individual magnetic monopole.

Such a magnetic monopole would presumably make an outward-going or inward-going magnetic field (which would vary like $1/r^2$), but such a pattern of magnetic field has never been observed (Figure 21.38). The nonobservation of a monopole field pattern can be expressed as a "Gauss's law" for magnetism—that the net flux of the magnetic field $\vec{B}$ over a closed surface is equal to the (nonexistent) "magnetic charge" inside:

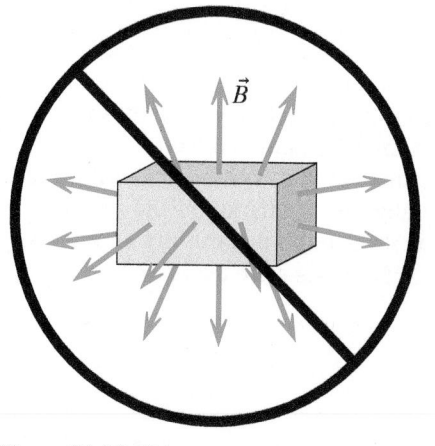

Figure 21.38 This pattern of magnetic field has never been observed.

GAUSS'S LAW FOR MAGNETISM

$$\oint \vec{B} \bullet \hat{n}\, dA = 0$$

As we saw in Chapter 17, a magnet can be thought of as consisting of atomic current loops (or electrons that act like current loops). If you cut a magnet in two, you don't get two magnetic monopoles—you just get two magnets!

21.6 PATTERNS OF MAGNETIC FIELD: AMPERE'S LAW

There is a relationship called Ampere's law that is an alternative version of the Biot–Savart law, just as Gauss's law is an alternative version of Coulomb's law. Ampere's law refers to a closed path rather than a closed surface.

We will study Ampere's law for the additional insights it provides about the relationship between current and magnetic field, and for certain important results that are most easily obtained using Ampere's law. Moreover, Ampere's law is relativistically correct, as is Gauss's law. Even more important, in Chapter 23 we will study Maxwell's extension of Ampere's law that makes a connection between time-varying electric and magnetic fields, which leads to understanding the nature of light.

Patterns of Magnetic Field

Moving charges produce magnetic fields, and we know how to calculate or estimate the magnitude and direction of the magnetic field due to a particular distribution of moving charges or currents. Sometimes, however, it is useful to reason in the other direction: from an observed pattern of magnetic field it may be possible to infer what currents are responsible for this pattern.

Suppose that you can't look inside the circular region shown in Figure 21.39, but you can measure the magnetic field all along the boundary of the region.

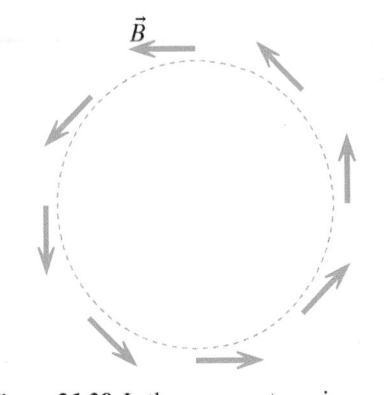

Figure 21.39 Is there current passing through this region? Does conventional current run into or out of the page?

QUESTION Can you conclude that there is some current inside the region? Does conventional current go into the page or out of the page?

A likely possibility is that a current-carrying wire is located at the center of the circle, with conventional current headed out of the page. This would produce the observed pattern of magnetic field.

Again, suppose that you can't look inside the circular region shown in Figure 21.40, but you can measure the magnetic field all along the boundary of the region.

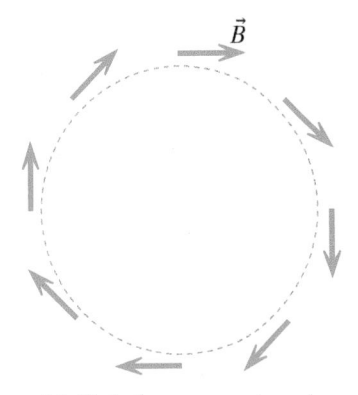

Figure 21.40 Is there current passing through this region? Does conventional current run into or out of the page?

QUESTION Can you conclude that there is some current inside the region? Does conventional current go into the page or out of the page?

A wire carrying conventional current into the page at the center of the circle would produce the observed pattern of magnetic field.

Ampere's law is a quantitative relationship between measurements of magnetic field along a closed path and the amount and direction of the currents passing through that boundary. Note the close parallel with Gauss's law, which relates measurements of electric field on a closed surface to the amount of charge inside that closed surface.

Quantifying the Magnetic Field Pattern

It is clear that there is some connection between the pattern of magnetic field along a closed path and the amount of current passing through the region bounded by that path, but how can we turn this into a quantitative relationship? We need to define a quantitative measure of the amount and direction of magnetic field along an entire closed path. Let's consider the magnetic field of a very long current-carrying wire a distance r from the wire (Figure 21.41), which we calculated using the Biot–Savart law in Chapter 17:

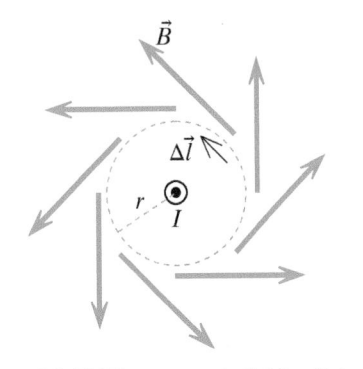

Figure 21.41 The magnetic field a distance r from a very long straight wire.

$$B_{\text{wire}} \approx \frac{\mu_0}{4\pi}\frac{2I}{r} \quad \text{(a distance } r \text{ from a very long wire)}$$

Given the curly character of the magnetic field around a wire, it is probably useful to integrate $\int \vec{B} \bullet d\vec{l}$ along a round-trip circular path around a wire and see what we get:

$$\oint \vec{B} \bullet d\vec{l} = \left(\frac{\mu_0}{4\pi} \frac{2I}{r} \right) \oint dl$$

because $\vec{B} \| d\vec{l}$ all along the path, and $r =$ constant,

$$\oint \vec{B} \bullet d\vec{l} = \left(\frac{\mu_0}{4\pi} \frac{2I}{r} \right) (2\pi r) = \mu_0 I$$

This is Ampère's law: the path integral of the magnetic field is equal to μ_0 times the current passing through the region enclosed by the path. (This is similar to Gauss's law, in which the surface integral of the electric field is equal to $1/\varepsilon_0$ times the charge inside the closed surface.)

In order to obtain a general proof of this quantitative relationship between current and magnetic field, we need to show that we can follow any path that surrounds the current, and that currents outside the path contribute zero to the path integral.

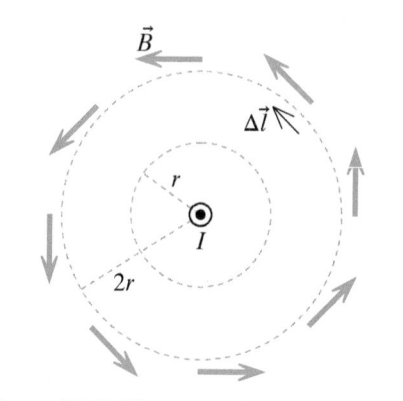

Figure 21.42 The magnetic field a distance $2r$ from a very long straight wire.

A Circular Path with Twice the Radius

It is easy to see that our result is independent of the radius r of the path we choose. For example, suppose that we choose a path along a larger circle of radius $2r$ (Figure 21.42).

QUESTION Why is the path integral the same along this longer path?

We know that the magnetic field near a very long wire is proportional to $1/r$, where r is the perpendicular distance to the wire. Therefore on a circle of twice the radius (and twice the circumference) the magnetic field is half as big, making the path integral be the same as before. The path integral of the magnetic field is independent of how big the circle is. This is similar to Gauss's law, where if you double the radius of a Gaussian sphere surrounding a charge, the surface area of the sphere goes up by a factor of four, but the electric field drops by a factor of four.

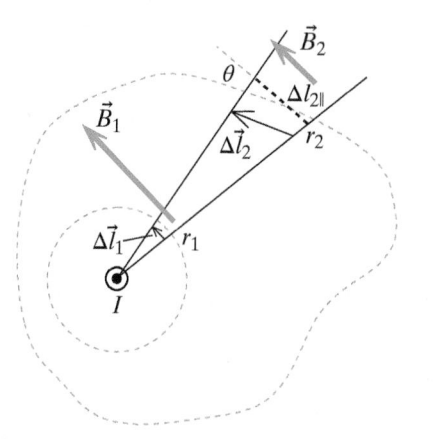

Figure 21.43 Comparing the path integral of magnetic field on a noncircular path to the path integral on a circular path.

A Noncircular Path

Suppose that we evaluate the path integral along a noncircular path. Focus on the contribution $\vec{B}_1 \bullet \Delta \vec{l}_1$ along a short arc of the circle and the contribution $\vec{B}_2 \bullet \Delta \vec{l}_2$ along the corresponding section of the noncircular path (Figure 21.43).

In Figure 21.43 the parallel projection $\Delta l_{2\|}$ of $\Delta \vec{l}_2$ along the direction of $\vec{B}_2$ is equal to $\Delta l_2 \cos \theta$, and by similar triangles it is also true that

$$\frac{\Delta l_{2\|}}{r_2} = \frac{\Delta l_1}{r_1}$$

Therefore

$$\vec{B}_2 \bullet \Delta \vec{l}_2 = B_2 \Delta l_2 \cos \theta = B_2 \Delta l_{2\|} = B_2 (r_2/r_1) \Delta l_1$$

However, since the magnetic field near a long straight wire is inversely proportional to the distance from the wire, we have $B_2/B_1 = r_1/r_2$, so

$$\vec{B}_2 \bullet \Delta \vec{l}_2 = B_1 \Delta l_1 = \vec{B}_1 \bullet \Delta \vec{l}_1$$

Since this is true for any arc, it must be that along the noncircular path $\oint \vec{B} \bullet d\vec{l}$ has the same value it has along the circular path, which is $\mu_0 I$. We have almost enough to prove Ampère's law. All that remains is to show that any current-carrying wires outside the closed path don't contribute.

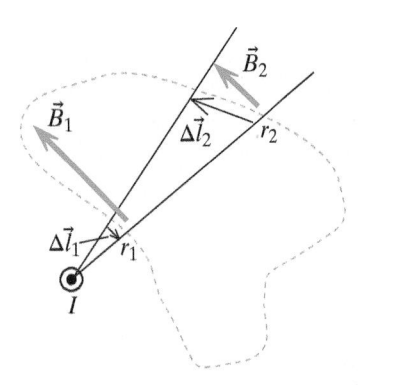

Figure 21.44 A current outside the path makes a zero net contribution to the path integral of the magnetic field.

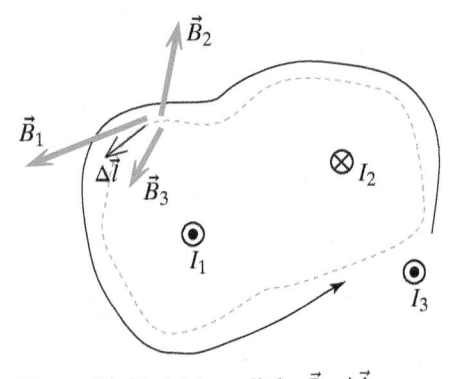

Figure 21.45 Add up all the $\vec{B} \bullet \Delta \vec{l}$ contributions along this counterclockwise path.

Effect of Currents Outside the Path

Consider Figure 21.44, in which there is a current outside the path.

> QUESTION Briefly explain why $\vec{B}_2 \bullet \Delta \vec{l}_2 = -\vec{B}_1 \bullet \Delta \vec{l}_1$ as we go around counterclockwise, and why $\oint \vec{B} \bullet d\vec{l} = 0$ in this case.

If $\Delta \vec{l}_1$ and $\Delta \vec{l}_2$ are short enough that they approximate arc segments on the two circles, then the length $\Delta l_{2\perp} = \Delta l_2 \cos \theta$ goes up by r but B goes down by $1/r$, so the magnitude of the product $B \Delta l_\perp$ is the same at both locations. As we go around the path counterclockwise, $\vec{B}_1 \bullet \Delta \vec{l}_1$ is negative but $\vec{B}_2 \bullet \Delta \vec{l}_2$ is positive, so $\vec{B}_1 \bullet \Delta \vec{l}_1 + \vec{B}_2 \bullet \Delta \vec{l}_2 = 0$. Adding up all such pairs, we find that a current that is outside the chosen closed path adds zero to the path integral $\oint \vec{B} \bullet d\vec{l}$.

Let's consider a specific example of three current-carrying wires (Figure 21.45). We draw a mathematical boundary that encloses two of the wires, and we walk counterclockwise along this path. For each individual current we have

$$\oint \vec{B}_1 \bullet d\vec{l} = \mu_0 I_1 \quad \text{(since } I_1 \text{ heads out of the page)}$$

$$\oint \vec{B}_2 \bullet d\vec{l} = -\mu_0 I_2 \quad \text{(since } I_2 \text{ heads into the page)}$$

$$\oint \vec{B}_3 \bullet d\vec{l} = 0 \quad \text{(since } I_3 \text{ is outside the boundary)}$$

If we add these up, we have

$$\oint (\vec{B}_1 + \vec{B}_2 + \vec{B}_3) \bullet d\vec{l} = \mu_0 (I_1 - I_2)$$

However, $(\vec{B}_1 + \vec{B}_2 + \vec{B}_3)$ is the net magnetic field $\vec{B}$ due to *all* the currents, both inside and outside the boundary, and it is this net field that we would measure along the boundary. Note that we count a current positive or negative depending on whether it makes a magnetic field in the direction of our counterclockwise walk along the path, or opposite to our walk, and this is equivalent to specifying the sign of the current depending on whether the current comes out of the page or goes into the page.

Putting it all together, we see that the path integral of magnetic field is equal to μ_0 times the sum of the enclosed currents (plus or minus), and currents outside the path contribute zero to the sum. This relationship is called Ampère's law:

AMPÈRE'S LAW

$$\oint \vec{B} \bullet d\vec{l} = \mu_0 \sum I_{\text{inside path}}$$

All the currents in the universe contribute by superposition to the magnetic field, yet Ampère's law refers only to those currents inside the chosen path because the outside currents contribute zero to the path integral. There is a strong parallel with Gauss's law.

Despite its very different form, Ampère's law is essentially equivalent to the Biot–Savart law from which it was derived. However, in some cases it is much easier to use Ampère's law than the Biot–Savart law. Moreover, Ampère's law is relativistically correct, whereas the Biot–Savart law is not, since it doesn't include retardation. This is just like the relationship between Gauss's law and Coulomb's law.

Note that Ampère's law involves the component of $\vec{B}$ parallel to a path, whereas Gauss's law involves the component of $\vec{E}$ perpendicular to a surface.

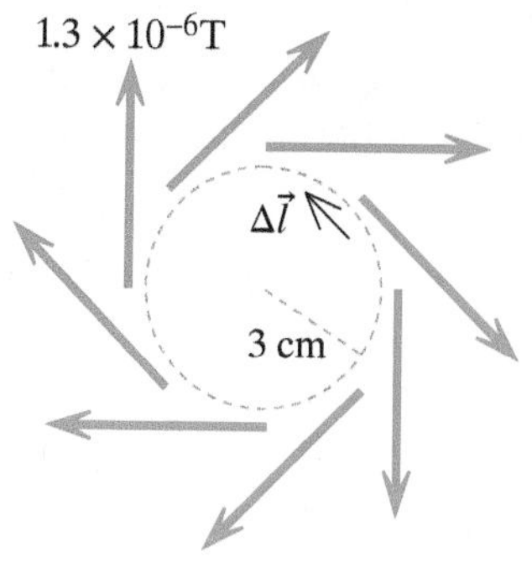

1.3 × 10⁻⁶T

1.3×10^{-6}T

$\Delta \vec{l}$

3 cm

Figure 21.46 A magnetic field of 1.3×10^{-6} T is measured everywhere on a circular path of radius 3 cm.

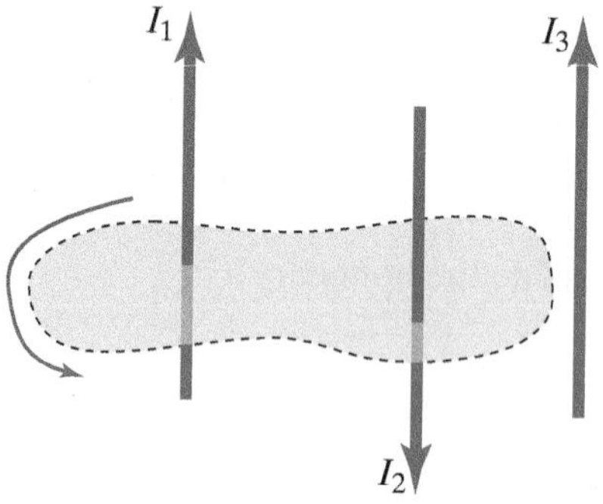

I_1

I_3

I_2

Figure 21.47 Curl the fingers of your right hand in the direction of the counterclockwise path. Piercings in the direction of your thumb (I_1) count positive; piercings in the opposite direction count negative (I_2). Don't count I_3, because it doesn't pierce the soap film.

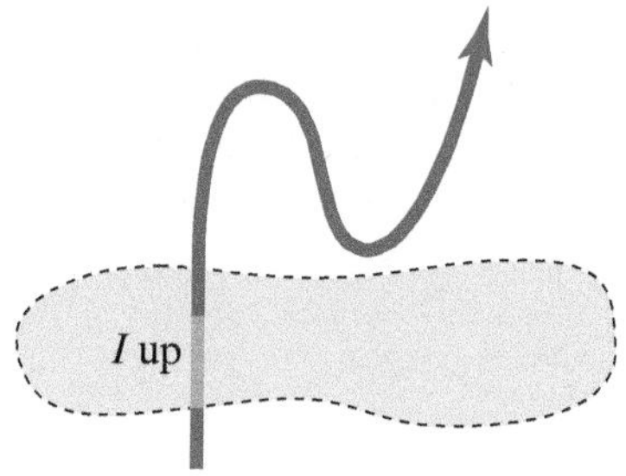

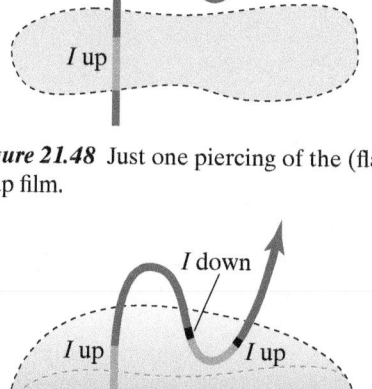

I up

Figure 21.48 Just one piercing of the (flat) soap film.

I down

I up

I up

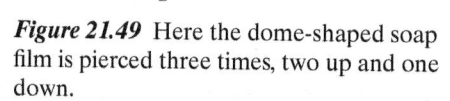

Figure 21.49 Here the dome-shaped soap film is pierced three times, two up and one down.

Checkpoint 2 Along the path shown in Figure 21.46 the magnetic field is measured and is found to be uniform in magnitude and always tangent to the circular path. If the radius of the path is 3 cm and B along the path is 1.3×10^{-6} T, what are the magnitude and direction of the current enclosed by the path?

Which Currents Are "Inside" the Path?

Before we apply Ampere's law to some important situations, we should pause to be more precise about how to decide what currents are "inside" a closed boundary. In the simple case of a single wire it is easy to say, but in more complicated cases this can be a bit tricky. We offer a procedure that is easy to follow and that moreover leads in a later chapter to an important extension of Ampere's law, made by Maxwell.

Just as every application of Gauss's law must start with choosing a closed surface to which to apply the law, so every application of Ampere's law must start with choosing a closed path that is the boundary of the region of interest.

After choosing a closed path, imagine stretching a soap film over your chosen path. You calculate the net current through the boundary by counting the current-carrying wires that pierce your soap film. You walk counterclockwise around the path, adding up $\oint \vec{B} \bullet d\vec{l}$ as you go. You count as positive any currents that pierce the soap film coming up through the film. To be more precise, let the fingers of your right hand curl counterclockwise around the path, and count a current as positive if it goes in the direction of your thumb. Any currents that pierce the film going in the opposite direction are counted as negative.

In Figure 21.47 we have $\oint \vec{B} \bullet d\vec{l} = \mu_0 \sum I_{\text{inside path}} = \mu_0(I_1 - I_2)$, since I_1 pierces the imaginary soap film in the upward direction, I_2 in the downward direction, and I_3 doesn't pierce the film.

Though we will not carry out the proof (which is based on dividing up the soap film into little areas and applying Ampere's law to the boundary of each area, then adding up the results), it can be shown that Ampere's law is still valid even if the closed boundary path does not lie in a plane, or even if the soap film that is anchored on the boundary is deformed into a nonflat shape. Certainly in the case of the straight current-carrying wires shown above, deforming the soap film upward or downward would make no difference in counting the piercings.

For a more interesting case, consider a wire with an S-shaped bend in it, with a closed boundary drawn around the wire as shown in Figure 21.48. How much current is "inside" the boundary? If we count the current by counting the piercings of the soap film, we can show that we get the same (correct) answer for any and all deformations of the soap film, as long as the film stays anchored at its edges to the boundary path.

In Figure 21.48, there is just one piercing of the imaginary soap film, and $\sum I = +I$. In Figure 21.49, there are three piercings of the imaginary soap film, but when we add up all the currents we again get the result that $\sum I = (+I) + (-I) + (+I) = +I$. Therefore, deforming the imaginary soap film does not affect the accounting for the amount of current enclosed by the boundary of that film (our chosen closed path).

Applications of Ampere's Law

We'll apply Ampere's law to some situations in which there are lots of currents distributed over a region. Here is a summary of the key steps in applying Ampere's law:

1. Choose a (mathematical) closed path as a boundary.
2. Stretch an imaginary soap film over the boundary.

3. Walk around the boundary counterclockwise, integrating $\oint \vec{B} \bullet d\vec{l}$.

4. Add up the positive and negative currents that pierce the soap film; this is $\sum I_{\text{inside path}}$. Count as positive those currents that pierce the imaginary soap film coming out of the film (that is, in the direction of your right thumb with the fingers of your right hand curling around in the direction of your walk); currents that pierce the film going into the film are counted as negative.

5. Apply Ampere's law, equating the two sums:

$$\oint \vec{B} \bullet d\vec{l} = \mu_0 \sum I_{\text{inside path}}$$

We'll apply this procedure to find the magnetic field due to a long thick wire, a solenoid, and a toroid.

A Long Thick Wire

Consider the problem of determining the magnetic field near a long *thick* wire (Figure 21.50). If we use the Biot–Savart law, we face the difficult problem of setting up and evaluating an integral not only along the length of the wire but also across the thickness of the wire, with a varying distance to those current elements. However, the result is very easy to get using Ampere's law.

Draw a circular path of radius r around the thick circular wire, assuming by symmetry that the magnetic field B is tangent to the path and constant in magnitude. Stretch an imaginary soap film over this circular path.

QUESTION Using Ampere's law, calculate the magnitude of the magnetic field at a distance r from the center of a long thick wire.

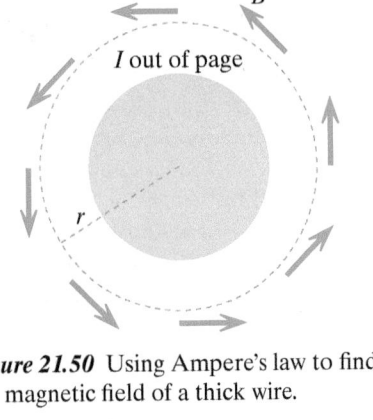

Figure 21.50 Using Ampere's law to find the magnetic field of a thick wire.

The path integral of the magnetic field is simply $B(2\pi r)$, since B is by symmetry constant in magnitude and is always parallel (tangential) to the circular path. The amount of current piercing the imaginary soap film is $\mu_0 I$, so we have this:

$$B(2\pi r) = \mu_0 I$$
$$B = \frac{\mu_0 I}{2\pi r} = \frac{\mu_0}{4\pi}\frac{2I}{r} \quad \text{(thick wire)}$$

This simple result (equivalent to the magnetic field of a thin wire) would have been very difficult to obtain using the Biot–Savart law directly. Even though we derived Ampere's law from the Biot–Savart law, there are symmetrical situations for which it is much easier to use Ampere's law.

A Solenoid

A long coil of wire with a small diameter is called a "solenoid." A simple electromagnet made by winding a wire on a nail is an iron-filled solenoid. We can calculate the magnetic field along the axis of a solenoid consisting of lots of circular loops. It turns out that the magnetic field is nearly the same in magnitude and direction throughout the interior of the solenoid, as long as you don't get too near the ends. One of the uses of solenoids is to make a uniform magnetic field. Solenoids also play an important role in some kinds of electronic circuits.

Figure 21.51 shows a portion of a very long solenoid. We assume that if we are far from the ends, the magnetic field inside the solenoid is fairly uniform and the magnetic field outside is very small due to near cancellation of the contributions of the near and far sides of the loops. (See the analysis using the Biot–Savart law, at the end of Chapter 17.)

We choose a rectangular path as shown in Figure 21.51, and we stretch an imaginary soap film over the rectangular frame. If there are N loops of wire

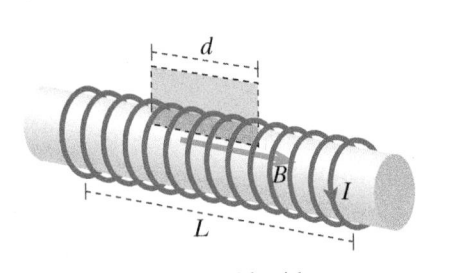

Figure 21.51 A solenoid, with a rectangular path chosen in such a way that the surface bounded by the path is pierced by some of the current-carrying wires.

along the entire length L of the solenoid, there are $(N/L)d$ wires that pierce the soap film, coming out of the page.

> QUESTION Use Ampere's law to show that the magnetic field inside the solenoid is $B \approx \mu_0 NI/L$, which agrees with the much more difficult analysis at the end of Chapter 17 based on the Biot–Savart law.

The path integral is just Bd, because along the two ends of the rectangular path the magnetic field is perpendicular to the path, and along the outside portion of the path the magnetic field is assumed to be very small. The current piercing the soap film is the number of piercings times I, or $[(N/L)d]I$. Ampere's law says that

$$Bd = \mu_0[(N/L)d]I$$

from which we easily conclude this:

$$B = \frac{\mu_0 NI}{L} \quad \text{(solenoid)}$$

The length d does not appear in the result, because the magnetic field cannot depend on the length of the mathematical path that we chose.

No matter where we choose to place that part of the path that is inside the solenoid, we would get this same result. That means that we expect the magnetic field to be quite uniform inside the solenoid, as long as we are far from the ends. This is confirmed by a numerical computation using the Biot–Savart law, as shown in Figure 21.52 (see the computational problem at the end of Chapter 17). The computation includes locations outside the solenoid, but the arrows representing the field there are too small to be seen. Clearly, our assumption is justified that the field is small outside the solenoid, far from the ends.

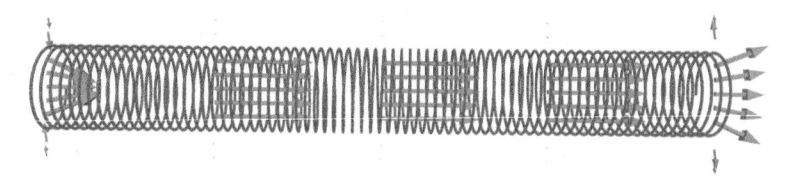

Figure 21.52

A Toroid

Figure 21.53 shows a toroid, a "doughnut" wrapped with wire that is essentially a solenoid bent around to make the ends meet. By symmetry, the magnetic field on a circular path inside the toroid is tangent to the path and constant in magnitude, as shown in a top view in Figure 21.54.

We draw a closed path in the form of a circle of radius r inside the windings. If there are N loops of wire wound on the solenoid, there are N wires that pierce a soap film stretched over that closed path.

> QUESTION Use Ampere's law to show that the magnetic field inside the toroid, a distance r from the center, is $B = (\mu_0 NI)/(2\pi r)$.

The path integral of the magnetic field is simply $B(2\pi r)$, since $\vec{B}$ is constant in magnitude and always tangential (parallel to $\Delta \vec{l}$). The current piercing the soap film is NI, so Ampere's law is this:

$$B(2\pi r) = \mu_0 NI$$

From this we find the magnetic field:

$$B = \frac{\mu_0 NI}{2\pi r} \quad \text{(toroid)}$$

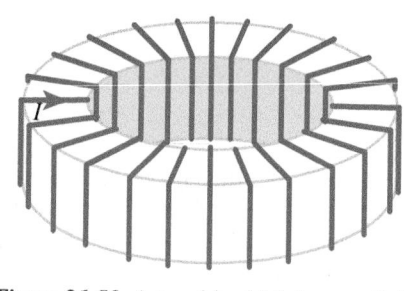

Figure 21.53 A toroid, which is essentially a solenoid bent around to make the ends meet.

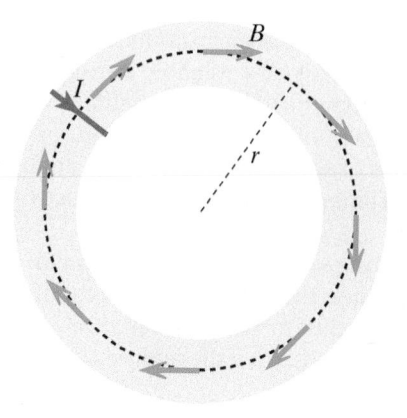

Figure 21.54 Top view of the toroid.

Note that B is proportional to $1/r$, and the magnetic field varies across the cross section of the toroid. We could not have easily used the Biot–Savart law to obtain this result, because the integration over all the current elements would have been extremely difficult. Yet Ampere's law gives us the result right away, with little calculation.

21.7 MAXWELL'S EQUATIONS

In this chapter we have developed three new equations: Gauss's law for electricity, Gauss's law for magnetism, and Ampere's law for magnetism. In a way, we already have an "Ampere's law for electricity." The round-trip path integral of the electric field produced by stationary or slowly moving point charges is zero:

$$\oint \vec{E} \bullet d\vec{l} = 0 \quad \text{since round-trip potential difference is zero}$$

The four equations taken together are the basis for the four equations known collectively as "Maxwell's equations," though two of the equations are incomplete. What we have just called "Ampere's law for electricity" when completed is called "Faraday's law" and is the subject of Chapter 22. What we have called "Ampere's law" when completed is called the "Ampere–Maxwell law" and is discussed in Chapter 23. In summary, here are the four equations:

MAXWELL'S EQUATIONS (INCOMPLETE)

$$\oint \vec{E} \bullet \hat{n}\, dA = \frac{\sum q_{\text{inside}}}{\varepsilon_0} \qquad \text{Gauss's law for electricity}$$

$$\oint \vec{B} \bullet \hat{n}\, dA = 0 \qquad \text{Gauss's law for magnetism}$$

$$\oint \vec{E} \bullet d\vec{l} = 0 \qquad \text{Incomplete version of Faraday's law}$$

$$\oint \vec{B} \bullet d\vec{l} = \mu_0 \sum I_{\text{inside path}} \qquad \text{Ampere's law (incomplete version of Ampere–Maxwell law)}$$

Note the structure of these laws. Two of the equations involve integrals over a surface (for electric field or magnetic field), and two involve integrals along a path (for electric field or magnetic field). The incompleteness in the last two laws comes from effects we have not yet studied. The complete forms of these laws include terms involving the time derivatives of electric and magnetic fields.

21.8 SEMICONDUCTOR DEVICES

On the student web site for this textbook, www.wiley.com/college/chabay, there is available at no charge an optional Supplement S2, "Semiconductor Devices." You now have the tools and concepts you need to be able to learn about these elements of modern electronic devices.

21.9 *THE DIFFERENTIAL FORM OF GAUSS'S LAW

This section deals with an advanced topic, which is not required as background for later chapters. It includes the important issue of the relativistic correctness of Gauss's law.

We have pointed out several times that unless charges are motionless or moving very slowly compared to the speed of light, there is a flaw in an attempt to calculate electric field at some location by using Coulomb's law. The problem is relativistic retardation. The electric field depends on where fast-moving particles were at some time in the past, not where they are now. Therefore, Coulomb's law is not consistent with the theory of special relativity. Gauss's law, however, *is* consistent with relativity. In this section we will see why.

One way we could avoid problems with retardation would be to come up with some physical law that would relate a property of the electric field at some location and time (x, y, z, t) to source charges at the *same* location and time (x, y, z, t). If we could write our law in this form, relating source and fields at the same location and time, retardation wouldn't be a problem, and our law could be consistent with special relativity, unlike Coulomb's law.

In brief, we're looking for

- a property of the electric field $\vec{E}$ at some (x, y, z, t)
- that is related to source charges at the *same* (x, y, z, t).

There is a difficulty in this scheme with point charges, since the electric field is infinite at the location of a point charge. After we discuss the general idea, we'll explain how point charges are dealt with.

Divergence

We will now introduce an appropriate property of the electric field, called the "divergence," which has the characteristics we want: we can relate the divergence to the charges at the same location and time.

Consider a very small Gaussian surface around a location in space, enclosing a very small volume ΔV. (Here V is volume, not potential!) In this region there is a charge density ρ measured in coulombs per cubic meter (ρ is lowercase Greek rho). The charge inside the tiny volume is $\rho \Delta V$ (multiply coulombs per cubic meter times volume in cubic meters). From Gauss's law, we know that the net flux on the surface of the tiny volume is:

$$\oint \vec{E} \bullet \hat{n}\, dA = \frac{\rho \Delta V}{\varepsilon_0}$$

Divide by the tiny volume ΔV:

$$\frac{\oint \vec{E} \bullet \hat{n}\, dA}{\Delta V} = \frac{\rho}{\varepsilon_0}$$

On the left side of this equation we have the net flux per unit volume. In the limit as the volume becomes arbitrarily small, $\Delta V \to 0$, this ratio is called the "divergence" (div) of the electric field at this location:

$$\operatorname{div}(\vec{E}) \equiv \lim_{\Delta V \to 0} \frac{\oint \vec{E} \bullet \hat{n}\, dA}{\Delta V}$$

(Note that this is a scalar, not a vector.) This is the property we are looking for, because we now have an equation in a form that doesn't have retardation problems:

$$\operatorname{div}(\vec{E}) = \frac{\rho}{\varepsilon_0}$$

Why is this property of the electric field called "divergence"? Look again at the pictures at the beginning of this chapter and at Figure 21.55. If there is positive charge inside a (vanishingly small) volume, the electric field directions "diverge," the net electric flux is positive, and $\operatorname{div}(\vec{E})$ is positive. If there is negative charge inside the tiny volume, the electric field directions converge ("antidiverge"), the net electric flux is negative, and $\operatorname{div}(\vec{E})$ is

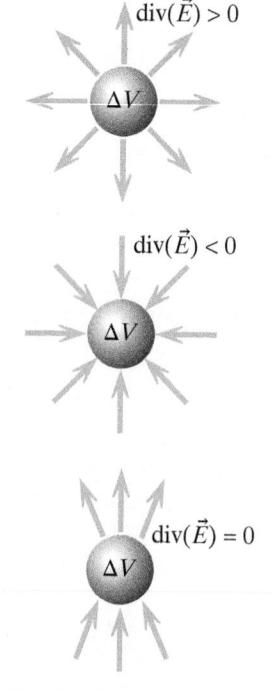

Figure 21.55 The divergence of the electric field at a location enclosed by a tiny volume $\Delta V \to 0$.

negative. If there is no charge inside the volume, the electric field looks as though it simply flows through the volume at this location with neither net divergence nor net convergence, the net electric flux is zero, and $\text{div}(\vec{E})$ is zero. In this introduction to the topic, we write the divergence as $\text{div}(\vec{E})$, an older notation that reminds us of the definition. Later we will introduce a different notation, $\vec{\nabla} \bullet \vec{E}$, that is normally used for this property.

Divergence Is Relativistically Correct

The divergence (net electric flux per unit volume, in the limit as the volume goes to zero) is a measure of an important property of the electric field at some location and time, and this is proportional to the charge density at that *same* location and time. This is called a "local" relationship between charge and field, because the charge and the divergence property are evaluated at the same location and time, rather than at different locations and times. This avoids the problem of relativistic retardation and forms the basis for what is called a "local field theory" that is consistent with special relativity.

In the beginning of this chapter we developed what is called the "integral form" of Gauss's law, involving an integral of electric flux over a (possibly large) closed surface. What we have just developed is the "differential form" of Gauss's law:

DIFFERENTIAL FORM OF GAUSS'S LAW

$$\text{div}(\vec{E}) = \frac{\rho}{\varepsilon_0}$$

where the divergence is defined as

$$\text{div}(\vec{E}) \equiv \lim_{\Delta V \to 0} \frac{\oint \vec{E} \bullet \hat{n}\, dA}{\Delta V}$$

which is a scalar quantity.

You can show that the units of the divergence of the electric field are the same as the units of ρ/ε_0.

Checkpoint 3 A lead nucleus is spherical with a radius of about 7×10^{-15} m. The nucleus contains 82 protons (and typically 126 neutrons). Because of their motions the protons can be considered on average to be uniformly distributed throughout the nucleus. Based on the net flux at the surface of the nucleus, calculate the divergence of the electric field as electric flux per unit volume. Repeat the calculation at a radius of 3.5×10^{-15} m. (You can use Gauss's law to determine the magnitude of the electric field at this radius.) Also calculate the quantity ρ/ε_0 inside the nucleus.

Why the Integral Form of Gauss's Law Is Relativistically Correct

In the beginning of this chapter we derived the integral form of Gauss's law by starting from Coulomb's law for stationary charges. The differential form was in turn derived from the integral form, yet the differential form of Gauss's law has the important property of being consistent with the theory of relativity, so it is actually more general than the Coulomb's law that we started from. Now we start from the differential form of Gauss's law, which is relativistically correct, and show that the integral form is also relativistically correct.

There is a purely mathematical theorem, the "divergence theorem," that shows that the volume integral of the divergence of a vector field,

$$\int \text{div}(\vec{E})\, dV$$

throughout a large volume at a particular time is equal to a familiar surface integral of the perpendicular component of that vector field,

$$\oint \vec{E} \bullet \hat{n}\, dA$$

on the closed bounding surface of the volume. In other words, the mathematical "divergence theorem" gives this result:

$$\int \text{div}(\vec{E})\, dV = \oint \vec{E} \bullet \hat{n}\, dA$$

However, $\text{div}(\vec{E}) = \rho/\varepsilon_0$, so

$$\int \text{div}(\vec{E})\, dV = \int \frac{\rho}{\varepsilon_0}\, dV = \frac{Q_{\text{inside}}}{\varepsilon_0}$$

and the volume integral of the charge density ρ is the charge contained inside the bounding surface at this moment in time. You can think of this as simultaneously calculating the divergence of every tiny volume ΔV inside the surface to get the amount of charge in that tiny volume, then adding up all these charges to get the total charge inside the surface.

Combining these results we obtain the integral form of Gauss's law, starting from the differential form, and using only a purely mathematical theorem in the derivation:

$$\oint \vec{E} \bullet \hat{n}\, dA = \frac{Q_{\text{inside}}}{\varepsilon_0}$$

At a fixed moment in time we have integrated the differential form of Gauss's law, which is relativistically valid, and we have recovered the integral form of Gauss's law. This is why even the integral form of Gauss's law, unlike Coulomb's law, is consistent with the special theory of relativity (see the earlier discussion in Section 21.3).

Dealing with Point Charges

When calculating the divergence by taking the limit as the volume goes to zero, $\Delta V \to 0$, if ΔV shrinks down onto the exact location of a point charge such as an electron, the limiting process doesn't work—the flux per unit volume becomes infinite. (At a location arbitrarily close to the point charge, there is no difficulty, and the charge density at that location is zero.)

The mathematical difficulty with point charges is handled formally through the use of the "delta function." The delta function is defined to be zero everywhere except at one point, where it is infinite, and its integral is exactly 1. In other words, the delta function is an infinite spike with zero width, constructed in such a way that the product of the infinite height and the zero width is finite (Figure 21.56). This peculiar entity is not a true function but is an example of a class of mathematical objects called "generalized functions" or "distributions."

An equivalent way of dealing with the mathematical problem is to approximate the point charge as a sphere of charge with a very small radius and a very large charge density. In that case the divergence limit is well defined

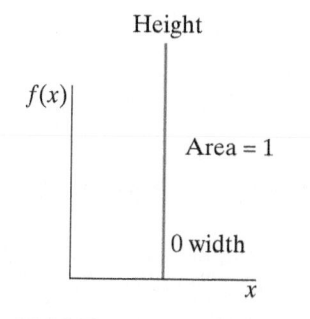

Figure 21.56 The one-dimensional delta function.

throughout the tiny sphere of charge, as it was in your calculations of the divergence inside a lead nucleus (Checkpoint 3).

In the practical case of the interior of ordinary atomic matter, we can stop shrinking the volume at a size that, though tiny, is still large enough to include lots of atoms. In this case we are averaging the divergence and the charge density over a tiny volume containing many protons and electrons. This is similar to what we did at the end of Chapter 16, where we found an average electric field inside an insulator, averaged over many atoms.

Another Way to Calculate the Divergence

It is possible to calculate the divergence directly from the definition, as a limit as $\Delta V \to 0$. In this section we present another way to calculate the divergence. Consider the tiny box-shaped surface shown in Figure 21.57. In this tiny region the electric field is in the x direction, but varying in magnitude. Let's use the definition of divergence to calculate this property at a location inside the surface, in the limit that the volume goes to zero:

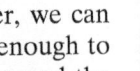

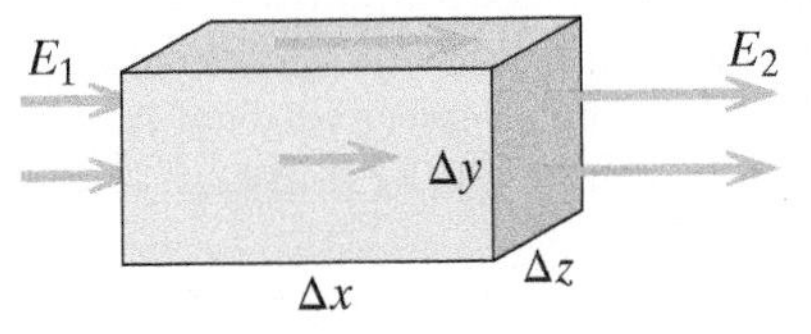

E_1 E_2

Δy

Δx Δz

Figure 21.57 Calculating the divergence of the electric field on the closed surface of a tiny volume.

$$\text{div}(\vec{E}) \equiv \lim_{\Delta V \to 0} \frac{\oint \vec{E} \cdot \hat{n}\, dA}{\Delta V} = \lim_{\Delta V \to 0} \frac{(E_2 - E_1)\Delta y \Delta z}{\Delta x \Delta y \Delta z}$$

$$\text{div}(\vec{E}) = \lim_{\Delta x \to 0} \frac{(E_2 - E_1)}{\Delta x}$$

However, this limit is the definition of the derivative of the electric field with respect to x:

$$\text{div}(\vec{E}) = \frac{dE}{dx}$$

More precisely, the divergence in this case is equal to the "partial derivative" of E_x with respect to x, where the special partial derivative notation indicates that we hold y and z constant while taking the derivative:

$$\text{div}(\vec{E}) = \frac{\partial E_x}{\partial x}$$

More generally, there may be contributions to the divergence involving E_y or E_z, which we would calculate in the same way. Adding together all these contributions to the divergence, we find the following:

DIFFERENTIAL FORM OF GAUSS'S LAW

$$\text{div}(\vec{E}) = \frac{\partial E_x}{\partial x} + \frac{\partial E_y}{\partial y} + \frac{\partial E_z}{\partial z} = \frac{\rho}{\varepsilon_0}$$

Electric Potential and Gauss's Law

Remember that electric field can be expressed as the negative gradient of the electric potential (just as force can be expressed as the negative gradient of potential energy). Writing V for electric potential (not volume!), we have this:

$$E_x = -\frac{\partial V}{\partial x}, \quad E_y = -\frac{\partial V}{\partial y}, \quad E_z = -\frac{\partial V}{\partial z}$$

Putting these gradient forms into the differential form of Gauss's law gives the following important equation:

$$\frac{\partial^2 V}{\partial x^2} + \frac{\partial^2 V}{\partial y^2} + \frac{\partial^2 V}{\partial z^2} = -\frac{\rho}{\varepsilon_0}$$

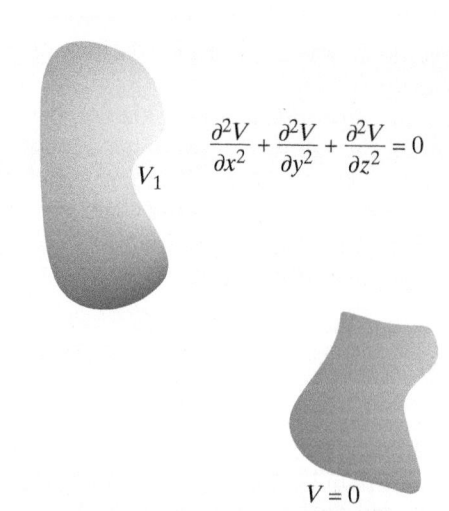

$$\frac{\partial^2 V}{\partial x^2} + \frac{\partial^2 V}{\partial y^2} + \frac{\partial^2 V}{\partial z^2} = 0$$

V_1

$V = 0$

Figure 21.58 A boundary value problem. One piece of metal is maintained at a fixed potential V_1 relative to a potential defined to be $V = 0$ for the other piece. In the empty space Gauss's law gives a differential equation to be solved for $V(x, y, z)$.

Here we have a differential equation relating the electric potential at some location to the charge density at that same location. A frequent use of this equation is in determining the electric potential in empty space (where the charge density ρ is zero) near some charged pieces of metal that are maintained at known potentials (a "boundary value" problem, Figure 21.58). If we knew the charge distribution on these pieces of metal, we could calculate the electric potential and electric field anywhere, through superposition. Often, however, we know only the fixed potentials of the pieces of metal, not the charge distribution. In such cases we can use the following differential equation, called "Laplace's equation":

$$\frac{\partial^2 V}{\partial x^2} + \frac{\partial^2 V}{\partial y^2} + \frac{\partial^2 V}{\partial z^2} = 0 \quad \text{(in empty space)}$$

Throughout the empty space between the pieces of metal, the electric potential must be a solution of this differential equation. There exist both analytical and numerical methods for solving Laplace's equation. At locations very near the pieces of metal, the solution of Laplace's equation in the empty space must equal the known potentials. This is an example of what is called a "boundary value problem," which is an important class of problem in science and engineering.

Checkpoint 4 In a certain region of space, the electric potential depends on position in the following way, where V_0 and a are constants: $V = V_0 + a(x^2 + y^2 + z^2)$. What is the charge density in this region? In a different region of space, the electric potential depends on position in the following way, where V_0 and b are constants: $V = V_0 + b(3x + 4y - 2z)$. What is the charge density in this region?

Standard Notation

To help introduce the topic, we have written the divergence as $\text{div}(\vec{E})$. The standard notation normally used for divergence is this:

$$\vec{\nabla} \bullet \vec{E} = \frac{\partial E_x}{\partial x} + \frac{\partial E_y}{\partial y} + \frac{\partial E_z}{\partial z} = \frac{\rho}{\varepsilon_0}$$

The $\vec{\nabla}$ (or "del") operator is expressed in the following form:

$$\vec{\nabla} = \left\langle \frac{\partial}{\partial x}, \frac{\partial}{\partial y}, \frac{\partial}{\partial z} \right\rangle$$

An operator operates on some quantity to its right. In this case the operation involves partial differentiation. The usual rules of calculating the dot product give the following for the quantity $\vec{\nabla} \bullet \vec{E}$:

$$\left\langle \frac{\partial}{\partial x}, \frac{\partial}{\partial y}, \frac{\partial}{\partial z} \right\rangle \bullet \langle E_x, E_y, E_z \rangle = \frac{\partial E_x}{\partial x} + \frac{\partial E_y}{\partial y} + \frac{\partial E_z}{\partial z}$$

Also, note that the del operator can be used to express the concept of gradient:

$$\vec{E} = -\left(\frac{\partial V}{\partial x}\hat{\imath} + \frac{\partial V}{\partial y}\hat{\jmath} + \frac{\partial V}{\partial z}\hat{k} \right) = -\vec{\nabla}V$$

We can rewrite the differential form of Gauss's law like this:

$$\vec{\nabla} \bullet (\vec{\nabla}V) = \frac{\partial^2 V}{\partial x^2} + \frac{\partial^2 V}{\partial y^2} + \frac{\partial^2 V}{\partial z^2} = -\frac{\rho}{\varepsilon_0}$$

This combination, the divergence of a gradient, is called the "Laplacian" and is written like this:

$$\nabla^2 V = \frac{\partial^2 V}{\partial x^2} + \frac{\partial^2 V}{\partial y^2} + \frac{\partial^2 V}{\partial z^2} = -\frac{\rho}{\varepsilon_0}$$

In regions where there are no charges, we have Laplace's equation:

$$\nabla^2 V = \frac{\partial^2 V}{\partial x^2} + \frac{\partial^2 V}{\partial y^2} + \frac{\partial^2 V}{\partial z^2} = 0 \quad \text{(in empty space)}$$

21.10 *THE DIFFERENTIAL FORM OF AMPERE'S LAW

We have pointed out several times that unless charges are moving very slowly compared to the speed of light, there is a flaw in trying to calculate magnetic field at some location by using the Biot–Savart law. The problem is relativistic retardation. The magnetic field depends on where fast-moving particles were at some time in the past, not where they are now, so the Biot–Savart law is not consistent with the theory of special relativity. Ampere's law, however, is consistent with relativity. In this section we will see why.

The argument is very similar to the one presented in connection with Gauss's law. We look for

- a property of the magnetic field $\vec{B}$ at some (x,y,z,t)
- that is related to moving source charges at the *same* (x,y,z,t).

Curl

We introduce an appropriate property of the magnetic field, called the "curl." Consider a very small Amperian path around a location in space, enclosing a very small area ΔA. In this region there is a current density $\vec{J}$ measured in amperes per square meter. The direction of the vector $\vec{J}$ is the direction of the conventional current at that location. The current passing through the tiny area is $\vec{J} \bullet \hat{n} \, \Delta A$ (multiply amperes per square meter times area in square meters), where $\hat{n}$ is a "normal" unit vector perpendicular to the area. From Ampere's law, we know that the path integral of magnetic field around the perimeter of the tiny area is this:

$$\oint \vec{B} \bullet d\vec{l} = \mu_0 \vec{J} \bullet \hat{n} \, \Delta A$$

Divide by the tiny area ΔA:

$$\frac{\oint \vec{B} \bullet d\vec{l}}{\Delta A} = \mu_0 \vec{J} \bullet \hat{n}$$

On the left side of this equation we have the path integral of magnetic field per unit area. In the limit as the area becomes arbitrarily small, $\Delta A \to 0$, this ratio is called the component in the $\hat{n}$ direction of the "curl" of the magnetic field at this location:

$$[\text{curl}(\vec{B})] \bullet \hat{n} \equiv \lim_{\Delta A \to 0} \frac{\oint \vec{B} \bullet d\vec{l}}{\Delta A}$$

This is the property we are looking for. We obtain an equation in a form that doesn't have retardation problems:

$$\text{curl}(\vec{B}) = \mu_0 \vec{J} \quad \text{(note that this is a } \textit{vector} \text{ equation)}$$

Why is this property of the magnetic field called the curl of the magnetic field? Look at Figure 21.59. If there is a current crossing through a (vanishingly small) area, the magnetic field curls around the current.

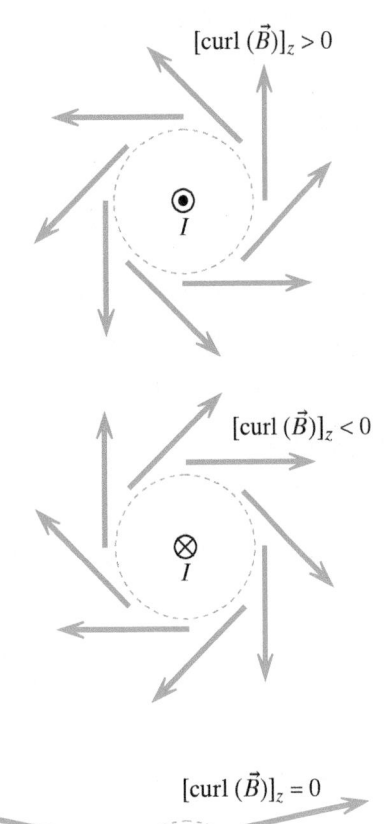

$[\text{curl} \, (\vec{B})]_z > 0$

$[\text{curl} \, (\vec{B})]_z < 0$

$[\text{curl} \, (\vec{B})]_z = 0$

Figure 21.59 The curl of the magnetic field at a location surrounded by a tiny $\Delta A \to 0$.

Curl Is Relativistically Correct

The curl (net path integral of magnetic field per unit area, in the limit as the area goes to zero) is a measure of an important property of the magnetic field at some location and time, and this is proportional to the current density at that *same* location and time. This is called a "local" relationship between current and field, because the current and the curl property are evaluated at the same location and time, rather than at different locations and times. This avoids the problem of relativistic retardation and forms the basis for what is called a "local field theory" that is consistent with special relativity.

Earlier in this chapter we developed what is called the "integral form" of Ampere's law, involving a path integral of magnetic field over a (possibly long) closed path. What we have just developed is the "differential form" of Ampere's law:

DIFFERENTIAL FORM OF AMPERE'S LAW

$$\text{curl}(\vec{B}) = \mu_0 \vec{J}$$

where the component of the curl in the $\hat{n}$ direction, perpendicular to a tiny area ΔA, is defined as

$$[\text{curl}(\vec{B})] \bullet \hat{n} \equiv \lim_{\Delta A \to 0} \frac{\oint \vec{B} \bullet d\vec{l}}{\Delta A}$$

Using the del operator, Ampere's law is written like this:

$$\vec{\nabla} \times \vec{B} = \mu_0 \vec{J}$$

The differential form of Ampere's law is correct in linking current density to magnetic field at the same location and time. However, we will see later that Ampere's law is incomplete.

> **Checkpoint 5** Based on its definition as path integral of magnetic field per unit area, what are the units of the curl of the magnetic field? Show that these units are the same as the units of $\mu_0 J$.

SUMMARY

Fundamental Principles

Gauss's law

$$\oint \vec{E} \bullet \hat{n} \, dA = \frac{\sum q_{\text{inside}}}{\varepsilon_0}$$

Gauss's law for magnetism

$$\oint \vec{B} \bullet \hat{n} \, dA = 0$$

Ampere's law

$$\oint \vec{B} \bullet d\vec{l} = \mu_0 \sum I_{\text{inside path}}$$

Alternative Notations for Gauss's Law

Finite sum of electric flux

$$\sum_{\text{closed surface}} \vec{E} \bullet \hat{n} \, \Delta A = \frac{\sum q_{\text{inside}}}{\varepsilon_0}$$

$E_\perp = E$ perpendicular to surface:

$$\oint E_\perp \, dA = \frac{\sum q_{\text{inside}}}{\varepsilon_0}$$

$A_\perp = A$ perpendicular to electric field:

$$\oint E \, dA_\perp = \frac{\sum q_{\text{inside}}}{\varepsilon_0}$$

$d\vec{A} = \hat{n}\,dA$ for representing area and normal:

$$\oint \vec{E} \bullet d\vec{A} = \frac{\sum q_{\text{inside}}}{\varepsilon_0}$$

The circle superimposed on the integral sign means "integrate over the entire closed surface."

Electric flux on a surface (not necessarily closed)

$$\Phi_{\text{el}} = \int \vec{E} \bullet \hat{n}\,dA$$

Results
We used Gauss's law to find the electric field of a plate and a sphere. We also proved some important properties about

metals, especially that excess charge is on the surface and that there is no field in an empty hole in a metal in equilibrium. We used Ampere's law to find the magnetic field of a thick wire, a solenoid, and a toroid.

Advanced *Differential Form of Gauss's Law

$$\vec{\nabla} \bullet \vec{E} = \frac{\partial E_x}{\partial x} + \frac{\partial E_y}{\partial y} + \frac{\partial E_z}{\partial z} = \frac{\rho}{\varepsilon_0}, \text{ or}$$

$$\nabla^2 V = \frac{\partial^2 V}{\partial x^2} + \frac{\partial^2 V}{\partial y^2} + \frac{\partial^2 V}{\partial z^2} = -\frac{\rho}{\varepsilon_0}$$

Advanced *Differential Form of Ampere's Law

$$\vec{\nabla} \times \vec{B} = \mu_0 \vec{J}$$

QUESTIONS

Q1 The electric field on a closed surface is due to all the charges in the Universe, including the charges outside the closed surface. Explain why the total flux nevertheless is proportional only to the charges that are inside the surface, with no apparent influence of the charges outside.

Q2 In Chapter 15 we calculated the electric field at a location on the axis of a uniformly charged ring. Without doing all those calculations, explain why we can't use Gauss's law to determine the electric field at that location.

Q3 Explain why you are safe and unaffected inside a car that is struck by lightning. Also explain why it might not be safe to step out of the car just after the lightning strike, with one foot in the car and one on the ground. (Note that it is current that kills; a rather small current passing through the region of the heart can be fatal.)

Q4 Based on its definition as electric flux per unit volume, what are the units of the divergence of the electric field?

Q5 In the region shown in Figure 21.60, the magnetic field is vertical and was measured to have the values shown on the surface of a cylinder. Why should you suspect something is wrong with these measurements?

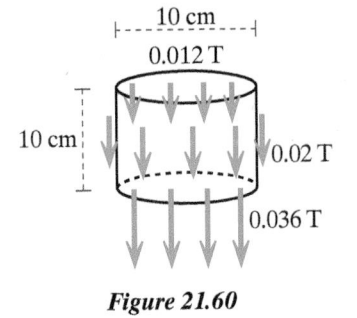

Figure 21.60

PROBLEMS

Section 21.2

•**P6** Figure 21.61 shows a disk-shaped region of radius 2 cm on which there is a uniform electric field of magnitude 300 V/m at an angle of $30°$ to the plane of the disk. Assume that $\hat{n}$ points upward, in the $+y$ direction. Calculate the electric flux on the disk, and include the correct units.

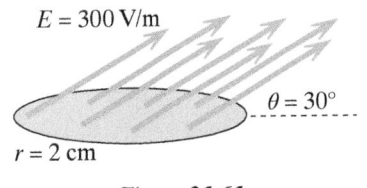

Figure 21.61

Section 21.3

•**P7** Figure 21.62 shows a box on whose surfaces the electric field is measured to be horizontal and to the right. On the left face

(3 cm by 2 cm) the magnitude of the electric field is 400 V/m, and on the right face the magnitude of the electric field is 1000 V/m. On the other faces only the direction is known (horizontal). Calculate the electric flux on every face of the box, the total flux, and the total amount of charge that is inside the box.

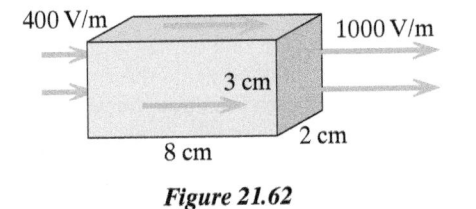

Figure 21.62

•**P8** A Gaussian cylinder is drawn around some hidden charges, as shown in Figure 21.63. Given this pattern of electric field on the surface of the cylinder, how much net charge is inside? Invent a possible charge distribution inside the cylinder that could give this pattern of electric field.

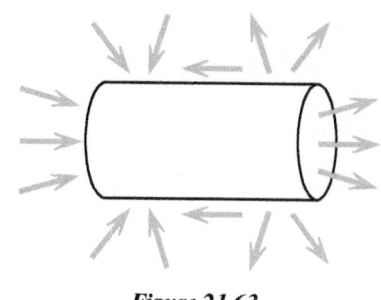

Figure 21.63

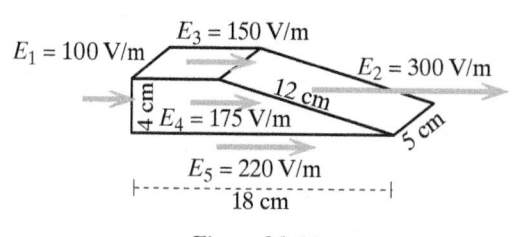

is $E_5 = 220$ V/m. How much charge is inside the box? Explain briefly.

Figure 21.66

••P9 The electric field has been measured to be vertically upward everywhere on the surface of a box 20 cm long, 4 cm high, and 3 cm deep, as shown in Figure 21.64. All over the bottom of the box $E_1 = 1500$ V/m, all over the sides $E_2 = 1000$ V/m, and all over the top $E_3 = 600$ V/m. What can you conclude about the contents of the box? Include a numerical result.

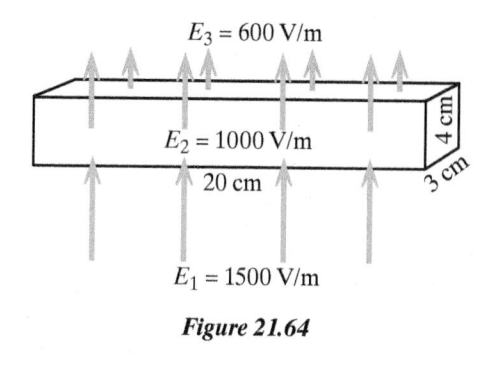

Figure 21.64

••P10 The electric field is horizontal and has the values indicated on the surface of a cylinder as shown in Figure 21.65. What can you deduce from this pattern of electric field? Include a numerical result.

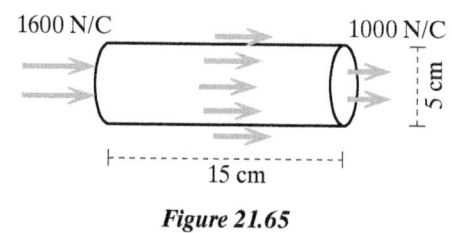

Figure 21.65

••P11 Consider a cube with one corner at the origin and with sides of length 10 cm positioned along the *xyz* axes. There is an electric field $\vec{E} = \langle 50, 200y, 0 \rangle$ N/m throughout the region that has a constant *x* component and a *y* component that increases linearly with *y*. **(a)** Draw a diagram showing $\hat{n}$ and $\vec{E}$ for every side of the cube. **(b)** Determine how much charge is inside the cube.

••P12 The electric field has been measured to be horizontal and to the right everywhere on the closed box as shown in Figure 21.66. All over the left side of the box $E_1 = 100$ V/m, and all over the right (slanting) side of the box $E_2 = 300$ V/m. On the top the average field is $E_3 = 150$ V/m, on the front and back the average field is $E_4 = 175$ V/m, and on the bottom the average field

••P13 The electric field is measured at locations on every surface of a rectangular box oriented with its long side on the *x* axis, as shown in Figure 21.67. The width *w* of the box is 6 cm, the height *h* of the box is 4 cm, and the depth *d* of the box is 2 cm. On the bottom of the box the electric field $\vec{E}_1$ is found to be $\langle 50, 40, 0 \rangle$ N/C. On every other surface of the box (front, back, top, left end, right end) the electric field $\vec{E}_2$ is found to be $\langle 120, 96, 0 \rangle$ N/C. (For clarity, the electric field on the front, back, and left faces has not been shown in the diagram.) **(a)** For every face of the box, specify $\hat{n}$. **(b)** Calculate the electric flux on every face of the box. **(c)** What is the net charge inside the box?

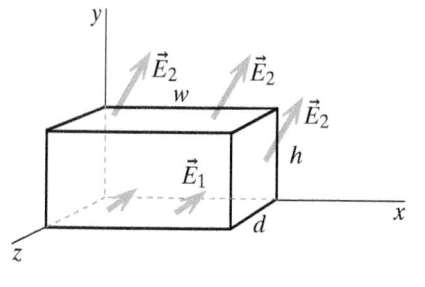

Figure 21.67

••P14 The electric field is measured all over a cubical surface, and the pattern of field detected is shown in Figure 21.68. On the right side of the cube and the bottom of the cube, the electric field has the value $\langle 370, -370, 0 \rangle$ N/C. On the top of the cube and the left side of the cube, the electric field is zero. On half of the front and back faces, the electric field has the value $\langle 370, -370, 0 \rangle$ N/C; on the other half of the front and back faces, the electric field is zero. One edge of the cube is 55 cm long. **(a)** What is the net electric flux on this cubical surface? **(b)** What is the net charge inside this cubical surface? **(c)** Can you think of a charge distribution that could generate this pattern of electric field?

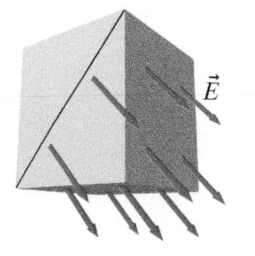

Figure 21.68

••P15 The center of a thin paper cube in outer space is located at the origin. Each edge is 10 cm long. The only other objects in the neighborhood are some small charged particles whose charges and positions at this instant are the following: $+9$ nC at $\langle 4,2,-3\rangle$ cm, -6 nC at $\langle 1,-3,2\rangle$ cm, $+7$ nC at $\langle 15,0,4\rangle$ cm, $+4$ nC at $\langle -1,3,2\rangle$ cm, and -3 nC at $\langle -2,12,-3\rangle$ cm. The electric flux on five of the six faces of the cube totals 564 V·m. What is the flux on the other face?

Section 21.4

••P16 A negative point charge $-Q$ is at the center of a hollow insulating spherical shell, which has an inner radius R_1 and an outer radius R_2. There is a total charge of $+3Q$ spread uniformly throughout the volume of the insulating shell, not just on its surface. Determine the electric field for **(a)** $r < R_1$, **(b)** $R_1 < r < R_2$, and **(c)** $R_2 < r$.

••P17 You may have seen a coaxial cable connected to a television set. As shown in Figure 21.69, a coaxial cable consists of a central copper wire of radius r_1 surrounded by a hollow copper tube (typically made of braided copper wire) of inner radius r_2 and outer radius r_3. Normally the space between the central wire and the outer tube is filled with an insulator, but in this problem assume for simplicity that this space is filled with air. Assume that no current runs in the cable.

Suppose that a coaxial cable is straight and has a very long length L, and that the central wire carries a charge $+Q$ uniformly distributed along the wire (so that the charge per unit length is $+Q/L$ everywhere along the wire). Also suppose that the outer tube carries a charge $-Q$ uniformly distributed along its length L. The cylindrical symmetry of the situation indicates that the electric field must point radially outward or radially inward. The electric field cannot have any component parallel to the cable. In this problem, draw mathematical Gaussian cylinders of length d (with d much less than the cable length L) and appropriate radius r, centered on the central wire. **(a)** Use a mathematical Gaussian cylinder located inside the central wire $(r < r_1)$ and another Gaussian cylinder with a radius in the interior of the outer tube $(r_2 < r < r_3)$ to determine the exact amount and location of the charge on the inner and outer conductors. (*Hint:* What do you know about the electric field in the interior of the two conductors? What do you know about the flux on the ends of your Gaussian cylinders?) **(b)** Use a mathematical Gaussian cylinder whose radius is in the air gap $(r_1 < r < r_2)$ to determine the electric field in the gap as a function of r. (Don't forget to consider the flux on the ends of your Gaussian cylinder.) **(c)** Use a mathematical Gaussian cylinder whose radius is outside the cable $(r > r_3)$ to determine the electric field outside the cable. (Don't forget to consider the flux on the ends of your Gaussian cylinder.)

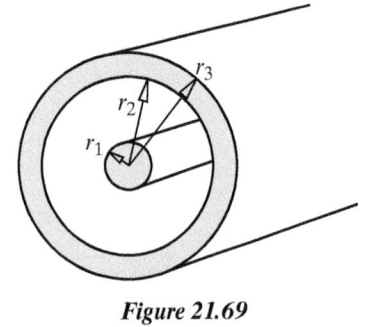

Figure 21.69

••P18 A straight circular plastic cylinder of length L and radius R (where $R \ll L$) is irradiated with a beam of protons so that

there is a total excess charge Q distributed uniformly throughout the cylinder. Find the electric field inside the cylinder, a distance r from the center of the cylinder far from the ends, where $r < R$.

••P19 Figure 21.70 shows a close-up of the central region of a capacitor made of two large metal plates of area A, very close together and charged equally and oppositely. There are $+Q$ and $-Q$ on the inner surfaces of the plates and small amounts of charge $+q$ and $-q$ on the outer surfaces.

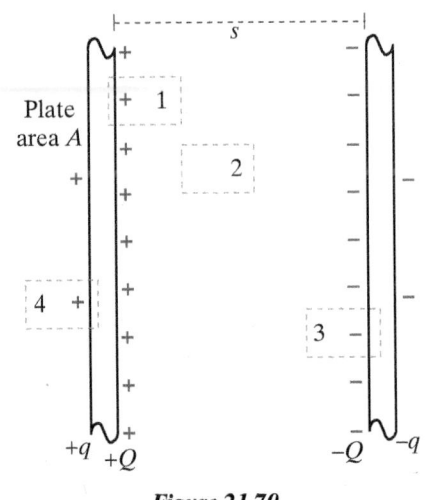

Figure 21.70

(a) Knowing Q, determine E: Consider a Gaussian surface in the shape of a shoe box, with one end of area A_{box} in the interior of the left plate and the other end in the air gap (surface 1 in the diagram). Using only the fact that the electric field is expected to be horizontal everywhere in this region, use Gauss's law to determine the magnitude of the field in the air gap. Check that your result agrees with our earlier calculations in Chapter 15. (Be sure to consider the flux on all faces of your Gaussian box.) **(b)** Proof by contradiction: Consider Gaussian box 2, with both ends in the gap. Use Gauss's law to prove that the electric field is constant in magnitude throughout the gap. **(c)** Knowing Q, determine E: Consider Gaussian box 3, with its left end in the gap and its right end in the interior of the right plate. Show that Gauss's law applied to this Gaussian box gives the correct magnitude of the electric field in the gap. **(d)** Knowing E, determine q/A: Finally, consider Gaussian box 4, with its left end outside the capacitor and its right end in the interior of the left plate. In Chapter 15 we determined that the fringe field is approximately given by

$$E_{fringe} \approx \frac{Q/A}{2\varepsilon_0} \frac{s}{R}$$

where s is the gap width and R is the radius of the circular plates. Use this information and Gaussian box 4 to determine the approximate amount of charge q on the outer surface of the plate.

•••P20 In Chapter 20 we showed that the nonuniform direction of the magnetic field of a magnet is responsible for the force $F = I(2\pi R)B\sin\theta$ exerted by the magnet on a current loop. However, we didn't know the angle θ, which is a measure of the nonuniformity of the direction of the magnetic field (Figure 21.71). Gauss's law for magnetism can be used to determine this angle θ. **(a)** Apply Gauss's law for magnetism to a thin disk of radius R and thickness Δx located where the current loop will be placed. Use Gauss's law to determine the component $B_3 \sin\theta$ of the magnetic field that is perpendicular to the axis of the magnet at a radius R from the axis. Assume that R is small enough that the x component of magnetic field is approximately uniform everywhere on the left face of the disk and also on the right face of the disk (but with a smaller value). The magnet has a magnetic dipole moment μ.

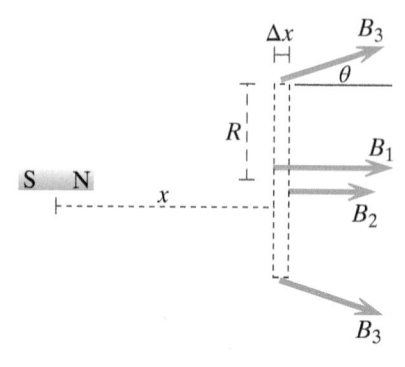

Figure 21.71

(b) Now imagine placing the current loop at this location, and show that $F = I(2\pi R)B_3 \sin\theta$ can be rewritten as $F = \mu_{loop}|dB/dx|$, which is the result we obtained from potential-energy arguments in Chapter 20.

•••P21 A solid chunk of metal carries a net positive charge and is also polarized by other charges that are not on the metal, as shown in Figure 21.72. The electric field very near the surface of a metal in equilibrium is perpendicular to the surface: any component of the electric field parallel to the surface would drive currents in the metal. (Along a rectangular path, part inside and part outside the metal, the round-trip potential difference is zero, so the parallel component of electric field is the same just inside and just outside the surface.) Let S be the "local" surface charge density—the surface charge per unit area (C/m^2) at a particular location on the surface. **(a)** Calculate the magnitude E of the electric field in the air very near a location on the surface of the metal, in terms of the surface charge density S at that location. Use a cylindrical Gaussian surface as shown, with one face just inside the metal and one face just outside. Don't skip steps in your analysis!

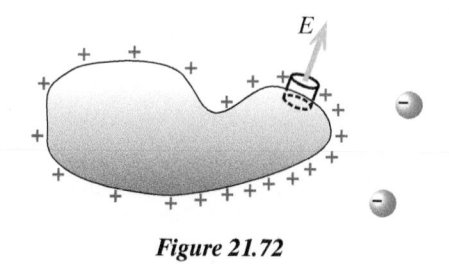

Figure 21.72

(b) Show that nearby charges contribute half of E, and distant charges contribute the other half.

Section 21.6

•P22 The magnetic field has been measured to be horizontal everywhere along a rectangular path 20 cm long and 4 cm high,

as shown in Figure 21.73. Along the bottom the average magnetic field $B_1 = 1.5 \times 10^{-4}$ T, along the sides the average magnetic field $B_2 = 1.0 \times 10^{-4}$ T, and along the top the average magnetic field $B_3 = 0.6 \times 10^{-4}$ T. What can you conclude about the electric currents in the area that is surrounded by the rectangular path?

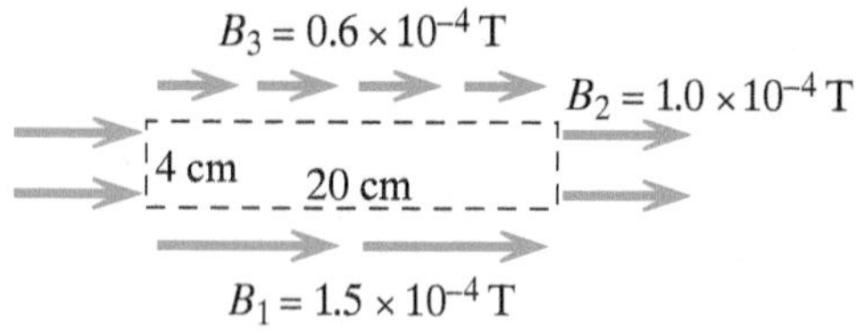

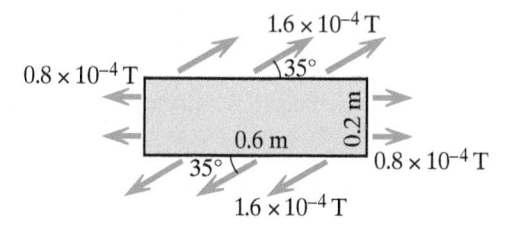

Figure 21.73

••P23 Figure 21.74 shows a measured pattern of magnetic field in space. How much current I passes through the shaded area? In what direction?

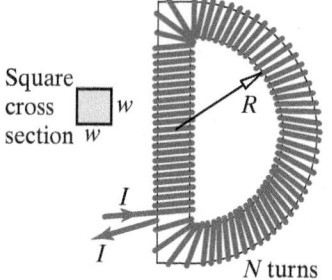

Figure 21.74

••P24 A D-shaped frame is made out of plastic of small square cross section and tightly wrapped uniformly with N turns of wire as shown in Figure 21.75, so that the magnetic field has essentially the same magnitude throughout the plastic. (R is much bigger than w.)

With a current I flowing, what is the magnetic field inside the plastic? Show the direction of the magnetic field in the plastic at several locations.

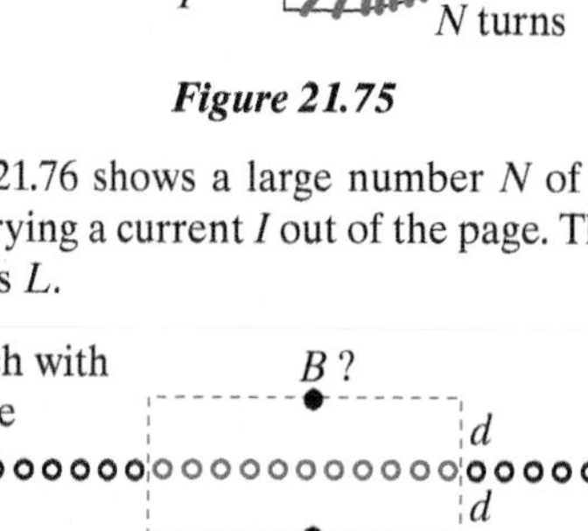

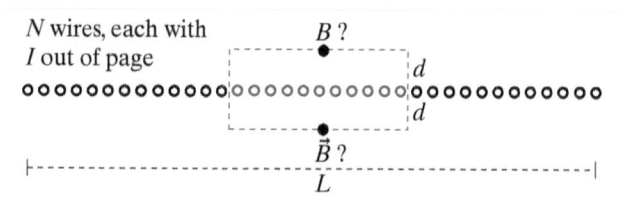

Figure 21.75

••P25 Figure 21.76 shows a large number N of closely packed wires, each carrying a current I out of the page. The width of this sheet of wires is L.

N wires, each with I out of page

Figure 21.76

(a) Show (and explain) the direction of the magnetic field at the two indicated locations, a distance d from the middle wire, where

d is much less than $L (d \ll L)$. **(b)** Using the rectangular path shown on the diagram, calculate the magnitude of the magnetic field at the indicated locations, in terms of the given physical quantities N, I, and L. **(c)** Using your result from part (b), find $\vec{B}$ in the region between two parallel current sheets with equal currents running in opposite directions. **(d)** What is $\vec{B}$ outside both sheets?

••P26 A long plastic rod with rectangular cross section $h \times w$ is wound with a coil of length d consisting of N closely packed turns of wire with negligible resistance. A current I runs through the coil.

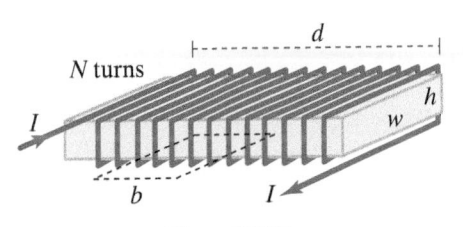

Figure 21.77

Using the rectangular path of length b shown on the diagram in Figure 21.77, prove that the magnetic field inside the coil is $B = \mu_0 NI/d$, and describe the direction of $\vec{B}$ inside the coil.

••P27 A coaxial cable consists of an inner metal wire of radius r_1 and an outer metal cylinder of radius r_2 (Figure 21.78). There is constant current I_1 to the right in the wire and constant current $I_2 < I_1$ to the left in the cylinder. Consider a circular Amperian path of radius R centered on the wire and perpendicular to the wire. Determine the magnitude and direction of the magnetic field at a location R above the wire, far from the ends of

the coaxial cable. Explain in detail, including all assumptions, directions, calculations, and so on. Don't skip steps.

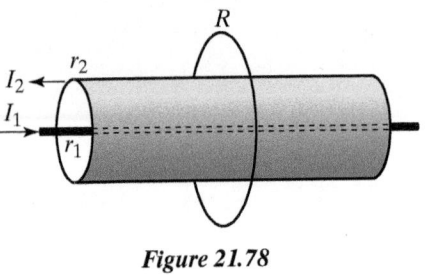

Figure 21.78

••P28 A thick wire of radius R carries a constant current I (Figure 21.79). Find the magnetic field inside the wire, a distance r from the center of the wire and far from the ends of the wire, where $r < R$ and R is much less than the length of the wire.

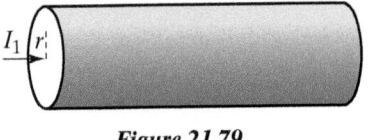

Figure 21.79

Section 21.9

•P29 A lead nucleus is spherical with a radius of about 7×10^{-15} m. The nucleus contains 82 protons (and typically 126 neutrons). Because of their motions the protons can be considered on average to be uniformly distributed throughout the nucleus. Based on the net flux at the surface of the nucleus, calculate the divergence of the electric field as electric flux per unit volume.

COMPUTATIONAL PROBLEM

•••P30 A point charge of 2×10^{-9} C is located at $\langle -0.03, 0, 0 \rangle$ m. Write a program to calculate the net electric flux on a cubical Gaussian surface centered at $\langle 0.03, 0, 0 \rangle$ m. The length of a side of the cube is 0.01 m. **(a)** Begin by considering each face of the cube as a single area element ΔA, and using the value of $\vec{E}$ at the center of each face. How does the net flux compare to $\sum q_{inside}/\varepsilon_0$? **(b)** Refine the calculation by dividing each face

of the cube into four smaller squares. How does your result change? **(c)** Continue to subdivide each square into smaller squares until your result no longer changes. Now how does the net flux compare to $\sum q_{inside}/\varepsilon_0$? **(d)** Move the charged particle to location $\langle 0.035, 0, 0 \rangle$ m, inside the cube, and recalculate the total flux. How does the net flux compare to $\sum q_{inside}/\varepsilon_0$?

ANSWERS TO CHECKPOINTS

1 1.73×10^{-3} V·m
2 0.195 A, into the page
3 1.03×10^{36} N/C/m; 1.03×10^{36} N/C/m; 1.03×10^{36} N/C/m

4 $-6\varepsilon_0 a$; 0
5 T/m

Faraday's Law

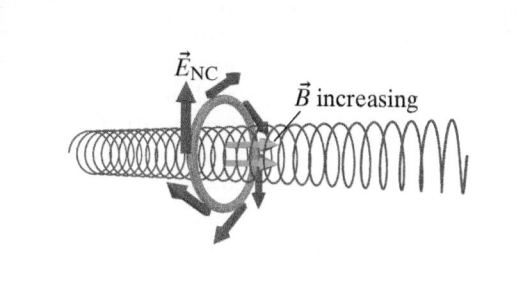

OBJECTIVES

After studying this chapter you should be able to

- Calculate the magnitude and direction of the electric field associated with a time-varying magnetic field.

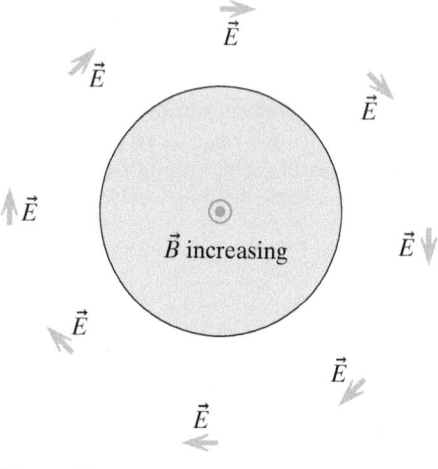

Figure 22.1 In a region where there is a time-varying magnetic field, there is also a curly electric field.

In previous chapters we have considered electric and magnetic fields as separate entities. Stationary charges make electric fields, and electric fields affect other charges. Moving charges make magnetic fields as well as electric fields, and magnetic fields affect other moving charges. In this chapter, however, we will see that there is a connection between electric and magnetic fields that we have not previously suspected. We find experimentally that if there is a time-varying magnetic field in a region of space, there is also a curly electric field throughout that region and surrounding regions (Figure 22.1). In Chapter 16, by considering the round-trip integral of electric field, we proved that a curly pattern of electric field cannot be made by a collection of stationary charges, so this is a particularly interesting phenomenon. No matter how an electric field comes into being, it has the same effect on a charge q (that is, $\vec{F} = q\vec{E}$).

The historical term "magnetic induction" is often used to describe this phenomenon, and one says that the time-varying magnetic field "induces" the curly electric field. This is somewhat misleading. It is more correct to say that anywhere we observe a time-varying magnetic field, we also observe a curly electric field. Faraday's law relates these observations quantitatively.

22.1 CURLY ELECTRIC FIELDS

In Chapter 17 using the Biot–Savart law and in Chapter 21 using Ampere's law we found that with a current I_1 in a solenoid, the magnetic field B_1 inside a tightly wound solenoid (not too near the ends) is $B_1 = \mu_0 N I_1 / d$, where N is the number of turns of wire and d is the length of the solenoid. The magnetic field outside the solenoid (not too near the ends) is very small and for a very long solenoid can be taken to be zero.

If the current is constant, then B_1 is constant in time. In that case a charge that is in motion somewhere outside the solenoid will not experience an electric or magnetic force, because the electric and magnetic fields outside the solenoid are essentially zero. However, something remarkable happens if we vary the current, so that the magnetic field B_1 inside the solenoid varies with time (Figure 22.2). There is still almost no magnetic field outside the solenoid, but we observe a curly electric field both inside and outside of the solenoid. This unusual electric field curls around the axis of the solenoid (end view; Figure 22.3). The electric field is proportional to dB_1/dt, the rate of change of the magnetic field.

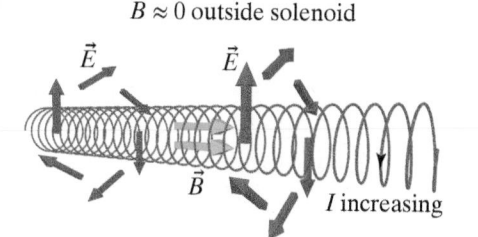

Figure 22.2 With a time-varying current in the solenoid there is a curly electric field.

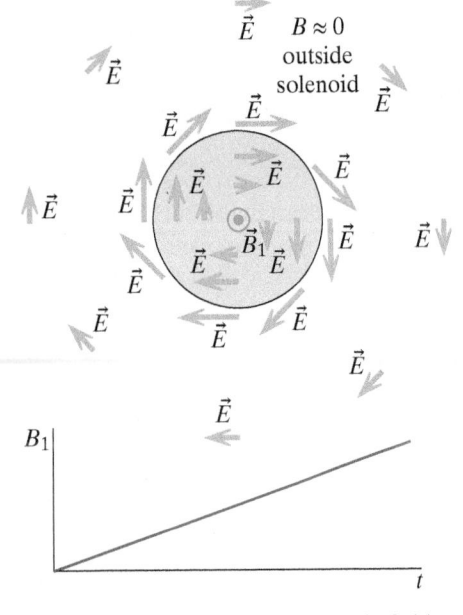

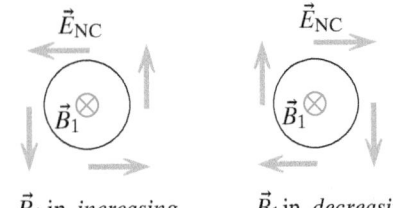

Figure 22.3 There is a curly electric field in the presence of a time-varying magnetic field (top). In this case B_1 is increasing with time (bottom).

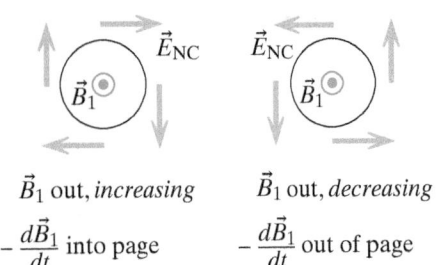

$\vec{B}_1$ out, *increasing*
$-\dfrac{d\vec{B}_1}{dt}$ into page

$\vec{B}_1$ out, *decreasing*
$-\dfrac{d\vec{B}_1}{dt}$ out of page

$\vec{B}_1$ in, *increasing*
$-\dfrac{d\vec{B}_1}{dt}$ out of page

$\vec{B}_1$ in, *decreasing*
$-\dfrac{d\vec{B}_1}{dt}$ into page

Figure 22.4 Four cases: magnetic field out or in, increasing or decreasing.

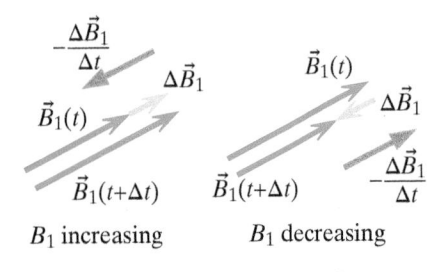

B_1 increasing B_1 decreasing

Figure 22.5 Find the change in the magnetic field as a basis for determining the direction of $-d\vec{B}_1/dt$.

Inside the solenoid the electric field is proportional to r, the distance from the axis (smaller near the axis). Outside the solenoid the curly electric field is proportional to $1/r$; the electric field gets smaller as you go farther away from the solenoid.

The curly electric field has the usual effect on charges: a charge q experiences a force $\vec{F} = q\vec{E}$. For example, a proton placed above the solenoid in Figure 22.3 will be initially pushed to the right. Because this curly electric field is associated with the time-varying current and magnetic field rather than with stationary charges, we call such an electric field a "non-Coulomb" field, $\vec{E}_{NC}$.

QUESTION Explain how you know that this pattern of curly electric field cannot be produced by an arrangement of stationary charges.

Traveling in a complete loop around the solenoid, $\oint \vec{E}_{NC} \bullet d\vec{l} \neq 0$. Since the round-trip integral of the electric field due to stationary charges is always zero, this pattern of electric field cannot be produced by stationary charges.

NON-COULOMB ELECTRIC FIELD

- A *Coulomb* electric field is produced by charges according to Coulomb's law:

$$\vec{E} = \frac{1}{4\pi\varepsilon_0}\frac{q}{r^2}\hat{r}$$

- A *non-Coulomb* electric field $\vec{E}_{NC}$ is associated with time-varying magnetic fields $d\vec{B}/dt$. Outside of a long solenoid inside of which the magnetic field is B_1, the non-Coulomb electric field is proportional to dB_1/dt and decreases with distance like $1/r$.

Figure 22.4 shows what is observed experimentally in four different cases, where the magnetic field points out of or into the page, and increases or decreases with time. From these results you can see that it is not the direction of $\vec{B}_1$ that determines the direction of $\vec{E}_{NC}$, but rather the direction of the rate of change of $\vec{B}_1$. The following right-hand rule summarizes the experimental observations:

DIRECTION OF THE CURLY ELECTRIC FIELD

With the thumb of your right hand pointing in the direction of $-d\vec{B}_1/dt$, your fingers curl around in the direction of $\vec{E}_{NC}$.

Finding the Direction of $-d\vec{B}/dt$

In order to use this right-hand rule, you need to be able to determine the direction of the vector quantity $-d\vec{B}_1/dt$. It helps to think about this quantity in the form $-\Delta\vec{B}_1/\Delta t$, where Δt is a small, finite time interval. From this form you can see that the direction of $-d\vec{B}_1/dt$ is the same as the direction of $-\Delta\vec{B}_1$, the negative of the change in direction of the magnetic field during a short time.

A good way to find the direction of $-d\vec{B}_1/dt$ is to draw the magnetic field $\vec{B}_1(t)$ at a time t, and the magnetic field $\vec{B}_1(t + \Delta t)$ at a slightly later time $t + \Delta t$, and observe the change $\Delta\vec{B}_1$ (Figure 22.5). Then $-\Delta\vec{B}_1$ is the direction of $-d\vec{B}_1/dt$.

Checkpoint 1 A magnetic field near the floor points up and is increasing. Looking down at the floor, does the non-Coulomb electric field curl clockwise or counterclockwise? A magnetic field near the ceiling points down and is decreasing. Looking up at the ceiling, does the non-Coulomb electric field curl clockwise or counterclockwise?

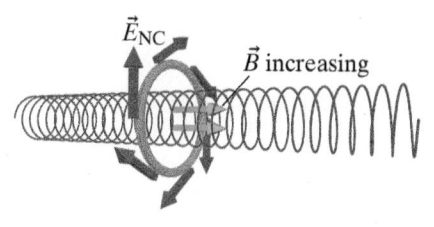

Figure 22.6 A metal ring is placed around the solenoid.

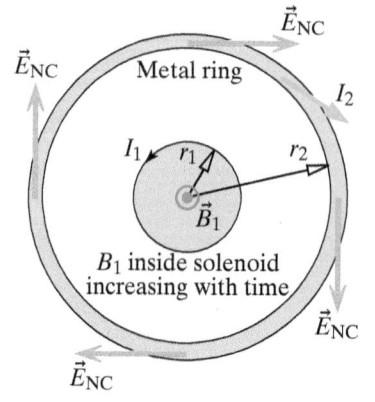

Figure 22.7 End view: The non-Coulomb electric field drives a current I_2 in the ring.

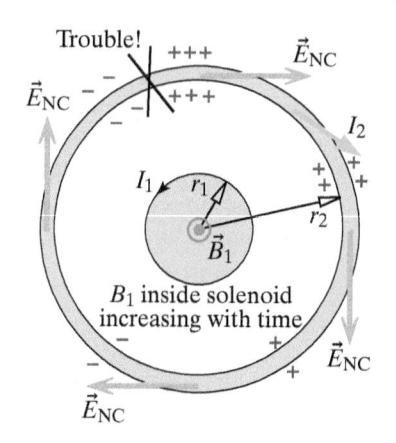

Figure 22.8 This pattern of surface charge is impossible, because it would imply a huge E at the marked location, and in the wrong direction!

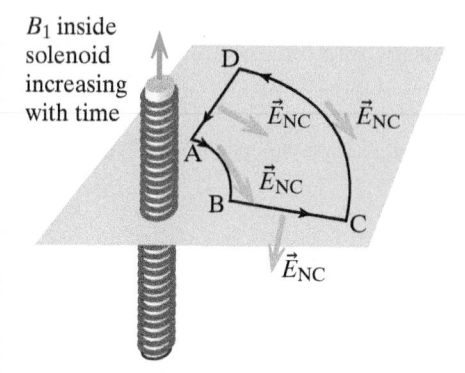

Figure 22.9 A path that does not encircle the solenoid.

Driving Current with a Non-Coulomb Electric Field

Suppose that we place a circular metal ring of radius r_2 around a solenoid (Figure 22.6), with the magnetic field in the solenoid increasing with time (Figure 22.7). The non-Coulomb electric field inside the metal will drive conventional current clockwise around the ring ($-d\vec{B}_1/dt$ points into the page; point your right thumb into the page and see how your fingers curl clockwise). The technological importance of this effect is that it can make current run in a wire just as though a battery were present.

The current I_2 in the ring is proportional to the electric field E_{NC} inside the metal, as in an ordinary circuit. However, in ordinary circuits there are charges on the surface of the wires that, together with charges on the battery, produce the electric field inside the metal that drives the current.

> QUESTION Think about a possible pattern of surface charge on this ring. Why is it impossible to draw a plausible gradient of surface charge on the ring in this situation?

The symmetry of the ring makes it impossible to have a surface charge gradient along the ring. We reason by contradiction: If you pick one point and draw positive surface charge there, then gradually decrease the amount of positive charge and increase the amount of negative surface charge, you find that when you get back to the starting point there is a huge change in surface charge (from $-$ to $+$), which would produce a huge E, in the wrong direction (Figure 22.8). However, there is nothing special about the point you picked, so it can't have a different electric field from all other points on the ring, and there cannot be a varying surface charge around the ring. (You may have realized, however, that there must be a tiny pileup of electrons on the outside of the ring, which then provides the radially inward force that turns the electron current; the pileup is extremely small because the radial acceleration $\bar{v}^2/r$ is very small due to the low drift speed $\bar{v}$ of the mobile electrons.)

The emf is the (non-Coulomb) energy input per unit charge. The (non-Coulomb) force per unit charge is the (non-Coulomb) field E_{NC}.

> QUESTION Therefore, what is the emf in terms of E_{NC} for this ring, if the ring has a radius r_2?

$$\text{emf} = \oint \vec{E}_{NC} \bullet d\vec{l} = E_{NC}(2\pi r_2)$$

since E_{NC} is constant and parallel to the path. The current in the ring of radius r_2 is $I_2 = \text{emf}/R$, where R is the resistance of the ring. It is as though we had inserted a battery into the ring.

> QUESTION If the metal ring had a radius r_2 twice as large, what would be the emf, since $\text{emf} = \oint \vec{E}_{NC} \bullet d\vec{l}$? Remember that the experimental observations show that the non-Coulomb electric field outside the solenoid is proportional to $1/r$.

Double the radius implies half the electric field, so the product $E_{NC}(2\pi r_2)$ stays the same. Apparently the emf in a ring encircling the solenoid is the same for any radius. (In fact, we get the same emf around any circuit surrounding the solenoid, not just a circular ring.)

> QUESTION Consider a round-trip path that does not encircle the solenoid (Figure 22.9). The electric field is shown all along the path A-B-C-D-A. Use the fact that $E_{NC} = \text{emf}/(2\pi r)$ to explain why we have the result $\text{emf} = \oint \vec{E}_{NC} \bullet d\vec{l} = 0$ around this path that does not encircle the solenoid.

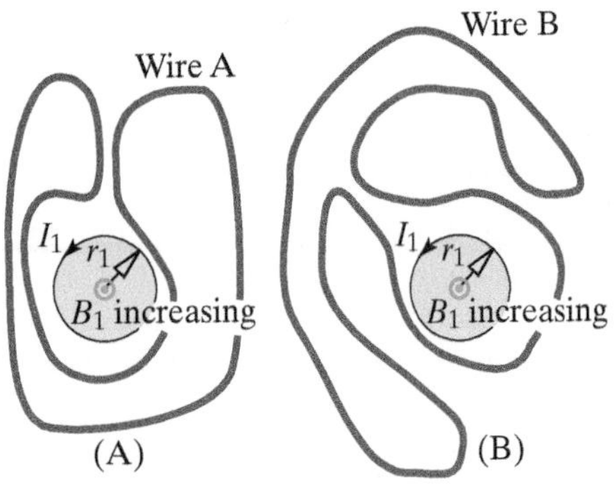

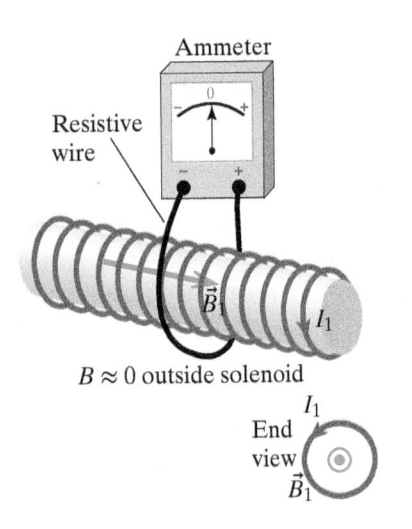

Figure 22.10 Will current run in these purple wires?

As you make a round trip around the path A-B-C-D-A, the contribution to $\oint \vec{E}_{NC} \bullet d\vec{l}$ is positive along A-B, zero along B-C (the parallel component of electric field is zero), negative along C-D, and zero along D-A. Moreover, the magnitude of the contribution along C-D is equal to that along A-B, because the longer path is compensated by the smaller electric field (which is proportional to $1/r$). Therefore we have a zero emf along this path. The non-Coulomb electric field would polarize a wire that followed this path but would not drive current around the loop.

We see that in order to drive current in a wire the wire must encircle a region where the magnetic field is changing. If the wire doesn't encircle the region of changing magnetic field, there is no emf.

> **Checkpoint 2** **(a)** In Figure 22.10, will current run in wire A? **(b)** In wire B? On a circular path of radius 10 cm in air around a solenoid with increasing magnetic field, the emf is 30 V. **(c)** What is the magnitude of the non-Coulomb electric field on this path? **(d)** A wire with resistance 4 Ω is placed along the path. What is the current in the wire?

22.2 FARADAY'S LAW

So far we have a right-hand rule for determining the direction of the curly electric field, but we don't have a way to calculate the magnitude of the non-Coulomb electric field. Faraday's law states a quantitative relationship between the rate of change of the magnetic field and the magnitude of the non-Coulomb electric field. To establish this quantitative relationship, in principle we could vary the magnetic field and measure E_{NC} by observing the effect of the curly electric field on an individual charged particle, but it can be difficult to track the path of an individual charged particle. Alternatively, we can construct a circuit following a path where we expect a non-Coulomb electric field and measure the current in the circuit as a function of dB/dt. We will describe circuit experiments that lead to Faraday's law.

Observing Current Caused by Non-Coulomb Electric Fields

In Figure 22.11, if we vary the magnetic field B_1 by varying I_1 and use the ammeter to measure the current in the loop around the solenoid, we can infer something about the resulting non-Coulomb electric field E_{NC} from its integral around that loop, which is the emf. Unfortunately you can't observe the phenomenon with simple equipment such as a compass; you need a sensitive ammeter. Perhaps your instructor will demonstrate the effects or arrange for you to experiment with appropriate equipment.

Initially the solenoid current I_1 is constant, so B_1 is constant, and no current runs through the resistive wire and the ammeter. Next we vary the current I_1 in the long solenoid, thus causing the magnetic field B_1 to vary, and we observe the current I_2 in the outer wire. In the solenoid we first increase the current rapidly, then hold the current constant, and then slowly decrease the current, at half the rate we used at first (Figure 22.12). If we know the resistance R of the circuit containing the ammeter we can determine the emf from the ammeter reading, since emf $- RI_2 = 0$; the wire acts as though a battery were inserted.

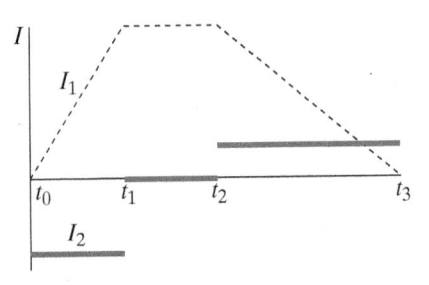

Figure 22.11 An ammeter measures current in a loop surrounding the solenoid. Initially I_1 is constant, so B_1 is constant, and no current runs through the ammeter. (The power supply and connections to the solenoid are not shown.)

Figure 22.12 Vary the solenoid current I_1 and observe the current I_2 that runs in the outer wire, through the ammeter.

1. While the solenoid current I_1 is increasing from t_0 to t_1 (Figure 22.12), B_1 is increasing, and the current I_2 runs clockwise in the circuit, out of the "+" terminal of the ammeter. The ammeter is observed to read a negative current. Remember that conventional current flowing into the positive terminal of an ammeter gives a positive reading.

If $dB/dt \neq 0$, there is a curly E_{NC}.

2. While the solenoid current I_1 is held constant from t_1 to t_2 (Figure 22.12), the magnetic field in the solenoid isn't changing, and there is no emf in the circuit. The ammeter reading is zero.

No dB/dt, no curly E_{NC}, even if there is a large (constant) B.

3. While the solenoid current I_1 is decreasing from t_2 to t_3 (at half the initial rate; see Figure 22.12), B_1 is decreasing, and the current I_2 runs counterclockwise in the circuit, into the $+$ terminal of the ammeter. The ammeter is observed to read a positive current. The current I_2 is observed to be half what it was during the first interval.

E_{NC} (and associated emf) outside the solenoid are proportional to dB/dt inside the solenoid.

4. There is one more crucial experiment. If we use a solenoid with twice the cross-sectional area but the same magnetic field, we find that the current I_2 is twice as big (which means that the emf is twice as big).

E_{NC} (and associated emf) outside the solenoid are proportional to the cross-sectional area of the solenoid.

Magnetic Flux

Putting together these experimental observations and being quantitative about the emf, we find experimentally that the magnitude of the emf is numerically equal to this:

$$|\text{emf}| = \left| \frac{d}{dt}(B_1 \pi r_1^2) \right|$$

The quantity $(B_1 \pi r_1^2)$ is called the "magnetic flux" Φ_{mag} on the area encircled by the circuit (Φ is the uppercase Greek letter phi). Magnetic flux is calculated in the same way as the electric flux introduced in Chapter 21 on Gauss's law.

The magnetic flux on a small area ΔA is $\vec{B} \bullet \hat{n} \Delta A = B_\perp \Delta A$, where $\hat{n}$ is a dimensionless unit vector perpendicular to the area ΔA ($\hat{n}$ is called the "normal" to the surface); $B_\perp$ is the perpendicular component of magnetic field. We add up all such contributions over an extended surface to get the magnetic flux over that surface (Figure 22.13):

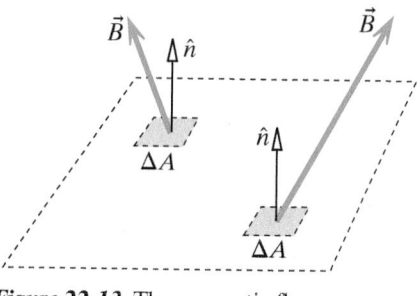

Figure 22.13 The magnetic flux on an area is the sum of the magnetic flux on each small subarea.

DEFINITION OF MAGNETIC FLUX

$$\Phi_{mag} = \int \vec{B} \bullet \hat{n} \, dA = \int B_\perp \, dA$$

Note that magnetic flux can be positive, negative, or zero, depending on the orientation of the magnetic field $\vec{B}$ relative to the normal $\hat{n}$.

> **Checkpoint 3** A uniform magnetic field of 3 T points 30° away from the perpendicular to the plane of a rectangular loop of wire 0.1 m by 0.2 m (Figure 22.14). What is the magnetic flux on this loop?

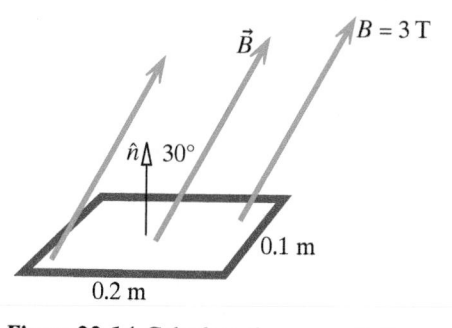

Figure 22.14 Calculate the magnetic flux on the area enclosed by this loop of wire.

Faraday's Law — Quantitative

The experimental fact that in the presence of a time-varying magnetic field there is an emf whose magnitude is equal to the rate of change of magnetic flux is called "Faraday's law" (discovered by the British scientist Michael Faraday in 1831). Faraday's law summarizes a great variety of experimental data, not just the data for the experiments we have discussed. Faraday's law is a major physical law concerning the relationship between electric and magnetic fields. Unlike the motional emf we studied in Chapter 20, Faraday's law cannot be derived from any of the other fundamental principles we have studied so far.

FARADAY'S LAW

$$\text{emf} = -\frac{d\Phi_{\text{mag}}}{dt}$$

where $\text{emf} = \oint \vec{E}_{\text{NC}} \bullet d\vec{l}$ and $\Phi_{\text{mag}} \equiv \int \vec{B} \bullet \hat{n} dA.$

In words: The emf along a round-trip path is equal to the rate of change of the magnetic flux on the area encircled by the path.

Direction: With the thumb of your right hand pointing in the direction of $-d\vec{B}/dt$, your fingers curl around in the direction of $\vec{E}_{\text{NC}}$.

The meaning of the minus sign: If the thumb of your right hand points in the direction of $-d\vec{B}/dt$ (that is, the opposite of the direction in which the magnetic field is increasing), your fingers curl around in the direction along which the path integral of electric field is positive:

$$\oint \vec{E}_{\text{NC}} \bullet d\vec{l} = -\frac{d}{dt}\left[\int \vec{B} \bullet \hat{n} dA\right] \quad \text{(sign given by right-hand rule)}$$

For most of our work it is simplest to calculate the magnitude of the effect, ignoring signs and directions, and then give the appropriate sign or direction based on the right-hand rule. We will include the minus sign in Faraday's law, as a reminder of what we need to do to get directions.

What area exactly do we use when calculating the magnetic flux in Faraday's law? Imagine a soap film stretched over the closed path around which we are calculating $\oint \vec{E}_{\text{NC}} \bullet d\vec{l}$. We calculate Φ_{mag} on the area covered by that soap film.

Putting these pieces together, we have the following form:

FORMAL VERSION OF FARADAY'S LAW

$$\oint \vec{E}_{\text{NC}} \bullet d\vec{l} = -\frac{d}{dt}\int \vec{B} \bullet \hat{n} dA \quad \text{(sign given by right-hand rule)}$$

Faraday's law summarizes the experiments we've described so far:

- A bigger emf is associated with a faster rate of change of magnetic flux (if the magnetic flux is constant there is no emf).
- If there is a bigger area with the same perpendicular magnetic field there is a bigger emf.

> **Checkpoint 4** A wire of resistance $10\,\Omega$ and length 2.5 m is bent into a circle that is concentric with and encircles a solenoid in which the magnetic flux changes from $5\,\text{T}\cdot\text{m}^2$ to $3\,\text{T}\cdot\text{m}^2$ in 0.1 s. What is the emf in the wire? What is the non-Coulomb electric field in the wire? What is the current in the wire?

The Coulomb Electric Field Can Be Included in Faraday's Law

We will usually write $\text{emf} = \oint \vec{E}_{\text{NC}} \bullet d\vec{l}$ as a reminder that it is the curly non-Coulomb electric field that has a nonzero round-trip path integral, which we call the emf around that path. However, we could also calculate the emf in terms of the net electric field $\vec{E} = \vec{E}_{\text{C}} + \vec{E}_{\text{NC}}$, where $\vec{E}_{\text{C}}$ is the Coulomb electric field due to charges:

$$\text{emf} = \oint \vec{E} \bullet d\vec{l}$$

This is true because we have

$$\text{emf} = \oint \vec{E} \cdot d\vec{l} = \oint (\vec{E}_C + \vec{E}_{NC}) \cdot d\vec{l} = 0 + \oint \vec{E}_{NC} \cdot d\vec{l}$$

since $\oint \vec{E}_C \cdot d\vec{l} = 0$ (the round-trip integral of a Coulomb electric field is zero).

We can apply Faraday's law to analyze quantitatively the situation of a circuit with an ammeter surrounding a solenoid (Figure 22.15; a cross section of the solenoid is shown). There is magnetic field $\vec{B}_1$ only over the circle of radius r_1 (the solenoid), and it points in the same direction as $\hat{n}$, the unit vector perpendicular to the surface.

EXAMPLE

A Circuit Surrounding a Solenoid

In Figure 22.15 the magnetic field in the solenoid increases from 0.1 T to 0.7 T in 0.2 s, and the area of the solenoid is 3 cm². **(a)** Calculate the magnetic flux Φ_{mag} on the area enclosed by the solenoid, the flux on the other portions of the area encircled by the circuit, and the total flux on the area encircled by the circuit. **(b)** What is the emf around the circuit? **(c)** If the resistance of the wire plus ammeter is 0.5 Ω, what current will the ammeter display?

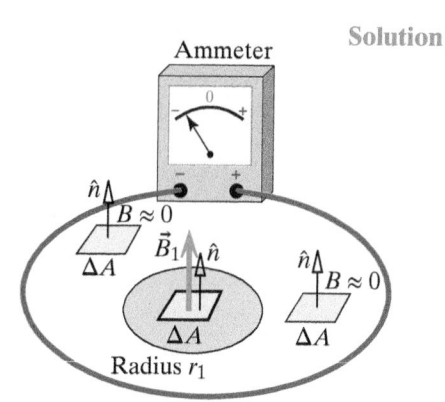

Ammeter

$\hat{n}$

$B \approx 0$

ΔA

$\vec{B}_1$ $\hat{n}$

$\hat{n}$

$B \approx 0$

ΔA

ΔA

Radius r_1

Figure 22.15 A cross section through the solenoid; calculate the flux enclosed by the wire.

Solution **(a)** On the area enclosed by the solenoid, $\Phi_{mag} = B_1(\pi r_1^2)$, because $B_\perp$ is constant and equal to B_1 throughout the cross-section of the solenoid. On every small area outside the solenoid, the magnetic field is nearly zero, so $\Phi_{mag} = 0$. Therefore the total flux through the outer wire is $\Phi_{mag} = B_1(\pi r_1^2)$.

What counts is the magnetic flux encircled by the circuit, not the magnetic flux on a closed surface such as a box or sphere. Unlike electric flux, the total magnetic flux on a closed surface is always zero, because "magnetic monopoles" seem not to exist.

(b) Experimentally we find that the emf around the circuit is numerically equal to $d\Phi_{mag}/dt$, as predicted by Faraday's law. Faraday's law predicts the following average emf:

$$\text{emf} = \frac{\Delta \Phi_{mag}}{\Delta t} = \frac{(0.7 \text{ T})(3 \times 10^{-4} \text{ m}^2) - (0.1 \text{ T})(3 \times 10^{-4} \text{ m}^2)}{(0.2 \text{ s})} = 9 \times 10^{-4} \text{ V}$$

(c) The circuit acts as though a battery were inserted, $\text{emf} - RI = 0$, so

$$I = \frac{\text{emf}}{R} = \frac{(9 \times 10^{-4} \text{ V})}{(0.5 \text{ }\Omega)} = 1.8 \times 10^{-3} \text{ A}$$

One more experiment: let's reconnect the wire and ammeter so that they don't encircle the solenoid (Figure 22.16).

QUESTION As we increase the current (and magnetic field) in the solenoid, what would you predict from Faraday's law about the ammeter reading? Why?

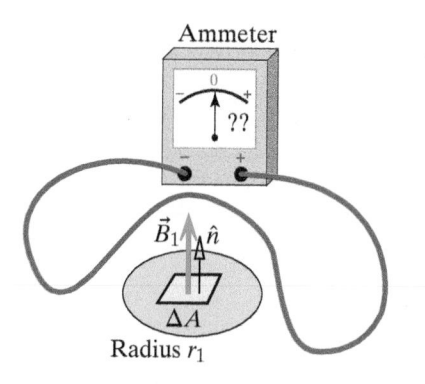

Ammeter

0

??

$\vec{B}_1$ $\hat{n}$

ΔA

Radius r_1

Figure 22.16 What happens if the circuit does not encircle the solenoid?

Within the area bounded by the wire outside the solenoid, there is practically no magnetic field, so no magnetic flux Φ_{mag}, and no rate of change of magnetic flux $d\Phi_{mag}/dt$. When we try the experiment, we do indeed find that the ammeter shows little or no current.

Voltmeter Readings

Since a voltmeter acts like an ammeter with a large resistance in series, a voltmeter may give a puzzling reading in the presence of time-varying magnetic flux. Consider the two voltmeters shown in Figure 22.17. You normally expect that when voltmeter leads are connected to each other, the voltmeter must read zero volts.

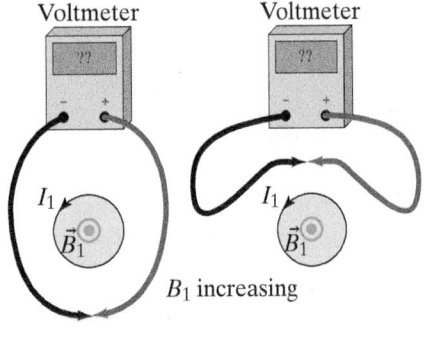

Figure 22.17 What do these voltmeters read?

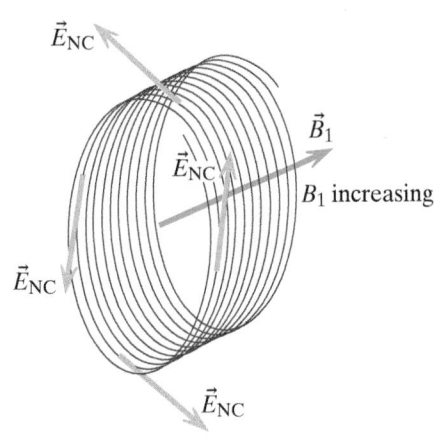

Figure 22.18 The emf of a thin coil is approximately N times the emf of one loop.

Figure 22.19 Move coil 1 toward coil 2, and there is a time-varying magnetic field inside coil 2.

Figure 22.20 Moving a magnet toward coil 2 creates a time-varying magnetic field inside the coil.

Figure 22.21 Rotating a bar magnet (or coil 1) produces a time-varying magnetic field inside coil 2.

QUESTION Do these voltmeters read zero?

The voltmeter leads on the left of Figure 22.17 encircle a region of changing magnetic field, so the voltmeter will read an emf equal to $d\Phi_{mag}/dt$. The leads of the other voltmeter don't encircle a region of changing magnetic field, so the voltmeter reads zero.

Because of the effect of time-varying magnetic fields, you have to be a bit careful in interpreting a voltmeter reading when there are varying magnetic fields around. For example, when there are sinusoidally alternating currents (AC) there are time-varying magnetic fields due to those time-varying currents. If the leads of an AC voltmeter happen to surround some AC magnetic flux, this will affect the voltmeter reading.

The Emf for a Coil with Multiple Turns

In many devices, instead of just one loop of wire surrounding a time-varying flux there is a coil with many turns, which increases the effect: the emf in a coil containing N turns is approximately N times the emf for one loop of the coil. This point needs some discussion.

In the space around the increasing flux in the coil, there is a curly pattern of non-Coulomb electric field. If the loops are tightly wound so that they are very close together, E_{NC} is about the same in each loop (Figure 22.18). The emf from one end of the coil to the other is the integral of the non-Coulomb field:

$$\text{emf} = \int_{N\text{ turns}} \vec{E}_{NC} \bullet d\vec{l} = N(E_{NC}L_{\text{one turn}}) = N(\text{emf}_{\text{one turn}})$$

Because of this, Faraday's law for a coil is written in the following form:

FARADAY'S LAW FOR A COIL

$$\text{emf} = -N\frac{d\Phi_{mag}}{dt} \quad \text{(sign given by right-hand rule)}$$

The emf in a coil of N turns is equal to N times the rate of change of the magnetic flux on one loop of the coil.

Faraday's Law and Moving Coils or Magnets

A time-varying magnetic field is associated with a curly electric field. One way to create a time-varying magnetic field is by varying the current in a stationary coil, but changes in the position of currents can also produce a time-varying magnetic field and the associated curly electric field. With a steady current in one coil in Figure 22.19, you can move that coil closer to a second coil. This increases the magnetic field (and magnetic flux) inside the second coil, and while the magnetic field is increasing there is an emf that can drive a current in the second coil.

You will also find an emf if you move a bar magnet toward or away from coil 2, since this creates a time-varying magnetic field (and magnetic flux) inside the coil (Figure 22.20).

You can also get an emf by rotating the first coil (or a bar magnet; Figure 22.21), since this changes the flux through the second coil as long as you are rotating and therefore changing the magnetic field (and magnetic flux) in the region of space surrounded by the second coil.

These various experiments give additional confirmation that the emf in one or more loops of wire is associated with a time-varying magnetic field, and numerically equal to the rate of change of the magnetic flux enclosed by the loops:

$$\text{emf} = -N\frac{d\Phi_{mag}}{dt} = -N\frac{d}{dt}\left(\sum \vec{B}_1 \bullet \hat{n}\, \Delta A_2\right)$$

An Interesting Complication

The current I_1 in a coil of wire makes a magnetic field B_1. If I_1 varies with time, there is a time-varying magnetic field dB_1/dt in a second coil and an emf and current I_2 in the second coil. This current I_2 *also* makes a magnetic field B_2, an effect that we have ignored until now. If I_2 is constant (due to a constant dB_1/dt), the additional magnetic field B_2 is constant and so does not contribute any additional non-Coulomb electric field.

However, if $d^2B_1/dt^2 \neq 0$ (there is a time-varying rate of change), then I_2 isn't constant, and there is a time-varying magnetic field dB_2/dt and a non-Coulomb electric field and emf of its own. In many cases this effect is small compared to the main effect and can be ignored. If the effect is sizable, you can see that a full analysis could be rather difficult, because you have a changing I_1 making a changing I_2 that contributes to the change, which ... Whew!

We will mostly avoid trying to analyze this complicated kind of situation in this introductory textbook, though we will see an example in the discussion of superconductors later in this chapter. A related issue is self-inductance, which we will also study later in this chapter.

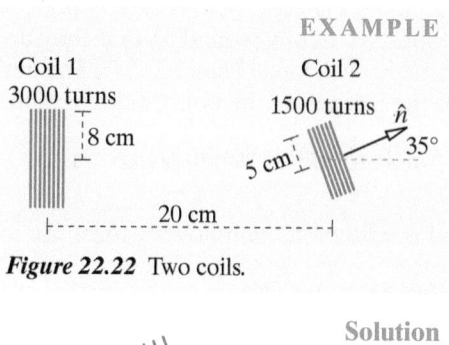

Coil 1
3000 turns
8 cm

Coil 2
1500 turns $\hat{n}$
5 cm
35°

20 cm

Figure 22.22 Two coils.

EXAMPLE

Two Coils

Two coils of wire are near each other, 20 cm apart, as shown in Figure 22.22. Coil 1 has a radius 8 cm and contains 3000 loops of wire. It is connected to a power supply whose output voltage can be changed, so that the current I_1 in coil 1 can be varied. Coil 2 contains 1500 loops of wire and has radius 5 cm, and is rotated 35° from the axis of coil 1, as shown in Figure 22.22. The current in coil 1 changes from 0 to 3 A in 4 ms. Calculate the emf in coil 2.

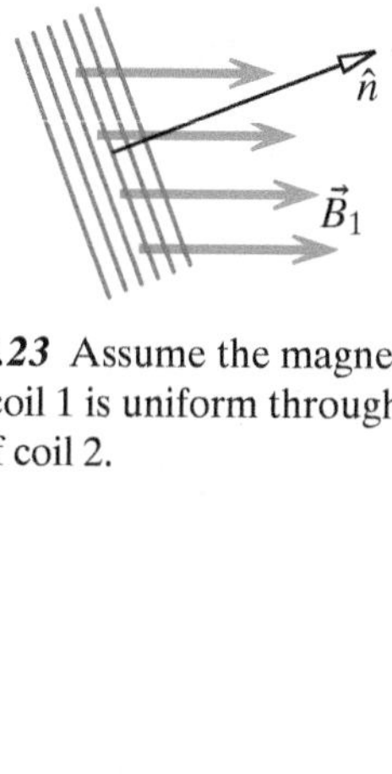

Figure 22.23 Assume the magnetic field made by coil 1 is uniform through the interior of coil 2.

Solution

Start from fundamental principles. For a given current I_1 in the first coil, we can find B_1, the magnetic field that this current makes at the location of coil 2. We know (dI_1/dt), so we can find $(d\Phi_{\text{mag}}/dt)$ in coil 2, which gives us the emf in coil 2.

Approximations: Assume that the coils are far enough apart that the magnetic field of the first coil is approximately uniform in direction and magnitude throughout the interior of the second coil (Figure 22.23), and can be approximated as a magnetic dipole field. Also assume that the length of the coils is small compared to the distance between them (essentially treating all loops as if they were located at the center of the coil).

Work out everything symbolically first; plug in numbers at the end.

Magnetic field B_1 at the location of coil 2:

$$B_1 \approx N_1 \frac{\mu_0}{4\pi} \frac{2I_1(\pi r_1^2)}{z^3}$$

Magnetic flux through one loop of coil 2:

$$\Phi_{\text{mag}} \approx B_1(\pi r_2^2)\cos\theta$$

$$= N_1 \frac{\mu_0}{4\pi} \frac{2I_1(\pi r_1^2)}{z^3}(\pi r_2^2)\cos\theta$$

I_1 is the only quantity that is changing, so the emf around one loop of coil 2 is:

$$\text{emf} = \left|\frac{d\Phi_{\text{mag}}}{dt}\right| \approx N_1 \frac{\mu_0}{4\pi} \frac{2(\pi r_1^2)(\pi r_2^2)}{z^3}\cos\theta \frac{dI_1}{dt}$$

Total emf for all of coil 2:

$$\text{emf} = N_2 \left| \frac{d\Phi_{\text{mag}}}{dt} \right| \approx N_1 N_2 \frac{\mu_0}{4\pi} \frac{2(\pi r_1^2)(\pi r_2^2)}{z^3} \cos\theta \frac{dI_1}{dt}$$

$$\frac{(1500)(3000)(1\times10^{-7}\,\text{T}\cdot\text{m/A})(2)(0.08\,\text{m})^2(\pi^2)(0.05\,\text{m})^2\cos 35°(3\,\text{A})}{(0.20\,\text{m})^3(4\times10^{-3}\,\text{s})}$$

$$= 10.9\,\text{V}$$

EXAMPLE   **Wire Loop and a Solenoid**

A length of wire whose total resistance is R is made into a loop with two quarter-circle arcs of radius r_1 and radius r_2 and two straight radial sections as shown in Figure 22.24. A very long solenoid is positioned as shown, going into and out of the page. The solenoid consists of N turns of wire wound tightly onto a cylinder of radius r and length d, and it carries a counterclockwise current I_s that is increasing at a rate dI_s/dt.

What are the magnitude and direction of the magnetic field at C, at the center of the two arcs? The "fringe" magnetic field of the solenoid (that is, the magnetic field produced by the solenoid outside of the solenoid) is negligible at point C. Show all relevant quantities on a diagram.

Solution   As usual, start from fundamentals. Initially the only sources of field are the moving charges in the solenoid (which make B_s in the solenoid) and dB_s/dt in the solenoid (with an associated curly non-Coulomb electric field around the solenoid). The curly electric field constitutes an emf that drives current I_{loop} in the surrounding wire. The current-carrying wires produce a magnetic field at location C.

Minor point: Because the wire is not circular, pileup leads to tiny amounts of surface charge on the wire. The net electric field in the wire (the superposition of the non-Coulomb and Coulomb fields) follows the wire and is uniform in magnitude throughout the wire.

(a) Inside the solenoid, $B_s = \dfrac{\mu_0 N I_s}{d}$ out of the page. See Figure 22.25.

(b) Flux through *outer* loop =

$$B_s A_s \cos(0°) + (0)A_{\text{rest of loop}} = \frac{\mu_0 N I_s}{d}\pi r^2$$

(c) $-d\vec{B}/dt$ into page, so $\vec{E}_{\text{NC}}$ and current I_{loop} are clockwise as shown.

$$|\text{emf}| = \frac{d}{dt}\left(\frac{\mu_0 N I_s}{d}\pi r^2 \right) = \frac{\mu_0 N}{d}\pi r^2 \frac{dI_s}{dt}$$

so $I_{\text{loop}} = \dfrac{\text{emf}}{R} = \dfrac{\mu_0 N}{d}\dfrac{\pi r^2}{R}\dfrac{dI_s}{dt}.$

(d) $B_{\text{at }C}$ due to straight segments, inner wire, and outer wire:

$$B_{\text{straight segments}} = 0 \text{ because } d\vec{l}\times\hat{r} = 0 \quad \text{(see diagram)}$$

$$B_{\text{inner wire}} = \left| \int_{\text{quarter loop}} \frac{\mu_0}{4\pi} \frac{I_{\text{loop}}\, d\vec{l}\times\hat{r}}{r_1^2} \right| = \frac{\mu_0}{4\pi}\frac{I_{\text{loop}}}{r_1^2}\int_{\text{quarter loop}} dl\sin(90°)$$

$$B_{\text{inner wire}} = \frac{\mu_0}{4\pi}\frac{I_{\text{loop}}}{r_1^2}\left(\frac{2\pi r_1}{4} \right) = \frac{\mu_0}{4\pi}\frac{\pi I_{\text{loop}}}{2r_1}, \text{ and } B_{\text{outer wire}} = \frac{\mu_0}{4\pi}\frac{\pi I_{\text{loop}}}{2r_2}$$

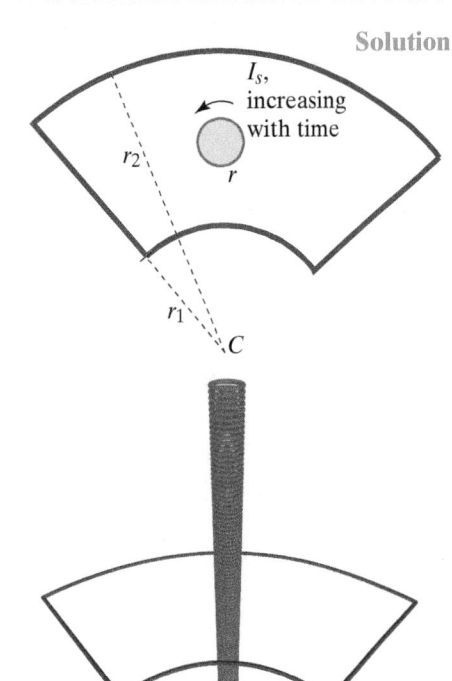

I_s, increasing with time

r_2

r

r_1

C

Figure 22.24 Upper: A long solenoid passes through a loop of wire. Lower: A 3D view.

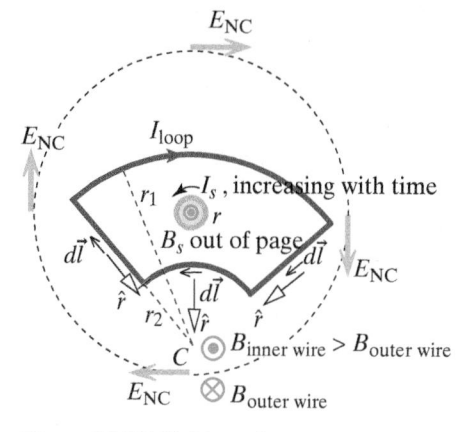

Figure 22.25 Fields and currents.

$B_{\text{inner wire}}$ is out of the page, and $B_{\text{outer wire}}$ is into the page at location C. Since B is proportional to $1/r$, $B_{\text{inner wire}}$ is larger than $B_{\text{outer wire}}$. The net field is out of the page, with this magnitude:

$$B_{\text{net at C}} = B_{\text{inner wire}} - B_{\text{outer wire}} = \frac{\mu_0}{4\pi} \frac{\pi I_{\text{loop}}}{2} \left(\frac{1}{r_1} - \frac{1}{r_2} \right)$$

$$B_{\text{net at C}} = \frac{\mu_0}{4\pi} \left(\frac{\pi}{2} \right) \left(\frac{\mu_0 N}{d} \frac{\pi r^2}{R} \frac{dI_s}{dt} \right) \left(\frac{1}{r_1} - \frac{1}{r_2} \right)$$

QUESTION Suppose that in this example the solenoid had been filled with an iron core. How would this have affected the emf in the wire?

Recall from Chapter 17 that iron has a multiplier effect. When you apply a magnetic field to iron, magnetic domains that are nearly aligned with the applied magnetic field tend to grow at the expense of domains that are not aligned, and within a domain the individual magnetic dipoles can rotate to be more aligned to the applied magnetic field. The new magnetic field due to the aligned magnetic dipoles in the iron can be much larger than the applied magnetic field, and as long as not all domains are fully aligned, this larger magnetic field will be approximately proportional to the applied field.

To be concrete, suppose that the magnetic field of a solenoid without an iron core varies from 0 to 0.1 T in 0.1 s, so that $dB/dt = 0.1/0.1 = 1$ T/s. Suppose that with an iron core present the magnetic field is 10 times greater and varies from 0 to 1 T in 0.1 s, so that $dB/dt = 1/0.1 = 10$ T/s. With 10 times the rate of change of magnetic field, the emf generated around the solenoid will be 10 times greater. This is the reason that you will often see coils of wire wrapped around iron cores in various kinds of electrical devices.

22.3 FARADAY'S LAW AND MOTIONAL EMF

In Chapter 20 we studied motional emf: when a wire moves through a magnetic field, magnetic forces drive current along the wire. We calculated the motional emf in a moving bar of length L as $(q|\vec{v} \times \vec{B}|)L/q$, where $\vec{v}$ is the velocity of the bar. If $\vec{v}$ is perpendicular to the magnetic field, the motional emf is simply vBL. There is another way to calculate motional emf that is often mathematically easier, in terms of magnetic flux.

As shown in Figure 22.26, in a short time Δt the bar moved a distance $\Delta x = v\Delta t$, and the area surrounded by the current-carrying pieces of the circuit increased by an amount $\Delta A = L\Delta x = Lv\Delta t$. There is an increased magnetic flux through the circuit of amount

$$\Delta \Phi_{\text{mag}} = B_\perp \Delta A = B(Lv\Delta t)$$

Dividing by Δt we find this:

$$\frac{\Delta \Phi_{\text{mag}}}{\Delta t} = BLv$$

However, BLv is equal to the emf vBL associated with the magnetic force, and in the limit of small Δt we have the following:

MOTIONAL EMF = RATE OF CHANGE OF MAGNETIC FLUX

$$|\text{emf}| = \left| \frac{d\Phi_{\text{mag}}}{dt} \right| \quad \text{(sign given by direction of magnetic force)}$$

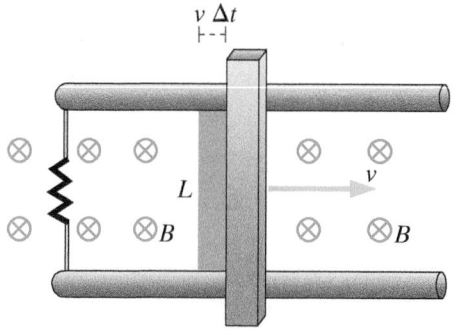

Figure 22.26 There is increased flux through the circuit as the bar moves.

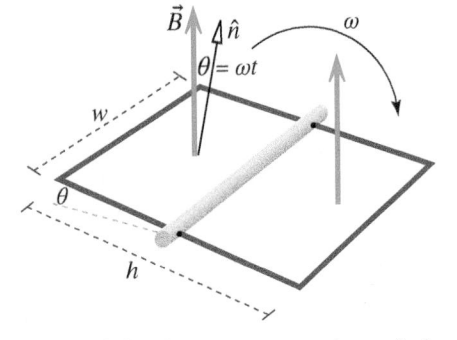

Figure 22.27 We can compute the emf of a generator from the rate of change of flux.

Although we have proved this result only for a particular special case, it can be shown that this result applies in general to calculating motional emf—that is, an emf associated with the motion of a piece of a circuit in a magnetic field.

As an example, consider a generator (Figure 22.27). When the normal to the loop is at an angle $\theta = \omega t$ to the magnetic field, the flux through the loop is $Bhw \cos(\omega t)$, so emf $= |d\Phi/dt| = \omega Bhw \sin(\omega t)$, which is the same result we obtained in Chapter 20 when we calculated the emf directly in terms of magnetic forces acting on electrons in the moving wires (motional emf).

> **Checkpoint 5** A uniform, non-time-varying magnetic field of 3 T points $30°$ away from the perpendicular to the plane of a rectangular loop of wire 0.1 m by 0.2 m (Figure 22.28). The loop as a whole is moved in such a way that it maintains its shape and its orientation in the uniform magnetic field. What is the emf around the loop during this move? In 0.1 s the loop in Figure 22.28 is stretched to be 0.12 m by 0.22 m while keeping the center of the loop in one place. What is the average emf around the loop during this time?

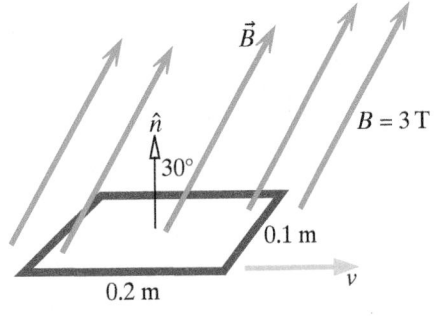

Figure 22.28 Move this rectangular circuit around in a uniform, constant magnetic field.

The Two Pieces of the Flux Derivative

A time-varying magnetic field is associated with a curly (non-Coulomb) electric field. We can quantitatively predict the behavior of circuits enclosing a time-varying magnetic field by calculating the rate of change of magnetic flux on the surface bounded by the circuit.

However, in the case of the sliding bar we just saw that there is another way that magnetic flux can change. If the magnetic field is constant but the shape or orientation of the loop changes so as to enclose more or less flux, the flux changes with time even though the magnetic field doesn't, and this is the phenomenon of motional emf associated with magnetic forces.

Any change in magnetic flux, whether due to a change in magnetic field or a change in shape or orientation, produces an emf equal to the rate of change of flux. Both effects can be present simultaneously. In Figure 22.29, suppose that as you drag a metal bar along rails the magnitude of the uniform magnetic field is also increasing with time. The flux is $B_\perp A$, and the rate of change of the flux is due in part to the change in the magnetic field, and in part to the change in the area enclosed by the changing path:

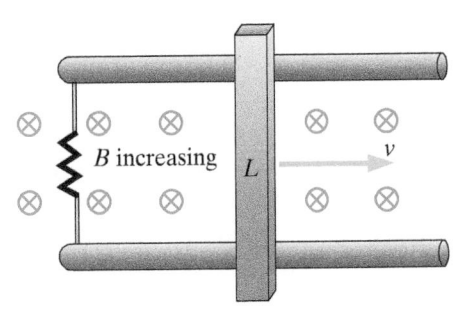

Figure 22.29 Because B is increasing there is a curly electric field $\vec{E}_{NC}$, and because the bar is moving there is also motional emf.

$$\frac{d\Phi_{mag}}{dt} = \frac{d}{dt}(B_\perp A)$$
$$= \frac{dB_\perp}{dt}A + B_\perp \frac{dA}{dt}$$

The first term, $(dB_\perp/dt)A$, is nonzero if $B_\perp$ varies with time (Faraday's law). The second term, $B_\perp dA/dt$, is nonzero if there is a change in the area enclosing magnetic flux (motional emf).

Evaluating these two terms for the case of a bar moving along rails, with the magnetic field changing in time as well as there being a change in the area enclosing magnetic flux, we have this:

$$\frac{d\Phi_{mag}}{dt} = \frac{dB_\perp}{dt}A + B_\perp \frac{dA}{dt}$$
$$= \frac{dB_\perp}{dt}A + B_\perp \frac{Ldx}{dt}$$
$$= \frac{dB_\perp}{dt}A + B_\perp Lv$$

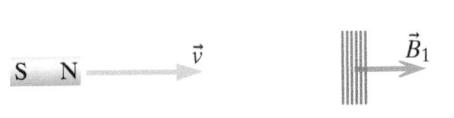

Figure 22.30 Move a magnet toward a stationary coil; there is a changing magnetic field in the coil, and a curly electric field $\vec{E}_{NC}$.

In some but not all cases the two effects can be related through a change of reference frame. If you move a magnet toward a stationary coil (Figure 22.30),

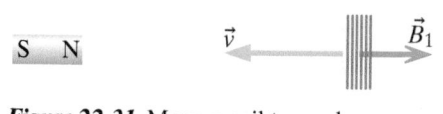

Figure 22.31 Move a coil toward a stationary magnet; there is no changing magnetic field at any fixed location in space, but there is motional emf in the coil.

the magnetic field in the coil is increasing with time, and there is a curly non-Coulomb electric field that drives current in the coil.

If on the other hand you move a coil toward a stationary magnet (Figure 22.31), there is no fixed location in space where the magnetic field is changing with time, but the coil is moving in a magnetic field, so there is a magnetic force on the mobile charges in the wire, and motional emf. (It is true that the magnetic field at the location of the moving coil is changing with time, but this is not a fixed location; there is no location where $dB/dt \neq 0$.)

In both situations you can calculate the same emf from the same rate of change of flux, and you observe the same current to run in the coil, though the mechanism looks different in the two cases. You can shift between one situation and the other simply by changing your reference frame (move with the magnet, or move with the coil). The principle of relativity implies that you should predict the same emf and the same current in either frame of reference, and you do.

It is not always possible to relate the two kinds of effects simply by changing reference frame. For example, consider a bar sliding along rails. In the reference frame of the bar, the bar is at rest but the resistor is moving. In neither reference frame is there a time-varying magnetic field.

(A full analysis in the reference frame of the bar is complicated. If the bar is at rest in this frame, why is it polarized? We saw in Chapter 20 that the answer is given by special relativity: when you transform to the frame of the bar in the presence of a uniform magnetic field, a transverse electric field of magnitude vB appears, and in this reference frame it is this new electric field that polarizes the bar.)

Summary: Non-Coulomb Fields and Forces

If the magnetic field changes ($d\vec{B}/dt$ nonzero), there is a non-Coulomb electric field $\vec{E}_{NC}$ that curls around the region of changing flux, and emf $= \oint \vec{E}_{NC} \bullet d\vec{l}$ is nonzero. In purely motional emf, however, where the path changes but the magnetic field doesn't change, the non-Coulomb forces are magnetic forces, not electric.

The non-Coulomb electric field $\vec{E}_{NC}$ affects stationary as well as moving charges, but the (non-Coulomb) magnetic forces only affect moving charges, such as electrons in the moving metal bar. It is somewhat odd that the magnitude of the emf in both of these situations is $d\Phi_{mag}/dt$, but that's how it works.

We have identified three different kinds of electric and magnetic forces on the electrons inside a wire:

1. Forces from Coulomb electric fields due to surface charges: superposition of contributions of the form $\vec{E} = \dfrac{1}{4\pi\varepsilon_0}\dfrac{q}{r^2}\hat{r}$

2. Forces from non-Coulomb electric fields associated with time-varying magnetic fields, $\dfrac{d\vec{B}}{dt}$

3. Magnetic forces $q\vec{v} \times \vec{B}$ if the wire is moving (yielding a motional $|\text{emf}| = \left|\dfrac{d\Phi_{mag}}{dt}\right|$ due to change of path)

All of these effects can be present simultaneously. For example, if the magnetic field is changing while you pull a bar along rails, there is a Coulomb electric field due to the charges that build up on the surfaces of the circuit elements, there is a non-Coulomb electric field due to the changing magnetic field, and there are non-Coulomb magnetic forces on electrons inside the moving bar. The round-trip integral of the Coulomb electric field is zero. The round-trip integral of the non-Coulomb electric field and the non-Coulomb magnetic force per unit charge together gives the emf, and the current I is given by $I = \text{emf}/R$, where $|\text{emf}| = |d\Phi_{\text{mag}}/dt|$.

22.4 MAXWELL'S EQUATIONS

In Chapter 21 we summarized the four Maxwell's equations that describe the patterning of electric and magnetic fields. We can update Maxwell's equations now that we know that the round-trip integral of electric field need not be zero (Faraday's law):

MAXWELL'S EQUATIONS

$$\oint \vec{E} \bullet \hat{n} dA = \frac{1}{\varepsilon_0} \sum Q_{\text{inside the surface}} \qquad \text{Gauss's law}$$

$$\oint \vec{B} \bullet \hat{n} dA = 0 \qquad \text{Gauss's law for magnetism}$$

$$\oint \vec{E} \bullet d\vec{l} = -\frac{d}{dt} \int \vec{B} \bullet \hat{n} dA \qquad \text{Faraday's law}$$

$$\oint \vec{B} \bullet d\vec{l} = \mu_0 \sum I_{\text{inside path}} \qquad \text{Ampere's law (incomplete)}$$

In Chapter 23 we will see that Ampere's law needs to be modified, at which point Maxwell's equations will be complete.

Maxwell's equations are a summary of a large number of diverse experiments concerning the possible patterns of and relationships between $\vec{E}$ and $\vec{B}$ in three-dimensional space. No violation of these relations has ever been observed.

You might wonder, "Where is Coulomb's law in all this? Where is the Biot–Savart law?" The equation for the electric field of a stationary point charge can be deduced from Gauss's law. Draw a spherical Gaussian surface around the charge and assume by spherical symmetry that on that surface $\vec{E}$ is everywhere radial and everywhere has the same magnitude. Then the electric flux over the whole spherical surface is $E(4\pi r^2)$, which Gauss's law says should be equal to q/ε_0. Solving for the field, we find the familiar result:

$$\vec{E} = \frac{1}{4\pi\varepsilon_0} \frac{q}{r^2} \hat{r}$$

Moreover, Faraday's law assures us that $\oint \vec{E} \bullet d\vec{l} = 0$ (when there is no time-varying magnetic field), and this is consistent with the outward-going or inward-going pattern of electric field around a stationary point charge, with no curly electric field.

Similarly, starting from Ampere's law it is possible to derive the Biot–Savart law (the magnetic field made by moving charges) but the mathematics is much more difficult and beyond the scope of this introductory course. The curly nature of the magnetic field deduced from Ampere's law is consistent with Gauss's law for magnetism, which says that magnetic field cannot have an outward-going or inward-going pattern but must be purely curly. (Another way of saying this is that no one has ever found a magnetic monopole, whose magnetic field would be outward-going or inward-going, and Gauss's law for magnetism embodies this experimental null result.)

Maxwell's equations also can be used to predict the electric and magnetic fields of accelerated charges or time-varying currents. In this chapter we have seen that there is a curly electric field in the presence of a time-varying current, and this fact of nature is embedded in Faraday's law, which relates the path integral of electric field to the rate of change of magnetic field. In Chapter 23 we will see additional consequences of the fields of accelerated charges.

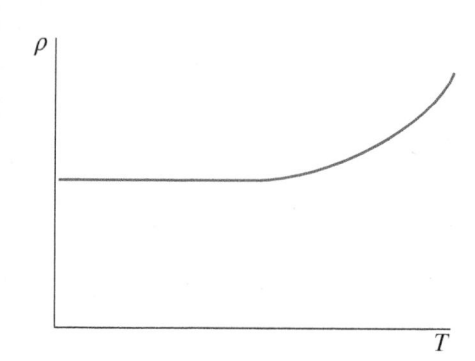

Figure 22.32 Resistivity vs. temperature for an ordinary metal.

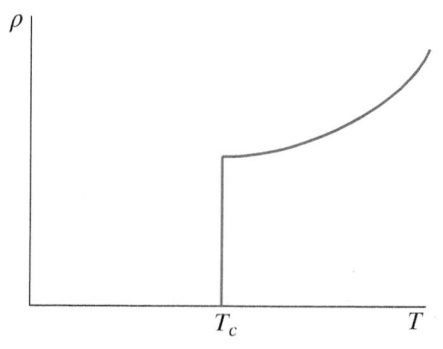

Figure 22.33 Resistivity vs. temperature for a superconductor.

I runs for *years*!

Figure 22.34 A current in a superconducting ring persists for years.

22.5 SUPERCONDUCTORS

As you lower the temperature, the resistivity of an ordinary metal decreases but remains nonzero even at absolute zero (Figure 22.32), whereas some materials lose all resistance to electric current. Such materials are called "superconductors." It isn't that the resistance merely gets very small. Below a critical temperature T_C that is different for each superconducting material, the resistance vanishes completely, as indicated in Figure 22.33. To put it another way, the mobility becomes infinite. A nonzero electric field in a superconductor would produce an infinite current! As a result, the electric field inside a superconductor must always be zero, yet a nonzero current can run anyway.

In a ring made of lead that is kept at a temperature below 7.2 K (that is, only 7.2 kelvins above absolute zero), one can observe a current that persists undiminished for years (Figure 22.34)!

> QUESTION How can you check that the current in the lead ring hasn't changed since last month, without inserting an ammeter into the ring or touching the ring in any way?

Since the current loop produces a magnetic field, you could use a compass or other more sensitive device to measure the magnetic field produced by the current.

> QUESTION Does this persistent current violate the fundamental principle of conservation of energy? Why or why not?

No energy is being dissipated in the ring. Since its resistance is zero, the power used up is zero (power $= RI^2$). Energy conservation is not violated.

> QUESTION Do the persistent atomic currents in a bar magnet violate the principle of conservation of energy? Why or why not?

The "currents" in a bar magnet are due to the spin and orbital motion of electrons in the atoms of the magnet. The energy of these electrons remains constant—no energy is used up, so no energy input is required. Energy conservation is not violated.

A fruitful way to think about a superconductor is to say that the persistent current is similar to the persistent atomic currents in a magnet, but the diameter of a persistent current loop in a superconductor can be huge compared to an atom. In both the magnet and the superconductor the persistent currents can be explained only by means of quantum mechanics. In both cases the explanation hinges on the fact that these systems have discrete energy levels. Since the energy levels are separated by a gap, there can be no small energy changes, and hence there is no mechanism for energy dissipation.

High-Temperature Superconductors

The lack of resistance in a superconductor offers obvious advantages for power transmission and in electromagnets. However, for a long time after 1911, when the Dutch physicist Kamerlingh Onnes discovered superconductivity in mercury below 4.2 K, all superconductors required very low temperatures. Recently materials have been discovered that are superconducting at temperatures above the boiling point of liquid nitrogen (77 K, −196 °C). Because liquid nitrogen is rather inexpensive to produce, this opens up new opportunities for the use of superconductors. There is even hope that some material might be found that would be a superconductor at room temperature.

Magnetic Flux through a Superconducting Ring

The complete lack of electric resistance has some odd consequences for the effect of magnetic fields on a superconductor. Suppose you try to change

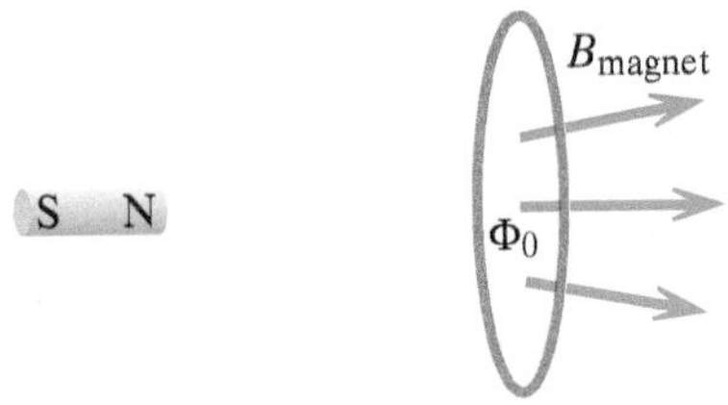

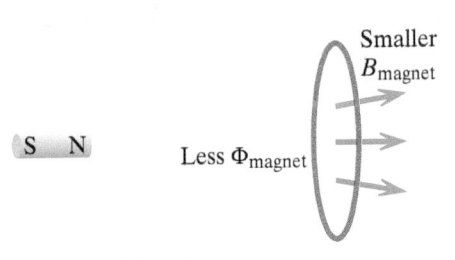

Figure 22.35 Start with a ring with initial flux Φ_0, then cool down the ring to make it become superconducting.

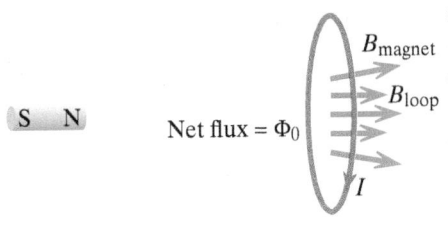

Figure 22.36 At low temperature, move the magnet away from the ring, reducing the flux in the ring contributed by the magnet.

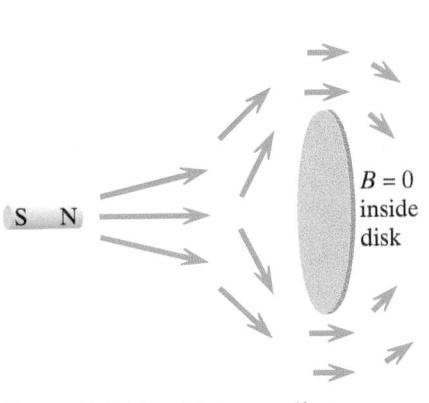

Figure 22.37 Current runs in the superconducting ring in such a way that the net change in the flux in the ring is zero.

the flux through a superconducting metal ring. This changing flux would be associated with a curly electric field (whose round-trip integral is an emf equal to $d\Phi_{mag}/dt$). This curly electric field would drive an infinite current in the ring, because there is no resistance! Because an infinite current is impossible, we conclude that it is not possible to change the flux through a superconducting ring, and this is what is observed.

What does happen when you try to change the flux through the ring by an amount $\Delta\Phi_{mag}$? Since the flux cannot change, what happens is that a (noninfinite) current runs in the ring in such a way as to produce an amount of flux $-\Delta\Phi_{mag}$, so that the net flux change is zero! We'll consider a specific experimental situation to illustrate the phenomenon.

Start with a ring and a magnet at room temperature, so that there is a certain amount of flux Φ_0 through the ring (Figure 22.35), then cool the ring below the critical temperature at which it becomes superconducting (obviously the ring has to be made of a material that does become superconducting at some low temperature).

Next move the magnet away from the ring, thus decreasing the flux made by the magnet inside the ring (Figure 22.36); there is a change in the flux due to the magnet, of some amount $-\Delta\Phi$.

In a nonsuperconducting material, there would be an emf in the ring, with a current emf $= RI$. However, since the resistance R of the superconducting ring is zero, the emf must be zero (0 times I is 0). Assuming that Faraday's law applies to superconductors, it must be that just enough current runs in the superconductor, in the appropriate direction, to increase the flux by an amount $+\Delta\Phi$. The net change in the flux is zero, and the flux remains equal to the original flux Φ_0.

Experimentally, that's exactly what happens (Figure 22.37). Enough current runs in the ring to make exactly the right amount of magnetic flux to make up for the decrease in the flux due to moving the magnet farther away, so that the flux through the ring remains constant. Even though a current runs, the electric field inside the superconductor is zero.

Checkpoint 6 When the magnet is moved very far away, how much flux is inside the ring compared to the original flux Φ_0? How much of this flux is due to the current in the ring?

The Meissner Effect

For many years it was assumed that a similar effect would happen with a solid disk instead of a hollow ring. It was expected that as you cooled the disk below its critical temperature the magnetic flux through the disk would stay the same, and that when you moved the magnet away there would be a current in the disk that would maintain the original flux. In 1933 physicists were astonished to discover that something else happened, something quite dramatic. As soon as the disk was cooled down below the critical temperature for the onset of superconductivity, the magnetic field throughout the interior of the solid disk suddenly went to zero!

This peculiar phenomenon is called the Meissner effect, and it applies to a particular class of materials called "Type I" superconductors. The configuration of net magnetic field (due to the magnet plus the currents inside the superconductor) looks something like Figure 22.38, with no net magnetic field inside the superconductor. Persistent currents are created in the superconductor, which make a magnetic field that exactly cancels the magnetic field due to the magnet everywhere inside the superconductor. Note that the currents in Figure 22.38 are oriented so that the disk and magnet repel each other.

Figure 22.38 The Meissner effect: Currents run in the solid superconducting disk in such a way that the net magnetic field is zero inside the disk.

The Meissner effect was totally unexpected, and it cannot be explained simply in terms of the lack of electric resistance in a superconductor (plus Faraday's law) as we did for the hollow ring. The Meissner effect is a special quantum-mechanical property of superconductors, and its explanation along with the explanation of other properties of superconductors was a triumph of the Bardeen–Cooper–Schrieffer (BCS) theory published in 1957.

Given that both the electric field and the magnetic field must be zero inside the superconductor, by using Gauss's law and Ampere's law it can be shown that the superconducting currents can run only on the surface of the superconductor, not in the interior. In particular, current in the interior would produce magnetic field throughout the material.

The electric field in an ordinary conductor in equilibrium goes to zero due to polarization. However, the fact that electric and magnetic fields must be zero in a superconductor even when not in equilibrium is quite a different, quantum-mechanical effect.

Magnetic Levitation

The Meissner effect is the basis for a dramatic kind of magnetic levitation. A magnet can hover above a superconducting disk, because no matter how the magnet is oriented, currents run in the superconducting disk in such a way as to repel the magnet. One design for maglev trains uses this superconducting effect.

This is in contrast to the situation with ordinary magnets. If you try to balance a magnet above another magnet, with the slightest misalignment the upper magnet flips over and is attracted downward rather than repelled upward.

The Levitron® toy achieves levitation with ordinary magnets by giving the upper magnet a high spin, so that the magnetic torque makes the spin direction precess rather than flip over. When a superconductor supports the upper magnet, gyroscopic stabilization is not needed.

22.6 INDUCTANCE

Changing the current in a coil can "induce" an emf in a second coil. A related effect happens even with a single coil: an attempt to change the current in a coil leads to an additional emf in the coil itself, because the coil surrounds a region of time-varying magnetic field (dB/dt), produced by itself. We can show that this self-induced emf acts in a direction to oppose the change in the current. As a result, there is a kind of sluggishness in any coil of wire: it is hard to change the current, either to increase it or to decrease it. It is not difficult to calculate this self-induction effect quantitatively for a solenoid.

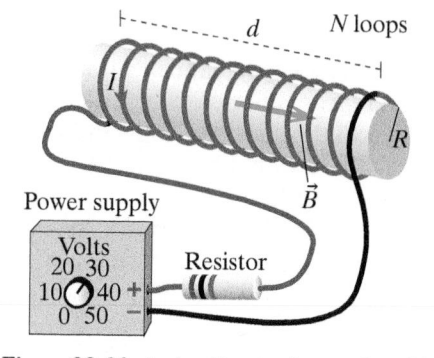

Figure 22.39 A circuit contains a solenoid.

A very long plastic cylinder with radius R and length $d \gg R$ is wound with N closely packed loops of wire with negligible resistance. The solenoid is placed in series with a power supply and a resistor, and current I runs through the solenoid in the direction shown in Figure 22.39. This current makes a magnetic field B inside the solenoid that points to the right.

If the current is steady, the electric field inside the metal of the solenoid coil is nearly zero (very small resistance of this wire). Now suppose that you try to increase the current by turning up the voltage of the power supply. If the current increases with time, there will be an increasing magnetic flux enclosed by each loop of the solenoid. In Chapters 17 and 21 we found that the magnetic field inside a solenoid of length d is $B = \mu_0 NI/d$.

> QUESTION How much emf is generated in one loop of the coil if the current changes at a known rate dI/dt?

We have this:

$$\text{emf} = \left|\frac{d\Phi_{mag}}{dt}\right| = \frac{d}{dt}\left[\frac{\mu_0 NI}{d}\pi R^2\right] = \frac{\mu_0 N}{d}\pi R^2\frac{dI}{dt} \quad \text{(one loop)}$$

QUESTION There are N loops, and each loop encloses approximately the same amount of time-varying magnetic flux. Therefore, what is the net emf along the full length of the solenoid?

Evidently we add up all the individual one-loop emfs in series to obtain the emf from one end of the coil to the other:

$$\text{emf} = N\left[\frac{\mu_0 N}{d}\pi R^2\frac{dI}{dt}\right] = \frac{\mu_0 N^2}{d}\pi R^2\frac{dI}{dt} \quad \text{(entire solenoid)}$$

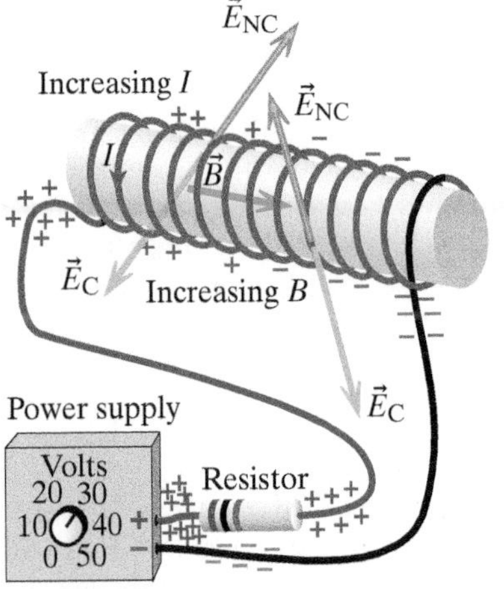

QUESTION If we increase the current I, what is the direction of the associated curly electric field, clockwise or counterclockwise as seen from the right end of the solenoid? Does this curly electric field act to assist or oppose the change in I?

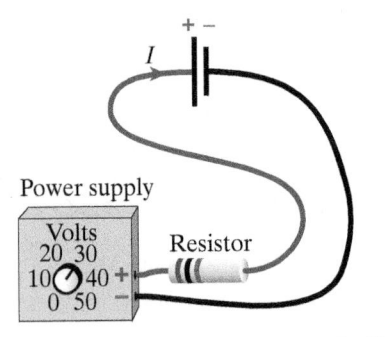

Figure 22.40 The non-Coulomb electric field opposes the increase in the current, and polarizes the solenoid.

If the current is increasing, the non-Coulomb electric field $\vec{E}_{NC}$ curls clockwise around the solenoid, opposing the increase in the current and polarizing the solenoid (Figure 22.40). The new surface charges produce a Coulomb electric field $\vec{E}_C$ that follows the wire and points opposite to the non-Coulomb field. If the resistance of the solenoid wire is very small, the magnitude of the two electric fields is nearly equal ($E_C \approx E_{NC}$).

Note the similarity to a battery: a non-Coulomb emf_{ind} maintains a charge separation, and there is a potential drop ΔV_{sol} along the solenoid that is numerically equal to emf_{ind}. If the wire resistance r_{sol} is significant, we have $\Delta V_{sol} = \text{emf}_{ind} - r_{sol}I$, just like a battery with internal resistance. You can think of this self-induced emf as making the solenoid act like a battery that has been put in the circuit "backwards," opposing the change in the current (Figure 22.41).

It is standard practice to lump the many constants together and write that the self-induced emf is proportional to the rate of change of the current, with proportionality constant L:

$$|\text{emf}_{ind}| = L\left|\frac{dI}{dt}\right|$$

Figure 22.41 The coil acts temporarily like a battery inserted to oppose the change.

The proportionality constant L is called the "inductance" or "self-inductance" of the device, which is called an "inductor."

QUESTION Use the results above to express the inductance L in terms of the properties of a solenoid.

Evidently the (self-)inductance of a solenoid is this:

$$L = \frac{\mu_0 N^2}{d}\pi R^2$$

The unit of inductance (V·s/A) is called the "henry" (H) in honor of the 19th-century American physicist Joseph Henry, who simultaneously with Michael Faraday in 1831 discovered the effects of time-varying magnetic fields.

QUESTION What if you decrease the current through the solenoid? What happens to the emf? Does it assist or oppose the change?

If you go through the previous analysis but with a decreasing current, you find that the emf goes in the other direction, tending to drive current in the original current direction and therefore opposing the decrease. The solenoid introduces some sluggishness into the circuit: it is more difficult to increase the current or to decrease the current. The sign of the effect should not be surprising, since if the self-induced emf assisted rather than opposed an increase in the current, the current would grow without limit!

> **Checkpoint 7** To get an idea of the order of magnitude of inductance, calculate the self-inductance in henries for a solenoid with 1000 loops of wire wound on a rod 10 cm long with radius 1 cm. If the solenoid were filled with iron so that the actual magnetic field were 10 times larger for the same current in the solenoid, what would be the inductance?

Transformers

Electric energy is moved long distances with potential differences of hundreds of thousands of volts, but home appliances need around a hundred volts. Conversion from high voltage to low voltage can be accomplished with transformers in which a primary coil with many turns makes an emf in a secondary coil with few turns. Suppose that you make a solenoid by wrapping $N_1 = 100$ turns around a hollow cylinder $d = 0.1$ m long for the primary coil, and you wrap $N_2 = 20$ turns around the outside for the secondary coil (Figure 22.42). Connect the 100-turn primary coil to an AC power supply whose voltage is $\Delta V_{max} \cos(\omega t)$. What emf will develop along the secondary coil?

The magnetic field made by the primary coil is $B = \mu_0 N_1 I/d$, and if the cross-sectional area of the solenoid is A, the emf in one turn of the secondary coil is $A dB/dt$. The total emf in the secondary coil is $N_2 = 20$ times the emf in one turn, so the potential difference across the secondary coil is $N_2 A(\mu_0 N_1/d)dI/dt$.

What is the potential difference across the primary coil? We just saw that it is $L dI/dt$, where $L = A \mu_0 N_1^2/d$ (since $A = \pi R^2$), so the potential difference across the primary coil is $A(\mu_0 N_1^2/d)dI/dt$.

Comparing $\text{emf}_2 = N_2 A(\mu_0 N_1/d)dI/dt$ with $\text{emf}_1 = A(\mu_0 N_1^2/d)dI/dt$, we see that $\text{emf}_2 = (N_2/N_1)\text{emf}_1$, or $20/100 = 0.2$ times the emf in the primary coil. The ratio of the number of turns determines the change in voltage. If ΔV_{max} is 500 V across the primary coil, it will be only 100 V across the secondary coil.

Because energy is conserved and power is $I\Delta V$, the smaller voltage in the secondary coil is accompanied by a larger current. This two-coil device, a "transformer," has reduced the voltage by a factor of 0.2 and increased the current by a factor of 5, assuming that the process is fully efficient.

We've just described briefly a "step-down" transformer, which reduces the voltage (and increases the current). There are also "step-up" transformers, in which the primary has few turns and the secondary many; the effect is to increase the voltage and decrease the current.

In actual practice the inside of the coils is normally iron rather than air, which has the effect of making the voltage across the secondary coil be less affected by the type of circuit elements connected to the secondary coil.

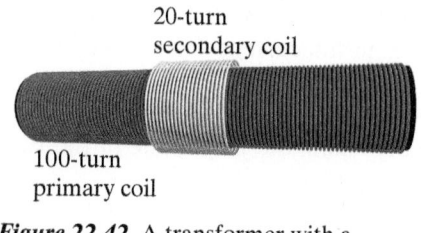

20-turn
secondary coil

100-turn
primary coil

Figure 22.42 A transformer with a 100-turn primary coil and a 20-turn secondary coil.

> **Checkpoint 8** A transformer has 500 turns in its primary coil and 20,000 turns in its secondary coil. An AC voltage with maximum voltage of 100 V is applied to the primary coil. What is the maximum voltage across the secondary coil?

Energy Density

In Chapter 16 we showed that there is an energy density associated with electric field, by expressing the energy in a capacitor in terms of the electric field in the capacitor:

$$\frac{\text{Electric energy}}{\text{Volume}} = \frac{1}{2}\varepsilon_0 E^2$$

Now we can show that there is energy density associated with magnetic field, by expressing the energy in an inductor in terms of the magnetic field in the inductor. We obtain an important result about magnetic energy that is quite general, despite the fact that the derivation is for the specific case of an inductor.

It is easy to show that the magnetic energy stored in an inductor is $\frac{1}{2}LI^2$. The power going into an inductor is $I\Delta V$ as for any device, and

$$P = I\Delta V = I(\text{emf}) = I\left(L\frac{dI}{dt}\right)$$

Integrating over time, we have

$$\text{Energy input} = \int (I\Delta V)dt = L\int_{t_i}^{t_f} I\frac{dI}{dt}dt = L\int_{I_i}^{I_f} I\,dI$$

$$= L\left[\frac{1}{2}I^2\right]_{I_i}^{I_f} = \Delta\left(\frac{1}{2}LI^2\right)$$

Since $B = \dfrac{\mu_0 NI}{d}$ and $L = \dfrac{\mu_0 N^2}{d}\pi R^2$, we have

$$\text{Magnetic energy} = \frac{1}{2}\left(\frac{\mu_0 N^2}{d}\pi R^2\right)\left(\frac{Bd}{\mu_0 N}\right)^2$$

$$= \frac{1}{2}\frac{1}{\mu_0}(\pi R^2 d)B^2$$

The magnetic field of the solenoid is large in the volume $(\pi R^2 d)$. Therefore we have the following result for the electric and magnetic energy densities:

ELECTRIC AND MAGNETIC ENERGY DENSITIES

$$\frac{\text{Energy}}{\text{Volume}} = \frac{1}{2}\varepsilon_0 E^2 + \frac{1}{2}\frac{1}{\mu_0}B^2$$

measured in J/m^3.

Although we have calculated the magnetic energy density for the specific case of a long solenoid, this is a general result. The interpretation is that where there are electric or magnetic fields in space, there is an associated energy density, proportional to the square of the field (E^2 and/or B^2).

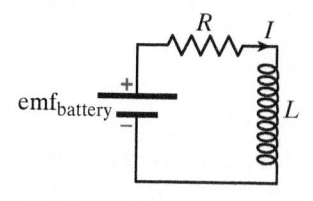

Figure 22.43 An RL circuit containing a resistor and an inductor.

22.7 *INDUCTOR CIRCUITS

Current in an RL Circuit

A circuit containing a resistor R and an inductor L is called an RL circuit. Figure 22.43 shows a series RL circuit some time after a switch was closed. The energy conservation loop rule for this series circuit is this:

$$\Delta V_{battery} + \Delta V_{resistor} + \Delta V_{inductor} = 0$$

$$emf_{battery} - RI - L\frac{dI}{dt} = 0$$

The term $-LdI/dt$ in the loop equation has a minus sign because, as we have seen, the emf of the inductor opposes the attempt to increase the current when we close the switch. The properties of this differential equation are similar to those of the differential equation of a resistor-capacitor (RC) circuit analyzed in Chapter 19.

Let the time when we closed the switch be $t = 0$. The current was of course zero just before closing the switch, and it is also zero just after closing the switch, since the sluggishness of the inductor does not permit an instantaneous change from $I = 0$ to $I = $ nonzero. In fact, such an instantaneous change would mean that dI/dt would be infinite, which would require an infinite battery voltage to overcome the infinite self-induced emf in the inductor.

QUESTION Prove that

$$I = \frac{emf_{battery}}{R}\left[1 - e^{-(R/L)t}\right]$$

by substituting I and dI/dt into the energy conservation equation. What is the final current in the circuit (that is, after a very long time)? Does that make sense?

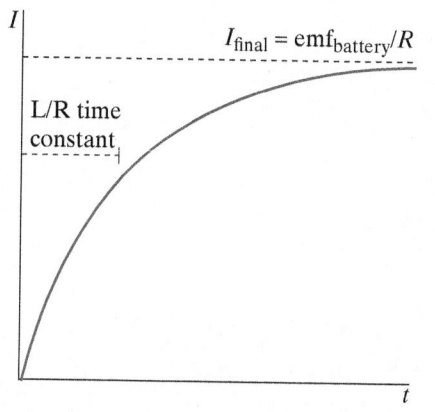

Figure 22.44 A graph of current vs. time in the RL circuit. The switch was closed at $t = 0$.

For large t, the exponential goes to zero, and we have $I = emf_{battery}/R$. This makes sense, because after a long time there is a steady-state current, and the flux does not change any more. With no change in the flux, no emf is generated in the inductor, and the current has the value that it would have if the inductor were not in the circuit. The current builds up slowly to this value. Figure 22.44 shows a graph of the current vs. time. One measure of the time it takes for the current to build up is the "time constant" L/R; at this time the exponential factor has dropped to $1/e$:

$$e^{-(R/L)t} = e^{-(R/L)(L/R)} = e^{-1} = \frac{1}{e} = 0.37$$

You see that having an inductor in a circuit makes the circuit somewhat sluggish. It takes a while for the current to reach the value that it would have reached almost immediately if the circuit had consisted just of a battery and a resistor. (Note, however, that any circuit has some inductance and some sluggishness, because changing the current from zero to nonzero involves a time-varying magnetic flux.)

QUESTION It is interesting to see what happens when you try to open the switch after the steady current has been attained. If the current were to drop from I to zero in a short time Δt, explain why the emf along the solenoid would be very large, and could be much larger than the battery emf.

The emf depends on dB/dt, which is proportional to dI/dt. If the time interval is very short, then dI/dt is very large, and the emf is very large. As a result of this large emf, you may see a spark jump across the opening switch. This phenomenon makes it dangerous to open a switch in an inductive circuit if there are explosive gases around.

Current in an LC Circuit

We have studied many examples of equilibrium and steady-state currents. We have also seen examples of a slow approach to equilibrium (an RC circuit) or to a steady-state current (the RL circuit we just examined).

There is another possibility: a circuit might oscillate, with charge sloshing back and forth forever—the system never settles down to a final equilibrium or a steady state. Of the systems we've analyzed or experimented with, none of them oscillated continuously, because there was always some resistance in the system that would have damped out any such oscillatory tendencies. It is possible that during the first few nanoseconds when surface charge is rearranging itself in a circuit, there may be some sloshing of charge back and forth, but this oscillation dies out due to dissipation in the resistive wires.

A circuit containing an inductor L and a capacitor C is called an LC circuit (Figure 22.45). Such a circuit can oscillate if the resistance is small. In this circuit, the connecting wires are low-resistance thick copper wires. Suppose that the capacitor is initially charged, and then you close a switch, connecting the capacitor to the inductor.

At first it is difficult for charge on the capacitor to flow through the inductor, because the inductor opposes attempts to change the current (and the initial current was zero). However, the inductor can't completely prevent the current from changing, so little by little there is more and more current, which of course drains the capacitor of its charge.

We'll see that just at the moment when the capacitor runs out of charge, there is a current in the inductor that can't change to zero instantaneously, due to the sluggishness of the inductor. So the system doesn't come to equilibrium but overshoots the equilibrium condition and pours charge into the capacitor. When the capacitor gets fully charged (opposite in sign to the original condition), the capacitor starts discharging back through the inductor.

This oscillatory cycle repeats forever if there is no resistance in the circuit. If there is some resistance, the oscillations eventually die out, but the system may go through many cycles before equilibrium is reached.

The energy conservation loop rule for this circuit is

$$\Delta V_{\text{capacitor}} + \Delta V_{\text{inductor}} = \frac{1}{C}Q - L\frac{dI}{dt} = 0$$

where Q is the charge on the upper plate of the capacitor, and I is the conventional current leaving the upper plate and going through the inductor.

QUESTION Can you explain why $I = -dQ/dt$?

dQ/dt is the amount of charge flowing off of the capacitor plate per second. This is the same thing as the current. Because charge is leaving the plate, dQ is negative, so $I = -dQ/dt$. Since $I = -dQ/dt$, we can rewrite the energy conservation equation as

$$\frac{1}{C}Q + L\frac{d^2Q}{dt^2} = 0$$

QUESTION Show that

$$Q = Q_i \cos\left(\frac{1}{\sqrt{LC}}t\right)$$

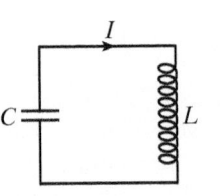

Figure 22.45 An LC circuit consists of an inductor and a capacitor.

is a possible solution of the rewritten energy conservation equation, by substituting Q and its second derivative into the equation.

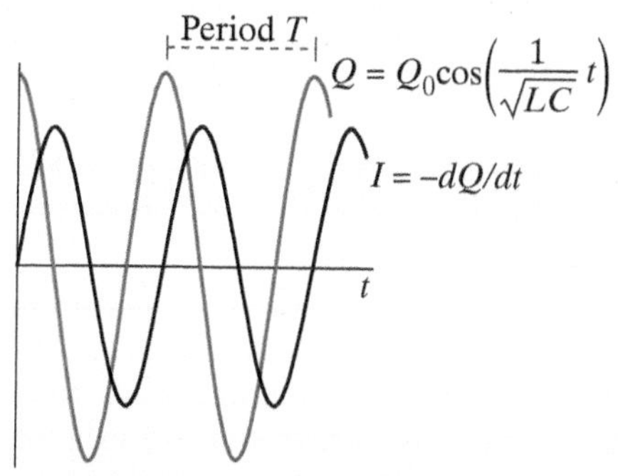

Moreover, this solution fits the initial conditions at $t = 0$, just after closing the switch, if Q_i is the initial charge on the upper plate, since this expression reduces to $Q = Q_i \cos(0) = Q_i$. (For other initial conditions, the correct solution is a sine, or a combination of sine and cosine.)

> QUESTION Now that you know Q on the upper plate of the capacitor as a function of time, calculate the current I through the inductor as a function of time.

The current is given by the following:

$$I = -\frac{dQ}{dt} = \frac{Q_i}{\sqrt{LC}} \sin\left(\frac{1}{\sqrt{LC}}t\right)$$

Now that we have expressions for Q and I as a function of time, we can graph both quantities (Figure 22.46). This is very special: charge oscillates back and forth in the circuit forever (if there is negligible resistance). There is no equilibrium, and there is no steady state. No battery or other source of energy input is needed, because there is no dissipation. The charge just keeps going back and forth on its own. If there is some resistance, the oscillations slowly die out, as shown in Figure 22.47 (the loop equation must be modified to contain the term $-RI$).

Note that the current I does not become large instantaneously (due to the sluggishness of the inductor). Also note that the current, which is the rate of depletion of the capacitor charge, reaches a maximum just when Q goes to zero. This maximum current starts charging the bottom plate of the capacitor positive, which makes the current decrease. When the current becomes zero, the system is in a state very similar to its original state, but inverted (the bottom plate is positive). Now the system runs the other way and eventually gets back after one complete cycle to a state in which the top plate is again positive, and the current is momentarily zero. Then the process repeats.

After one complete cycle, $(1/\sqrt{LC})t = 2\pi$, so the period of the oscillation is $T = 2\pi\sqrt{LC}$. The frequency $f = 1/T = 1/(2\pi\sqrt{LC})$.

> **Checkpoint 9** What is the oscillation frequency of an LC circuit whose capacitor has a capacitance of 1 μF and whose inductor has an inductance of 1 mH? (Both of these are fairly typical values for capacitors and inductors in electronic circuits. See Checkpoint 7 for a numerical example of inductance.)

Energy in an LC Circuit

Another way to talk about an LC circuit is in terms of stored energy. The original energy was the electric energy stored in the capacitor, equal to $Q^2/(2C)$; see Chapter 19. At the instant during the oscillation when the charge on the capacitor is momentarily zero, there is no energy stored in the capacitor; all the energy at that moment is stored in the inductor, in the form of magnetic energy. As the system oscillates, energy is passed back and forth between the capacitor and the inductor.

Earlier in this chapter we showed that the magnetic energy stored in an inductor is $\frac{1}{2}LI^2$. At $t = 0$, the current is zero, and all the energy in the system is the electric energy of the capacitor. At the instant when $Q = 0$ on the capacitor, the current is a maximum, and all the energy in the system is the magnetic energy of the inductor.

Period T

$$Q = Q_0 \cos\left(\frac{1}{\sqrt{LC}}t\right)$$

$I = -dQ/dt$

Figure 22.46 Capacitor charge and inductor current in an LC circuit with no resistance.

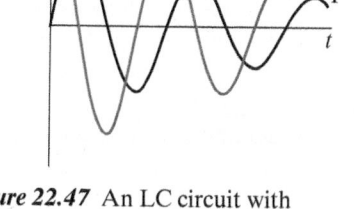

Figure 22.47 An LC circuit with resistance.

QUESTION If Q_i is the initial charge on the capacitor, use an energy argument to find the maximum current I_{max} in the circuit.

Initially, all of the stored energy is electric:

$$U = U_{electric} + U_{magnetic} = \frac{1}{2}\frac{Q_i^2}{C}$$

When $I = I_{max}$, $Q = 0$, so all of the stored energy is magnetic:

$$U = U_{electric} + U_{magnetic} = \frac{1}{2}LI_{max}^2$$

Since the total energy in the circuit does not change, because the circuit has no resistance and no energy is dissipated as heat, the following must be true:

$$\frac{1}{2}LI_{max}^2 = \frac{1}{2}\frac{Q_i^2}{C}, \quad \text{and} \quad I_{max} = \frac{Q_i}{\sqrt{LC}}$$

QUESTION Show that this result is consistent with $Q = Q_i \cos(t/\sqrt{LC})$ by calculating $I = -dQ/dt$ and seeing what the maximum I is.

Starting from the charge as a function of time, we have this:

$$I = -\frac{dQ}{dt} = -\frac{Q_i}{\sqrt{LC}}\sin\left(\frac{t}{\sqrt{LC}}\right) \quad \text{so that} \quad I_{max} = \frac{Q_i}{\sqrt{LC}}$$

This is the same result we obtained by using an energy argument.

We have been discussing "free" oscillations, in which the oscillations (perpetual or dying out) proceed with no inputs from the outside. If you try to force the oscillations by applying an AC voltage from the outside (with an AC power supply), you find that it is difficult to get much current to run unless you nearly match the free oscillation frequency $f = 1/(2\pi\sqrt{LC})$. This is an example of what is called a "resonance" phenomenon (see the end of Chapter 7), in which a system (the LC circuit in this case) responds significantly only when the forcing of the system is done at a frequency close to the free-oscillation frequency.

AC Circuits

For completeness, it should be mentioned that an important use of inductors is in AC circuits (sinusoidally alternating current). You have just seen that LC circuits are characterized by sinusoidally alternating currents, as are generators with loops rotating in a magnetic field. Whenever there is an AC current passing through an inductor, there is an AC emf generated in the inductor due to the time-varying flux. If the current is a sine function, the emf is a cosine function, and vice versa.

There is an elaborate mathematical formalism for dealing with AC circuits, including accounting for the fact that the current and voltage in an AC circuit need not reach their maxima at the same time (a "phase shift," which was 90° in the LC circuit we just analyzed). We will not go into these complications here, but it is worth noting that an important element in understanding such circuits is to recognize the role of the self-induced emf in an inductor, and the fact that in an inductor a sinusoidal current is associated with a cosinusoidal voltage.

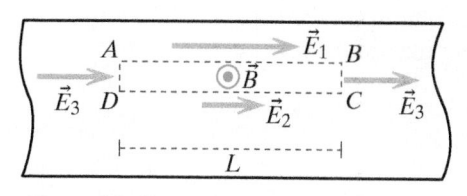

Figure 22.48 An AC current produces a time-varying magnetic field in a wire, and $E_1 > E_2$.

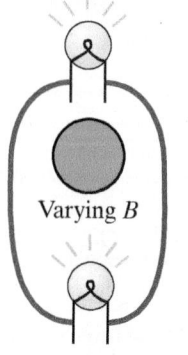

Figure 22.49 Two light bulbs connected around a long solenoid with varying B.

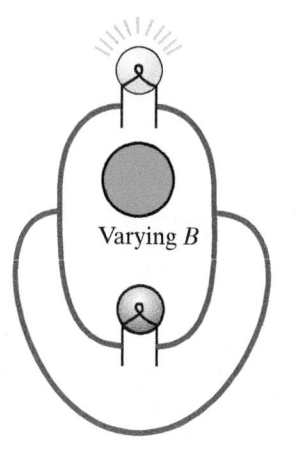

Figure 22.50 Add a thick copper wire to the two-bulb circuit.

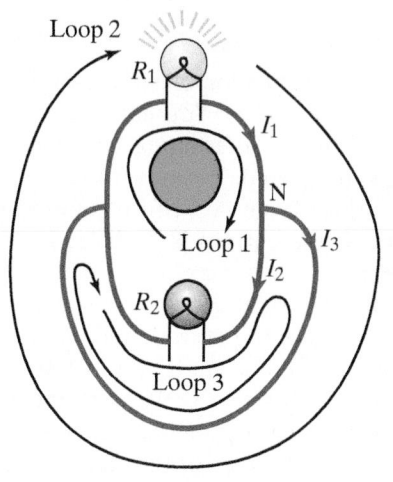

Figure 22.51 Write loop and node equations for this circuit.

Skin Effect

In Chapter 18 we showed that in a DC circuit the electric field was uniform across the cross section of a wire, because the round-trip path potential difference ΔV_{ABCDA} in Figure 22.48 had to be zero, so that $\vec{E}_1$ had to be equal to $\vec{E}_2$. This is not true in an AC circuit, because the magnetic field made by the AC current varies sinusoidally in time. The magnetic field inside the loop $ABCDA$ is out of the page. When the current is increasing, the magnetic field is increasing out of the page, and there will be a clockwise curly non-Coulomb electric field. Going around the path $ABCDA$, the path integral $\int \vec{E} \cdot d\vec{l} = E_1 L - E_2 L = d\Phi/dt$, which is positive. Therefore $E_1 > E_2$. The bigger $d\Phi/dt$, the larger the difference between E_1 and E_2.

The higher the AC frequency of the current, the bigger is $d\Phi/dt$, so that at very high frequencies E_2 is very small compared to E_1, and there is significant electric field only very near the surface of the wire. Since $J = I/A = \sigma E$, there is almost no current density anywhere in the interior of the wire: all of the current runs very near the surface of the wire. This high-frequency phenomenon is called "skin effect."

22.8 *SOME PECULIAR CIRCUITS

This section offers examples of surprising aspects of circuits when there is a time-varying magnetic field.

Two Bulbs Near a Solenoid

We can show you a very simple circuit that behaves in a rather surprising way. Consider two light bulbs connected in series around a solenoid with a varying magnetic field (Figure 22.49). The solenoid carries an alternating current (AC) in order to provide a time-varying magnetic flux through the circuit, so that there is an emf to light the bulbs. If the flux varies as $\Phi_{\text{mag}} = \Phi_0 \sin(\omega t)$, in the bulb circuit we have a varying emf $= \text{emf}_0 \cos(\omega t)$, since the emf is proportional to the rate of change of the magnetic flux. If each bulb has a resistance R, there is an alternating current through the bulbs of this amount:

$$I = \frac{\text{emf}}{2R} = \frac{\text{emf}_0}{2R} \cos(\omega t)$$

Notice that a uniformly increasing flux ($\Phi_{\text{mag}} = \Phi_0 t$) would yield a constant emf, but a sinusoidally varying flux yields a cosinusoidally varying emf. It is very convenient to use AC currents to study the effects of time-varying magnetic fields, because both the initiating current and the induced current vary as sines and cosines, and AC voltmeters and ammeters can be used to measure both parts of the circuit.

Now we alter the circuit by connecting a thick copper wire across the circuit (Figure 22.50). You may find it surprising that the top bulb gets much brighter and the bottom bulb no longer glows. Let's try to understand how this happens.

As usual, we need to write charge-conservation node equations and energy conservation loop equations, so we need to label the nodes and loops as in Figure 22.51 (any two of these three loops would be enough, but we'd like to show how to handle all of them).

Here is the key to analyzing this peculiar circuit. The round trip around loop 1 or loop 2 takes in an emf (just as though there were a battery in the loop). However, the round trip around loop 3 does not enclose any time-varying flux and so has no emf. Therefore the loop and node rules are the following, where R_1 and R_2 are the resistances of the hot top bulb and cold bottom bulb, and the resistance of the added copper wire is essentially zero:

$$\text{loop 1: emf} - R_1 I_1 - R_2 I_2 = 0 \qquad \text{node N: } I_1 = I_2 + I_3$$
$$\text{loop 2: emf} - R_1 I_1 = 0$$
$$\text{loop 3: } R_2 I_2 = 0 \text{ (no flux enclosed)}$$

From the loop 3 equation we see that $I_2 = 0$, which is why the bottom bulb doesn't glow.

With $I_2 = 0$, the node N rule says that $I_1 = I_3$: all the current in the top bulb goes through the copper wire (and none through the bottom bulb). Both the loop 1 and loop 2 equations reduce to $R_1 I_1 = $ emf, so the current through the top bulb is $I_1 = $ emf$/R_1 = (\text{emf}_0/R_1)\cos(\omega t)$, which is nearly twice what it was before we added the extra copper wire ($R_1 > R$ due to higher temperature).

Checkpoint 10 Figure 22.52 shows four different ways to connect the copper wire. Based on the analysis we have just carried out, involving identifying whether or not there is a battery-like emf in a loop, what is the brightness of both bulbs in circuits 1, 2, 3, and 4?

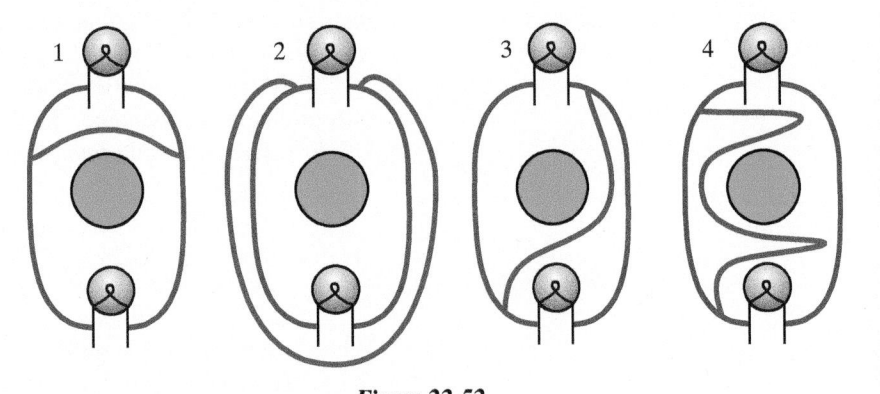

Figure 22.52

Coulomb and Non-Coulomb Electric Fields in a Nonuniform Ring

In a circular ring encircling a solenoid with a time-varying magnetic field, there is current driven by a non-Coulomb electric field. There is no gradient of surface charge along the ring, although there is a small transverse polarization required to turn the electrons in a circle (top of Figure 22.53).

If there is a thin section of the ring, however, charge will pile up at the ends of the thin section, and there will be surface charges all along the ring (bottom of Figure 22.53). These surface charges will produce a Coulomb electric field $\vec{E}_C$, which adds vectorially to the non-Coulomb electric field $\vec{E}_{NC}$ that is associated with the changing magnetic flux. The net electric field $\vec{E}_{NC} + \vec{E}_C$ is what drives the electrons around the ring.

Inside the thin section the net electric field is larger than it was for the uniform ring, and in the thick sections the net electric field is smaller than it was for the uniform ring.

> QUESTION If both rings are made out of the same material, how do you know that the current must be smaller in the ring with the thin section than it was in the uniform ring?

In the thick section of the nonuniform ring, the only thing that changed is that the net electric field is smaller, so the current in the thick section must be

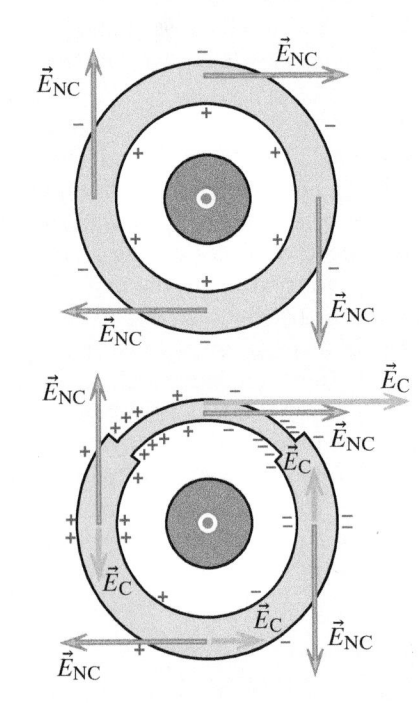

Figure 22.53 A uniform metal ring, and a metal ring with a thin section.

smaller. (Of course this same smaller current runs through the thin section of this ring.)

Note that the Coulomb electric field $\vec{E}_C$ is large and goes clockwise in the thin section, but $\vec{E}_C$ is small and goes counterclockwise in the thick section. Since $\vec{E}_C$ is due to point charges, the round-trip integral of $\vec{E}_C$ is zero.

On the other hand, the non-Coulomb electric field $\vec{E}_{NC}$ due to the changing magnetic flux goes clockwise all around the ring, and its round-trip integral is emf $= d\Phi_{mag}/dt$, which is not zero.

Two Competing Effects in a Shrinking Ring

Suppose that we stretch a springy metal ring of radius R and place it in a region of uniform magnetic field B pointing out of the page, which is increasing at a rate of dB/dt (Figure 22.54). When we let go of the ring, it contracts at a rate dR/dt (which is a negative number). What is the emf around the metal ring? Assume that the ring has high resistance and therefore carries only a small current, so we can neglect any magnetic field contributed by current in the ring.

In terms of flux, there are two competing effects here. The increasing magnetic field B tends to increase the flux, but the decreasing area tends to decrease the flux. Whether or not a current will flow depends on the relative magnitudes of these two contributions to the net flux.

Since the magnetic field is the same throughout the ring and is perpendicular to the page, the flux through the ring at any instant is $\Phi_{mag} = B(\pi R^2)$. The emf is the rate of change of the flux:

$$\text{emf} = \frac{d\Phi_{mag}}{dt} = \frac{d}{dt}[B(\pi R^2)] = \frac{dB}{dt}(\pi R^2) + B\frac{d(\pi R^2)}{dt}$$
$$= \frac{dB}{dt}(\pi R^2) + B(2\pi R)\frac{dR}{dt}$$

What is the physical significance of these two contributions to the emf? In the presence of a changing magnetic field, there is a clockwise non-Coulomb electric field in the wire. The first term, involving dB/dt, is associated with this non-Coulomb electric field in the wire.

Because the wire is moving inward, there is a magnetic force on the electrons in the wire. In this case, $\vec{v} \times \vec{B}$ (the magnetic force per unit charge) is counterclockwise in the ring. The second term, involving the negative quantity dR/dt, is associated with the magnetic force on the moving pieces of the ring (motional emf). Note that $|dR/dt|$ is the speed of pieces of ring toward the center, and $2\pi r$ is the circumference, so the contribution to the emf is similar to emf $= BLv$ for a rod of length L sliding on rails.

Note that if $(dB/dt)(\pi R^2) = B(2\pi R)|(dR/dt)|$, the emf would be zero even though we had both a changing magnetic field and a changing area. In this case the magnetic force on an electron inside the ring is equal and opposite to the non-Coulomb electric force, due to the non-Coulomb electric field.

There is no surface charge gradient around this symmetrical ring, and no Coulomb electric field around the ring. The forces on electrons in the metal ring are due to the non-Coulomb electric field and to the non-Coulomb magnetic force on the moving pieces of the ring.

22.9 *THE DIFFERENTIAL FORM OF FARADAY'S LAW

As with Gauss's law and Ampere's law, there is a differential form of Faraday's law, valid at a particular location and time. Using properties of vector calculus

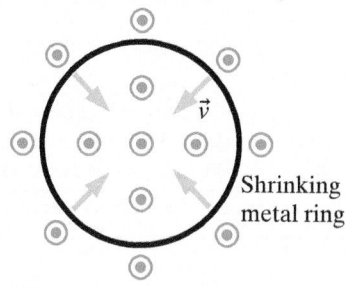

B out of page and increasing

Figure 22.54 A springy metal ring that had been stretched and is now shrinking.

one can show that the integral form of Faraday's law is equivalent to the following, where the derivative is a "partial derivative" with respect to time (holding position constant):

DIFFERENTIAL FORM OF FARADAY'S LAW

$$\text{curl}(\vec{E}) = -\frac{\partial \vec{B}}{\partial t} \quad \text{or} \quad \vec{\nabla} \times \vec{E} = -\frac{\partial \vec{B}}{\partial t}$$

22.10 *LENZ'S RULE

With the thumb of your right hand pointing in the direction of $-d\vec{B}_1/dt$, your fingers curl around in the direction of $\vec{E}_{NC}$. Another rule, called "Lenz's rule," can also be used to get the direction of the non-Coulomb electric field. We describe Lenz's rule briefly, although it is unnecessary to learn and use Lenz's rule in this introductory textbook on electricity and magnetism, for reasons that we will give below.

Imagine that you place a wire around the changing flux, so that the non-Coulomb electric field drives current in the wire:

LENZ'S RULE

The non-Coulomb electric field would drive current in a direction to make a magnetic field that attempts to keep the flux constant.

To see how this works, consider the first example in Figure 22.55, where $\vec{B}_1$ points out of the page and is increasing, and the non-Coulomb electric field curls clockwise around the solenoid. If a wire encircles the solenoid, conventional current runs clockwise in the wire, and this current produces an additional magnetic field in the region, pointing into the page.

This additional magnetic field is in the opposite direction to the change in $\vec{B}_1$, so we say that the additional magnetic field attempts to keep the magnetic flux constant despite the increase in B_1. The induced current does not succeed at keeping the net flux constant (unless the wire is made of superconducting material). In particular, if B_1 is increasing at a constant rate, the current I_2 in the ring and the additional magnetic field are constant by Faraday's law, so the net flux does increase.

If on the other hand $\vec{B}_1$ points out of the page and is decreasing (the second example in Figure 22.55), there is a counterclockwise non-Coulomb electric field that would drive conventional current counterclockwise in a wire, which would make a magnetic field out of the page, in the direction to attempt to keep the magnetic flux constant despite the decrease in B_1. (Again, the net flux will nevertheless decrease, unless the ring is superconducting.)

QUESTION Go through this analysis for the third and fourth examples shown in Figure 22.55 and convince yourself that Lenz's rule does correctly summarize the experimental data for the direction of the non-Coulomb electric field.

An interesting property of Lenz's rule is that it also correctly gives the direction of current in the case of motional emf, or in situations where both a time-varying magnetic field and motion contribute to a change in the flux enclosed by a part of a circuit.

We have not emphasized Lenz's rule for two reasons. We have emphasized time-varying magnetic field and curly non-Coulomb electric field (rather than just time-varying flux and emf), to give a stronger sense of mechanism and to distinguish the effects of time-varying magnetic fields from the effects of the motion of a wire in a magnetic field.

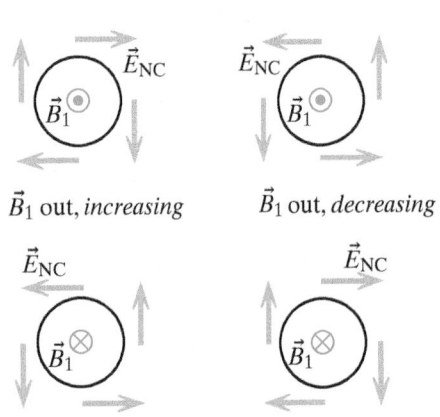

$\vec{B}_1$ out, *increasing* $\vec{B}_1$ out, *decreasing*

$\vec{B}_1$ in, *increasing* $\vec{B}_1$ in, *decreasing*

Figure 22.55 Using Lenz's rule to determine the direction of the non-Coulomb electric field.

Also, it is very easy to make serious conceptual mistakes using Lenz's rule when first studying the effect of time-varying magnetic fields, due to a natural tendency to try to use it for predicting quantitatively the magnitude of the induced current flow, not just its direction. In particular, except for superconductors, it is not true that "enough current runs to cancel the change in the magnetic field."

SUMMARY

Faraday's Law
The new phenomenon introduced in this chapter is that a time-varying magnetic field is accompanied by a non-Coulomb electric field that curls around the region of varying field. Unlike the Coulomb electric field produced by point charges, the non-Coulomb electric field has a nonzero round-trip path integral if the path encircles a time-varying flux.

Faraday's Law

$$\text{emf} = -\frac{d\Phi_{mag}}{dt}$$

where $\text{emf} = \oint \vec{E}_{NC} \bullet d\vec{l}$ and $\Phi_{mag} \equiv \int \vec{B} \bullet \hat{n} dA$.

In words: The emf along a round-trip path is equal to the rate of change of the magnetic flux on the area encircled by the path.

Direction: With the thumb of your right hand pointing in the direction of $-d\vec{B}/dt$, your fingers curl around in the direction of $\vec{E}_{NC}$.

Formal version of Faraday's law

$$\oint \vec{E}_{NC} \bullet d\vec{l} = -\frac{d}{dt} \int \vec{B} \bullet \hat{n} dA$$

(with the sign given by the right-hand rule).

Faraday's law also covers the case where the flux changes due to changes in the path ("motional emf"), and the non-Coulomb force in this case is a magnetic force.

In general, there can be any combination of three effects acting on electrons in a circuit:

1. Coulomb electric field due to surface charges
2. Non-Coulomb electric field associated with time-varying magnetic field
3. Magnetic forces on moving portions of the circuit (motional emf)

Superconductors
Superconductors have zero resistance, and a consequence of this is that a superconducting ring maintains a constant flux in the area enclosed by the ring. The magnetic field inside a superconductor is always zero (the Meissner effect), and this important property cannot be explained merely in terms of zero resistance.

Inductance
When you vary the current in a coil, the varying flux enclosed by the coil is associated with an emf in the *same* coil, which makes it difficult to change the current. This sluggishness plays a role in RL and LC circuits, and it is quantified by the "inductance" L of the coil. The emf is proportional to the rate of change of the magnetic flux, which is proportional to the rate of change of the current:

$$|\text{emf}_{ind}| = L\left|\frac{dI}{dt}\right|$$

We calculated the inductance in one particular case, that of a long solenoid:

$$L_{solenoid} = \frac{\mu_0 N^2}{d}\pi R^2$$

Electric and Magnetic Energy Densities

$$\frac{\text{Energy}}{\text{Volume}} = \frac{1}{2}\varepsilon_0 E^2 + \frac{1}{2}\frac{1}{\mu_0}B^2$$

measured in J/m^3.

RL (Resistor–Inductor) Circuits
The current in an RL series circuit varies with time:

$$I = \frac{\text{emf}_{battery}}{R}\left[1 - e^{-(R/L)t}\right]$$

LC (Inductor–Capacitor) Circuits
An LC circuit oscillates sinusoidally with a frequency $f = 1/(2\pi\sqrt{LC})$:

$$Q = Q_i \cos\left(\frac{1}{\sqrt{LC}}t\right)$$

If there is no resistance, an LC circuit never reaches a final equilibrium or a final steady state.

*The Differential Form of Faraday's Law

$$\text{curl}(\vec{E}) = -\frac{\partial \vec{B}}{\partial t} \quad \text{or} \quad \vec{\nabla} \times \vec{E} = -\frac{\partial \vec{B}}{\partial t}$$

QUESTIONS

Q1 In Figure 22.56 a bar magnet with its S pole on top moves upward toward the origin, $\langle 0,0,0 \rangle$.

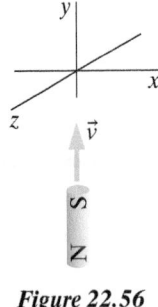

Figure 22.56

(a) At the origin, what is the direction of $\vec{B}$ at this instant, time t_1? **(b)** At the origin, what is the direction of $\vec{B}$ at a slightly later time t_2, when the bar magnet is still underneath the origin but closer to it? **(c)** At the origin, is the magnitude of the magnetic field $|\vec{B}|$ larger at time t_1 or time t_2? **(d)** At the origin, what is the direction of $\Delta \vec{B}$, the vector change in the magnetic field from time t_1 to time t_2? **(e)** At the origin, what is the direction of $-\Delta \vec{B}/\Delta t$? **(f)** Looking down at the origin from above (along the $+y$ axis), is the pattern of curly electric field clockwise or counterclockwise? **(g)** What is the direction of $\vec{E}_{NC}$ at the location $\langle 1,0,0 \rangle$?

Q2 Two coils of wire are near each other, positioned on a common axis (Figure 22.57). Coil 1 is connected to a power supply whose output voltage can be adjusted by turning a knob, so that the current I_1 in coil 1 can be varied, and I_1 is measured by ammeter 1.

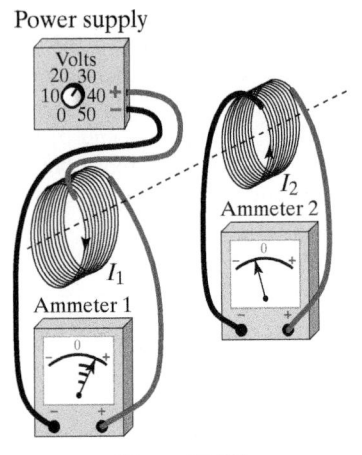

Figure 22.57

Current I_2 in coil 2 is measured by ammeter 2. The ammeters have needles that deflect positive or negative depending on the direction of current passing through the ammeter, and ammeters read positive if conventional current flows into the + terminal. Figure 22.58 is a graph of I_1 vs. time. Draw a graph of I_2 vs. time over the same time interval. Explain your reasoning.

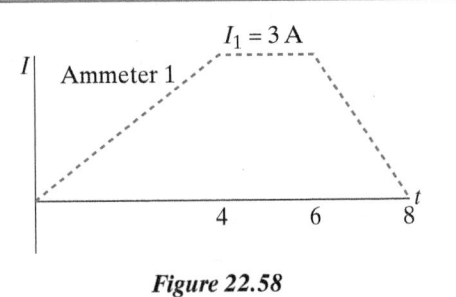

Figure 22.58

Q3 Two metal rings lie side-by-side on a table (Figure 22.59). Current in the left ring runs clockwise and is increasing with time, so a current runs in the right ring. Does this current run clockwise or counterclockwise? Explain, using a diagram. (*Hint:* Think carefully about the direction of magnetic field in the right ring produced by the left ring, taking into consideration what sections of the left ring are closest.)

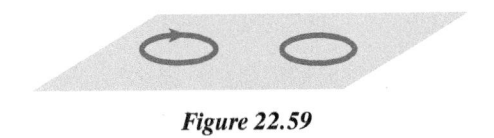

Figure 22.59

Q4 A very long straight wire (essentially infinite in length) carries a current of 6 ampere (Figure 22.60). The wire passes through the center of a circular metal ring of radius 4 cm and resistance 2 Ω that is perpendicular to the wire. If the current in the wire increases at a rate of 0.25 A/s, what is the current in the ring? Explain carefully.

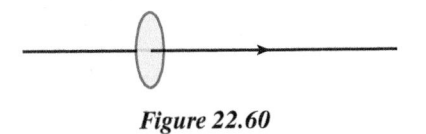

Figure 22.60

Q5 A bar magnet is dropped through a vertical copper tube and is observed to fall very slowly, despite the fact that mechanical friction between the magnet and the tube is negligible (Figure 22.61). Explain carefully, including adequate diagrams.

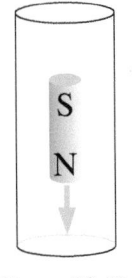

Figure 22.61

Q6 Suppose that you move a bar magnet toward the coil in Figure 22.62, with the S end of the bar magnet closest to the coil. **(a)** Will the ammeter read positive or negative? **(b)** Now move the bar magnet away from the coil, with the S end still closest to the coil. Will the ammeter read + or −?

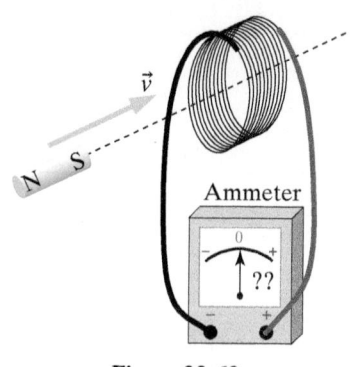

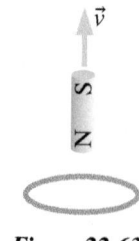

Figure 22.62

Q7 A bar magnet is held vertically above a horizontal metal ring, with the south pole of the magnet at the top (Figure 22.63). If the magnet is lifted straight up, will current run clockwise or counterclockwise in the ring, as seen from above?

Figure 22.63

Q8 One of the methods physicists have used to search for magnetic monopoles is to monitor the current produced in a loop of wire. Draw graphs of current in the loop vs. time for an electrically uncharged magnetic monopole passing through the loop, and for an electrically uncharged magnetic dipole (such as a neutron) passing through the loop with its north end head-first. Don't worry too much about the details of the exact moment when the particle goes through the plane of the loop; concentrate on the times just before and just after this event. Explain the differences in the two graphs.

(A signal corresponding to a magnetic monopole was seen once in such an experiment, but no one has been able to reproduce this result, and most physicists seem to believe that the supposed event was due to extraneous noise in the system or other malfunction of the apparatus.)

Q9 A conventional current I runs through a coil in the direction shown in Figure 22.64. A single loop of copper wire is near the coil. The loop and the coil are stationary.

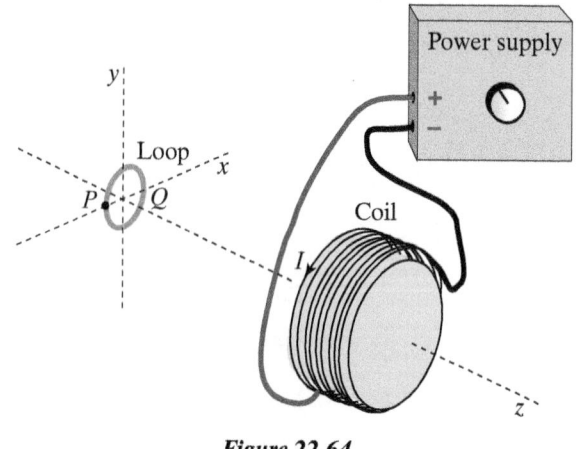

Figure 22.64

(a) In this initial state (constant current in coil), what is the direction of the magnetic field at the center of the copper loop, due to the current in the coil? (b) In this initial state, what is the direction of the electric field at location P inside the copper loop? (c) What is the direction of the electric field at location Q inside the copper loop?

Next the power supply is adjusted so the current in the coil *decreases* with time. (d) Now, what is the direction of $d\vec{B}/dt$ at the center of the copper loop? (e) What is the direction of $-d\vec{B}/dt$ at the center of the copper loop? (f) What is the direction of the electric field at location P in the copper wire? At location Q? (g) At this moment, is the magnitude of the magnetic flux inside the copper loop increasing, decreasing, or staying constant?

Q10 Would the inductance of a solenoid be larger or smaller if the solenoid is filled with iron? Explain briefly.

Q11 (a) When an RL circuit has been connected to a 1.5 V battery for a very *short* time, what is the potential difference from one end of the resistor to the other? (b) When an RL circuit has been connected to a 1.5 V battery for a very *long* time, what is the potential difference from one end of the resistor to the other? (c) Explain briefly the difference between equilibrium, steady-state current, and the behavior of an LC circuit.

PROBLEMS

Section 22.2

•**P12** In Figure 22.65 the solenoid radius is 2 cm and the ring radius is 10 cm. $B = 0.5$ T inside the solenoid, and $B \approx 0$ outside the solenoid. What is the magnetic flux through the outer ring?

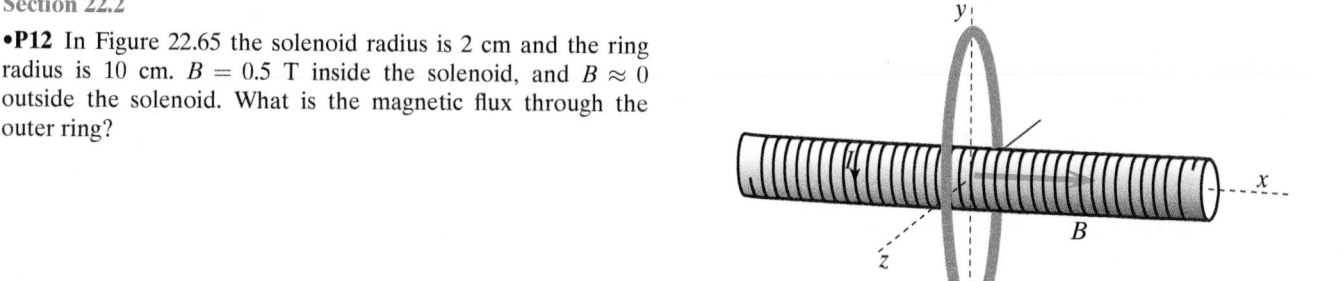

Figure 22.65

•P13 In Figure 22.66 the north pole of a bar magnet points toward a thin circular coil of wire containing 40 turns. The magnet is moved away from the coil, so that the flux through one turn inside the coil decreases by $0.3 \, \text{T} \cdot \text{m}^2$ in 0.2 s. What is the average emf in the coil during this time interval? Viewed from the right side (opposite the bar magnet), does the induced current run clockwise or counterclockwise? Explain briefly.

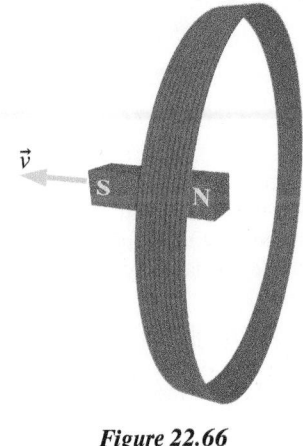

Figure 22.66

•P14 The magnetic field in a solenoid is $B = \mu_0 NI/d$. A circular wire of radius 10 cm is concentric with a solenoid of radius 2 cm and length $d = 1$ m, containing 10,000 turns. The current increases at a rate of 50 A/s. What is the non-Coulomb electric field in the wire?

•P15 In Figure 22.67 there is a circular region of radius r_1 in which there is a uniform magnetic field B pointing out of the paper (the magnetic field is essentially zero outside this region). The magnetic field is *decreasing* at this moment at a rate $|dB/dt|$. A wire of radius r_2 and resistance R lies entirely *outside* the magnetic-field region (where there is no magnetic field). In which direction does conventional current I flow in the ring? What is the magnitude of the current I? Show the pattern of (non-Coulomb) electric field in the ring. What is the magnitude E_{NC} of the non-Coulomb electric field?

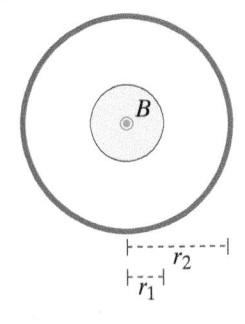

Figure 22.67

•P16 A thick copper wire connected to a voltmeter surrounds a region of time-varying magnetic flux, and the voltmeter reads 10 V. If instead of a single wire we use a coil of thick copper wire containing 20 turns, what does the voltmeter read? A thin

Nichrome wire connected to an ammeter surrounds a region of time-varying magnetic flux, and the ammeter reads 10 A. If instead of a single wire we use a coil of thin Nichrome wire containing 20 turns, what does the ammeter read?

••P17 In Figure 22.64 a coil of wire with 300 turns is connected to a power supply, and a current runs in the coil. A single loop of copper wire is located near the coil, with its axis on the same line as the axis of the coil. The loop and the coil are stationary. The radius of the coil is 9 cm, and the radius of the loop is 4 cm. The center of the loop is 22 cm from the center of the coil. At time t_1 the conventional current I in the coil is 5 A, and it is not changing. **(a)** What is the absolute value of the magnetic flux through the loop at time t_1? **(b)** What approximations or assumptions did you make in calculating your answer to part (a)? **(c)** What is the direction of the electric field at location P in the copper wire of the loop at time t_1? (Remember that at this time the current in the coil is constant.)

At a later time t_2 the current in the coil begins to decrease. **(d)** Now what is the direction of the curly electric field in the loop?

At time t_2 the current in the coil is decreasing at a rate $dI/dt = -0.3$ A/s. **(e)** At this time, what is the absolute value of the rate of change of the magnetic flux through the loop? **(f)** At this time, what is the absolute value of the emf in the loop? **(g)** What is the magnitude of the electric field at location P, which is inside the wire? **(h)** Now the wire loop is removed. Everything else remains as it was at time t_2; the current is still changing at the same rate. What is the magnitude of the electric field at location P?

••P18 In Figure 22.68 the magnetic field is uniform and out of the page inside a circle of radius R, and is essentially zero outside the circular region. The magnitude of the magnetic field is changing with time; as a function of time the magnitude of the magnetic field is $(B_0 + bt^3)$, where $B_0 = 1.5$ T, $b = 1.4$ T/s^3, $r_1 = 3.6$ cm, $r_2 = 51$ cm, $t = 1.3$ s, and $R = 17$ cm.

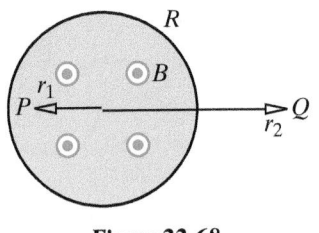

Figure 22.68

(a) What is the direction of the electric field at location P, a distance r_1 to the left of the center ($r_1 < R$)? **(b)** What is the magnitude of the electric field at location P? (*Hint:* Remember that "emf" is the integral of the non-Coulomb electric field around a closed path.) **(c)** What is the direction of the electric field at location Q, a distance r_2 to the right of the center ($r_2 > R$)? **(d)** What is the magnitude of the induced electric field at location Q?

••P19 In Figure 22.69 a thin circular coil of radius r_1 with N_1 turns carries a current $I_1 = a + bt + ct^2$, where t is the time and a, b, and c are positive constants. A second thin circular coil of radius r_2 with N_2 turns is located a long distance x along the axis of the first coil. The second coil is connected to an oscilloscope, which has very high resistance.

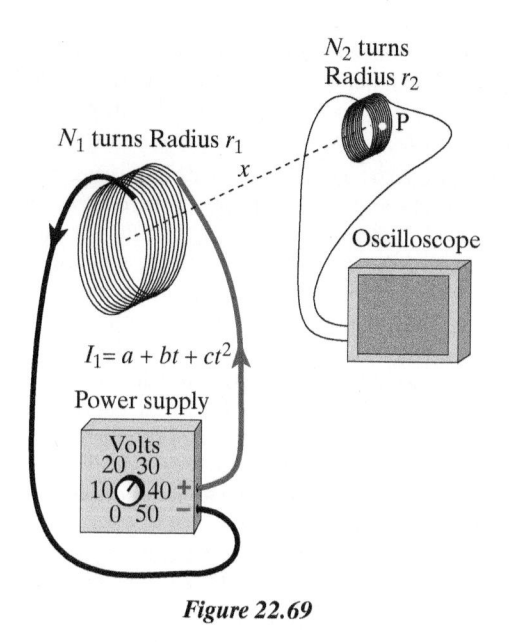

$I_1 = a + bt + ct^2$

Power supply

Figure 22.69

(a) As a function of time t, calculate the magnitude of the voltage that is displayed on the oscilloscope. Explain your work carefully, but you do not have to worry about signs. **(b)** At point P on the drawing (on the right side of the second coil), draw a vector representing the non-Coulomb electric field. **(c)** Calculate the magnitude of this non-Coulomb electric field.

••P20 In Figure 22.70 a long straight wire carrying current I is moving with speed v toward a small circular coil of radius r containing N turns, which is attached to a voltmeter as shown. The long wire is in the plane of the coil. (Only a small portion of the wire is shown in the diagram.)

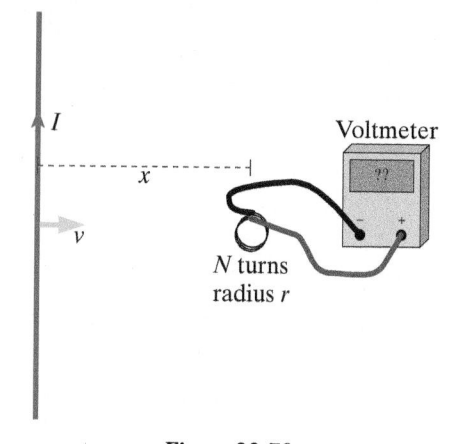

Figure 22.70

The radius of the coil is 0.02 m, and the coil has 11 turns. At a particular instant $I = 3$ A, $v = 3.2$ m/s, and the distance from the wire to the center of the coil is $x = 0.13$ m. **(a)** What is the magnitude of the rate of change of the magnetic field inside the coil? You will need to calculate this symbolically before you can get a number. Write an expression for the magnetic field due to the wire at the location of the coil. Use the approximate equation, since the wire is very long. Remember the chain rule, and remember that $v = dx/dt$. **(b)** What is the magnitude of the voltmeter reading? Remember that this includes all 11 turns of the coil. **(c)** What is the direction of $d\vec{B}/dt$? **(d)** What is the direction of the curly electric field in the coil?

••P21 A coil with 3000 turns and radius 5 cm is connected to an oscilloscope (Figure 22.71). You move a bar magnet toward the coil along the coil's axis, moving from 40 cm away to 30 cm away in 0.2 s. The bar magnet has a magnetic moment of 0.8 A·m².

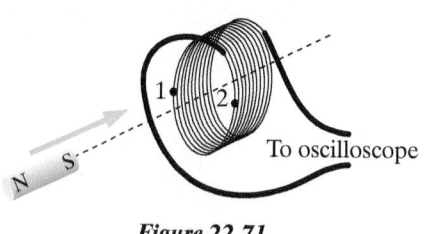

Figure 22.71 .

(a) On the diagram, draw the non-Coulomb electric field vectors at locations 1 and 2. Briefly explain your choices graphically. **(b)** What is the approximate magnitude of the signal observed on the oscilloscope? **(c)** What approximations or simplifying assumptions did you make?

••P22 In Figure 22.72 a toroid has a rectangular cross section with an inner radius $r_1 = 9$ cm, an outer radius $r_2 = 12$ cm, and a height $h = 5$ cm, and it is wrapped around by many densely packed turns of current-carrying wire (not shown in the diagram). The direction of the magnetic field inside the windings is shown on the diagram. There is essentially no magnetic field outside the windings. A wire is connected to a sensitive ammeter as shown. The resistance of the wire and ammeter is $R = 1.4$ ohms.

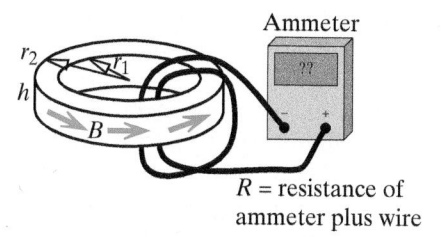

R = resistance of ammeter plus wire

Figure 22.72

The current in the windings of the toroid is varied so that the magnetic field inside the windings, averaged over the cross section, varies with time as shown in Figure 22.73:

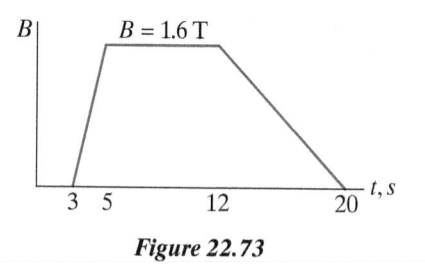

Figure 22.73

Make a careful graph of the ammeter reading, including sign, as a function of time. Label your graph, and explain the numerical aspects of the graph, including signs.

••P23 In Figure 22.74 a thin coil of radius $r_1 = 4$ cm, containing $N_1 = 5000$ turns, is connected through a resistor $R = 100$ Ω to an AC power supply running at a frequency $f = 2500$ H, so that the current through the resistor (and coil) is $I_1 \sin(2\pi \times 2500t)$. The voltage across the resistor triggers an oscilloscope that also displays this voltage, which is 10 V peak-to-peak and therefore has an amplitude of 5 V (and therefore the amplitude of the current is $I_1 = 0.05$ A).

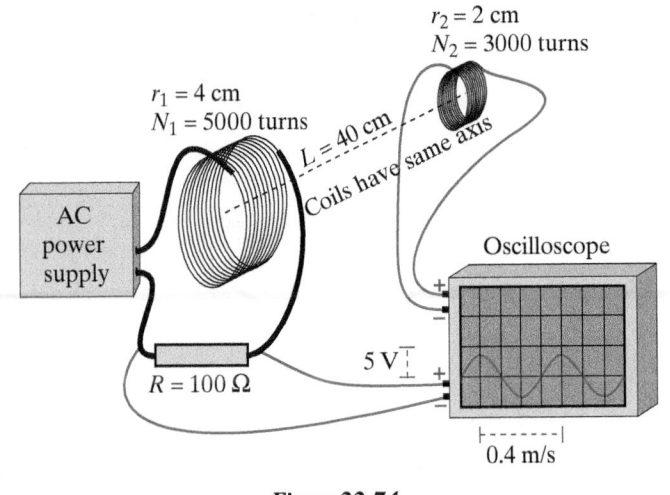

Figure 22.74

A second thin coil of radius $r_2 = 2$ cm, containing $N_2 = 3000$ turns, is a distance $L = 40$ cm from the first coil. The axes of the two coils are along the same line. The second coil is connected to the upper input of the oscilloscope, so that the voltage across the second coil can be displayed along with the voltage across the resistor. **(a)** Assume you have adjusted the VOLTS/DIVISION knob on the upper input so that you can easily see the signal from the second coil. Sketch the second coil voltage along with the resistor voltage on the oscilloscope display, paying careful attention to the shape and positioning of the two voltage curves with respect to each other. Explain briefly. **(b)** Calculate the amplitude (maximum voltage) of the second coil voltage. If you must make simplifying assumptions, state clearly what they are.

••P24 A thin rectangular coil lies flat on a low-friction table (Figure 22.75). A very long straight wire also lies flat on the table, a distance z from the coil. The wire carries a conventional current I to the right as shown, and this current is decreasing: $I = a - bt$, where t is the time in seconds, and a and b are positive constants. The coil has length L and width w, where $w \ll z$. It has N turns of wire with total resistance R.

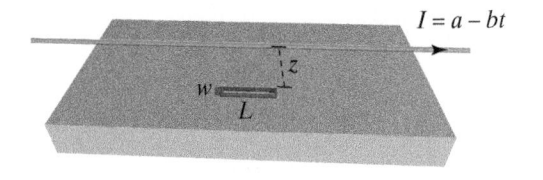

Figure 22.75

What are the initial magnitude and direction of the nonzero net force that is acting on the coil? You can neglect friction. Explain in detail. If you must make simplifying assumptions, state clearly what they are, but bear in mind that the net force is not zero.

••P25 There are a lot of numbers in this problem. Just about the only way to get it right is to work out each step symbolically first, and then plug numbers into the final symbolic result.

Two coils of wire are aligned with their axes along the z axis, as shown in Figure 22.76. Coil 1 is connected to a power supply and conventional current flows clockwise through coil 1, as seen from the location of coil 2. Coil 2 is connected to a voltmeter. The distance between the centers of the coils is 0.14 m.

Coil 1 has $N_1 = 570$ turns of wire, and its radius is $R_1 = 0.07$ m. The current through coil 1 is changing with time. At $t = 0$ s, the current through coil 1 is $I_0 = 18$ A. At $t = 0.4$ s, the current through coil 1 is $I_{0.4} = 6$ A.

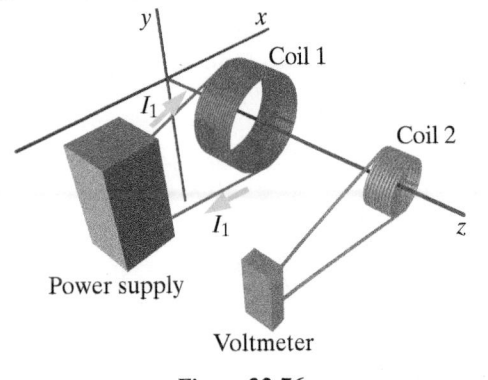

Figure 22.76

Coil 2 has $N_2 = 275$ turns of wire, and its radius is $R_2 = 0.03$ m. **(a)** Inside coil 2, what is the direction of $d\vec{B}/dt$ during this interval? **(b)** What is the direction of the electric field inside the wire of coil 2, at a location on the top of coil 2? **(c)** At time $t = 0$, what is the magnetic flux through one turn of coil 2? Remember that all turns of coil 1 contribute to the magnetic field. Note that the coils are not very far apart (compared to their radii). **(d)** What approximations did you make in calculating the flux through one turn of coil 2? **(e)** At $t = 0.4$ s, what is the magnetic flux through one turn of coil 2? **(f)** What is the emf in one turn of coil 2 during this time interval? **(g)** The voltmeter is connected across all turns of coil 2. What is the reading on the voltmeter during this time interval? **(h)** During this interval, what is the magnitude of the non-Coulomb electric field inside the wire of coil 2? Remember that the emf measured by the voltmeter involves the entire length of the wire making up coil 2. **(i)** At $t = 0.5$ s, the current in coil 1 becomes constant, at 5 A. Which of the following statements are true? (1) The voltmeter now reads 0 V. (2) The electric field inside the wire of coil 2 now points in the opposite direction. (3) The voltmeter reading is about the same as it was at $t = 0.4$ s. (4) The electric field inside the wire of coil 2 is now 0 V/m.

••P26 In Figure 22.77 a very small circular metal ring of radius $r = 0.5$ cm and resistance $x = 5$ Ω is at the center of a large concentric circular metal ring of radius $R = 50$ cm. The two rings lie in the same plane. At $l = 3$ s, the large ring carries a *clockwise* current of 3 A. At $l = 3.2$ s, the large ring carries a *counterclockwise* current of 5 A.

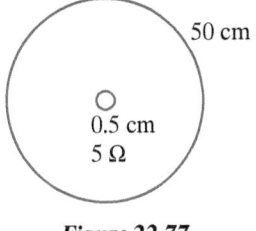

Figure 22.77

(a) What is the magnitude of the emf in the small ring during this time? **(b)** What are the magnitude and direction of the average current in the small ring? **(c)** What is the average electric field in the small ring during this time? Give both the magnitude and direction of the electric field. Draw a diagram, showing all relevant quantities.

••P27 A very long, tightly wound solenoid has a circular cross section of radius 3 cm (only a portion of the very long solenoid is shown in Figure 22.78). The magnetic field outside the solenoid is negligible. Throughout the inside of the solenoid the magnetic field B is uniform, to the left as shown, but varying with time t: $B = (0.07 + 0.03t^2)$ T. Surrounding the circular solenoid is a loop of 4 turns of wire in the shape of a rectangle 10 cm by 15 cm. The total resistance of the 4-turn loop is 0.1 ohm.

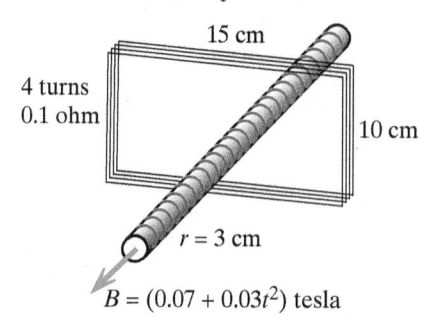

Figure 22.78

(a) At $t = 2$ s, what is the direction of the current in the 4-turn loop? Explain briefly. **(b)** At $t = 2$ s, what is the magnitude of the current in the 4-turn loop? Explain briefly.

••P28 Tall towers support power lines 50 m above the ground and 20 m apart that run from a hydroelectric plant to a large city, carrying 60 H alternating current with amplitude 1×10^4 A (Figure 22.79). That is, the current in both of the power lines is $I = (1 \times 10^4 \text{A}) \sin(2\pi \cdot 60 \text{ H} \cdot t)$.

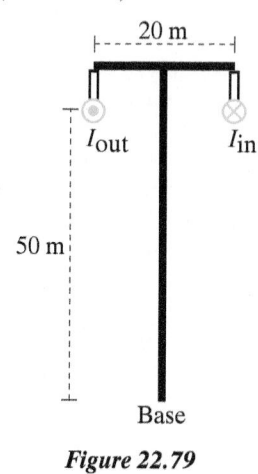

Figure 22.79

(a) Calculate the amplitude (largest magnitude) and direction of the magnetic field produced by the two power lines at the base of the tower, when a current of 1×10^4 A in the left power line is headed out of the page, and a current of 1×10^4 A in the right power line is headed into the page. **(b)** This magnetic field is not large compared to the Earth's magnetic field, but it varies in time and so might have different biological effects than the Earth's steady field. For a person lying on the ground at the base of the tower, approximately what is the maximum emf produced around the perimeter of the body, which is about 2 m long by half a meter wide?

••P29 In Figure 22.80 a very long solenoid of length d and radius r_1 is tightly wound uniformly with N turns of wire. A variable power supply forces a current to run in the solenoid

of amount $I_1 = p - kt$, where p and k are positive constants (so that the current is initially equal to p but decreases with time). A circular metal ring of radius r_2 and resistance R is centered on the solenoid and located near the middle of the solenoid, very far from the ends of the solenoid (so that the solenoid contributes essentially no magnetic field outside the solenoid).

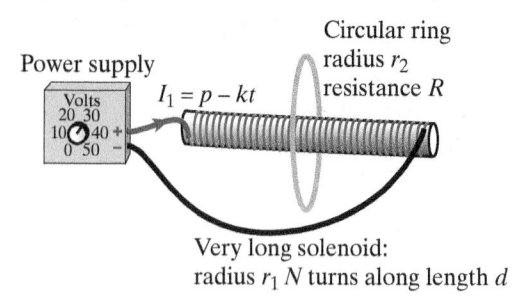

Figure 22.80

(a) What are the magnitude and the direction of the magnetic field at the center of the ring, due to the solenoid and the ring, as a function of the time t? Explain your work; part of the credit will be given for the clarity of your explanation, including clarity of appropriate diagrams. **(b)** Qualitatively, how would these results have been affected if an iron rod had been inserted into the solenoid? Very briefly, explain why.

••P30 You throw a bar magnet downward with its south end pointing down (Figure 22.81). The bar magnet has a magnetic dipole moment of 1.2 A·m². Lying on the table is a nearly flat circular coil of 1000 turns of wire, with radius 5 cm. The coil is connected to an oscilloscope, which has a very large resistance.

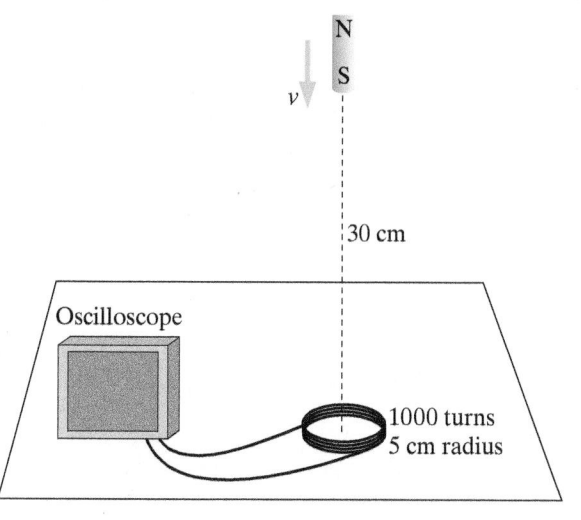

Figure 22.81

(a) On a diagram, show the pattern of non-Coulomb electric field in the coil. Explain briefly. **(b)** At the instant when the magnet is 30 cm above the table, the oscilloscope indicates a voltage of magnitude 2 mV. What is the speed v of the magnet at that instant?

••P31 Suppose that you have an electron moving with speed comparable to the speed of light in a circular orbit of radius r in a large region of uniform magnetic field B. **(a)** What must be the relativistic momentum p of the electron? **(b)** Now the uniform magnetic field begins to increase with time: $B = B_0 + bt$, where B_0 and b are positive constants. In one orbit, how much does the energy of the electron increase, assuming that in one orbit the radius doesn't change very much? (This effect was exploited in the "betatron," an electron accelerator invented in the 1940s.) *Note:* It turns out that the electron's energy increases by less than the amount you calculated, for reasons that will become clear in Chapter 23.

••P32 A thin coil is located at the origin; its radius is 3 cm and its axis lies on the x axis. It has 30 low-resistance turns and is connected to a 75 Ω resistor. A second thin coil is located at $\langle 15,0,0 \rangle$ cm and is traveling toward the origin with a speed of 8 m/s; it has 50 low-resistance turns, its axis lies on the x axis, its radius is 4 cm, and it has a current of 12 A, powered by a battery. What is the magnitude of the current in the first coil?

••P33 A bar magnet whose dipole moment is $\langle 0,0,6 \rangle$ A·m² has a constant velocity of $\langle 0,0,5 \rangle$ m/s. When the center of the magnet is at location $\langle 1,5,3 \rangle$ m, what is the (vector) electric field at location $\langle 1.04,5,2 \rangle$ m?

••P34 A very long wire carries a current I_1 *upward* as shown in Figure 22.82, and this current is *decreasing with time*. Nearby is a rectangular loop of wire lying in the plane of the long wire, and containing a resistor R; the resistance of the rest of the loop is negligible compared to R. The loop has a width w and height h, and is located a distance d to the right of the long wire.

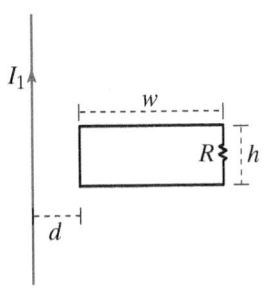

Figure 22.82

(a) Does the current I_2 in the loop run clockwise or counterclockwise? Explain, using a diagram. **(b)** Show that the magnitude of the current I_2 in the loop is this:

$$I_2 = \frac{\mu_0}{4\pi} \frac{(2h)\ln[(d+w)/d]}{R} \left| \frac{dI_1}{dt} \right|$$

(*Hint:* Divide the area into narrow vertical strips along which you know the magnetic field.)

••P35 In Figure 22.83 a single circular loop of wire with radius R carries a large clockwise constant current $I_{\text{loop}} = I_0$, which constrains a proton of mass M and charge e to travel in a small circle of radius r at constant speed around the center of the loop, in the plane of the loop. The orbit radius r is much smaller than the loop radius R: $r \ll R$.

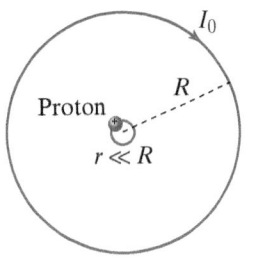

Figure 22.83

(a) Draw a diagram of the proton orbit, indicating the direction that the proton travels, clockwise or counterclockwise. Explain

briefly. **(b)** What is the speed v of the proton, in terms of the known quantities I_0, R, r, M, and e? Explain your work, including any approximations you must make. **(c)** The current was constant for a while, but at a certain time $t = t_0$ it began to decrease slowly, so that after t_0 the current was $I_{\text{loop}} = I_0 - k(t - t_0)$. On your diagram of the proton orbit, draw electric field vectors at four locations (one in each quadrant) and explain briefly. **(d)** When the current decreases, does the proton speed up or slow down?

Section 22.3

••P36 A copper bar of length h and electric resistance R slides with negligible friction on metal rails that have negligible electric resistance (Figure 22.84). The rails are connected on the right with a wire of negligible electric resistance, and a magnetic compass is placed under this wire (the diagram is a top view). The compass needle deflects to the right of north, as shown on the diagram.

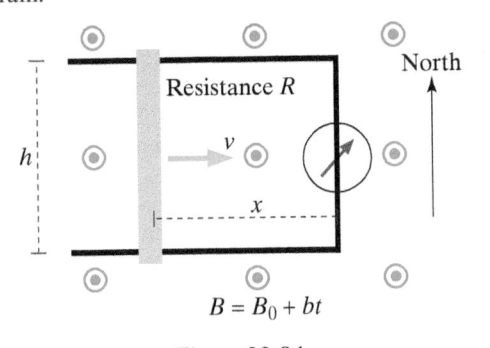

Figure 22.84

Throughout this region there is a uniform magnetic field B pointing out of the page, produced by large coils that are not shown. This magnetic field is increasing with time, and the magnitude is $B = B_0 + bt$, where B_0 and b are constants, and t is the time in seconds. You slide the copper bar to the right and at time $t = 0$ you release the bar when it is a distance x from the right end of the apparatus. At that instant the bar is moving to the right with a speed v. **(a)** Calculate the magnitude of the initial current I in this circuit. **(b)** Calculate the magnitude of the net force on the bar just after you release it. **(c)** Will the bar speed up, slow down, or slide at a constant speed? Explain briefly.

Section 22.5

••P37 A superconducting ring of radius R has a permanent conventional current I in the direction shown in Figure 22.85 and is oriented with its axis along the x axis. The ring is moving to the right along the x axis with uniform speed v. Calculate the electric field (magnitude and direction) at location $\langle x,y,0 \rangle$ relative to the center of the ring. If you must make any simplifying assumptions or approximations, state them explicitly.

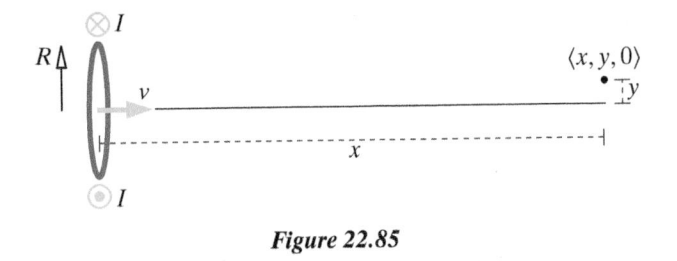

Figure 22.85

Section 22.6

•P38 A thin coil has 12 rectangular turns of wire. When a current of 3 A runs through the coil, there is a total flux of 1×10^{-3} T·m² enclosed by one turn of the coil (note that $\Phi = kI$, and you can calculate the proportionality constant k). Determine the inductance in henries.

•P39 A transformer has a primary coil with 100 turns and a secondary coil of 350 turns. The AC voltage across the primary coil has a maximum of 120 V and the AC current through the primary coil has a maximum of 4 A. What are the maximum values of the voltage and current for the secondary coil?

••P40 A 1000-turn solenoid has a radius of 1.4 cm and a length of 25 cm. The current in the solenoid is 8 A. **(a)** What is the inductance of this solenoid? **(b)** Inside the solenoid, what is the magnetic energy density (J/m³) far from the ends of the solenoid? **(c)** What is the total magnetic energy, in the approximation that there is little magnetic field outside the solenoid and the magnetic field is nearly uniform inside the solenoid? **(d)** Show that the result in part (c) is equal to $\frac{1}{2}LI^2$.

••P41 It is now possible to buy capacitors that have a capacitance of one farad. **(a)** Design a solenoid so that when it is connected to a charged 1 F capacitor, the circuit will oscillate with a period of one second. Give all the relevant parameters of the solenoid (length, etc.) so that someone could build the solenoid from your design specifications. Assume that there is wire available with low enough resistance that the resistance of the solenoid is negligible, although this may be difficult to achieve in practice unless the wire is superconducting. **(b)** If the 1 F

capacitor is initially charged to 3 V, what is the maximum current that will run through the inductor?

••P42 A cylindrical solenoid 40 cm long with a radius of 5 mm has 300 tightly wound turns of wire uniformly distributed along its length (Figure 22.86). Around the middle of the solenoid is a 2-turn rectangular loop 3 cm by 2 cm made of resistive wire having a resistance of 150 Ω.

One microsecond after connecting the loose wire to the battery to form a series circuit with the battery and a 20 Ω resistor, what is the magnitude of the current in the rectangular loop and its direction (clockwise or counterclockwise in the diagram)?

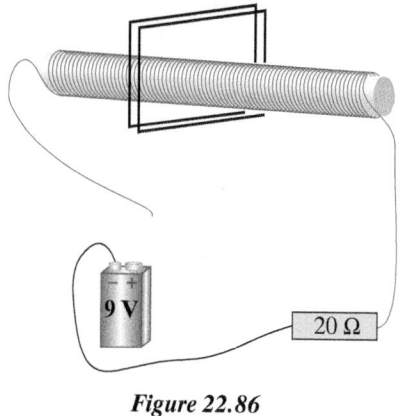

Figure 22.86

ANSWERS TO CHECKPOINTS

1 Clockwise; counterclockwise
2 (a) No current in wire A. **(b)** Current will run in wire B; **(c)** 47.7 V/m; **(d)** 7.5 A
3 0.052 T·m²
4 20 V; 8 V/m; 2 A
5 0 (there is no flux change); 0.166 V
6 The flux must still be equal to Φ_0. Since the magnet is too far away to contribute, all the flux must be due to the current in the ring.
7 4×10^{-3} H; 4×10^{-2} H
8 4000 V

9 5033 Hz
10 In the first three cases, for the bulb that is bright a loop can be drawn that consists of just that bulb plus a wire that encircles the varying magnetic flux. For the bulb that is dark a loop can be drawn that consists of just that bulb plus a wire that does not encircle the varying magnetic flux. (1) Lower bulb bright, upper dark; (2) upper bulb bright, lower dark; (3) upper bulb bright, lower dark. For the fourth case, both bulbs are in a loop that encircles the varying magnetic flux, so they both glow. The current is smaller than in the other circuits, because the emf drives two bulbs in series.

Electromagnetic Radiation

OBJECTIVES

After studying this chapter you should be able to

- Calculate the 3D radiative electric and magnetic fields produced by an accelerated charge.
- Mathematically relate the period, wavelength, and speed of a sinusoidal electromagnetic wave, and relate energy flux to amplitude.
- Explain physical phenomena involving the re-radiation of electromagnetic radiation.
- Mathematically relate index of refraction, frequency, speed of propagation, and angle of deflection of light in a medium.

We are approaching the climax of our study of electricity and magnetism: the classical model of electromagnetic radiation, or light. In Volume 1 we studied aspects of the quantum mechanical model of light, which involves particles called photons that have energy and momentum but zero rest mass. However, there are important properties of light, and aspects of the interaction of light and matter, that we cannot explain with our simple particle model. These phenomena, which have to do with the wavelike nature of light, can be explained by the classical model of electromagnetic radiation, which we will explore in this chapter. Aspects of this classical model are important to a fuller understanding of the quantum mechanical model of light, as well.

23.1 MAXWELL'S EQUATIONS

The four Maxwell equations provide a complete description of possible spatial patterns of electric and magnetic field in space. So far we have complete versions of three of these equations: Gauss's law, Gauss's law for magnetism, and Faraday's law. However, we have an incomplete version of the fourth equation, Ampere's law. Before we can go further, we need to complete this equation. Once we have all of Maxwell's equations, we will be able to show that some surprising configurations of electric and magnetic fields in space are not only possible but extremely important.

The Ampere–Maxwell Law

As we saw in Chapter 22, Faraday's law expresses a connection between a time-varying magnetic field in a region of space and a curly (non-Coulomb) electric field throughout space. You may have wondered whether a time-varying electric field in a region might be associated with a magnetic field throughout space. In fact, this does turn out to be the case. We will see that Ampere's law needs to be modified to include this relation.

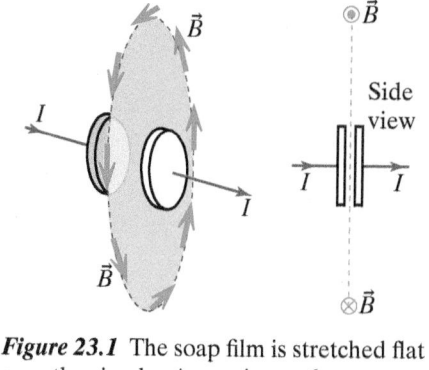

Figure 23.1 The soap film is stretched flat over the circular Amperian path.

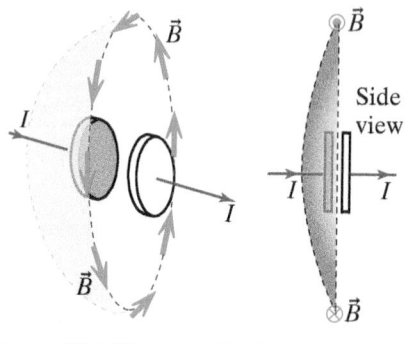

Figure 23.2 The soap film bulges out and is pierced by the current on the left.

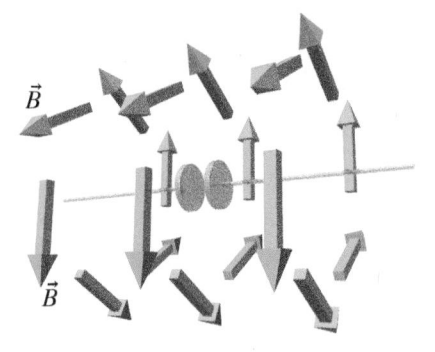

Figure 23.3 Probable magnetic field near the capacitor.

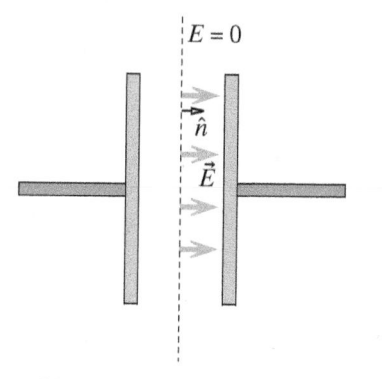

Figure 23.4 Calculating flux through a surface between the plates of the capacitor (side view).

If we try to apply Ampere's law to a circuit containing long straight wires carrying current to charge a capacitor, we will find what appears to be a contradiction. We choose a circular "Amperian" path around a charging capacitor, and stretch an imaginary soap film over the chosen path. Now consider two different soap film positions.

In Figure 23.1 the path does not enclose any current because the soap film is flat and passes between the capacitor plates. We must therefore conclude that $\oint \vec{B} \cdot d\vec{l}$ is zero along this circular path.

However, in Figure 23.2 the path does enclose the current I because the soap film bulges out and is pierced by the long straight wire on the left. In this case we conclude that $\oint \vec{B} \cdot d\vec{l}$ is equal to $\mu_0 I$. This contradicts our previous result.

QUESTION Which result seems more reasonable? Would you expect the magnetic field along the chosen path to be zero or nonzero?

Presumably the magnetic field along the path is really nonzero, because the current in the wire contributes to the magnetic field at all locations. We would expect the magnetic field around the capacitor to be nearly the same as the magnetic field around a nearby section of the wire, as indicated in Figure 23.3.

In Chapter 17 we found by applying the Biot–Savart law applied to a long wire that the magnetic field near the wire was

$$B = \frac{\mu_0}{4\pi} \frac{2I}{r} \quad \text{(a distance } r \text{ from the center of a long wire)}$$

This should be approximately the magnetic field along the circular path in Figure 23.1, because there is only a very short piece of wire missing from the long straight wire. Therefore around the Amperian path in Figure 23.1 we should have

$$\oint \vec{B} \cdot d\vec{l} \approx \frac{\mu_0}{4\pi} \frac{2I}{r} 2\pi r = \mu_0 I$$

This dilemma was resolved by James Clerk Maxwell, a Scottish physicist. Knowing of Faraday's discovery that a time-varying magnetic field is accompanied by an electric field, he guessed that a time-varying electric field might be accompanied by a magnetic field. Between the plates of the capacitor there is a time-varying electric field, because the electric field between the plates is growing as the capacitor becomes more and more charged. Maxwell guessed that the time-varying electric field could account for the magnetic field around the capacitor.

Let's explore this idea quantitatively. By analogy to Faraday's law, we will guess that the time rate of change of the electric flux in a region is related to the integral of the magnetic field along a path surrounding the region.

First we need an expression for the electric flux through the surface shown in Figure 23.4. As we saw in Chapter 21,

$$\Phi_{\text{electric}} = \int \vec{E} \cdot \hat{n} dA$$

As shown in Figure 23.4, inside the capacitor the electric field is parallel to $\hat{n}$ and has magnitude $(Q/A)/\varepsilon_0$. Outside the capacitor the electric field is very small if the plates are large and close together. Therefore

$$\Phi_{\text{electric}} = \left(\frac{Q}{A\varepsilon_0}\right) A \cos 0° = \frac{Q}{\varepsilon_0}$$

QUESTION What is the time derivative of the electric flux in this situation?

In a time Δt the amount of charge ΔQ that flows onto the capacitor is $\Delta Q = I\Delta t$. Thus, the rate of change of Q, the amount of charge on the capacitor, is

$$\frac{dQ}{dt} = I$$

Therefore, the time rate of change of electric flux is

$$\frac{d\Phi_{\text{electric}}}{dt} = \frac{d}{dt}\left(\frac{Q}{\varepsilon_0}\right) = \frac{I}{\varepsilon_0}$$

The quantity $\varepsilon_0 \dfrac{d\Phi_{\text{electric}}}{dt}$ has units of amperes, so we will combine it with I in Ampere's law. We will call this more general form of Ampere's law the "Ampere–Maxwell law":

THE AMPERE–MAXWELL LAW

$$\oint \vec{B} \bullet d\vec{l} = \mu_0 \left[\sum I_{\text{inside path}} + \varepsilon_0 \frac{d\Phi_{\text{electric}}}{dt}\right]$$

$$\Phi_{\text{electric}} = \int \vec{E} \cdot \hat{n}\, dA$$

If our imaginary soap film is outside the capacitor, there is just the term involving I but no term involving changing electric flux. If our imaginary soap film goes between the capacitor plates, there is just the term involving changing electric flux but no term involving I. The Ampere–Maxwell law handles both cases.

It also handles the case of a bulge in the soap film near the center of the capacitor plate (Figure 23.5). In this case the film doesn't enclose all of the electric flux. The soap film is pierced not only by the incoming current I but also by a lesser amount of radial current I_2 flowing outward from the center of the capacitor plate toward the edge of the plate, so that the net current piercing the soap film is $(I - I_2)$. In this case the Ampere–Maxwell equation contains both a current term and a flux term, adding up to $\mu_0 I$.

We interpret the Ampere–Maxwell law as saying that a time-varying electric field is always associated with a magnetic field. It is quite difficult to verify Maxwell's ingenious guess by a direct experiment, because in order to get a changing electric field you have to run a current to deposit a growing amount of charge, and this unavoidable current may be the source of the magnetic field you observe. Unless you are very clever in your geometrical arrangement, the real currents may mask the effects of the time-varying electric field.

There has in fact been direct experimental verification of the electric flux term, but the most compelling confirmation of Maxwell's inspired guess is indirect, in the explanation he was able to provide for the nature of light and radio waves, as we will soon see.

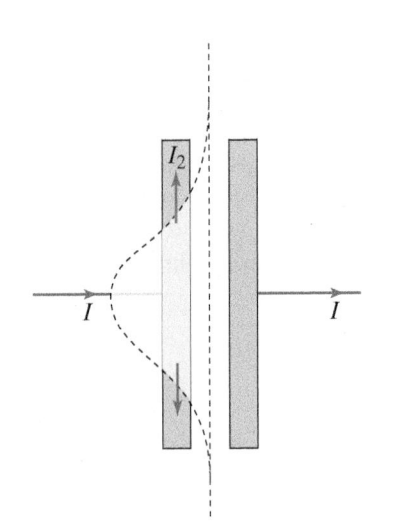

Figure 23.5 The soap film has a bulge near the center of the plate.

Maxwell's Equations (Integral Form)

We can now list the four equations that summarize all electromagnetic effects and are collectively called "Maxwell's equations" (although Maxwell himself did not write them in this modern form). Note that the first three are familiar equations from previous chapters:

MAXWELL'S EQUATIONS

$$\oint \vec{E} \bullet \hat{n} dA = \frac{1}{\varepsilon_0} \sum Q_{\text{inside the surface}} \qquad \text{Gauss's law}$$

$$\oint \vec{B} \bullet \hat{n} dA = 0 \qquad \text{Gauss's law for magnetism}$$

$$\oint \vec{E} \bullet d\vec{l} = -\frac{d}{dt} \int \vec{B} \bullet \hat{n} dA \qquad \text{Faraday's law}$$

$$\oint \vec{B} \bullet d\vec{l} = \mu_0 \left[\sum I_{\text{inside path}} + \varepsilon_0 \frac{d}{dt} \int \vec{E} \bullet \hat{n} dA \right] \quad \text{Ampere–Maxwell law}$$

These are the "integral" versions of Maxwell's equations. The "differential" versions of Maxwell's equations are given at the end of this chapter.

In addition to Maxwell's equations, to complete our list of fundamental relations of electricity and magnetism we need to add the equation for the net electric and magnetic force, the Lorentz force, which defines what we mean by electric and magnetic fields in terms of their effects on charges:

THE NET ELECTRIC AND MAGNETIC FORCE

$$\vec{F} = q\vec{E} + q\vec{v} \times \vec{B} \qquad (d\vec{F} = Id\vec{l} \times \vec{B} \text{ for currents})$$

23.2 FIELDS TRAVELING THROUGH SPACE

A time-varying magnetic field is associated with an electric field, and a time-varying electric field is associated with a magnetic field. Might it be possible for time-varying electric and magnetic fields in empty space to exist without any charges or currents nearby?

If we can show that this is possible (that is, if this would not be inconsistent with one or more of Maxwell's equations), we will still need to answer the question of how such a process might get started. Eventually we will show that such a configuration of fields can be produced by an accelerated charge, but first we will concentrate on establishing that such a configuration of electric and magnetic fields is indeed possible.

Solving Maxwell's Equations

In principle, a straightforward way to show how this could work would be to solve the partial differential equations represented by Maxwell's equations in their differential form, but the required mathematics is beyond the scope of this introductory textbook. However, we can do what we did in finding a solution to the differential equation for an RC circuit in Chapter 19. In that case we guessed that an exponential function would be a solution for the current as a function of time. We then showed that this function was a solution of the differential equation, by taking derivatives of the guessed function and plugging them into the differential equation. This is one of the most common schemes for solving a differential equation, and it can be used to solve integral equations as well.

Solution Plan

We will take the following approach to the problem:

- Propose a particular configuration of electric and magnetic fields in space and time.
- Show that this is consistent with all four Maxwell equations, and is therefore a possible solution.
- Show that an accelerated charge produces such fields.
- Identify the effects that such fields would have on matter.
- Analyze a variety of phenomena involving such fields.

A Simple Configuration of Traveling Fields

We could consider various configurations of time-varying electric and magnetic fields, in the absence of any charges, and we would find that most of them do not satisfy Maxwell's equations and hence are not possible. However, if we add the assumption that the region of nonzero field is "traveling" through space at a velocity $\vec{v}$, we can find a solution that works. This idea—that the region where the fields are nonzero must move through space—is the key to finding time-varying electric and magnetic fields that satisfy all of Maxwell's equations.

You may recognize this phenomenon—a disturbance moving through space—as a wave. To someone familiar with differential equations, the differential form of Maxwell's equations suggests a wave as a possible solution. Many waveforms are possible, including the familiar sinusoidal wave. To simplify our analysis, we will first consider the simplest possible wave: a single propagating pulse, as shown in Figure 23.6.

A Moving Slab (or Pulse)

Consider a rectangular "slab" of space, of thickness d and very large height and depth, filled with a uniform electric field $\vec{E}$ pointing up and a uniform magnetic field $\vec{B}$ pointing out of the page (Figure 23.6). The electric and magnetic fields are zero outside the slab. We imagine the field-filled slab moving to the right with an unknown speed v, by which we mean that at a later time the slab-shaped region of space that contains fields is somewhere to the right of the present position (and at that time the fields are zero in the original position).

Time-Varying Fields

If you stand at a location to the right of this region, as shown in Figure 23.6, for a while you don't experience any electric or magnetic fields. Then, as shown in Figure 23.7, the fields at your location change. For a short time you can detect an upward-pointing electric field, and a magnetic field coming out of the page. After the slab passes by, you again experience no electric or magnetic fields. You might describe your experience by saying that a short "pulse" of fields passed by. As the edges of the pulse pass a location, there are time-varying fields at that location. (In a real pulse, the magnitude of the field would not change instantaneously, so dE/dt and dB/dt at the edges of the pulse would not be infinite.)

Propagation

We say that in our example the fields "propagate" to the right, or that the "direction of propagation" is to the right. "Propagation" means the spreading of something (in this case, electric and magnetic fields) into a new region. Strictly speaking, the fields don't move. Rather, it's just that at one instant in time there are fields in a particular slab-shaped region of space, and at a later time there are fields in a different slab-shaped region of space, to the right of the original region (Figure 23.8). At a particular instant in time, the fields vary with location. At a particular location in space, the fields vary with time (Figure 23.7). Mathematically, we can say that the region of space in which $\vec{E}$ and $\vec{B}$ are nonzero is propagating to the right with velocity $\vec{v}$.

We will show that this configuration of electric and magnetic fields is consistent with all four of Maxwell's equations. Of course, we have chosen to consider these particular directions for $\vec{E}$, $\vec{B}$, and $\vec{v}$ because we know that they will work. However, if we try other sets of relative directions, say with $\vec{E}$ and $\vec{B}$ both pointing down and $\vec{v}$ up, we find that these directions are not consistent with all four of Maxwell's equations. Only the arrangement we treat here turns

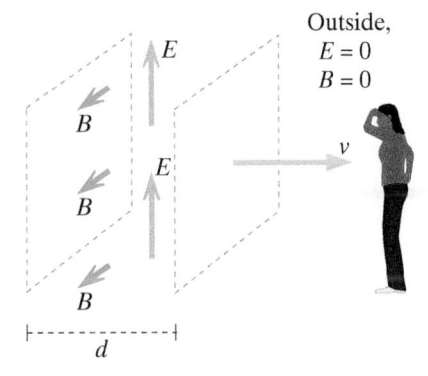

Figure 23.6 A slab of space contains electric and magnetic fields, with no fields outside the slab. The slab moves to the right with speed v.

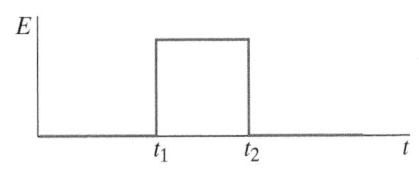

Figure 23.7 Magnitude of electric field observed at a particular location in space, as a function of time, as the slab passes. Note that at times t_1 and t_2, the electric field at this location is changing ($dE/dt \neq 0$).

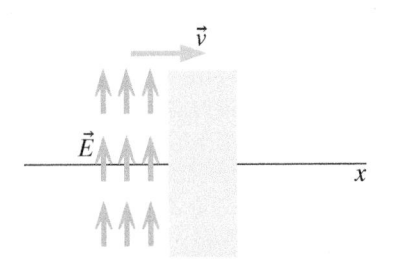

Figure 23.8 At a given instant the electric field is nonzero in the region shown. In the next instant, the region of nonzero field will propagate to the shaded area. For clarity, magnetic field is not shown here.

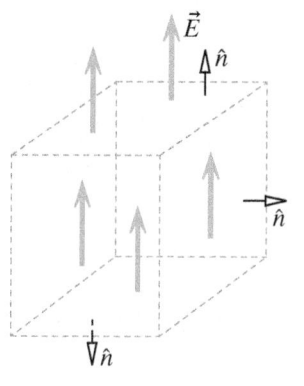

Figure 23.9 Applying Gauss's law to a box-shaped region within the slab, we see that the net electric flux on the box is 0. $\vec{B}$ is not shown.

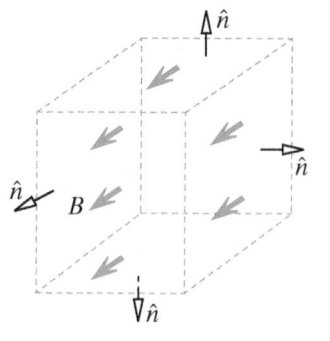

Figure 23.10 Applying Gauss's law for magnetism to a box-shaped surface within the slab, we find that the net magnetic flux on the box is 0. $\vec{E}$ is not shown.

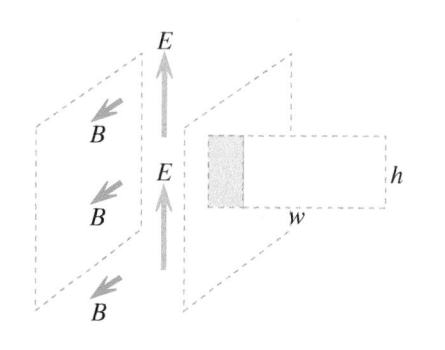

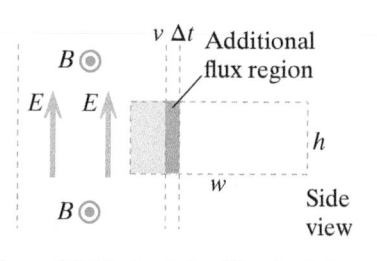

Figure 23.11 Applying Faraday's law to a rectangular loop partly inside the slab. The lower diagram shows a side view.

out to satisfy Maxwell's equations, with $\vec{E}$, $\vec{B}$, and $\vec{v}$ perpendicular to each other as shown in Figure 23.6. (Of course, $\vec{v}$ can be in any direction; it is the relative orientation of $\vec{E}$, $\vec{B}$, and $\vec{v}$ that is important.)

Gauss's Law

First we'll consider Gauss's law,

$$\oint \vec{E} \bullet \hat{n} dA = \frac{1}{\varepsilon_0} \sum Q_{\text{inside the surface}}$$

We need to show that our chosen configuration of electric fields is permissible in a region of space where there are no charged particles. We choose an imaginary closed box, fixed in space, through which the field-filled slab is moving (Figure 23.9). The positive flux on the top of the box is exactly canceled by an equal negative flux on the bottom. On the sides $\vec{E}$ is perpendicular to $\hat{n}$, so the flux on the sides is zero. Since the net flux is zero, there must be no charge in the box, as we expected.

Gauss's Law for Magnetism

Next we consider Gauss's law for magnetism,

$$\oint \vec{B} \bullet \hat{n} dA = 0$$

Using the same box-shaped surface (Figure 23.10), we find that the positive magnetic flux on the front of the box is canceled by an equal amount of negative flux on the back of the box. The other sides contribute no flux, because $\vec{B}$ is perpendicular to $\hat{n}$. Thus, this configuration of magnetic field can exist only in a region where there are no magnetic monopoles, which is fine, because as far as we know there are no magnetic monopoles anywhere!

Faraday's Law

Faraday's law relates a changing magnetic flux to a non-Coulomb electric field,

$$\oint \vec{E} \bullet d\vec{l} = -\frac{d}{dt} \int \vec{B} \bullet \hat{n} dA$$

Because the region of space in which $\vec{E}$ and $\vec{B}$ are nonzero is "moving," the magnetic field at a particular location in space changes with time, and this time-varying magnetic field is associated with a curly electric field. To consider Faraday's law, we draw a vertical rectangular path, h high and w wide, partly inside the slab and partly outside as shown in Figure 23.11, which includes a side view. This rectangular path is fixed in space, whereas the slab "moves" to the right. Let's apply Faraday's law to the rectangular path. First we will consider the right-hand side of the equation.

QUESTION If the slab is moving to the right at speed v, calculate the rate of change of magnetic flux enclosed by the path in terms of B, by calculating how much additional flux there is in a time Δt, then dividing by Δt. Note in the side view in Figure 23.11 the additional region of magnetic flux that appears in an amount of time Δt.

In a time Δt the area of magnetic field inside our chosen path increases by an amount $(v\Delta t)h$, and since the magnetic field B is uniform within the slab, the magnetic flux increases by an amount $B(v\Delta t)h\cos 0 = Bvh\Delta t$. Therefore the rate of change of magnetic flux inside our chosen path is this:

$$\frac{d\Phi_{\text{magnetic}}}{dt} = Bvh$$

By Faraday's law, this rate of change of magnetic flux is equal to the emf around the path, which is a path integral of the electric field (the left-hand side of the equation).

> QUESTION Keeping in mind the fact that part of the path is outside the slab, what is the path integral of the electric field in terms of the electric field E?

The only portion of the path along which there is a parallel component of nonzero electric field is along the piece of length h, so we have this:

$$|\text{emf}| = \left| \oint \vec{E} \bullet d\vec{l} \right| = Eh\cos 0 = Eh$$

> QUESTION Therefore Faraday's law establishes an algebraic relation between E and B. State this relation.

The emf is equal to the rate of change of magnetic flux, $Eh = Bvh$, so we have the following relationship between B and E in the slab:

$$E = vB$$

> QUESTION Explain how the direction of $\vec{E}$ is consistent with Faraday's law.

In the region ahead of the slab, the magnetic field changes from being zero to pointing out of the page, so $\Delta \vec{B}$ points out of the page and $-d\vec{B}/dt$ points into the page (Figure 23.12). With the right thumb pointing in the direction of $-d\vec{B}/dt$, the fingers of the right hand curl clockwise, which is the sense of the electric field around the path.

We have shown that our chosen configuration of moving fields is consistent with Faraday's law if $E = vB$.

The Ampere–Maxwell Law

The Ampere–Maxwell law relates a changing electric flux to a magnetic field:

$$\oint \vec{B} \bullet d\vec{l} = \mu_0 \left[\sum I_{\text{inside path}} + \varepsilon_0 \frac{d\Phi_{\text{electric}}}{dt} \right]$$

In this situation there are no currents. Because the field-containing slab is "moving," the electric field at a location in space changes with time, and by the Ampere–Maxwell law this changing electric field must be associated with a curly magnetic field. To consider the Ampere–Maxwell law, we draw a rectangular path, fixed in space, that is horizontal (so this rectangle is perpendicular to the electric field). Figure 23.13 includes a view from above, looking down on the upward-pointing electric field. First we will consider the right-hand side of the equation:

> QUESTION Let's apply the Ampere–Maxwell law to this path. If the slab is moving to the right at speed v, calculate the rate of change of electric flux enclosed by the path in terms of E, just as you calculated the rate of change of magnetic flux in the preceding section. Note in the view from above in Figure 23.13 the additional region of electric flux that appears in an amount of time Δt.

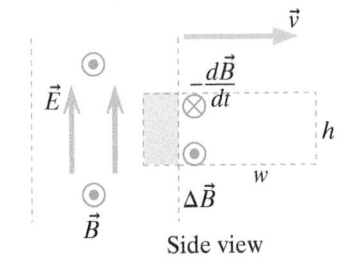

Side view

Figure 23.12 The direction of the electric field is consistent with Faraday's law.

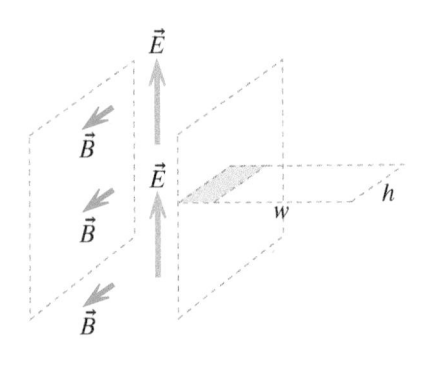

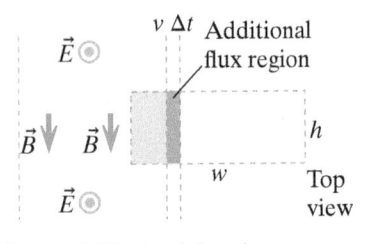

Top view

Figure 23.13 Applying the Ampere–Maxwell law to the slab. The lower diagram shows a view from above.

In a time Δt the area of electric field inside our chosen path increases by an amount $(v\Delta t)h$, and since the electric field E is uniform within the slab, the electric flux increases by an amount $E(v\Delta t)h\cos 0 = Evh\Delta t$. Therefore the rate of change of electric flux inside our chosen path is this:

$$\frac{d\Phi_{electric}}{dt} = Evh$$

QUESTION Keeping in mind the fact that part of the path is outside the slab, what is the path integral of the magnetic field (the left-hand side of the equation) in terms of B?

The only portion of the path along which there is a parallel component of nonzero magnetic field is along the piece of length h, so we have this:

$$\oint \vec{B} \bullet d\vec{l} = Bh\cos 0 = Bh$$

QUESTION Therefore application of the Ampere–Maxwell law establishes another algebraic relation between E and B. State this relation.

There is no current I, so the path integral of the magnetic field is equal to $\mu_0\varepsilon_0$ times the rate of change of electric flux, $Bh = \mu_0\varepsilon_0 Evh$, so we have the following relationship between B and E in the slab:

$$B = \mu_0\varepsilon_0 vE$$

Speed of Propagation

Since we have shown that our chosen configuration of fields is possible if the region of nonzero field is traveling with speed v, we are very interested in finding out what v must be. We have two different algebraic relations involving E and B, both of them involving the unknown speed v with which the pulse of electric and magnetic fields is traveling through space:

$$E = vB$$
$$B = \mu_0\varepsilon_0 vE$$

QUESTION Solve for v and evaluate the result numerically to see with what speed the fields must propagate in order that this pattern of fields be consistent with all four of Maxwell's equations.

By combining $E = vB$ and $B = \mu_0\varepsilon_0 vE$, we find that

$$v^2 = \frac{1}{(\mu_0\varepsilon_0)}$$

Evaluating this numerically we obtain this:

$$v = \sqrt{\frac{1}{\mu_0\varepsilon_0}} = \sqrt{\left(\frac{1}{4\pi\varepsilon_0}\right)\left(\frac{4\pi}{\mu_0}\right)}$$

$$v = \sqrt{(9\times 10^9 \ \text{N}\cdot\text{m}^2/\text{C}^2)(1\times 10^7 \ \text{A/T}\cdot\text{m})} = 3\times 10^8 \ \text{m/s}$$

An electromagnetic wave propagates at the speed of light! This is a spectacular result: the pulse is predicted to advance at a speed that is the same as the known speed of light. At the time that Maxwell went through a similar line of reasoning in the mid-1800s, no one had any idea as to the real nature of light, but the speed of light was known to be 3×10^8 m/s. Maxwell went through a theoretical analysis leading to the possibility of "moving" fields, and he predicted the speed of propagation of such fields by using values for μ_0 and ε_0 that were determined from simple experiments on electricity and magnetism.

When the predicted speed turned out to be the known speed of light, Maxwell concluded that light must be a combination of time-varying electric and magnetic fields that can propagate through otherwise empty space, far from any charges or currents. This theoretical analysis was brilliantly confirmed in the period 1885–1889 by Heinrich Hertz, who accelerated charges in a spark and observed the first radio waves, which also propagate at the speed of light.

It is useful to note that the direction of propagation $\vec{v}$ of the pulse is the same as the direction of the cross product $\vec{E} \times \vec{B}$ (Figure 23.14). We can summarize the relation between E and B in a pulse:

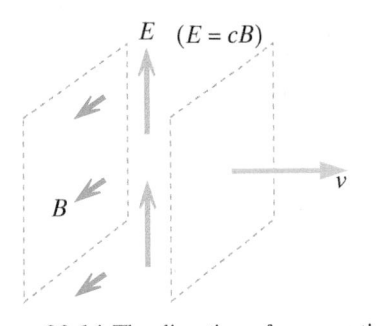

Figure 23.14 The direction of propagation of the electromagnetic pulse.

RELATION OF E TO B IN AN ELECTROMAGNETIC WAVE

$$E = cB$$

where c is the speed of light. The direction of $\vec{v}$ is the direction of $\vec{E} \times \vec{B}$.

The fact that $E = cB$ is easy to remember if you note that electric force is $q\vec{E}$ while magnetic force is $q\vec{v} \times \vec{B}$, so cB has the same units as E.

> **Checkpoint 1** A pulse of radiation propagates with velocity $\vec{v} = \langle 0, c, 0 \rangle$. The electric field in the pulse is $\vec{E} = \langle 0, 0, 1 \times 10^6 \rangle$ N/C. What is the magnetic field (vector) in the pulse?

Electromagnetic Radiation

The pulse of propagating electric and magnetic fields is called "electromagnetic radiation"—"electromagnetic" because it involves both electric fields and magnetic fields, and "radiation" because it "radiates" outward from a source. Visible light—electromagnetic radiation that can be detected by the human eye—makes up only part of the spectrum of electromagnetic radiation. When we say "light," we often mean not only visible light, but also any kind of electromagnetic radiation: microwaves, radio waves, infrared, ultraviolet, x-rays, or gamma-rays. Later in this chapter we will discuss the electromagnetic spectrum in more detail.

In any case, there is only one arrangement of propagating electric and magnetic fields that satisfies Maxwell's equations in empty space: $\vec{E}$ and $\vec{B}$ at right angles to each other, with $E = cB$, and the region containing nonzero fields propagating in the direction of $\vec{E} \times \vec{B}$ at the speed of light. Now that we have successfully shown that these propagating fields are consistent with Maxwell's equations, our next task is to investigate how such fields are produced.

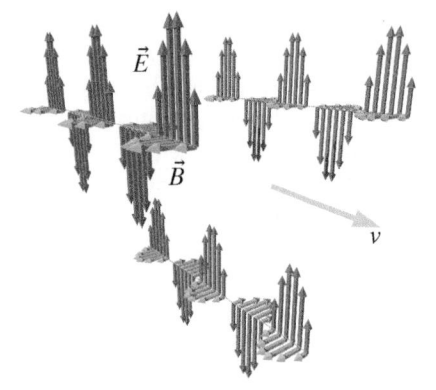

Figure 23.15 A sinusoidal electromagnetic wave.

Other Waveforms

The simple pulse of radiation that we've been considering is not the only possible kind of electromagnetic wave. In fact, any waveform that can be made up from successive short pulses of different magnitudes would also satisfy Maxwell's equations. For example, a square wave or a sinusoidal wave (Figure 23.15) would also be possible. We will see later that sinusoidal electromagnetic waves are particularly common.

23.3   ACCELERATED CHARGES PRODUCE RADIATION

We have seen that a pulse of electromagnetic radiation propagating with the speed of light is consistent with Maxwell's equations. The next question we need to answer is how such a configuration of propagating fields could be

Figure 23.16 Directions of electric field surrounding a stationary positive charge.

produced. It turns out that the way to produce electromagnetic radiation is to accelerate a charge.

Consider a positive charge q sitting at rest (Figure 23.16). The direction of the electric field produced by this charge is radially outward in all directions, which we indicate with "electric field lines" showing the direction (but not the magnitude) of the electric field at each point.

We kick the charge downward, so that in a very short time it picks up a little bit of speed and then coasts downward at that slow speed (there is no friction). Some time after the kick, we take a snapshot of the direction of electric field throughout this region, and the snapshot has the surprising and peculiar appearance shown in Figure 23.17. We will discuss what is observed by two people stationed at location 1 and location 2. Observer 1 sees an electric field that points away from the present location of the slowly moving charge.

QUESTION Explain why at this instant in time observer 2 sees an electric field that points away from the original position of the charge rather than the present location of the charge.

The key issue is retardation. Information about the change in the electric field has not yet reached observer 2. Since changes in the electric field propagate at the speed of light, it takes a while for the new field to appear at observer 2's position.

Between observer 1 and observer 2 there is a kink in the electric field direction. At some later time this kink will be at location 2 (Figure 23.18).

QUESTION If the distance from the original location of the charge to observer 2 is r, how long after we kicked the charge does observer 2 detect a sideways (transverse) electric field?

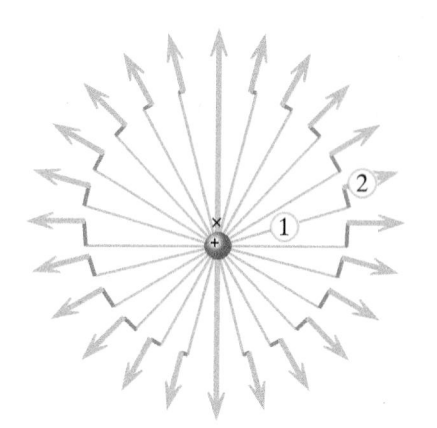

Figure 23.17 Directions of electric field shortly after the charge was kicked. The × marks the original location of the charge.

Since the transverse electric field propagates at the speed of light, we have $r = ct$ and $t = r/c$. If observer 2 is very far away from the original position of the charge, the radius of curvature r of the propagating radiation is very large, and observer 2 experiences a brief pulse of transverse electric field that looks very much like the transverse electric field associated with the flat moving slab we discussed earlier. An accelerated charge produces electromagnetic radiation. Of course, these flat diagrams show only a cross-sectional slice through a 3D pattern of radiation, shown in Figure 23.19.

It is clear from our application of Faraday's law and the Ampere–Maxwell law to the moving slab that a pulse of radiation must contain both electric and magnetic fields, so that the pulse can move forever through otherwise empty space, far from any charges. The transverse electric field is accompanied by a transverse magnetic field perpendicular to the electric field, with $\vec{E} \times \vec{B}$ in the direction of propagation, which means that the radiative magnetic field must point out of the page on the right of Figures 23.18 and 23.19, and into the page on the left of these figures.

Outside the expanding sphere of radiation there is no magnetic field because the charge was initially at rest. Inside the expanding sphere there is a charge moving downward with constant velocity, and it makes an ordinary magnetic field (by the Biot–Savart law) that is out of the page on the right and into the page on the left. In this case the Biot–Savart magnetic field is in the same direction as the radiative magnetic field, but this need not be the case. If the charge's acceleration is opposite to its velocity (slowing down rather than speeding up), the two kinds of magnetic fields point in opposite directions.

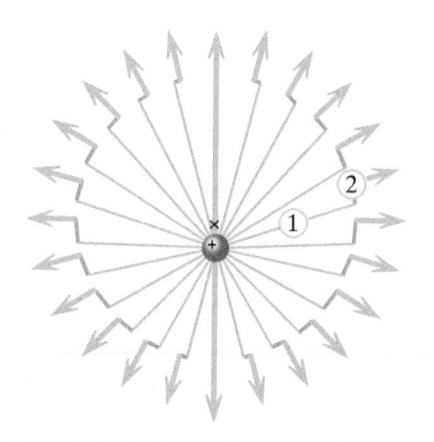

Figure 23.18 Information about the change has reached observer 2.

QUESTION A charge is initially at rest. Then you move it to another position, leaving it again at rest. Is there any radiation?

Yes. You have to accelerate the charge to get it moving, then decelerate it later to bring it again to rest, so there will be at least two pulses of radiation.

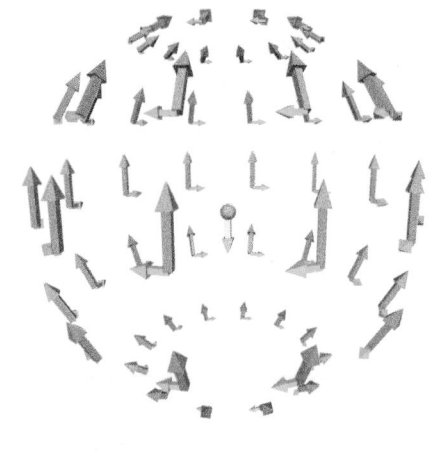

Figure 23.19 Radiation expands in a spherical pattern from a charge (center) that has been briefly accelerated downward. The orange arrows represent electric field, and the cyan arrows represent the accompanying magnetic field.

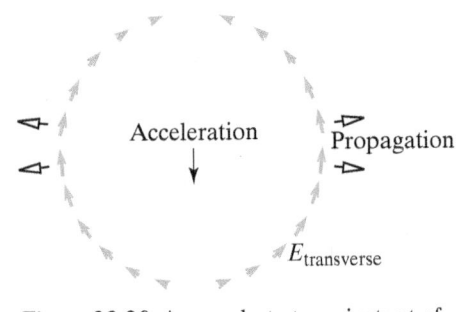

Figure 23.20 A snapshot at one instant of the transverse electric field, which propagates away from the accelerated charge. Magnetic field is not shown.

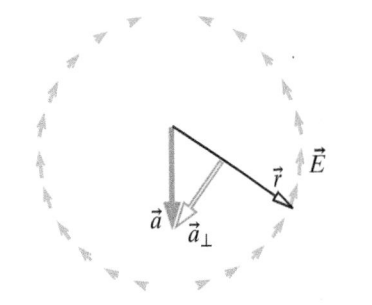

Figure 23.21 The transverse electric field is proportional to the transverse component of acceleration.

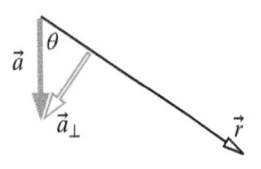

Figure 23.22 The magnitude of $\vec{a}_\perp$ is $a\sin\theta$.

Direction of the Transverse Electric Field

If we draw only the transverse electric field at some time after the kick, we see a pattern like that shown in Figure 23.20. Bear in mind that Figure 23.20 is a two-dimensional slice through the three-dimensional phenomenon that is shown in Figure 23.19, with the pulse headed out of the page and into the page as well as moving in the plane of the diagram.

QUESTION In what directions is the electromagnetic radiation most intense? In what directions is there no electromagnetic radiation?

It is most intense to the sides; there is zero radiation in the direction of the acceleration (and the opposite direction).

Magnitude of the Transverse Electric Field

The qualitative argument we have just made gives the direction of the radiative fields. This argument can be made quantitative by an application of Gauss's law, to derive expressions for the magnitude and direction of the radiative electric and magnetic fields. This derivation, due originally to Purcell, is given in Section 23.11. The results of that derivation give the following description of the radiative fields.

The transverse electric field is associated with the kink in the electric field direction that is associated with accelerated motion of the source charge, and from the diagrams of this process it is plausible that the magnitude of the radiative field is proportional to $-q\vec{a}_\perp$, where $\vec{a}_\perp$ is the "projected" acceleration of the charge (projected onto a plane perpendicular to the line of sight, as in Figure 23.21). Note that the magnitude of $\vec{a}_\perp$ is just $a\sin\theta$ (Figure 23.22).

The direction of propagation of the radiation is outward from the source charge, in the direction of the unit vector $\hat{r}$, drawn as usual from the source charge toward the observation location. This is also the direction of $\vec{E}\times\vec{B}$.

The radiative (transverse) electric field is proportional to the charge q and the projected acceleration $\vec{a}_\perp$, and it turns out to be inversely proportional to the distance r. The full relationship is the following:

PRODUCING A RADIATIVE ELECTRIC FIELD

$$\vec{E}_{\text{radiative}} = \frac{1}{4\pi\varepsilon_0}\frac{-q\vec{a}_\perp}{c^2 r}$$

There are two important points to notice in this equation:

- The direction of the transverse electric field is opposite to $q\vec{a}_\perp$.
- The field falls off with distance at the rate of $1/r$.

This is very different from the $1/r^2$ behavior of the ordinary electric field of a stationary charge or of a charge moving at a constant velocity. It is this much slower fall-off with distance that makes it possible for electromagnetic radiation to affect matter that is very far from the accelerated charges that produced the radiation. We see extremely distant stars because of this slow fall-off with distance.

In earlier chapters we showed that there is energy density associated with electric and magnetic fields, proportional to the squares of the field magnitudes. Similarly, later in this chapter we will show that the radiative power flow in watts per square meter is proportional to the square of the electric field, so the power flow per square meter falls off like $1/r^2$. Energy outflow over larger and larger spherical surfaces is constant, because the surface area $4\pi r^2$ grows proportional to r^2.

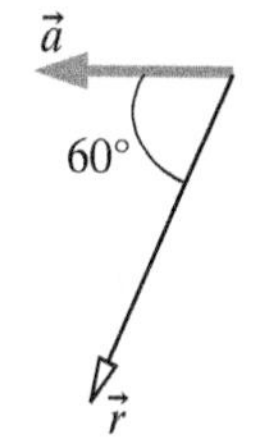

Figure 23.23 An electron is accelerated in the direction shown (Checkpoint 2).

Note that if the accelerated object is an electron or other negative charge, all the radiative electric field vectors in Figure 23.21 will point in the opposite direction, because the electric field is proportional to the charge q.

> **Checkpoint 2** An electron is briefly accelerated in the direction shown in Figure 23.23. The resultant electromagnetic radiation is observed at a location indicated by the vector $\vec{r}$ in the diagram. On a copy of the diagram, draw two vectors: $\vec{a}_\perp$, and $\vec{E}$ at the observation location.

Stability of Atoms

Classically (before quantum mechanics), any accelerated charge must emit electromagnetic radiation. Since electromagnetic radiation can accelerate other charges, there must be a flow of energy associated with the radiation. This implies that an accelerated charge loses some mechanical energy simply by virtue of emitting radiation.

Before quantum mechanics, this posed a puzzle for the stability of atoms. If we think of an atom as a miniature solar system, with electrons orbiting the nucleus, the electrons are undergoing accelerated motion (radial acceleration $= v^2/r$) and ought to lose energy by radiation and spiral into the nucleus. When one calculates how long an atom should last, it is a tiny fraction of a second! This paradox also applies to the persistent atomic currents in a magnet or the persistent currents in a superconducting ring.

This paradox was resolved by quantum mechanics. The basic idea is that the stable energies of a system are discrete, and in some cases (atoms, magnets, superconductors) there is no lower-energy state to which the system can go, and so there is no way to radiate energy and in the process drop to a lower-energy state. On the other hand, an atom that is in a state with energy above the "ground state" can and does drop to a lower-energy state and emits energy in the form of light.

EXAMPLE   **Electron Enters Region of Electric Field**

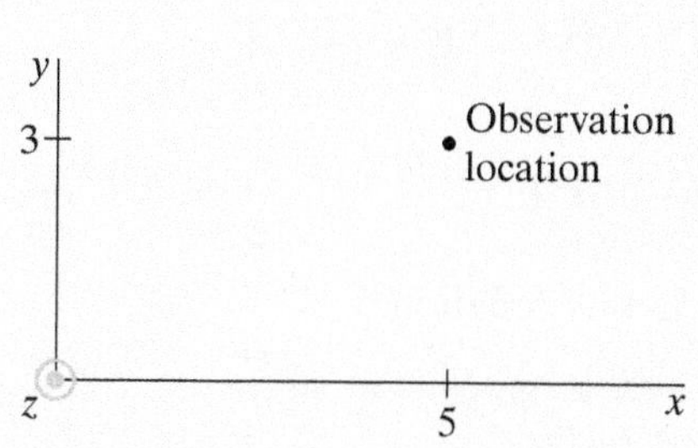

Figure 23.24 An electron enters a region of electric field.

An electron is traveling at nonrelativistic speed along the z axis in the $+z$ direction in a region of nearly zero fields (Figure 23.24). At $t = 0$ it reaches the origin $\langle 0,0,0 \rangle$ and enters a region where the electric field is $\langle 0,0,2 \times 10^{16} \text{ V/m} \rangle$. You observe fields at location $\langle 5,3,0 \rangle$ m. In the following, explain your work fully. **(a)** When do you first observe radiative fields? Explain. **(b)** On the diagram, show the direction of the initial radiative electric field at your location. Explain. **(c)** On the diagram, show the direction of the initial radiative magnetic field at your location. Explain. **(d)** What is the magnitude of the initial radiative electric field at your location? **(e)** What is the magnitude of the initial radiative magnetic field at your location?

Solution   **(a)** It takes $\dfrac{\sqrt{(5 \text{ m})^2 + (3 \text{ m})^2}}{3 \times 10^8 \text{ m/s}} = 1.9 \times 10^{-8}$ s for light to propagate from the origin to the observation location.

(b) See Figure 23.25. The electron undergoes an acceleration (deceleration) in the $-z$ direction, so $\vec{a}_\perp$ is in the $-z$ direction, and $-q\vec{a}_\perp = -(-e)\vec{a}_\perp$ is in the $-z$ direction. This produces a radiative electric field in the $-z$ direction, as shown.

(c) The direction of propagation is given by the direction of $\vec{E} \times \vec{B}$, which requires that $\vec{B}$ point in the direction shown.

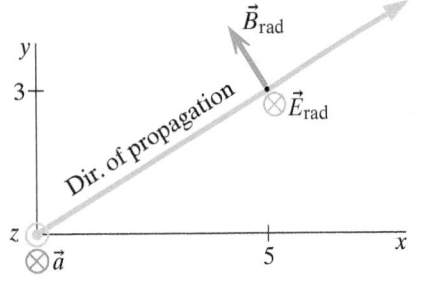

Figure 23.25 Directions of radiative fields.

It is important to note that the direction of the magnetic field predicted by the Biot–Savart law is in the opposite direction, because it depends on the velocity of the electron (in the $+z$ direction), not the acceleration (in the $-z$ direction). At a far distance the radiative magnetic field $(1/r)$ is much larger than the Biot–Savart magnetic field $(1/r^2)$.

(d) The speed is nonrelativistic, so $a_\perp = a = eE/m$, and

$$E_{\text{rad}} = \frac{1}{4\pi\varepsilon_0} \frac{e(eE/m)}{c^2 r}$$

$$E_{\text{rad}} = \left(9\times 10^9 \frac{\text{N}\cdot\text{m}^2}{\text{C}^2}\right) \frac{(1.6\times 10^{-19}\text{ C})^2(2\times 10^6\text{ V/m})}{(3\times 10^8\text{ m/s})^2(\sqrt{(5\,\text{m})^2+(3\,\text{m})^2})(9\times 10^{-31}\text{ kg})}$$

$$E_{\text{rad}} = 9.8\times 10^{-10}\text{ V/m}$$

(e) $B_{\text{rad}} = \dfrac{E_{\text{rad}}}{c} = \dfrac{9.8\times 10^{-10}\text{ V/m}}{3\times 10^8\text{ m/s}} = 3.2\times 10^{-18}\text{ T}$

Summary: Production of Electric and Magnetic Fields

Here is a summary of the most basic properties of how electric and magnetic fields are produced:

FIELDS MADE BY CHARGES

- A charge at rest makes a $1/r^2$ electric field but no magnetic field.
- A charge moving with constant velocity makes a $1/r^2$ electric field and a $1/r^2$ magnetic field.
- An accelerated charge in addition makes electromagnetic radiation, with a $1/r$ electric field and an accompanying $1/r$ magnetic field.

Remember that so far, no one has found an isolated "magnetic monopole," but if there were such objects, they would presumably produce fields as follows:

FIELDS MADE BY (HYPOTHETICAL) MAGNETIC MONOPOLES

- A magnetic monopole at rest would make a $1/r^2$ magnetic field, but no electric field.
- A magnetic monopole moving with constant velocity would make a $1/r^2$ magnetic field and a $1/r^2$ electric field.
- An accelerated magnetic monopole would in addition make electromagnetic radiation, with a $1/r$ magnetic field and an accompanying $1/r$ electric field.

23.4 SINUSOIDAL ELECTROMAGNETIC RADIATION

If a charge is given just one kick to get it moving, during the brief acceleration the charge emits a single brief pulse of electromagnetic radiation. However, the charge emits continuous radiation if it is moved sinusoidally, with the position of the charge described by

$$y = y_{\text{max}} \sin(\omega t)$$

where ω is the "angular frequency" in radians per second. Because the electric interaction between the electron cloud in an atom and the nucleus is spring-like (in the sense that the force is proportional to displacement), electrons in atoms can be usefully modeled as charges on springs; see Problem P62 in Chapter 15.

Ordinary matter emits excess energy in the form of sinusoidal electromagnetic radiation.

QUESTION If the charge moves sinusoidally, its y position is described by $y = y_{max} \sin(\omega t)$. Why will the electromagnetic radiation emitted also be sinusoidal?

$$v_y = \frac{dy}{dt} = \omega y_{max} \cos(\omega t)$$

$$a_y = \frac{dv_y}{dt} = -\omega^2 y_{max} \sin(\omega t)$$

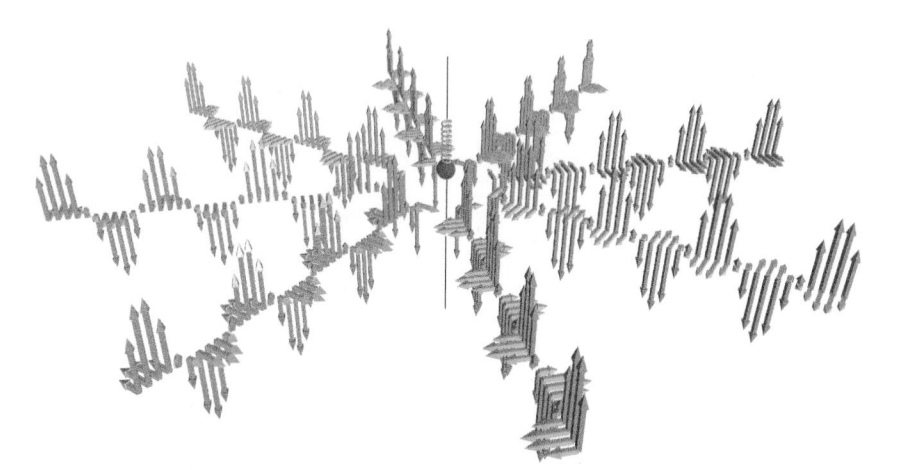

Figure 23.26 A charge moving sinusoidally in the $\pm y$ direction emits sinusoidal electromagnetic radiation, which propagates outward from the source. For clarity, this radiation is shown only in the xz plane, and the $1/r$ falloff in magnitude is not shown. As usual electric fields are represented by orange arrows, magnetic fields by cyan arrows.

The y component of the acceleration of the charge is also sinusoidal. The sinusoidal motion involves a sinusoidal acceleration which produces sinusoidal electromagnetic radiation. This radiation propagates away from the accelerated charge in all directions except $\pm y$ (Figure 23.26). For waves propagating in the xz plane, we have:

$$E_y = \frac{1}{4\pi\epsilon_0} \frac{-qa_y}{c^2 r}$$

$$E_y = \frac{1}{4\pi\epsilon_0} \frac{q\omega^2 y_{max}}{c^2 r} \sin(\omega t)$$

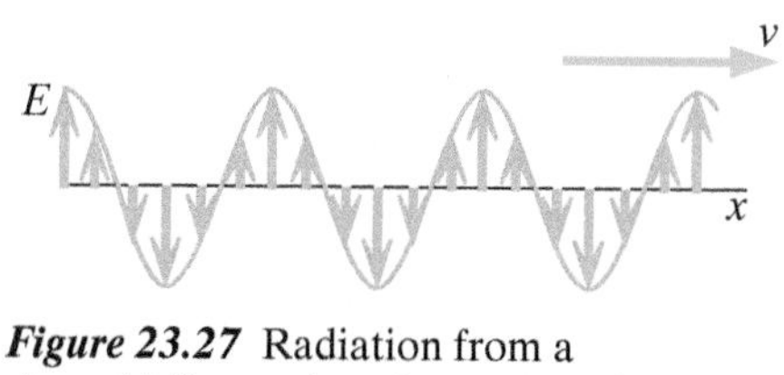

A snapshot of such a wave at a single instant in time is shown in Figure 23.27. Often only the outline of the wave is shown in diagrams (Figure 23.28).

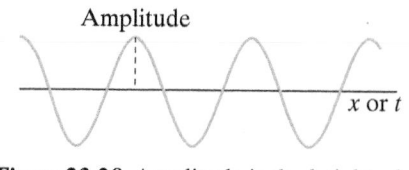

Figure 23.27 Radiation from a sinusoidally moving charge, at a single instant in time. Only the electric field is shown in this snapshot. Often only the outline of the wave is shown in diagrams (solid line).

Amplitude

The *amplitude* of a sinusoidal wave is the height of a peak in the wave, measured from the zero line, as shown in Figure 23.28. In the case of an electromagnetic wave, the amplitude is the maximum magnitude of the electric field.

Figure 23.28 Amplitude is the height of a wave crest.

Period

Imagine that you are standing at a particular location, observing a sinusoidal wave passing by. The *period* of the wave is the time between crests at your

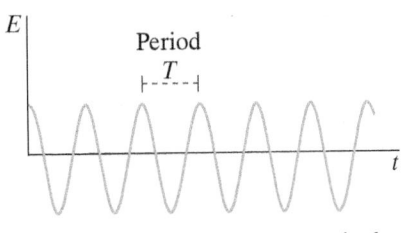

Figure 23.29 A plot of E vs. t at a single location in space. The period of a wave is the time between arrival of successive crests at a single location.

location. You could imagine measuring the period of a wave using a stopwatch: starting the timer when a crest (maximum amplitude) reaches your location, and stopping the timer when the next crest reaches your location. Figure 23.29 shows a plot of E vs. t at a specific location; the period is indicated on the graph.

Frequency

The frequency f of the sinusoidal electromagnetic wave is the number of peaks per second that pass a given location. It is the inverse of the period:

$$f = \frac{1}{T}$$

The units of frequency are s^{-1}, or hertz (abbreviated Hz). Frequency f is related to angular frequency ω in radians per second by

$$f = \frac{\omega}{2\pi}$$

Wavelength

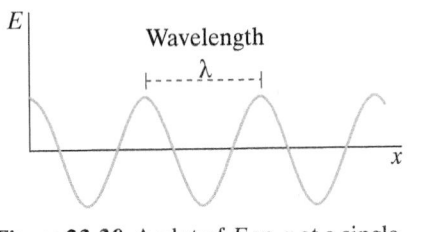

Figure 23.30 A plot of E vs. x at a single instant in time. The wavelength λ is the distance between two maxima.

Sinusoidal electromagnetic radiation is often described in terms of wavelength λ instead of frequency f, and $c = \lambda f$, because the wave advances a distance $\lambda = cT$ in one period $T = 1/f$, as illustrated in Figure 23.30. Electromagnetic radiation of different wavelength ranges has been given different names, such as *x-rays* or *microwaves*. Radiation with wavelengths of about 400–700 nm (1 nm = 1×10^{-9} m) is special because it can be detected by molecules in the human eye; we refer to this as *visible light*.

It should also be evident that electromagnetic radiation need not be sinusoidal. The moving slab that we treated earlier represented a single pulse and was not at all like a sinusoid. However, sinusoidal radiation is of great practical importance because so many different kinds of systems emit sinusoidal radiation, from atoms to radio stations.

Speed

How might we measure the speed of propagation of an electromagnetic wave? One can think of two different approaches:

(a) Follow a wave crest: If you watch one particular wave crest, you will see that it travels a distance λ (one wavelength) in a time T (one period). Therefore the *speed* of the crest is

$$v = \frac{\lambda}{T}$$

(b) Time the arrival of a radiative electric field: One could imagine a different way of measuring the speed of an electromagnetic wave. Suppose that you and a friend synchronize your clocks, then travel to locations that are a distance d apart. Your friend aims a laser at your location, and precisely at time t_1, turns on the laser. You record the time t_2 at which you first detect the radiative electric field in the laser light, and knowing the distance between the locations and the elapsed time $\Delta t = t_2 - t_2$, you calculate the speed at which the laser light traveled toward you:

$$v = \frac{d}{\Delta t}$$

In a vacuum, these two ways of measuring the speed of a sinusoidal electromagnetic wave will give the same answer: 3×10^8 m/s. However, this will not necessarily be the case if part or all of the space through which the light wave travels is filled with a medium such as water, glass, or even air. In this case, method 2 (measuring the time required for information about a *change* in the electromagnetic field to travel a given distance) will still give 3×10^8 m/s. However, method 1 (timing the interval between crests in a steady state electromagnetic wave inside the medium) will give a different answer, which will almost always be less than 3×10^8 m/s. In Section 23.8 we consider the reason for this discrepancy.

Combining the equations defining speed, wavelength, and frequency, we get these relations:

$$v = \frac{\lambda}{T} \text{ and } v = \lambda f$$

Checkpoint 3 The wavelength of green light from a laser pointer is 532 nm. What is the frequency of this light?

Polarized Radiation

The term *polarized* refers to a beam of electromagnetic radiation in which the electric field is aligned only along one axis. For example, in the sinusoidal radiation illustrated in Figure 23.27, the electric field is either in the $+y$ direction or in the $-y$ direction. Likewise, if you imagine standing at a location in the xz plane and observing the radiation emitted by the oscillating charge shown in Figure 23.26, you would detect an electric field in either the $+y$ or $-y$ direction.

How can radiation be unpolarized? A collection of oscillating source charges, in which the charges oscillate along different directions, will produce unpolarized radiation (Figure 23.31). Direct sunlight, or light from an ordinary light bulb, is unpolarized. It is possible to construct devices called polarizers, which allow only radiation with an electric field in a particular direction to pass through. Polarizers are discussed in Section 23.6.

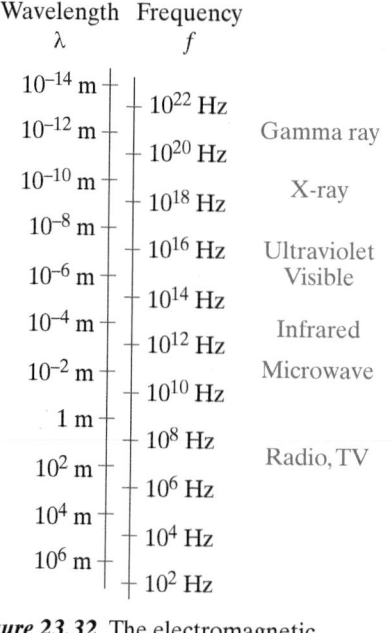

Figure 23.31 Radiation emitted by a collection of charges oscillating in different directions is not polarized; the electric field is not aligned in a single plane.

Transmitters

How can we produce electromagnetic radiation of a desired frequency? Radio transmitters have an LC circuit connected to the transmitting antenna, and either the inductance L or the capacitance C can be adjusted to choose the desired frequency (see Chapter 22). This circuit accelerates electrons in the antenna, and the accelerated electrons radiate.

Visible light is often produced by making some material so hot that the thermal oscillation of the atoms reaches frequencies in the visible region. This is the situation in the filament of a flashlight bulb. In this case there is a broad mixture of different frequencies that we perceive as white light.

Energy transitions between two quantized atomic levels produce light of a single frequency, as in a laser.

The Electromagnetic Spectrum

Figure 23.32 is a summary of the "electromagnetic spectrum," consisting of the ranges of wavelengths and frequencies that are characteristic of various mechanisms for producing and detecting electromagnetic radiation. The names given to radiation of different wavelengths are historical and were assigned before it was clear that all these entities were fundamentally the same. We still find these names useful for referring to electromagnetic radiation in different frequency or wavelength ranges.

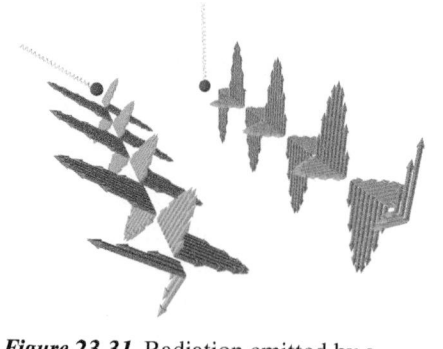

Figure 23.32 The electromagnetic spectrum: wavelength and frequency (logarithmic scale).

23.5 ENERGY AND MOMENTUM IN RADIATION

In Chapter 8 we saw that according to the particle model of light, photons have both energy and momentum, which they can transfer to matter. In the classical model that we are now studying, it is also the case that electromagnetic radiation carries both momentum and energy, and can impart both energy and momentum to matter.

Energy in Electromagnetic Radiation

How would a pulse of electromagnetic radiation interact with ordinary matter? Since electromagnetic radiation is composed of electric and magnetic fields, we can analyze its effects straightforwardly, in terms of the effects of electric and magnetic fields on charged particles. We can see the main effects by examining what would happen to a single charge exposed to the pulse of electromagnetic radiation. We'll hang a small ball with a positive charge $+q$ from a spring, so that the ball can move up and down in the direction of the electric field, and we'll watch the ball to see what happens when the pulse of radiation goes by (Figure 23.33).

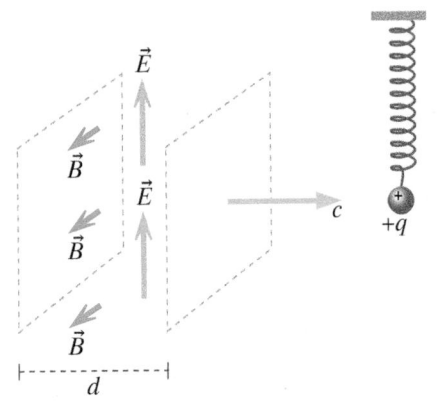

Figure 23.33 A hanging charged ball is affected by electromagnetic radiation.

A Charged Ball on a Spring

Of course nothing happens until the pulse reaches the ball, at which time there is a short pulse of electric and magnetic fields that act on the ball. If the slab containing the fields has a thickness d, the pulse lasts for a short time d/c (c is the speed of light).

For example, if $d = 30$ cm (1 ft), the pulse duration is only one nanosecond. During that brief time, the charged ball experiences an upward electric force $F = qE$, where E is the magnitude of the electric field in the electromagnetic pulse. This brief impulse $F\Delta t$ gives the ball some upward momentum:

$$\Delta p = p - 0 = F\Delta t = (qE)\left(\frac{d}{c}\right)$$

The interaction is over so quickly that the ball hardly has time to move a significant distance upward during the short duration of the pulse, but because the ball acquires some upward momentum, it will now oscillate up and down on the spring for a while after the pulse has passed by (until mechanical friction damps out the oscillations).

Energy Transfer

Since the kinetic energy of the ball has to come from somewhere, it must be the case that there is energy carried by electromagnetic radiation, and that when the ball is accelerated some of this energy is transferred from the radiation to the charged ball. Since the momentum of the ball is proportional to the magnitude of the electric field in the pulse,

$$\Delta K = K - 0 = \frac{p^2}{2m} = \left(qE\frac{d}{c}\right)^2\left(\frac{1}{2m}\right)$$

assuming the speed of the ball remains low compared to the speed of light during this brief interaction. This implies that the amount of energy in a pulse of electromagnetic radiation is proportional to the square of the electric field:

Radiative energy $\propto E^2$

Conservation of Energy

Since the ball has gained energy, the radiation should have lost an equal amount of energy. Apparently the electric field E should be smaller after the pulse has accelerated the ball, and later in the chapter we will see how this can come about.

Energy Density

We can be quantitative about the amount of energy carried by electromagnetic radiation. Earlier we found that there is an energy density (energy per unit volume) associated with electric and magnetic fields:

ENERGY DENSITY IN ELECTRIC AND MAGNETIC FIELDS

$$\frac{\text{Energy}}{\text{Volume}} = \frac{1}{2}\varepsilon_0 E^2 + \frac{1}{2}\frac{1}{\mu_0}B^2, \quad \text{measured in joules/m}^3$$

For the particular case of electromagnetic radiation, $E = cB$, so

ENERGY DENSITY IN ELECTROMAGNETIC RADIATION

$$\frac{\text{Energy}}{\text{Volume}} = \frac{1}{2}\varepsilon_0 E^2 + \frac{1}{2}\frac{1}{\mu_0}\left(\frac{E}{c}\right)^2 = \frac{1}{2}\varepsilon_0 E^2\left(1 + \frac{1}{\mu_0\varepsilon_0 c^2}\right) = \varepsilon_0 E^2$$

since $\mu_0\varepsilon_0 = 1/c^2$; the electric and magnetic energy densities are equal.

The Poynting Vector

Energy density is related to energy flux in joules per second per square meter, or watts per square meter (in this case, flux denotes a flow of energy through a bounded surface). Consider a cross-sectional area A of an advancing slab of electromagnetic radiation (Figure 23.34). In a time Δt, a volume $A(c\,\Delta t)$ of electric and magnetic fields passes through the area A. The amount of energy passing through area A in the time Δt is $\varepsilon_0 E^2(Ac\,\Delta t)$, and therefore the energy flux is $\varepsilon_0 E^2 c$ in W/m^2. Since $E = cB$, and $\mu_0\varepsilon_0 = 1/c^2$, we can write this:

$$\text{energy flux} = \varepsilon_0 EBc^2 = \frac{1}{\mu_0}EB$$

Since the direction of the energy flow is given by $\vec{E} \times \vec{B}$, and since $\vec{E}$ and $\vec{B}$ are perpendicular to each other, the magnitude $|\vec{E} \times \vec{B}| = EB$. We define the Poynting vector $\vec{S}$, whose magnitude is the rate of energy flux in watts per square meter and whose direction is the direction of propagation of the electromagnetic radiation:

ENERGY FLUX (THE POYNTING VECTOR)

$$\vec{S} = \frac{1}{\mu_0}\vec{E} \times \vec{B}, \quad \text{in W/m}^2$$

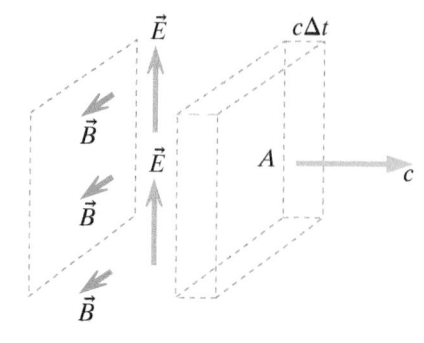

Figure 23.34 The advancing slab of radiation carries an energy flux in W/m^2.

Checkpoint 4 In the vicinity of Earth's orbit around the Sun, the energy intensity of sunlight is about 1400 W/m^2. What is the approximate magnitude of the electric field in sunlight? (What you calculate is actually the root-mean-square or rms magnitude of the electric field, because in sunlight the magnitude of the electric field at a fixed location varies sinusoidally, and the intensity is proportional to E^2.)

Momentum in Electromagnetic Radiation

The classical model of electromagnetic radiation can also account for the transfer of momentum from a pulse of radiation to an object.

In the previous section we saw that the electric field in a pulse of radiation exerted a force on a charged object—starting it moving in a direction perpendicular to the direction of propagation of the pulse. While the main effect of electromagnetic radiation is this sideways ("transverse") kick of the electric field, the magnetic field also has an interesting effect, although it is very small and hard to observe. Consider again the charged ball hanging from the spring (Figure 23.35). As soon as the electric field imparts some upward momentum Δp to the ball, the magnetic field will exert a force on the moving ball, which has a speed v upward.

QUESTION What is the direction of the magnetic force on the ball, which carries a positive charge $+q$?

The direction of $+q\vec{v} \times \vec{B}$ is to the right.

QUESTION Now consider what happens if the ball carries a negative charge $-q$ (Figure 23.36). What is the direction of the magnetic force on the negatively charged ball?

The direction of $-q\vec{v} \times \vec{B}$ is again to the right, because the charge is negative (q is a positive value) and the velocity of the charge is downward. Thus the direction of the magnetic force is the same whether the ball is positively charged or negatively charged. This effect that electromagnetic radiation has on matter is called "radiation pressure." It even acts on neutral matter, because it affects the protons and electrons in the material in the same way. It is difficult to observe because it is a small effect.

We can estimate the magnitude. A charge q acquires a speed v due to the passage of the electromagnetic pulse. The average speed is $v/2$, and the average magnetic force is $|q\vec{v} \times \vec{B}| = q(v/2)B\sin(90°) = q(v/2)(E/c)$ since $B = E/c$ in the slab of radiation. Thus the ratio of the average magnetic force to the electric force is this:

$$\frac{\overline{F}_m}{F_e} = \frac{q(v/2)(E/c)}{qE} = \frac{1}{2}\frac{v}{c}$$

Normally the speed v acquired due to the passage of an electromagnetic pulse is much less than the speed of light, so the effect of the magnetic field is very small compared to the effect of the electric field.

Electromagnetic Radiation Carries Momentum

A charged particle exposed to sinusoidal electromagnetic radiation (Figure 23.37) would gain no net momentum in a direction perpendicular to the direction of propagation of the radiation, since it would alternately be accelerated in opposite directions by the alternating electric field, and the average electric force would be zero. However, the average magnetic force would not be zero, so the particle's momentum in the direction of $\vec{v}$ would increase.

Recall that relativistic momentum p and energy E are related to each other for a particle of mass m by the equation $E^2 = (pc)^2 + (mc^2)^2$ (here E is energy, not electric field!). In the quantum view of light as photons, light can be considered to have particle-like properties with zero mass, in which case $E = pc$. This relation between energy E and momentum p is also

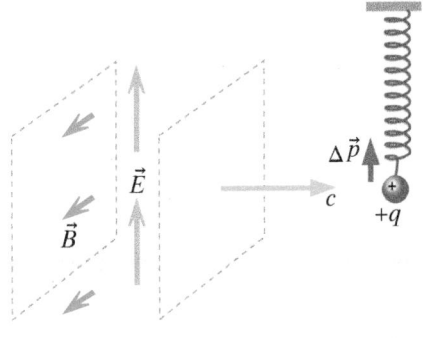

Figure 23.35 What is the direction of the magnetic force on the positively charged ball?

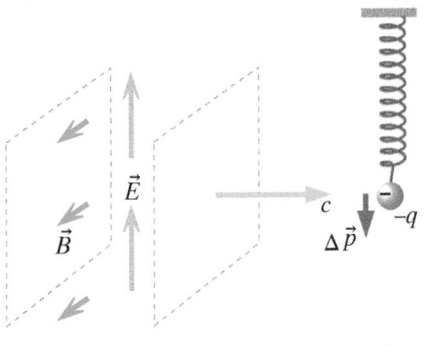

Figure 23.36 What is the direction of the magnetic force on the negatively charged ball?

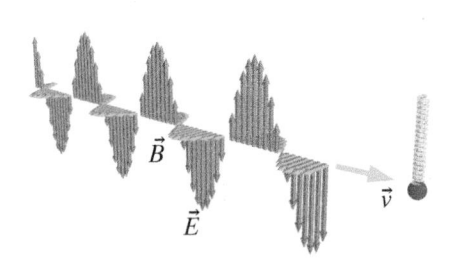

Figure 23.37 Sinusoidal radiation incident upon a charged particle.

valid in Maxwell's classical theory of light, and the energy and momentum carried by electromagnetic radiation are both proportional to the square of the electric field. The momentum flux is $1/c$ times the energy flux given by the Poynting vector:

MOMENTUM FLUX

$$\frac{\vec{S}}{c} = \frac{1}{\mu_0 c}\vec{E} \times \vec{B}, \quad \text{in N/m}^2$$

The units of the Poynting vector are W/m^2, and dividing by c means that the momentum flux is in $W \cdot s/m^3$, which is J/m^3, or $N \cdot m/m^3$, which is N/m^2. Hence the momentum flux has dimensions of pressure.

Conservation of Momentum

If electromagnetic radiation imparts momentum to a particle, the radiation should lose an equal amount of momentum. Again, we see that the electric and magnetic fields should be smaller in magnitude after an interaction with matter. In a moment we will see how this can come about through re-radiation.

Radiation Pressure

When electromagnetic radiation strikes neutral matter, both the protons and the electrons experience a force in the direction of propagation of the radiation (radiation pressure). Despite the fact that radiation pressure is a small effect, there have been serious proposals to build spaceships with huge solar sails, many kilometers in diameter, that could move around the solar system thanks to the radiation pressure of sunlight on the sails. The effect is doubled if the sails are highly reflective, and the plans call for sails made of very thin aluminized plastic.

> **Checkpoint 5** Since force is dp/dt, the force due to radiation pressure reflected off of a solar sail can be calculated as 2 times the radiative momentum striking the sail per second. In the vicinity of Earth's orbit around the Sun, the energy intensity of sunlight is about 1400 W/m^2. What is the approximate magnitude of the pressure (force per square meter) on the sail? For comparison, atmospheric pressure is about 1×10^5 N/m^2.

Re-radiation or "Scattering"

Consider once again the charged ball hanging from the spring, exposed to electromagnetic radiation and therefore oscillating at the radiation frequency (Figure 23.37).

Wait a minute! Won't the ball radiate? After all, the oscillating ball is a sinusoidally moving (and sinusoidally accelerating) charged object, so shouldn't it emit electromagnetic radiation? Yes!

The oscillating ball will re-radiate in (nearly) all directions (see Figure 23.19). The re-radiation by the charged ball redistributes the energy in new directions. This re-radiation is often called "scattering" because part of the original energy moving to the right has been "scattered" into other directions. Figure 23.38 illustrates the effect of the superposition of incident radiation and re-radiation by the charged ball.

QUESTION Explain why scattering (re-radiation) is never observed in the direction of or directly opposite to the direction of the original transverse electric field (that is, up or down in Figure 23.38).

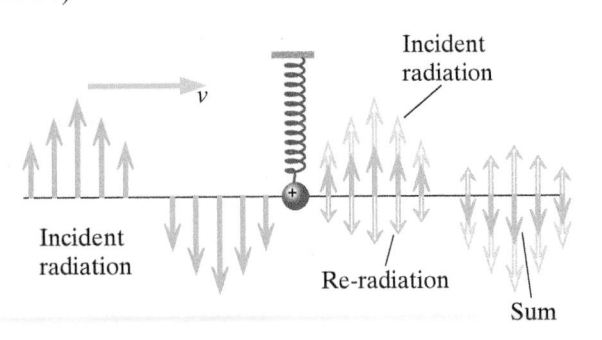

Figure 23.38 Re-radiation by the accelerated charged ball. The accelerated charge re-radiates in almost all directions; for clarity only re-radiation in the $+x$ direction has been shown in this diagram. The outgoing radiation is the superposition of the incident radiation and the re-radiated fields. E outgoing is smaller than E incoming because some energy has been scattered in other directions.

Straight above or below the accelerated charge, the projected acceleration $\vec{a}_\perp$ is zero, so there is no re-radiation in those directions.

We saw earlier that incoming radiation loses energy by giving kinetic energy to a charged ball, proportional to E^2. The re-radiation by the accelerated charge provides a mechanism for decreasing the energy of the original radiation. To the right of the charged ball, the outgoing electromagnetic radiation is the superposition of the original fields plus the fields radiated by the accelerated ball. If an incoming electric field points upward, it accelerates a positively charged ball upward, and to the right of the ball the re-radiated electric field points downward.

As a result, the net outgoing electric field to the right of the ball in Figure 23.38 is smaller than the original incoming electric field. The energy in the electromagnetic radiation, which is proportional to E^2, is smaller after giving kinetic energy to the ball.

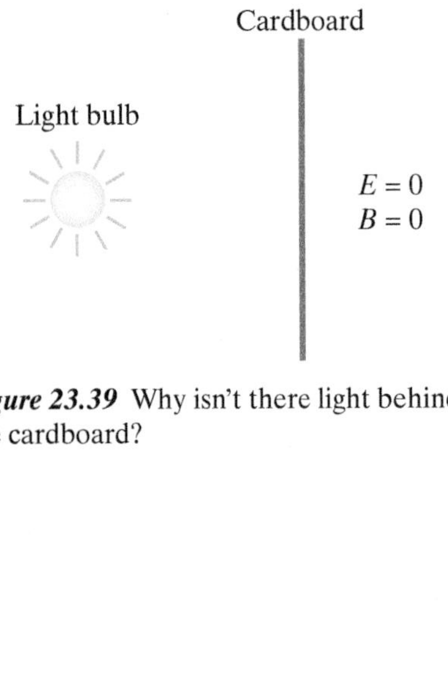

> **Checkpoint 6** Why doesn't light go through a piece of cardboard (Figure 23.39)? According to the superposition principle, the electric and magnetic fields radiated by the accelerated charges in the light bulb are definitely present to the right of the cardboard, so why isn't there any light there? What must be true of the light re-radiated by charges in the cardboard?

Cardboard

Light bulb

$E = 0$
$B = 0$

Figure 23.39 Why isn't there light behind the cardboard?

23.6 EFFECTS OF RADIATION ON MATTER

Because radiation pressure is such a small effect, the most important aspect of the interaction between electromagnetic radiation and matter is the effect that the electric field has on the charged particles inside atoms and molecules. For the rest of the chapter, we will concentrate on the electric field in electromagnetic radiation, and neglect the magnetic field. Of course the magnetic field must be present in electromagnetic radiation, but usually only the electric field need be considered when we try to understand the main effects of electromagnetic radiation on matter, in terms of our model of electromagnetic radiation.

In the following sections we will consider a wide range of important and interesting phenomena in terms of what we have learned about how to produce electromagnetic radiation consisting of propagating electric and

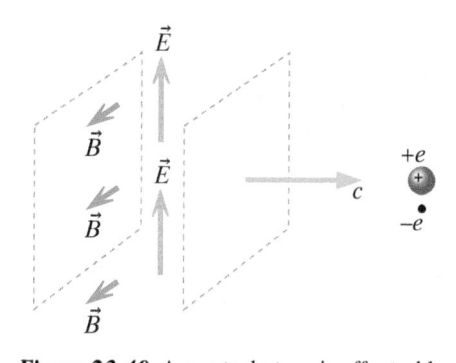

Figure 23.40 A neutral atom is affected by electromagnetic radiation.

Figure 23.41 A high-frequency radio transmitter and receiver.

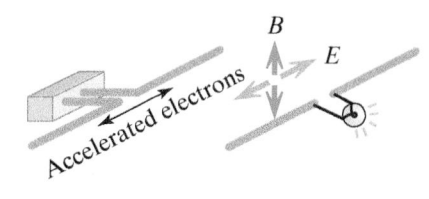

Figure 23.42 The radiative electric field drives current through the bulb.

Figure 23.43 When the distance is greater, the bulb is dimmer.

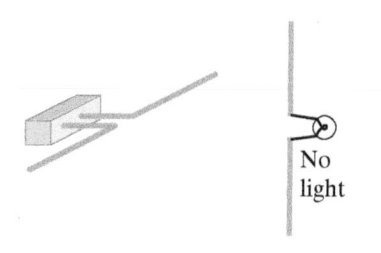

Figure 23.44 With the receiving antenna vertical, the bulb doesn't light.

magnetic fields, and what we know about the effects that electric and magnetic fields have on the charged particles in matter.

The Effect of Radiation on a Neutral Atom

The main effect that a pulse of radiation has on matter is that it gives a brief electric kick to charged objects. This kick is sideways or transverse to the direction of propagation of the fields. What would a pulse of electromagnetic radiation do to a neutral atom or molecule? A simple model of an atom or molecule is adequate to get the main idea. Consider a positive core (nucleus plus tightly bound inner electrons) and a single, loosely bound outer electron (Figure 23.40).

> QUESTION What happens to this atom when the pulse of electromagnetic radiation goes by? Which part of the atom picks up more speed: the electron or the positive core?

The electric field polarizes the atom. During this process the electron moves farther than the much more massive ion (nucleus plus inner electrons); the magnitude of the force eE is the same on both objects. For many purposes the interaction of electromagnetic radiation with an atom can be analyzed by modeling an outer electron as being connected to the rest of the atom by a spring; see Problem P62 in Chapter 15.

According to our model of electromagnetic radiation, this is the basic mechanism for the interaction of light and matter: the electric field of the light exerts forces that move charged particles inside atoms and molecules. Although electromagnetic radiation is rather unfamiliar compared to topics we studied earlier, there is nothing unfamiliar about the effect of the associated electric field on charged particles, which is simply $\vec{F} = q\vec{E}$.

Polarizers

Figure 23.41 illustrates a demonstration of electromagnetic radiation that is sometimes presented in physics lectures. A radio transmitter consists of a high-frequency AC voltage supply (frequency around 300 MHz) connected to a transmitting antenna in the form of two metal rods, and a radio receiver is made of two metal rods connected by a flashlight bulb. When positioned as shown, the bulb lights although it is not connected to the circuit!

The high-frequency AC voltage drives the mobile electrons back and forth in the transmitting antenna, and the sinusoidal motion of these electrons implies a sinusoidal acceleration, which produces electromagnetic radiation (Figure 23.42). For the radiation that propagates to the right, the oscillating electric field is horizontal, and it can accelerate the mobile electrons in the receiving antenna, driving them back and forth through the flashlight bulb, and lighting the bulb.

If you move the receiver and transmitter farther apart (Figure 23.43), the bulb gets dimmer because of the $1/r$ fall-off of the magnitude of the radiative electric field.

If you turn the receiving antenna vertical, the bulb does not light (Figure 23.44).

> QUESTION Explain why the bulb does not light when the receiving antenna is vertical.

The emitted radiation is polarized along the axis of the transmitting antenna, but the radiated electric field is perpendicular to the receiving antenna. The electrons in the antenna are accelerated back and forth across the width of the wire (not its length), so the current does not go through the bulb.

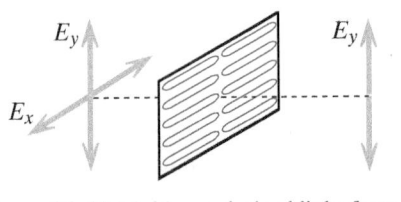

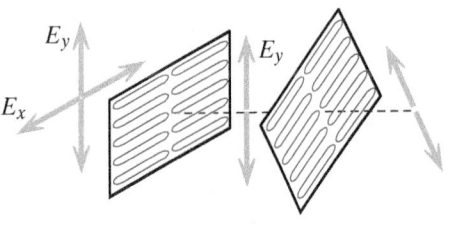

Figure 23.45 Making polarized light from unpolarized light. The incident beam contains unpolarized sinusoidal radiation consisting of waves with electric fields in all directions perpendicular to the direction of propagation.

Figure 23.46 A second polarizer rotates the polarization direction.

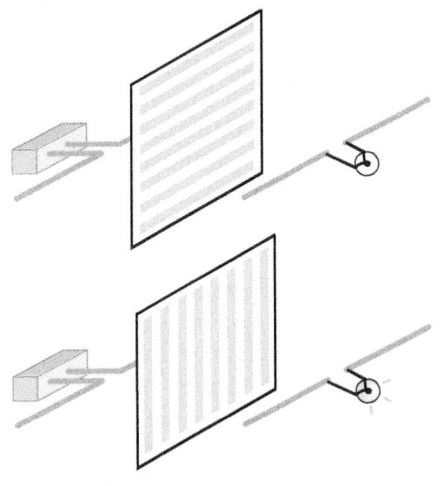

Figure 23.47 A polarizer for radio waves.

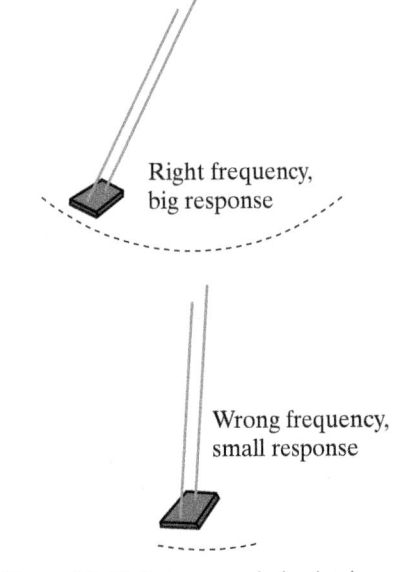

Right frequency, big response

Wrong frequency, small response

Figure 23.48 Resonance behavior in pushing a swing.

You can produce polarized visible light by passing unpolarized light through a sheet of a special plastic whose long molecules are aligned with each other, such as the material in Polaroid® sunglasses (Figure 23.45). Light that is polarized in the same direction as these long molecules is able to drive electrons along the length of the molecule and lose energy to this relatively large-scale motion. Light that is polarized perpendicular to these long molecules can't move the electrons very far and consequently loses much less energy in the plastic. For light that is polarized at some other angle only the component of electric field that is perpendicular to the long molecules manages to show up after the polarizer. The net effect is that polarized light is produced.

If you place a second polarizer after the first one, its orientation determines whether all, some, or none of the polarized light is seen after the polarizer, and if any component of the light is seen after the second polarizer it is polarized in alignment with the second polarizer (Figure 23.46).

QUESTION Polaroid® sunglasses are useful at the beach because unpolarized sunlight that glances off the water happens to become polarized in the process, with the electric field horizontal. In order to block this glare but let other (unpolarized) aspects of the scene come through, should the long molecules in the plastic lenses be horizontal or vertical?

Since we want to diminish the horizontally polarized radiation (glare), the molecules should also be horizontal. Radiation with a horizontal electric field will be absorbed and re-radiated, with only a fraction seen in the original direction. Vertically polarized light will be seen unaffected.

You can observe these effects with radio waves, too. For a polarizer, use a piece of cardboard that has a series of parallel metal foils taped to the cardboard. When the foils are aligned with the antennas, the bulb does not glow; but when the foils are aligned perpendicular to the antennas, the bulb does glow (Figure 23.47).

Checkpoint 7 Briefly explain the observations in the previous paragraph (Figure 23.47). (*Hint:* Think about the re-radiation that may occur from the wires.)

Effects that Depend on Frequency

Let us again suspend a charged ball from a spring and watch what happens as this sinusoidally varying electromagnetic radiation goes by (Figure 23.37). In the vertical direction the electric force on the ball varies like $qE_{max} \sin{(\omega t)}$, and this forces the ball to oscillate up and down at the same angular frequency ω. How large an oscillation we get depends on how close ω is to the angular frequency at which the spring–ball system would freely oscillate in the absence of the radiation. If the radiation frequency nearly matches the free-oscillation frequency, the ball will oscillate up and down with a large motion. If on the other hand the radiation frequency is very different from the free-oscillation frequency, the ball will move up and down very little, even if the radiative electric field is quite large.

Resonance

This phenomenon is called "resonance." It is easy to observe resonance when you push someone on a swing (Figure 23.48). If you give little pushes at exactly the same frequency as the swing's free-oscillation frequency (for example, every time the swing comes all the way back), even these little pushes add up to produce a very large motion, and very small pushes are sufficient to maintain this large motion. However, if you push at a frequency different from the free-oscillation frequency, the swing is forced to oscillate at your frequency but will hardly move at all, because your inputs don't reinforce each other at the right times.

There is an extremely important consequence of the fact that systems respond strongly to sinusoidal radiation only if the frequency of the radiation closely matches the frequency at which these systems would oscillate freely: Different frequencies of sinusoidal electromagnetic radiation affect matter differently. For example:

- Sinusoidal electromagnetic radiation with a very high frequency, $f = \omega/2\pi$ of about 1×10^{15} Hz, has a big effect on the organic molecules in your retina, and you detect "visible light." However, a radio does not respond to visible light.
- Sinusoidal electromagnetic radiation with a frequency of about 1×10^6 hertz (1000 kHz) has a big effect on the mobile electrons in the metal of a radio antenna, and the radio receiver detects "radio waves." You can tune a radio to be sensitive to electromagnetic radiation at 1020 kHz and be insensitive to all other radio frequencies, so that you hear just one station at a time. However, your eye does not detect radio waves.
- The very high frequency of the sinusoidal electromagnetic radiation known as x-rays has very little effect on the atoms of your body and so mostly passes right through with no interaction; there is just enough interaction that shadows of the denser parts of your body appear on film sensitive to x-rays.

For many purposes involving the classical interaction of light and matter, a neutral atom can be modeled as though an electron in the atom acts like a mass on a spring, whose free-oscillation frequency determines the resonance response to electromagnetic radiation. See Problem P62 in Chapter 15.

Color Vision

In your retina there are three kinds of organic molecules, which we might call R, G, and B. Type R responds strongly to electromagnetic radiation near 560 nm (0.56×10^{-6} m, or 5.4×10^{14} Hz); type G responds strongly to electromagnetic radiation near 530 nm (0.53×10^{-6} m, or 5.7×10^{14} Hz); and type B responds strongly to electromagnetic radiation near 420 nm (0.42×10^{-6} m, or 7.1×10^{14} Hz). Each of these molecules responds less or not at all to the wavelengths favored by the other molecules (Figure 23.49). Your brain interprets signals from R molecules as indicating red light; signals from G molecules indicate green light; and signals from B molecules indicate blue light.

Combinations of these responses are interpreted as all of the other colors that you can perceive, many of which are not in the rainbow. For example, a mixture of electromagnetic radiation at 700 nm (red) and at 400 nm (blue) is interpreted by the brain as magenta. In a rainbow, red and blue are separated from each other, so magenta doesn't occur. Equal stimulation of the R, G, and B molecules is interpreted by the brain as white light, another color that is not in the rainbow.

There is some overlap in wavelengths in the sensitivity of the R, G, and B receptors. For example, light containing the single wavelength of 588 nm (the yellow light of burning sodium) stimulates both the R and the G receptors, so you also perceive yellow with a suitable mixture of two wavelengths, one red and one green. Color television and computer screens use closely spaced red and green dots to fool the brain into seeing a yellow dot.

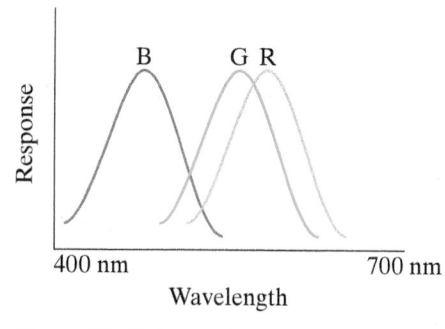

Figure 23.49 Approximate response of light-sensitive molecules in the human eye to visible light as a function of wavelength.

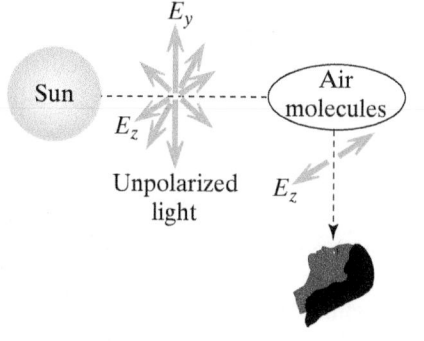

Figure 23.50 Looking up at the sky—why is there any light? (Unpolarized light is indicated schematically here.)

Why the Sky Is Blue (and Polarized)

Why is the light we see coming from the sky blue (and polarized)? Suppose that you look straight up at the blue sky, in the early morning or late afternoon, so that the Sun is near the horizon (Figure 23.50).

QUESTION First, why do you see any light at all coming from the sky overhead? After all, if you stand on the Moon and look up, the sky is black (except for stars).

The reason you see any light at all coming down from straight overhead on the Earth is that electrons in the air molecules all through the atmosphere above your head are accelerated by the sunlight and re-radiate. The Moon has no atmosphere to scatter the sunlight. Interestingly, it is also critical that the air molecules are moving around randomly, so that in any cubic centimeter of air the number of molecules continually fluctuates. It can be shown that without this fluctuation the re-radiations from the fixed array of molecules would cancel each other out. This is discussed in Supplement S3, in the section "Further Applications of the Wave Model."

Second, the sky light coming from overhead is polarized, despite the fact that light from the Sun is unpolarized due to being emitted by many atoms at random orientations. This polarization is easy to observe for yourself. While wearing Polaroid® sunglasses on a cloudless day, look straight overhead in early morning or late afternoon and turn your head to vary the orientation of the polarizer. You will see a marked change in the intensity of the sky light.

QUESTION Second, explain why the sky light is polarized, and state what the direction of polarization is.

The electric field of light from the Sun is perpendicular to the direction of propagation. The light from the Sun is not polarized, so the electric field can point anywhere in the yz plane (Figure 23.50). However, only charges in air molecules that are accelerated in the $\pm z$ direction will re-radiate in the $-y$ direction, downward toward you, so the sky light is polarized with the electric field in the $\pm z$ direction.

QUESTION Third, why is the sky light blue?

Light from the Sun is a mixture of many different frequencies of sinusoidal electromagnetic radiation. The spectrum of sunlight runs from red to blue, and the mixture is perceived by you as nearly white (but somewhat yellowish). So why does the light from the sky look blue instead of yellow? After all, the electrons are forced to oscillate at the same frequency as the light that hits them.

The electrons in the atom are shaken by the oscillating electric field. If the position of the electron can be represented by $x = A\cos(\omega t)$, then the acceleration is $a = -\omega^2 A\cos(\omega t)$. The magnitude of the re-radiated electric field is proportional to the acceleration, and the power (energy/second) of the re-radiation is proportional to the square of the acceleration. We expect the re-radiation that comes to your eye to be proportional to ω^4. This implies that re-radiation of the high-frequency blue component of sunlight will be much greater than the re-radiation of the low-frequency red component of sunlight. The ratio of blue to red frequencies is nearly a factor of 2, so the ratio of re-radiated intensities is about a factor of $2^4 = 16$.

Why Distant Mountains Appear Bluish

On a very clear day, with little dust or haze in the air, distant mountains nevertheless look bluish. An indication that there is little haze is the fact that distant mountain ridges appear with crisp and sharp lines, not blurred as they would look in haze or fog. The explanation is essentially the same as the explanation of why the sky is blue! When you look up at the sky and see a brilliant blue color, you are looking through the equivalent of only a few kilometers of sea-level air (the atmosphere extends out to many tens of kilometers, but the air has exponentially decreasing density as you go higher).

When you look at a distant mountain through many kilometers of air, sunlight coming from the side scatters off that column of air into your eyes, adding a bluish light to the image of the mountain. You need a dark mountain as a background in order to be able to notice this relatively low-intensity bluish light.

23.7 LIGHT PROPAGATION THROUGH A MEDIUM

The propagation of electromagnetic radiation through matter is more complicated than the propagation of radiation through vacuum, because the electric and magnetic fields in the incident radiation affect charged particles in the medium. The affected particles may be accelerated and re-radiate. The net field is the superposition (sum) of the original field and the re-radiated field. In the following sections we will explore the optical consequences of this superposition process.

Wavefronts

In the sinusoidal electromagnetic radiation emitted by an oscillating charge, a *wavefront* is considered to be a set of peaks (amplitude maxima) all emitted at the same time. In Figure 23.51, the circles mark successive wavefronts in the radiation emitted by a sinusoidally oscillating charge. To simplify diagrams, often only the circles representing the wavefront are shown, as in Figure 23.52.

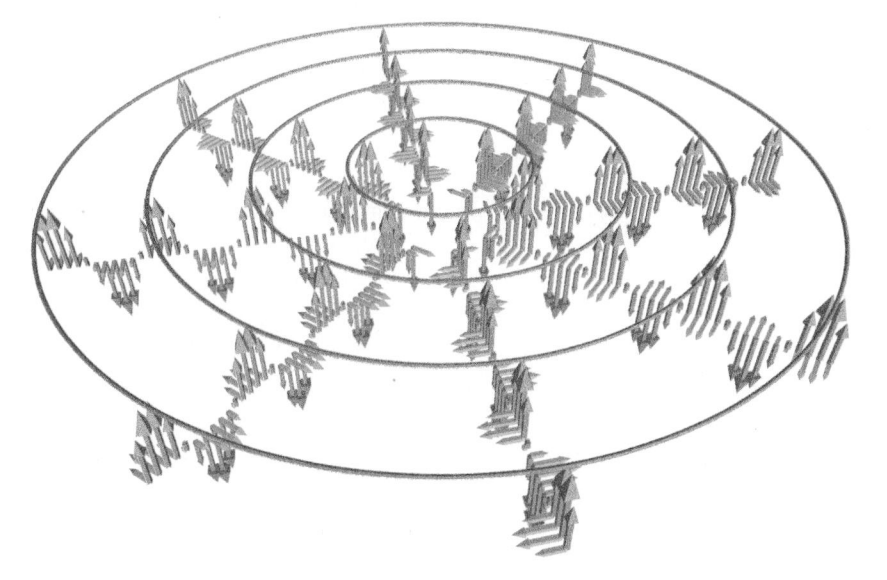

Figure 23.51 The circles mark successive wavefronts in the radiation emitted by a charge at the center (not shown) oscillating along the y axis. For clarity, only radiation emitted in the xz plane is shown (radiation is in fact emitted in all directions except the $\pm y$ direction).

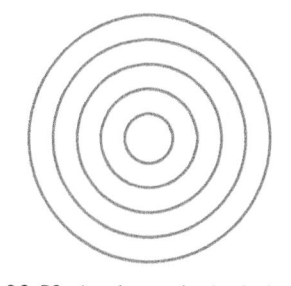

Figure 23.52 A schematic depiction of the wavefronts spreading out from an oscillating source charge.

Figure 23.53 Far from the source, segments of wavefronts appear nearly planar, and parallel to each other.

Plane Waves

The wavefronts emanating from an oscillating charge are nearly spherical, as indicated in Figure 23.52. However, as the wave propagates outward, the radius of curvature increases so that to an observer far from the source (where *far* means a distance large compared to the wavelength of the radiation), successive wavefronts appear nearly planar, and parallel to each other, as indicated in Figure 23.53. In this situation, we describe the wave as a *plane wave*. The plane referred to is the plane defined by the $\vec{E}$ and $\vec{B}$ vectors, as indicated in

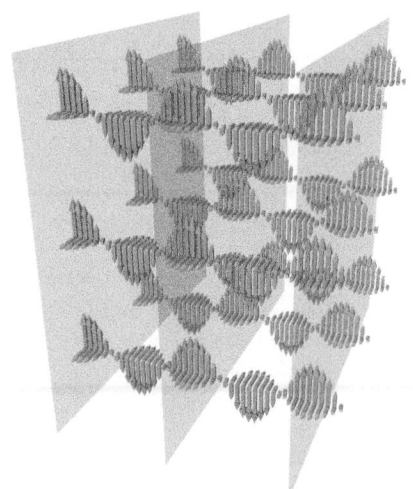

Figure 23.54 A plane wave. The gray planes indicate the location of parallel (or nearly parallel) wavefronts.

Figure 23.55 A line called a *ray* (the red line) is often used to describe a section of a wavefront propagating in a given direction. Rays are drawn perpendicular to wavefronts.

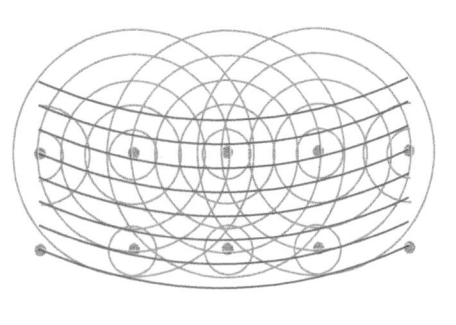

Figure 23.56 Gray dots represent atoms in a solid medium. The blue curves represent wavefronts in the incident radiation. The light red circles represent wavefronts in the re-radiation from charges in the atoms that were accelerated by the incident radiation. The net field at any location is the sum of all fields at that location.

Figure 23.54; this plane is perpendicular to the direction of propagation of the wave.

The radiation traveling in a particular direction of propagation is often described as a *ray* and represented by a single line in the direction of propagation, perpendicular to wavefronts (Figure 23.55). Although there are not really discrete rays—the radiation emitted by an oscillating charge is spatially continuous—imagining particular rays can be useful in understanding the geometric behavior of light as it travels through objects such as lenses or prisms.

Re-radiation and Superposition

When electromagnetic radiation travels through a region filled with a medium such as glass, air, water, metal, plastic, or anything other than a vacuum, it may interact with charged particles in the medium. As discussed previously, the electric field in the incoming radiation accelerates charged particles in the medium, causing these accelerated charges to emit electromagnetic radiation as well. (The magnitude of the electric force on a proton is the same as that on an electron, but the acceleration of a low-mass electron is much larger, so it is the electrons that contribute almost all of the re-radiation.) Therefore, at any location inside a medium, the net radiative electric and magnetic fields are the superposition of the incident (initial) sinusoidally oscillating fields and the re-radiated oscillating fields. As indicated schematically in Figure 23.56, this sum can be quite complex, especially in a dense medium such as a solid material, which is composed of many charged particles. The behavior of light in a material depends very much on the structure of the material at the atomic level.

One possible result of superposition is that the incident wave and re-radiated waves may superimpose in just such a way that troughs in one wave occur at the same locations as crests in the other wave. In this case, the net amplitude is zero, and no electromagnetic radiation emerges from the other side of the medium. This is called *destructive interference*. In this case the material is *opaque*. Opacity depends on the wavelength of the incident radiation. For example, cardboard can be opaque to visible light, but not to microwave radiation—a useful property if you wish to use a microwave oven to reheat leftover food in a cardboard box.

In a material that is transparent to electromagnetic radiation of a particular wavelength, incident radiation and re-radiation add *constructively*, so the amplitude of the waves inside the material is not zero. However, this addition process can result in changes in the wavelength of the net waves and in the speed at which a particular wave crest travels through the medium.

Index of Refraction

Most commonly, in ordinary solid materials, the incident wave and re-radiated waves combine to create a resultant net wave whose crests travel through the medium at a speed slower than c. Because the frequency f of the re-radiation is the same as that of the incident radiation, the wavelength inside the medium is less than the wavelength in vacuum: $v = \lambda/T = \lambda f$, so $\lambda = v/f$, and smaller v means shorter λ (because f is unchanged).

The *index of refraction* of a medium is defined as the ratio of c, the speed of a wavefront in a vacuum to v, the speed of a wavefront in the medium, both measured by the time for one crest to travel a given distance:

$$n = \frac{c}{v}$$

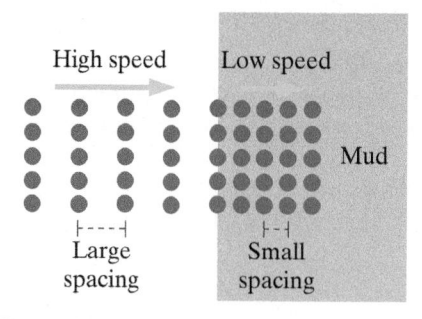

High speed   Low speed

Mud

Large
spacing

Small
spacing

Figure 23.57 When members of a
marching band encounter mud and slow
down, the ranks get closer together.

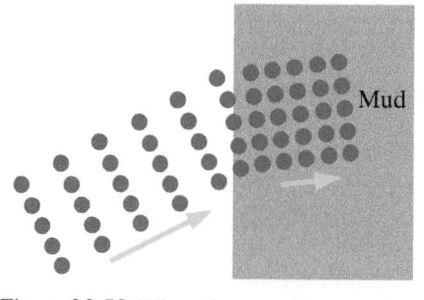

Mud

Figure 23.58 When the marching band
hits the mud at an angle, the direction of
march changes.

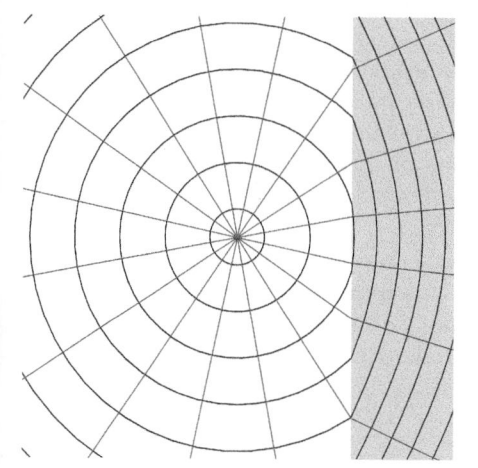

Figure 23.59 Light spreads out in air from
a sinusoidally oscillating point source and
encounters a slab of glass. The red rays
indicate the directions of propagation, and
the curving arcs are the wavefronts,
perpendicular to the rays.

Here are indices of refraction for some typical materials, for a wavelength of
589 nm (yellow sodium light):

Material	n
Vacuum	1.0000
Air	1.00029
Water	1.33
Fused quartz	1.46
Plexiglas	1.51
Zircon	1.92

Checkpoint 8 Light propagates through gallium phosphide at a speed of
8.57×10^7 m/s. What is the index of refraction of gallium phosphide?

23.8 REFRACTION: BENDING OF LIGHT

The fact that the speed of a wavefront is different in different materials leads
to bending of a ray when a wave moves from one medium into another.

A Marching Band Changes Direction

There is a useful analogy that helps explain what happens to wavefronts and
rays of light at the surface between the two materials. Think of a marching band
that is marching quickly across grass. When it comes to a muddy part of the field
it has to slow down. In Figure 23.57 all members of each marching rank reach
the mud at the same time, and the pileup on reaching the mud makes the ranks
(wavefronts) closer together in the mud than on the grass. This is analogous to
the decreased wavelength when light passes from air into glass.

In Figure 23.58 the band approaches the mud at an angle. When the
first band members in a rank encounter the mud and slow down, there are
other marchers in their rank who are still on the grass and still moving quickly.
If the marchers are required to maintain the same spacing within their rank as
before, the ranks (wavefronts) in the mud end up being at a different angle than
the ranks still on grass. The direction of march changes; the outgoing "ray" is
in a different direction than the incoming "ray." We'll look at this phenomenon
more quantitatively in the case of light.

Bending of Light

In Figure 23.59 light from a sinusoidally oscillating point source in air
encounters the surface of a slab of glass (the gray area). The wavefronts,
indicated by the curving arcs, get closer together in the glass, as a result
of the superposition of the fields of the point source and the additional
fields reradiated by accelerated charges in the glass. The wavelength λ of the
radiation is shorter inside the glass than it was in the air. Because the frequency
and period of the oscillation don't change, the speed of a wavefront in the
glass is slower than it was in air (speed = λ/T). The red rays represent
the directions of propagation, which are perpendicular to the wavefronts. The
rays bend inward at the surface between the air and the glass. The bending of
rays of light at an interface between two materials is called "refraction."

Reflection

In addition to the refraction (bending of light), there is also some reflection of
the light in situations like that shown in Figure 23.59. The accelerated electrons
in the glass radiate in all directions and, in addition to making the wavelength
shorter in the glass, they produce light in backward directions. It looks as

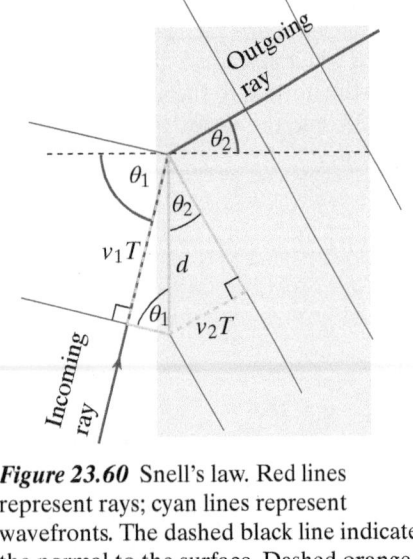

Figure 23.60 Snell's law. Red lines represent rays; cyan lines represent wavefronts. The dashed black line indicates the normal to the surface. Dashed orange lines indicate the distance traveled by a wavefront in time T. The gray medium (glass) has a higher index of refraction.

though this "reflected" light bounces off the surface, but in fact this new light is emitted by all of the accelerated charges throughout the glass. The reflected light may be quite weak; you have probably noticed that reflections off a sheet of glass can be quite dim.

Snell's Law

We can derive a relationship called "Snell's law" between the bending of the light rays at an interface and the speed of wavefronts in the two media. Zoom in close enough that the wavefronts can be approximated by parallel straight lines (Figure 23.60). It is customary to express the directions of the incoming and outgoing rays as angles between the rays and "normals" to the surface (lines that are perpendicular to the surface).

In Figure 23.60 the incoming ray is at an angle θ_1 to the normal and the outgoing ray is at an angle θ_2 to the normal. The same angles occur in the orange triangles, since if the sides of an angle are perpendicular to the sides of another angle, these two angles are either equal to each other or their sum is 180° (in either case their sines will be equal).

The wavefront speed in air is v_1 and the wavefront speed in the glass is v_2. In a time T (the period), a wavefront in air travels a distance $v_1 T$ and a wavefront in glass travels a distance $v_2 T$, indicated by dashed orange lines in the diagram. The two orange triangles share a side of length d.

In the triangle on the left (in air):

$$\sin\theta_1 = \frac{v_1 T}{d}$$

In the triangle on the right (in glass):

$$\sin\theta_2 = \frac{v_2 T}{d}$$

Since $\frac{T}{d}$ is the same for both triangles,

$$\frac{\sin\theta_1}{v_1} = \frac{\sin\theta_2}{v_2}$$

It is customary to express this result in terms of the index of refraction $n = c/v$, which can be written as $v = c/n$:

$$\frac{\sin\theta_1}{c/n_1} = \frac{\sin\theta_2}{c/n_2}$$

$$\frac{n_1 \sin\theta_1}{c} = \frac{n_2 \sin\theta_2}{c}$$

Canceling the common factor c, we obtain Snell's law:

SNELL'S LAW

$$n_1 \sin\theta_1 = n_2 \sin\theta_2$$

where n_1 and n_2 are the indices of refraction of the two media, and θ_1 and θ_2 are the angles the incident and transmitted rays, respectively, make with the normal to the surface.

For concreteness we've talked about air and glass, but this result is valid for any two materials whose indices of refraction are n_1 and n_2.

EXAMPLE   **Ray Travels from Water into Glass**

In Figure 23.61 a beam of light travels through water and hits a glass surface. The angle θ_1 between the incident beam and the normal to the glass surface is 23°. The index of refraction of water is 1.33, and the index of refraction of this type of glass is 1.65. What is the angle θ_2?

Solution   From Snell's law:

$$n_1 \sin\theta_1 = n_2 \sin\theta_2$$

$$\sin\theta_2 = \frac{n_1 \sin\theta_1}{n_2}$$

$$\theta_2 = \arcsin\frac{1.33\sin 23°}{1.65} = 18.4°$$

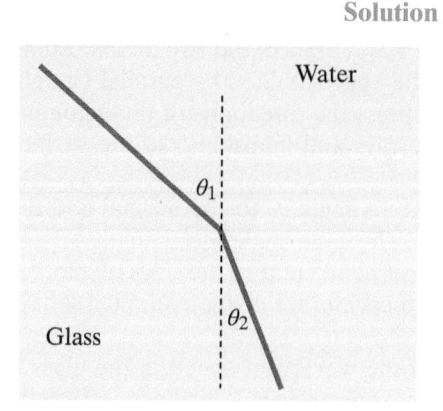

Figure 23.61 A beam of light travels from water into glass.

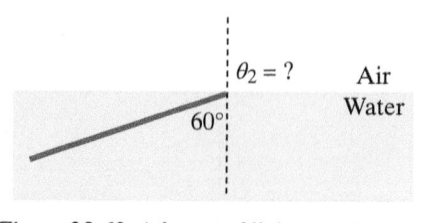

Figure 23.62 A beam of light travels through water to the water–air interface, at an angle of 60° to the normal.

Total Internal Reflection

We've been considering light that passes from air to glass, with a decrease in the wavefront speed associated with an increase in the index of refraction. Something unusual happens in the other direction, going from a higher index of refraction to a lower index of refraction. For example, suppose that light goes from water (with an index of refraction of 1.33) into air, whose index of refraction is very close to 1.0. Consider a ray in the water at an angle of 60° to the normal, as shown in Figure 23.62. Let's use Snell's law to predict the angle the ray will make to the normal in the air:

$$1.33\sin 60° = 1.0\sin\theta_2$$

$$\sin\theta_2 = 1.33\sin 60° = 1.15$$

QUESTION   Try taking the arcsine of 1.15 on your calculator.

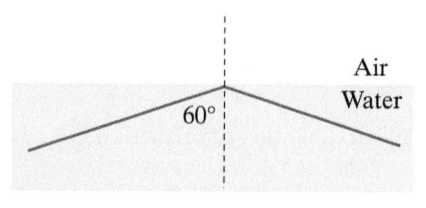

Figure 23.63 Total internal reflection.

What happens? You get an error! The sine function varies between −1 and +1, and there is no angle whose sine is 1.15. Snell's law tells us that there is no possible angle for this ray to make in the air. What happens if a ray of light at an angle of 60° to the vertical tries to emerge from a pool of water into the air above? It doesn't emerge but reflects back down into the water, as indicated in Figure 23.63. This phenomenon is called "total internal reflection."

Figure 23.64 shows rays of light emitted from a source that is underwater, emerging into the air above. The largest possible angle from the vertical (the normal to the surface) for a ray in the air is 90°.

QUESTION   What is the angle to the vertical for a ray in the water that gives a ray in the air at 90° to the vertical? This is a horizontal ray that just skims the surface. (*Hint:* Use Snell's law.)

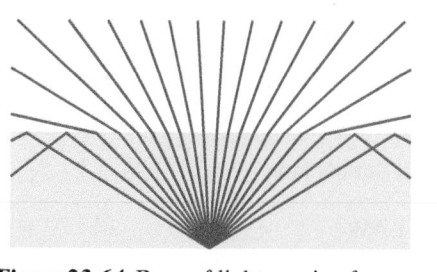

Figure 23.64 Rays of light coming from a source underwater bend outward when passing into the air, where the index of refraction is smaller. Underwater rays whose refracted angle would be greater than 90° cannot escape the water and are reflected downward ("total internal reflection").

Snell's law says that $1.33\sin\theta_1 = 1.0\sin 90°$, so $\sin\theta_1 = 1.0/1.33 = 0.75$, and $\theta_1 = 49°$. Any underwater rays at more than 49° to the vertical cannot escape into the air but are "totally internally reflected" back down into the water, as you can see in Figure 23.64. If you look up at the surface while swimming underwater, at a large angle to the vertical you see bright reflections off the surface due to this effect.

Fiber Optics

An important application of the phenomenon of total internal reflection is the use of "fiber optics" for long-distance high-speed internet connections. It is possible to transmit much more data over long distances by means

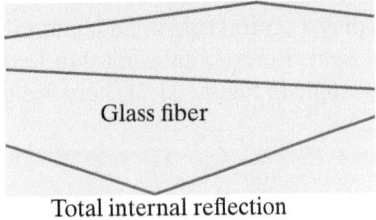

Figure 23.65 In glass fibers used for long-distance high-speed Internet communication, total internal reflection keeps rays from escaping.

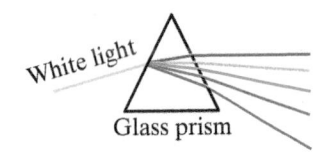

Figure 23.66 Because the index of refraction of some substances depends strongly on wavelength, light of different wavelengths is bent through significantly different angles when passing through a prism.

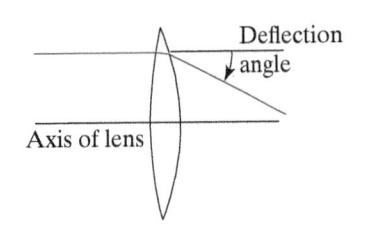

Figure 23.67 Two prisms can be used to make light rays converge or diverge.

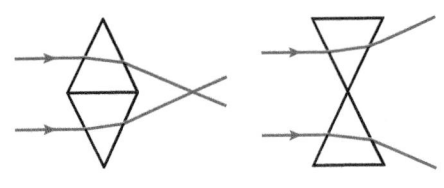

Figure 23.68 A converging lens deflects the incoming ray toward the axis of the lens.

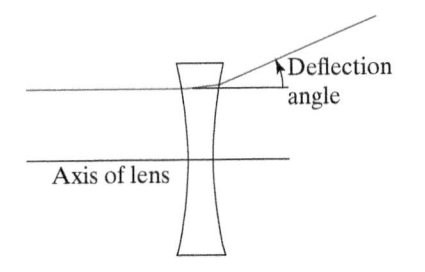

Figure 23.69 A diverging lens deflects the incoming ray away from the axis of the lens.

of light propagating in glass fibers than by means of electrical signals propagating in copper wires, but there are two major design issues. The glass must be extremely pure in order to be extraordinarily transparent, so that signal strength is maintained over long distances. The second issue is that the glass fibers must be designed in such a way that light rays traveling through the glass undergo total internal reflection at the surface of the fiber rather than escaping. This is achieved by using glass with an appropriate index of refraction. In some designs there is a variation in the index of refraction as one approaches the surface. Figure 23.65 shows the basic idea: rays that are not quite along the axis are kept in the glass fiber by total internal reflection.

Checkpoint 9 A certain glass has an index of refraction of 1.6 and is in air. What is the largest angle from the vertical for which a ray originating in the glass can emerge into the air? (For larger angles there will be total internal reflection.)

23.9 LENSES

Prisms: Refraction at Two Surfaces

The index of refraction of most materials is not exactly the same for light of different wavelengths. This variation is due to differences in how light of different wavelengths interacts with the charged particles that compose the material. As a result of the wavelength dependence of index of refraction, a glass prism can be used to show that white light in fact contains a mixture of all colors of light, as shown schematically in Figure 23.66, where a few of the colors in white light are shown. Since the index of refraction depends on the wavelength, light of different wavelengths is bent by different amounts as it passes from air into glass, and again when it passes from glass back into the air.

Two Prisms Are Like a Lens

If two prisms are placed base to base as shown at the left of Figure 23.67, light rays incident on the glass are bent toward the center line: they converge. If instead the two prisms are placed tip to tip as shown at the right of Figure 23.67, incoming rays are bent away from the center line: they diverge. These stacks of prisms, whose function is to bend incoming light rays, are like very simple lenses.

Lenses: Converging and Diverging

Figure 23.68 shows the cross section of a lens made of transparent glass or plastic, whose index of refraction is greater than that of the air. The surfaces are spherical. A line through the center of the lens, perpendicular to the lens, is called the axis of the lens (or the "optic axis" of the lens). An incoming ray is refracted at the first spherical surface, and then again at the second spherical surface. The net effect is that the ray is deflected from its original direction toward the axis of the lens. We say this is a "converging" lens, because incoming rays "converge" toward the axis. Converging lenses are thicker in the middle than at the edge.

Figure 23.69 shows a "diverging" lens. Incoming rays are deflected away from the axis of the lens. Diverging lenses are thinner in the middle than at the edge. Figure 23.70 shows two more examples of lenses, one converging and one diverging.

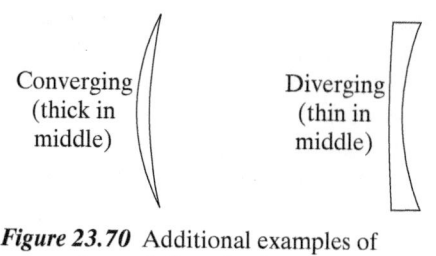

Figure 23.70 Additional examples of converging and diverging lenses.

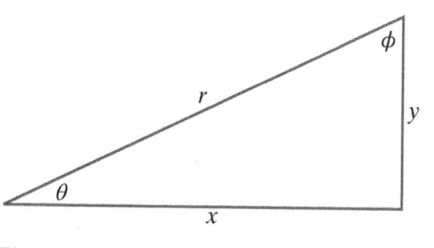

Figure 23.71 A right triangle.

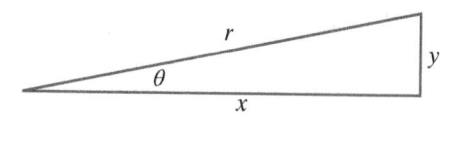

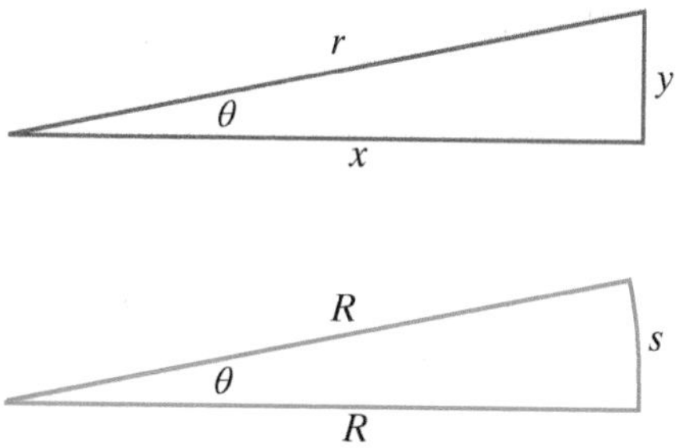

Figure 23.72 A right triangle (red) with a small angle θ, and a sector of a circle (green) with the same small angle θ.

Brief Review of Triangles

Sines, cosines, and tangents of small angles play a central role in understanding the details of how thin lenses bend rays of light. Before analyzing thin lenses we offer a very brief review. For the right triangle in Figure 23.71, here are the key relationships:

$$\sin\theta = \frac{y}{r} \qquad\qquad y = r\sin\theta$$

$$\cos\theta = \frac{x}{r} \qquad\qquad x = r\cos\theta$$

$$\tan\theta = \frac{y}{x} = \frac{\sin\theta}{\cos\theta} \qquad \theta + \phi = 90°$$

The Small-Angle Approximation

Measuring angles in terms of radians plays an important role, especially when dealing with small angles. If an arc of a circle of radius R has a length s, the angle θ measured in radians is defined to be $\theta = s/R$. A full circle has an arc length of the circumference, $s = 2\pi R$, and the angle in radians for a full circle is $s/R = 2\pi$, or 6.28 rad in a circle. To convert radians to degrees, multiply the number of radians by $(360°)/(2\pi \text{ rad})$. For example, the number of degrees in one radian is

$$(1 \text{ rad})\frac{(360°)}{(2\pi \text{ rad})} = \frac{360°}{2\pi} = 57.3°$$

In Figure 23.72 are two figures that look very similar: a right triangle (red) with a small angle θ, and a sector of a circle (green) with the same small angle θ. Evidently $s \approx y$ and $R \approx r$, so the sine, y/r, or tangent, y/x, of a small angle is nearly equal to the angle measured in radians, (s/R):

$$\tan\theta \approx \theta \quad \text{measured in radians}$$

$$\sin\theta \approx \theta \quad \text{measured in radians}$$

$$\cos\theta \approx 1 \quad \text{and} \quad x \approx R \text{ and } x \approx r$$

It is easy to verify the validity of these approximations using your calculator. For example:

> QUESTION 5° is $5(2\pi)/360° = 0.08727$ rad. Calculate $\tan 5°$ and $\sin 5°$.

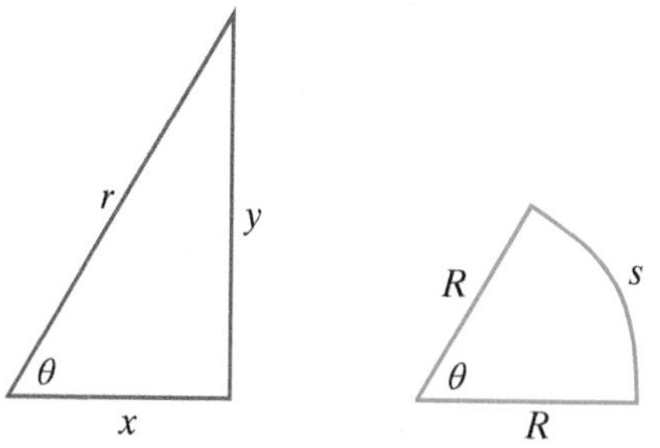

You should find that $\tan 5°$ is 0.08749 and $\sin 5°$ is 0.08716; both of these numbers are approximately equal to 0.08727 rad. This approximation is called the "small-angle" approximation. As you can see, it is a very good approximation for small angles.

In contrast, in the large-angle triangle and sector shown in Figure 23.73, s and y are quite different in length, so $\theta = s/R$ is quite different from $\tan\theta = y/x$. Also, $\cos\theta$ is not approximately equal to 1, so x is not approximately equal to r, nor is $\sin\theta \approx \theta$.

Thin Lenses

If the y position of a light ray does not change very much as the ray travels through the lens, we call the lens a *thin lens*. Figure 23.74 is the output of a computer program that traced two rays through a thin converging lens, using Snell's law at both spherical surfaces. Notice that:

Figure 23.73 A right triangle (red) with a large angle θ, and a sector of a circle (green) with the same large angle θ.

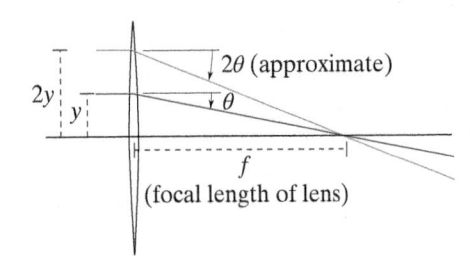

Figure 23.74 A ray that is twice as far from the axis is deflected twice as much (small-angle approximation).

- To a good approximation, the ray that is twice as far from the axis is deflected twice as much: Deflection is proportional to distance from the axis.
- Both rays converge to nearly the same point on the axis, a distance f from the center of the lens.

These results apply to all rays parallel to the axis that pass through thin lenses, not just the two rays computed in Figure 23.74.

The distance f is called the "focal length of the lens" and is a property of the radii of curvature of the two spherical surfaces and of the index of refraction of the material of the lens.

Deflection Angles for Thin Lenses

In Figure 23.74, consider the right triangle whose horizontal base is f and height is y. The angle opposite y is θ, so we have

$$\tan\theta = \frac{y}{f}$$

Now consider a second right triangle with the same base but with height $2y$; the angle opposite $2y$ is approximately 2θ:

$$\tan 2\theta = \frac{2y}{f}$$

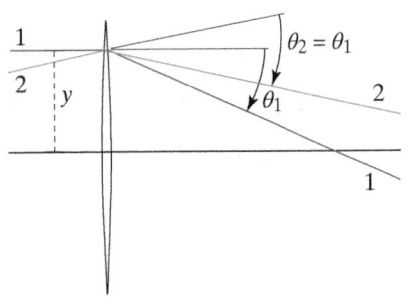

Figure 23.75 In the small-angle thin-lens approximation, all rays that hit the lens at a distance y from the axis are deflected through the same angle $\theta_1 y/f$.

Measuring angles in radians, if θ and 2θ are small:

$$\theta \approx \frac{y}{f}$$

$$2\theta \approx \frac{2y}{f}$$

We see that in the small-angle approximation, the deflection angle is simply $\theta \approx y/f$, where y is the distance from the axis where the ray hits the lens. Moreover, it doesn't matter what the angle of the incoming ray is. In Figure 23.75 two rays approach the same point on the lens but arrive along different angles to the axis. The computer ray tracing using Snell's law shows that the deflection angles are the same for these two rays. The approach angle doesn't affect the deflection angle, which is determined solely by the distance from the axis where the ray hits the lens.

THIN-LENS DEFLECTION ANGLES

For small angles, the deflection angle for any ray passing through a thin lens is y/f, where y is the distance from the axis where the ray hits the lens, and f is the focal length of the lens.

To keep track of signs and directions, one says that a diverging lens has a negative focal length. Then the deflection angle $\theta \approx y/f$ is negative, and a negative deflection angle is interpreted as diverging away from the axis.

The Thin-Lens Equation

In the preceding discussion we noted that:

- A thin lens changes the angle of a ray by an amount proportional to how far from the axis the ray hits the lens.
- The deflection angle is y/f, where f is the focal length of the lens and y is the distance from the axis where the ray hits the lens.

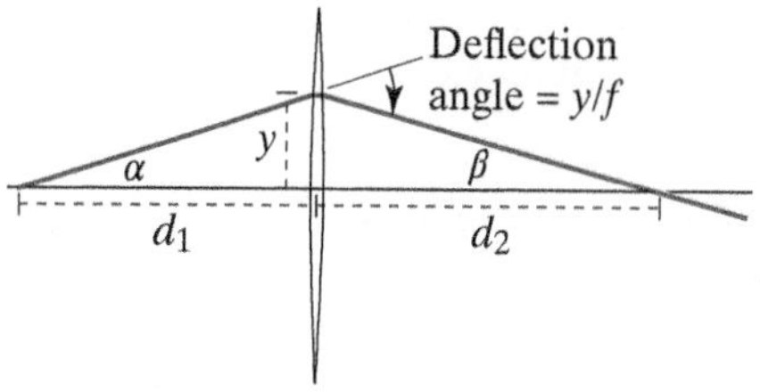

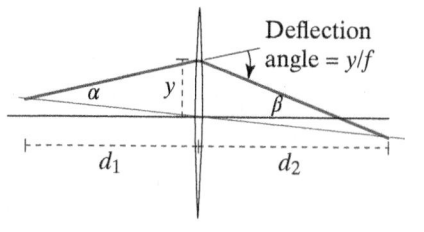

Figure 23.76 Proving the thin-lens equation.

From this basic property we can derive a "lens equation" that will be particularly useful in analyzing images, to be discussed in the next numbered section. In Figure 23.76 we consider a ray that starts from (or goes through) a point on the axis, a distance d_1 to the left of the (thin) lens and is deflected toward the axis, crossing the axis a distance d_2 from the lens. For clarity we have exaggerated the vertical dimension so that some angles look rather large, but in the small-angle approximation we have

$$\alpha \approx \tan \alpha = \frac{y}{d_1}$$

$$\beta \approx \tan \beta = \frac{y}{d_2}$$

The angles of the large triangle whose base is $d_1 + d_2$ add up to 180°, from which we deduce that the deflection angle y/f must be equal to $\alpha + \beta$. Using the results we just obtained for α and β, we have this:

$$\frac{y}{f} = \frac{y}{d_1} + \frac{y}{d_2}$$

Dividing through by y, we obtain the thin-lens equation:

THE THIN-LENS EQUATION

$$\frac{1}{f} = \frac{1}{d_1} + \frac{1}{d_2}$$

When Rays Don't Start from the Axis

If we rotate the rays in Figure 23.76 through a small angle, as in Figure 23.77, the lens equation is still valid (in the small-angle approximation). The deflection angle is still the same, because that depends only on y, the distance from the axis where the ray hits the lens. Also, because the cosine of a small angle is nearly equal to 1, the distances along the axis of the lens hardly change when the source is a little off the axis. The result is that we can use the lens equation to find locations along the axis even for rays that don't start or end on the axis itself, as long as the angles are small.

Figure 23.77 The thin-lens equation is also valid for rays that don't start on the axis.

Mathematical Proof: Two Approximations

We omit a mathematical proof that the deflection angle is in fact approximately equal to y/f, because it is algebraically and geometrically somewhat complicated. There are two approximations that must be made. One is the small-angle approximation. Another is the thin-lens approximation: the lens must be thin compared to its focal length. For clarity, in some of the figures the lenses are shown fatter than a "thin" lens would be.

> **Checkpoint 10** A small light bulb is 0.3 m to the left of a converging lens whose focal length is 0.2 m. How far to the right of the lens will rays from the bulb cross the axis?

23.10   IMAGE FORMATION

In this section we will discuss images that are made by lenses. An image is an *apparent* source of light. It is not a real light source, because there is no actual object at the location of an image.

Real Images

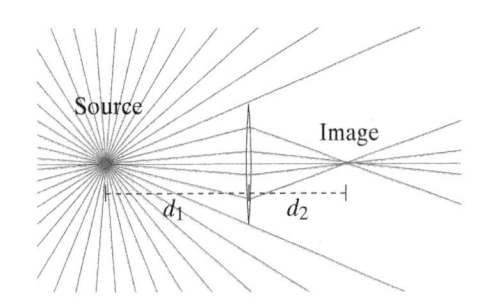

Figure 23.78 Forming a real image.

In Figure 23.78 we see an example of the formation of what is called a "real image" of a point source (which could be a small light bulb). On the left, a small point-like lamp emits light in all directions. Those rays that hit the lens are deflected toward the axis; the other rays simply keep going straight. It is because the deflection angle y/f of a ray is proportional to how far from the axis the ray hits the lens (y) that those rays that hit the lens all pass (nearly) through the same point on the axis, a distance d_2 to the right of the lens.

The thin-lens equation

$$\frac{1}{f} = \frac{1}{d_1} + \frac{1}{d_2}$$

implies that all rays that start at a distance d_1 to the left of the lens will cross the axis at the same distance d_2 to the right of the lens. Of course these rays keep going, and they seem to be coming from this new point. In many ways this point acts just like a real source. If you stand farther to the right and look toward the lens, you will see what looks like a lamp at the new location.

We call this an "image," by which we mean an apparent source of light—apparent because there is no actual object at the location of the image. If you place a sheet of film at the location of the image, or an array of digital light sensors such as you find in a modern digital camera, you'll detect a spot of light at the location of the real image, because rays of light actually hit that spot. This kind of image is called a "real image" in contrast to a "virtual image," which we'll discuss soon.

A real image acts in some ways just like a real source, but there are differences. The real source emits light in all directions, but the image on the right emits light headed only to the right, in a cone whose angular size is determined by the diameter of the lens.

Aberrations: Limitations of the Simple Thin-Lens Model

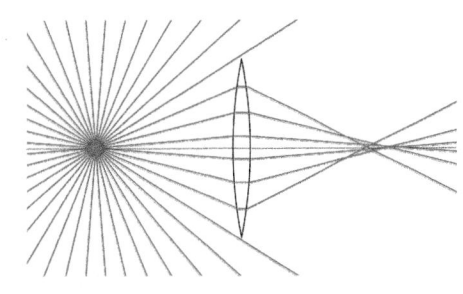

Figure 23.79 If the assumptions of a thin-lens and small angles are not valid, the predictions of the thin-lens model are inadequate. Here we see that not all rays converge to quite the same point. This is "spherical aberration."

If we take a closer look, we can see limitations of the thin-lens model, due to the approximations of a thin lens and small angles. Figure 23.79 is again a computer calculation using Snell's law at the two spherical lens surfaces, with a somewhat thick lens and somewhat large angles. The rays do come to nearly the same point, but not quite. We say that this lens has significant "aberration" in its image formation. Note that it is the largest angles that deviate the most, reflecting the role of the small-angle approximation in our simplified model. It is a challenge to designers of optical systems to reduce aberrations to a minimum. The aberration associated with spherical lens surfaces is called "spherical aberration," and this particular kind of aberration can be minimized by making lens surfaces that are more complicated than spherical ones.

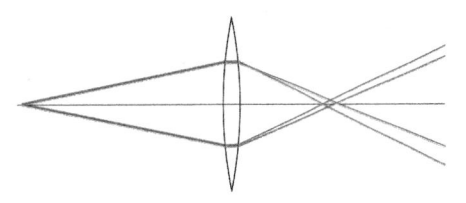

Figure 23.80 Because the index of refraction is typically larger for blue light than red light, a lens bends blue light more than red light, giving rise to "chromatic aberration."

Another kind of aberration is due to the fact that the index of refraction of a lens varies with the wavelength of the light, so that the focal length f of the lens is different for red light and blue light. The index of refraction in glass for blue light is larger than the index for red light, so blue light is bent more than red light. If a source emits a mixture of red and blue light the blue image will be closer to the lens than the red image, as shown in Figure 23.80. A white-light source, which emits a wide range of colors between red and blue, will form images that lie between the red and blue images.

This aberration associated with color is called "chromatic aberration." It can be minimized by combining two lenses made of different kinds of glass (with different indices of refraction), with different focal lengths, which partially compensate for each other's aberrations. Such a lens is called "achromatic" (Greek for "without color"), though the correction isn't perfect. One of the advantages of curved mirrors used in large telescopes is that angles of reflection depend only on the shape of the mirror, not on the color of the light, so reflecting telescopes are truly achromatic.

Virtual Images

In Figure 23.81 we've traced rays through a weaker lens (it has a smaller index of refraction, though still larger than that of air) so that there isn't enough bending of the rays to bring them to a point on the axis to the right of the lens. If we follow these rays back they *seem* to come from a point to the left of the actual source, but they don't really come from there. If you go to that place, you won't find any light headed to the right from there. Contrast this virtual image with the real image in Figure 23.78. There are real rays of light emanating from a real image, and if you go to that place, you'll find those rays of light. If, however, you put detectors at the location of the virtual image in Figure 23.81 you won't detect a spot of light, because there are no actual rays of light converging to that point or emanating from it. In fact, in the situation shown, where the lamp emits light in all directions, your entire detector will be illuminated rather than one point.

This is not to say that virtual images are somehow not important; they are. In many optical instruments from magnifying glasses to microscopes, it is often a virtual image that we observe with our eyes. However, if you follow the rays all the way backward, there is no actual source of light at the point from which the rays seem to emanate.

In Figure 23.82 you can see another example of a virtual image. The rays of light emanating from an object underwater seem to be coming from a location above the actual source of the light. There is a virtual image of the object at this higher location. Your eyes are fooled by the fact that the light rays you see seem to be radiating from the image point rather than from the actual point. This effect can be seen by looking at a coin at the bottom of a pan full of water; the coin seems to be higher than expected.

If you look carefully at Figure 23.82 you can see that the imaging isn't perfect: the rays seem to come from a small region rather than from a point. This is another example of aberrations. Typically, the wider the range of ray angles in an optical system, the worse the aberrations and the more difficult it is to compensate for the aberrations in a more complicated design. In Figure 23.83 you can see that if we include an even wider range of rays the virtual image is even less point-like. Also, you can see that if you look at a coin in a pan of water from off to the side, the bundle of rays that pass through your eye seem to come from a higher location than when you look straight down at the coin. This effect is easily observable.

Images of Extended Objects

So far we've been analyzing the images of single point sources. Next we'll consider the image of an extended object, a short stick that has a red lamp at one end and a blue lamp at the other. Figure 23.84 traces rays from the two ends of the stick.

You see two real images corresponding to the two lamps, one red and one blue. Note that there is an inversion: the upper (blue) source forms the lower image. If you place an array of digital light detectors, like those found in a camera, where the real images are you will get two spots, with the blue spot below the red spot. The inversion is typical of real images. If you have a converging lens available, hold it away from your eye and look through it at a distant object. As long as the distance from the lens to your eye is longer than the focal length of the lens, you'll see an inverted image.

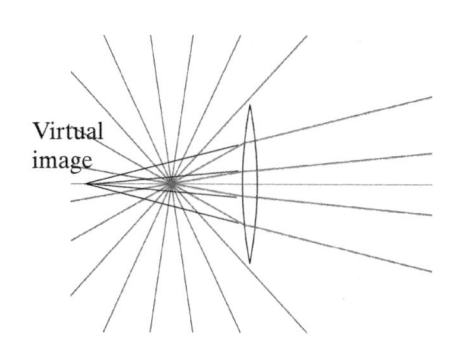

Figure 23.81 Forming a virtual image. To an observer on the right side of the lens, light appears to be coming from the location of the virtual image.

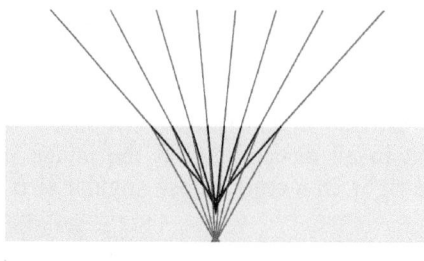

Figure 23.82 Rays emerging from the water seem to be coming from a location above the actual source. This location is a virtual image.

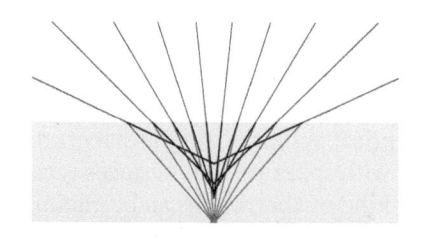

Figure 23.83 The wider the range of angles of the rays, the less point-like the virtual image.

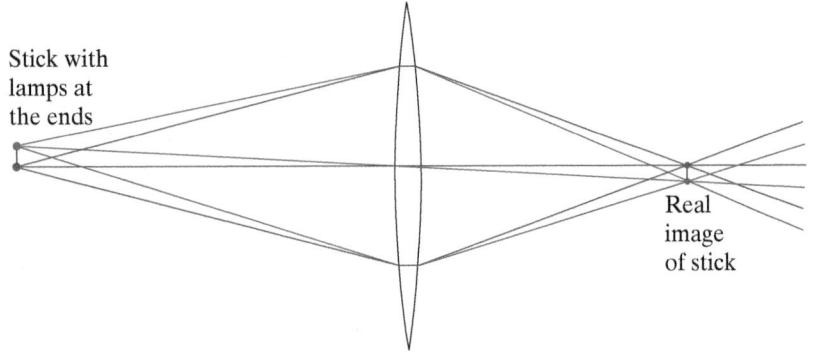

Figure 23.84 Two point sources of light at the end of a stick, a red lamp and a blue lamp, produce two real images (only a few rays are shown for clarity). Note the inversion: the upper, blue source forms the lower image and the lower, red source forms the upper image.

Focal Point

Let us apply the thin-lens equation $1/f = 1/d_1 + 1/d_2$ to the situation of a source far from the lens.

> QUESTION If a source is very far away, rays hit the lens nearly parallel to the axis. In this case (source is very far away), approximately how big is the term $1/d_1$?

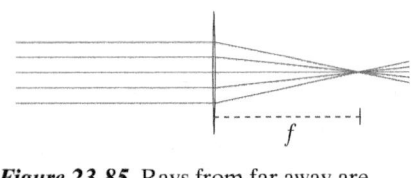

Figure 23.85 Rays from far away are nearly parallel and pass through the focal point.

If d_1 is very large, $1/d_1$ is nearly zero, and the lens equation reduces to $1/f = 1/d_2$, so $d_2 = f$. Rays parallel to the axis, or nearly parallel when coming from a very distant source, are deflected in such a way as to pass through a point a distance f beyond a converging lens; this point is called a focal point of the lens (Figure 23.85).

DEFINITION OF FOCAL POINT

> The two points a distance f along the axis from the lens are called the "focal points" of a lens. They are the points on the axis through which initially parallel rays pass after being deflected by the lens.

Ray Tracing: Determining Image Locations Graphically

Since the red lamp in Figure 23.84 lies on the axis, we know that its image will also lie on the axis—but what about the blue image?

We'll show how to determine the location of the blue image using graphical analysis. If we trace the paths of two or more different rays as they leave the source and pass through the lens, these deflected rays will intersect at the location of the image. Of the many rays we might draw in a diagram, there are two rays that are particularly easy to draw.

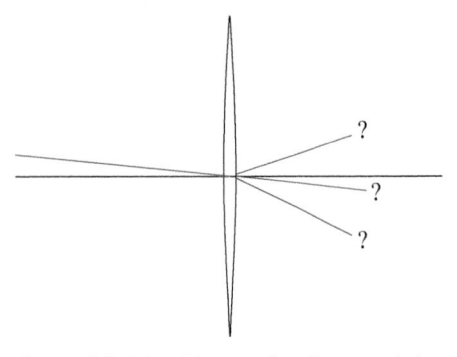

Figure 23.86 A blue ray heads toward the center of a lens. How will it be deflected?

(a) First, draw a ray from the blue lamp toward the center of the lens (Figure 23.86).

> QUESTION How much will this ray be deflected in passing through the lens? How do you know?

Since the deflection angle for a ray that hits the lens at a perpendicular distance y from the axis is y/f, and since $y = 0$ for the ray you drew, the deflection will be zero. In the small-angle, thin-lens approximation, a ray that hits the center of a lens goes straight through without deflection, no matter what the angle of the ray is.

(b) Second, draw a ray that starts out parallel to the axis.

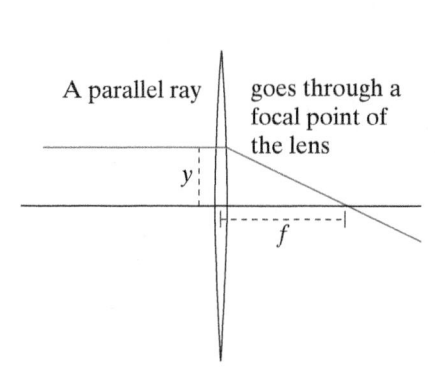

Figure 23.87 A parallel ray is deflected to cross the axis at a distance *f* from the lens. This is one of the two "focal points" of the lens.

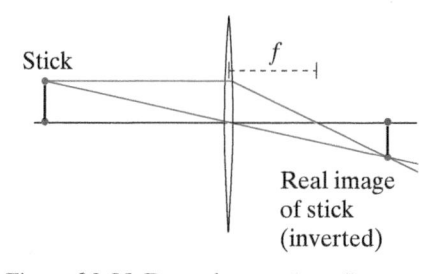

Figure 23.88 Draw the two "easy" rays from the blue lamp through the lens. The intersection is the location of the real image of the blue lamp.

The path of this ray is also easy to determine. A ray that starts out parallel to the axis is deflected by a converging lens in such a way as to pass through a point on the axis to the right of the lens that is a distance *f* from the lens. This is shown in Figure 23.87, the key point being that the deflection of the parallel ray is y/f (small-angle approximation). The triangle whose base is *f* and height is *y* has an angle whose tangent (and, approximately, the angle in radians) is y/f, equal to the deflection angle.

The two rays that we drew intersect at the image location (Figure 23.88). We drew only two of the rays that leave the blue lamp. All other rays from the lamp that go through the lens will also go through the location of the image of the lamp, in the thin-lens, small-angle approximation. To the extent that these approximations are not quite valid, the many rays will pass nearly but not exactly through the same point, and the image of the lamp won't be a nice clean, compact point; it will be somewhat smeared out.

Determining Image Locations Algebraically

How can we calculate the locations of the red and blue real images quantitatively? Use the lens equation, $1/f = 1/d_1 + 1/d_2$, to find the distance along the axis of these images. The lens equation is based on small-angle and thin-lens approximations, where all angles are assumed to be small compared to 1 rad. For that reason, the height of the blue lamp above the axis of the lens doesn't affect our use of the lens equation; we simply use the distances along the axis for d_1 and d_2. If we have a lens with known focal length *f*, and the red and blue lamps are a distance d_1 from the lens, we can calculate how far along the axis the images are from the lens: $1/d_2 = 1/f - 1/d_1$.

You know that all the rays from the blue lamp that hit the lens get deflected in such a way as to pass through the (real) image. One of those rays on its way to the image passes through the center of the lens, without deflection. You also know the distance along the axis d_2 of the blue image. This is enough information to determine the location of the blue image.

EXAMPLE   **Using Algebra to Locate an Image**

Suppose that a converging lens is located at the origin $\langle 0,0,0 \rangle$ and has a focal length of 12 cm, with its axis pointing along the *x* axis. There is a stick 12.5 cm to the left of the lens. At one end of the stick is a red lamp, located at $\langle -12.5,0,0 \rangle$ cm, and at the other end is a blue lamp 1 cm above the red lamp, at $\langle -12.5,1,0 \rangle$ cm (Figure 23.89). Find the locations of the images of the red and blue lamps; the complete image of the stick will lie between the images of the two lamps that are at the ends of the stick.

Solution   For the *x* coordinate of the image, use the lens equation:

$$\frac{1}{12} = \frac{1}{12.5} + \frac{1}{d_2}$$
$$d_2 = 300 \, \text{cm}$$

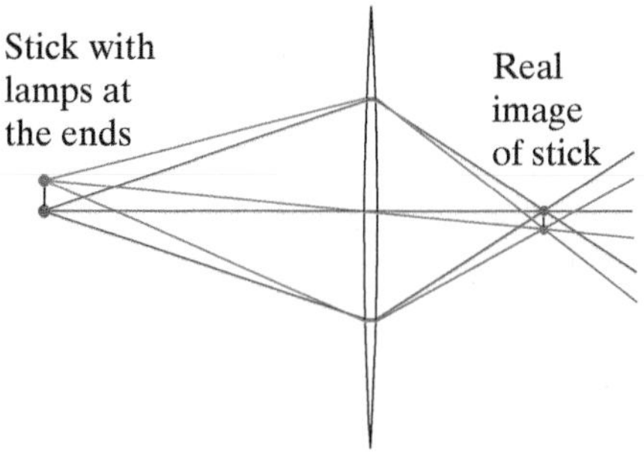

For the *y* coordinate of the image, we already know that the image of the red lamp lies on the *x* axis. The image location of the red lamp is $\langle 300,0,0 \rangle$ cm.

For the blue lamp, the ray passing through the center of the lens is undeflected. Therefore the slope of the ray is the same to the left and right of the lens:

$$\frac{-1}{12.5} = \frac{y_2}{300}$$
$$y_2 = -24 \, \text{cm}$$

Figure 23.89 We will calculate the location of the images of the red and blue lamps, at the ends of the stick.

The blue lamp was only 1 cm above the axis and 12.5 cm along the axis from the lens, but its real image is 24 cm below the axis, 300 cm along the axis from the lens. As a vector, the image location of the blue lamp is $\langle 300, -24, 0 \rangle$ cm. The complete image of the stick lies between $\langle 300, 0, 0 \rangle$ cm and $\langle 300, -24, 0 \rangle$ cm. The image is 24 times as big as the actual object. This factor is called the "magnification."

We didn't need to convert to meters in doing the analysis, since all the quantities were given in centimeters.

When the distance of a lamp from a lens is only slightly larger than the focal length f of the lens, as it is in this case (12.5 cm is only slightly larger than 12 cm), the image is very far from the lens (300 cm or 3 m in this case). Evidently Figure 23.89 is not to scale!

Computer and Movie Projectors

The effect we've just seen, that a source near the lens can make a real image far from the lens that is greatly enlarged, is the basis for computer and movie projection. A small glowing or illuminated object is placed a distance slightly greater than f from a converging lens, and an image of the object is displayed on a screen that in this case must be placed 300 cm from the lens. If the screen is not accurately placed 300 cm from the lens, the red and blue rays won't converge to a point on the screen, and there will be big blobs of red light and blue light on the misplaced screen. To put it in more practical terms, if the screen is placed 300 cm away but the stick and the lens aren't accurately positioned 12.5 cm apart, the image will be out of focus, with points turning into blobs. Projectors have a lens held in a screw mount to make it easy to focus the image on the screen by changing the object-to-lens distance simply by turning the lens in the screw mount.

Each illuminated point in the object is represented by a dot of light on the screen, making a complete picture of the object. The image is inverted, so the object must be mounted upside down in order to make a screen display that is right side up. In Figure 23.90 a lens makes a real image of an illuminated R on a sheet of ground glass. Rays are drawn from four numbered corners of the R through the center of the lens to the image. Study the Figure carefully to understand why the image is inverted.

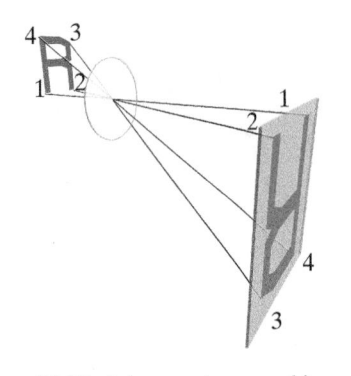

Figure 23.90 A lens makes a real image of an illuminated R on a sheet of ground glass. Rays from four numbered corners of the R are drawn through the center of the lens to show why the image is inverted.

> **Checkpoint 11** An object 5 cm tall is 20 cm to the left of a converging lens, which forms a real image of the object on a screen 500 cm to the right of the lens. **(a)** How tall is the image on the screen? **(b)** What is the magnification? **(c)** What is the focal length of this lens?

A Lens Gathers Light

In addition to forming images by bending rays, a major function of a lens is to gather light. In Figure 23.90, imagine replacing the lens by a sheet of cardboard with a pinhole at its center.

> QUESTION Would there still be an (inverted) image of an R on the ground glass?

Yes, because the rays traced in Figure 23.90 would still go through the pinhole to the same places on the glass as before. However, the image would be very dim, because very little light from any source point could pass through the pinhole. Figure 23.91 shows an image made by a pinhole. For clarity the inverted R is shown much brighter than it would really be.

In contrast, with the lens in place, a large bundle of rays from one source point is focused to one corresponding point on the ground glass, so the image

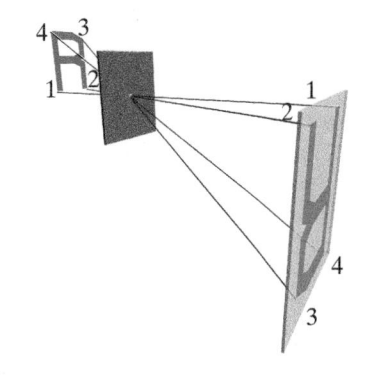

Figure 23.91 A pinhole makes a real image. The image would actually be very dim.

is much brighter. Remember that in Figure 23.90 we show only those single rays from the four source points that go through the center of the lens; a large bundle of rays from each source point passes through the full area of the lens, but for clarity these rays are not shown in the figure.

The larger the lens, the brighter the image, though there is a practical limit, because the larger the lens, the larger the angles of the rays, and at some point the aberrations for a spherical lens would distort the image too much because the small-angle approximation would no longer be valid. (Moreover, the larger the lens, the more expensive it is to manufacture.)

There is another difference between a pinhole and a lens. With a pinhole, there will be a (dim) image on the ground glass no matter how far from the pinhole the glass is positioned. In contrast, a lens will focus a bundle of rays from one source point to one corresponding point on the glass only if the glass is placed at the correct distance from the lens, which we have calculated using the thin-lens equation.

The light-gathering property of a lens is indicated by the "f-number," which is the ratio of the focal length to the diameter of the lens, f/D; a larger diameter means a smaller f-number. The amount of light collected is proportional to the area of the lens and so is proportional to the diameter squared and to the inverse square of the f-number. Compare f-numbers of 2.8 and 4.0: a lens with an f-number of 2.8 will collect $(4.0/2.8)^2 = 2$ times as much light as a lens with an f-number of 4.0.

Cameras often have an adjustable circular aperture, called a "diaphragm," for reducing the effective size of the lens when the scene is too bright and one needs to reduce the brightness of the image. One sets the diaphragm by specifying the effective f-number. Reducing the effective size of the lens by making the opening in the diaphragm smaller (increasing the f-number) also reduces aberrations by reducing the maximum angles of rays that are accepted, thereby improving image quality.

Parallel Beams of Light

An important special case is the image made by an object that is very far away from the lens. In the lens equation $1/f = 1/d_1 + 1/d_2$, if d_1 is very large, $1/d_1$ is very small compared to the other terms, so approximately we have $1/f = 1/d_2$, or $d_2 = f$; all the rays from the distant object pass through a point a distance f beyond the lens. The incoming rays are nearly parallel. Figure 23.92 shows this situation. A ray that hits the lens at a distance y from the axis is deflected through an angle y/f, and since the incoming ray is parallel to the axis, it will cross the axis a distance f beyond the lens.

A practical application of this is the concentration of solar energy. The Sun is so far away that rays from the center of the Sun are nearly parallel as they hit a lens, and they get concentrated at the focal point, a distance f beyond the lens. Rays from one edge of the Sun are also nearly parallel but at a slightly different angle (Figure 23.93). The radius of the Sun is 7×10^8 m, and its distance from the Earth is 1.5×10^{11} m, so the angular width corresponding to the radius of the Sun is $7 \times 10^8 / 1.5 \times 10^{11} = 4.7 \times 10^{-3}$ radians (about one-quarter of a degree).

Suppose that the focal length of the lens is $f = 20$ cm. Place a screen 20 cm behind the lens, so that the incoming rays will focus onto the screen. A ray that goes from the edge of the Sun through the center of the lens will be displaced $(4.7 \times 10^{-3})(20) = 9.4 \times 10^{-2}$ cm from the axis of the lens when it hits the screen. That is, the (real) image of the Sun on the screen will have a radius of 9.4×10^{-2} cm. If the lens has a radius of 5 cm, the effect is to concentrate solar energy from an area of $\pi r^2 = \pi(5)^2$ cm^2 into an area of $\pi y^2 = \pi(9.4 \times 10^{-2})^2$ cm^2. The energy concentration factor is $(5/(9.4 \times 10^{-2})^2 \approx 3000$! The spot on the screen gets very hot.

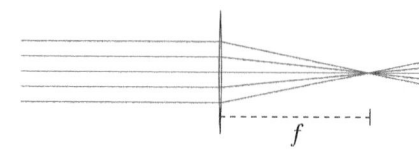

Figure 23.92 Parallel rays converge at the focal point, a distance f to the right of the lens.

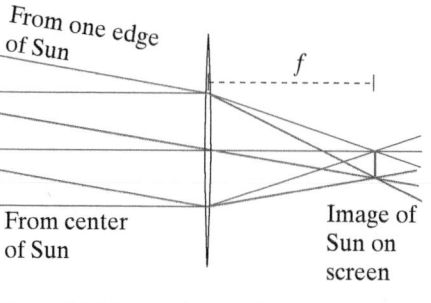

Figure 23.93 Nearly parallel rays arriving from the center of the Sun, and another set of nearly parallel rays arriving from the edge of the Sun. A real image of the Sun is formed on a screen.

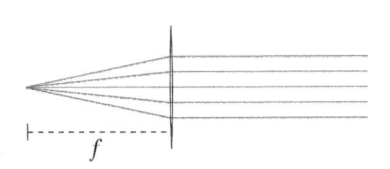

Figure 23.94 Rays emitted from a focal point of a converging lens produce a parallel beam.

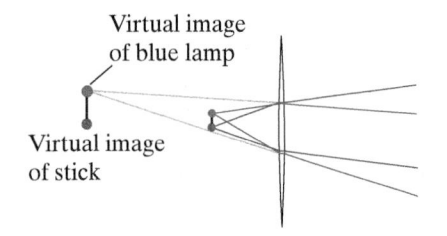

Figure 23.95 A virtual image of a stick, which is larger than the actual stick and right side up.

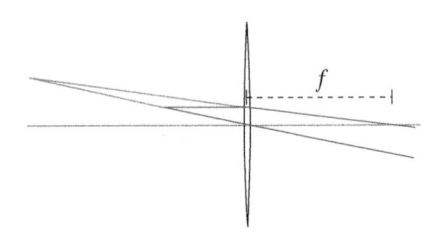

Figure 23.96 "Easy" rays emitted by the blue lamp: one that goes through the center of the lens, undeflected, and one that goes parallel to the axis then through the focal point.

A converging lens can also be used to make a parallel beam of light from a point source. In Figure 23.94 we see that if we place a point source at a focal point, a distance f from a converging lens, all rays that hit the lens will emerge parallel to the axis, because the deflection angle is y/f and the angle of the incoming ray is also y/f.

It's interesting to see what the lens equation predicts for this case, where d_1 is equal to the focal length f: $1/f = 1/f + 1/d_2$, so $1/d_2$ is zero or approximately zero.

QUESTION If $1/d_2$ is nearly zero, what does this say about d_2?

For $1/d_2$ to be very small, d_2 must be very large. If we place the source close to f from the lens, the place where the rays cross after passing through the lens will be at a very great distance. In the limit, those rays will cross only at infinity. That is, these rays are parallel.

Analyzing Virtual Images

We've discussed how to predict where a real image will be formed. The prediction of the location of a virtual image is a little trickier, because no rays actually go through the virtual image. Rather, the lens deflects rays in such a way that those rays *seem* to come from an image, as shown in Figure 23.95. The object is again a short stick with a red lamp at one end and a blue lamp at the other. The virtual image of the stick is larger than the actual stick, and right side up.

First we'll analyze this situation graphically. As before, there are two rays whose deflections are easy to determine. A ray from the blue lamp that hits the center of the lens will pass straight through, with zero deflection (Figure 23.96). A ray from the blue lamp that starts out parallel to the axis is deflected by the converging lens in such a way as to pass through a point on the axis to the right of the lens that is a distance f from the lens. Draw these two particularly "easy" rays, the one through the center of the lens and the one that starts parallel to the axis and crosses the axis a distance f to the right of the lens. Follow these rays back to the point where they intersect; this is the location of the virtual image of the blue lamp. All other rays (that don't miss the lens) will also seem to come from the location of the image, in the thin-lens, small-angle approximation. The virtual image is upright, not inverted.

Next, let's calculate quantitatively the location of the virtual image of the stick:

The converging lens is located at the origin $\langle 0,0,0 \rangle$ and has a focal length of 12 cm, with its axis pointing along the x axis. The red lamp is located at $\langle -7,0,0 \rangle$ cm and the blue lamp is 1 cm above the red lamp, at $\langle -7,1,0 \rangle$ cm. Find the locations of the images of the red and blue lamps; the complete image of the stick will lie between the images of the two lamps that are at the ends of the stick.

1. Lens equation: $1/12 = 1/7 + 1/d_2$, which yields $d_2 = -16.8$ cm.

QUESTION Can you guess the physical significance of this distance turning out to be negative?

It means that the location along the axis is to the *left* of the lens, on the same side as the source. To the right of the lens, rays seem to be diverging from a location to the left of the actual source of light. The location of the virtual image of the red lamp is $\langle -16.8,0,0 \rangle$ cm.

2. Ray through center of lens: On the lamp side of the lens the slope of this ray coming from the blue lamp is $-1/7$, and on the right side of the lens the slope is the same (no deflection). This is one of the rays that seem to be coming from the virtual source, so we follow it back to a location whose distance

along the axis is 16.8 cm to the left of the lens. The distance from the axis is 16.8 times the slope, which is $(16.8)(1/7) = 2.4$ cm above the axis. The vector position of the virtual image of the blue lamp is $\langle -16.8, 2.4, 0 \rangle$ cm.

A Magnifying Glass

This particular arrangement we've just discussed is an example of a magnifying glass. If you look at a stick 1 cm long from a distance of 30 cm (a typical viewing distance), the angular width (the angle "subtended" by the stick) is 1/30 rad, or about 0.03 rad. Place the lens 7 cm in front of the stick, and the angular width is now $1/7 = 0.14$ rad. That is, the difference in angle of the central ray coming from one end of the stick and the central ray coming from the other end of the stick is 1/7, or 0.14 rad. Without the lens, you would have had to be $0.14/0.03 = 4.3$ times closer for the stick to appear to have that angular width in your field of view, and you might not be able to focus on the object if your eye were that close to it. We say that the magnification of the object is 4.3. The virtual image of the stick appears longer.

If you have a converging lens available, hold it away from your eye and look through it at a nearby object. As long as the distance from the lens to the object is less than the focal length of the lens, you'll see a virtual image, which is right side up, as is typical of virtual images. Figure 23.97 shows the situation in 3D. Rays from the corners of the illuminated R that go through the center of the lens seem to be coming from the corners of the large right side up virtual image of the R.

Figure 23.97 A converging lens makes an enlarged, right-side-up virtual image of an illuminated R. Rays from four corners of the R are drawn through the center of the lens to show why the image is right side up and larger.

Diverging Lenses

Another way to make a virtual image is with a diverging lens. In fact, because a diverging lens cannot make rays converge toward the axis, this kind of lens cannot make real images. In contrast, we've seen that a converging lens can produce either real or virtual images, depending on where the source is located.

In analyzing the effects of a diverging lens, two of the rays from a source are again particularly easy to draw. A ray that hits the center of the lens passes through without deflection. A ray that starts out parallel to the axis and hits the lens at a distance y from the axis is deflected *outward*, *away* from the axis, through a deflection angle $y/|f|$. To avoid having to use absolute-value notation, it is standard practice to say that a diverging lens has a negative focal length. That is, f is a negative number. Note that the virtual image of the stick is smaller than the stick and is right side up; if you have a diverging lens you can see these two effects simply by looking at an object through the lens.

In Figure 23.98 we see that for an initially parallel ray we draw the outgoing ray as though it came from a location a distance $|f|$ to the left of the diverging lens, whereas with a converging lens a parallel ray goes through the axis at a distance f to the right of the lens. Equivalently, we can say that for both kinds of lens an initially parallel ray crosses (or seems to cross) the axis at a location f to the right of the lens, where f has a negative value for a diverging lens and hence is to the left of the lens.

This sign convention carries over to the lens equation, which we continue to write as $1/f = 1/d_1 + 1/d_2$, but where f is a negative number for a diverging lens. We already saw that d_1 or d_2 could also be negative numbers.

Figure 23.98 "Easy" rays emitted by the blue lamp: one that goes through the center of the lens, undeflected, and one that goes parallel to the axis, then is deflected outward through an angle that looks like it is coming from the focal point on the left.

Sign Conventions

Let's summarize the sign conventions that are implicit in the thin-lens equation $1/f = 1/d_1 + 1/d_2$:

- f is positive for a converging lens, negative for a diverging lens.

- We measure d_1 from the source to the lens and we measure d_2 from the lens to the image.
- The distances d_1 and d_2 are considered positive if they are in the same direction that the light is traveling and negative otherwise. Both d_1 and d_2 are positive for the simple case of a distant source and converging lens.

With these sign conventions the thin-lens equation applies to all combinations of thin lenses (and spherical mirrors) and all combinations of source or image locations. Implicit in these conventions is that if the source is to the left of the lens and d_2 is positive, the image is to the right of the lens and is real and inverted, whereas if d_2 is negative, the image seems to be to the left of the lens and is virtual and right side up.

What does it mean for d_1 to be negative? Suppose that lens 1 would make a real image 20 cm to the right of lens 2 if lens 2 were not there. The effect of lens 2 can be calculated by setting d_1 to -20 cm: the "source" can be considered to be a real image that would be to the right of lens 2.

EXAMPLE

Sign Conventions: A Converging Lens

A point source of light is located to the left of a converging lens of focal length $+10$ cm. **(a)** If the source is 12 cm to the left of the lens, where is the image? **(b)** Is the image real or virtual? **(c)** The source is moved to be 8 cm to the left of the lens; now where is the image? **(d)** Is this image real or virtual?

Solution   **(a)** The thin-lens equation is $1/10 = 1/12 + 1/d_2$, where we have used centimeters for f and d_1, so our answer will be in centimeters. Solving, $d_2 = 60$ cm.

(b) Because d_2 is positive, rays have converged to a point 60 cm to the right of the lens, where there is a real image from which light rays originate. If a screen is placed 60 cm to the right of the lens an inverted image will be seen on the screen.

(c) $1/10 = 1/8 + 1/d_2$, so $d_2 = -40$ cm.

(d) Because d_2 is negative, the image is to the left of the lens and is a virtual image from which light rays *seem* to originate, but in fact no rays come from there. If a screen is placed 40 cm to the left of the lens, no image will be seen on the screen.

In part (a) the source is to the left of the left focal point (12 cm to the left of the lens, 2 cm to the left of the focal point) and a real image is formed. In part (b) the source is to the right of the left focal point (8 cm to the left of the lens, 2 cm to the right of the focal point) and there is a virtual image. This is a general result that emerges from the thin-lens equation: sources farther than f from a converging lens form real images. Sources closer to a converging lens than f form virtual images.

EXAMPLE

Sign Conventions: A Diverging Lens

A point source of light is located to the left of a diverging lens of focal length -10 cm. **(a)** If the source is 12 cm to the left of the lens, where is the image? **(b)** Is the image real or virtual? **(c)** The source is moved to be 8 cm to the left of the lens; now where is the image? **(d)** Is this image real or virtual?

Solution   **(a)** The thin-lens equation is $1/(-10) = 1/12 + 1/d_2$. Solving, $d_2 = -5.45$ cm.

(b) Because d_2 is negative, the image is to the left of the lens and is a virtual image from which light rays *seem* to originate, but in fact no rays come from there. If a screen is placed −5.45 cm to the left of the lens, no image will be seen on the screen. Looking through the lens from the right, you would see a (virtual) image that seems to be 5.45 cm behind the lens.

(c) $1/(-10) = 1/8 + 1/d_2$, so $d_2 = -4.44$ cm.

(d) Because d_2 is negative, there is again a virtual image to the left of the lens, at a location not very different from what we found in part (a).

A single diverging lens always forms a virtual image and cannot form a real image, no matter where the source is placed.

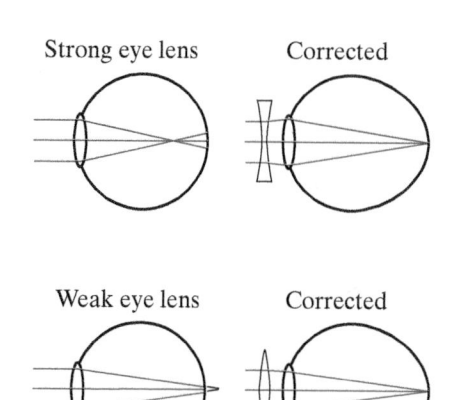

Figure 23.99 An eye lens that is too strong or too weak fails to focus rays onto the retina at the back of the eye. A diverging or converging lens can correct this.

Eyeglasses and Contact Lenses

An important use of diverging lenses is to correct defective vision (Figure 23.99). If the lens in the eye is too strong and bends rays too much, rays coming from a point source will form a (real) image in front of the retina, and the retina will detect a blob of light rather than a point. A diverging lens bends rays away from the axis and can compensate for too much bending toward the axis by the eye lens. The effect is to move the real image from in front of the retina onto the retina.

In contrast, if the eye lens is too weak and doesn't bend rays enough, rays coming from a point source would form a (real) image behind the retina, and the retina will detect a blob of light rather than a point. A converging lens, which bends rays toward the axis, can add to the bending by the eye lens and move the image from behind the retina onto the retina.

Eyeglass and contact lens properties are given in terms of "diopters," which is the inverse of the focal length in meters. A lens with +2.5 diopters has a focal length $f = 1/2.5 = 0.4$ m, or 40 cm. A lens of −2.5 diopters is a diverging lens with focal length −40 cm.

Multiple Lenses

Many optical instruments including binoculars, telescopes, and microscopes contain two or more lenses. In a telescope, the front lens produces a real image of distant objects, which you examine with another lens (the eyepiece) that acts as a magnifying glass. The front lens is made much larger than the lens in your eye, so it captures much more light than your own eye can, making the image much brighter than it would otherwise be.

The design and analysis of multilens systems depend on the fact that light coming from a real or virtual image acts like the light coming from an actual object (except for being restricted in the range of angles of the rays), so in analyzing what a second lens will do you can pretend that there is an actual light-emitting object at the location of the real or virtual image made by the first lens.

Mirrors

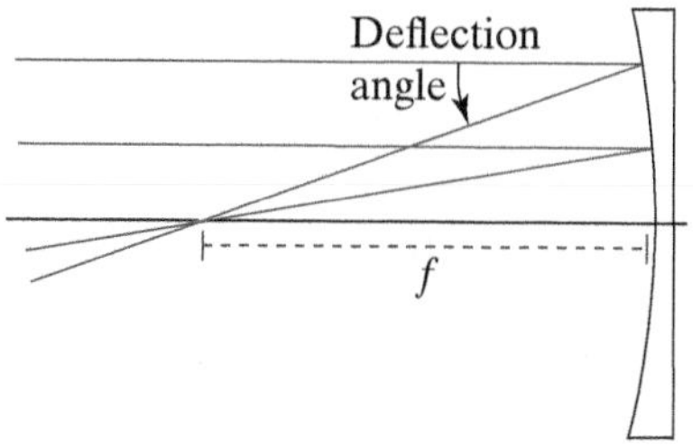

The effect of a spherical mirror is essentially the same as that of a lens with spherical surfaces (Figure 23.100). The thin-lens equation still applies, with the same sign conventions. For example, with a distant source to the left of the mirror in Figure 23.100, what is the sign of d_2? We measure d_2 from the lens (or mirror in this case) to the image, and it is positive because that is the direction in which the light is propagating.

For rays at large angles, spherical mirrors exhibit the same kind of spherical aberration that is characteristic of spherical lenses. There is, however, a very important advantage of a mirror: all colors are reflected in the same way, so there is no chromatic aberration, whereas the variation of the index of refraction in a lens causes the refraction of a ray to differ for different colors.

Figure 23.100 A spherical mirror acts in a way that is similar to a spherical lens.

A practical advantage for large astronomical telescopes is that a mirror can be very thin and weigh much less than a lens of the same diameter and focal length. The largest telescope lens ever made is only 1 m in diameter, but there are now telescope mirrors over 10 m in diameter. Moreover, a lens can be supported only around its edge, but a large mirror can be supported all over its back surface.

Many telescope mirrors are made with a paraboloid surface rather than a spherical surface, because incoming rays that are parallel to the axis are all reflected through the focal point no matter how far from the optical axis they are (or how large the deflection angle is). This is approximately true for a spherical lens (or mirror) only in the small-angle approximation, which is equivalent to the rays being near the optical axis.

23.11 *THE FIELD OF AN ACCELERATED CHARGE

One can use Gauss's law to determine the equation for the electric field of an accelerated charge. We will construct a Gaussian surface near to but not enclosing a charge that is briefly accelerated. Since no charge is enclosed by the surface, the net flux on the surface is zero. We can find the flux on the left and right portions of the surface, and knowing this we can draw conclusions about the magnitude and direction of the field on the side portion of the surface, which includes the radiative electric field.

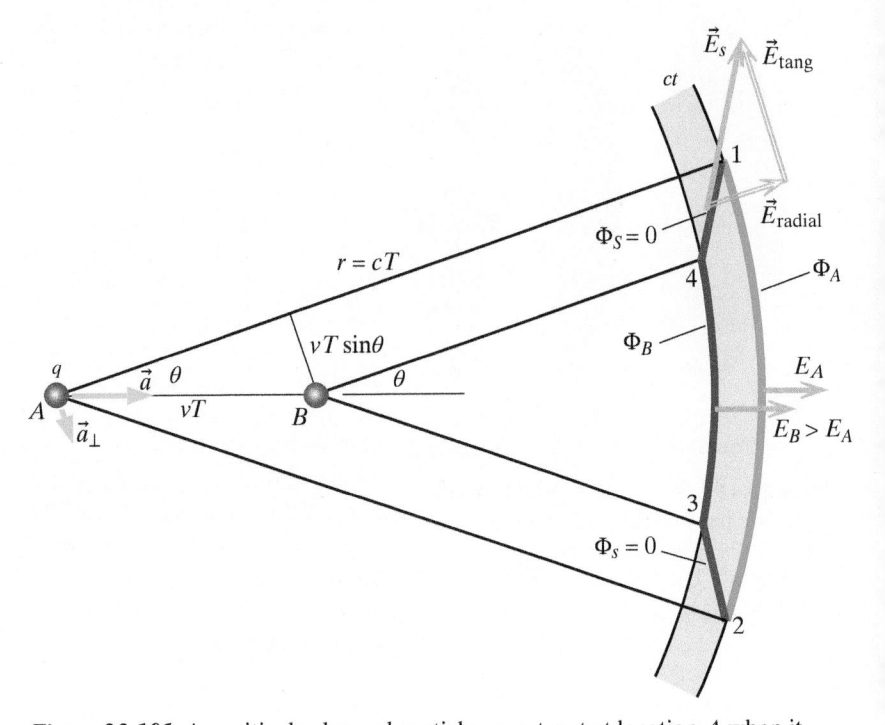

Figure 23.101 A positively charged particle was at rest at location A when it experienced a brief acceleration to the right and then coasted at constant speed. The diagram is not to scale.

In Figure 23.101 a positively charged particle was at rest at location A when it experienced a brief acceleration a to the right, lasting a very short time t. It then coasted at a constant low speed $v = at$, passing location B at time T, having gone a distance vT. The diagram is a slice through a 3D distribution of electric field. For clarity, the diagram is not drawn to scale: the

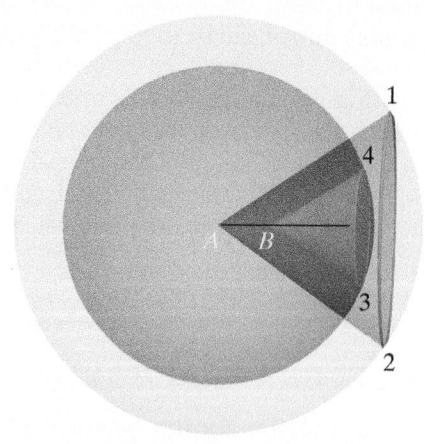

Figure 23.102 A 3D view of the geometry used to find the radiative field of an accelerated charge.

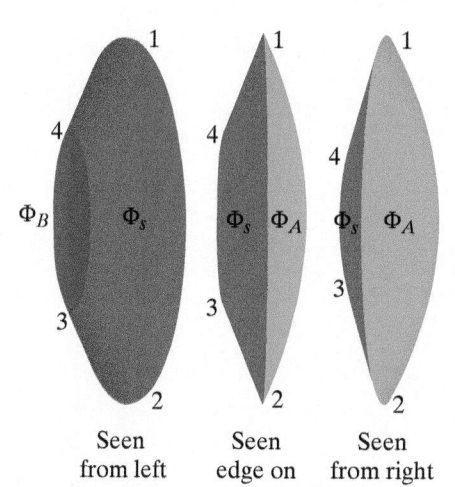

Figure 23.103 3D views of the Gaussian surface we have chosen. The blue portion on the left is concave, the green portion on the right is convex, and the slanted side portion is red.

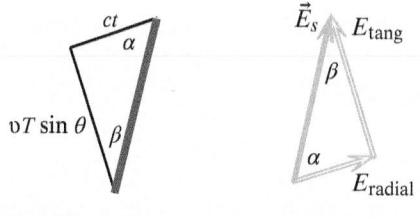

Figure 23.104 Two similar triangles. The triangle on the left is enlarged from the upper slanted surface in Figure 23.101.

distance vT is actually much shorter than the distance cT since $v \ll c$, and the acceleration time t is much smaller than the time T so the distance ct is much smaller than cT.

Observers at a distance greater than $r = cT$ still see the electric field originally associated with the stationary charge at location A, but other observers see the electric field of a slowly moving charge, which at time T is at location B. The cyan area is an expanding thin spherical shell whose thickness ct corresponds to the brief time t of the acceleration.

Figure 23.102 shows a 3D view of the situation. The cone with its apex at location A intersects the outer surface of the spherical shell at the locations 1 and 2 that are shown in Figure 23.101, and the cone with its apex at location B intersects the inner surface of the spherical shell at locations 3 and 4 that are shown in Figure 23.101.

We choose a Gaussian surface marked with red, green, and blue lines on Figure 23.101 (again, the diagram is a slice through the 3D closed Gaussian surface). This Gaussian surface does not contain any charge, so the total electric flux over this surface is zero. Three views of this Gaussian closed surface are shown in Figure 23.103. Locations 1, 2, 3, and 4 are shown. Also shown are the surfaces on which we will determine the fluxes Φ_A, Φ_B, and Φ_s.

The field E_A on the outside of the spherical shell is what it was when the charge was still at location A, but the field E_B on the inside of the expanding spherical shell is larger, because E_B is the field of the charge at location B, which is closer than location A.

Suppose that the size of the angle θ is such that the right (green) portion of our Gaussian surface represents one-tenth of the total area of a sphere around location A. Then the flux Φ_A on the right portion of our Gaussian surface is $+0.1q/\varepsilon_0$, since the total flux on a complete sphere would be $+q/\varepsilon_0$. We have the same angle θ at location B, so the flux on the left (blue) side of our Gaussian surface is $-0.1q/\varepsilon_0$, and $|\Phi_A| = |\Phi_B|$. Because B is closer than A, $E_B > E_A$, but the right side of our Gaussian surface has more area than the left side. These two effects compensate each other to make the magnitude of the flux on the left side equal to the magnitude of the flux on the right side.

Since $\Phi_A + \Phi_B = 0$, it must be that the flux Φ_s on the other, ring-shaped (red) portion of the Gaussian surface is also zero, and given the symmetry of the situation it must be that the electric field $\vec{E}_s$ on that surface is everywhere parallel to the surface. Note in Figure 23.101 that $\vec{E}_s$ has a tangential component $\vec{E}_{\text{tang}}$ that is parallel to $-q\vec{a}_\perp$, which is what we expect for a radiative electric field. The triangle in Figure 23.101 for which the upper red line is the hypotenuse is shown enlarged in Figure 23.104, and this triangle is similar to the triangle formed by $\vec{E}_s$ and its components. From $\tan \alpha$ in Figure 23.104 we obtain a relationship between the tangential and radial components of $\vec{E}_s = \vec{E}_{\text{tang}} + \vec{E}_{\text{radial}}$ that we'll need in a moment:

$$\tan \alpha = \frac{E_{\text{tang}}}{E_{\text{radial}}} = \frac{vT\sin\theta}{ct}$$

$$E_{\text{tang}} = E_{\text{radial}}\frac{vT\sin\theta}{ct}$$

Next, we can determine E_{radial} and from that determine E_{tang}. In Figure 23.105 is a close-up of the expanding spherical shell. We choose another Gaussian surface, this time one that is just inside and just outside the inner surface of the shell, with very small areas and therefore negligible flux on the top and bottom in the diagram. Because this Gaussian surface includes no charge, the magnitude of the flux $E_{\text{radial}}\Delta A$ on the left and right sides of this Gaussian

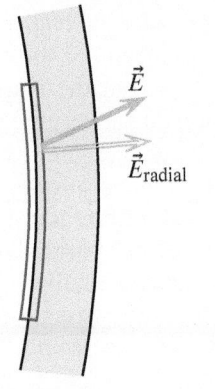

Figure 23.105 A close-up look at a small section of the expanding spherical shell.

surface must be equal, which proves that the radial component of the electric field doesn't change as we cross from just inside the inner surface of the shell to just outside the inner surface of the shell. In particular, the radial component of $\vec{E}_s$ must be the same as the electric field of a charge at location B (and we use the approximation that since $vT \ll cT$, the distance from B to the spherical shell is approximately r):

$$E_{\text{radial}} = \frac{1}{4\pi\varepsilon_0}\frac{q}{r^2}$$

Substitute this result into our earlier equation and simplify, noting that $r = cT$ (so $T = r/c$) and v/t is the acceleration a:

$$E_{\text{tang}} = \frac{1}{4\pi\varepsilon_0}\frac{q}{r^2}\frac{vT\sin\theta}{ct}$$
$$= \frac{1}{4\pi\varepsilon_0}\frac{qa\sin\theta}{c^2 r}$$

Note that $a\sin\theta$ is the perpendicular component of the vector acceleration $\vec{a}$. Evidently $\vec{E}_{\text{tang}}$ is the transverse radiative field, and $\vec{E}_{\text{radial}}$ is the ordinary Coulomb electric field. Because the Coulomb electric field falls off as $1/r^2$, at large distances only the $1/r$ radiative field is significant. The radiative electric field is accompanied by a magnetic field whose magnitude is $B = E_{\text{rad}}/c$.

Taking into account the direction of $\vec{E}_{\text{tang}}$ as well as its magnitude, we have proved that the transverse radiative electric field is

$$\vec{E}_{\text{radiative}} = \frac{1}{4\pi\varepsilon_0}\frac{-q\vec{a}_\perp}{c^2 r}$$

This derivation of the radiative electric field is based on that given in Appendix B of E. M. Purcell, *Electricity and Magnetism*, Berkeley Physics Course Vol. 2, 2nd edition, 1985, McGraw-Hill.

23.12 *DIFFERENTIAL FORM OF MAXWELL'S EQUATIONS

Here are the differential forms of Maxwell's equations, where $\vec{J}$ is the current density in amperes per square meter:

MAXWELL'S EQUATIONS—DIFFERENTIAL FORMS

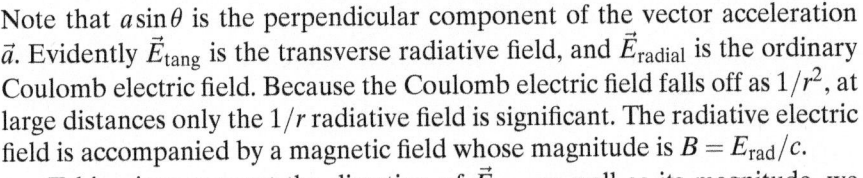

$\text{div}(\vec{E}) = \vec{\nabla} \bullet \vec{E} = \dfrac{\rho}{\varepsilon_0}$ ⠀⠀⠀ Gauss's law

$\text{div}(\vec{B}) = \vec{\nabla} \bullet \vec{B} = 0$ ⠀⠀⠀ Gauss's law for magnetism

$\text{curl}(\vec{E}) = \vec{\nabla} \times \vec{E} = -\dfrac{\partial \vec{B}}{\partial t}$ ⠀⠀⠀ Faraday's law

$\text{curl}(\vec{B}) = \vec{\nabla} \times \vec{B} = \mu_0\left(\vec{J} + \varepsilon_0\dfrac{\partial \vec{E}}{\partial t}\right)$ ⠀⠀⠀ Ampere–Maxwell law

SUMMARY

Maxwell's Equations

$$\oint \vec{E} \bullet \hat{n} dA = \frac{1}{\varepsilon_0} \sum Q_{\text{inside surface}} \quad \text{Gauss's law}$$

$$\oint \vec{B} \bullet \hat{n} dA = 0 \quad \text{Gauss's law for magnetism}$$

$$\oint \vec{E} \bullet d\vec{l} = -\frac{d}{dt} \int \vec{B} \bullet \hat{n} dA \quad \text{Faraday's law}$$

Ampere–Maxwell law:

$$\oint \vec{B} \bullet d\vec{l} = \mu_0 \left[\sum I_{\text{inside path}} + \varepsilon_0 \frac{d}{dt} \int \vec{E} \bullet \hat{n} dA \right]$$

Magnetic and electric flux terms

$$\Phi_{\text{mag}} = \int \vec{B} \bullet \hat{n} dA \qquad \Phi_{\text{el}} = \int \vec{E} \bullet \hat{n} dA$$

The force due to electric and magnetic interactions

$$\vec{F} = q\vec{E} + q\vec{v} \times \vec{B} \qquad (d\vec{F} = Id\vec{l} \times \vec{B} \text{ for currents})$$

Producing a radiative electric field (Figure 23.106)

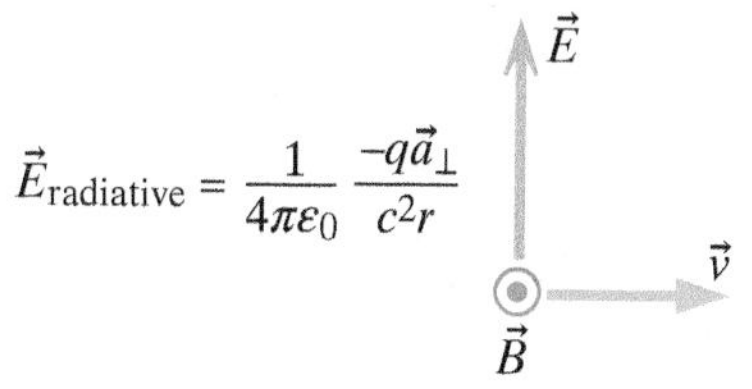

$$\vec{E}_{\text{radiative}} = \frac{1}{4\pi\varepsilon_0} \frac{-q\vec{a}_\perp}{c^2 r}$$

Figure 23.106

The direction of $\vec{v}$ is the direction of $\vec{E} \times \vec{B}$, $E = cB$ in electromagnetic radiation.

$$v = \frac{1}{\sqrt{\mu_0\varepsilon_0}} = c = 3 \times 10^8 \text{ m/s}$$

Effects of electromagnetic radiation on matter

The main effect that electromagnetic radiation has on matter is due to $\vec{F} = q\vec{E}_{\text{radiative}}$, but the magnetic field is responsible for radiation pressure, a very small effect.

If $\vec{E}_{\text{radiative}}$ makes a charge accelerate, that charge radiates in turn ("re-radiation" or "scattering").

Different systems respond differently to sinusoidal electromagnetic radiation of a particular frequency $f(c = f\lambda)$. Systems that freely oscillate sinusoidally with a frequency f respond very strongly to sinusoidal electromagnetic radiation of the same frequency f ("resonance" phenomenon).

Energy flux of electromagnetic radiation

$$\vec{S} = \frac{1}{\mu_0}\vec{E} \times \vec{B}, \text{ in W/m}^2 \quad \text{(the "Poynting vector")}$$

Momentum flux of electromagnetic radiation

$$\frac{\vec{S}}{c} = \frac{1}{\mu_0 c}\vec{E} \times \vec{B}, \quad \text{in N/m}^2$$

The sky is blue and polarized due to re-radiation of sunlight, with blue light re-radiated more strongly than red light.

Geometric Optics

Snell's law and the index of refraction $n = c/v$:
$$n_1 \sin\theta_1 = n_2 \sin\theta_2$$
Converging and diverging lenses

The thin-lens equation: $\dfrac{1}{f} = \dfrac{1}{d_1} + \dfrac{1}{d_2}$

Real and virtual images

QUESTIONS

Q1 Match the physical situation with the Maxwell's equation that would be useful in analyzing the situation:

(a) Gauss's law	1. Find the magnetic field inside a current-carrying coaxial cable.
(b) Gauss's law for magnetism	2. Predict what an AC voltmeter reads whose leads encircle a coil.
(c) Faraday's law	3. Find the amount of charge inside a box.
(d) Ampere–Maxwell law	4. Relate the magnetic flux on one face of a cube to the magnetic flux on the other faces.

Q2 At a time t_0 an electric field is detected in the region shown on the left diagram in Figure 23.107. The electric field is zero at all other locations. At a later time t_1 an electric field is detected in the region shown on the right diagram; the electric field is zero elsewhere. You conclude that an electromagnetic wave is passing through the region.   **(a)** What is the direction of propagation of the electromagnetic wave? **(b)** If you were to measure the magnetic field in the region where the electric field is nonzero, what would be the direction of the magnetic field?

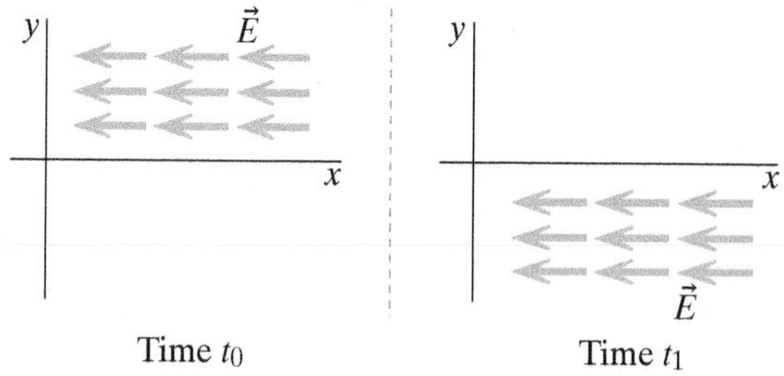

Figure 23.107

Q3 A pulse of electromagnetic radiation is detected at a particular location. In the pulse, the electric field is in the $+y$ direction and the magnetic field is in the $-z$ direction. What is the direction of propagation of this pulse of radiation?

Q4 Show that a moving region of space in which $\vec{E} = \langle 0,E,0 \rangle$, $\vec{B} = \langle 0,-B,0 \rangle$, and $\vec{v} = \langle -c,0,0 \rangle$ does not satisfy Maxwell's equations, and therefore is not a possible configuration of traveling fields.

Q5 A proton is accelerated in the direction shown in Figure 23.108. **(a)** Which arrow $(a–j)$ best indicates the direction of propagation of radiation reaching observation location P? **(b)** Which arrow $(a–j)$ best indicates the direction of $\vec{a}_\perp$ at observation location P? **(c)** Which arrow $(a–j)$ best indicates the direction of the radiative electric field observed at location P? **(d)** Which arrow $(a–j)$ best indicates the direction of the radiative electric field observed at location S?

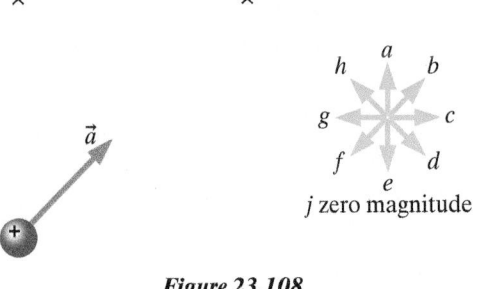
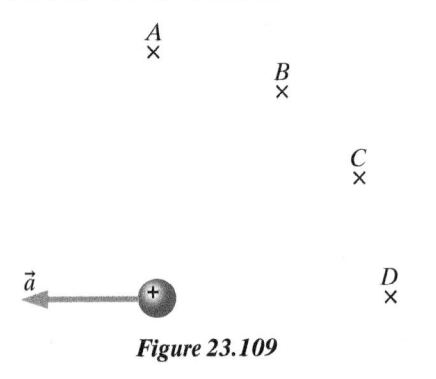

Figure 23.108

Q6 A proton is accelerated in the direction shown in Figure 23.109. At which location(s) $(A–D)$ will electromagnetic radiation be detected a short time later?

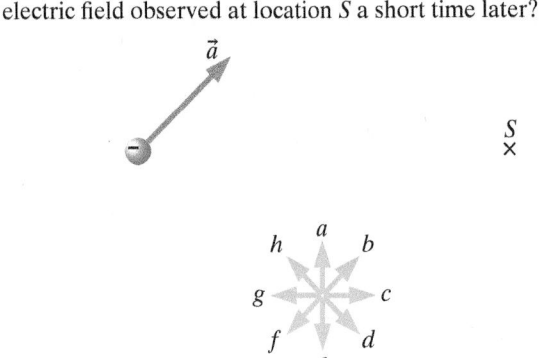

Figure 23.109

Q7 An electron is briefly accelerated in the direction shown in Figure 23.110. What will be the direction $(a–j)$ of the radiative electric field observed at location S a short time later?

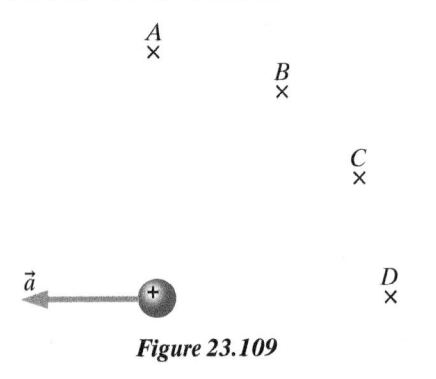

Figure 23.110

Q8 When informally discussing the effect of electromagnetic radiation on matter, physicists often talk about the effects that electromagnetic radiation has on the electrons in the matter. Why

would they omit commenting on the effects on the nucleus, which is also charged?

Q9 Explain why electromagnetic radiation moving to the right applies radiation pressure to the right on a neutral grain of dust in the air, even though the grain has zero net charge.

Q10 Figure 23.111 shows several different orientations (A, B, and C) of the receiving antenna discussed earlier. In each case, predict the brightness of the bulb and explain briefly.

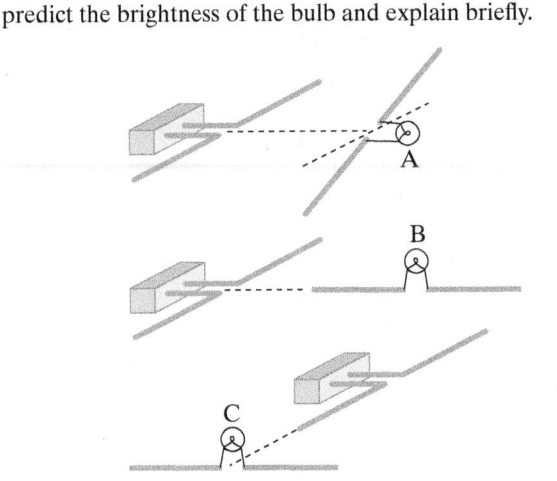

Figure 23.111

Q11 You may have noticed that television antennas on housetops have metal rods mounted horizontally (Figure 23.112), at least in the United States (we're not talking about satellite dishes). What does this imply about the construction of the transmitting antennas used by television stations? Why?

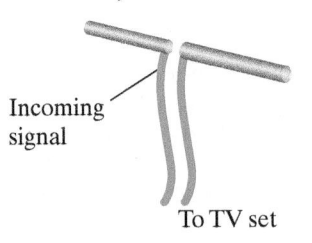

Figure 23.112

Q12 Television antennas often have one or more horizontal metal bars mounted behind the receiving antenna and insulated from it, as shown in Figure 23.113. Explain qualitatively why adding a second metal bar can make the television signal either stronger or weaker, depending on the distance between the front receiving antenna and the second bar.

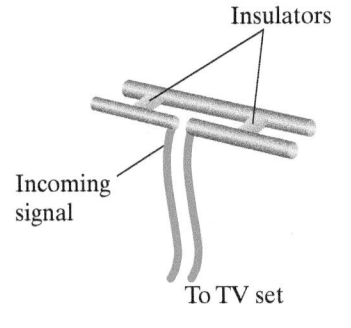

Figure 23.113

Q13 At dawn, with the Sun just rising in the east, you face the Sun and bend your head back to look straight up, and you examine the blue sky light with a Polaroid® filter. Why is the light polarized? What is the direction of the electric field, east–west or north–south? Explain carefully.

Q14 On a cloudless day, when you look away from the Sun at the rest of the sky, the sky is bright and you cannot see the stars. On the Moon, however, the sky is dark and you *can* see the stars even when the Sun is visible. Explain briefly.

Q15 If a beam of light from a medium with a higher index of refraction emerges into a medium with a lower index of refraction, what happens? (1) The emerging beam bends toward the normal (smaller angle). (2) The emerging beam bends away from the normal (larger angle). (3) The emerging beam does not bend at all.

PROBLEMS

Section 23.1

••P16 N closely spaced turns of wire are wound in the direction indicated on a hollow plastic ring of radius R, with circular cross section (Figure 23.114).

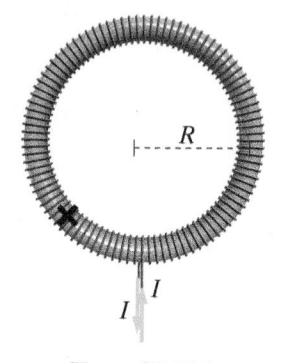

Figure 23.114

(a) If the current in the wire in Figure 23.114 is I, determine the magnetic field B at the location indicated by the ×, at the center of the cross section of the ring, and indicate the direction by drawing a vector at that location. Give a complete derivation, not just a final result. Use a circular path. (b) Throughout this region there is a uniform electric field E into the paper. This electric field begins to increase at a rate dE/dt, and there continues to be a current I in the wire. Now what is the magnitude of the magnetic field at the indicated location?

••P17 A rectangle 4 cm by 8 cm is drawn in empty space, in a region where there is no matter present, and the magnetic field is measured at all points along the rectangle as shown (Figure 23.115). What can you conclude about the region enclosed by the rectangle? Be as quantitative as you can.

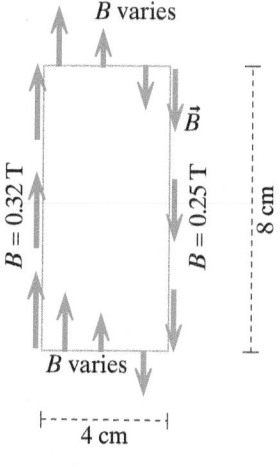

Figure 23.115

Section 23.2

•P18 If the magnetic field in a particular pulse has a magnitude of 1×10^{-5} T (comparable to the Earth's magnetic field), what is the magnitude of the associated electric field?

•P19 In Figure 23.116 electromagnetic radiation is moving to the right, and at this time and place the electric field is horizontal and points out of the page. The magnitude of the electric field is $E = 3000$ N/C. What are the magnitude and direction of the associated magnetic field at this time and place?

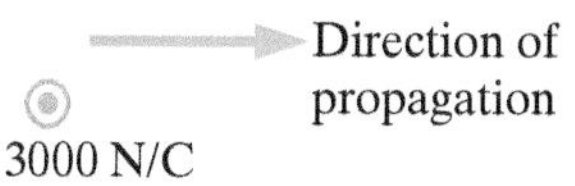

⊙   ⟶ Direction of
$E = 3000$ N/C   propagation

Figure 23.116

•P20 A pulse of electromagnetic radiation is propagating in the $+y$ direction. You have two devices that can detect electric and magnetic fields. You place detector 1 at location $\langle 0, -3, 0 \rangle$ m and detector 2 at location $\langle 0, 3, 0 \rangle$ m. (a) At time $t = 0$, detector 1 detects an electric field in the $-x$ direction. At that instant, what is the direction of the magnetic field at the location of detector 1? (b) At what time will detector 2 detect electric and magnetic fields?

•P21 A pulse of radiation propagates with velocity $\langle 0, 0, -c \rangle$. The electric field in the pulse is $\langle 7.2 \times 10^6, 0, 0 \rangle$ N/C. What is the magnetic field in the pulse?

Section 23.3

•P22 An electric field of 1×10^6 N/C acts on an electron, resulting in an acceleration of 1.8×10^{17} m/s² for a short time. What is the magnitude of the radiative electric field observed at a location a distance of 2 cm away along a line perpendicular to the direction of the acceleration?

•P23 A proton is briefly accelerated in the direction of the arrow labeled $\vec{a}$ in Figure 23.117. The vector $\vec{r}$ indicates the location of an observation location relative to the initial position of the proton. The vector $\vec{a}_\perp$ for this observation location is indicated on the diagram. (a) If θ is 40°, and the magnitude of the acceleration is 1.8×10^{17} m/s², what is the magnitude of $\vec{a}_\perp$? (b) What is the magnitude of the radiative electric field at the indicated observation location, if the magnitude of $\vec{r}$ is 0.011 m? (c) At the observation location, what is the direction of the radiative electric field? (1) The same as the direction of $\vec{a}$, (2) Opposite to the direction of $\vec{a}$, (3) The same as the direction of $\vec{a}_\perp$, (4) Opposite to the direction of $\vec{a}_\perp$, (5) The same as the direction of $\vec{r}$, (6) Opposite to the direction of $\vec{r}$, (7) Into the page, (8) Out of the page. (d) At the observation location, what is the direction of the ordinary Coulomb electric field due to the proton? (Same choices as above.) (e) At the observation location, what is the direction of the radiative magnetic field? (Same choices as above.)

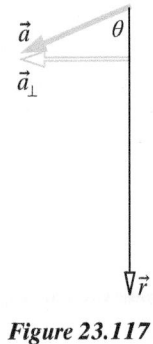

Figure 23.117

•**P24** A proton is accelerated in the direction shown by the arrow labeled $\vec{a}$ in Figure 23.118. Which of the arrows labeled A–F correctly shows the direction of the vector $\vec{a}_\perp$ at the observation location indicated by the vector $\vec{r}$?

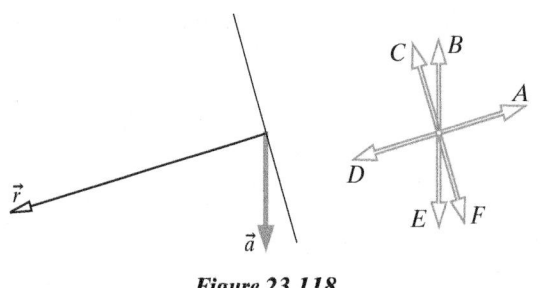

Figure 23.118

•**P25** A proton located at the origin is accelerated in the $-y$ direction for a brief time. **(a)** How much time passes before a detector located at $\langle 0.15,0,0 \rangle$ m detects a radiative electric field? **(b)** What is the direction of the radiative electric field observed at this location? **(c)** If the accelerated particle had been an electron instead of a proton, what would have been the direction of the radiative electric field at this location?

•**P26** At time $t = 0$ an electron at the origin is subjected to a force that briefly accelerates it in the $+z$ direction, with an acceleration of 8×10^{17} m/s^2. Before trying to answer the following questions, draw a clear diagram. Location D is $\langle -26,0,0 \rangle$ m, and location H is at $\langle 0,16,0 \rangle$ m. **(a)** At what time is a radiative electric field first detected at location D? **(b)** What is the direction of propagation of electromagnetic radiation that is detected at location D? **(c)** What is the direction of the radiative electric field observed at location D? **(d)** What is the magnitude of the radiative electric field observed at location D? **(e)** What is the direction of the radiative magnetic field observed at location D? **(f)** What is the magnitude of the radiative magnetic field observed at location D? **(g)** At what time is a radiative electric field first detected at location H? **(h)** What is the direction of propagation of electromagnetic radiation that is detected at location H? **(i)** What is the direction of the radiative electric field observed at location H? **(j)** What is the magnitude of the radiative electric field observed at location H? **(k)** What is the direction of the radiative magnetic field observed at location H? **(l)** What is the magnitude of the radiative magnetic field observed at location H?

•**P27** If the electric field inside a capacitor exceeds about 3×10^6 V/m, the few free electrons in the air are accelerated enough to trigger an avalanche and make a spark. In the spark shown in Figure 23.119, electrons are accelerated upward and positive ions are accelerated downward.

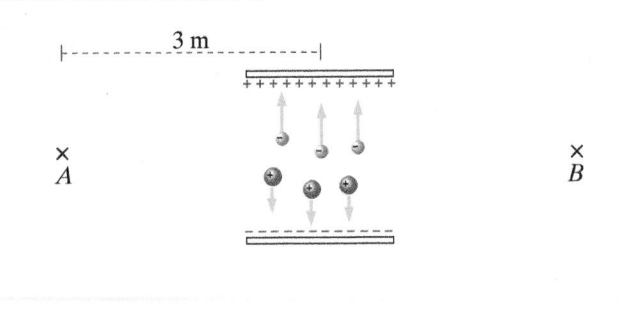

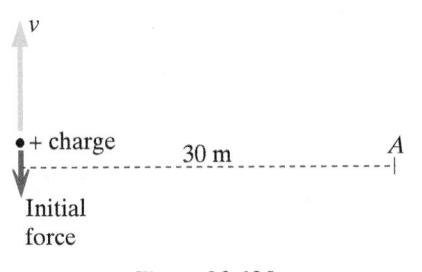

Figure 23.119

(a) Qualitatively, explain the directions and relative magnitudes of the radiative electric fields at location A, to the left of the capacitor, due to the indicated motions of the electrons and of the ions (actually, in a spark there are also accelerations in the opposite direction when the electrons and ions collide with air molecules). **(b)** Repeat the analysis for locations B (to the right) and C (to the right and down a bit). **(c)** If you are at location A, 3 m to the left of the capacitor, how long after the initiation of the spark could you first detect a magnetic field? What is the direction of the radiative magnetic field? (Before the spark occurs there is no magnetic field anywhere in this region.) **(d)** In addition to producing electromagnetic radiation, the moving charges in the spark produce electric fields according to Coulomb's law and magnetic fields according to the Biot–Savart law. Explain why these fields are much smaller than the radiative fields, if you are far away from the capacitor.

••**P28** A positive charge coasts upward at a constant velocity for a long time. Then at $t = 0$ (Figure 23.120) a force acts downward on it for 1 ns (1×10^{-9} s). After this force stops acting, the charge coasts upward at a smaller constant speed for 1 ns; then a force acts upward for 1 ns and it resumes its original speed. The new position reached at $t = 3$ ns is much less than a millimeter from the original position.

Figure 23.120

You stand at location A, 30 m to the right of the charge (Figure 23.120), with instruments for measuring electric and magnetic fields. What will you observe due to the motion of the positive charge, at what times? You do not need to calculate the magnitudes of the electric and magnetic fields, but you do need to specify their directions, and the times when these fields are observed.

••**P29** An electron is initially at rest. At time $t_1 = 0$ it is accelerated upward with an acceleration of 1×10^{18} m/s^2 for a very short time (this large acceleration is possible because the electron has a very small mass). We make observations at location A, 15 m from the electron (Figure 23.121).

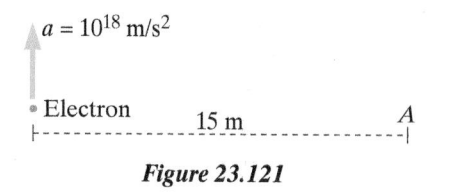

$a = 10^{18}$ m/s^2

• Electron   15 m A

Figure 23.121

(a) At time $t_2 = 1$ ns (1×10^{-9} s), what are the magnitude and direction of the electric field at location A due to the electron? (b) At what time t_3 does the electric field at location A change? (c) What is the direction of the radiative electric field at location A at time t_3? (d) What is the magnitude of this electric field? How does it compare to the magnitude of the field at location A at time t_2? (e) Just after time t_3, what is the direction of the magnetic force on a positive charge that was initially at rest at location A? Explain with a diagram.

••**P30** At time $t = 0$, electrons in a vertical copper wire (Figure 23.122) are accelerated downward for a very short time Δt (by a power supply that is not shown). A proton, initially at rest, is located a very large distance L from the wire.

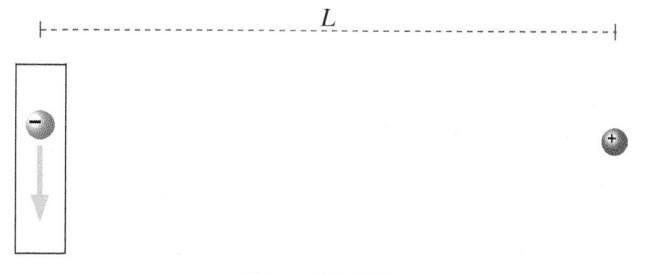

Figure 23.122

Explain in detail what happens to the proton (neglect gravity), and at what time.

••**P31** A slab (pulse) of electromagnetic radiation that is 15 cm thick is propagating downward (in the $-y$ direction) toward a short horizontal copper wire located at the origin and oriented parallel to the z axis, as shown in Figure 23.123. (a) The direction of the electric field inside the slab is out of the page (in the $+z$ direction). On a diagram, show and describe clearly the direction of the magnetic field inside the slab.

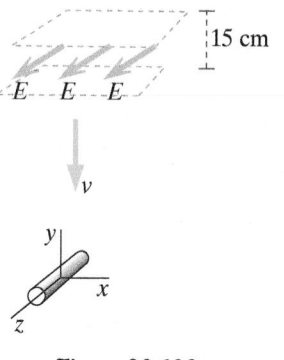

15 cm

E E E

v

Figure 23.123

(b) You stand on the x axis at location $\langle 12, 0, 0 \rangle$ m, at right angles to the direction of propagation of the pulse. Your friend stands on the z axis at location $\langle 0, 0, 12 \rangle$ m. The pulse passes the copper wire at time $t_1 = 0$, and at a later time t_2 you observe new nonzero electric and magnetic fields at your location, but your friend does not. Explain. What is t_2? (Give a numerical answer.) How long a time do these new fields last for you? (c) On a diagram, show and describe clearly the directions of these new nonzero fields ($\vec{E}$ and $\vec{B}$) at your location. Explain briefly but carefully.

Section 23.4

•**P32** A particular AM radio station broadcasts at a frequency of 990 kHz. (a) What is the wavelength of this electromagnetic radiation? (b) How much time is required for the radiation to propagate from the broadcasting antenna to a radio 3 km away?

•**P33** What is the wavelength of ultraviolet light whose frequency is 5×10^{15} Hz?

•**P34** The wavelength of red light is about 695 nm. What are the frequency and period of this radiation?

•**P35** Calculate the wave length for several examples of sinusoidal electromagnetic radiation:

radio, 1000 kHz, $\lambda = ?$

television, 100 MHz, $\lambda = ?$

red light, 4.3×10^{14} Hz, $\lambda = ?$

blue light, 7.5×10^{14} Hz, $\lambda = ?$

(Note for comparison that an atomic diameter is about 1×10^{-10} m.)

Section 23.5

•**P36** A small laser used as a pointer produces a beam of red light 5 mm in diameter and has a power output of 5 mW. What is the magnitude of the electric field in the laser beam?

••**P37** A 100 W light bulb is placed in a fixture with a reflector that makes a spot of radius 20 cm. Calculate approximately the amplitude of the radiative electric field in the spot.

Section 23.6

•**P38** A sinusoidal beam of electromagnetic radiation that is only 2 cm high propagates in the $+x$ direction, with the electric field in the $\pm z$ direction. The beam strikes a metal rod that is centered on the origin and aligned along the z axis, as shown in Figure 23.124. You place detectors sensitive to electromagnetic radiation at location $\langle 0, -17, 0 \rangle$ m (detector A) and location $\langle 0, 0, 17 \rangle$ m (detector B). Answer the following questions for detector A and for detector B: Will the detector detect electromagnetic radiation? If so: (a) What is the direction of propagation of the electromagnetic radiation detected? (b) What is the direction of the electric field in the radiation? (c) What is the direction of the magnetic field in the radiation?

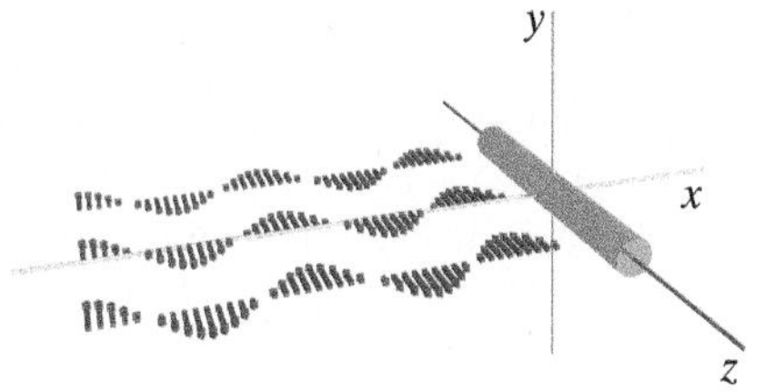

Figure 23.124

Section 23.8

•**P39** The index of refraction of water is 1.33. At what speed would a wave crest in a beam of light travel through this medium?

•**P40** If a diver who is underwater shines a flashlight upward toward the surface, at an angle of 29° from the normal, at what angle does the light emerge from the water? The indices of refraction are: water, 1.33; air, 1.00029.

•**P41** To get total internal reflection at the interface of water (refractive index 1.33) and a plastic whose refractive index is 1.46: **(a)** Which material must the light start in? (1) It doesn't matter, (2) Water, (3) Plastic, **(b)** What is the critical angle?

•**P42** For a particular liquid–air interface the critical angle (for total internal reflection) is 59.8°. The index of refraction of air is 1.00029. What is the index of refraction of the liquid?

••**P43** A spotlight mounted beside a swimming pool sends a beam of light at an angle into the pool, as shown in Figure 23.125. The light hits the water 2.4 m from the edge of the pool. The pool is 2 m deep and the light is mounted 1.2 m above the surface of the water. (The diagram is not drawn to scale.) The indices of refraction are: water, 1.33; air, 1.00029.

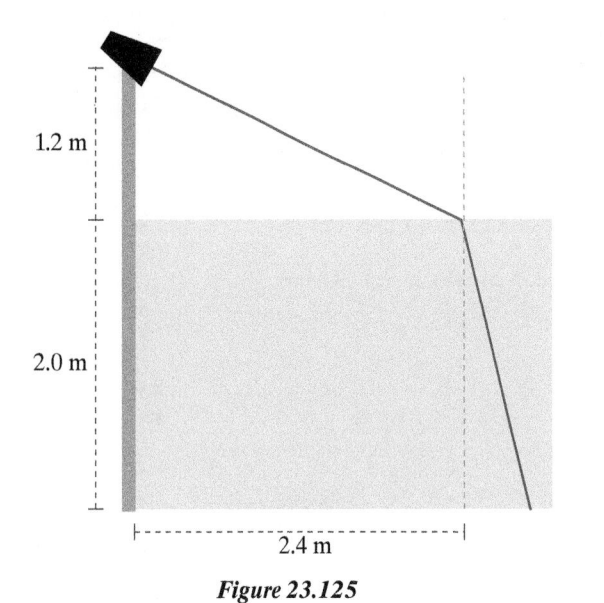

Figure 23.125

(a) What is the angle of the beam from the normal in the air?
(b) What is the angle of the beam from the normal in the water?
(c) How far from the edge of the pool is the spot of light on the bottom of the pool?

Section 23.10

•**P44** A point source of green light is placed on the axis of a thin converging lens, 15 cm to the left of the lens. The lens has a focal length of 10 cm. Where is the location of the image of the source? Is it a real or a virtual image? If you placed a sheet of paper at the location of the image, what would you see on the paper?

•**P45** A point source of green light is placed on the axis of a thin converging lens, 7 cm to the left of the lens. The lens has a focal length of 10 cm. Where is the location of the image of the source? Is it a real or a virtual image? If you placed a sheet of paper at the location of the image, what would you see on the paper?

•**P46** A thin diverging lens of focal length 25 cm is placed 18 cm to the right of a point source of blue light on the axis of the lens. Where is the image of the source? Is it a real or a virtual image? If you placed a sheet of paper at the location of the image, what would you see on the paper?

•**P47** A thin converging lens of focal length 20 cm is located at the origin and its axis lies on the x axis. A point source of red light is placed at location $\langle -25, 1, 0 \rangle$ cm. Where is the location of the image of the source? Is it a real or a virtual image? It helps to draw a diagram and trace the "easy" rays.

•**P48** A thin converging lens of focal length 20 cm is located at the origin and its axis lies on the x axis. A point source of red light is placed at location $\langle -14, 1, 0 \rangle$ cm. Where is the location of the image of the source? Is it a real or a virtual image? It helps to draw a diagram and trace the "easy" rays.

•**P49** You have a thin converging lens whose focal length is 0.3 m and a sheet of white paper on which to display a real image of a small tree that is 2 m tall and 40 m away. How far in back of the lens should you place the paper in order to get a sharp image of the tree? How tall is the image of the tree? Is it inverted or right side up?

••**P50** You have a small but bright computer display that is only 4 cm wide by 3 cm tall. You will place a lens near the display to project an image of the display about 2.5 m tall onto a large screen that is 6 m from the lens of the projector. **(a)** Should you use a converging lens or a diverging lens? **(b)** Approximately, what should the focal length of the lens be? **(c)** Suppose that you choose a lens that has the focal length that you calculated in part (b). What exactly should be the distance from the computer display to the lens? **(d)** In order that the screen display be right side up, should the computer display be inverted or right side up? **(e)** On another occasion the screen is moved close, so that it is only 4 m from the lens. To focus the image, you have to readjust the distance between the computer display and the lens. What is the new exact distance from the computer display to the lens? **(f)** How tall is the image on the screen, now that the screen is only 4 m from the lens?

COMPUTATIONAL PROBLEMS

•••**P51** **(a)** Write a computer program that displays a sinusoidal electromagnetic wave propagating through space. Make the wavelength 600 m (corresponding to a frequency of about 500 kHz, which is in the AM radio frequency band) and the amplitude of the electric field 1.0×10^4 V/m. (This is roughly the amplitude that would be measured a few meters from a 50,000 W radio transmitter. On average, the amplitude of the radiative field from the Sun at Earth's orbital radius is about 700 V/m.)

Animate the wave as a function of time, making it propagate in the positive x direction, with the electric field polarized in the y direction. Display both the electric and magnetic field vectors, and make sure they have the correct directions relative to each other. Display at least three full wavelengths, with enough observation locations per wavelength that you can clearly see the sinusoidal character of the wave. Think carefully about scaling. The length of the arrow objects must be scaled such that both the arrows and the wavelength can be seen. Also, to see smooth wave-like motion, the time step must be a small fraction of the period of the wave.

(b) Place a positron initially at rest in the presence of the electromagnetic wave from part (a). (Using a positron instead of an electron makes it easier to think about signs and directions.) Modify your program to model the motion of the positron due to its interaction with the electromagnetic wave. Leave a trail.

This is a relativistic situation since the instantaneous speed of the particle can get quite high. To accurately model the positron's motion, you'll need to use the relativistic relationship between momentum and velocity:

$$\vec{p} = \frac{m\vec{v}}{\sqrt{1-(v^2/c^2)}} \quad \text{and therefore} \quad \vec{v} = \frac{\vec{p}}{m}\left(\frac{1}{\sqrt{1+(p/(mc))^2}}\right)$$

ANSWERS TO CHECKPOINTS

1 $\vec{B} = \langle 3.3 \times 10^{-3}, 0, 0 \rangle$ T

2 An electron accelerated to the left produces a radiative electric field as shown in Figure 23.126.

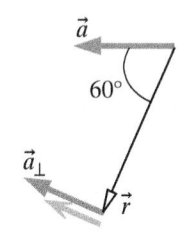

Figure 23.126

3 5.6×10^{14} Hz

4 726 V/m

5 9.3×10^{-6} N/m²

6 Since electric fields are not blocked by matter, the radiated fields must be present on the other side of the cardboard. Yet there is clearly no light there, so the net field must be zero there. There must be an additional source of electric and magnetic field somewhere, producing a field that adds to the radiative field to produce a zero net field. The electrons in the atoms of the cardboard must be accelerated by the electric field of the incoming radiation in such a way that the re-radiated field is equal and opposite to the incident field. The net field on the other side of the cardboard is the superposition of the two fields, and is zero.

7 Top: Horizontally polarized radiation accelerates electrons back and forth along the horizontal metal foils. The electrons re-radiate, with much of the radiation going off in other directions. The net field reaching the antenna does accelerate electrons along the antenna, but does not produce enough current to light the bulb. Bottom: Horizontally polarized radiation across the narrow width of the foils is not very effective in causing a large response in the foils. There is minimal re-radiation, so most of the original radiation accelerates electrons in the receiver along its length, thus lighting the bulb.

8 $n = c/v = 3.5$

9 $38.7°$

10 0.6 m

11 (a) 125 cm; **(b)** 25; **(c)** 19.23 cm

Answers to Odd-Numbered Problems

Chapter 1

P13 2.15×10^7 m/s   **P15** **(a)** $\vec{a} = \langle -4, -3, 0 \rangle$ **(b)** $\vec{b} = \langle -4, -3, 0 \rangle$
(c) True **(d)** $\vec{c} = \langle 4, 3, 0 \rangle$ **(e)** True **(f)** False   **P17** $\langle 0.04, -3.4, 60.0 \rangle$
P19 $\langle 0.58, 0.58, 0.58 \rangle$; $\langle 0.58, 0.58, 0.58 \rangle$ **P21** 458.26 m/s^2
$\langle 0.872872, 0.436436, -0.218218 \rangle$
P23 **(a)** $\langle 3 \times 10^{-10}, -3 \times 10^{-10}, 8 \times 10^{-10} \rangle$ m
(b) 9.1×10^{-10} m **(c)** $\langle 0.33, -0.33, 0.88 \rangle$
P25 **(a)** $\langle -5.5, -20, 0 \rangle$ m **(b)** 20.74 m
P27 **(a)** $\langle -10 \times 10^{10}, -17 \times 10^{10}, 0 \rangle$ m
(b) $\langle 10 \times 10^{10}, 17 \times 10^{10}, 0 \rangle$ m   **P29** $\langle x_p - x_e, y_p - y_e, z_p - z_e \rangle$;
$\langle x_e - x_p, y_e - y_p, z_e - z_p \rangle$
P31 **(a)** $\langle -2.01 \times 10^5, 5.2 \times 10^4, -1 \times 10^3 \rangle$ m/s
(b) 2.08×10^5 m/s   **P33** **(a)** $\langle 974, 0, 684 \rangle$ m
(b) $\langle 1.50, 0, 1.05 \rangle$ m/s   **P35** 3.0 s;
$\langle 5.33 \times 10^2, -7.33 \times 10^2, 5.67 \times 10^2 \rangle$ m
P37 **(a)** $\langle 4.4, -6.4, 0 \rangle$ m/s **(b)** $\langle 4, -7.7, 0 \rangle$ m/s **(c)** The time
interval from $t = 6.3$ s to 6.8 s **(d)** $\langle 0.132, -0.192, 0 \rangle$ m
P39 **(a)** $\langle 0, 9 \times 10^5, -4 \times 10^5 \rangle$ m/s **(b)** $\langle 0.02, 6.34, -2.86 \rangle$ m
P41 147 m   **P43** **(a)** $\langle 22.3, 26.1, 0 \rangle$ m/s **(b)** $\langle 44.6, 52.2, 0 \rangle$ m
(c) $\Delta t = 1.0$ s was too big.   **P45** 6.2 kg $\cdot$ m/s
P47 2.24×10^5 kg $\cdot$ m/s   **P49** 6.7 kg $\cdot$ m/s   **P51** **(a)** $\langle -2mv_x, 0, 0 \rangle$
(b) 0 **P53** $\langle 0, 4.1, 0 \rangle$ kg $\cdot$ m/s   **P55** 0; 500 kg $\cdot$ m/s
P57 $\langle 20, 50, -10 \rangle$ m/s   **P59** $\langle 0, 0, -11 \rangle$ m
P61 7.35×10^{-19} kg $\cdot$ m/s   **P63** 1.93×10^{-20} kg $\cdot$ m/s   **P65** 1250
P67 $0.65c$

Chapter 2

P9 $\langle -9510, 0, 0 \rangle$ N   **P11** $\langle 180, -180, 700 \rangle$ N
P13 **(a)** $\langle 0.011, -0.005, -0.003 \rangle$ kg $\cdot$ m/s
(b) $\langle 0.003, -0.049, -0.003 \rangle$ kg $\cdot$ m/s **(c)** $\langle -0.004, -0.022, 0 \rangle$ N
P15 $\langle 91, 106, 80 \rangle$ m/s   **P17** **(a)** 12 m/s **(b)** 0.06 s **(c)** 1.0×10^6 N
(d) 41 **(e)** $\vec{v}_{avg} = (\vec{v}_i + \vec{v}_f)/2$ was valid   **P19** Depends on
what approximations you make.
P21 $\langle 0.15, 2.102, 0 \rangle$ m, $\langle 3, 2.04, 0 \rangle$ m/s;
$\langle 0.3, 2.106, 0 \rangle$ m, $\langle 3, 0.08, 0 \rangle$ m/s;
$\langle 0.45, 2.012, 0 \rangle$ m, $\langle 3, -1.88, 0 \rangle$ m/s   **P23** $\langle -3.83, 0, 2.79 \rangle$ m
P25 **(a)** 4 m/s **(b)** 12 m   **P27** **(a)** 90 mi **(b)** 37.5 mi/h **(c)** 45
mi/h   **P29** **(a)** none **(b)** 6 **(c)** 3 **(d)** 2 **(e)** none **(f)** none **(g)** 5
(h) 1 **(i)** 4 **P31** **(a)** 6 **(b)** 7 **(c)** 5 **P33** assuming the flower pot
fell from rest, it started from 5.5 m above the top of the window
P35 **(a)** $\langle -11, 11.1, -6 \rangle$ m/s **(b)** arithmetic
(c) $\langle -11, 13.6, -6 \rangle$ m/s **(d)** $\langle 3.5, 6.8, -9 \rangle$ m **(e)** 0
(f) $0 = 16$ m/s $+ (-9.8$ N/kg$)\Delta t$ **(g)** 1.63 s **(h)** 13 m
P37 **(a)** $\Delta t = \sqrt{2h/g}$ **(b)** $v_{fy} = -g\sqrt{2h/g}$ **P39** **(a)** 1 **(b)** $9/8$
(c) $3/4$ **P41** 23 cm   **P43** $\langle 0, 0.2807, 0 \rangle$ m   **P45** At $t = 0.04$ s,
$s = 0.0157$ m and $L = 0.0843$ m. At $t = 0.08$ s, $s = 0.0466$ m and
$L = 0.0554$ m   **P47** **(a)** $\Delta t \approx 6.15 \times 10^{-10}$ s **(b)** 18 cm

Chapter 3

P11 $3/16$ **P13** **(a)** $\langle 2.8 \times 10^8, 0, -2.8 \times 10^8 \rangle$ m **(b)** 4.0×10^8 m
(c) $\langle 0.7, 0, -0.7 \rangle$ **(d)** $\langle -4.9 \times 10^{28}, 0, 4.9 \times 10^{28} \rangle$ N
P15 1.12×10^{24} N   **P17** **(a)** 2×10^{22} N **(b)** 2×10^{22} N
P19 **(a)** $\langle -7 \times 10^{11}, 5 \times 10^{11}, 0 \rangle$ m **(b)** 8.6×10^{11} m
(c) $\langle -0.81, 0.58, 0 \rangle$ **(d)** 1.8×10^{21} N **(e)** 1.8×10^{21} N
(f) $\langle -1.5 \times 10^{21}, 1.0 \times 10^{21}, 0 \rangle$ N
(g) $\langle 1.5 \times 10^{21}, -1.0 \times 10^{21}, 0 \rangle$ N   **P21** about 1×10^{-6} N and
about 10 N   **P23** 3.2×10^4 m   **P25** mg
P27 **(a)** $\langle 1.39 \times 10^{-4}, -2.32 \times 10^{-4}, -1.86 \times 10^{-4} \rangle$ N
(b) $\langle 8.4 \times 10^{-4}, 1.40 \times 10^{-3}, 1.12 \times 10^{-3} \rangle$ kg $\cdot$ m/s
P29 **(a)** $\langle 3.04 \times 10^3, 1.496 \times 10^4, 0 \rangle$ m/s **(b)** $\langle 3.004 \times 10^{12}$,
$4.01 \times 10^{12}, 0 \rangle$ m **(c)** Force and velocity are not constant, so a
large time interval will give inaccurate results.   **P31** **(b)** about
5 kg $\cdot$ m/s **(c)** about 500 m   **P33** To the right   **P35** c
P37 **(a)** $\langle -0.9, 0.6, 0 \rangle$ m **(b)** 1.08 m **(c)** $\langle -0.833, 0.556, 0 \rangle$
(d) 2.29×10^{-16} N **(e)** $\langle 1.91 \times 10^{-16}, 1.27 \times 10^{-16}, 0 \rangle$ N
(f) 6.17×10^{-8} N **(g)** $\langle -5.14 \times 10^{-8}, 3.43 \times 10^{-8}, 0 \rangle$ N
(h) 2.69×10^8 **(i)** 2.69×10^8 **P39** 2.3×10^{-8} N, 5.4×10^{-51} N;
will repel   **P41** **(a)** $\langle 2.2, -0.4, 2.6 \rangle$ kg $\cdot$ m/s **(b)** $\langle 0, -7.84, 0 \rangle$ N
(c) $\langle 2.2, -1.2, 2.6 \rangle$ kg $\cdot$ m/s   **P43** 4.6×10^6 m, so inside the Earth
P45 **(a)** $\langle 1.875, 0.5, 0 \rangle$ kg $\cdot$ m/s **(b)** $\langle 5.36, 1.43, 0 \rangle$ m/s
P47 $\langle -2, 14, 7 \rangle$ kg $\cdot$ m/s   **P49** **(a)** $\langle 0, 0.09, -0.18 \rangle$ kg $\cdot$ m/s
(b) $\langle 0, 0.09, -0.18 \rangle$ kg $\cdot$ m/s **(c)** $\langle 0, 1.5, -3 \rangle$ m/s   **P51** **(a)** $m|\vec{v}|$
(b) $|\vec{F}|$ **(c)** $m|\vec{v}|$ **(d)** $|\vec{v}_{mosquito}| \approx (M/m)|\vec{v}_{car}|$
P53 **(a)** $\langle 25, 1, 0 \rangle$ kg $\cdot$ m/s **(b)** approximately zero
(c) $\langle 25, 1, 0 \rangle$ kg $\cdot$ m/s **(d)** $\langle 7, -4, 0 \rangle$ kg $\cdot$ m/s
P55 $\langle 3138, -1750, 4200 \rangle$ m/s   **P57** $\langle 13.4, -1.4, 0 \rangle$ m/s
P59 $v_{fx} = [m(v_1x - v_2x) + Mv]/M$; $v_{fy} = [m(v_1y - v_2y)]/M$
P61 $-(m/M)\langle v \cos \theta, v \sin \theta, 0 \rangle$
P63 about 1×10^{-13} m/s

Chapter 4

P21 1.7×10^4 kg/m^3 **P23** **(a)** 1.06×10^{-25} kg/atom
(b) 2.02×10^8 atoms **(c)** 0.870 kg   **P25** 1.35×10^4 N/m
P27 450 N/m   **P29** 1950 N/m   **P31** 27 N/m   **P33** 46 N/m
P35 **(a)** 5×10^{-5} m^2 **(b)** 7.2×10^{-5} m^2 **(c)** 3.6×10^{-5} m^2
P37 0.21 mm   **P39** 2.0×10^{11} N/m^2
P41 **(a)** $-F/(m_1 + m_2 + m_3)$ **(b)** $-m_3F/(m_1 + m_2 + m_3)$;
$-(m_2 + m_3)F/(m_1 + m_2 + m_3)$
(c) $\langle -m_3F/(m_1 + m_2 + m_3), 0, 0 \rangle$ **P43** 3 N   **P45** $\langle 10.2, 0, 0 \rangle$ m;
$\langle 4.23, 0, 0 \rangle$ m/s   **P47** depends on your data   **P49** $3Mg$
P51 -0.105 m   **P53** 1.6 s   **P55** 0.4 s   **P57** **(a)** 0.94 m/s
(b) 15 m/s^2 **P61** **(a)** $dp/dt = -mg \sin(s/L)$
(b) $d^2s/dt^2 + (g/L)s = 0$ **(c)** $2\pi\sqrt{(L/g)}$ **P63** 2710 m/s
P65 about 1×10^{19} molecules   **P67** 20.4 m; 19.8 m

Chapter 5

P7 (a) $\vec{0}$ (b) 3 (c) -392 N (d) 392 N (e) 497.4 N (f) 306.2 N
(g) -306.2 N **P9** (a) $\vec{0}$ (b) $\vec{0}$ (c) rope, floor (d) 195.3 N
(e) -195.3 N (f) Earth, floor, rope (g) 163.9 N (h) 294 N
(i) 130 N **P11** (a) 6098 N, 3980 N, 7840 N (b) 0.049, 0.032, 0.063
P13 (a) 1040 N, 1420 N (b) 528 N, 726 N **P15** (a) 2.95 m/s^2
(b) 39.3 N **P17** 2.24 m/s, tangential **P19** (a) $(1.8 \times 10^{23}$ N$)\hat{p}$
(b) 3.13×10^{29} kg$\cdot$m/s **P21** $\langle -4.97 \times 1022, 1.88 \times 10^{22}, 0 \rangle$ N,
$\langle 6.47 \times 10^{22}, 1.71 \times 10^{23}, 0 \rangle$ N **P23** (a) 3.67 m/s (b) tangent to
path (c) inward, toward center **P25** (a) d (b) b
(c) 8.74×10^{-15} N **P27** $\vec{0}$, 0.349 m/s, 14.0 kg$\cdot$m/s, 0.977 N
P29 1.76 m/s, 66.3 N **P31** 8.78 N **P33** (a) a (b) $\vec{0}$ (c) 492 N
(d) $+x$ (e) $\langle 0, 492, 0 \rangle$ N (f) $\langle 296, 369, 0 \rangle$ N **P35** $\sqrt{Rg}$
P37 (a) $\vec{0}$ (b) 8.71 N (c) 871 N (d) $\langle 0, 1130, 0 \rangle$ N
(e) 1.4×10^4 N/m **P39** 6.65×10^{12} kg **P41** 1.46 m/s
P43 (a) 13.7 m/s (b) direction changes (c) spring exerts a force
(d) 300 N (e) 32.8 kg **P45** $\sqrt{4\pi^2 mR/(k_s(R-L))}$
P47 3.0×10^4 m/s **P49** (a) $\langle 0, 200, 0 \rangle$ kg$\cdot$m/s^2
(b) $\langle 0, -549, 0 \rangle$ N (c) $\langle 0, 749, 0 \rangle$ N (d) $\langle 0, -200, 0 \rangle$ kg$\cdot$m/s^2
(e) $\langle 0, -549, 0 \rangle$ N (f) $\langle 0, 349, 0 \rangle$ N (g) heavier (h) lighter
P51 2.69 m/s; 1.21 s **P53** (a) 4.24×10^7 m (b) 0,48 s (c) 85 min
$= 1.4$ hr (d) 7.9×10^3 m/s (e) 3250 s $= 0.9$ hr
P55 (a) 1.60×10^6 m/s(0.005c) (b) yes (c) 1.11×10^{37} kg
(d) 5.6 million solar masses

Chapter 6

P9 6 **P11** 0.99999985c; 1.5×10^{-10} J **P13** (a) 5.76×10^{-10} J
(b) no (c) 1.48×10^{-10} J (d) C **P15** 81.6 J **P17** 0
P19 104 J **P21** 585 J **P23** $\langle 0, -4.9, 0 \rangle$ N; 4.9 J
P25 -8.6 J **P27** 2.5×10^{-19} J
P29 189 J **P31** 2.58×10^{-9} J; 4.59×10^{-10} J; 2.12×10^{-9} J;
7.28×10^{-9} J; 4.59×10^{-10} J; 6.82×10^{-9} J
P33 10.2 m/s **P35** 4.5 m/s **P37** 9.81 m/s **P39** 1×10^{-13} N
P41 a, b, e, and f; 0.99 **P43** 0.5 MeV
P45 (a) 3.499767×10^{-8} J (b) 3.499708×10^{-8} J (c) decreased
(d) 3.72 MeV **P47** (b) 1.194×10^{-12} J (c) 2.2×10^{-14} J
(d) $\gamma = 1.002$ **P49** 19.8 m/s **P51** -2200 J **P53** 6.3 m/s
P55 (a) 9.8 J (b) 6.3 m/s **P57** 102 m **P59** (a) 2970 m/s
(b) 2835 m/s **P61** 25.5 m/s **P63** 4.26×10^3 m/s **P65** 0.99c
P67 $mgR/2$, half **P69** (a) $mgL(1 - \cos\theta)$ (c) $-mg\sin\theta$
(d) $\sqrt{4gL}$ **P71** (b) $\sqrt{1 - m^2/M^2}c$, high speed (c) 0.99c
(d) 8.4×10^{-18} m **P73** (a) $\langle 0, 13, 0 \rangle$ m(b) $\langle 0, -784, 0 \rangle$ N
(c) -1.02×10^4 J (d) 0 (e) -1.02×10^4 J (f) 0 (g) -1.02×10^4 J
(h) 0 (i) -1.02×10^4 J (j) -1.02×10^4 J
(k) all the same **P75** (a) 1.26×10^7 m/s (b) 1.57×10^{-14} m
(c) 6.38×10^{-15} m; gap is only 3×10^{-15} m (d) 1.9×10^{13} J

Chapter 7

P17 0.5 m/s **P19** (a) Mg/k_s; $-\frac{1}{2}(Mg)^2/k_s$ (b) $-2Mg/k_s$
(c) $\sqrt{k_s s^2/M - 2g(L-s)}$ **P21** 2.37 m **P23** 0.0746 m
P25 (a) -0.2 eV (b) 0.2 eV (c) $6a/r^7$ **P27** 80.5°C **P29** -516 J
|31| (a) 58.9°C (b) energy transfer between system and
surroundings is negligible; heat capacities independent of
temperature (c) 1.81×10^4 J **P33** 490 J; 2040 kg/s **P35** 3 W
P37 (a) 9 J (b) 9 J (c) 3 m/s (d) 0.2 s (e) 0 (f) -9 J (g) 15 m/s
(h) 18 m/s (i) 225 J (j) 325 J (k) 99 J (l) 3.3 m (m) 99 J (n) 3 m
(o) 90 J (p) -9 J **P39** 31.9 m; too large **P41** (a) 353 N (b) 696
N **P43** (a) same as hand (b) small (c) large (d) reduced

Chapter 8

P9 13.6 eV **P11** 6190 atoms **P13** 1.4 eV **P15** (a) excited states
of 1.91 eV and 2.48 eV above the ground state; excited states of
0.57 eV and 2.48 eV above the ground state (b) Bombard with
electrons whose kinetic energy is greater than 0.57 eV and less
than 1.91 eV and see whether there is a dark line. **P17** (a) 0.8,
1.9, 1.1, 3.8, 3.0, and 1.9 eV (note two transitions of 1.9 eV)
(b) 1.9, 3.0, and 3.8 eV **P19** 11, 5, 2, 9, 3, and 6 eV; 6, 9, and 11
eV **P21** (a) 3, 2.5, 1.9, 0.5, 1.1, and 0.6 eV (b) 1.9, 2.5, and 3.0
eV **P23** (a) one possible scheme: ground state, and 0.3, 0.8,
and 2.8 eV above the ground state (b) no (c) 0.3, 0.8, and 2.8 eV
P25 (a) 0.023 eV (b) 4.1×10^{-38} kg (c) 0.023, 0.046, and 0.069
eV **P27** (a) 0.015 eV (b) 1.8 eV (c) 120 (d) 0.03 and 0.045 eV

Chapter 9

P9 4.6×10^6 m from center of Earth **P11** $\langle L/4, L/4, 0 \rangle$ from
lower left; $\langle -3L/4, L/4, 0 \rangle$ from lower right **P13** 4.7×10^4 J
P15 (a) $\langle 25, 97, 0 \rangle$ kg$\cdot$m/s (b) $\langle 3.125, 12.125, 0 \rangle$ m/s (c) 809 J
(d) 627.125 J (e) 181.875 J (f) 68.203 J and 113.672 J, which add
to 181.875 J **P17** (a) center of mass moves with constant
velocity (b) the centers of the two orbits coincide at the center
of mass; M_1 must be greater than M_2 (c) 7.5×10^{29} kg and
5×10^{29} kg **P19** 14.3 J **P21** 395 J **P23** 10.5 rad/s
P25 (a) 2.99×10^4 m/s (b) 2.686×10^{33} J (c) 7.27×10^{-5} rad/s
(d) 2.60×10^{29} J (e) 2.69×10^{33} J **P27** (a) 0.007 J (b) 0.90 m/s
(c) 0.207 J (d) 313 rad/s **P29** (a) 6.75 m/s (b) 225 J (c) 65 J
P31 typical results are 1100 N, -350 J, 0.2 s **P33** (a) $\sqrt{gh + v_i^2}$

(b) $\left((M+m)gh + (M+\frac{1}{2}m)v_i^2)/(M+\frac{1}{2}m)\right)^{\frac{1}{2}}$

P35 (a) $\sqrt{Fb/M}$ (b) $F(d - b)$ **P37** (a) $\left(2Fw/M + v_i^2\right)^{\frac{1}{2}}$

(b) $\left((2Fd)/(3mr^2) + \omega_i^2\right)^{\frac{1}{2}}$ **P39** (a) 24.2 J (b) 25.0 J

P41 $\sqrt{2Mg(y - y_0)/I + \omega_i^2}$ **P43** (a) $\sqrt{2Fd/(M + 4m)}$

(b) $\left(Fw/(2mb^2 + \frac{1}{4}MR^2)\right)^{\frac{1}{2}}$ **P45** (a) $\sqrt{2Fd/(M + 2m)}$

(b) $Fs - \frac{1}{4}\left(MR^2 + mL^2\right)\omega^2$

Chapter 10

P15 $v_{1,f} = v_{1,i}(m_1 - m_2)/(m_1 + m_2)$; $v_{2,f} = 2m_1 v_{1,i}/(m_1 + m_2)$
P17 (a) $\langle 312, -398, 185 \rangle$ m/s (b) 4.3×10^6 J (c) 1
P19 (a) $\langle 2500, -3500, -1500 \rangle$ kg$\cdot$m/s (b) 8.02×10^7 J (c) 0
(d) 6.82×10^7 J (e) 1.20×10^7 J (f) 4.8×10^7 J (g) 0 **P21** (a) car
+ truck; Momentum Principle (b) 0 (c) 2.73×10^6 J (d) inelastic
P23 (a) $\langle -1.40 \times 10^{-19}, 0, 0 \rangle$ kg$\cdot$m/s
(b) $\langle 2.86 \times 10^{-19}, 0, 0 \rangle$ kg$\cdot$m/s (c) 9.17 MeV (d) 0.78 MeV
(e) 2.28×10^{-14} m **P25** $|\vec{p}_n| = |\vec{p}_{\pi-}|$; $E_n + E_{\pi-} = 1196$ MeV;
$E_n^2 - (p_n c)^2 = (939 \text{ MeV})^2$; $E_{\pi-} - (p_{\pi-} c)^2 = (140 \text{ MeV})^2$
P27 (a) 0.5c (b) 1.87 m **P29** (a) 3 m/s (b) 4.8×10^{-8} eV;
negligible compared to 10.2 eV **P31** 259 MeV; 8×10^7 m/s
P33 (answers given for x components) (a) $mv_1/(m + M)$
(b) $Mv_1/(m + M)$; $-mv_1/(m + M)$ (c) The momenta have
equal magnitudes before the collision, and after the collision
they also have to have equal magnitudes in order for the total
momentum to remain zero. If a magnitude changes, that would
mean the energy changes, so the Energy Principle would be
violated. (d) $-(M - m)v/(M + m)$; $2mv/(M + m)$

Chapter 11

P13 $22\,\hat{k}$ **P15** $\langle 0,0,144 \rangle$ kg·m^2/s **P17 (a)** out of page
(b) $d_1 p_1 \sin\alpha$ **(c)** out of page **(d)** $d_2 p_2$ **P19** 7.1×10^{33} kg·m^2/s
P21 34.1 J **P23 (a)** 0.18 kg·m^2 **(b)** 14.2 J **(c)** 2.26 kg·m^2/s
P25 (a) 0.0765 kg·m^2 **(b)** 0.91 kg·m^2/s **(c)** 6.28 J
P27 (a) $\langle 29.4,0,0 \rangle$ kg·m/s **(b)** $\langle 49,0,0 \rangle$ m/s
(c) $\langle 0,0,-25.9 \rangle$ kg·m^2/s **(d)** $\langle 0,0,4.95 \rangle$ kg·m^2/s
(e) $\langle 0,0,-30.9 \rangle$ kg·m^2/s **(f)** $\langle 32.9,0,0 \rangle$ kg·m/s
P29 (a) 1.44 kg·m^2/s **(b)** 9.72 kg·m^2/s, into page
(c) 11.16 kg·m^2/s, into page **P31** $\langle 4,3.8,0 \rangle$ kg·m^2/s
P33 $d/dt\langle 0,0,xmv_y \rangle = \langle 0,0,-xmg \rangle$ **P35 (a)** -0.323 kg·m^2/s
(b) -0.323 kg·m^2/s **(c)** -0.138 kg·m^2/s **(d)** 1 **P37 (a)** 1, 3, 4,
and 5 **(b)** 0.189 m/s **P39 (a)** 2.01 s **(b)** 1.64 s **P41 (a)** 7.42 N
(b) 0.177 m **P43 (a)** about 0.5 s **(b)** about 100 J
P45 (a) about 0.2 s **(b)** about 800 J
P47 (a) $2m_1 v_1 + (-m_2 v_2 \cos\theta_2) = m_1 v_3 \cos\theta_3 + m_2 v_4 \cos\theta_4$
(b) $-m_2 v_2 \sin\theta_2 = 2m_1 v_3 \sin\theta_3 - m_2 v_4 \sin\theta_4$
(c) $2m_1(L/2)^2 \omega_1 + (-bm_2 v_2 \cos\theta_2) =$
$2m_1(L/2)^2 \omega_2 + bm_2 v_4 \cos\theta_4$ **P49 (a)** $\langle 2582,-9.6,0 \rangle$ m/s,
7.375 rad/s out of page **(b)** 1.55×10^7 J **P51 (a)** $\langle 8.68,1.3,0 \rangle$
m/s **(b)** 166 rad/s **(c)** 803 J **P53 (a)** 4.37×10^4 m/s
(b) increases 1 hour **P55 (a)** $\langle 0,0,-0.0256 \rangle$ kg·m^2/s
(b) $L_z = -0.186$ kg·m^2/s **(c)** $\omega_z = -145$ rad/s **P57 (b)**
147 N·m **(c)** 1 and 3 **P59** 3.67 rad/s **P61 (b)** $\frac{1}{2}g\sin\theta$; $\frac{2}{3}g\sin\theta$;
$\frac{5}{7}g\sin\theta$ **(c)** same time; sphere
(d) $\frac{1}{2}\left(M + I_{CM}/R^2\right)v_{CM}^2 = Mg\sin(\theta)\Delta x$;
$\frac{1}{2}Mv_{CM}^2 = \left(Mg\sin\theta - I_{CM}/R^2\,(dv_{CM}/dt)\right)\Delta x$; time derivatives
agree with force and torque analyses **P63** 12 rad/s
P65 (a) 20.9 rad/s **(b)** 104.7 rad **(c)** 6000° **P67 (a)** 4 rad/s^2
(b) 30 rad/s **(c)** 90 rad **(d)** 5160° **P69 (a)** 62.8 rad/s **(b)** 377 rad
(c) 69.8 rad/s **(d)** 332 rad **(e)** 99.5 m **P71** 3.55 rad, 203°
P73 0.543L **P75** 3.15×10^{-34} J·s **P77** -54.4 eV, -13.6 eV,
40.8 eV, all 4 times the values for hydrogen
P79 (a) 2.38×10^{-54} kg·m^2 **(b)** 4.7×10^{-15} m **(c)** 7×10^{-15} m;
nucleus may not be spherical **P83 (a)** clockwise **(b)** 32 s
P85 (a) clockwise as seen from above **(b)** slower precession
(c) same

Chapter 12

P17 252; 1 **P19** 18000 **P21** 8.04×10^{-22} J
P23 5.4×10^{88} year, or about 5×10^{78} times the age of the
Universe **P25 (a)** 5.65×10^{-23} J/K **(b)** 6.62×10^{-23}
J/K **(c)** 1.23×10^{-23} J/K **P27 (a)** 7.86 J/K **(b)** -7.74 J/K
(c) 0.12 J/K **(d)** 0 **P29** $E = \frac{1}{4}a^2 T^2$ **P31** $C = \frac{1}{2}a^2 T/N_{atoms}$
(J/K/atom) **P33 (a)** 215.55 K **(b)** 273.02 K **(c)** 3.48×10^{-23}
J/K/atom **P35 (a)** 3.61006×10^{-21} J **(b)** temperatures 118.7 K
and 121.0 K **(c)** 2.62×10^{-23} J/K/atom **P37** about 250 K
P39 25.1°C **P41 (a)** $v_i + F\Delta t/(M+m)$
(b) $\omega_i + 2(F+f)\Delta t/(MR)$ **(c)** $F(d-x) - \frac{1}{4}MR^2(\omega_f^2 - \omega_i^2)$
(d) 500 K **P43** 4.04×10^{-21} J, 1/40 eV **P45** 8300 m **P47** 1350
m/s; 510 m/s **P49** similar to Figure 12.54, but with different
break points **P51** about 2 cm **P53** about 600 m/s; rms speed
is larger; roughly one-tenth **P55 (a)** about 0.004 **(b)** 1×10^{17}
(c) 120 km **P57 (a)** helium **(b)** 8 K **P59** 108 K

Chapter 13

P17 $\langle 0,-4.48 \times 10^{-17},0 \rangle$ N **P19** $\langle 3.2 \times 10^{-15}, 3.2 \times 10^{-15},0 \rangle$
N **P21 (a)** negative **(b)** down and to the left
(c) $\langle 0.707,0.707,0 \rangle$ **(d)** $\langle -1.4 \times 10^{-5}, -1.4 \times 10^{-5},0 \rangle$ N

(e) $\langle -0.707,-0.707,0 \rangle$ **P23** $\langle 0,0,-9.13 \times 10^4 \rangle$ N/C
P25 9.38×10^3 N/C **P27 (a)** $\langle -0.6,-0.7,-0.2 \rangle$ m
(b) $\langle 0.5,-0.1,-0.5 \rangle$ m **(c)** $\langle 1.1,0.6,-0.3 \rangle$ m **(d)** 1.29 m
(e) $\langle 0.854,0.466,-0.233 \rangle$ **(f)** 48.8 N/C
(g) $\langle 41.7,22.7,-11.4 \rangle$ N/C **P29** $\langle -2550,-2550,0 \rangle$ N/C
P31 1.13×10^4 N/C **P33** $\langle -225,0,0 \rangle$ N/C
P35 (a) $\langle 0.8,0.7,-0.8 \rangle$ m **(b)** $\langle 0.5,1,-0.5 \rangle$ m **(c)** $\langle -0.3,0.3,0.3 \rangle$
m **(d)** 0.520 m **(e)** $\langle -0.577,0.577,0.577 \rangle$ **(f)** -5.33×10^{-9} N/C
(g) $\langle 3.08 \times 10^{-9}, -3.08 \times 10^{-9}, -3.08 \times 10^{-9} \rangle$ N/C
P37 $\langle 3,0,0 \rangle$ m **P39 (a)** $\langle 0,-5.92 \times 10^{-7},0 \rangle$ m
(b) $\langle 0,5.92 \times 10^{-7},0 \rangle$ m **P41** $\langle -960,-960,0 \rangle$ N/C
P43 $\langle 0,3 \times 10^{-6},0 \rangle$ m **P45 (a)** $+x$ **(b)** 7.31×10^5 N/C
(c) -7.69×10^{-8} m to the left
P47 (a) $\langle -1.15 \times 10^7, 8.64 \times 10^6,0 \rangle$ N/C **(b)** $\langle 0,-3 \times 10^7,0 \rangle$
N/C **(c)** $\langle -1.15 \times 10^7, -2.13 \times 10^7,0 \rangle$ N/C **(d)** $\langle 23.0,$
$42.6,0 \rangle$ N/C **(e)** $\langle 0,4 \times 10^7,0 \rangle$ N/C **(f)** $\langle -8.64 \times 10^6,$
$-6.48 \times 10^6,0 \rangle$ N/C **(g)** $\langle 1.125 \times 10^7,0,0 \rangle$ N/C
(h) $\langle 2.61 \times 10^6, 3.35 \times 10^7,0 \rangle$ N/C **(i)** $\langle -7.83 \times 10^{-3},$
$-0.101,0 \rangle$ N/C **P49 (a)** $\langle 1.92 \times 10^5,0,0 \rangle$ N/C
(b) $\langle -1.27 \times 10^5,0,0 \rangle$ N/C **(c)** $\langle -3.07 \times 10^{-14},0,0 \rangle$ N
P51 (a) $\vec{0}$ **(b)** 3.38×10^4 N/C **(c)** 2.22×10^3 N/C
P53 1.07×10^4 N/C **P55** $\langle 0,-4.43 \times 10^{-15},0 \rangle$ N **P57**
(a) 64.8 N/C **(b)** 32.4 N/C **P59** $\langle 0,105,0 \rangle$ N/C **P61 (a)** right
end **(b)** 8.3×10^{-8} C

Chapter 14

P27 4×10^{-10} C **P29** 2 and 3 **P31** 1 **P33** 1/243
P35 (a) 2 **(b)** g **(c)** c **(d)** g **P37 (a)** 894 m/s^2 **(b)** 1/32 **P41** c
P43 1.25×10^{-4} m/s **P45** 3 **P47** b **P49** j **P51 (a)** 1 **(b)** 3
(c) 2 **P53 (a)** B **(b)** 3 **P55** 4.08×10^{-4} m/s
P57 (a) 675 N/C **(b)** 9.1×10^{-3} N/C **(c)** increase **(d)** no **P59**
(a) $(1/(4\pi\epsilon_0)\langle 2p/(L+R+b)^3 - Q/(R+b)^2, -p/(L+R+b)^3, 0\rangle$
(b) $L \gg s$ **(c)** $\vec{E}_4$ points to the right and downward; $\vec{E}_{ball}$ points
to the left and upward **(d)** positive surface charge on the lower
right, negative surface charge on the upper left **(e)** $\vec{0}$ **P61** 1, 2,
and 4 **P63 (a)** 2.5 nC **(b)** 3

Chapter 15

P21 (a) $-Q/(2A)$ **(b)** $-Qdx/(2A)$ **(c)** $\langle -x,y,0 \rangle$ **(d)** $\sqrt{x^2+y^2}$
(e) x **P23 (a)** 0.024 m **(b)** -4×10^{-10} C **(c)** Q/N **(d)** L/dL
(e) $(Q/L)dL$ **P25 (a)** $-(1/(4\pi\epsilon_0))2Q\,|\vec{p}|\,/(Lx^2)$ **(b)** 2×10^4
m/s^2 **P27 (a)** each piece is an arc of radius R and angle $\Delta\theta$
(b) $(1/(4\pi\epsilon_0))(|Q|/(\alpha R^2))\Delta\theta\,\langle \cos\theta, \sin\theta,0 \rangle$
(c) $(1/(4\pi\epsilon_0))(|Q|/(\alpha R^2))\,(1 - \cos\alpha)$ **(d)** units correct; field
points toward rod midpoint (symmetry); for small α, field
approaches that of a point charge **P29 (a)** small-radius rings of
charge; contribution of each ring is to the right
(b) $(1/(4\pi\epsilon_0))(Q/L)\Delta x/(L+d-x)^2\,\langle 1,0,0 \rangle$
(c) $(1/(4\pi\epsilon_0))Q/(d(L+d))$ **(d)** for $d \gg L$, field approaches that
of a point charge **P31 (a)** 3.048×10^4 N/C toward the
negative ring **(b)** 2.743×10^{-4} N toward the positive ring
P33 0 **P35 (a)** 2.031×10^6 N/C **(b)** 2.012×10^6 N/C **(c)** differs
by less than 1% **P37 (a)** 2.795×10^6 N/C **(b)** 2.760×10^6 N/C
(c) differs by only 1.3% **P39 (a)** negative on left face, positive
on right face **(b)** 0 **(c)** 2.66×10^{-9} C **P41** 1.84×10^{-5} C
P43 1.01×10^{-4} C **P45 (a)** curves downward between plates,
then moves in a straight line down and to the right
(b) 1.8×10^{16} m/s^2 **(c)** 3.18×10^{-9} C; negative
P47 $Q(s/R)/(2 - s/R - 2t/R)$ **P49 (a)** metal ball polarizes
with right side more negative than left side; molecules in plastic

polarize with positive ends pointing toward metal ball **(b)** 7780 N/C to the left, ignoring the field produced by the polarized molecules in the plastic **(c)** 0 **(d)** 7780 N/C to the right **P51** 4.59×10^4 N/C inward; 3.51×10^4 N/C outward **P53 (a)** $\langle -5 \times 10^{-5}, 0, 0 \rangle$ N/C **(b)** 0 **(c)** left side more positive than right side **P55 (a)** dipoles are radial with negative ends pointing toward center **(b)** no polarization **(c)** same **(d)** negative on inner surface, positive on outer surface **(e)** $-Q$ on inner surface, $+Q$ on outer surface **(f)** same **P57 (a)** 249 N/C; 0 **(b)** 364 N/C; -182 N/C **(c)** -5.82×10^{-17} N; 2.91×10^{-17} N **P59 (a)** $(1/(4\pi\epsilon_0))qs/L^3$ to the right **(b)** 0; right side of ball more positive than left side; dipole field points to the right and field of charges on ball points to left **P61** $(1/(4\pi\epsilon_0))^2 (4\alpha Q^2/L)/|\vec{d}|^4$

Chapter 16

P21 3.11×10^6 m/s **P23 (a)** 1.64×10^{-25} J **(b)** 3.01×10^{-20} J **P25** -7.52×10^{-23} J **P27** 3.6×10^4 V/m **P29 (a)** $\langle -1, 0, 0 \rangle$ m **(b)** 750 V **(c)** 1.2×10^{-16} J = 750 eV **(d)** -1.2×10^{-16} J = -750 eV **P31 (a)** $\langle 0.7, 0, 0 \rangle$ m **(b)** 595 V **(c)** 9.52×10^{-17} J = 595 eV **(d)** -9.52×10^{-17} J = -595 eV **P33** 7860 V/m **P35** 9×10^3 V **P37** 1500 V **P39** B; 150 V/m to the left **P41 (a)** to the right **(b)** for example, $R = 0.1$ m, $Q_A = 2.78 \times 10^{-8}$ C and $Q_B = -2.78 \times 10^{-8}$ C **P43** -1200 V **P45** 1.92×10^{-19} J **P47 (a)** -205 V **(b)** 205 V **(c)** -3.28×10^{-17} J **(d)** 4.64×10^{-17} J **P49 (a)** $(q^2/(4\pi\epsilon_0))(1/((x-s/2)^2 + y^2)^{1/2} - 1/((x+s/2)^2 + y^2)^{1/2})$ **(b)** $(1/(4\pi\epsilon_0))px/r^3$ **(c)** $(1/(4\pi\epsilon_0))2p/x^3$ **(d)** E is in x direction **P51** $(Q_2 b - Q_1 a)/(A\epsilon_0)$ **P53** $(1/(4\pi\epsilon_0))2(Q/L)b/r$ **P55** $(1/(4\pi\epsilon_0))Qe(1/R - 1/(R+h))$ **P57 (a)** $(Q_1/(4\pi\epsilon_0))(1/L - 1/(L+R_2))$ **(b)** $(Q_2/(4\pi\epsilon_0))(1/(R_2+d) - 1/R_2) + (Q_1/(4\pi\epsilon_0))(1/(L+R_2) - 1/(L+R_2+d))$ **(c)** decrease **P59 (a)** 37 V **(b)** 3214 V **P61** $(2Q/(4\pi\epsilon_0))(1/(R+d+L) - 1/(R+d))$ **P63** $(1/(4\pi\epsilon_0))(Q/R)\ln(d/(d-h)) + (Q/(2\epsilon_0))(h/A)$ **P65** 8 V **P67 (a)** 0 **(b)** 0 **P69** $(1/(4\pi\epsilon_0))(q_1/r_{1,A} - q_2/r_{2,A})$ **P71** -450 V **P73** 608 V/m **P75** $-(1/(4\pi\epsilon_0))Q/(a^2 + h^2)^{1/2}$ **P77** $(11/8)(1/(4\pi\epsilon_0))(Q/R)$ **P79 (a)** $Q_2 = -Q$; $Q_3 = 5Q$ **(b)** $(1/(4\pi\epsilon_0))5Q/r_3$; $(1/(4\pi\epsilon_0))5Q/r_3$; $(1/(4\pi\epsilon_0))5Q/r_3$; $(1/(4\pi\epsilon_0))(5Q/r_3 - Q/r_2 + Q/r_1)$; $(1/(4\pi\epsilon_0))(5Q/r_3 - Q/r_2 + Q/r_1)$ **P81 (a)** no polarization **(b)** $-(1/(4\pi\epsilon_0))Q/R$ **P83** 250 V; 100 V; 250 V; 600 V **P85** $(1/(4\pi\epsilon_0))Qz/(z^2 + R^2)^{3/2}$ **P87 (a)** 0 **(b)** $-(1/(4\pi\epsilon_0))qs(1/a^2 - 1/b^2)$ **(c)** $(1/(4\pi\epsilon_0))eqs(1/a^2 - 1/b^2)$

Chapter 17

P15 $\langle 0, 0, 7.5 \rangle$; $\langle 0, 0, -7.5 \rangle$ **P17** $\langle 15, 15, 6 \rangle$ **P19** 0 at 2 and 5; $\langle 0, 0, -1.28 \times 10^{-17} \rangle$ T at 1 and 6; $\langle 0, 0, 1.28 \times 10^{-17} \rangle$ T at 3 and 4 **P21** 0 at 2 and 5; $\langle 0, 0, -5.28 \times 10^{-18} \rangle$ T at 3 and 4; $\langle 0, 0, 5.28 \times 10^{-18} \rangle$ T at 1 and 6 **P23 (a)** electric $\langle -1, 0, 0 \rangle$; magnetic $\langle 0, 0, 1 \rangle$ **(b)** electric 5.76×10^9 N/C; magnetic 0.66 T **P25 (a)** 7.5×10^{19} electrons/s **(b)** 3.15×10^{-5} m/s **P27** 8.98×10^{28} m^{-3} **P29 (a)** 3.5×10^{-6} T **(b)** north **P31** 5.14×10^{-6} T **P33 (a)** $+x$ **(b)** 0.1625 m **(c)** 0.1625 m **(d)** $\langle 0.1625, 0, 0 \rangle$ m **(e)** $\langle -0.08125, 0, 0 \rangle$ m **(f)** $\langle 0.16225, 0.178, 0 \rangle$ m **(g)** $\langle 0.674, 0.739, 0 \rangle$ **(h)** $\langle 0, 0, 0.120 \rangle$ m **(i)** $\langle 0, 0, 2.95 \times 10^{-6} \rangle$ T **P35 (a)** conventional current down and to the left; magnetic field down and to the right **(b)** 0.088 A **P37 (a)** 0 **(b)** $\langle 0, 0, hdx \rangle$ **(c)** $+x$ **P39 (a)** 0.0917 A **(b)** $\langle 0, 0, -1.63 \times 10^{-6} \rangle$ T **(c)** assumed $r \ll L$; neglected curved part of wire **P41 (a)** 1.11×10^{-7} T

(b) 1.11×10^{-5} T **P43** 1.5 A **P45** $(\mu_0/(4\pi))I\pi(1/R_1 + 1/R_2)$; direction $\langle 0, 0, -1 \rangle$ **P49 (a)** into the page **(b)** $(\mu_0/(4\pi))I(\pi/R + 2/h)$ **P51** $(\mu_0/(4\pi))I\pi(1/R + 3/r)$; $\langle 0, 0, -1 \rangle$ **P53** 4.3 A·m^2 **P55** 0.5 A·m^2 **P57 (a)** S end **(b)** 1 and 5 **(c)** 12

Chapter 18

P21 1, 3, 6, 8, and 9 **P23** $i_A = i_B + i_C$, $i_B + i_C = i_D$, $i_A = i_D$; i_D **P25 (a)** 11 A **(b)** yes **(c)** 5 A **(d)** no **P27** 0.016 V/m **P29** 2, 5, 8, and 9 **P31** 6.24×10^{19} electrons/s **P33** 1; 6×10^{18} electrons/sl 0.26 V/m **P35 (a)** downward **(b)** upward **(c)** 19.8 V/m; to the left **P37** 5 V/m; same **P39** 3.33 V/m; same **P41 (a)** 3.5 V **(b)** C **P43 (a)** 2 **(b)** 1.3 V $+ V_E - V_G + 1.3$ V $+ V_B - V_D = 0$ **(c)** 5 V/m **(d)** 6.73×10^{18} s^{-1} **(e)** 5 V/m **(f)** (c) **P45 (a)** 1, 2, and 4 **(b)** 1.3 V $+ 2(V_A - V_C) + V_C - V_E = 0$ **(c)** $i_B = i_D = i_F$ **(d)** 21.9 V/m; 0.925 V/m **P47 (a)** 1.2×10^{18} s^{-1} **(b)** neglect connection wires; filaments same lengths; ideal battery (no internal resistance) **(c)** high positive density near $+$ terminal of battery; small gradient along wires; large gradient across bulbs (largest along thin-filament bulb) **P49 (a)** 3×10^{17} s^{-1} **(b)** 4.5×10^{17} s^{-1} **(c)** do the experiment **(d)** 6×10^{17} s^{-1} **P51 (a)** 2.27 V/m **(b)** 2.27 V/m **(c)** 3 **P53** 18.7°

Chapter 19

P29 d **P31** c **P33** d **P35** 1, 3, and 4 **P37 (a)** 1, 2, 3, and 5 **(b)** 1, 2, 3, and 5 **(c)** c **(d)** a **P39** 1.85×10^{-8} F **P41** 8×10^{-10} C **P43** battery: $-2E$, 0; bulb: 0, 0; capacitor: $+E$, $-E$; surroundings: $+E$, $+E$ **P45** 4.06×10^7 $(\Omega \cdot m)^{-1}$ **P47** 40 Ω **P49 (a)** 0.83 Ω, 3.6 A **(b)** 4.2×10^{-4} Ω, 15.0015 V **P51** 240 Ω; 280 Ω **P53** bulb resistance increases with temperature **P55** 3 C; 1.9×10^{19} electrons **P57** 5×10^{-3} V/m **P59 (a)** field is small in thick wires, large in thin wires; field goes counterclockwise **(b)** roughly, a small gradient along the thick wires, large along the thin wires **(c)** $n(1/4)\pi d_2^2 uK / \left((d_2/d_1)^2(2L_1 + L_3) + 2L_2 \right)$ **(d)** $n(1/4)\pi d_2^2 uK / (2L_2)$ **(e)** $K/4$ **P61 (a)** E decreases with x **(b)** mobility increases with x (lower temperature) **(c)** $I/(ne(wh)(u_0 + kx))$ **(d)** negative **(e)** $(I/(ne(wh)k))\ln((u_0 + kd)/u_0)$ **(f)** $(1/(ne(wh)k))\ln((u_0 + kd)/u_0)$ **P63 (a)** 3.0 V $- I_4(15\,\Omega) - I_3(30\,\Omega + 20\,\Omega) = 0$; 3.0 V $- I_4(15\,\Omega) - I_1(20\,\Omega) = 0$; $I_4 = I_3 + I_1$ **(b)** compass 1 deflects 3° toward NE; compass 2 deflects 4.2° toward NW **(c)** 126 V/m **(d)** $E_1 > E_2$ **(e)** 4.6 mm **(f)** -2.4 V **(g)** 4.8×10^{-19} J **(h)** 0.153 W **P67** 4 A **P69 (a)** 34.3 Ω **(b)** 0 **(c)** 0.26 A **P71 (a)** 15 A **(b)** 0 **P73** 10 V **P75 (a)** $C \times$emf **(b)** field: polarized molecules reduce fringe field, so current runs again; potential: gap voltage reduced, so current runs again **(c)** $(\text{emf}/R)(1 - 1/K)$ **(d)** $KC \times$emf **P77 (a)** $E = 3600$ V/m; too small to make a spark **(b)** 3.83×10^{-7} C **(c)** 6.12 V **(d)** 3.02 V **P79 (a)** $ABCEFGA$: $20 - 10I_1 - 15I_4 - 12I_6 - 20I_1 = 0$; $DECD$: $5 + 15I_4 - 20I_2 = 0$; $DFED$: $-30I_3 + 12I_6 - 5 = 0$; C: $I_1 = I_2 + I_4$; D: $I_2 = I_5 + I_3$; F: $I_3 + I_6 = I_1$ **(b)** do the check **(c)** 33.2 V **(d)** 1.63 W **(e)** 1730 V/m

Chapter 20

P25 $\langle 0, 0, 1.92 \times 10^{-14} \rangle$ N **P27** $\vec{0}$ **P29** 1.64 T **P31 (a)** left accelerating plate is positive; top deflection plate is positive; B into page **(b)** 1600 V; 320 V **(c)** 1371 V; 274 V **P33 (a)** 8.9° **(b)** 4×10^{-17} N **P35 (a)** CCW **(b)** 9.2 T **P37** $\langle 0, 0, 0.243 \rangle$ N

P39 **(a)** 5.52 A **(b)** $-y$ **P41** **(a)** $\frac{1}{4}(\mu_0/(4\pi))2\pi I(1/a - 1/b)$; $\langle 0,0,1\rangle$ **(b)** $\frac{1}{4}(\mu_0/(4\pi))2\pi Iev(1/a - 1/b)$; $\langle 0,1,0\rangle$ **P43** 1.14 T into the page **P45** 9500 m/s **P47** $\langle 0,0,-1863\rangle$ V/m
P49 **(a)** $-z$ **(b)** $-y$ **(c)** 1.08×10^{-4} T **(d)** 61.6 V/m **P51** **(a)** B **(b)** upward **(c)** upward **(d)** 3.30×10^7 m/s; 2.40×10^{-3} T **(e)** 2, 5, 6 **(f)** 650 V **P53** $(5/4)(\mu_0/(4\pi))2\pi IV/R$, $-y$
P55 **(a)** positive **(b)** 4.8×10^{-3} m/s **(c)** 9.8×10^{-4} (m/s)/(V/m) **(d)** 4×10^{23} m^{-3} **(e)** 2.4 Ω **P57** **(a)** If you connect the + lead to the bottom of the bar (with the − lead straight above it) and the voltmeter reads positive, conventional current in the long wire goes to the right. **(b)** $nedh\Delta V/((\mu_0/(4\pi))2I_{ammeter})$
P59 1.77×10^{-5} V; 0.022 A **P61** 2 **P63** **(a)** 1 **(b)** 0 **(c)** slightly less than 4.48×10^{-19} N **(d)** 4.48×10^{-19} N **(e)** 1.26 V **(f)** to the left **P65** **(a)** 0; 0 **(b)** vBh/R, vB^2h^2/R **(c)** 0, 0 **(d)** vBh/R, vB^2h^2/R **(e)** 0, 0 **P67** west end has excess electrons; about 2500 electrons **P69** **(a)** negative below, positive above **(b)** 0.96A, CCW **(c)** 2 **(d)** 0.035 N to the right **P71** $\langle 0,1.2,0\rangle$ N **P73** 120 J **P75** **(a)** 24 V **(b)** 11.5 W **P77** **(a)** $+1.6 \times 10^{-19}$ C **(b)** 14.4 V **(c)** 0 **(d)** -14.4 V **(e)** 14.4 eV **(f)** 14.4 eV **P79** 7.5×10^{-6} C

Chapter 21

P7 left -0.24 V·m; right 0.60 V·m; all other faces 0; total 0.36 V·m; 3.19×10^{-12} C **P9** contains -5×10^{-11} C
P11 contains 1.8×10^{-12} C **P13** total flux 0.0672 V·m; contains 5.95×10^{-13} C **P15** -227 V·m **P17** **(a)** $+Q$ on surface of inner wire; $-Q$ on inner surface of outer tube **(b)** $(Q/L)/(2\pi\epsilon_0 r)$ **(c)** 0 **P19** **(a)** $(Q/A)/\epsilon_0$ **(b)** assume E varies and show this is inconsistent with Gauss's law **(c)** $(Q/A)/\epsilon_0$ **(d)** $Qs/(2R)$ **P21** **(a)** S/ϵ_0 **(b)** the field of a large sheet would be $(1/2)S/\epsilon_0$ **P23** 125 A, out of page **P25** **(a)** left above, right below **(b)** $\mu_0NI/(2L)$ **(c)** μ_0NI/L **(d)** 0 **P27** $(\mu_0/(4\pi))(I_1 - I_2)/R$, out of page **P29** 1×10^{36} N/C/m

Chapter 22

P13 1.5 V; CCW **P15** CCW; $|dB/dt|\pi r_1^2/R$; tangent, CCW; $|dB/dt|r_1^2/(2r_2)$ **P17** **(a)** 2.6×10^{-6} T·m^2 **(b)** treat coil as magnetic dipole **(c)** there is no electric field **(d)** $-y$ **(e)** 2.16×10^{-7} V **(f)** 2.16×10^{-7} V **(g)** 8.59×10^{-7} V/m **(h)** 8.59×10^{-7} V/m **P19** **(a)** $(\mu_0/(4\pi))(2N_1N_2\pi^2 r_1^2 r_2^2/x^3)(b + 2ct)$ **(b)** downward **(c)** $(\mu_0/(4\pi))(N_1N_2\pi r_1^2 r_2/x^3)(b + 2ct)$ **P21** **(a)** upward at 1, downward at 2 **(b)** 0.404 mV **(c)** B uniform across coil; used average dB/dt **P23** **(a)** sine wave in coil 1, $-$cosine wave in coil 2 **(b)** 0.233 V; assume $z \gg r_1$ and coils are thin **P25** **(a)** $+z$ **(b)** $+x$ **(c)** 2.33×10^{-5} T·m^2 **(d)** coils are thin; B uniform across coil 2 **(e)** 7.76×10^{-6} T·m^2 **(f)** 3.88×10^{-5} V **(g)** 10.7 mV **(h)** 2.06×10^{-4} V/m **(i)** 1 and 4 **P27** **(a)** CCW **(b)** 13.6 mA **P29** **(a)** $(\pi r_1^2/R)|(\mu_0N(-k))/d|$; to the right

(b) much larger current in ring **P31** **(a)** eBR **(b)** $\pi r^2 be$
P33 1.13×10^{-6} V/m **P35** **(a)** CCW **(b)** $(e/m)(\mu_0/(4\pi))2I_0Ar/R^3$ **(c)** tangent to orbit, CW **(d)** slows down **P37** $(\mu_0/(4\pi))3\pi R^2 Iyv/x^4$ **P39** 420 V, 1.14 A **P41** **(a)** radius = 0.01 m, length = 0.04 m, 1600 turns **(b)** 18.8 A

Chapter 23

P17 $dE/dt = 1.3 \times 10^{18}$ V/m/s inside the loop **P19** 1×10^{-5} T in $-y$ direction **P21** 0.024 T in $-y$ direction **P23** **(a)** 1.16×10^{17} m/s^2 **(b)** 1.69×10^{-7} V/m **(c)** 4 **(d)** 5 **(e)** 8 **P25** **(a)** 5×10^{-10} s **(b)** $+y$ **(c)** $-y$ **P27** **(a)** electrons have low mass so experience larger acceleration than the ions; electric field at A is in $+y$ direction **(b)** electric field at B is in $+y$ direction; electric field at C is up and to the right **(c)** 1×10^{-8} s; into page **(d)** proportional to $1/r^2$ instead of $1/r$
P29 **(a)** 6.4×10^{-12} V/m to the left **(b)** 50 ns **(c)** upward **(d)** 1.07×10^{-9} V/m; much larger **(e)** to the right **P31** **(a)** to the left **(b)** 4×10^{-8} s; 5×10^{-10} s **(c)** E out of page, B up **P33** 60 nM **P35** 300 m; 3 m; 698 nm; 400 nm **P37** 550 V/m **P39** 2.66×10^8 m/s **P41** **(a)** 3 **(b)** 65.6° **P43** **(a)** 63.4° **(b)** 42.3° **(c)** 4.2 m **P45** 0.23 m to the left of the lens; virtual; no image **P47** $\langle 100, -4, 0\rangle$ cm; real **P49** 0.302 m; 1.5 cm; inverted

Supplement S1

P3 2.4×10^{22} molecules/s **P5** **(b)** down to 1 percent in 4.6 min **P7** about 2 hours **P9** **(a)** 1.0×10^5 N/m^2 **(b)** 1300 m/s **(c)** 4×10^{20} atoms **(d)** helium leaves faster **(e)** decrease **P11** 3.17 **P13** **(a)** 2910 J **(b)** -2910 J **(c)** 0 J **P15** **(a)** 214 K **(b)** 313.6 J **(c)** 2784 N **(d)** 2652 N **(e)** 2514 N **P17** 25 s; increase **P19** 0.05 J **P21** **(a)** $NkT_H \ln(V_2/V_1)$ **(b)** $(T_H/T_L)^{3/2} = V_3/V_2$ **(c)** $NkT_L \ln(V_4/V_3)$ **(d)** $(T_H/T_L)^{3/2} = V_3/V_2 = V_4/V_1$; putting the pieces together, one finds that Q_H/T_H does indeed equal Q_L/T_L **P23** 5.9 mm

Supplement S2

P1 5×10^{-8} m

Supplement S3

P13 7.5×10^{14} Hz; 1.3×10^{-15} s **P15** 1.33 m
P17 **(a)** reradiation; electric field in and out of page, magnetic field in the xy plane, perpendicular to the electric field **(b)** 23.6°; 11.5° **P19** 23.6°; 53.1° **P21** 1.2×10^{-10} m; 1.6×10^{-10} m
P23 212 nm **P25** **(a)** 28 cm **(b)** wider **(c)** narrower **P27** 9.4 N **P29** 0.02 m; 0.6 m; 7.5 ms; 1.67 m^{-1}; 838 rad/s; 133 Hz **P31** $0.04\cos((2\pi/0.2)(x + 45t))$ **P33** 8.2×10^{-14} m; extremely small compared to 1×10^{-10} m **P35** **(a)** 89.6 m/s **(b)** 128 Hz, 70 cm; 192 Hz, 46.7 cm; 256 Hz, 35 cm; 320 Hz, 28 cm **P37** 365 nm **P39** 774 V/m; 2.8×10^{20} photons/s

Index

The text of Supplements 1–3 can be found at www.wiley.com/college/chabay

PERIODIC TABLE OF THE ELEMENTS

Atomic number →
Symbol →
Name (IUPAC) →
Atomic mass →

6
C
Carbon
12.011

IUPAC recommendations →
Chemical Abstracts Service group notation →

1 IA	2 IIA	3 IIIB	4 IVB	5 VB	6 VIB	7 VIIB	8 VIIIB	9 VIIIB	10 VIIIB	11 IB	12 IIB	13 IIIA	14 IVA	15 VA	16 VIA	17 VIIA	18 VIIIA
1 **H** Hydrogen 1.0079																	2 **He** Helium 4.0026
3 **Li** Lithium 6.941	4 **Be** Beryllium 9.0122											5 **B** Boron 10.811	6 **C** Carbon 12.011	7 **N** Nitrogen 14.007	8 **O** Oxygen 15.999	9 **F** Fluorine 18.998	10 **Ne** Neon 20.180
11 **Na** Sodium 22.990	12 **Mg** Magnesium 24.305	3	4	5	6	7	8	9	10	11	12	13 **Al** Aluminum 26.982	14 **Si** Silicon 28.086	15 **P** Phosphorus 30.974	16 **S** Sulfur 32.065	17 **Cl** Chlorine 35.453	18 **Ar** Argon 39.948
19 **K** Potassium 39.098	20 **Ca** Calcium 40.078	21 **Sc** Scandium 44.956	22 **Ti** Titanium 47.867	23 **V** Vanadium 50.942	24 **Cr** Chromium 51.996	25 **Mn** Manganese 54.938	26 **Fe** Iron 55.845	27 **Co** Cobalt 58.933	28 **Ni** Nickel 58.693	29 **Cu** Copper 63.546	30 **Zn** Zinc 65.409	31 **Ga** Gallium 69.723	32 **Ge** Germanium 72.64	33 **As** Arsenic 74.922	34 **Se** Selenium 78.96	35 **Br** Bromine 79.904	36 **Kr** Krypton 83.798
37 **Rb** Rubidium 85.468	38 **Sr** Strontium 87.62	39 **Y** Yttrium 88.906	40 **Zr** Zirconium 91.224	41 **Nb** Niobium 92.906	42 **Mo** Molybdenum 95.94	43 **Tc** Technetium (98)	44 **Ru** Ruthenium 101.07	45 **Rh** Rhodium 102.91	46 **Pd** Palladium 106.42	47 **Ag** Silver 107.87	48 **Cd** Cadmium 112.41	49 **In** Indium 114.82	50 **Sn** Tin 118.71	51 **Sb** Antimony 121.76	52 **Te** Tellurium 127.60	53 **I** Iodine 126.90	54 **Xe** Xenon 131.29
55 **Cs** Cesium 132.91	56 **Ba** Barium 137.33	57-71 lanthanoids	72 **Hf** Hafnium 178.49	73 **Ta** Tantalum 180.95	74 **W** Tungsten 183.84	75 **Re** Rhenium 186.21	76 **Os** Osmium 190.23	77 **Ir** Iridium 192.22	78 **Pt** Platinum 195.08	79 **Au** Gold 196.97	80 **Hg** Mercury 200.59	81 **Tl** Thallium 204.38	82 **Pb** Lead 207.2	83 **Bi** Bismuth 208.98	84 **Po** Polonium (209)	85 **At** Astatine (210)	86 **Rn** Radon (222)
87 **Fr** Francium (223)	88 **Ra** Radium (226)	89-103 actinoids	104 **Rf** Rutherfordium (267)	105 **Db** Dubnium (268)	106 **Sg** Seaborgium (269)	107 **Bh** Bohrium (270)	108 **Hs** Hassium (269)	109 **Mt** Meitnerium (278)	110 **Ds** Darmstadtium (281)	111 **Rg** Roentgenium (281)	112 **Cn** Copernicum (285)	113	114 **Fl** Flerovium (289)	115	116 **Lv** Livermorium (293)		118

57 **La** Lanthanum 138.91	58 **Ce** Cerium 140.12	59 **Pr** Praseodymium 140.91	60 **Nd** Neodymium 144.24	61 **Pm** Promethium (145)	62 **Sm** Samarium 150.36	63 **Eu** Europium 151.96	64 **Gd** Gadolinium 157.25	65 **Tb** Terbium 158.93	66 **Dy** Dysprosium 162.50	67 **Ho** Holmium 164.93	68 **Er** Erbium 167.26	69 **Tm** Thulium 168.93	70 **Yb** Ytterbium 173.04	71 **Lu** Lutetium 174.97
89 **Ac** Actinium (227)	90 **Th** Thorium 232.04	91 **Pa** Protactinium 231.04	92 **U** Uranium 238.03	93 **Np** Neptunium (237)	94 **Pu** Plutonium (244)	95 **Am** Americium (243)	96 **Cm** Curium (247)	97 **Bk** Berkelium (247)	98 **Cf** Californium (251)	99 **Es** Einsteinium (252)	100 **Fm** Fermium (257)	101 **Md** Mendelevium (258)	102 **No** Nobelium (259)	103 **Lr** Lawrencium (266)

Numbers in parentheses are the masses of the longest-lived isotopes of elements that have no stable isotopes.

Greek Alphabet and Its Uses in this Textbook

Alpha	A		α	alpha particle (helium-4 nucleus)	Nu	N			ν	neutrino; frequency $(=f)$
Beta	B		β		Xi	Ξ			ξ	
Gamma	Γ		γ	relativistic factor, photon	Omicron	O			o	
Delta	Δ	change of; small quantity of	δ		Pi	Π			π	circumference/ diameter of a circle
Epsilon	E		ϵ	a small quantity	Rho	P			ρ	density, resistivity
Zeta	Z		ζ		Sigma	Σ	sum		σ	conductivity
Eta	H		η		Tau	T			τ	torque
Theta	Θ		θ	angle	Upsilon	Υ			υ	
Iota	I		ι		Phi	Φ	flux		ϕ	phase angle
Kappa	K		κ		Chi	X			χ	
Lambda	Λ		λ	wavelength	Psi	Ψ			ψ	
Mu	M		μ	micro (1×10^{-6}), muon, coefficient of friction	Omega	Ω	# of microstates of a system, ohm, precession angular frequency		ω	angular frequency, angular speed

Units of Measurement

kg = kilogram   m = meter   s = second   C = coulomb
N = newton $(\text{kg} \cdot \text{m/s}^2)$ J = joule $(\text{N} \cdot \text{m})$ W = watt (J/s)
A = ampere (C/s)   V = volt (J/C)   T = tesla (N/C)/(m/s)   eV = electron volt $= 1.6 \times 10^{-19}$ J

Important Prefixes

pico (p) $= 1 \times 10^{-12}$ nano (n) $= 1 \times 10^{-9}$ micro $(\mu) = 1 \times 10^{-6}$ milli (m) $= 1 \times 10^{-3}$
centi (c) $= 1 \times 10^{-2}$ kilo (k) $= 1 \times 10^{3}$ mega (M) $= 1 \times 10^{6}$ giga (G) $= 1 \times 10^{9}$

Unit Conversions

1 kg = 2.2 pounds   1 inch = 2.54 cm   1 foot = 30.5 cm   1 mile = 1.6 km
1 liter = 1.1 quart $T_{\text{Kelvin}} = T_{\text{Celsius}} + 273.15$ $T_{\text{Fahrenheit}} = (9/5)T_{\text{Celsius}} + 32$
1 cal = 4.2 J   1 kcal $= 4.2 \times 10^{3}$ J

1 kcal is 1 food calorie, also called 1 "large calorie"

Physical Constants

$G = 6.7 \times 10^{-11} \dfrac{\text{N} \cdot \text{m}^2}{\text{kg}^2}$; universal gravitational constant

$c = 3 \times 10^8$ m/s; speed of light

Avogadro's number $= 6.02 \times 10^{23} \dfrac{\text{molecules}}{\text{moles}}$

$e = 1.6 \times 10^{-19}$ coulomb; proton charge

1 eV $=$ electron volt $= 1.6 \times 10^{-19}$ J

$h = 6.6 \times 10^{-34}$ joule·s; Planck's constant

$\hbar = \dfrac{h}{2\pi} = 1.05 \times 10^{-34}$ joule·s

$k_B = 1.38 \times 10^{-23}$ J/K; Boltzmann's constant $R = 8.3$ J/K; gas constant

$\dfrac{1}{4\pi\varepsilon_0} = 9 \times 10^9 \dfrac{\text{N} \cdot \text{m}^2}{\text{C}^2}$

$\varepsilon_0 = 8.85 \times 10^{-12} \dfrac{\text{C}^2}{\text{N} \cdot \text{m}^2}$

$\dfrac{\mu_0}{4\pi} = 1 \times 10^{-7} \dfrac{\text{T} \cdot \text{m}}{\text{A}}$

Masses and Distances

1 u $\approx m_{\text{proton}} \approx m_{\text{neutron}} \approx m_{\text{hydrogen atom}} \approx \dfrac{1 \times 10^{-3} \text{ kg/mole}}{6.02 \times 10^{23} \text{ atoms/mole}} = 1.7 \times 10^{-27}$ kg

$m_{\text{electron}} = 9 \times 10^{-31}$ kg   Radius of atom $\approx 1 \times 10^{-10}$ m

$M_{\text{Earth}} = 6 \times 10^{24}$ kg   Radius of Earth $= 6.4 \times 10^6$ m   Height of atmosphere ≈ 50 km

$M_{\text{Moon}} = 7 \times 10^{22}$ kg   Radius of Moon $= 1.75 \times 10^6$ m   Distance from Earth to Moon $= 4 \times 10^8$ m

$M_{\text{Sun}} = 2 \times 10^{30}$ kg   Radius of Sun $= 7 \times 10^8$ m   Distance from Earth to Sun $= 1.5 \times 10^{11}$ m

$m_{\text{neutron}} = 939.6$ MeV/c^2 $m_{\text{proton}} = 938.3$ MeV/c^2 $m_{\text{electron}} = 0.511$ MeV/c^2

Miscellaneous

$g = +9.8$ N/kg; magnitude of the gravitational field near the Earth's surface

Breakdown strength of air $\approx 3 \times 10^6$ N/C

Horizontal component of Earth's magnetic field $\approx 2 \times 10^{-5}$ T in much of the United States.

Circumference of circle $= 2\pi r$ Area of circle $= \pi r^2$

Surface area of sphere $= 4\pi r^2$ Volume of sphere $= \frac{4}{3}\pi r^3$